Cell Growth

NATO ADVANCED STUDY INSTITUTES SERIES

A series of edited volumes comprising multifaceted studies of contemporary scientific issued by some of the best scientific minds in the world, assembled in cooperation with NATO Scientific Affairs Division.

Series A: Life Sciences

Recent Volumes in this Series

Volume 30 – Human Physical Growth and Maturation
edited by Francis E. Johnston, Alex F. Roche, and Charles Susanne

Volume 31 – Transfer of Cell Constituents into Eukaryotic Cells
edited by J. E. Celis, A. Graessmann, and A. Loyter

Volume 32 – The Blood – Retinal Barriers
edited by Jose G. Cunha-Vaz

Volume 33 – Photoreception and Sensory Transduction in Aneural Organisms
edited by Francesco Lenci and Giuliano Colombetti

Volume 34 – Lasers in Biology and Medicine
edited by F. Hillenkamp, R. Pratesi, and C. A. Sacchi

Volume 35 – Environmental Physiology of Fishes
edited by M. A. Ali

Volume 36 – The Prostaglandin System: Endoperoxides, Prostacyclin, and Thromboxanes
edited by F. Berti and G. P. Velo

Volume 37 – Active Defense Mechanisms in Plants
edited by R. K. S. Wood

Volume 38 – Cell Growth
edited by Claudio Nicolini

Volume 39 – Drug Development, Regulatory Assessment, and Postmarketing Surveillance
edited by William M. Wardell and Giampaolo Velo

This series is published by an international board of publishers in conjunction with NATO Scientific Affairs Division

A Life Sciences B Physics	Plenum Publishing Corporation London and New York
C Mathematical and Physical Sciences	D. Riedel Publishing Company Dordrecht, Boston, and London
D Behavioral and Social Sciences E Applied Sciences	Sijthoff & Noordhoff International Publishers Alphen aan den Rijn, The Netherlands, and Germantown, U.S.A.

Cell Growth

Edited by

Claudio Nicolini

Temple University Health Sciences Center
Philadelphia, Pennsylvania
and National Research Council
Genoa, Italy

PLENUM PRESS • NEW YORK AND LONDON
Published in cooperation with NATO Scientific Affairs Division

Library of Congress Cataloging in Publication Data

NATO Advanced Study Institute on Cell Growth (1980 : Erice, Italy)
Cell growth.

(NATO advanced study institutes series. Series A, Life sciences ; v. 38)
"Proceedings of a NATO Advanced Study Institute on Cell Growth, held October 18-31, 1980 in Erice, Sicily"–T.p. verso.
Bibliography: p.
Includes index.
1. Cell proliferation–Congresses. 2. Cell cycle–Congresses. 3. Cancer cells–Congresses. I. Nicolini, Claudio A. II. Title. III. Series. [DNLM: 1. Cell cycle–Congresses. QH 605 N279c 1980]
QH605.N28 1980 574.87'62 81-15732
ISBN 978-1-4684-4048-5 ISBN 978-1-4684-4046-1 (eBook) AACR2
DOI 10.1007/978-1-4684-4046-1

Proceedings of a NATO Advanced Study Institute on Cell Growth,
held October 18-31, 1980, in Erice, Sicily

Softcover reprint of the hardcover 1st edition 1982
A Division of Plenum Publishing Corporation
233 Spring Street, New York, N.Y. 10013

To My Sons
Davide and Peter Christian

PREFACE

During October 18-31, 1980, the first course of the International School of Pure and Applied Biostructure, a NATO Advanced Study Institute was held at the "Ettore Majorana Center for Scientific Culture" in Erice, Sicily, co-sponsored by national and international agencies. The subject of the course was "Cell Growth", with participants (from 16 different countries) selected worldwide.

The study of cell growth has been one of humanity's most challenging problems and it has been approached from many different points of view, such as biochemistry, genetic engineering, cell biology, zoology, oncology, immunology, biophysics and a few other fields. It has been very difficult to keep such varied points of view all in one room and in one audience, because of the heterogeneity of background and inherent difficulty of communication, with occasional nominalistic rather than factual debates. This Institute aimed to bypass those limitations by approaching in a structured and tutorial fashion the problem of cell growth in three dimensions: (1) in terms of the various disciplines involved, from molecular to cellular biology, from genetic engineering to clinical oncology, from biophysics to immunology; (2) in terms of the system studied, from prokaryotes to eukaryotes and cancer cells; (3) in terms of the various levels of macromolecular organization, from membrane to cytoskeleton and chromatin. The emphasis has been placed in the basic sciences, which in the long term constitute the only route to a complete understanding of the molecular-cellular mechanisms regulating growth, differentiation and ultimately cancer.

This book is the result of this Advanced Study Institute and aims to present a structured and widely interdisciplinary view of the current knowledge on normal and abnormal cell growth, up to the very recent findings. Limited space is also left to the presentation of controversial opinions (not necessarily shared by the editor) to permit an unbiased update of the state of the art. The book has been edited in a tutorial format, with the

cooperation of several leading scientists. We wish to express our gratitude to Frank Kendall and Wolfried Linden for their invaluable and critical cooperation prior, during and after the Institute and publication of this volume.

Claudio Nicolini

INTRODUCTION

A Perspective View of the Cell Cycle

Daniel Mazia

Hopkins Marine Station of Stanford University

Pacific Grove, California 93950, U.S.A.

The intensive study of the Cell Cycle has a 30-year history, during which it has grown from an activity of a few cell biologists into an industry. Having learned a good deal by pursuing some rather simple assumptions, the field is ready to replace answers with questions, even to go so far as to ask: "What are the real problems of the Cell Cycle?" One felt this in discussions that filled our days on a mountain-top in ancient Sicily.

The modern history of the field began with and remains dominated by a single discovery, made around 1950. A fact simple in itself - that DNA doubled between cell divisions - gave us the S-phase. It followed from early results that entrance into the S-phase was a commitment to future division. The standardization of the autoradiographic technique, especially after the introduction of tritiated thymidine, defined and conventionalized the phases of the cycle, although they had no inherent content but merely said that in some cells (which became "typical" for no other reason) something happened between the previous division and the beginning of S and something happened between the end of S and the next division. The central problem of the Cell Cycle became the search for the events (so far eluding us) which command or allow the cell to enter S. This emphasis is justified by the best of reasons: it promises a key to the understanding of order or anarchy in the production of cells in higher organisms. However, cell biologists who want to know how one cell becomes two cells have to deal with problems other than the initiation of the replication of DNA and may even find fruitful approaches to the latter problem in considering the cell cycle as a whole.

In the oldest thought on cell reproduction, the greatest emphasis was placed on the Cell Cycle as a sequence of doubling and halving of cell size. The prevailing view was that the completion of the growth of a cell at a "critical mass" was the trigger to cell division. The old idea is easily dismissed by appeal to exceptions but embodies some deep questions of the Cell Cycle: how do we explain the upper limit of cell mass that can be serviced by one nucleus; how can a cell sense its size; what are the special events that take place at the boundary between interphase and mitosis? Current work, especially on yeast cells, reopens the question of a causal relation between cell size and the onset of division. Other work, of which much more is needed, searches for molecules which are synthesized at the end of G2 (when there is a G2 phase).

Now some advances in the rapidly growing field of the structure of chromatin are very promising; we can expect to know the difference between chromatin in interphase and chromatin in mitosis; after all, the condensation of chromatin is the first sign of mitosis. Thus, the early history of the study of the Cell Cycle gave us the two revolutionary events which define the cycle: the initiation of chromosome replication and the initiation of mitosis.

In the discussions at Erice, we were prepared to question the reality of the conventional phases of the Cell Cycle. Considering all of the events by which one cell becomes two, there is no rule that requires that certain ones must take place between division and the beginning of S and others must take place between the end of S and the onset of mitosis. One can purify the paradigm of the Cell Cycle by focussing on what may be called the Reproductive Wheel. In one turn of the Reproductive Wheel, one nucleus makes two equivalent and separated nuclei and the cell is divided. We would say that one cell has made two cells whether or not the cell has grown. We would then see growth in the usual cycle (doubling and halving) as the augmentation of cellular structure and biochemical machinery that supports the turn of the Reproductive Wheel. If a cell has a G1 period, it would only be saying that some events needed for continuing into chromosome replication take place - in that kind of cell - during a period following division. If there is no G1 period, the activities required for starting S took place in the previous generation. If there is no G2, all of the events necessary for entrance into mitosis were completed before the end of S. The events included in cell growth impose the limits on the reproductive events, but they need not follow similar schedules in different kinds of cells.

It is instructive to examine a special kind of cell in which the cell cycle is liberated from restraints of growth: the egg cell. In such cells, it may take as little as 10 minutes to double the number of nuclei. The Reproductive Wheel can turn that rapidly when it is fully supplied with all that it needs. We do not, however, say that such a cycle is independent of events which, in other cells, fall within G1 and G2. On the contrary, we recognize that the maturation of the egg, which may have taken weeks or months, has "fattened" the huge cell with those products which other cells need time to make within a period between divisions, some before the onset of S, others between the end of S and the start of mitosis. What we want to know is the events or molecules which determine the turning of the Reproductive Wheel. The conventional phases of the cycle can help us in working on particular types of cells but will disappoint us if we generalize them.

The Reproductive Wheel expresses itself above all in the Chromosome Cycle. What we can determine easily is that the chromosomes decondense at the time of division; replicate between divisions; begin to condense at the beginning of mitosis; split (separating sister chromosomes) at the end of metaphase; move to the poles and decondense for the next cycle. Some of the chapters in this volume touch on an idea that links the Chromosome Cycle even more closely to the Cell Cycle as a whole. There is evidence that chromosomes decondense gradually (in cells which have a G1) as a cell passes from division into the next interphase. It is possible to speculate that the chromosomes must reach a certain degree of decondensation before they can replicate and that they cannot condense again until they have replicated. If we could consider the Chromosome Cycle to be the foundation of the Cell Cycle, it would provide an intelligible focus for the study of the infinity of events which make up the description of the life history of the cell.

The reader will find in this volume a great wealth of new knowledge about the growing and reproducing cell, viewed at all levels from minutely molecular analysis to the consideration of cell populations. The organizers of the School at Erice aimed beyond the communication of the results of single research efforts toward criticism and judgment. One hopes that the spirit of the School will come through in these pages.

CONTENTS

SECTION I: WHAT IS A CELL?

SECTION II: CELL PROBES

SECTION I:
WHAT IS A CELL?

STRUCTURE OF THE EUKARYOTIC CELL

Nanne Nanninga

Department of Electron Microscopy and Molecular Cytology

University of Amsterdam, Amsterdam, The Netherlands

INTRODUCTION

The purpose of this tutorial review is to present some recent advances on the structure and origin of the eukaryotic cell. There are many types of eukaryotic cells and, therefore, some abstraction is unavoidable. In addition, nowadays cell biologists have a clear preference for animal cells as compared to plant cells. Within the group of animal cells again a limited number of cell types are being studied. These limitations should be kept in mind.

I have attempted to avoid merely cataloguing a number of cellular structures. When looking back into the past, i.e. the last twenty five years it is possible to find a common denominator. In many instances the picture one has in mind of the eukaryotic cell is largely determined by the electron microscopic technique used. As will be shown below, a technique that allows in particular the visualization of membranes will lead to a cell concept in which the occurrance of membrane-bounded compartments is stressed. By contrast, a technique that preserves the structures inbetween the membranes will lead to emphasis on the non-membranous cellular skeleton. It is within this framework that I would like to present the material.

HISTORICAL ASPECTS

A fascinating period in the study of cells has been the passing of the border between microscopic and submicroscopic structures.

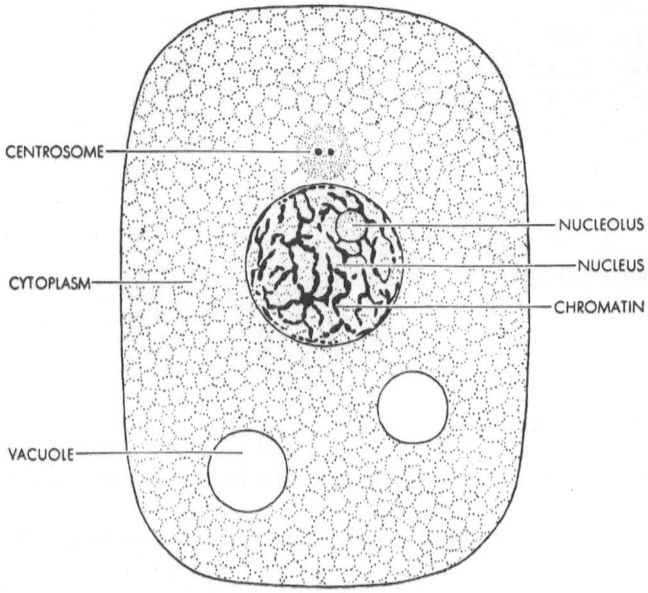

Fig. 1. Cell structure typical for the pre-electron microscopic era (1922). From "The Living Cell" by J. Brachet, Scientific American, September 1961.

This can be clearly seen from the pioneering work of Frey-Wyssling (1, 2). An illustrative example is the evolution (or involution) of a concept like groundplasm: "What is left over after removal of all known particulate bodies of the cytoplasm is a homogeneous mass as seen in the electronmicroscope which we call groundplasm" (2). An old scheme of the cell shows an ill-defined texture between nucleus and cell boundary (Fig. 1). Various investigators have stressed the fibrillar, granular and/or reticular aspect of the groundplasm. Progress was hampered by the limited resolution of the light microscope and the often unindentified occurrance of preparation artefacts. This state of affairs was continued with the emergence of electron microscopy, though at a higher resolution. Typical examples are the following. Bretschneider (3) advocated a detailed reticular concept (Fig. 2a), Frey-Wyssling (1) considered globular macromolecules as elementary units. These would associate (theory of junctions; in German "Hafpunkt-Theorie") to form fibrillar elements, macromolecular films or three-dimensional porous textures of the plasma gel (Fig. 2b). A now strikingly modern concept was obtained by Haguenau and Bernhard (4) in 1952 (Figs. 3a and b).

Improved electron microscopic techniques brought into focus the lamellar aspect of the ground and cytoplasm. Important advances were achieved by Sjöstrand and coworkers (e.g., ref. 5).

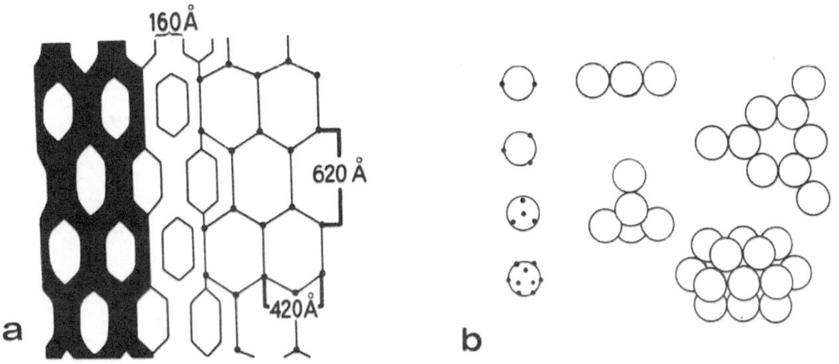

Fig. 2. Hypothetical structure of the groundplasm. (a) detailed proposition of Bretschneider (3) as based on chemical experiments; (b) theory of junctions of Frey-Wyssling (1). Globular macromolecules aggregate by junctions. From top till bottom: coordination number 2, beaded chain; coordination number 3, porous film; coordination number 4, tetrahedral space group; coordination number 12, close packing.

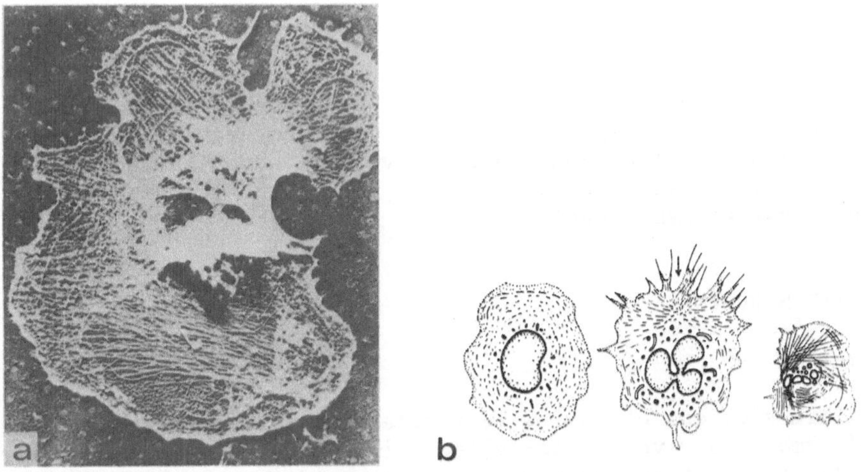

Fig. 3. (a) Fibrillar structure of the groundplasm in a blood platelet (4). Fixation with chromic acid; (b) arrangement of submicroscopic fibrils in (from left to right) a leukocyte, a multinuclear granulocyte and a blood platelet.

The lamellar or membranous aspect of the cell was especially prominent after application of potassium permanganate as a fixative. Membranes became visible as triple-layered structures (unit membrane

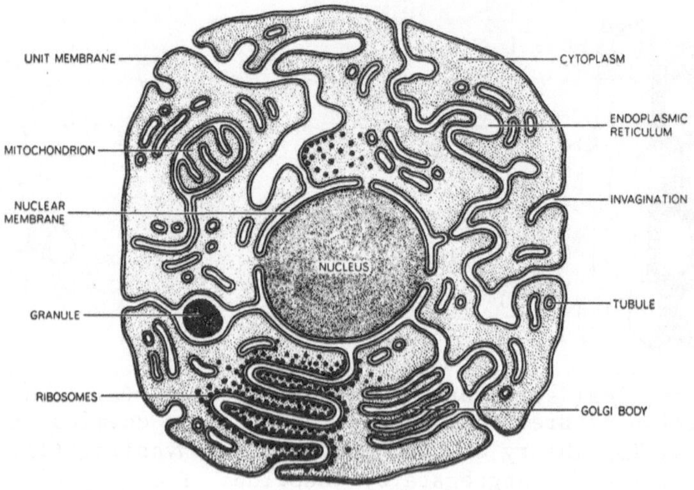

Fig. 4. A cell composed of membrane-bounded compartments (6). Diagram from "The Membrane of the Living Cell" by J.D. Robertson, Scientific American, April 1962.

concept) and the cell was conceived of as an entity divided by membranes (Robertson (6); Fig. 4). This model has been extremely stimulating for further research on cell structure. Subsequent investigations clarified the structure of the unindentified spaces between the membranes. This development depended heavily on the introduction of glutaraldehyde as a fixative (7). Note that potassium permanganate is very destructive with respect to cellular preservations (except membranes).

MEMBRANE STRUCTURE

Asymmetry

Thin sections of osmium tetroxide or potassium permanganate fixed membranes reveal a so-called triple-layered structure: electron dense, electron translucent, electron dense. It has often been observed that the two electron dense layers differ in appearance. This points to a difference in chemical composition. Morphologically, this is most clearly observed in freeze-fractured membranes. With this technique (Fig. 5) frozen membranes are split along the hydrophobic interior (8), revealing two complementary leaflets (Fig. 6; e.g. ref. 9). The nature of the particles is generally proteinaceous, whereas it is far less clear whether an observed particle is fractured and/or plastically deformed. However, the distribution of the particles over the two leaflets clearly reflects the asymmetry of the membrane.

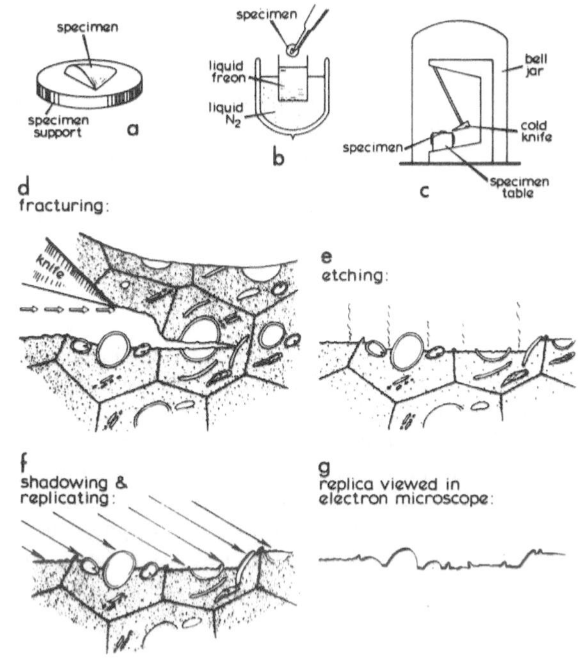

Fig. 5. Pictorial representation of the freeze-fracture technique (12).

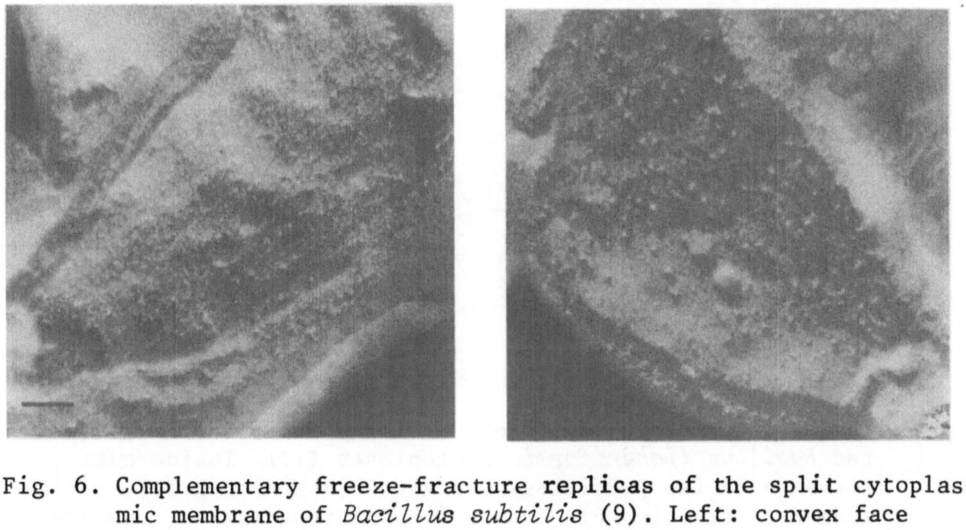

Fig. 6. Complementary freeze-fracture replicas of the split cytoplasmic membrane of *Bacillus subtilis* (9). Left: convex face with many particles; right: concave face with few particles. Bar: 0.1 µm.

Inside-out

The freeze-fracturing technique has also been useful in demonstrating the existence of inside-out vesicles. Originally this was shown for fragmented erythrocyte membranes (10). An example of bacterial membranes (11) is shown in Fig. 7. Inside-out vesicles have a convex face with few particles and a concave face with numerous particles. It has often been noticed that the convex fracture face carries more particles than the complementary concave one (12).

Phase separation

The mobility of the fatty acid chains in membrane phospholipids can be reduced by lowering the temperature. Eventually the chains may become tightly packed while pushing the proteins away. The random distribution of particles as seen in freeze-fracture faces is changed into a pattern with smooth phospholipid areas surrounded by particles (Fig. 8; ref. 13). The separation in phases shows the lateral mobility of membrane components. This temperature dependant phenomenon should not be confused with patching and capping (see below).

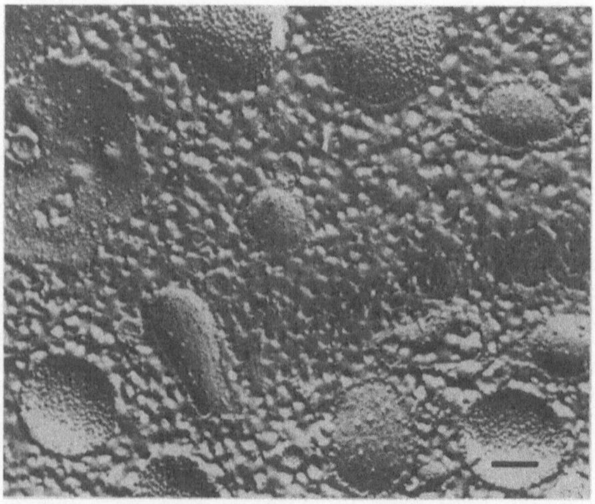

Fig. 7. Inside-out and right side-out vesicles derived from fragmented *Bacillus licheniformis* protoplasts (11). Inside-out: convex with few particles, concave with many particles; right side-out (cf. Fig. 6): convex with many particles, concave with few particles. Bar: 0.1 μm.

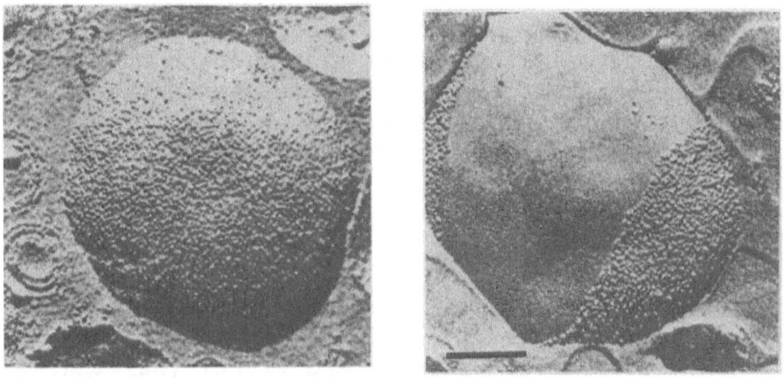

Fig. 8. Phase separation in the cytoplasmic membrane of *Escherichia coli* (13). Left: membrane at 45°C; right membrane at 0°C. Bar: 0.2 µm.

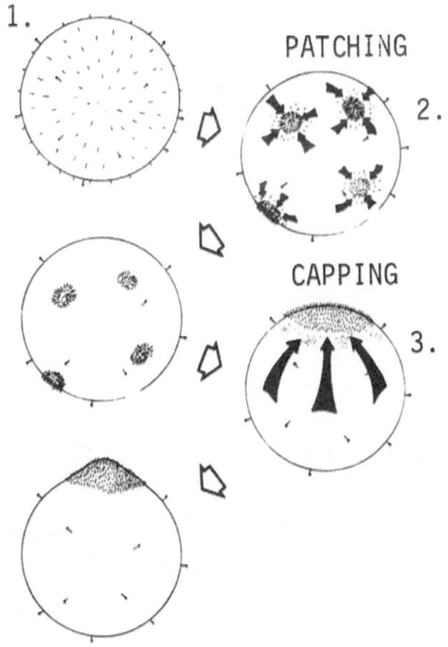

Fig. 9. Lectin-induced lateral movement of plasma membrane components. Component (1) retains its original distribution, whereas component (2) coalesces to form patches and eventually one polar cap (diagram from ref. 14).

Patching and Capping

Plasma membranes contain proteins which carry sugar chains (glycoproteins). Generally such sugar chains occur on the cell exterior. Lectins may bind to the sugars. Because of the lateral mobility of the glycoprotein lectin-induced patches arise. Eventually the patches may fuse to form a cap (Figs. 9 and 10).

Cell Fusion

Another example of membrane plasticity is cell fusion. Cells from different species may be fused with the aid of Sendai virus or simply in the presence of polyethylene glycol (Baserga, this volume). Cell surface markers of the respective membranes intermingle quickly afterwards.

Fluid Mosaic Membrane

The above mentioned aspects fit into a model in which proteins are virtually floating into a bilayer of phospholipids (ref. 15; Fig. 10). A useful research object has been and still is the easily accessible erythrocyte (15). With respect to the proteins the operational distinction between integral (intrinsic) and peripheral (extrinsic) components has been made. Integral proteins can be removed with detergents, whereas peripheral are easily detached by salt and/or EDTA treatment. An integral protein like glycophorin

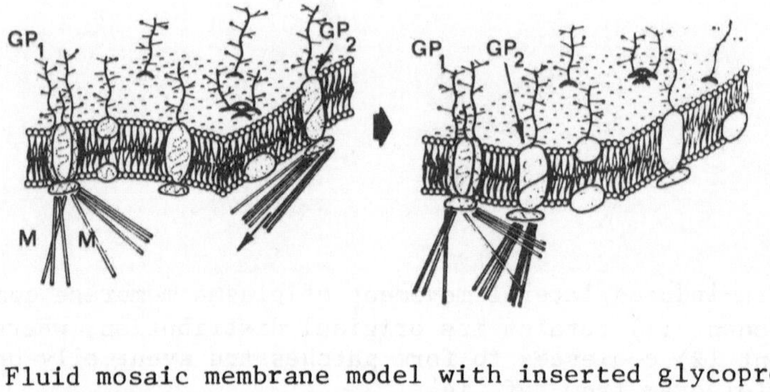

Fig. 10. Fluid mosaic membrane model with inserted glycoprotein molecules (GP) and association with microfilament-microtubule cytoskeletal assemblage (M). Shown is the lateral mobility of GP2 (from ref. 15).

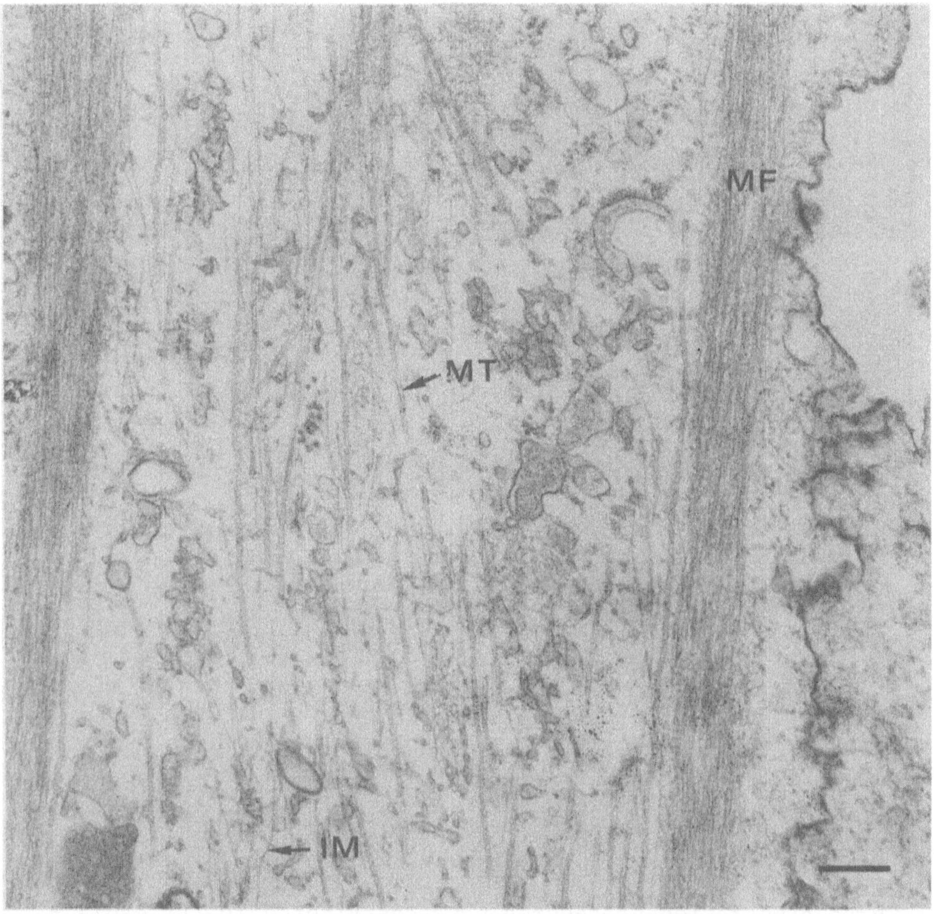

Fig. 11. Cytoskeletal components in a cultured 3T3 fibroblast. Fixation with glutaraldehyde, osmium tetroxide and tannin. MF, microfilament; IM, intermediate filament; MT, microtubule. Bar: 0.2 μm. (Courtesy D. Mesland).

projects into the cell environment with sugar chains attached (Fig. 10). At the cytoplasmic part of the molecule contact is presumably made with a contractile cytoplasmic skeleton (cytoskeleton; see below). The model thus suggests a phospholipid framework into which proteins are inserted. Yet an alternative view may be considered. When phospholipids are extracted with Triton X-100 from fibroblasts the overall cell shape is maintained (17). Close at the original location of the cell membrane a presumably proteinaceous lamina seems to be present. This leads one to the idea of a protein

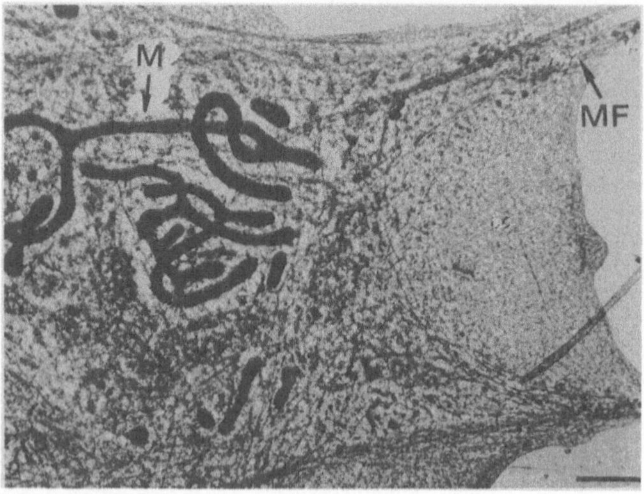

Fig. 12. Whole mount critical point-dried 3T3 fibroblast showing cellular filaments. MF, microfilaments. Note thread-like mitochondria (M). From ref. 21. Bar: 1 μm.

matrix into which phospholipids are organized as a bilayer structure. For some recent reviews see e.g. refs. 18 and 19.

CYTOSKELETON

Three preparation procedures have contributed to the evolvement of the concept of the cytoskeleton. The use of glutaraldehyde for fixation led to the detection of filamentous structures of various thicknesses in the thin-sectioned cytoplasm (Fig. 11). In whole mount freeze-dried (20) or critical point dried cells (Fig. 12; ref. 21) such structures have likewise been observed. Last but not least an impressive contribution has been made on the light microscopic level through the visualization of fluorescently labelled cytoplasmic filaments. In the latter method cells are made permeable for fluorescently labelled antibodies which bind to particular components in the cell. In the case of actin the antibody may be replaced by heavy meromyosin or its subfragment. These latter two components are obtained by proteolytic digestion of myosin. Alternatively actin filaments may be seen by electron microscopy for the binding of heavy meromyosin or its subfragment to actin produces a typical arrow head appearance (decoration; Fig. 13; ref. 22).

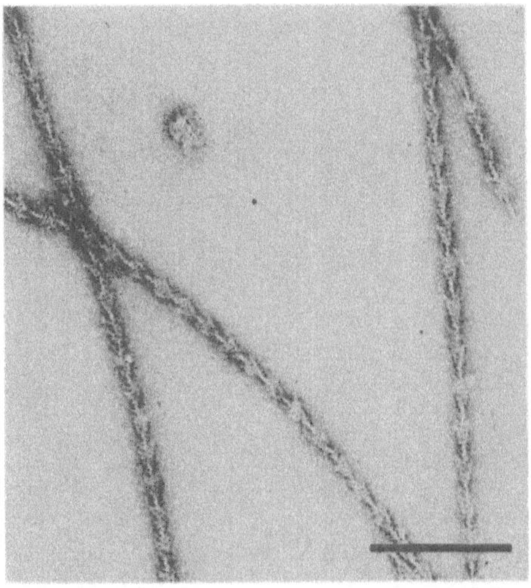

Fig. 13. Actin filaments decorated with a subfragment of heavy meromyosin. Negative staining with 1% uranyl acetate. From ref. 22. Bar: 0.1 μm.

At present three categories of constituants of the cytoskeleton can be distinguished: microfilaments, intermediate filaments and microtubules. The various structures differ with respect to diameter, chemical composition and function.

Microfilaments

Microfilaments have a thickness in the order of 7 nm and are mainly composed of polymerized actin with a molecular weight of 42,000 daltons (for a review, e.g. ref. 23). Though originally considered as a specific component of muscle, actin has now been demonstrated at many cellular sites (Fig. 11). The distribution of microfilaments is especially well visible by immunofluorescence of whole cells (Fig. 14). This allows one to follow the microfilament pattern when cells proceed through the division cycle or when they move (Fig. 15). Though one should be careful to extrapolate to other cell types it is to be expected that a general picture on the dynamics of microfilament distribution will emerge in the near future.

Intermediate filaments

These filaments have a thickness in the order of 10 nm (Fig.

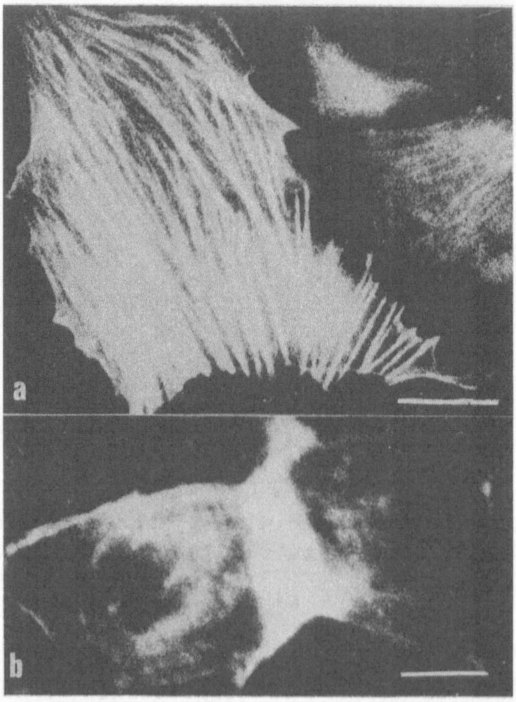

Fig. 14. (a) Actin filaments decorated with fluorescently-labelled heavy meromyosin. Interphase chick fibroblast cell;(b) telophase cell of a rat kangaroo. Actin is concentrated in the cleavage furrow. Decoration with fluorescently-labelled heavy meromyosin subfragment. Bars represent 10 and 5 µm, respectively. From ref. 23.

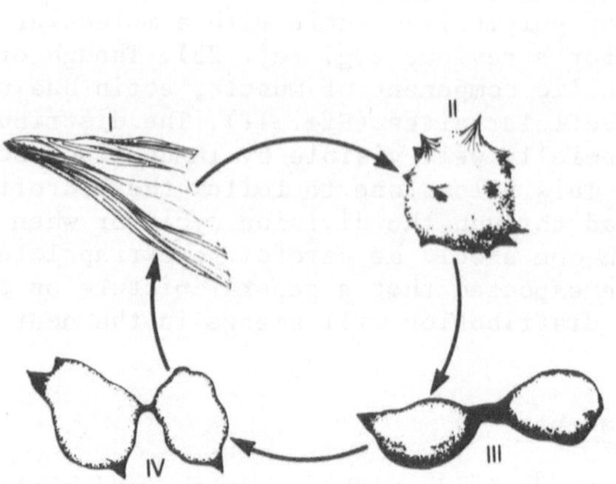

Fig. 15. Shape alterations of the cell and distribution of actin during the division cycle of chick fibroblasts (23).

11). There are now several proteins known with molecular weights roughly in the order of 50,000 daltons which may form intermediate filaments. As a group the intermediate filaments seem far less homogeneous than the microfilaments. The distribution of for instance prekaratine has been studied by indirect immunofluorescence in PtK2 cells (ref. 24; Fig. 16). The localization of desmin, α-actinin and vitmentin has been studied in chicken skeletal muscle (25). Vimentin filaments have also been observed as perinuclear rings in chicken fibroblasts (26). In the above mentioned experiments immunofluorescence on the light microscopic level played a vital role. From a general point of view intermediate filaments may be considered as "mechanical integrators of cellular space" (27).

Microtubules

Microtubules are hollow tubes with a diameter of about 25 nm. They are composed of dimers of α- and β-tubulin, each having a molecular weight of about 120,000 daltons (for reviews see, e.g. refs. 23 and 28). They are the main component of the mitotic spindle and centriole. They occur in basal bodies and cilia and also run through the cytoplasm. Microtubules are typical structures which were detected after better preservation of the cytoplasm by glutaraldehyde (Fig. 11). As with the microfilaments and intermediate filaments the

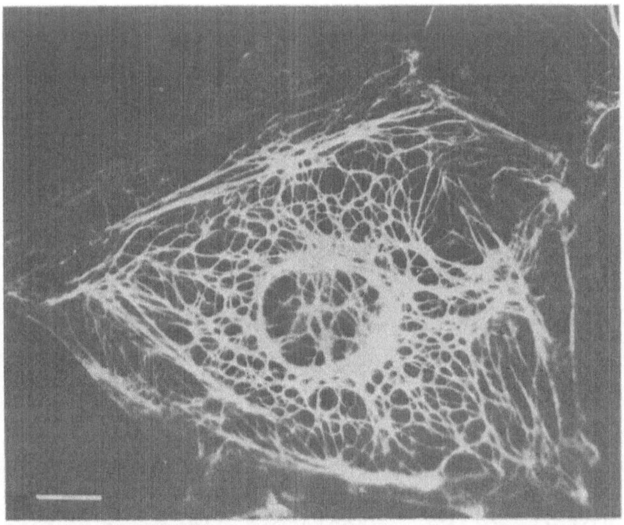

Fig. 16. Immunofluorescence microscopy of prekaratin arrangement in a cultured rat kangaroo PtK2 cell. Antibodies had been raised against cow hoof prekaration. From ref. 24. Bar: 10 μm.

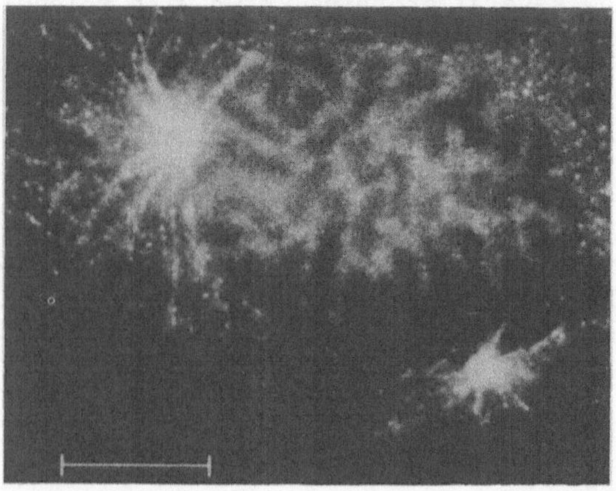

Fig. 17. Immunofluorescence microscopy with antitubulin antibody in a prophase 3T3 cell. Two MTOCs are visible. From ref. 29. Bar: 10 μm.

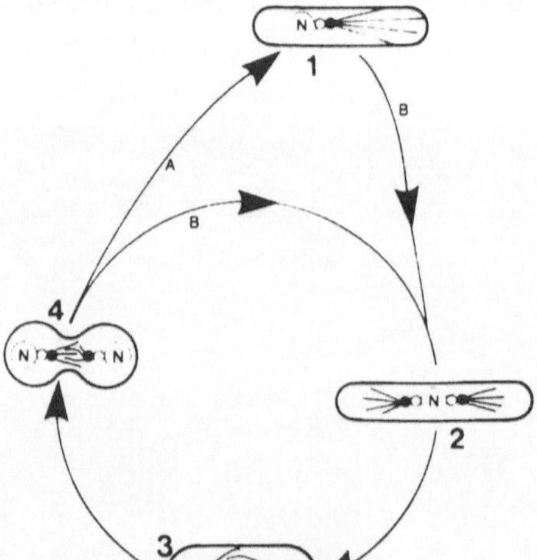

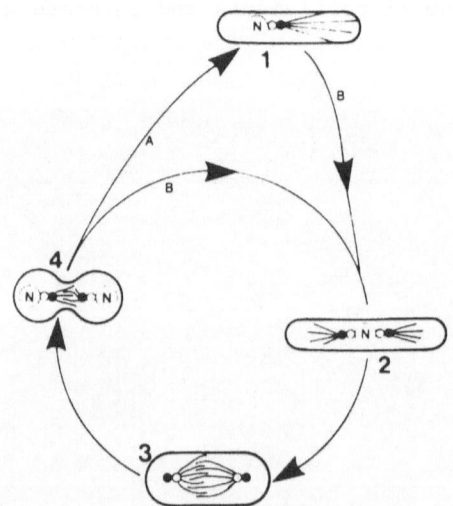

Fig. 18. Diagram to illustrate the microtubule cycle. (1) Cytoplasmic microtubules; (2) commitment to divide (B) involves replication (activation of MTOCs (oo); (3) spindle component (o) is active, cytoplasmic component (o) is inactive during nuclear division; (4) cytoplasmic microtubules participate in cytoplasmic division. Daughter cells continue to grow (A), or re-enter the division cycle (B). From ref. 30.

cellular arrangement could be especially deduced by application of fluorescently labelled antibodies (Fig. 17). Special attention was given to microtubule organizing centers (MTOC's; Fig. 18). Upon mitosis microtubules radiate from the MTOC to and along the metaphase chromosomes. After mitosis they disappear and subsequently radiate into the cytoplasm in the opposite direction.

In summary, the cytoskeleton emergences as a highly dynamic structure. Comparision with Fig. 4 shows how the groundplasm has been filled in.

MITOCHONDRIA AND CHLOROPLASTS

Though it is not my aim to give a complete list of cytoplasmic inclusions mitochondria deserve some attention. Their name describes them as threadlike structures. However, under the influence of their image seen in electron microscopic thin sections a very schematic picture came to the foreground: a comparatively short cylindrical body with hemispherical caps. Serial sections from which mitochondrial shapes were reconstructed led to the dramatic conclusion that different sectioned mitochondrial profiles belonged to one and the

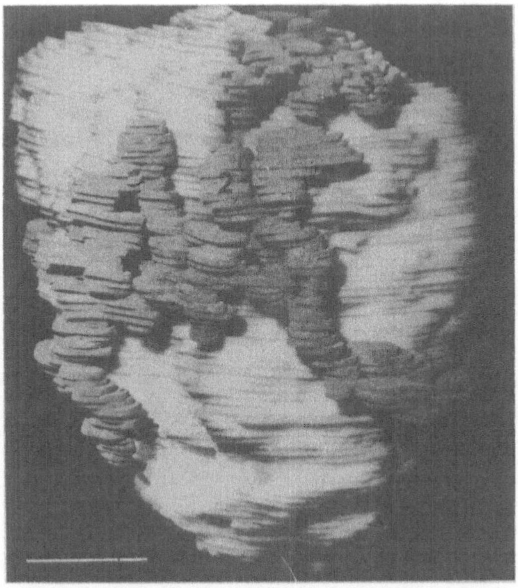

Fig. 19. Tubular mitochondrial network surrounding the nucleus of a mouse lymphocyte. The model is based on ultrathin serial sections. From ref. 31. Bar: 1 μm.

same organelle (Figs. 19 and 20). A similar conclusion has been drawn for chloroplasts (32). In addition to serial sectioning the spatial organization of mitochondria could also be followed in whole mount preparations (Fig. 12). This insight reconciled the early light microscopic observation with modern electron microscopic ones. Finally some remarks on ribosomes. These ubiquitous cytoplasmic ribonucleoprotein particles have been studied extensively (for a review see, e.g. ref. 33). However, relatively little attention has been paid to their arrangement inside the cell. Recently, the association of polyribosomes with a skeletal framework has been stressed (34). Older literature can be found in ref. 35.

THE NUCLEUS

The nucleus now attracts the attention of many investigators. Analysis of its enigmatic organization received a new impetus from the discovery of the nucleosome (36). The nucleosome may be considered as a basic building stone of chromatin, i.e. the nucleoprotein

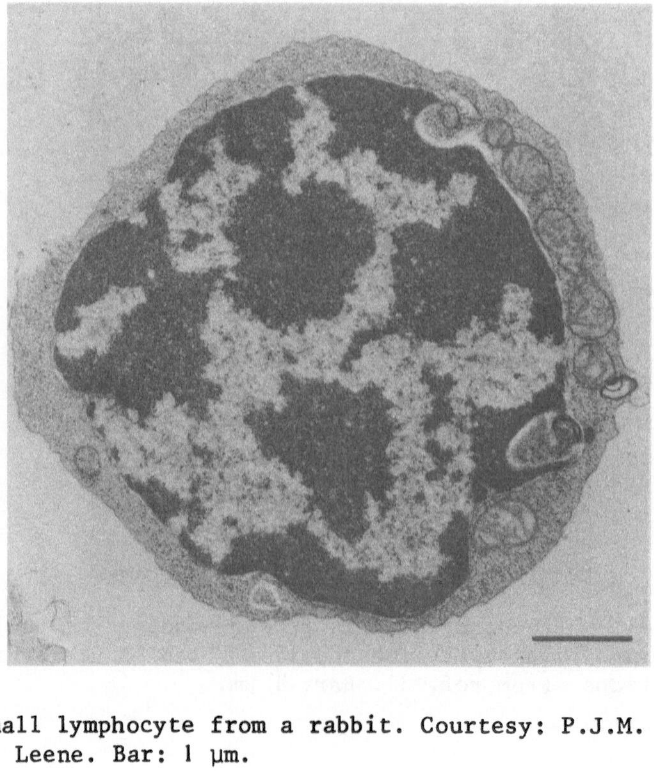

Fig. 20. Small lymphocyte from a rabbit. Courtesy: P.J.M. Roholl and W. Leene. Bar: 1 μm.

structure of the nucleus (for details see elsewhere in this volume). To which extent nucleosome structure is maintained during DNA-replication or transcription is likely to emerge in the very near future.

Nucleosomes are packed into higher orders of organization, the exact nature of which has still to be clarified. Examples of such organizational levels are euchromatin, heterochromatin and metaphase chromosomes. The loosely organized euchromatin is supposed to be active in transcription, whereas heterochromatin is not. The latter is often observed as compact electron dense patches alongside the nuclear membrane (Fig. 20). A specialized region in the nucleus is the nucleolus; here ribosomal RNA is transcribed. Well known is the arrangement of transcribed RNA as feather-like structures, first described by Miller et al. (37) for spread preparations.

Metaphase chromosomes

Apart from the distinction loose euchromatin and compact heterochromatin, there is the dispersed interphase chromosome organization as contrasted to the condensed well delineated metaphase chromosome.

Thin sections of artificially swollen chromosomes reveal a central axis from which fibrils radiate (Fig. 21; ref. 38). The length and thickness of the fibrils depend at least in part on the ionic environment. When histones are removed a recognizable skeleton of the metaphase chromosome remains (Fig. 22; ref. 39). The skeleton predominantly contains non-histon protein and DNA can be seen radiating out. When the DNA is more extensively spread loops are observed with a length in the order of 10 to 30 μm. It remains to be established whether or not this size fortuitously resembles that of a replicon (40). The above observations have been fitted into a radial-loop model of the metaphase chromosome (38, 39). See further the other contributions in this volume.

THE NUCLEAR MATRIX

When isolated nuclei are subsequently extracted in the presence of low magnesium, 2M NaCl, detergent and nucleases a histon, phospholipid and nucleic acid depleted structure remains. This structure still possesses the dimensions of the nucleus (Fig. 23; ref. 41). The nuclear matrix may be considered as the nucleoplasmic counterpart of the cytoskeleton. So far, however, characteristic protein constituants have not been purified. A skeleton that remains after extraction might represent a real cytological feature. On the other hand, the skeleton might be composed of substances the location of which was originally determined by the extracted components. The same reasoning applies to the metaphase chromosomal skeleton. Nevertheless, whether significant or not, the two skeletons provide concepts stimulating for further research.

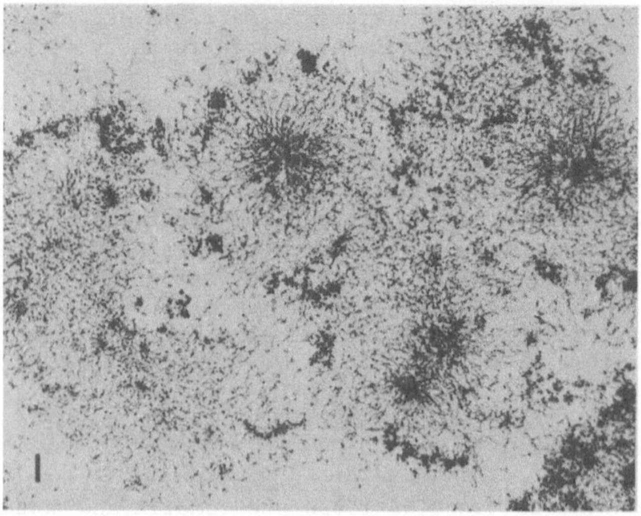

Fig. 21. Cross sections through EDTA swollen HeLa metaphase chromosomes. From ref. 38. Bar: 0.2 μm

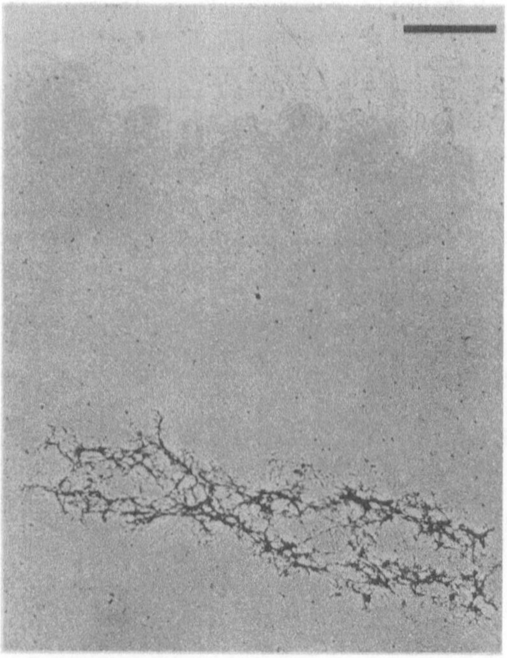

Fig. 22. Histon depleted HeLa metaphase chromosomes. The chromosomes have been treated with a mixture of dextran sulfate and heparin. From ref. 39. Bar: 2 μm.

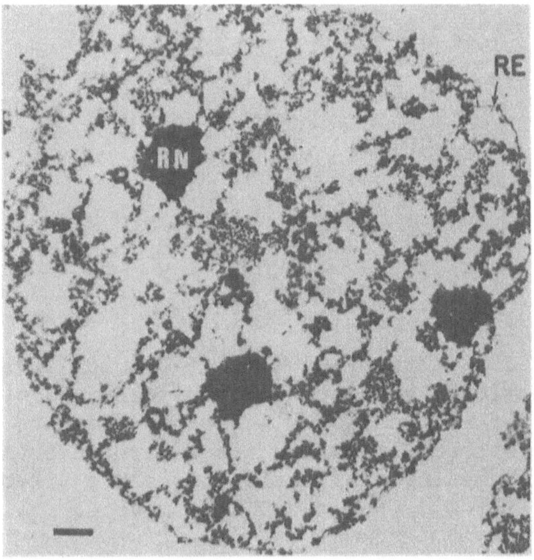

Fig. 23. Nuclear matrix derived from extracted rat liver nuclei. RE, residual envelope; RN, residual nucleolus. From ref. 41. Bar: 0.5 µm.

REPLICON ORGANIZATION

It is well known that DNA replication starts at many sites (origins) and proceeds bidirectionally (40). Multiple replication sites become visible after disruption of chromosomal structure. A sequence of linearly organized replication sections (replicons) with moving replication apparatuses (Fig. 24a) might be for this reason a too simple abstraction. An alternative is to assume fixed replication sites (42) which are grouped together (43) as indicated in Fig. 24b. Such an arrangement in which DNA is the moving agent might favor the coordination between the various replication sections.

CONCLUDING REMARKS

In the foregoing I have drawn some lines from past to present on eukaryotic cell structure. The presentation started with a description of the enigmatic groundplasm. Being elusive to the light microscope, with electron microscopic techniques still in their infancy, ample opportunities for speculation existed (cf. Figs. 2a and 2b). Further improvements in chemical fixation, embedding and thin

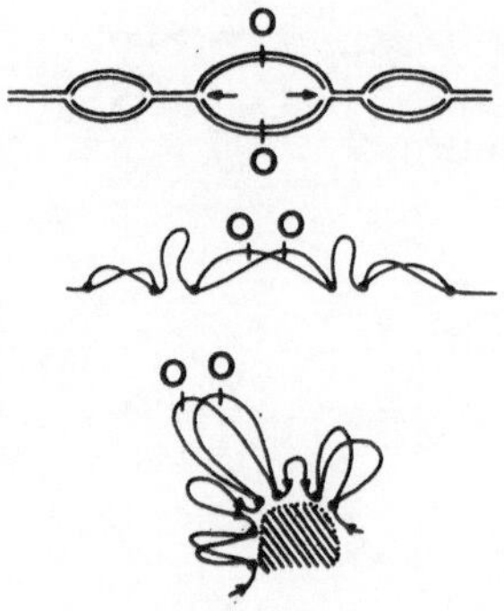

Fig. 24. (a) Model of Huberman and Riggs (40) with several DNA replication sections; each section replicating bidirectionally. O, origin of replication. The arrows indicate moving replication forks; (b) schematic aid to arrive at structure (c). In this scheme the replication forks are fixed and grouped together (42, 43). Now the DNA is moving.

sectioning brought the lamellar aspect of the groundplasm to the foreground. This culminated in a cell model dominated by membrane-bounded compartments (Fig. 4). Subsequently, the spaces between the membranes have been filled in (cytoskeleton). Remarkable is the come back of the light microscope in the field of immunofluorescence microscopy.

I am indebted to C.L. Woldringh for reading the manuscript, to J. Woons for preparing the figures and to C.E.A. van Wijngaarden-Pater for help in preparing the manuscript.

1. Frey-Wyssling, A. (1953) Submicroscopic Morphology of Protoplasm (Elsevier Publishing Company, Amsterdam).
2. Frey-Wyssling, A.(1957) Macromolecules in Cell Structure (Harvard University Press, Cambridge, Mass.).
3. Bretschneider, L.H. (1950) Proc. Acad. Sci. Amsterdam C53, 1476-1489.
4. Haguenau, F. & Bernhard, W. (1952) Exp. Cell Res. 3, 629-639.
5. Sjöstrand, F.A. & Hanzon, V. (1954) Exp. Cell Res. 7, 393-346.
6. Robertson, J.D. (1959) in The Structure and Function of Subcellular Components, ed. Crook, E.M. (Cambridge University Press, Cambridge), Vol. 16, pp. 1-43.
7. Sabatini, D.D., Bensch, K. & Barnett, R.J. (1963) J. Cell Biol. 17, 19-58.

8. Branton, D. (1966) Proc. Natl. Acad. Sci. USA 55, 1048-1056.
9. Nanninga, N. (1971) J. Cell Biol. 49, 564-570.
10. Steck, T.L., Weinstein, R.S., Straus, J.H. & Wallach, D.F.H. (1970) Science 168, 225-257.
11. Wientjes, F.B., van 't Riet, J. & Nanninga, N. (1979) Biochim. Biophys. Acta 553, 213-223.
12. Branton, D. & Deamer, D.W. (1972) Membrane Structure (Springer Verlag, Wien).
13. Letellier, L. & Schechter, E. (1976) J. Microsc. Biol. Cell. 25, 191-195.
14. Hopkins, C.R. (1978) Structure and Function of Cells (W.B. Saunders Company Ltd., London).
15. Nicolson, G.L. (1976) Biochim. Biophys. Acta 457, 57-108.
16. Steck, Th. L. (1974) J. Cell Biol. 62, 1-19.
17. Ben-Ze'ev, A., Duerr, A., Solomon, F. & Penman, S. (1979) Cell 17, 859-865.
18. Wickner, W. (1979) Annu. Rev. Biochem. 48, 23-45.
19. Op den Kamp, J.A.F. (1979) Annu. Rev. Biochem. 48, 47-71.
20. Heuser, J.E. & Kirchner, M.W. (1980) J. Cell Biol. 86, 212-234.
21. Temmink, J.H.M. & Spiele, H. (1980) J. Cell Science 41, 19-32.
22. Spudich, J.A., Huxley, H.E. & Finch, J.T. (1972) J. Mol. Biol. 72, 619-632.
23. Adelstein, R.S., Scordilis, S.P. & Trotter, J.A. (1979) Meth. Achiev. Exp. Pathol. 8, 1-41.
24. Weber, K., Osborn, M. & Franke, W.W. (1980) Eur. J. Cell Biol. 23, 110-114.
25. Granger, B.L. & Lazarides, E. (1978). Cell 15, 1253-1268.
26. Blose, S. (1979) Proc. Natl. Acad. Sci. USA 76, 3372-3376.
27. Lazarides, E. (1980) Nature 283, 249-255.
28. Sanger, J.W. & Sanger, J.M. (1979) Meth. Achiev. Exp. Pathol. 8, 110-142.
29. Tucker, R.W., Pardee, A.B. & Fujiwara, K. (1979) Cell 17, 527-535.
30. Lloyd, C. (1979) Trends in Biochem. Sciences 4, 187-189.
31. Rancourt, M.W., McKee, A.P. & Pollack, W. (1975) J. Ultrastruct. Res. 51, 418-424.
32. Pellegrini, M. (1980) J. Cell Sci. 43, 137-166.
33. Wool, I.G. (1979) Annu. Rev. Biochem. 48, 719-754.
34. Fulton, A.B., Wan, K.M. & Penman, S. (1980) Cell 20, 849-857.
35. Nanninga, N (1973) Int. Rev. Cytol. 35, 135-188.
36. Olins, A.L. & Olins D.E. (1974) Science 183, 330-332.
37. Miller, O.L. Jr. & Beatty, B.R. (1969) Science 164, 955-957.
38. Marsden, M.P.F. & Laemmli, U.K. (1979) Cell 17, 849-858.
39. Paulson, J.R. & Laemmli, U.K. (1977) Cell 12, 817-828.
40. Huberman, J.A. & Riggs, A.D. (1968) J. Mol. Biol. 32, 327-341.
41. Berezny, R. & Coffey, D.S. (1977) J. Cell Biol. 73, 616-637.
42. Dingman, C.W. (1974) J. Theor. Biol. 43, 187-195.
43. Pardoll, D.M., Vogelstein, B. & Coffey, D.A. (1980) Cell 19, 527-536.

CELL AND CONTRACTILE PROTEIN EVOLUTION

P. Omodeo
Istituto di Biologia Animale, Università di Padova (Italy)

E. Capanna
Istituto di Anatomia Comparata, Università di Roma (Italy)

V. Pallini
Istituto di Zoologia, Università di Siena (Italy)

This paper deals with the significance of the contractile proteins in the evolution of the eukaryotic cell and with their own evolutionary history.

First, some introductory remarks on the taxonomic position of some primitive organisms are necessary. Putting them into the right frame is useful in biological research and especially in cytology.

For instance, when Loewy (1) in 1952 demonstrated the presence of actin and myosin in *Physarum polycephalum*, many biochemists wondered what species of vegetable a slime mould was. A cytologist, encountering the assertion that the actin from the mixomycete *Dictyostelium discoideum* is related to the actin of *Acanthamoeba castellanii* and is like that of mammals would be similarly puzzled. And he would be shocked to learn that some recent taxonomists put *Dictyostelium* and *Amoeba* even in the same order while others assign them to two different kingdoms.

It is unnecessary to go over the many general classifications that have lately been proposed. It is sufficient to start from the five kingdoms of living organisms as arranged by Lynn Margulis (2) in 1976.

The Taxonomic Frame

The first kingdom is the Prokaryota with well defined boundaries and so requires no particular comment.

The second is the Protista comprehending the Protoza, the Algae, the Mixomycetes *sensu lato* (that is the slime moulds) and the fungi having flagellated gametes without cell wall.

The third kingdom accommodates the filamentous fungi devoid of a flagellated stage (amastigote) belonging to the Ascomycetes, Basidiomycetes and Zygomycetes.

On the fourth and fifth kingdoms, of higher plants (from mosses upward) and animals (Metazoa), no comment is necessary.

The classification proposed by Margulis is good and quite justified, but according to Omodeo (3) it may be improved by adding a fourth taxon to the kingdom Fungi.

In fact, red algae (Rhodophyceae), as some specialists have pointed out (4,5) are similar to Ascomycetes: the relevant characters of their cells correspond to those of the Fungi even in some lesser particulars (Fig. 2), apart from the type of polysaccharide forming their walls, and, of course, photosynthesis.

The second difference is not relevant because, as Margulis herself (6,7) and others (8,9) have maintained, the acquisition of chloroplasts, and thus of photosynthesis, came from endosymbiosis with a blue-green alga, and this event apparently occurred at least twice with two or more forms of blue-green algae (cf. Omodeo, 3).

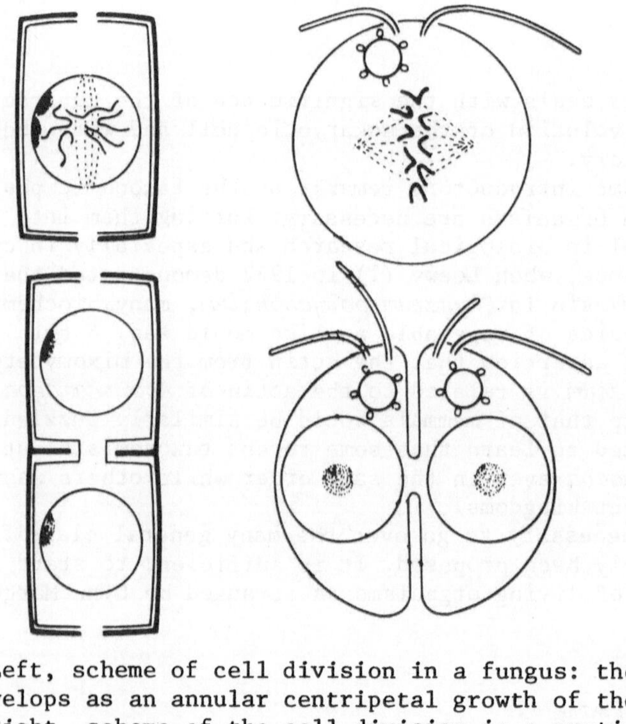

Fig. 1. Left, scheme of cell division in a fungus: the septum develops as an annular centripetal growth of the cell wall. Right, scheme of the cell division in a protist (monad): the cytokinesis is effected by the constriction of a ring of contractile fibres underlying the plasmalemma. Note in the fungal cell the absence of contractile vacuoles.

The first difference does not seem important because other surely related organisms differ in the nature of the glycan they are able to synthesize.

Thus it seems suitable to transfer these primitive algae to the kingdom Fungi.

The Place of Fungi in the Phylogenetic Tree

Now the question arises whether the Protists are more primitive than the Fungi (as above defined) or viceversa.

For a long time the filamentous Fungi were considered rather hi-highly evolved because of the complexity of their fruiting bodies, the mushrooms. The other fungi which do not form visible fruiting bodies and resemble the algae were considered "lower fungi" and called Phycomycetes.

With the advent of the electron microscope, the cell organization of these organisms became well known, so that this view could hardly be sustained, as Omodeo (10) and Taylor (11) have stressed. Indeed those who have studied unicellular Fungi, such as *Saccharomyces*, or unicellular red algae, as *Cyanidium*, agree that these have the simplest cells among the Eukaryotes, and there is no reason to believe that these free-living organisms underwent regressive evolution.

More generally, we emphasize that forms which are part of the kingdom Fungi, as above defined, occupy the lowest grade in the Eukaryote series for the following reasons:

1) their chromatin is more simply organized and their DNA content is the lowest among the Eukaryotes:

2) they are primitively devoid of flagella, basal bodies and amoeboid motility;

3) the mitosis is of primitive type and their cytokinesis is more like that of the prokaryotes than that of the eukaryotes;

4) their metabolic aptitudes are broader and rather different from those of the remaining Eukaryotes.

In the Fungi mitotis goes on with the nucleolus and nuclear envelope remaining and the endonuclear mitotic spindle consists of a few microtubules which arise from two centrosomal plaques (12). Cytokinesis is effected by annular ingrowth of the wall in relation to the waist of the cell, but the diaphragm so formed generally has a central hole which, in the pluricellular Rhodophyceae and in the Basidiomycetes, is partly or completely closed by a plug (Fig. 2). Only in yeasts and in unicellular red algae (Porphyridiales) is division complete, and the plasma membrane does not connect daughter cells. Therefore fungal cytokinesis is similar to that in the Prokaryota but wholly unlike that in the protists (as above defined) where it is accomplished by the contraction of a ring of fibres.

The fungal cell wall is a *passive* defence against endo-osmotic flux. In Protist cells living in water with fluctuating osmolarity the defense against endo-osmotic flux is, on the contrary, *active*, and water in excess is evacuated by contractile vacuoles lined by a felt of actomyosinic fibres. In these cells the same fibres provide for amoeboid movement, phagocytosis and macropinocytosis (13), activities completely unknown in the Fungi.

These organisms are devoid of any type of motility; even the "spermatia" (i.e. the gamete-like cells of red algae) are non-motile and are passively carried by the movement of the water toward the oogonia of the same or of another individual (14).

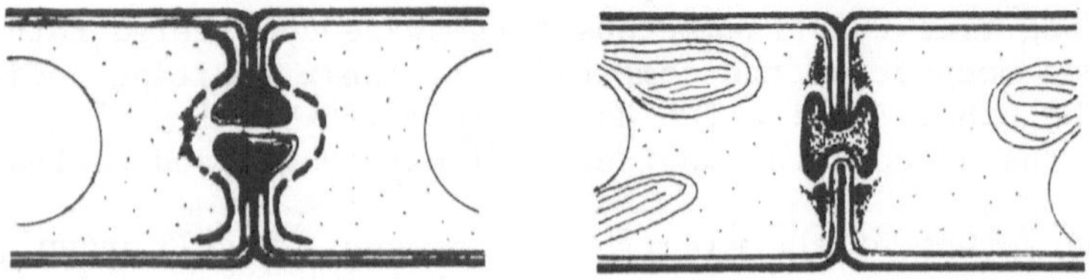

Fig. 2. Left, a perforate septum with septal pore cap in a Basidiomycete hypha. Right, a perforate septum between sister cells in a Rhodophycea; note the plug occluding the pore.

Contractile Proteins in the Lower Eukaryotes

As the fungal cell lacks any type of motility, it is tempting to conclude that it lacks also the proteins of the two contractile systems of higher eukaryotes (actomyosin and tubulin-dynein). But this is not the case, for there is direct and indirect evidence that the- these proteins are present in Fungi.

Actin has been isolated in baker's yeast (*Saccharomyces cerevisiae*) by Koteliansky *et al.* (15) and its gene has been isolated and sequenced by Ng and Abelson (16).

Tubulin has been isolated in the same yeast (17) and microtubules have been observed in the mitotic spindle of many fungi. Davidse and Flach (18) and Baum *et al.* (19) report that these microtubules have little or no affinity with the common agents which depolymerize the microtubules in higher eukaryotic organisms; their role in the morphogenesis of the ascomycete *Fusarium* (=*Gibberella*) has been analyzed by Howard and Haist (20).

Indirect evidence of dynein in the Fungi comes from the mechanics of their mitosis which is similar to that of some protists (12). It is obvious that energy expenditure is necessary for the sister chromatid separation and for their migration toward the cell poles. The point is: which contractile system functions as the force producer?

Most authors believe that the force is produced by the tubulin-dynein system, aided by the polymerization that lengthens some microtubule bundles. But Forer (21) maintains that it is the actomyosin system. As cytochalasin and antimyosin antibodies block actomyosin but do not inhibit mitosis (22, 23, 24) and as the microtubules are the more important component of the mitotic spindle which is rich in dynein, it is wiser to assume that the protagonist of mitotic movement is the tubulin-dynein system and that the actomyosin system is eventually a supporting actor (25, 26).

Studies on the proteins of the mitotic apparatus of the Fungi have still to be done, but it seems unquestionable that one of the two contractile systems must be present, and the preference goes to tubulin-dynein.

Other indirect evidence on contractile proteins concern the synaptonemal complex of prophase gametocytes.

Quite recently De Martino *et al.* (26) using the technique of fluorescent antibodies have detected smooth muscle myosin and actin in the meiotic spermatocyte of the mouse, rat and macaca. Actin shows as small spots during the whole meiotic prophase. Each spot corresponds to one attachment site of each bivalent to the nuclear envelope, but there are two additional spots which correspond to the attachment sites of the free ends of the heterochromosomes (Fig. 3).

Myosin is evident from the zygonema stage to the diplonema as filaments associated with the central area of the synaptonemal complex (Fig. 4).

After DNAse treatment the fluorescent anti-actin sera stain the central elements of the synaptonemal complex. Capanna *et al.* (27) suggest that actin plays a role in connecting bivalents to the nuclear envelope and that actin and myosin participate together in the process of pairing homologous chromosomes.

This suggestion seems reasonable and it has an interesting implication. As the synaptonemal complex is a very old and stable (though transient) structure of meiotic eucells from yeasts to man, these proteins, which are its constituents, must be as old. That means that action and myosin are present in the cell nucleus from the origin of eukaryotic organisms.

Such an assertion obviously needs some experimental control; one of us (E.C.) is working on it.

Contractile Systems in Phylogeny

At this point in our analysis we shall attempt a first conclusion.

In the filamentous amastigote Fungi and in the red algae cell motility and cytoplasmic contractility are very likely absent, but the contractile proteins are present and appear to have a fundamental role in the mitotic and meiotic processes; they may have also a role in chromatin condensation (28). Contractile proteins, moreover, constitute the bulk of nonhistone proteins of the eucell nucleus (29,30).

Such a situation suggests that the contractile proteins appeared at the onset of the eukaryote cell as a component of the chromatin. In the meantime (or later) tubulin and actin participated, together with the cell wall, in the morphogenesis of the cell (31, 32).

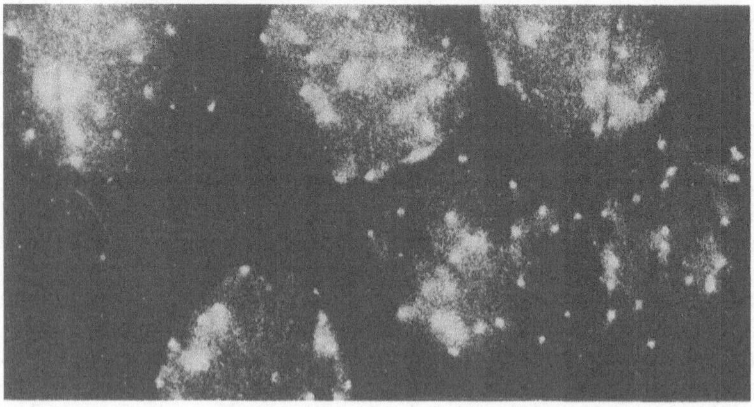

Fig. 3. Frozen section of lepto-zygotene spermatocytes of a mouse: the antiactin serum stains spots located on the nuclear envelope.

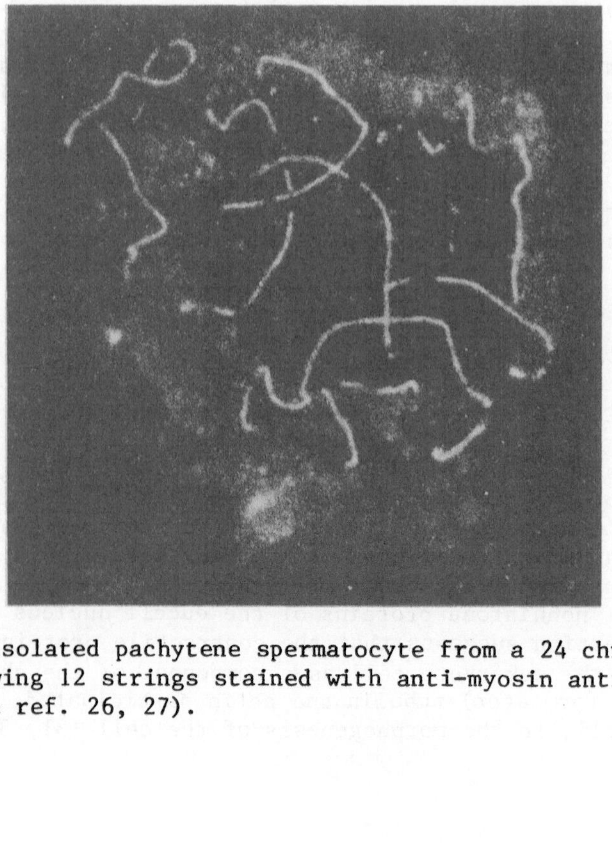

Fig. 4. An isolated pachytene spermatocyte from a 24 chromosome mouse, showing 12 strings stained with anti-myosin antiserum (see ref. 26, 27).

At an ensuing evolutionary level both the contractile proteins were engaged in cytoplasmic contractility, in cytokinesis and in cell motility. That happened, appartently, when the cell wall disappeared - at least during some stage in the life cycle (gametes?). In any case the event marks the transition from the Fungi (*sensu* Omodeo) to the Protista (*sensu* Margulis).

Flagella and cytoplasmic actomyosin remain the essential locomotory structures of protists, of plants, and of a few metazoa, but in most animals the actomyosin system evolves in the fibres of muscle cells which become much more important for locomotion. Nevertheless cytoplasmic actomyosin and the tubulin-dynein system persist in all eukaryotic organisms, and undergo further evolution.

Function and Evolution of Tubulin

The tubulin-dynein system is well known from the axoneme of the cilia and flagella, an amazingly constant structure from the lowest protists to the highest metazoans. But its functioning is not necessarily tied to axoneme geometry or to axoneme outer doublets: in the motile sperm tail of a dipteran insect (*Monarthropalpus*) we have an example of tubulin doublets irregularly scattered (33). The sperm tails of other insects (34), the mitotic spindle, the axostyle of the Trichomonadid and Hypermastigid protozoa (35) are examples of contractile structures where the dynein is bound to single microtubules.

The function of the tubulin-dynein system in the cytoplasm of non-dividing cells is not clear and its abundance in the vertebrate brain is surprising.

Microtubules participate not only in cytoplasmic and chromatin movements but also fundamentally in cell morphology and morphogenesis: there are few of them in the fungal cytoplasm which is modelled and sustained mainly by the rigid cell wall, but they are abundant in the naked cells where they form a cytoplasmic scaffolding that can be demonstrated by staining with fluorescent antibodies (36).

Tubulin molecules are the bricks forming the microtubules and some other similar structures. They are formed by two polypeptide chains, α and β, of apparent M.W. of about 55 and 53 kdal respectively.

A heterodimer formed by these two peptides polymerizes reversibly, both *in vitro* and *in vivo*, in microtubules; the process involves GTP and divalent cations.

On the evolutionary history of tubulin many interesting facts are known, but the general view is still incomplete.

It has been established that the tubulins of yeast and Microsporidia have a somewhat lower molecular weight than the tubulins of protists and metazoans. The fungal proteins, moreover, do not interact with the common antimicrotubular agents which are so effective in protists, plants and animals (18,19), while they bind to methyl benzimidazole-2-ylcarbamate. Lastly, the tubulin complementary DNA (cDNA) probes from chick brain are able under stringent conditions to hybridize to the DNA of all the vertebrates tested, and to sea urchin DNA, but they do not hybridize to yeast DNA (37). These statements indicate that the tubulins in Fungi are rather different from all other tubulins, and that fits in with the phylogenetic distance and amastigote conditions of these organisms.

A comparison between protist tubulins and those of animals and plants is nowadays possible but only on the basis of their interaction with drugs or substances of small molecules (38) ; this interaction is fairly uniform as is the centesimal aminoacidic content.

The primary structure, the best tool to assess the kinship between protein species, is known for some terminal sequences alone: comparison between aminoacid sequences of tubulins α and β from the chick brain and comparison between these and the α and β tubulins from the sea urchin sperm flagella show that the differences are very few (39). That implies that the two polypeptides derive from an ancestral polypeptide and that each is highly conservative. Ludueña (40) estimates that during tubulin history 0.45 point mutation per 100 residues per 100 million years has been accepted during evolution: "Such extreme conservatism is not surprising in view of the structural constraints imposed on tubulin as well as the fact that it must retain certain enzymatic activities intrinsic to the assembly process as well as binding sites for other proteins" (40, pg. 79).

In the eukaryotic plasma membrane a protein exists which has many properties in common with tubulin. According to Margulis *et al.* (41) another similar protein may be present in spirochaetes, but the point cannot be considered as definitively established (40).

On the number of genes that code for tubulins in different organisms we have only sporadic data. Cleveland *et al.* (37) with the method of cloned cDNA individuate in the chicken brain four genes for α polypeptide and as many for β polypeptide, which is evidently the minimum estimate for the species. Gozes *et al.* (42) by isoelectric focusing show that there are 5-6 "iso-tubulins" in the brain in the foetal rat and 9 in the adult: the minimum gene number involved is no less than 5.

Dynein

Dynein is a Mg^{++} or Ca^{++} activated ATPase having a Svedberg coefficient ranging from 11 to 31 S; it is formed by at least one polypeptide chain of exceptionally high molecular weight capable of interacting with microtubules.

The structure and localization of this protein have been studied by Gibbons and co-workers in sea urchin flagella (43). These flagella (Fig. 5) are composed mostly of tubulin, but they also contain polypeptide Aα and Aβ of slow electrophoretic migration corresponding to a M.W. of 300-340 kdal. The two large polypeptides, together with several components of lower molecular weight, form the isozyme 1 of dynein which is the main axonemal ATPase and has, in its native structure, a 21S coefficient, a weight of 1250 kdal, and forms the whole outer arm of the doublets (44,45,46). Isozyme 2 is constituted by polypeptide D, and its structure and localization have not yet been resolved in detail. As to polypeptides B and C, they have no ATPase activity but they may have a structural or regulatory function (47).

The basic phenomena underlying axonemal motility consist in the sliding of adjacent doublets produced by the interaction of dynein m molecules, located in the arms, with subfibres B of doublets (48,49). In the process dynein and tubulin play roles which are analogous respectively to those of myosin and actin in cytoplasmic and muscle contractility.

A complex axonemal structure and the presence of two enzymic forms of dynein are a widespread feature of protists (50), invertebrates (47), vertebrates (51) and plants (52) (from mosses to gymnosperms). However more that two forms of axonemal dynein are present in complex organisms.

To begin with, recent studies in mammals indicate that the cilia and flagella in the same organisms, although structurally very similar (but with a different mode of beating), contain different species of dynein (53). This is clearly shown in Figure 6. Bovine sperm flagella contain a group of high M.W. polypeptides which in electrophoresis migrate very close to the sea urchin sperm polypeptides. Similar components are detected in bovine tracheal and oviduct cilia, as is shown in the co-electrophoresis tracks E and F of Figure 6; the slowest migrating band I from bull sperm does not coincide with any of the ciliary bands, and the sperm band L does not match exactly with band 2 from the cilia.

More marked differences are found in the sedimentation properties of flagellar and ciliary dyneins: the sperm enzyme has a main peak at about 19 S, whereas the ciliary enzyme forms a main peak at about 11S (Fig. 7).

It is possible that these differences reflect functional adaptations to the different modes of beating in cilia and flagella. A parallel example is offered by the slow and fast types of striated muscle, which differ in their myosins more at the molecular than at the ultrastructural level.

The physico-chemical properties of dynein vary somewhat even in more or less related species, e.g. sea urchin and sea stars (47,54), bull and man (51,55), probably owing to the large size of its polypeptides (56).

Further kinds of dynein intervene in the motility of unspecialized cytoplasm: it is believed that microtubules play a role in cytoplasmic flow and chromosome movements, however the occurrence of dynein in cells without flagella has been demonstrated only very recently.

Brain tubulin preparations contain a Mg^{++} and Ca^{++} dependent ATPase, which, under the same ionic conditions used in the study of axonemal dyneins, sediments as two peaks at about 11 and 18S, forming a pattern somewhat similar to that of axonemal dyneins from the same animal species (Fig. 7). This enzyme co-sediments with a high M.W. polypeptide which in electrophoresis migrates very close to the sperm J polypeptide (Fig. 8).

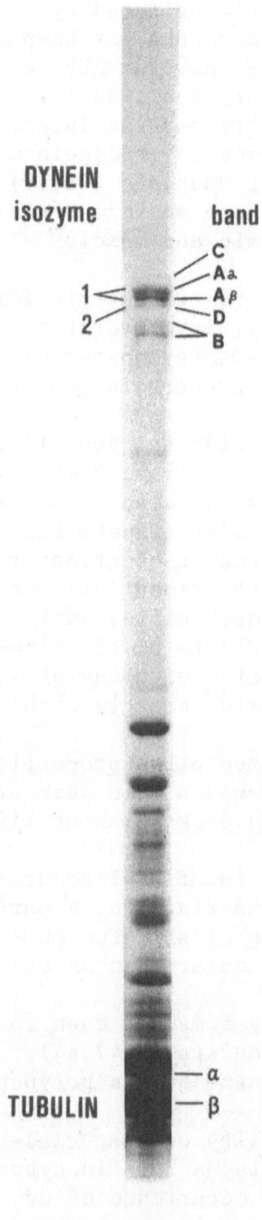

Fig. 5. Electrophoretic separation of flagellar polypeptides from spermatozoa of a Tyrrhenian sea urchin species. Supporting medium: a gel formed by a concentration gradient of polyacrylamide (3% at the top, 7% at the bottom), containing SDS and urea.
Electrophoretic migration (from top to bottom) is inversely proportional to the log of polypeptide molecular weight. Polypeptides below ca 50 kdal are not separated in this system and collect just below the tubulin bands.

It must be associated with brain cell microtubules since it tends to co-purify with tubulin through several cycles of polymerization-depolymerization and may well be defined as a dynein. However, some of its catalytic properties are different from those of axonemal dyneins: it is less sensitive to inhibition by Vanadate ions and it is not strictly specific for ATP, showing a high activity in the Mg^{++}-dependent hydrolysis of CTP (53).

A dynein-like protein resembling mammalian brain dynein in its nucleotide specificity and sensitivity to V inhibition has been recenttly isolated from the eggs of the fly *Ceratitis capitata.* This enzyme is formed by a polypeptide which resembles in its electrophoretic migration a sperm polypeptide from the same species (Fig. 9), and sediments at about 11 S (Pallini et al. unpublished data). Its association with tubulin has not yet been demonstrated, but it is quite likely, since insect eggs are known to be very rich in tubulin (57).

Insect egg dynein is related to sea urchin egg dynein, which also appears to be relatively insensitive to inhibition by V (58). It has been demonstrated that this sea urchin dynein becomes associated with the mitotic spindle of embryonic cells (59). A similar outcome may be assumed for insect egg dynein. Interestingly, the concentration of V required to inhibit these cytoplasmic dyneins agrees fairly well with those reported to inhibit the movements of chromosomes along the mitotic spindle *in vitro* (60). As to brain dynein, it may intervene, in the cytoplasmic transport phenomena or in the maintenance of the highly 'improbable' and precise forms of nerve cells.

In all cases, the cytoplasmic dyneins may produce a sliding of the cytoplasmic microtubules analogous to that of the axonemal outer

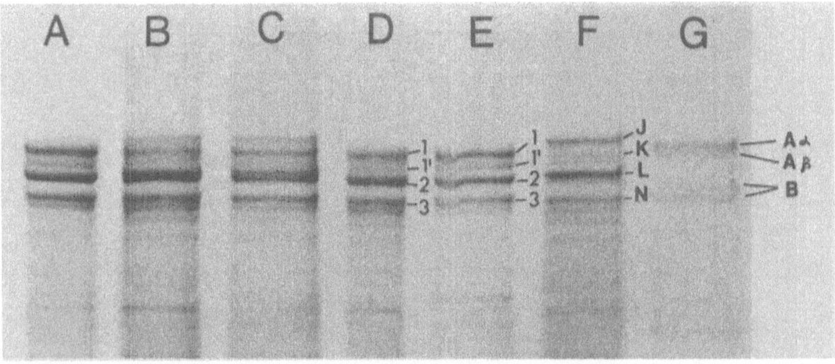

Fig. 6. Comparison of the heavy polypeptides of bovine cilia and flagella and of their nomenclature. Electrophoretic conditions as in Fig. 5. G, reference sea urchin polypeptides; F, bovine spermatozoa (bands indicated in ref. 51); E, oviduct cilia (bands indicated as in ref. 53); D, tracheal cilia (bands indicated as in ref. 53); C, coelectrophoresis of sperm and oviduct axoneme polypeptides; B, coelectrophoresis of sperm and tracheal axoneme polypeptide; A, coelectrophoresis of oviduct and tracheal axoneme polypeptides.

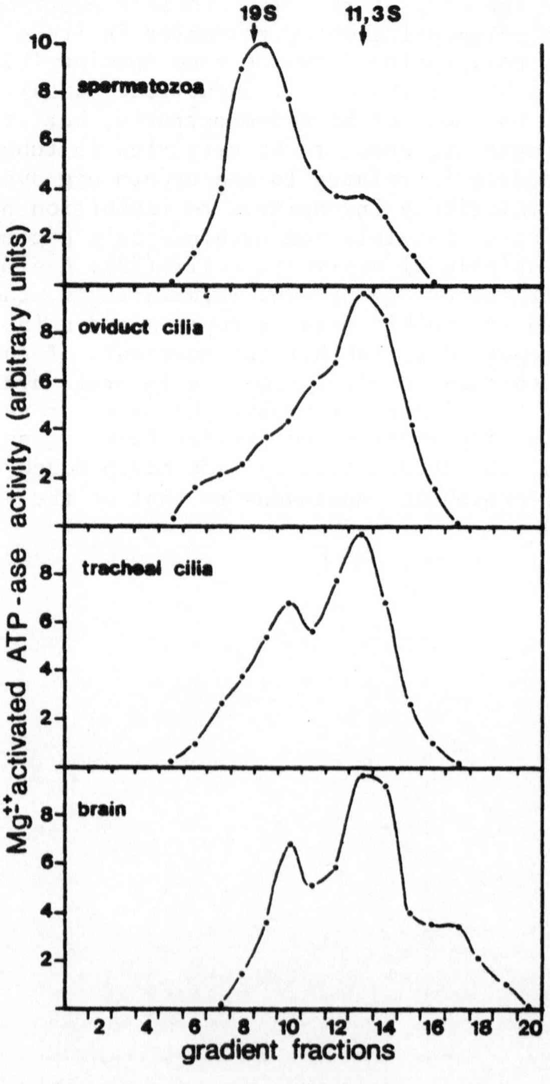

Fig. 7. Sedimentation behaviour of ATPases associated with axonemal and with brain cytoplasmic microtubules from the ox. ATPases had been made soluble by dialysis against 2 mM TRIS of pH 8.5, 0.5 mM EDTA (47). Sedimentation (from right to left) was performed on 5% - 25% sucrose gradient at 300 000 g (average) for 210 minutes. The arrows indicate the peak position of proteins (catalase and thyreoglobulin) of known sedimentation coefficient.

doublet microtubules, but other functions cannot be excluded. In fact, dynein may regulate the polymerization-depolymerization process of microtubules (61). Polymerization and depolymerization occur at the opposite ends of a microtubule (threadmilling) and may produce a transport of molecules or particles attached to the microtubular surface (61, 62)*.

Such data permit us to state that dynein evolved in divergent ways acquiring somewhat different functions, of which some are still unknown, yet from them it is not possible to state at all how these molecules are related and how they behaved during phylogeny.

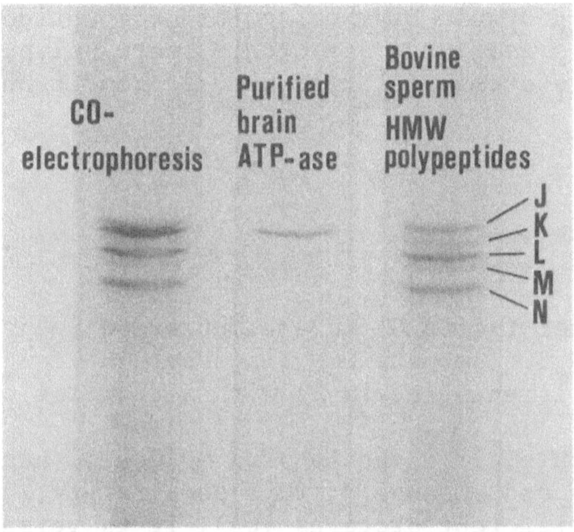

Fig. 8. Bovine brain and sperm dynein polypeptides. By coelectrophoresis it is clearly shown that the high molecular weight component of brain microtubules is very similar to the J band of spermatozoa (electrophoretic conditions as in Fig.5).

*Cytoplasmic dyneins may have a subtle role in morphogenesis. This is suggested by the abnormalities accompanying Kartagener's syndrome in man caused by the lack of arms in the axonemal outer doublets (68) and by abnormalities of the dynein heavy polypeptides (55). The disease is due to a recessive autosomal mutation affecting genes which control the production of dynein. Its pleiotropic phenotypic expressions include, in addition to male sterility and respiratory diseases arising from impaired ciliary function, the condition of "situs inversus viscerum" in about 50% of cases.

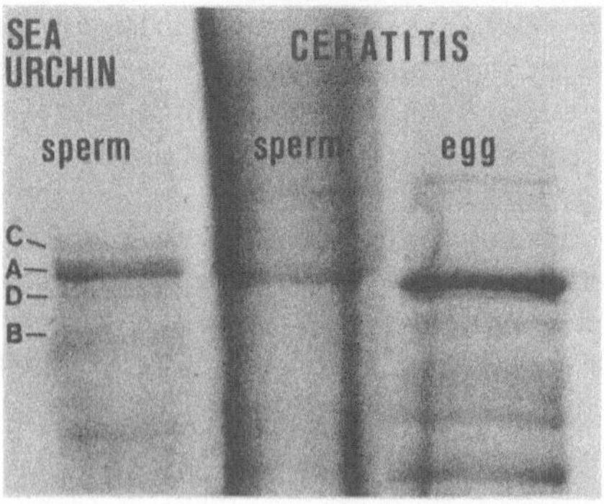

Fig. 9. Sperm and egg dynein high molecular polypeptides from the fly *Ceratitis capitata*. Sea urchin sperm polypeptides are shown as a reference (electrophoretic condition as in Figure 5).

Actin

The actin monomer (actin G) is a globular polypeptide with diameters of 5.5, 5 and 3.5 nm. Its M.W. is about 42 kdal. The actin polylymerizes in thin filaments (actin F) with a diameter of about 11 nm and length of up to 2 µm.

The actin molecule, like the tubulin molecule, has many binding sites on its surface and because of this peculiarity, which involves no less than eight different proteins (64), it is versatile, playing many important roles. Actin participates in different types of cytoplasmic contractility, morphogenesis, microanatomical strengthening (65,66), and even in mitosis. It is present in yeasts and plants and abundant in Protista and Metazoa (67).

The primary structure of the (nuclear?) actin of *Saccharomyces cerevisiae* has recently been established by mean of sequencing cloned DNA previously inserted in plasmids (16). The difference between yeast's actin and *Physarum*'s concerns 43 residues out of 374 (11.5%) and the differences from the β and γ cytoplasmic actins concern respectively 47 and 46 residues (12.3 and 12.6%). Curiously, the difference between the actin from yeast and that from mammalian muscle involves only 34 residues (9.1%). It is to be noted, however, that the residue substitutions, considered one by one, appear much more revolutionary than the mere percentage of substitutions suggests: the molecular form of yeast actin is surely different from that all other actins and thus its properties are different.

The primary structures of cytoplasmic non-muscle actins are known for a slime mould, *Physarum polycephalum* (68), for a small amoeba, *Acanthamoeba castellanii* (69), and for some mammals (70,71). These actins closely resemble each other: that of *Physarum* differs from that of the mammals in 16 residues only (5.8%). If we do not take into account the six first residues, which are highly variable in all actins so far studied, the difference amounts to only eleven residues out a total of 369 (3%) and the substitutions are conservative. Such a difference appears extremely low for organisms that began to diverge no less that 800 million years ago; it corresponds to 0 0.38 accepted point mutations per 100 residues per 100 million years. It is interesting to note that in all actins so far investigated (from amoebae to mammals) the histidine at position 73 is N^{γ} methylated.

In yeast and in the protists mentioned one actin species alone is known; Ng and Abelson (16) quote unpublished observations of R.A. Firtel, according to whom in *Dictyostelium discoideum* (a slime mould) there are about 17 genes for actin. It is not stated whether these genes code for identical polypeptides.

In *Drosophila* and in mammals there are six actin species, coded by as many genes (72): two species are in the cytoplasm, and four in the muscle. In the rabbit the actin of the cytoplasm differs from that of the skeletal muscle (73, 74) in 25 residues, this difference (9.1%) being almost twice that between the actin of *Physarum* and the actin of rabbit cytoplasm.

These facts indicate that in evolutionary history the ancestral gene coding for actin was duplicated five times and that some of the new genes were free to diverge from the original one which remained strictly conservative.

Actin seems to be related to a bacterial protein: the elongating factor of proteins synthesis; it may be related also to β actinin, a protein which regulates actin polymerization, having the same M.W. and a similar aminoacidic composition.

Myosin

Myosin is a large protein with ATPase activity stimulated by Ca^{++} and enhanced by actin. It is formed by two heavy polypeptides chains each weighing 200-215 kdal, by two 16-17 kdal light chains, and by two 17.5--20 kdal light chains. Myosin molecules self-assemble in bipolar filaments.

Myosin, like actin with which it interacts, seems to be present in all eukaryotic organisms where it is represented by many varieties (67) whose relationships are hard to establish, as is the way with the large proteins.

The comparative study of myosin extracted from different organisms is hindered by the lack of dara on their primary structure, inaccuracy in measurements and by the fact that the large molecule, although retaining the same aminoacidic sequence, may appear different because of postranslational modifications which differ in different organisms or even in cell types.

Myosin in the cytoplasm of *Physarum polycephalum* (1,75,76), *Dictyostelium discoideum* (77), *Acanthamoeba castellanii* (77) and *Amoeba proteus* (79) participates in amoeboid cell movements. In the first species a myosin type alone has been demonstrated, remarkably similar to those in vertebrate cells and fibres; its molecule is perhaps slightly longer and heavier (75,80),but on this point authors disagree (76). In *Acanthamoeba* three different myosin species have been found. One resembles the *Physarum* myosin, but the others are quite different, consisting of one heavy and two light peptides only which have unusual low M.W. and do not form bipolar filaments (81). The conventional myosin is localized in the inner cytoplasm while the unconventional one seem to be localized near the plasma membrane.

In the Vertebrates different myosins (isomyosins) have been found: in undifferentiated cytoplasm, in smooth, cardiac and skeletal muscle fibres. In skeletal muscle different myosin types are to be found, in conformity with the physiological role. According to a recent evaluation a mammal has no less than 12 myosin species, coded by as many loci.

The prokaryote protein most like myosin is an ATP-dependent DNA--unwinding enzyme (82), now renamed DNA-denaturing enzyme.

CONCLUSIONS

Above we stated that the comparative analysis of cell organization allows it to be said that in the lower eukaryotes (viz. amastigote filamentous Fungi) the contractile systems participate only in chromatin movements and that the acquisition of cytoplasmic contractility marks the transition from the Fungi to the Protista.

Comparisons of the biochemical structure properties of the actomyosin and tubulin-dynein systems are consistent with such assumption and make possible a further conclusion: the contractile proteins during their evolutionary history duplicated many times assuming different and important functions in the motility of the organism. With their new roles, their molecular structures diverged.

This history is like that of other proteins, yet it will be better understood only when the contractile systems in the lower Eukaryotes are better known.

ACKNOWLEDGEMENT

The authors thank Prof. R. Mitchell for help with the manuscript.

REFERENCES

1. Loewy A.G. (1952) *J. Cell. Comp. Physiol.* 40, 127-156.
2. Margulis L. (1976) *Taxon* 25, 391-403.
3. Omodeo P. (1981) *Boll. Zool.* 48, in press.
4. Dougherty E.C. & Allen M.B. (1960) in *Comparative Biochemistry of Photoreactive Systems*, ed. Allen M.B., Acad. Press,New York & London, pp. 129-144.

5. Cain R.F. (1972) *Mycologia* 64, 1-14.
6. Sagan L. (Margulis L.) (1967) *J. Theor. Biol.* 14: 225-274.
7. Margulis L. (1970) *Origin of Eukaryotic Cells*, Yale U.P., New Haven & London.
8. Omodeo P. (1969) *Boll. Zool.* 36, 291-310.
9. Taylor F.J.R. (1974) *Taxon* 23, 229-258.
10. Omodeo P. (1977) *Biologia*, UTET, Torino.
11. Taylor F.J.R. (1976) *Taxon* 25, 377-390.
12. Heath J.B. (ed.) (1978) *Nuclear Division in the Fungi*, Acad. Press, New York & London.
13. Allison A.C. (1976) in *Contractile systems in non-muscle tissue*, eds. Perry S.V., Margreth A., Adelstein R.S.; North Holland, Amsterdam.
14. Peel M.C. & Duckett J.G. (1975) in *The Biology of the Male Gamete*, eds. Duckett J.G. & Racey P.A., Linnean Society & Acad. Press, London, pp. 1-13.
15. Koteliansky V.E., Gluckhova M.A., Beniamin M.V., Surguchov A.P. & Smirnov V.N. (1979) *FEBS Lett.* 102, 55-58.
16. Ng R. & Abelson J. (1980) *Proc. Natl. Acad. Sci. USA* 77,3912-3996.
17. Haber J.E., Peloquin J.G., Halvorson H.O., Borisy G.G. (1972) *J. Cell. Biol.* 55, 355-367.
18. Davidse L.C. & Flack W. (1977) *J. Cell Biol.* 72, 174-193.
19. Baum P.J., Throner J., & Honig L. (1978) *Proc. Natl. Acad. Sci USA*, 75, 4962-4966.
20. Howard R.J. & Haist J.R. (1980) *J. Cell Biol.* 87, 55-64.
21. Forer A. (1977) in *Nuclear Division in the Fungi*, ed. I.B. Heath, Acad. Press, London.
22. Mabuchi I., Okuno M. (1977) *J. Cell Biol.* 74, 251-253.
23. Kiehart D.P., Inoué S., Mabuchi I. (1977) *J. Cell Biol.* 75, 258.
24. McIntosh J.R., Hepler P.K., Van Wie D.G. (1969) *Nature* 224: 659-663.
25. Mohri I. (1979) in *Cell Motility: Molecules and Organization*, eds. Hatano S., Ishikawa H., Sato H., University Park Press, Baltimore.
26. De Martino C., Capanna E., Nicotra M.R., Natali P.G. (1980) *Cell Tiss. Res.*, in press.
27. Capanna E., De Martino C., Natali P.G. (1980) *Boll. Zool.* 47, 233-239.
28. Rungger D., Rungger-Brandle E., Chaponnier C., Gabbiani G. (1979) *Nature* 282, 320-321.
29. Lesturgeon W.M., Forer A., Yang Y.-Z., Bertram J.S., Rusch H.P. (1975) *Bioch. Bioph. Acta* 379, 529-555.
30. Douvas A.S., Bakke A., Bonner J. (1977) in *The molecular biology of the mammalian genetic apparatus*, ed. Ts'o P.O.P., North--Holland, Amsterdam, Vol. 1, 143-163.
31. Howard R.J., Haist J.R. (1980) *J. Cell Biol.*87, 55-64.
32. Tucker J.B. (1979) in *Microtubules* eds. Robert K. & Hyams J.S. Acad. Press, London & New York, pp. 315-357.
33. Baccetti B., Dallai R. (1976) *J. Ultrastruct. Res.* 55, 50-69.

34. Dallai R. (1979), in *The Spermatozoon* eds. Fawcett D.W. & Bedford J.M., Urban & Schwartzenberg, Baltimore.
35. Cleveland L.R. (1965) *J. Cell Biol.* 24, 387-400.
36. Weber K. & Osborn M. (1979) in *Microtubules* eds. Roberts K. & Hyams J.S., Acad. Press, London & New York, pp. 279-313.
37. Cleveland D.W., Lopata M.A., MacDonald R.J., Cowan N.J., Rutter W.C. (1980) *J. Cell Biol.* 20, 95-106.
38. Dustin P. (1978) *The microtubules*, Springer Verlag, Heidelberg & New York.
39. Ludueña R.F. & Woodward D.O. (1975), *Ann. N. Y. Acad. Sci.* 253, 272-283.
40. Ludueña R.F. (1979) in *Microtubules* eds. Roberts K. & Hyams J.S., Acad. Press, London & New York, pp. 65-116.
41. Margulis L., Banerjee S., White T. (1969) *Science N.Y.*, 164, 1177-1178.
42. Gozes I., De Baetselier A., Littauer U.Z. (1980) *Eur. J. Biochem.* 103, 13-20.
43. Gibbons I.R. & Rowe A.J. (1965) *Science* 149, 424-426.
44. Gibbons I.R. & Fronk E. (1979) *J. Biol. Chem.* 254, 187-196.
45. Gibbons B.H. & Gibbons I.R. (1979) *J. Biol. Chem.* 254, 197-201.
46. Bell C.A., Fronk E. & Gibbons I.R. (1979) *J. Supramolecular Structure* 11, 311-317.
47. Gibbons I.R., Fronk E., Gibbons B.H. & Ogawa K. (1976) in *Cell Motility* eds. Goldman R.D., Pollard T.D. & Rosenbaum J.L., Cold Spring Harbor Lab., Vol. 3, pp. 915-932.
48. Satir P. (1968) *J. Cell Biol.* 39, 77-81.
49. Summers K.E. & Gibbons I.R. (1971) *Proc. Natl. Acad. Sci. USA* 68, 3092-3096.
50. Piperno G. & Luck D.J. (1979) *J. Biol. Chem.* 254, 3084-3090.
51. Pallini V. & Gibbons I.R., in preparation.
52. Norstog K. (1975) in *The Biology of the Male Gamete*, eds. Duckett J.G. & Racey P.A., Linnean Society & Acad. Press, London, pp. 135-142.
53. Pallini V., Bugnoli M., Mencarelli C., Scapigliati G. (1981) 35th Symp. Soc. Exp. Biol. Cambridge, in press.
54. Mabuchi I., Shimizu T., Mabuchi W. (1976) *Arch. Biochem. Bioph.* 176, 564-576.
55. Baccetti B., Burrini A.G., Pallini V. & Ranieri T. (1981) *J. Cell Biol.* 88, 102-107.
56. Chkraborty R. (1975) 14th Intern. Congress of Genetics, Moscow.
57. Green L.H., Brandis J.W., Turner F.R. & Raff R.A. (1975) *Biochemistry N.Y.*, 14: 4487-4491.
58. Pratt M.M. (1980) *Develop. Biol.* 74, 364-378.
59. Pratt M.M., Otter T. & Salmon E.D. (1980) *J. Cell Biol.* 86, 738-745.
60. Cande Z.W. & Wolniak S.M. (1978) *J. Cell Biol.* 79, 573-580.
61. Haimo L., Telzer B.R. & Rosenbaum J.L. (1979) *Proc. Natl. Acad. Sci. USA* 76, 5759-5763.
62. Margolis R.L., Wilson L., Kiefer B.I. (1978) *Nature* 272, 450-452.

63. Afzelius B.A. (1976) *Science* 193, 317-319.
64. Stossel T.P., Hartwig J.H., Yin H.L., Davies L.A. (1979) in *Cell Motility*, eds. Hatano S., Ishikawa H., Sato H., Univ. Park Press, Baltimore.
65. Pollard T.D., Fujiwara K., Niederman R., Maupin-Szamier P. (1976) in *Cell Motility*, eds. Goldman R., Pollard T.D., Rosenbaum J.L., Cold Spring Harbour Labor, Vol. 3, pp. 375-389.
66. Brinkley R., This issue.
67. Korn E.D. (1978) *Proc. Natl. Acad. Sci. USA* 75, 588-599.
68. Vandekerckhove J. & Weber K. (1978) *Nature* 276, 721-722.
69. Elzinga M. & Lu R.C. (1976) in *Contractile Systems in Non-Muscle Tissues*, eds. Perry S.V., Margreth A., Adelstein R., Elsevier North Holland, Amsterdam.
70. Vandekerckhove J. & Weber K. (1978) *Eur. J. Biochem.* 90, 451-462.
71. Vandekerckhove J. & Weber K. (1978) *Proc. Natl. Acad. Sci. USA* 75, 1106-1110.
72. Zulauf E., Tobin S.L., Sanchez F., McCarthy J. (1980) *Eur. J. Cell Biol.* 22, 13 (IInd Int. Congr."Cell Biology")
73. Lu R.C. & Elzinga M. (1977) *Biochemistry*, 10: 590-600.
74. Collins J.H. & Elzinga M. (1975) *J. Biol. Chem.* 250, 5915-5920.
75. Adelman M.R. & Taylor E.W. (1969) *Biochemistry* 8, 4964-4975, 4976-4988.
76. Nachmias V.T. (1979) in *Cell motility: molecules and organization*, eds. Hatano S., Hishikawa H. & Sato H., Univer. Park Press, Baltimore, pp. 33-57.
77. Mockrin S.C. & Spudich J.A. (1976) *Proc. Natl. Acad. Sci. USA* 73, 2321-2325.
78. Pollard T.D. & Korn D.E. (1973) *J. Biol. Chem.* 248, 4691-4697.
79. Condeelis J.S. (1977) *Anal. Biochem.* 78, 374-394.
80. Hatano S. & Takahashi K. (1971) *J. mechanochem. Cell Motility* 1, 7-14.
81. Godasi H. & Korn E.D. (1980) *Nature* 286: 452-456.
82. Abdel-Monem M., Lauppe H.-L., Kartenbeck J., Dürwald H., Hoffmann--Berling H. (1977) *J. Mol. Biol.* 110, 66)-685.

DISCUSSION (SECTION I)

BRADBURY: Is there any evidence these days that actin in the nucleus is involved in the separation of chromosomes?

NANNINGA: There is no evidence that actin is inside the mitotis spindle. In eucaryotic cells, there are microtubules pulling the metaphase spindles apart already. At first sight, actin may not be required, but this may not be a complete story and there may be a place for actin.

SELLS: Is there any information regarding the nuclear matrix with regard to the processing of hnRNA?

NANNINGA: If the nucleus is treated with low magnesium, 2M NaCl, detergent and nucleases; you end up with a reticulum structure. Whether this residue is related to original structure is not sure. Autoradiography has been done to see where newly formed DNA is located. The grains are found on the reticulum structure but this does not prove too much.

BAUER: The microfibres are linked to the cytoplasmic membrane and may be related to cell growth. Has it been shown that α-actinin is an anchor of microfilaments?

NANNINGA: I think it is true, but I cannot remember the experimental details.

NICOLINI: Could you please elaborate on the freeze fracture technique?

NANNINGA: It is a method of 1963 which comes from Moore and Muhlethaler from Frey-Wyssling's Lab. The principle is the following: Let's assume that we have a suspension of membrane-bounded round cells. The method is to freeze these organisms very quickly to a temperature or about - 150° C; the freezing occurs in thousanths of a second. Then the next step is that this frozen material is brought into a shadow-casting apparatus, where it is fractured at, for instance - 150° C with a cooled hammer. The fracture goes through the interior of the membrane in a continuous manner. Then the upper part of the fractured material is removed so that one ends up with a broken surface which is the frozen fractured surface of the biological material. The second step is to evaporate the heavy metal in such a way that you get a shadow effect and then the next step is to evaporate carbon on top of it so that you get a carbon layer. Both the platinum as a heavy metal as well as the carbon following exactly the contours of the speciment. Then you open the bell jar and put the replicated specimen into water and the replica is floated off. So, what you maintain is the replicated part of the biological structure which consists of the heavy metal showing the elevations and depressions which are reinforced by carbon. The fortunate part is that if you freeze a biological membrane and then fracture them, you split the membranes and then you are looking inside the membranes and the particle distribution on them. In short this is the freeze fracture technique.

You can do something else than that described in this presentation, where we fracture and what is broken falls off somewhere is the bell jar. One can also fracture in a controlled way where one can maintain the complementary replicas of the both fractured membrane and cell envelope.

NICOLINI: Could you study the chromatin structure with this technique?

NANNINGA: No, I do not think it is a very convenient method. You can study the nuclear envelopes and see the pores in the nuclear envelopes. I think you can not recognize the chromatin unless you chrystallize it.

WITTELESBERGER: What are the relative proportions of protein and lipid in the cell surface? Is there just a few proteins in a vast sea of lipid or are there more proteins than suggested by the Singer-Nicholson model?

NANNINGA: That model is correct as far as the drawings go but it is only a drawing. Probably there may be more matrix.

NICOLINI: Just to make justice to Stanley Bram, I saw some of his beautiful pictures of non crystallized material of high order chromatin structure. They were very clear and seemed very convincing. Knowing the high quality of his work, I would like to know on what grounds you are saying that one could not study the high structure of chromatin by freeze fracture?

NANNINGA: You can make crystals of nucleosomes in particular of core particles which has been done, for instance, in the lab of Klug in Cambridge. It may be that he was looking at crystallized cores.

BRADBURY: No, they were freeze fractured nuclei and he claimed that they were high order structures, coils and cores.

SELLS: I wanted to ask you about the concept of free and bound ribosomes in cells. There usually are ribosomes that are attached to the endoplasmic reticulum. Are there such things as free ribosomes or are they actually attached to the filaments or microtubules?

N. NANNINGA: It could be that so called polyribosomes are connected to microfilaments or intermediate filaments and I remember some fairly early publications a long time ago saying that they saw polyribosomes that were not attached to membranes but were associated as fibers, but I would not know whether one can generalize this.

SECTION II: CELL PROBES

IMMUNOCYTOLOGICAL METHODS

Stratis Avrameas

Unite d'Immunocytochimie
Departement de Biologie Moleculaire
Institut Pasteur
25, Rue du Docteur Roux
75724 PARIS CEDEX 15, France

INTRODUCTION

The localization of tissue or cellular antigens can be performed by using a suitable immunocytological technique. The immunocytological techniques combine the principles of cytology with the specificity of the immunological reaction and allow the localization and, in some instances, the relative quantitation of cellular antigenic constituents.

The basis of all immunocytological techniques is an antibody conjugate preparation with a marker substance visible by light and/or electron microscopy, which allows the localization of cellular constituents. The steps followed in all immunocytological techniques are the same: an antibody is first covalently linked with a marker substance and then this labelled antibody is allowed to react with the cellular antigen. Subsequently the excess labelled antibody is eliminated by washing and the marker substance attached to the cellular antigen is visualized by light or electron microscopy.

On the basis of the above procedure, introduced for the first time by Coons and co-workers 40 years ago (1) and termed the direct technique, several variants have been developed. One of the most commonly employed of these variants is the indirect or "sandwich" technique. In this variant, the tissue is incubated with an antiserum specific for the antigen under investigation and prepared in animal species A. The tissue is then washed and incubated with a labelled antibody directed against the immunoglobulins of species A.

The most widely used labels have been fluorescent (2) and radioactive (3) substances, ferritin (4) and enzymes (5, 6). Occasionally, viruses (7), hemocyanin (8) and heavy metals (9) have also been used to label the antibodies. Fluorescein-labelled antibodies are employed for the detection of tissue constituants with the light microscope while ferritin is used for electron microscopy. Radioisotope and enzyme-labelled antibodies can be employed for both light and electron microscopic studies. Some features are common to all immunocytological procedures while others are particular to a given method. We shall deal first with the common features and then with the particular features of each immunocytological procedure.

FEATURES COMMON TO ALL IMMUNOCYTOLOGICAL TECHNIQUES

Antisera and Antibodies

With all immunocytological procedures, the best results are obtained with antisera containing high titers of antibody. Generally, such antisera precipitate well with their antigen and the potency of an antiserum can therefore be evaluated by using several simple precipitin tests. Potent antisera containing several milligrams of antibody per ml are usually obtained by injecting animals several times with the antigen mixed with complete Freund's adjuvant. It should be stressed however that immunization of animals with antigens that have been preserved during evolution will never lead to the production of high titer antisera. Usually, such antisera do not precipitate or precipitate weakly with their antigen.

The marker substance used in a given immunocytological technique can be linked either to the γ globulin or to the IgG fraction of the antiserum or to the antibodies specifically isolated from this antiserum. The globulin fraction is prepared by salting-out procedures and IgG by salting out followed by ion-exchange chromatography (10). For the preparation of purified antibodies, specific immunoadsorbents are employed (11). Although good results are obtained with labelled immune γ globulins and IgG, it is preferable to use purified labelled antibodies when possible. The results obtained with purified labelled antibodies are much more reproducible, the specific staining is intense and the non-specific background is much less pronounced.

After the coupling reaction between the label and the antibody has occurred, the presence of labelled antibody in the reaction mixture is usually assessed by double diffusion and immunoelectrophoresis. Subsequently, in order to increase specific staining and decrease background staining unreacted antibody and label are removed from the labelled antibody preparation. This is generally accomplished by gel filtration or ion exchange chromatography.

The purified labelled antibody is then evaluated for antibody and label content. The ability of the labelled antibody to detect the desired antigen is determined by using a cellular preparation known to contain the antigen, and for comparison, if possible, with a previously evaluated conjugate known to give a positive reaction under the same conditions.

Tissue Processing and Immunocytological Reaction

The tissue containing the antigen under investigation should be processed in such a way as to have an acceptable tissue morphology, to avoid diffusion of cellular constituents and to preserve the antigenic determinants. The procedure selected for processing the cell preparation will depend on the characteristics of the antigen under consideration, whether the antigen is located on the cell surface or within the cell, and whether light or electron microscopic examination is planned. The detection of surface antigens on membranes for either light or electron microscopy does not raise any particular problems and it may be carried out on living cells by simply incubating the cell preparations with the labelled antibody. On the contrary, the detection of an intracellular antigen is often difficult to perform and requires rendering the antigen accessible to the labelled antibody. This can be accomplished either by sectioning the cell or by damaging the membrane through fixation or drying. Different conditions are employed depending on whether the intracellular antigen is to be visualized by light or electron microscopy.

Light microscopy. For light microscopy, the localization of antigens by the various immunocytological procedures is carried out on fixed or unfixed cell preparations.

In unfixed cell preparations, sections of frozen tissue, imprints or smears are air dried and then incubated with the labelled antibody. In general, simple drying seems to be less harmful than any fixative for the preservation of the cellular antigenic determinants. However, the use of dried cell preparations in immunocytological localization has two main disadvantages; first, tissue morphology is often poorly preserved and second, because drying only partially disrupts the membrane, difficulties of penetration of the labelled antibody into the interior of cells often occurs and consequently intracellular staining is faint.

For the localization of antigens on fixed cell preparations, two main procedures are used. In the first procedure, the tissue is fixed and then embedded in an adequate medium or frozen. From these preparations sections are obtained and incubated with the labelled antibody. In the second procedure, sections from frozen unfixed tissue are first prepared, fixed and then incubated with the labelled antibody. Fixation is usually carried out either with various aldehyde solutions or with several organic solvents. Fixation with aldehyde solutions will be considered later in the localization of antigens for ultrastructural study. With organic solvents, good results are obtained by fixing for 10 to 20 minutes at -20°C in absolute acetone, ethyl or methyl alcohol or mixtures of these. The main effect of these organic fixatives is to abruptly remove cellular water and hence to precipitate the cellular constituents which become irreversibly insolubilized. With some notable exceptions, fixation with organic solvents does not seem to alter most of the cellular antigenic determinants appreciably. These organic fixatives have the additional advantage of dissolving most of the cellular lipids, thus producing more complete disruption of the cell membranes and therefore easier penetration of the labelled antibody into the interior of the cells.

Electron microscopy. For immunocytological studies by electron microscopy, both satisfactory preservation of the tissue and occurrence of the immunological reaction must be reconciled. Organic solvents such those used for the detection of antigens by light microscopy cannot be employed because these fixatives completely destroy the ultrastructures. Good ultrastructural preservation is obtained only with aldehyde fixatives like formaldehyde or glutaraldehyde. These fixatives, especially glutaraldehyde, may hinder the accessibility of antigenic determinants to labelled antibodies by the introduction of cross-linkages among the various cellular macromolecules. Fixation with glutaraldehyde generally yields understained or negative preparations. Good results can be expected only when relatively light fixation with formaldehyde is carried out. The immunocytological reaction can be performed either on tissues after fixation and before embedding or after fixation and embedding. In the first case, good results are obtained with small blocks of tissue fixed at 4°C for up to 24 hours with 4% formaldehyde in 0.1 M phosphate or cacodylate buffer pH 7.2 - 7.4. After fixation, the blocks are washed for at least 24 hours at 4°C with several changes of the buffer solution; they are then reduced to small fragments and frozen. From these fragments 5-15 μ sections are cut and incubated for 2 hours with the labelled antibody. Subsequently, the sections are washed to remove excess conjugate, incubated with substrate if enzyme-labelled antibodies are used, washed, sometimes, post-fixed for 30 minutes with 2.5% glutaraldehyde, dehydrated, embedded and processed for electron microscopy. In the second case, after fixation and washing the blocks are dehydrated and embedded in plastic. Ultrathin sections of embedded tissue are cut, treated with a reagent allowing the partial removal of the plastic and then incubated with the labelled antibody. After washing (and if required enzymatic staining) the sections are processed for electron microscopy. Although with this procedure good results have been obtained with several antigens and especially with various hormones, it appears that in its present form, the procedure is not applicable to all antigens.

Controls for light and electron microscopy. With all immunocytological procedures, it is essential to run controls in parallel so as to distinguish between a specific immunological reaction and non-specific staining. The following controls are usually carried out: 1) the labelled antibody is adsorbed with its homologous antigen; this should prevent the specific immunological reaction from occurring on the cellular preparations; 2) the cellular preparation is treated with the specific but unlabelled antibody and then by labelled antibody; this should noticeably reduce the specific staining; 3) the cellular preparation is incubated with an unrelated labelled antibody; in this way only the non-specific staining should be visible. With all immunocytological techniques it is extremely difficult if not impossible to completely avoid non-specific staining. In general, less non-specific staining occurs when strong antisera are employed. The non-specific staining may be further reduced by extensive washing of the preparation after incubation with the labelled antibody.

Characteristics of Several Immunological Techniques

Immunofluorescence. Antigens are detected by immunofluorescence using antibodies labelled with fluorochromes. When fluorochromes are illuminated by light or short wave-length, they emit light of longer visible wavelength, which permits detection of the labelled antibody. Fluorescein and rhodamine are the two fluorochromes most often employed in immunofluorescence techniques. Fluorescein emits in the yellow green and rhodamine in the red wave-lengths of the visible spectrum. This permits double staining on the same cellular preparation, and even on the same cell. The two fluorochromes are coupled to antibodies via covalent bonds that are stable during the immunocytological procedures. Fluorescein and rhodamine are commercially available as the isothiocyanate derivatives. These active groups bind readily to free amino groups of proteins at alkaline pH.

Immunofluorescence is a very sensitive technique. In a cellular preparation a small quantity of antigen is sufficient to give a positive reaction. For example, 0.8×10^{-4} $\mu g/mm^2$ of pneumonococcal capsular polysaccharide is detectable by this technique. Immunofluorescence is certainly the most widely employed immunocytological technique and it has contributed enormously to our present knowledge in different areas of cell biology. Compared to the other immunocytological procedures, immunofluorescence is the simplest and most rapid technique. Its only limitation is that it cannot be employed for ultrastructural studies. Immunofluorescence is the most well documented immunocytological technique and many reviews on this subject have been published (2, 12, 13, 14).

Immunoferritin. The immunoferritin technique is applied for the localization of antigens at the ultrastructural level. Ferritin is a large protein molecule with a molecular weight of 700,000. It contains 23% iron by weight. This accounts for the electron density of ferritin and renders ferritin-labelled antibodies visible in the electron microscope.

Meta-xylene diisocyanate, p,p' difluoro-m,m' dinitrophenyl sulphone and glutaraldehyde have been employed for the conjugation of ferritin to antibodies. In order to avoid coupling procedures, a variant of the ferritin-labelled antibody technique has been developed. Hybrid antibodies possessing anti-immunoglobulin and anti-ferritin specificities were prepared and employed for the detection of cell surface antigen (15). The technique has been further developed by the preparation and use of a hybrid antibody in which the electron-dense marker is a virus (7).

The immuno-ferritin technique has been employed successfully for the ultrastructural localization of membrane or surface antigens. Although with this technique positive results have occasionally been reported for the intracellular detection of antigens, the use of ferritin-labelled antibodies for such localization is hampered by the high molecular weight of the ferritin. However, it should be noted that because of the high molecular weight of antibody-ferritin conjugates, probably due to steric hindrance, even cell surface antigens may not always be detected.

Compared to the other ultrastructural immunocytological techniques, immunoferritin allows the most precise localization of tissue antigens. It is probable, however, that each ferritin molecule does not detect only one antigenic determinant. Reviews on the immunoferritin technique have been published (14, 16, 17).

Immuno-autoradiography. In these procedures, antibodies labelled with radioactive tracers are employed for the light and electron microscopic detection of tissue antigens. Radioactive antibodies are first allowed to react with the antigen and the cellular preparation is subsequently processed for autoradiography.

Tritium and ^{125}I have been used as labels although in immunocytological procedures, the latter has been almost exclusively employed (18, 19). Tritium is introduced into antibodies by allowing tritiated acetic anhydride to react with antibody (18). ^{125}I is introduced in the tyrosine moities of antibodies by using the chloramine T or the lactoperoxidase method (20, 21).

Immuno-radioautography is potentially a very sensitive method and it has been applied successfully to the study of particular problems by light microscopy (22). However, the low resolution of light microscopy combined with the relatively low resolution inherent to radioautography precludes the use of immuno-radioautography for precise localization of antigens by light microscopy. With the electron microscope, antibodies labelled with ^{125}I permit resolution in the range of 600 to 1000 Å (23) and have been employed successfully in various immunocytological studies (19, 24, 25, 26). Studies performed with antibodies or their Fab fragments labelled with ^{125}I have shown that in cellular preparations, adequately fixed for electron microscopy, intracellular antigen cannot be detected with whole antibody because of lack of penetration. Iodinated Fab fragments, on the other hand, can penetrate the cell and be used for the ultrastructural intracellular detection of antigens. Consequently, immuno-radioautography using iodinated Fab fragments combined with electron microscopic peroxidase cytochemistry can be utilized for the simulations detection of two antigens (27).

Compared to the other immunocytological techniques, immuno-radioautography is probably the most difficult and time-consuming to carry out while giving the lowest resolution. However, it is probably the most sensitive and it gives the most reliable quantitative data.

Other immunocytological techniques. A few other immunocytological techniques have been reported but have not been widely employed to date. Thus, attempts have been made to render antibodies electron-dense by introducing heavy metals such as uranium and to use them for electron microscopy (28). Alternatively, antibodies have been allowed to adsorb to colloidal gold and the immuno-colloidal preparation obtained used for the identification of bacterial antigen and the localization of cell surface antigens (29). Finally, antibodies have been labelled with hemocyanin, the characteristic shape of which allowed localization of cell surface antigens by electron microscopy (8). Immunological techniques based on the use of enzyme-markers will be described in the following chapter.

REFERENCES

1. A.H. Coons, H.J. Creech and R.N. Jones, Immunological properties of an antibody containing a fluorescent group. Proc. Soc. Exp. Biol. and Med. (N.Y.) 47:200 (1941).
2. A.H. Coons, Histochemistry with labelled antibody. Int. Rev. Cytol. 5:1 (1956).
3. M.C. Berenbaum, The autoradiographic localization of intracellular antibody. Immunology 2:71 (1959).
4. S.J. Singer, Preparation of an electron-dense antibody conjugate. Nature (London) 183:1523 (1959).

5. S. Avrameas and J. Uriel, Methode de marquage d'antigenes et d'anticorps avec des enzymes et son application en immunodiffusion. C.R. Acad. Sci. (Paris) 262:2543 (1966).
6. P.K. Nakane and G.B. Pierce, Enzyme-labelled antibodies: preparation and application for the localization of antigens. J. Histochem. Cytochem. 14:929 (1966).
7. U. Hammerling, T. Aoki, H.A. Wood, L.J. Old, E.A. Boyse and E. DeHarven. New visual markers of antibody for electron microscopy. Nature (London) 223:1158
8. M.J. Karnovsky, E.R. Unanue and M. Lenenthal. Ligand-induced movement of lymphocyte membrane macromolecules. II. Mapping of surface moieties. J. Exp. Med. 136:907 (1972).
9. F.A. Pepe. The use of specific antibody in electron microscopy. I. Preparation of Mercury-labelled antibody. J. Biophys. Biochem. Cytol. 11:515 (1961).
10. J.L. Fahey and E.W. Terry. in: "Handbook of Experimental Immunology, Vol. 1 Ion Exchange Chromatography and Gel Filtration" D.M. Weir, editor, Alden and Mowbray Ltd.: Oxford, pp. 71 (1973).
11. S. Fuchs and M. Sela. in: "Handbook of Experimental Immunology, Vol. 1 Immunoadsorbents" D.M. Weir, editor, Alden and Mowbray Ltd.: Oxford, pp. 11 (1973).
12. R.C. Nairn, editor "Fluorescent Protein Tracing" Livingstone Ltd.: Edinburgh and London (1969).
13. G.D. Johnson and E.J. Holborow. in: "Handbook of Experimental Immunology, Vol. 1 Immunofluorescence" D.M. Weir, editor, Alden and Mowbray Ltd.: Oxford, pp. 18 (1973).
14. L.D. Sternberger, editor "Immunocytochemistry" Prentice-Hall Inc.: Englewood Cliffs (1974).
15. V. Hammerling, T. Aoki, E. DeHarven, E.A. Boyse and L. Old, Use of hybrid antibody with anti-γG and anti-ferritin specificities in locating cell surface antigens by electron microscopy.
16. G.A. Andres, K.G. Hsu and B.C. Seegal. in: "Handbook of Experimental Immunology, Vol. 2, Immunologic Techniques for the Identification of Antigens or Antibodies by Electron Microscopy" D.M. Weir, editor, Alden and Mowbray, Ltd.: Oxford, pp. 34 (1973).
17. C. Morgan, Int. Rev. Cytology, 32:291 (1972).
18. K. Ostrowski, E.A. Barnard, W. Sawicki, T. Chorzelski, A. Langner and A. Mikulski, Autoradiographic detection of antigens in cells using tritium-labeled antibodies. J. Histochem. Cytochem. 18:490 (1970).
19. C. Boosman and J.D. Feldman, The proportion and structure of cells forming antibody γG and γM immunoglobulins, and γG and γM antibodies. Cellular Immunol. 1:31 (1970).
20. F.C. Greenwood, W.M. Hunter and J.S. Glover, The preparation of ^{131}I-labelled human growth hormone of high specific radioactivity. Biochem. J. 89:114 (1963).

21. J.J. Marchalonis, An enzymic method for the trace iodination of immunoglobulins and other proteins. Biochem. J. 113: 299 (1969).
22. G.J.V. Nossal, N.L. Warner, H. Lewis and J. Sprent. Quantitative features of a sandwich radioimmunolabeling technique for lymphocyte surface receptors. J. Exp. Med. 135:405 (1972).
23. H.C. Fertuck and M.M. Salpeter. Localization of acetylcholine receptor by ^{125}I-labeled α-Bungaratoxin binding of mouse motor endplates. Proc. Nat. Acad. Sci. (USA) 71:1376 (1974).
24. G.L. Ada, J.H. Humphrey, B.A. Askonas, H.O. McDevitt and G.J.V. Nossal. Correlation of grain counts with radioactivity (^{125}I and tritium) in autoradiography. Exp. Cell Res. 41:557 (1966(.
25. W.D. Perkins, M.J. Karnovsky and E.R. Unanue. An ultrastructural study of lymphocytes with surface-bound immunoglobulin. J. Exp. Med. 135:267 (1972).
26. E.R. Unanue, W.D. Perkins and M.J. Karnovsky. Ligand-induced movement of lymphocyte membrane macromolecules. I. Analysis of immunofluorescence and ultrastructural radioautography. J. Exp. Med. 136:885 (1972).
27. N.K. Gonatas, A. Stieber, J. Gonatas, P. Gambetti, J.C. Antoine and S. Avrameas. Ultrastructural autoradiographic detection of intracellular immunoglobulins with iodinated Fab fragments of antibody: the combined use of ultrastructural autoradiography and peroxidase cytochemistry for the detection of two antigens: double labeling. J. Histochem. Cytochem. 22:999 (1974).
28. L.A. Sternberger. Electron microscopic immunocytochemistry: a review. J. Histochem. Cytochem. 15:139 (1967).
29. P.W. Faulk and M.G. Taylor. An immunocolloid method for the electron microscopy. Immunochemistry 8:1081 (1971).

QUALITATIVE AND QUANTITATIVE IMMUNOENZYMATIC TECHNIQUES

Stratis Avrameas

Unite d'Immunocytochimie
Department de Biologie Moleculaire
Institut Pasteur
25, Rue du Docteur Roux
75724 PARIS CEDEX 15, France

INTRODUCTION

Because an enzyme can be revealed and also accurately determined by appropriate procedures, enzyme-labelled antigens and antibodies can be used to localize, detect or measure constituents of biological interest (1,2). Such immunoenzyme techniques can be applied in several ways to both basic research and biomedical diagnosis. 1) Localization of cellular constituents; 2) quantitation of antigens or antibodies present in biological fluids or cells; 3) detection of small amounts of antigen-antibody immune precipitates.

The basic principles of all immunoenzyme procedures are the same: an antibody is first covalently linked to an enzyme, then the enzyme-labelled antibody is allowed to react with a cellular or otherwise immobilized antigen, and finally the sites of enzyme activity are measured or revealed by specific enzymatic staining. It is evident therefore that the most important factor in these procedures is the quality of the antibody-enzyme conjugate employed. Consequently the choice of the three components: 1) antibody, 2) enzyme, 3) conjugation procedure, is of prime importance. The conditions that the antibody should fulfill are the same for all immunoenzyme techniques and have already been mentioned in the chapter on immunocytological methods so they will not be considered in this chapter.

However, the requirements of the enzyme and conjugation procedure vary somewhat for each specific immunoenzymatic technique and hence they will be considered separately for each particular immunoenzymatic procedure.

LOCALIZATION OF CELLULAR CONSTITUENTS

The basic principles of these procedures are the same as those that apply to the detection of antigens by immunofluorescence but cytochemical staining is substituted for fluorescence visualization.

Enzyme

Mainly horseradish peroxidase, but also E. coli alkaline phosphatase and Aspergillus niger glucose oxidase-labelled antibodies have been used for the localization of antigens with the light and electron microscope. Peroxidase is at present the enzyme of choice because: a) its cytochemical staining both for the light and electron microscopes is easy to perform and it gives satisfactory and reproducible results; b) its relatively low molecular weight (40.000), which is less than that of all the other enzymes used in immunocytochemical methodology, allows peroxidase conjugates to penetrate more readily into the interior of fixed cells.

Conjugates

Enzyme conjugates are prepared by covalently coupling antigens or antibodies to enzymes. Various coupling procedures have been described but either glutaraldehyde (3, 4), or sodium m-periodate (5) coupling procedures have been most commonly used. The periodate procedure is based on the principle that active aldehyde groups are produced after periodate treatment of glycoproteic enzyme which can react with an antibody added subsequently. The technique has been used principally for the preparation of protein-peroxidase conjugates. The glutaraldehyde procedure is based on the principle that with this reagent cross-linkage is achieved between enzyme marker and proteins to be labelled through free e-amino groups. This reagent can be used in one step procedures for the preparation of almost all enzyme-protein conjugates useful in enzyme-immunoassays (3, 6) and in two-step procedures mainly for the preparation of peroxidase-protein conjugates (4). The main advantage of the latter procedure is that the peroxidase antibody conjugates obtained are more than 90% homogenous and are composed of one molecule of peroxidase coupled with one molecule of antibody. These homogenous and lowest possible molecular weight conjugates are the most suitable for use in immunocytological procedures. In addition to the use of conjugates prepared by covalently coupling enzymes to antigens or antibodies, it is also possible to make use of non-covalent interactions. The possibility of using the non-covalent interaction between an antibody and an enzyme has been described (7, 8, 9) and more recently the use of the strong interaction between avidin and biotin in immunoenzyme techniques has been also reported (10).

General Considerations

The enzyme-labelled antibody technique, or its variants, have been applied successfully to various immunocytological studies using the light microscope.

Because histochemical staining techniques give rise to final reaction products of different colors for different enzymes or even for the same enzyme, enzyme-labelled antibodies have been used for the simultaneous detection of two different cellular constituents. Alternatively, paired staining techniques combining autoradiography or immunofluorescence and immuno-enzymological methods have also been employed. Recently, enzyme immunocytochemical staining combined with phase contrast has been reported. This combination allowed the detection of fine morphological details not revealed or revealed with difficulty by all the other immunocytochemical staining techniques.

Satisfactory and reproducible results have been obtained with immunoenzyme techniques for the ultrastructural localization of cell surface antigens. Using the procedures described in the chapter on immunocytological methods for electron microscopy less satisfactory, but quite acceptable results were also obtained for the ultrastructural localization of intracellular antigens.

Comparison of peroxidase and fluorescein-conjugates antisera have shown that the sensitivities of both methods were similar. A comparison of ferritin-labelled and peroxidase-labelled antibody has indicated that the peroxidase-technique was several times more sensitive than the ferritin method (11).

QUANTITATION OF ANTIGENS AND ANTIBODIES BY ENZYME IMMUNOASSAYS

The basic principles of these procedures are the same as those developed for quantitative radio-immunoassays but the measurement of the enzyme activity replaces counting of the radioactivity (12, 13, 14, 15). The points discussed previously for the localization of cellular constituents are also important in enzyme immunoassay procedures. Thus, the sensitivity and reproducibility of the various immunoassays depend mainly upon the antibody, the enzyme and the enzyme-antibody conjugate employed. With these reagents, most of the above considerations concerning enzyme immunocytochemical staining are also applicable here and therefore only points not discussed previously will be added. An additional and important factor which is needed in enzyme immunoassays, i.e. the insoluble phase employed for the immobilization of the antigen or antibody will be discussed in some detail.

Enzyme

All the enzymes used in enzyme immunohistochemical procedures, i.e. peroxidase, alkaline phosphatase and glucose oxidase, have been also used in enzyme-immunoassays. In addition, since for the latter technique, unlike immunohistochemical staining, there are no particular problems of accessibility to the antigen by labelled antibody, even high molecular weight enzymes like E. coli β-galactosidase can be used (16).

Conjugates

The same considerations as those referred to the preparation of enzyme-antibody conjugates for use in immunocytological work are also applicable here. However, in enzyme-immunoassays, because there are no problems of penetration, high molecular weight conjugates can be used equally well, provided that there is no increase of the background values.

Solid Phase for the Immobilization of Antigen or Antibody

To immobilize antigens or antibody a number of solid phases have been employed. The adsorption of antibodies to plastic surfaces, a procedure introduced for radioimmunoassays (17), produces immobilized antibody preparations equally effective for enzyme-immunoassays. Covalent coupling of antibody to cellulose (18), agarose (19) and polyacrylamide (12) beads produces immobilized antibody preparations which also give reproducible results in enzyme-immunoassays. Recently, the use of magnetic solid phases has also been reported (20).

General Considerations

Over the past few years, various enzyme-immunoassays have been developed and extensively applied with success to the measurement of various constituents of biological interest.

As in radio-immunoassay, competitive and non-competitive procedures have been described in enzyme-immunoassays. The choice of the procedure will depend on the nature of the constituents to be determined. Non-competitive procedures, when applicable, have the advantage of requiring the use of labelled antibody rather than of labelled antigen. Labelled antibody (which has quite similar physiochemical characteristics in all mammalian species) is easier to prepare and standardize than labelled antigen. In assays involving measurements of antibody or of macromolecular antigens, non-competitive procedures offer more advantages than competitive ones.

Enzyme-immunoassays using peroxidase, glucose oxidase, alkaline phosphatase and β-galactosidase conjugates with the same antibody have been compared (20). Best results in terms of accuracy and reproducibility were obtained with alkaline phosphatase and β-galactosidase-labelled antibodies. With these two enzymes the sensitivity of the measurements were equal to that achieved with radioimmunoassays.

Using fluorescent, radioactive and light emission techniques, procedures have been developed which can measure extremely small quantities of enzyme. Such procedures, when applied to enzyme-immunoassays, allow measurement of very small quantities of antigen or antibody. Thus the use of radioactive substrates (21) has allowed the measurement of 10^{-17} grams of cholera toxin. Using 4-methyl-umbellyferyl-β-D-galactopyranoside, a fluorogenic substrate for β-galacotosidase, extremely small quantities of either humoral (16) or cellular constituents (22) have been measured and under certain conditions it was even possible to quantitate constituents on single cells (23, 24).

DETECTION OF IMMUNE PRECIPITATES

As with fluorescent and radio-labelled antibodies, enzyme-labelled antibodies can be employed to enhance the visualization of immune precipitates by enzymatic cytochemical staining (25). In this case, the basic principles of the procedures are the same as those developed for quantitative or qualitative immunodiffusion or immunofixation techniques, but enzyme staining is performed instead of protein staining.

For example, an enzyme-labelled antibody preparation can be employed for the immunological identification of electrophoretic bands. The slabs supporting the electrophoresed material are first incubated with an unlabelled antisera specific for the antigen to be identified. The slabs are then washed to eliminate excess antiserum and incubated with enzyme-labelled antibodies directed against the immunoglobulins of the first antiserum. After washing, the enzyme activity is revealed by using an appropriate substrate.

Such procedures increase the sensitivity of the detection of an immunoprecipitate 10 to 20-fold over the protein staining. When applied to quantitative single radial diffusion or electro-immunodiffusion techniques, concentrations of proteins ranging from 20 to 100 ng per ml can be measured (25).

All the considerations concerning immunocytochemical staining and enzyme-immunoassay are also applicable here. However, for immunodiffusion techniques, best results are obtained with glucose oxidase rather than with peroxidase or alkaline phosphatase-labelled antibodies.

CONCLUDING REMARKS

Compared to the other qualitative and quantitative immunological techniques based on the use of labelled antigen or antibody, the immunoenzymatic techniques present the following advantages: 1) the same enzyme-labelled antigen or antibody can be employed both for quantitative studies and qualitative work by light and electron microscopy; 2) enzyme-labelled antigens and antibodies stored under sterile conditions can be used for years without any appreciable loss of their enzymatic and immunological activities; 3) immunoenzymatic techniques entail a minimal risk of contamination or pollution; 4) there is no need for costly equipment and the procedure can be performed in simply equipped laboratories; 5) the possibility of using fluorogenic substrates makes enzyme-immunoassays a tool of tremendous potential.

REFERENCES

1. S. Avrameas and J. Uriel, Methode de marquage d'antigenes et d'anticorps avec des enzymes et son application en immunodiffusion. C.R. Acad. Sci. (Paris) 262:2543 (1966).
2. P.K. Nakane and G.B. Pierce, Enzyme-labelled antibodies. Preparation and application for the localization of antigens. J. Histochem. Cytochem. 14:929 (1966).
3. S. Avrameas, Coupling of enzymes to proteins with glutaraldehyde. Use of the conjugates for the detection of antigens and antibodies. Immunochemistry 6:43 (1969).
4. S. Avrameas and T. Ternynck, Peroxidase-labelled antibody and Fab conjugates with enhanced intracellular penetration. Immunochemistry 8:1175 (1971).
5. P.K. Nakane and A. Kawaoi, Peroxidase-labelled antibody. A new method of conjugation. J. Histochem. Cytochem. 22:1084 (1974).
6. S. Avrameas, T. Ternynck and J.L. Guesdon, Coupling of enzymes to antibodies and antigens. Scand. J. Immunol. 8:7 (1978).
7. S. Avrameas, Indirect immunoenzyme techniques for the intracellular detection of antigens. Immunochemistry 6:925 (1969).
8. T.E. Mason, R.F. Phifer, S.S. Spicer, R.A. Swallow and R.B. Dreskin, An immunoglobulin-enzyme bridge method for localizing tissue antigens. J. Histochem. Cytochem. 17:563 (1969).
9. L.A. Sternberger, P.H. Hardy, Jr., J.J. Cuculis and H.G. Meyer, The unlabelled antibody-enzyme method of immunohistochemistry. Preparation and properties of soluble antigen-antibody complex (horseradish peroxidase-anti-horseradish peroxidase) and its use in identification of spirochetes. J. Histochem. Cytochem. 18:315 (1970).

10. J.L. Guesdon, T. Ternynck and S. Avrameas, The use of avidin-biotin interaction in immunoenzymatic techniques. J. Histochem. Cytochem. 27:1131 (1979).
11. R. Bretton, T. Ternynck and S. Avrameas, Comparison of peroxidase and ferritin labeling of cell surface antigens. Exptl. Cell Res. 71:145 (1972).
12. S. Avrameas and B. Guilbert, Dosage enzymo-immunologique de proteines a l'aide d'immunoadsorbants et d'antigenes marques aux enzymes. C.R. Acad. Sci. (Paris) 272:2705 (1971b).
13. S. Avrameas and B. Guilbert, A method for quantitative determination of cellular immunoglobulins by enzyme-labelled antibodies. Europ. J. Immunol. 1:394 (1971a).
14. E. Engvall and P. Perlmann, Enzyme-linked immunoadsorbent assay (ELISA). Quantitative assay of immunoglobulin G. Immunochemistry 8:871 (1971).
15. B.K. Van Weemen and A.H.W.M. Schuurs, Immunoassay using antigen-enzyme conjugates. FEBS Lett. 15:232 (1971).
16. K. Kato, Y. Hamaguchi, M. Fukui and E. Ishikawa, Enzyme-linked immunoassay. I. Novel method for synthesis of the insulin-β-D-galactosidase conjugate and its applicability for insulin assay. J. Biochem. 78:235 (1975).
17. K. Catt and G.W. Tregear, Solid phase radioimmunoassay in antibody-coated tubes. Science 158:1570 (1967).
18. B. Ferrua, R. Maiolini and R. Masseyeff, Coupling of gammaglobulin to microcrystallin cellulose by periodate oxydation. J. Immunol. Meth. 25:49 (1979).
19. J.G. Streefkerk and A.M. Deelder, Serodiagnostic application of immunohistoperoxidase reactions on antigen-coupled agarose beads. J. Immunol. Meth. 7:225 (1975).
20. J.L. Guesdon and S. Avrameas, Magnetic solid phase enzyme-immunoassay. Immunochemistry 14:443 (1977).
21. C.C. Harris, R.H. Yolken, H. Krokan and I.C. Hsu. Ultrasensitive enzymatic radioimmunoassay: application to detection of cholera toxin and rotavirus. Proc. Natl. Acad. Sci. USA 76:5336 (1979).
22. D.J. Cameron and B.F. Erlanger, An enzyme-linked procedure for the detection and estimation of surface receptors on cells. J. Immunol. 16:1313 (1976).
23. P. Hosli, S. Avrameas, A. Ullman, E. Vogt and M. Rodrigot, Quantitative ultramicro-scale immunoenzymatic method for measuring Ig antigenic determinants in single cells. Clin. Chem. 24:1325 (1978).
24. S. Avrameas, P. Hosli, M. Stanislawski, M. Rodrigot and E. Vogt, A quantitative study at the single cell level of immunoglobulin antigenic determinants present on the surface of murine B and T lymphocytes. J. Immunol. 122:648 (1979).
25. J.L. Guesdon and S. Avrameas, An immunoenzymatic method for measuring low concentrations of antigens by single radial diffusion. Immunochemistry 11:595 (1974).

CELL FUSION AND THE INTRODUCTION OF NEW INFORMATION INTO TEMPERATURE-SENSITIVE MUTANTS OF MAMMALIAN CELLS

Renato Baserga, Christopher Potten, and P.M.L. Ming

Fels Research Institute and Department of Pathology
Temple University School of Medicine
Philadelphia, Pennsylvania 19140

INTRODUCTION

Cell fusion occurs in nature spontaneously. Barski et al. (1) were the first ones to demonstrate spontaneous fusion between mammalian cells in culture. However, it is likely that many multinucleated giant cells found in normal and pathological conditions (for instance, in measles) may be the result of cell fusion. In cell biology, though, cell fusion has been used to introduce new information, of either phenotypic or genotypic nature into viable mammalian cells. Cell fusion should therefore be discussed together with newer methods for introducing information into mammalian cells which include: manual microinjection, transfection, liposomes, loaded erythrocyte ghosts and permeabilization of membranes. Because of space limitations, we shall omit the last three from the discussion, although liposomes and permeabilized membranes have still unexplored potentials, and we shall take into consideration only: 1) cell fusion; 2) manual microinjection; and 3) transfection.

Cell Fusion

Techniques. Experimental cell fusion was first introduced in 1962 by Okada (2) who used inactivated Sendai virus to fuse mammalian cells. Sendai virus contains a fusogenic protein (3) that causes cellular membranes to fuse among themselves. Mammalian cells can be fused to each other or to chick erythrocytes or to fragments of cells, for instance, cytoplasts, that are chemically enucleated cells. Fusion by inactivated Sendai virus has been superseded in recent years by fusion with polyethylene glycol (PEG), which is more convenient, more reproducible and less toxic than fusion by Sendai virus. One usually uses 50% PEG of 1000 M.W., but the reader should consult the paper by Schneiderman et al. (4) for the detailed techniques.

When cytoplasts are fused to whole cells, the cytoplasts must be preapred by chemical enucleation, i.e. by treatment of cells with Cytochalasin B and centrifugation. With appropriate care (5) one can obtain a population of cytoplasts that is 99% pure, that is, where 99% of the cells are enucleated. One can even go further, but then cell loss becomes prohibitive.

When one fuses mammalian cells to chick erythrocytes, there is no problem is recognizing heterokaryons, because chick nuclei, even when reactivated, are considerably smaller than mammalian nuclei. When two mammalian cells are fused, or when a nucleated mammalian cells is fused with a cytoplast, problems arise in the proper identification of the fusion products, i.e. in the case of dikaryons (fusion products with two nuclei), the ability to distinguish homokaryons (two nuclei from the same type of cells) from heterokaryons (one nucleus from each type of cell). In the fusion products between cytoplasts and whole cells, one has to separate single cells from true cybridoids (resulting from fusion of a single cell with one or more cytoplasts). There are many ways to identify fusion products, but, in cell cycle studies, where one often uses incorporation of thymidine (^{3}H) as an end-point, the safest way is to prelabel different types of cells with latex beads of different sizes, for instance 0.8μ and 2μ. If a fusion product contains beads of a different size, it should be, in theory, a heterokaryon (or a cybridoid). In practice, certain stringent criteria have to be set, and these are explained in detail in the paper by Jonak and Baserga (6).

Applications. Several useful applications have come out of cell fusion, including: a) hybrid cell lines with special characteristics; b) complementation studies; c) mapping of genes on human chromosomes (or chromosomes of other species); and d) reactivation of chick erythrocyte nuclei. The first three introduce genotypic changes in mammalian cells and are based on the formation of hybrid cell lines. Hybrids can be obtained in reasonable numbers and in a reliable manner, if selective pressure to eliminate the parent cell lines can be applied. For instance, if one parent cell line is thermosensitive and does not grow at 39.5° and the other lacks the enzyme thymidine kinase, hybrids can easily be selected in HAT medium (TK$^-$ cells will not grow) at 39.5° (inhibitory for the temperature-sensitive cells). In some cases one of the parent cell lines can simply be a population of cells that do not grow in monolayers under standard conditions, for instance mouse peritoneal macrophages or lymphoblastoid cells in suspension. If selective pressure cannot be applied hybrids can be selected in a cell sorter.

Hybrid cell lines. These can be made to fit one's experimental design. A typical illustration is given in the chapter on "Protein and RNA Synthesis".

Complementation studies. If hybrids can be formed between two mutant cell lines, it means that they belong to two different complementation groups, i.e. that they are defective in different proteins or different subunits of the same protein. In cell cycle studies, good examples of complementation studies recently reported by our group (6,7) (to be discussed below), and by Prescott and collaborators in the genetic analysis of G_1 (8,9).

Mapping of genes on human chromosomes. By using karyotypic analysis of hybrid cells (and, occasionally, more sophisticated techniques of molecular biology), it has been possible to map more than 200 genes in human chromosomes (10). The method is simply based on the identification, in hybrids, between human and rodent cells of the human chromosome that can correct a given biochemical defect in the rodent cell line (usually mouse cells). Again in cell cycle studies, by using cell cycle-specific ts mutants of hamster cells (see below) hybridized to $HPGRT^-$ human fibroblasts, it has been possible to identify on separate human chromosomes genes that control certain required steps in the cell cycle. The results of these studies are summarized in Table 1. The significance of these findings lies in the fact that separate steps in cell proliferation can be mapped on genes located on five different human chromosomes and therefore, presumably, on different genes. The number of genes (five) is small when one considers that at least 32 genes have been identified as controlling the cell cycle of the less complicated yeast S. cerevisiae (11), but it is a first, necessary step toward a genetic analysis of the human cell cycle.

Reactivation of chick erythrocyte nuclei. These studies have given us valuable information on the mechanism of de-differentiation of cells. The chick erythrocyte is presumably a terminally differentiated cell, and yet, after fusion to a mammalian cell its nucleus is reactivated to synthesize both RNA and DNA and to code for chick-specific proteins (12). Reactivation of chick erythrocyte nuclei requires the migration of mammalian proteins of nuclear and nucleolar origin into chick nuclei (12). Interestingly, if one fuses serum-deprived ts mutants of G_1 with chick erythrocytes, the ts mutants of G_1 are incapable of reactivating chick erythrocyte nuclei when the fusion products are incubated at the temperature nonpermissive for the ts mutants (13). This indicates that some G_1 information must be present in order for the chick nuclei to be reactivated.

Use of Cell Fusion to Determine the Informational Content of Cell Cytoplasms. We will use as an illustration the experiments of Jonak and Baserga (6) with cytoplasts of G_1-specific ts mutants of the cell cycle. In these studies we used three different ts mutants, all of which arrest in G_1 at the nonpermissive temperature: AF8, ts13 and HJ-4 (see below). Cells were serum-deprived (G_0-G_1 cells) or serum-stimulated (S-phase cells), cytoplasts were prepared as described above, and then fused to whole cells of the same or of a different type. We then determined the ability of cytoplasts to complement the ts defect in the whole cell by studying the entry into S phase of the fusion products which we called cybridoid. The results are summarized in Table 2. Depending on the cell line used, and on the length of the period of serum deprivation, the information for certain mid- and late-G_1 functions may or may not be present in the cytoplasts of resting cells. However, S-phase cytoplasts always contain the full information for the complementation of the various ts functions (or the defective thymidine kinase in the TK^- cells). These results are the first rigorous demonstration that the nucleus is necessary to generate the information for the transition of mammalian cells from a resting to a growing stage. The results also confirm the earlier observations (14) that resting cells can go into deeper G_0 states with different informational content. In fact, the results with AF8 clearly raise the question of the definition of G_0 cells.

Transfection of Mammalian Cells with DNA (15)

This technique was first introduced by Graham and van der Eb who succeeded in transforming mammalian cells with fragments of Adenovirus DNA. The technique is deceptively simple and it

Table 1. Mapping on Human Chromosomes of Genes that are Defective in Some Temperature Sensitive Mutants of the Cell Cycle

Mutant Cell Line	Site of Arrest at the Nonpermissive Temperature	Complemented by Human Chromosome
AF8	G_1	3
K12	G_1	14
ts13	G_1	4
ts546	metaphase	6
H142	early S	9

From Ming et al (28, 29) and unpublished data.

Table 2. Effect of Fusion of Cytoplasts to Whole Cells Temperature-Sensitive for Entry into S Phase

Recipient Whole Cell	Donor Cytoplast from: S Phase Cells	Donor Cytoplast from: Cells Serum-deprived for (hrs)	% of Cybridoids entering S at the nonpermissive temperature
G_0AF8	ts13	--	35.0
G_0AF8	--	ts13	39.0
G_0AF8	--	AF8	8.0
G_0ts13	--	AF8 (72 hrs)	54.5
G_0ts13	--	AF8 (144 hrs)	19.5
G_0ts13	ts13	--	67.0
G_0ts13	--	--	4.0
G_0AF8	--	--	7.0

From Jonak and Baserga (6), and unpublished data. Cybridoids are fusion products that can be rigorously characterized as deriving from one whole cell fused to one or more cytoplast. These experiments were also carried out at permissive temperature simply to show that the fusion process is only slightly inhibitory (not shown). At the nonpermissive temperature, the fusion products were incubated in 10% serum.

consists of exposing cells to DNA under appropriate conditions which include the presence of calcium phosphate. The transfection technique has been used by Axel and co-workers who, in a series of brilliant experiments, have been able to transform TK^- cells into TK^+ cells by using the thymidine kinase gene of the Herpes virus (HSV-TK). We have used the same technique to correct the ts defect of two G_1-ts mutants, AF8 and ts13. Using 20-30 Kilo base pairs fragments of BHK DNA (BHK being the parent cell line of AF8 and ts13), we have been able to obtain from AF8 and ts13 a number of transformant clones that grow at the nonpermissive temperature. We have now repeated the experiment using BHK genome cloned in Charon phage. This is the first cloning of mammalian genes that are known to control the progression of cells through the cell cycle.

Manual Microinjection

This technique, pioneered by numerous investigators, has been made popular by the efforts of Graessmann and Graessmann (16) who have made the methodology fast, reliable and relatively uncomplicated. It consists of microinjecting directly into cells with the aid of a micropipette a solution containing the desired macromolecules: protein, RNA, DNA. About 2000 molecules per cell can be microinjected in a volume of 10^{-11} ml (these figures are a rough approximation), and the phenotypic changes induced by the microinjected macromolecules can be monitored by immunofluorescence, autoradiography or biochemical analysis (15). A microinjected gene, for instance the HSV-TK or the A gene of SV40, is promptly expressed and, in fact, the microinjected cells are phenotypically changed for at least several days. This technique is particularly useful for cell cycle studies in which phenotypic changes for limited periods of time are studied. Graessmann (16) and co-workers were the first ones to show that SV40 tantigen or cRNA from the SV40 A gene, when microinjected, can induce cellular DNA replication in resting mouse cells. In our laboratory we have microinjected a cloned SV40 A gene into G_1-ts mutants, i.e., ts13 and AF8 cells. The microinjected A gene caused cellular DNA replication in these cell lines at both permissive and nonpermissive temperatures. It seems therefore that the product(s) of the SV40 A gene can induce G_0 cells to enter S phase in the absence of certain ts G_1 functions that are required by serum-stimulated cells. Clearly, this is just a beginning - this technique can now be extensively used to identify genes and gene products that control the progression of cells through the cell cycle.

Conclusions

Transfer of genotypic or phenotypic information to mammalian cells can be obtained by a variety of techniques. Somatic cell fusion between two types of cells or between cytoplasts of one type and cells of another is still useful, especially for chromosome mapping, complementation studies and creation of particular hybrid cell lines (for instance, hybridomas). DNA transfection and manual microinjection introduce into cells specific macromolecules, which is a distinct advantage over somatic cell fusion. In the future, the use of liposomes entrapping nucleic acids or proteins may give us the desired goal of bulk introduction of specific macromolecules into viable mammalian cells.

ts Mutants of the Cell Cycle

In recent years mammalian cells with a stably altered phenotype have been extensively used in a variety of biochemical and genetic studies, extending also to cell proliferation and differentiation. Siminovitch (17) in an excellent review has marshalled the evidence that these stably altered phenotypes have, at least in some cases, a genetic basis; and, in agreement with him, we shall use here for these cells the term, mutant, although the term may turn out to be inappropriate in some instances. We are interested here in the mutants temperature-sensitive (ts) for growth and, more precisely, in cell cycle-specific ts mutants.

Cell cycle-specific ts mutants have been operationally defined as mutants that arrest at the nonpermissive temperature in a specific phase of the cell cycle (18). Several such mutants have been isolated and are listed by Thompson and Siminovitch (19). It is important to emphasize that cell cycle specific ts mutants must be differentiated from growth mutants. Growth mutants are simply mutants that do not grow at all at the nonpermissive temperature; however, when under non-restrictive conditions, they will stop at any point in the cell cycle, usually at the point in which they find themselves when the ts function is expressed. Cell cycle-specific ts mutants, instead, continue throughout the cell cycle, even under restrictive conditions, until they arrive at that specific phase of the cell cycle in which the ts function is needed. In this discussion we will be mostly interested in G_1 ts mutants, i.e., mutants operationally defined as mutants that arrest at the nonpermissive temperature in the G_1 phase of the cell cycle, while growing normally at the permissive temperature.

Production and Selection of ts Mutants

In general, the methods used to select ts growth mutants of animal cells have been based on the establishment of conditions for killing wild-type cells at the nonpermissive temperature, taking advantage of the fact that they multiply normally, while the ts mutants are arrested in growth and division (20). The method generally used consists in mutagenizing cells with N-methyl-N-nitrosoguanidine. After MNNG treatment the cells are allowed to recover from the toxicity and then they are shifted up to the nonpermissive temperature. The cultures are then exposed to agents that are lethal for dividing cells, for instance, 5-flurodeoxyuridine, or bromodeoxyuridine, followed by exposure to ultraviolet light. After the treatment the cells are shifted back to the permissive temperature to allow growth of the ts mutants. For more details one should consult the review by Basilico (20). These procedures are repeated two or three times, and growth mutants are thus selected. They are then characterized further to determine whether they are growth mutants or cell cycle specific ts mutants.

We know very little about the biochemical defects in ts mutants of the cell cycle. Only in one instance has some information been obtained on the biochemical basis of a cell cycle-specific ts mutation (see below). However, it is fair to say that temperature-sensitive mutations are usually due to point mutations that make the gene product nonfunctional at the restrictive temperature. While this has never been demonstrated in mammalian cells, it is likely, at least on the basis of what we know about temperature-sensitive mutations in bacteria.

Nature of the ts Mutation

There is one good illustration of a temperature-sensitive mutation in mammalian cells in which the evidence is substantial that the mutation is a point mutation. These are certain CHO mutants that are resistant to α-amanitin. α-Amanitin is a drug that specifically inhibits RNA polymerase II by binding to the α-amanitin binding subunit of RNA polymerase II (21). Ingles and co-workers (22) have developed a series of mutants that are highly resistant to α-amanitin and that can grow in the presence of high concentrations of α-amanitin. Some of these mutants are not only α-aminitin resistant but are also ts for growth, for instance, mutant AMA-R1 (22). Studies on the revertant frequency of these mutants have clearly indicated that the α-amanitin resistance and the temperature-sensitive mutation are on the same subunit and are a point mutation. It seems, therefore, that at least in this case, the ts mutation can be demonstrated to be due to a single point mutation in the α-amanitin binding subunit of RNA polymerase II. These experiments are also interesting in as much as they clearly indicate that α-amanitin has only one single point of action; namely, the α-amanitin binding subunit of RNA polymerase II. A point mutation in that subunit makes the cells α-amanitin resistant so that the cells can now grow in very high concentrations of the drug. This is an unusual situation as far as drugs are concerned, since most of the drugs have a variety of effects on cells. In the case of α-amanitin one can properly say that there is only one single point of action and that other effects that one may notice with α-amanitin are purely secondary to its effect on the α-amanitin binding subunit of RNA polymerase II.

Another inportant thing to define is the concept of the execution point. The execution point is defined (23) as the point in the cell cycle at which the ts function is no longer needed for progression. This should be made very clear. It is not the point at which the ts function is expressed. The ts function could have been expressed at earlier times and what the execution point simply says is that beyond that point the ts function is no longer needed. For instance, AF8 cells, whether stimulated by serum or collected after mitotic detachment have an execution point which is located about 8 hr before the beginning of S. This simply means that 7 hr before the beginning of S the ts function of tsAF8 is no longer needef for the progression through the cell cycle.

The definition of G_1 ts mutants implies also that: 1) when collected by mitotic detachment and plated at the nonpermissive temperature, the cells do not enter S; 2) when made quiescent by serum restriction and subsequently stimulated at the nonpermissive temperature, the cells do not enter S; 3) under both conditions the cells enter S at the permissive temperature; 4) the parental line is capable of entering S at both temperatures, and finally 5) that the cells are not arrested in other phases of the cell cycle when shifted up to the nonpermissive temperature. A few of the ts mutants do meet all of these criteria. It would have been desirable to make ts mutants from 3T3 cells which have been, and are, extensively used in studies on the cell cycle and in many other studies. Unfortunately, although there are some reports in the literature about ts mutants of the cell cycle derived from 3T3 cells, our own experience and the experience of several other individuals is that wild-type 3T3 cells do not do well at all once the temperature is about 39°. This makes it very difficult to select ts mutants and to handle them so as to clearly identify the arrest in the cell cycle progression as due to a temperature-sensitive function rather than to a general deterioration of cellular processes. For this reason, most of the studies on the cell cycle have been carried out with ts mutants of either Syrian hamster cells, especially BHK cells, or Chinese hamster cells. Wild-type Chinese hamster cells grow usually quite well at a temperature of 40°, while BHK, at least in our experience, grow just as well at 41° as at 37°. In the rest of the discussion we will concentrate on two mutants that Basilico and collaborators (20) derived from BHK cells, specifically tsAF8, and ts13.

AF8 and ts13 Mutants

AF8 grow normally at 34° and their nonpermissive temperature is 40°, at which temperature less than four percent of the cells may be found in S, either coming from mitosis or G_0. At the nonpermissive temperature they die in four to five days. The median execution point of the ts function is approximately 8 hr before S, regardless of previous growth conditions (24). Hybrids between AF8 and human cells containing all hamster chromosomes and human chromosome 3 can grow at 40°. AF8 and ts13 complement each other and presumably their biochemical defect is on a different gene, or at least on different subunits of the same enzyme. Figure 1 shows the different behavior of AF8 and ts13 in terms of cell cycle. While both cell types do not grow at 40°, and grow perfectly normal at 34°, there are some differences concerning their ability to enter S under certain conditions. In Fig. 1 the cells were made quiescent by serum deprivations for 48 hr in 0.5% serum. The cells were then stimulated in 10% serum at the nonpermissive temperature for 16 hr. They were then shifted down to 34° and replaced again in 0.5% serum. Figure 1 shows that ts13 cells entered DNA synthesis in great numbers while AF8 cells did not. Clearly, when the temperature is shifted down, but the cells are left in 10% serum, both ts13 and tsAF8 can enter S phase (not shown). It seems, therefore, that ts13 cells can complete all the serum-specified functions even at the nonpermissive temperature, so that, when the cells are brought down to the

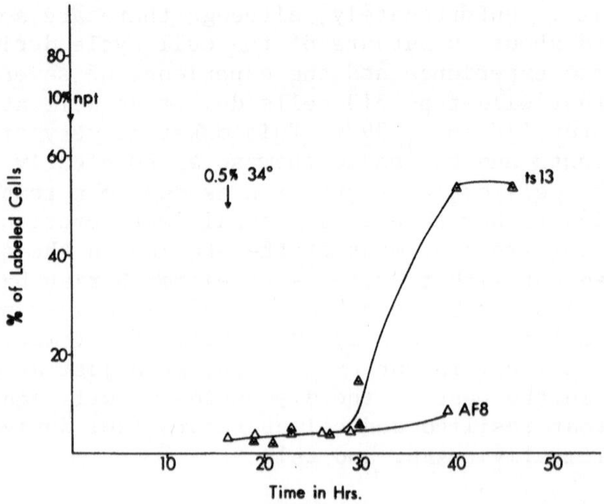

Figure 1. Entry into S phase of ts13 and tsAF8 Cells

Quiescent cells were stimulated with 10% serum at the nonpermissive temperature. After 16 hr, they were replaced in 0.5% serum, and shifted down to 34°. ▲—▲ ts13 cells; △—△ tsAF8 cells. Time (in hr) on the abscissa; cumulative labeling index on the ordinate.

permissive temperature even in 0.5% serum, they are capable of entering S phase normally. On the contrary, at least one of the serum-specified function, has not been carried out by AF8 cells at the nonpermissive temperature so that, if the cells are now brought down to 34° in low serum, they will not enter DNA synthesis at all. These experiments are interesting because they also clearly point out that the ts defects must be different; one, being in a function that is serum dependent, while the other is not.

We shall see below what is the biochemical defect in AF8 cells. ts13 cells are less characterized, but they are also known to arrest in G_1 with a median execution point about 3 hr before S (25). Like AF8, ts13 cells in G_0 cannot reactivate chick nuclei when the heterokaryons between quiescent cells and chick erythrocytes are incubated at the nonpermissive temperature. However, both ts mutants can reactivate the chick erythrocytes if incubated at the permissive temperatures.

Nature of the ts Defect in AF8 Cells

tsAF8 are the only ts mutants of the cell cycle for which some information is available about its biochemical defect. The first hint at the possible defect in AF8 came from the experiments of Rossini and Baserga (26) who found that RNA polymerase II activity in isolated nuclei decreased when AF8 cells were incubated at the nonpermissive temperature. RNA polymerase I activity was not affected and the cells remained viable for at least 60 hr, while RNA polymerase II activity was virtually gone by 24 hr. In the parent cell line, BHK, RNA polymerase II activity remained constant even at the temperature of 41°. In subsequent experiments Rossini et al. (27) found that the α-amanitin binding subunit of RNA polymerase II (and presumably the whole molecule) disappeared in tsAF8 at 40° with a half-life of $\sim$10 hr. Again, no such changes could be detected in tsAF8 at 34° or in BHK cells at either 34° or 41°. A rapid survey of protein synthesis indicated that protein synthesis was not affected for at least 30 hr at the nonpermissive temperature. The conclusion was reached that ts defect in AF8 had something to do with either the synthesis, the assembly or the stability of RNA polymerase II.

Further studies have confirmed this conclusion. Ingles et al. (personal communication) have found that AF8 do not complement with a ts mutant of CHO cells, that has a point mutation in the α-amanitin binding subunit of RNA polymerase II. In our laboratory we have shown that SV40 A gene and HSV-TK gene are not expressed when microinjected into tsAF8 at 40°, while normally expressed when microinjected in AF8 at 34°, or in other ts mutants at either temperature.

The implication of these findings is that RNA polymerase II is needed for the progression of cells in G_1, which in turn is strong evidence that unique copy genes must be transcribed in G_1. These results and those with cytoplasts previously described place, on a firm basis, the intuitive notion that transcription of nuclear genes is necessary for the transition of cells from the resting to the growing stage.

Cloning Wild Type Genes

Although it seems obvious that genes and their products are

necessary for cell proliferation, if one is asked to name one single gene whose expression has been rigorously demonstrated to be required for cell division, no answer is forthcoming. Very likely, the enzymes necessary for DNA replication are expressed during the transition from G_0 (or mitosis) to S, but even for these obvious gene products there is no demonstration that the genes are actually transcribed before S. It is therefore important to try to identify gene and gene products that control cell proliferation. For this purpose we have chosen two approaches: 1) the use of viral genes (especially SV40 and Adenovirus); and 2) gene cloning. We have therefore cloned the whole BHK genome in Charon phage, and we are now attempting to isolate the genes that are defective in AF8 and ts13 cells. Once these genes are isolated, they can be used as probes to study their expression during the cell cycle.

REFERENCES

1. G. Barski, S. Sorieul and F. Cornefert, J. Nat. Canc. Inst. 26:1269-1291 (1961).
2. Y. Okada, Exp. Cell Res. 26:98-107 (1962).
3. A. Scheid, M.C. Hsu and P. Schoppin, in: "Introduction of Macromolecules into Viable Mammalian Cells" R. Baserga, C. Croce and G. Rovera, eds., Alan Liss, New York, pp. 187-204 (1980).
4. S. Schneiderman, J.L. Farber and R. Baserga, Somantic Cell Genet. 5:263-269 (1979).
5. G. Veomet, J.W. Shay, P. Haugh and D.M. Prescott, Methods Cell Biol. 13:1-5 (1976).
6. G.J. Jonak and R. Baserga, Cell 18:117-123 (1979).
7. J. Floros and R. Baserga, Cell Biol. Int'l Rpts. 4:75-82 (1980).
8. R.M. Liskay, Exp. Cell Res. 114:69-77 (1978).
9. R.M. Liskay and D.M. Prescott, Proc. Nat. Acad. Sci. 75: 2873-2877 (1978).
10. Human Gene Mapping 4, Cytogenetics & Cell Genetics 22:1-730 (1978).
11. D.R.W. Edwards, J.B. Taylor, W.F. Wakeling, F.Z. Watts and I.R. Johnston, Cold Spring Harbor Symposium 43:577-586 (1978).
12. N.R. Ringertz and R.E. Savage, in: "Hybrids", Academic Press, New York, pp. 366 (1976).
13. R. Baserga, J. Cell. Physiol. 95:377-386 (1978).
14. L.H. Augenlicht and R. Baserga, Exp. Cell Res. 89:255-262 (1974).
15. R. Baserga, C. Croce and G. Rovera, eds. "Introduction of Macromolecules into Viable Mammalian Cells" Alan Liss, New York, pp. 357 (1980).

16. M. Graessmann and A. Graessmann, Proc. Nat. Acad. Sci. 73:366-370 (1976).
17. L. Siminovitch, Cell 7:1-11 (1976).
18. E.H.Y. Chu, J. Cell. Physiol. 95:365-366 (1978).
19. L. Thompson, L. Siminovitch, J. Cell. Physiol. 96:361-366 (1978).
20. C. Basilico, J. Cell. Physiol. 95:365-366 (1978).
21. M. Cochet-Meilhac and P. Chambon, Biochim. Biophys. Acta. 353:160-184 (1974).
22. C.J. Ingles, Proc. Natl. Acad. Sci. 75:405-409 (1978).
23. J.R. Pringle, J. Cell. Physiol. 95:393-406 (1978).
24. T. Ashihara, S.D. Chang and R. Baserga, J. Cell. Physiol. 96:15-22 (1978).
25. J. Floros, T. Ashihara and R. Baserga, Cell Biol. Int'l Rpts 2:259-269 (1978).
26. M. Rossini and R. Baserga, Biochemistry 17:858-863 (1978).
27. M. Rossini, S. Baserga, C.H. Huang, C.J. Ingles and R. Baserga, J. Cell Physiol. (in press) (1980).
28. P.M.L. Ming, H.L. Chang and R. Baserga, Proc. Nat. Acad. Sci. 73:2052-2055 (1976).
29. P.M.L. Ming, B. Lange and S. Kit, Cell Biol. Int'l Rpts 3:169-178 (1979).

VARIOUS AUTORADIOGRAPHIC METHODS AS A TOOL IN CELL GROWTH STUDIES

B. Maurer-Schultze

Institut für Medizinische Strahlenkunde
University of Würzburg, Germany

In studies in the field of cell biology the autoradiographic method demonstrates radioactivity in tissues and cells by means of a photographic emulsion. The biological specimens, histological sections, smears of cells or squashes of tissue, are covered with a particularly sensitive photographic emulsion. After the appropriate exposure time and photographic development silver grains are found in the emulsion above those cells or cell structures which are radioactively labeled (for the autoradiographic technique see Schultze, 1969; Amlacher, 1974; Rogers, 1979).

The photographic emulsion replaces the counter used in biochemical studies. The advantage of the autoradiographic method is to study the incorporation of labeled precursors into the undisrupted tissue or cell, while biochemical methods study the incorporation of labeled precursors per mg homogenized tissue or cells or cell constituents. The disadvantage of the autoradiographic method consists in the fact that it measures only the radioactive label; however, it gives no information about the chemical binding of this radioactivity. Therefore, additional biochemical studies are often necessary.

The sensitivity of the autoradiographic method is even higher than that of the counter. With suitable autoradiographic techniques, decay rates of one disintegration per day within one cell can be measured. Provided the autoradiographic method is appropriately applied it can well be used for quantitative studies as will be shown in the following.

One of the main fields of autoradiography in studies of cell biology is the investigation of metabolic processes in the cell.

In the following a brief review will be given on autoradiographic studies of the cellular protein synthesis with labeled amino acids, the RNA synthesis with 3-H-cytidine or -uridine and the special field of cell proliferation, studied with labeled thymidine (TdR).

PROTEIN SYNTHESIS

Fig. 1 shows autoradiographs of unstained sections of the stomach of the mouse 60 min after application of 10 different 3-H-labeled amino acids (Citoler et al., 1966; Schultze and Maurer, 1967; Maurer, 1972). It is striking that there is a great similarity in the relative grain density between the various cell types of the stomach after administration of the different amino acids. The chief cells of the stomach glands show the highest grain density, the submucous membrane and the smooth muscle a very low one. The surface epithelia have a slightly higher grain density. This similarity between the autoradiographs, i.e. the fact that the incoporation pattern does not depend on the amino acid used, shows that the amino acid incorporation represents the de-novo synthesis of protein molecules. This is true for other tissues, too. In addition, biochemical studies have shown that 97-100% of the incorporated amino acid activity was found to be peptide bound to protein (Schultze et al., 1960; Droz and Warshawsky, 1963).

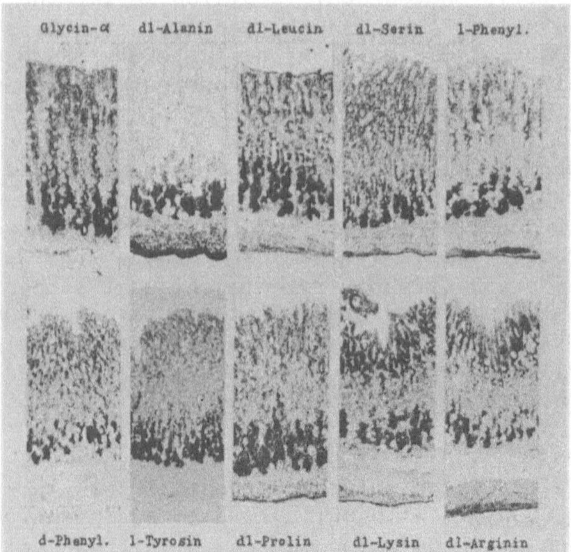

Fig. 1. Autoradiographs of unstained sections of the stomach of the mouse 60 min after application of 10 different 3-H-labeled amino acids (from Citoler et al., 1966).

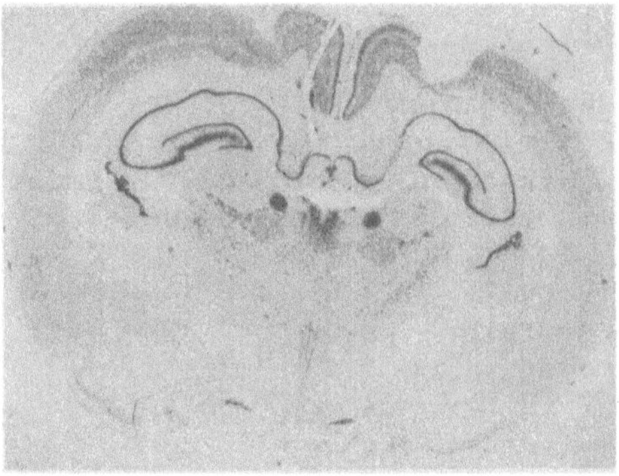

Fig. 2. Autoradiograph of an unstained section of the diencephalon of the rabbit 60 min after application of 35-S-thio-amino acids (from Oehlert et al., 1958).

Fig. 2 depicts an autoradiograph of an unstained cross section through the diencephalon of the rabbit 60 min after application of 35-S-thio-amino acids (Oehlert et al., 1958). The structure of the brain and the different nuclei of this region can be recognized even better than in cresyl violet stained sections. This is due to the high grain density above the neurons. The grain density above the white matter is about 50 times less. Because of the low amino acid incorporation into white matter, biochemical studies showed a very low amino acid incorporation rate per mg brain tissue. This is due to the fact that there are far fewer neurons than white matter per unit weight of brain tissue. Autoradiographic studies have shown that the neurons belong to the cells with the highest protein metabolism within the organism.

Autoradiographic studies of the protein synthesis of different tissues in mice, rats and rabbits with labeled amino acids show that there are certain cell types in the organism with a high amino acid incorporation rate, like neurons (irrespective of their localization in the organism), protein secreting cells (pancreas) as well as strongly proliferating cells (intestinal crypt cells). There are other cell types with a medium incorporation rate like liver parenchymal cells and others with a very low amino acid incorporation rate like muscle and connective tissue. The grain density between the first and the last group differs by a factor of 30-70 (Schultze et al., 1960; Citoler et al., 1966).

For a quantitative evaluation of the autoradiographic results, i.e. the interpretation of the grain densities on the autoradiographs as protein turnover rates, further knowledge about two factors is required:

1. the specific activity of the free amino acid
2. the amino acid composition of the protein in the different cell types.

1. Specific activity of the free amino acid: The labeled amino acid applied is rapidly distributed within the pool of corresponding free amino acid in the organism. This results in a certain specific activity of this free amino acid. The relationship between the specific activity of the free amino acid and the turnover rate of the protein bound amino acid is expressed as:

$$\text{Amino acid activity incorporated into protein during the time dt} = A_o \cdot dt \cdot s(t)$$

$$\text{Amino acid activity incorporated into protein between the time of injection (0) and sacrifice (T)} = A_o \cdot \int_o^T s(t) \cdot dt$$

$$= A_o \cdot s_{(mean)} \cdot T$$

A_o = amount of free amino acid incorporated into protein/unit time.
s_t = specific activity of the free amino acid at any given time
T = duration of the experiment.

A_o can be calculated, if the activity of the protein bound amino acid (between 0 and T) and the mean specific activity of the free amino acid is measured. As is shown by this formula, the activity incorporated into protein -this corresponds to the grain density in the autoradiographs- is not simply proportional to the turnover rate of the amino acid (A_o) but is also proportional to the mean specific activity of the free amino acid at the location of protein synthesis. Only if the mean specific activity is the same for the different tissues in the organism, the incorporated activity is proportional to the turnover rate, and only in this case the grain density represents the relative protein turnover rate of the different cell types.

For instance, if there is a blood brain barrier for certain amino acids (as found for alanine, serine, glycine and proline; Garweg, 1969; Schultze et al., 1972 b) the grain density above the neurons will be low after application of these amino acids, although the protein turnover rate of these neurons is high. This is due to the fact that the mean specific activity of these 4 amino acids in the brain is low because of the blood brain barrier.

2. Amino acid composition of the protein of different cell types: Even if the specific activity of the free amino acid is the same in the whole organism, the grain density depends on a second factor, namely the amino acid composition of the cellular protein. Provided the amino acid composition of the protein of two cell types is the same, but the protein turnover rate in the one cell type is 10 times higher than in the other, the turnover rate of the indivi-

dual amino acids will also be 10 times higher in the cell type with the higher protein turnover rate. The grain density on the autoradiographs will show a ratio of 10:1 after application of various amino acids.

Things are different, if the amino acid composition of the protein of two cell types is not the same. For instance, the relative grain density above the cells of the hair follicle in the rat skin is about 5 times higher after application of 35-S-thio-amino acids than after application of other labeled amino acids (Schultze et al., 1960). This is due to the greater amount of cystine incorporated into the protein of the hair follicle, since the keratin of the hair follicle contains about 7 times more cysteine, namely 17% cystine compared to 2.5% of the other amino acids. Thus, for determining the protein turnover rate from the grain density on the autoradiographs, the amino acid composition of the protein must be considered.

Duration of the experiment: Furthermore, the duration of the experiment also plays an important role in studies of protein synthesis. Only in short term experiments of about 1 h the incorporated amino acid activity represents the different turnover rates (provided the above mentioned prerequisites are met) because only in this case the catabolism of those proteins with a fast turnover rate is negligible. The mean life span of protein in neurons of the rat, for instance, is less than 1 day (6.6 h), that of the muscle cells about 50 days (Niklas et al., 1958). Furthermore, in long term experiments reutilization of labeled amino acids from catabolized proteins affects the results.

Concerning the protein synthesis in the different cell types of the mammalian organism, a rather interesting relationship was found between the protein synthetic rate of the entire cell and the proportion of nuclear volume to the cellular volume (Schultze et al., 1965; Citoler et al., 1966; Schultze and Maurer, 1967). The amount of labeled amino acid incorporated into the entire cell depends on the proportion of nuclear volume to the entire cell volume, i.e. on the nuclear-cell-volume ratio. There is a linear relationship between the protein turnover rate of a cell and its nuclear-cell-volume ratio. For instance, the neurons with a high nuclear-cell-volume ratio have a high protein turnover rate, while the protein turnover rate is very low in muscle cells with a very low nuclear-cell-volume ratio. This leads to the difference in grain density by a factor of 50 and more on the autoradiographs.

Particularly in the brain, labeled amino acids are also used for studies of neuronal connectivities and the determination of a fast and slow axonal flow (for review, e.g., see Cowan and Cuénod, 1975).

It should be mentioned that protein is synthesized during the whole cell cycle. Therefore, all cells are labeled after application of labeled amino acids. Only during metaphase the protein synthetic rate is somewhat reduced (for review, e.g., see Schultze, 1968).

RNA SYNTHESIS

Mainly 3-H-cytidine and -uridine are used as precursors in autoradiographic studies of RNA synthesis. However, these precursors are not exclusively incorporated into RNA, but to a small extent also into DNA. Only uridine, labeled in position 5, can be considered as a selective RNA label.

In autoradiographic studies of the RNA synthesis in cells and tissues the administration of 3-H-cytidine and -uridine leads to a strong labeling of the nucleolus within minutes. 30 min after labeling, the grain density above the nucleolus has further increased, however, there are also grains above the karyoplasm. 3 h after labeling, most of the 3-H-activity is in the cytoplasm. These autoradiographic findings represent the well known transport of RNA from the nucleus to the cytoplasm.

The quantitative evaluation of autoradiographs after application of RNA precursors is somewhat more complicated. In autoradiographic studies of RNA synthesis fixation of the tissue plays an important role. Depending on the fixative used, different amounts of labeled RNA are removed from the tissue (Schneider and Maurer, 1963, Antoni et al., 1965). As was shown in these experiments the amount of RNA which is removed by the various fixatives is even different for the cytoplasm and the nucleus. Thus, the choice of the fixative is important. Neutral formalin has proved to be satisfactory.

Influence of ß-selfabsorption: For a quantitative evaluation of autoradiographs in studies of the RNA synthesis the influence of ß-selfabsorption must be considered. Due to the short range of the tritium-ß-particles, ß-selfabsorption within the tissue greatly influences the autoradiographic effect. Because of the different mass density of the nucleolus, karyoplasm and cytoplasm, the ß-selfabsorption is quite different for these cell structures (Maurer and Primbsch, 1964). For instance, if the 3-H-activity within nucleolus, karyoplasm and cytoplasm of mouse esophagus epithelia (5 and 10 min after injection of 3-H-uridine and -cytidine) is calculated, based on grain counts on autoradiographs, a ratio of 25% : 70% : 5% is obtained. However, if these values are corrected for ß-selfabsorption of these cell structures, a different ratio must be derived from these experiments: 60% of the 3-H-activity are in the nucleolus, 36% in the karyoplasm and 4% in the cytoplasm (Schultze and Maurer, 1963).

Relationship between RNA- and protein synthesis

Comparative autoradiographic studies of the RNA- and protein synthesis of the different cell types in the mammalian organism have led to an interesting relationship (Schultze and Maurer, 1967): Apart from the different intracellular distribution of the grains, the mean grain density/unit area is very similar after application of protein and RNA precursors. That means cells with a high protein synthetic rate also have a high RNA synthetic rate and vice versa. This proportionality is represented by equation II.

$$\text{I} \quad \text{Protein turnover rate} = c_1 \cdot \text{RNA content}$$

$$\text{II} \quad \text{Protein turnover rate} = c_2 \cdot \text{RNA turnover rate}$$

$$\frac{\text{I}}{\text{II}} = \frac{\text{RNA content}}{\text{RNA turnover rate}} = \frac{c_2}{c_1} = ML_{RNA}$$

Due to the similarity of the autoradiographs in studies of the RNA and protein synthesis for the different cell types, "c_2" should have about the same value for the various kinds of cells.

On the other hand it is known from the work of Caspersson and Brachet that the protein turnover rate of a cell is proportional to its RNA content (equation I). That means that "c_1" should also be about the same for the different cell types. If equation I is divided by equation II, i.e. the amount of cellular RNA by the absolute RNA turnover rate in g/unit time, the quotient is constant. However, this quotient represents the mean life span of RNA related to the entire cellular RNA. Thus, the mean life span of RNA should be the same in different cell types.

Biochemical determinations of the RNA content and the RNA turnover rate for the liver of the mouse resulted in a mean life span of about 5 days for the entire cellular RNA and of about 1/2 a day for the nuclear RNA (Schultze, 1968). This corresponds well with the autoradiographic findings and the role of m-RNA in protein synthesis.

DNA SYNTHESIS

The investigation of DNA synthesis has developed into a special independent field of research, the field of cell kinetics. For studies of DNA synthesis a specific precursor is available, 3-H- or 14-C-labeled thymidine (TdR) which is practically exclusively incorporated into DNA. A further advantage of this precursor is its short availability time (for review, e.g., see Cleaver, 1967 and Schultze, 1969). After intravenous or intraperitoneal injection TdR rapidly diffuses from the blood stream and is available to most tissues.

The clearance from the serum occurs within minutes. However, this rapid decrease of free TdR in the serum allows no conclusions on the availability of TdR at the location of DNA synthesis, i.e. within the cell nucleus.

Availability time

The availability time at the location of DNA synthesis can be measured by counting the increase in the mean grain number/nucleus as a function of time after injection of labeled TdR. This availability time represents the time between the injection of labeled TdR and its incorporation into fixable DNA and includes all metabolic steps inbetween.

Fig. 3 gives an example of such a measurement of the availability time (Carmona et al., in prep.). Fig. 3b shows the increase in the mean grain number/nucleus of jejunal crypt cells of the mouse after a single injection of 3-H-TdR. For 3-H-TdR of high specific activity, commonly used in cell kinetic studies, the plateau of the mean grain number/nucleus is reached about 40-60 min after 3-H-TdR injection. That means from that time on 3-H-TdR is no longer incorporated into fixable DNA, since 3-H-TdR is no longer available in the nucleus. Thus, the maximum availability time is about 40-60 min. However, the majority of labeled TdR is already incorporated into fixable DNA within 20 min after injection. As shown in Fig. 3a the same is true for 14-C-TdR. In the case of 14-C-TdR the availability time was measured as mean grain count/nucleus or biochemically as specific activity of the DNA of the small intestine of the mouse.

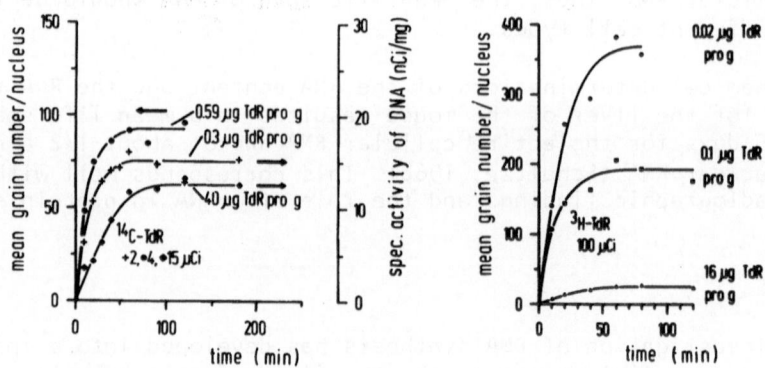

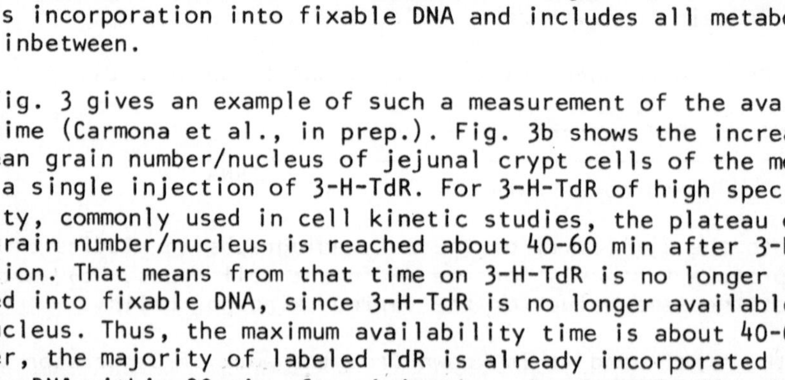

Fig. 3. Availability time of labeled thymidine (TdR). Single injection of a: 14-C-TdR and b: 3-H-TdR and different amounts of TdR.
The increase in labeling of the newly synthesized DNA in the crypt cells of the mouse was measured (from Carmona et al., in prep.).

This availability time is actually short compared to the duration of the S phase which is about 8 h for a great number of mammalian cell types (cf. Maurer and Schultze, 1968). That means that an injection of labeled TdR is more or less a pulse labeling which is a prerequisite for the application of labeled TdR in cell kinetic studies.

An injection of labeled TdR into an animal or the addition of labeled TdR to a cell culture leads to labeling of all cells that are in S phase at the time of application of the labeled TdR, i.e. of all cells which synthesize DNA. The ratio of labeled cells to all cells of a cell population, shortly after application of labeled TdR, represents the labeling index (LI).

In general 2 parameters can be derived from autoradiographs of cells or tissues after injection of labeled TdR:

1. the number of labeled cells or mitoses
2. the number of grains/nucleus.

Concerning the grain number/nucleus the question arises whether or not it represents the DNA synthetic rate. Some experimental data suggest that under physiological conditions the mean grain number/nucleus seems to represent the DNA synthetic rate (for review see Schultze, 1969). For instance, the S phase duration is about 8 h for most cell types of the mouse and the rat. On the other hand rat spermatogonia with an S phase which is about 3 times longer, also show a mean grain count/nucleus which is about 3 times smaller. However, strictly speaking the grain number/nucleus can only be taken as a measure for the DNA synthetic rate, if the portion of endogenous DNA synthesis, which is replaced by the exogenous TdR, is the same for the different cell types.

Percentage of endogenous DNA synthesis replaced by exogenous TdR

Fig. 4 contains this percentage of endogenous DNA synthesis which is replaced by exogenous TdR and it also contains the tracer dose for TdR (Lang et al., 1968; Maurer et al., in prep.). For these measurements the following experiments were carried out: One series of mice received 3-H-TdR with varying specific activity, i.e. the same 3-H-TdR activity was injected, however, the amount of TdR differed by a factor of 10 000. 1 h later the specific activity of the DNA of the small intestine was measured. The same was done for 14-C-TdR with another series of mice. As shown in Fig.4 the specific activity or the mean grain number/nucleus is the same up to about 0.5 μg TdR/g animal weight irrespective of the amount of TdR injected. That means that up to 0.5 μg/g TdR the 3-H- or 14-C-activity of the newly synthesized DNA does not depend on the amount of TdR applied; it only depends on the 3-H- or 14-C-activity administered. From 0.5 μg/g TdR on, the specific activity of the newly synthesized DNA strongly decreases. The 2 arrows depict the amount of TdR, usu-

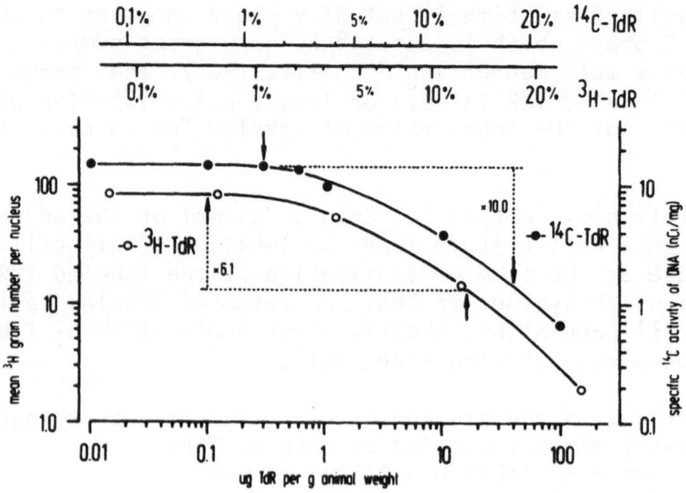

Fig. 4. Percent of endogenous DNA synthesis replaced by the exogenous 3-H- or 14-C-TdR.
The same activity but different amounts of TdR were applied. The specific activity or the mean grain count/nucleus of the jejunal crypt cells of the mouse were measured (from Maurer et al., in prep.)

ally applied in cell kinetic studies. In both cases the amount of TdR injected is within the tracer dose (0.5 µg/g).

The scales at the top of the curves represent the percentages of the entire endogenous DNA synthesis which is replaced by the exogenous TdR, if the corresponding amounts of TdR are injected according to the scale at the bottom. Fig. 4 shows that in the case of the amount of TdR commonly used in cell kinetic studies, only between 0.1 to 1.0% of the endogenous DNA synthesis is replaced by the injected labeled TdR. This is true for the jejunal crypt cells of the mouse, however, the same was found for other cell types (Lang et al., 1968).

These results clearly show that any alteration of the physiological conditions by experimental changes may strongly influence this small portion of exogenous TdR which is incorporated into DNA. Thus, the mean grain count/nucleus cannot be taken a priori as a measure for the DNA synthetic rate. For instance, a reduction of the mean grain count/nucleus after X-irradiation which is often interpreted as a decrease of the DNA synthetic rate, may well be due to an alteration of this small percentage of exogenous TdR. The mean grain count/nucleus may then be reduced, although the DNA synthetic rate might be the same.

Cell kinetic methods

At the beginning of cell kinetic studies the main interest was focussed on the measurement of the LI of the different cell types. However, the LI is only a rough estimate of the proliferative activity of a cell population.

Percent-labeled-mitoses-method (PLM): One of the main aims in cell kinetic studies is the determination of cell cycle parameters of different cell types in vivo and in vitro. For this purpose the PLM method introduced by Quastler and Sherman (1959) is used most frequently. With this method the passage of the labeled S phase population through the next and the following mitoses is observed by counting the percentage of labeled mitoses on autoradiographs as a function of time after a single injection of labeled TdR.

Fig. 8 (upper part) shows such a PLM curve for the jejunal crypt cells of the mouse after injection of 3-H-TdR. From those PLM curves a mean value for the duration of the S phase can be derived from the 50% level of the increasing and decreasing slope of the first peak or from the area under the first peak (Gerecke, 1970). The time difference between two successive waves of labeled mitoses represents the mean duration of the cycle. The time between the 3-H-TdR injection and the appearance of the first labeled mitoses is the minimum time for G_2, the time up to the 50% level of the increasing slope is the mean value for G_2+1/2 M. The duration of G_1 is obtained by subtracting S and G_2+M from the cycle time.

The PLM method is the most frequently applied method for determining cell cycle parameters. Such studies were carried out for a great number of cell types and have led to quite interesting general conclusions. For the majority of cell types studied - not only mammalian cells - the S phase duration is about 8 h. Furthermore, if the cells once have entered the S phase, they pass through S, G_2 and M in a rather constant time interval. This is true for many different cell types. The great difference in cycle time between the various cell types is due to the great difference in the duration of G_1 (Maurer and Schultze, 1968; Schultze, 1968).

Provided the cell cycle parameters are known, another important characteristic of cell proliferation can be determined, namely the growth fraction. In the majority of cell populations only a fraction of the cells proliferates. The portion of proliferating cells, related to all cells of a cell population, is designated as "growth fraction" (Mendelsohn, 1960; 1962). This growth fraction (GF) plays a role in many of the formulas used for calculating cell kinetic parameters. The only precise method to measure the GF consists in counting the LI on the autoradiographs and in determining the LI of the proliferating fraction only. The quotient represents the GF. The LI, counted on the autoradiographs, is related to all cells, whether they proliferate or not. It is not possible to distinguish histologically which of the unlabeled cells belong to the GF. The LI of the proliferating fraction can be obtained by dividing the duration of S by the cycle time, both values being derived from a PLM curve. This LI is related to proliferating cells only, since the PLM method is based on the observation of mitoses, i.e. of proliferating cells.

Continuous labeling method (CL): Apart from the commonly applied PLM method cell cycle parameters can also be determined by continuous labeling with labeled TdR (Fujita, 1962). When 3-H-TdR is continuously applied the percentage of labeled cells increases from the LI to 100%, if the GF is one. If the GF is less than one the percentage of labeled cells increases to a corresponding lower level. The time interval between the 3-H-TdR injection and the 100% level (or a lower one) represents the cycle time minus S. However, the CL method is not a suitable method for determining cell cycle parameters.

Grain count halving method: Both methods, the PLM and the CL method require serial sampling of tissues and quite high doses of labeled TdR. If the application of these methods is not possible a rough estimate of the cycle time can also be obtained by the grain count halving method (Alpen and Cranmore, 1959; Killmann et al., 1962; Baserga et al., 1963; Fried, 1968, 1970). With each cell division, i.e. after each cycle the labeled DNA of the mother cell is distributed more or less uniformly between the two daughter cells. Therefore, the mean grain number/nucleus is reduced roughly to half its value. Thus, from the decline of the mean grain count/nucleus as a function of time, following a single injection of 3-H-TdR, the cycle time can be derived. Strictly speaking this method can only be applied, if the cells grow exponentially. This method is not very precise. However, it provides a rough estimate of the cycle time with a few samples of the tissue studied.

Mode of growth: For most cell kinetic studies it is important to know whether the cell population studied grows exponentially or as a steady state system. In the case of steady state growth only one of the two daughter cells of a mitosis continues to proliferate while the other one leaves the proliferating pool. The number of cells remains constant. In an exponentially growing cell population both daughter cells of a mitosis continue to proliferate and the number of cells increases exponentially. Depending on the mode of growth the age distribution of the cells throughout the cycle is different. In the case of steady state growth the age distribution of the cells throughout the cycle is constant. In the case of exponential growth there are more younger cells than older ones within the cycle; every division must, on average, yield 2 cells and twice as many cells must leave the mitosis than enter it.

Depending on the mode of growth, the formulae for calculating cell kinetic parameters are different, since the relationship between the various parameters is different. For instance, for two cell populations with the same cycle time and the same duration of mitosis the mitotic index is different, depending on the mode of growth.

Proliferation of neural epithelial cells

In the following a brief example will be presented for how simple cell kinetic methods, as counting percentages of labeled cells and measuring the duration of S, enable us to obtain far reaching quantitative data on the proliferation of neural epithelial cells in the developing brain (Schultze et al., 1974).

For these experiments pregnant rats received a single 3-H-TdR injection between the 10th and 20th day of pregnancy. The percentage of labeled neurons of particular neuronal types were counted on autoradiographs of the brain of the 25 day old offspring. As can be seen in Fig.5 about 50-60% of the precursor cells of the 4 different cell types are labeled. These precursor cells stop proliferation between the 13th and 16th day of fetal development. Similar curves were obtained for other cell types.

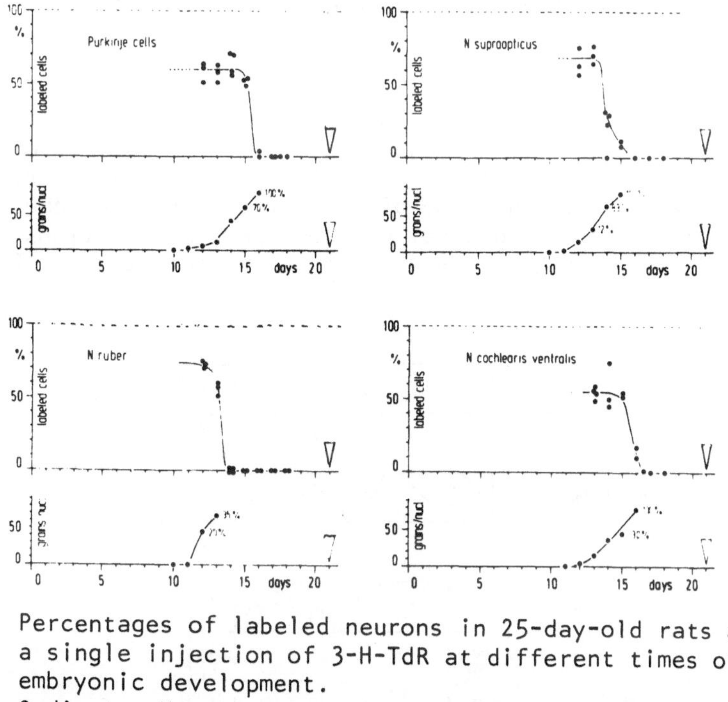

Fig. 5. Percentages of labeled neurons in 25-day-old rats after a single injection of 3-H-TdR at different times of embryonic development.
Ordinate: Upper curves, percentage of labeled neurons; Lower curves, mean grain number/nucleus. Abscissa: Time of prenatal 3-H-TdR injection (from Schultze et al., 1974).

These percentages of labeled neurons in the 25-day-old animals represent the labeling indices at the time of prenatal 3-H-TdR injection of those neural epithelial cells which later give rise to the particular neuronal type. From these labeling indices and the S phase duration the cell cycle time can be calculated for those neural epithelial cells which later give rise to particular neurons. A very short cycle time of about 10 h was found for the different cell types. Thus, the neural epithelial cells which later form different neurons divide every 10 h.

On the basis of the cycle time of these precursor cells and the cell number of a particular neuronal type, the "beginning of proliferation" of the corresponding neural epithelial cells can be determined. Surprisingly, the start of proliferation of the four different neuronal types in Fig. 5 was found to coincide with the formation of the neural plate in the rat at the 9th day of embryonic development. For details see Schultze et al. (1974).

Double labeling with 3-H- and 14-C-TdR

Most questions in cell kinetic studies can be answered by single labeling methods as mentioned above. However, if these methods cannot be applied or if additional information is needed, a very potent method can be used, namely the double labeling method with 3-H- and 14-C-TdR.

Principle of the method: The principle of the double labeling method is shown in Fig. 6. Animals receive a 1st injection of 3-H-TdR and, for instance, 1 h later a 2nd injection of 14-C-TdR. All cells which are in S at the time of the 1st injection become 3-H-labeled. During the time interval between the two injections the 3-H-labeled cells in late S leave the S phase and enter G_2. On the

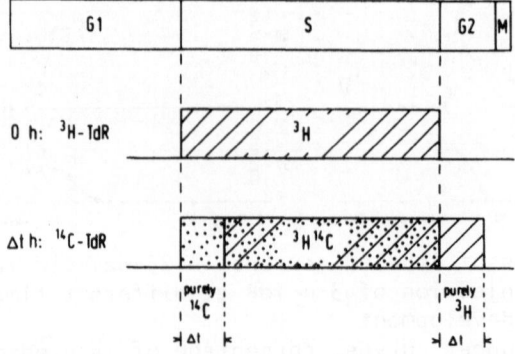

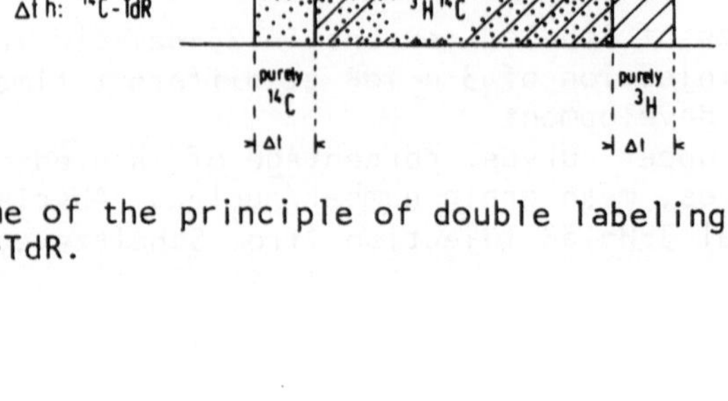

Fig. 6. Scheme of the principle of double labeling with 3-H- and 14-C-TdR.

other hand unlabeled G_1 cells enter S between the two injections. All cells which are in S at the time of the 2nd injection become 14-C-labeled. Thus, there are 3 groups of differently labeled cells: purely 3-H-labeled cells that have left the S phase between the two injections, purely 14-C-labeled cells that have entered S between the two injections and double labeled cells which were in S at both times of injection.

This double labeling method is frequently used for determining the duration of the S phase. However, this method also measures cell fluxes at the beginning and at the end of S and can therefore be used for the determination whether there is exponential or steady state growth (Maurer et al., 1972; Schultze et al., 1972a, Burholt et al., 1976).

Demonstration of 3-H- and 14-C-labeling by two emulsion layer autoradiography: The 3-H- and 14-C-labeled cells can be distinguished by the different ranges of their ß-particles which are separately registered by a two emulsion layer autoradiograph. For this purpose the labeled sample is covered by a thin emulsion layer of 1-2 µm (dipping technique) which registers the 3-H-ß-particles with their short range of 1-2 µm. This thin emulsion layer is exposed for the shortest possible time to get a sufficient number of 3-H- grains and the smallest number of spurious 14-C-grains. The thin emulsion layer is then developed in the usual way so that no further grains can be produced in it. The developed emulsion is covered by a gelatine layer of about 5 µm. This gelatine layer has the purpose to absorb all 3-H-ß-particles, also those with the maximum range of 0.5 mg/cm^2. A second, thick emulsion layer (about 20 µm) is then being brought onto the gelatine layer, it registers the 14-C-ß-particles as tracks of grains. This thick emulsion layer can be exposed as long as desirable for a sufficient number of 14-C-tracks. Fig. 7 shows an example for differently labeled jejunal crypt cells of the mouse.

Provided the 3-H-activity is much higher than the 14-C-activity all 3 categories of differently labeled cells can be quantitatively distinguished by such a two emulsion layer autoradiograph (Schultze et al., 1976). In the following, three examples of the multiple applicability of the double labeling method with 3-H- and 14-C-TdR, will be presented.

I. PLM method modified by double labeling with 3-H- and 14-C-TdR: As mentioned above the PLM method is the most frequently applied method for determining cell cycle parameters. Modification of the PLM method by double labeling leads to a much better resolution of the waves of the differently labeled mitoses and therefore allows the derivation of cycle parameters even in cases where the usual PLM method fails, i.e. if the duration of the cycle phases varies considerably. An example for such a modification for the jejunal

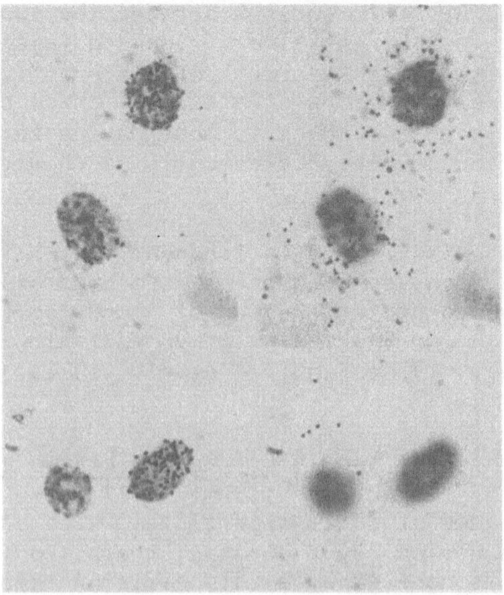

Fig. 7. Two emulsion layer autoradiograph of squashed jejunal crypt cells of the mouse.
Left: Thin emulsion layer in focus; right: thick emulsion layer in focus. At the top: double labeled cell nucleus, grains in the thin emulsion, tracks in the thick emulsion layer. Middle: purely 14-C-labeled cell nucleus, few spurious grains in the thin emulsion layer, tracks in the thick emulsion layer. Bottom: unlabeled and purely 3-H-labeled cell nuclei (grains in the thin, no tracks in the thick emulsion layer).

crypt cells of the mouse is shown in Fig. 8 (Schultze et al., 1979). For this purpose double labeling experiments were carried out, as described above, with a time interval of 2 h between the 2 injections. This leads to a 2-h-wide subpopulation of purely 3-H-labeled cells in G_2 and to another 2-h-wide subpopulation of purely 14-C-labeled cells at the beginning of S. The passage of these two subpopulations through the next and the following mitoses was followed by counting the percentages of purely 3-H- and purely 14-C-labeled mitoses on two emulsion layer autoradiographs. The percentage of purely 3-H-labeled mitoses (solid line) shows a first symmetrical peak which is only 2-h-wide and has an area of 2 h, as expected for a 2-h-wide subpopulation of cells. The passage through the next mitosis results in a flatter peak which is broadened, however, sharply limited to both sides. From the time difference of these two peaks a more exact cycle time can be derived than from the upper curve after single labeling. The resolution is much better. The relative broadening of the second peak compared to the first one provides the variance of the cycle time.

Compared to the purely 3-H-labeled cells, the purely 14-C-labeled cells must first pass through the entire S phase (Fig. 8). One S phase duration after the first peak of purely 3-H-labeled mitoses a flatter peak of purely 14-C-labeled mitoses appears (dashed line). From the distance of these two peaks a more exact value for the duration of S can be derived. Furthermore, the broadening of the second peak, compared to the first one, provides the variance of the S phase duration. Thus, for the first time the variance of the duration of S can be measured experimentally.

Two important conclusions can be drawn from these results: As shown in Fig. 9 there is a linear relationship between the variances of the transit times of the cells through the different waves of labeled mitoses and the corresponding transit times. This means that the variances of two successive cycle phases are additive. This leads to the important conclusion that the transit times of cells through two successive cycle phases are uncorrelated. This lack of correlation, however, is an implicit assumption in most of the

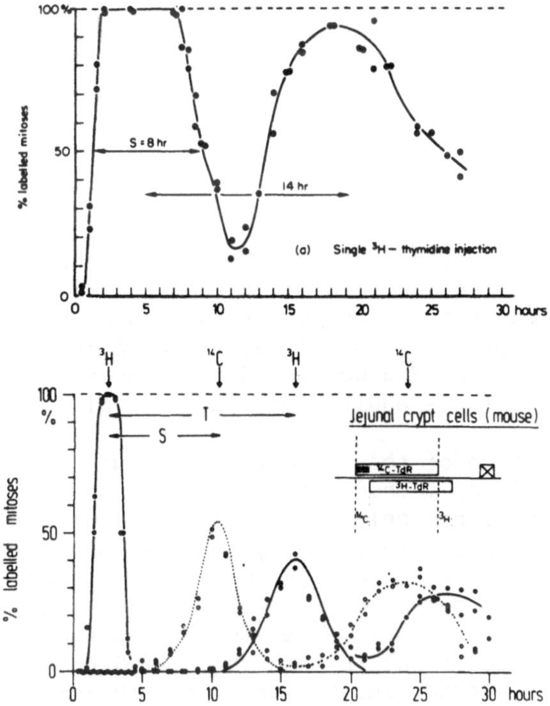

Fig. 8. Percent labeled mitoses curve for the jejunal crypt cells of the mouse. Upper curve after a single 3-H-TdR injection; lower curve after double labeling with 3-H- and 14-C-TdR injection (from Schultze et al., 1979).

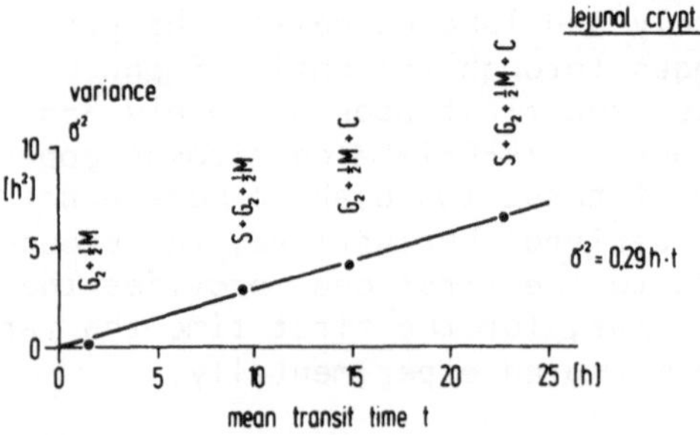

Fig. 9. Relation between the variances of the transit times of the cells through the different waves of labeled mitoses and the corresponding transit times (from Schultze et al., 1979).

mathematical models applied to the computer analysis of PLM curves. For the first time this assumption could be confirmed experimentally. This linearity also means that the variances of the transit times through the cycle phases are proportional to the duration of these phases. The common assumption that the variance of G1 is greater than that of other cycle phases is not true - at least not for the crypt cells.

II. Double labeling experiments in studies of glial cell proliferation: In studies of glial cell proliferation in the adult animal the double labeling method has led to results that would never have been obtained by other methods. Because of the few proliferating glial cells in the adult animal no cell kinetic parameters were known until some years ago. The PLM method cannot be applied because of the few mitoses present. Therefore, the duration of S was determined by double labeling and was found to be about 10 h for glial cells in the brain of the adult mouse (Korr et al., 1973).

The determination of the cycle time was much more difficult. The grain count halving method led to a surprisingly short cycle time of about 20 h. Since only a rough estimate of the cycle time can be obtained by this method the cycle time was measured by an independent method which in analogy to the percent-labeled-mitoses-method can be called "method-of-labeled-S-phases" (Korr et al., 1975). With this method the passage of the labeled S phase cells, not through the next mitosis but through the next S phase, is studied. This S phase can be recognized by labeling with 14-C-TdR. Since this method is not based on mitoses but on interphase cells, and since there are about 10-20 times more labeled interphase cells than mitoses, it can be applied to glial cell proliferation. However, a 10-h-wide S phase population and a window of the same width would not lead to distinct peaks. Therefore, in order to get a more narrow cell population, a trick was used: a 4-h-wide subpopulation of purely 3-H-labeled cells was cut off from the 10-h-wide S-phase population by double labeling.

Fig. 10 contains in the upper part a scheme of the experimental procedure. Mice received a first injection of 3-H-TdR and 4 h later a second injection of 14-C-TdR. This leads to a 4-h-wide subpopulation of purely 3-H-labeled cells in G_2. The passage of these cells through the next S phase was followed. For recognizing this S phase (hatched area) 14-C-TdR was injected at different time intervals after double labeling. The animals were always killed 1 h later. As soon as the purely 3-H-labeled cells reach the next S phase they become double labeled by the 14-C-TdR injection. In this case the ratio of purely 3-H- to double labeled cells (plotted on the ordinate) decreases and should be zero, if all purely 3-H-labeled cells have entered S. The ratio increases again as soon as the purely 3-H-labeled cells are no longer in S. The solid line in Fig. 2 represents the theoretical curve for the ratio: purely 3-H- to double labeled cells under the assumption of an S phase duration of 10 h, of G_2+M of 5 h (both measured values) and a cycle time of 20 h (derived from the grain count halving method). The measured values agree quite well with the theoretical curve. If such a curve is calculated for other cycle times (16 or 24 h) no agreement at all can

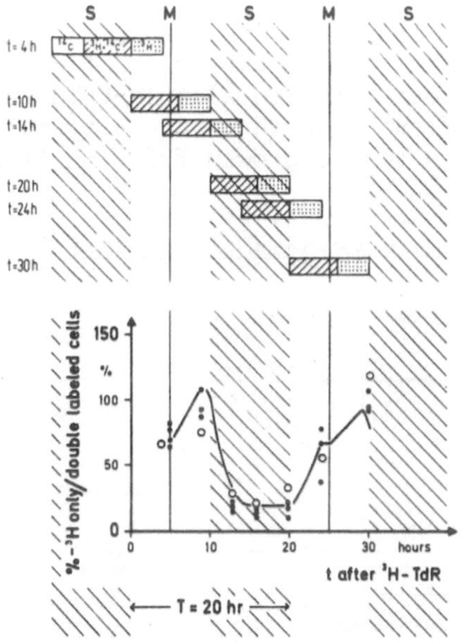

Fig. 10. Scheme and results for the determination of the mean cycle time of neuroglia by the "method of labeled S phases" (from Korr et al., 1975).

be achieved. This shows that the cycle time of glial cells is indeed about 20 h.

Glial cells in the brain of the adult untreated mouse have a LI of only 0.2% and a GF of only 0.4%; 99.6% are non-proliferating cells. But the few cells which proliferate divide every 20 h.

Furthermore, these experiments show that not all purely 3-H-labeled cells enter the next S phase; otherwise the measured values should decrease to zero. However, there are still about 20% purely 3-H-labeled cells. This is to be expected for a cell population in a steady state system.

On the other hand we were able to show that non-proliferating glial cells enter the proliferating pool.Fig. 11 shows schematically the corresponding experiments (Korr, 1978, 1980). Mice received a continuous infusion of 14-C-TdR for 24 h in order to label all proliferating cells. One day after the end of the 14-C-TdR infusion a single dose of 3-H-TdR was injected. Prior to the 3-H-TdR injection all proliferating glial cells are 14-C-labeled. These cells either continue to proliferate or stop proliferation, as mentioned above. But even these cells remain 14-C-labeled. Those cells which continue to proliferate become double labeled, if they are in S at the time of 3-H-TdR injection, and they remain purely 14-C-labeled, if they are not in S at that time. Thus, only purely 14-C- and double labeled cells should be observed. However, surprisingly enough quite a number of purely 3-H-labeled cells were seen. These cells must have entered the growth fraction during the time interval between the end of the continuous infusion with 14-C-TdR and the 3-H-TdR injection. This is the first time that such a transfer of cells from the non-growth fraction to the growth fraction is shown experimen-

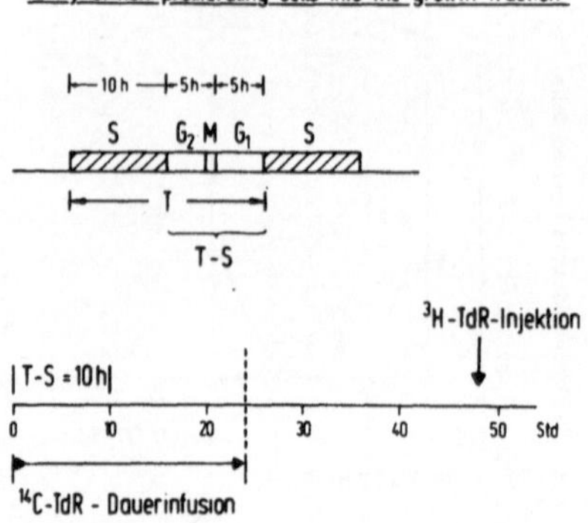

Fig. 11. Experimental design for demonstrating the passage of glial cells from the non-growth fraction to the growth fraction.

tally. That means that glial cells proliferate with a permanent exchange of cells between the growth fraction and the non-growth fraction.

III. Effect of cytotoxic drugs on cells in different cycle phases: The double labeling method enables us to study the effect of cytotoxic drugs on cells in different cycle phases also under in vivo conditions. In the following this will be demonstrated by the example of vincristine (VCR). VCR plays a role in the so-called synchronization therapy of malignant tumors. This therapy is based on the fact that most cytotoxic drugs exhibit their strongest cytocidal effect on cells in a specific cycle phase. Consequently one tried to increase the therapeutic effect by collecting cells in the drug-sensitive cycle phase, i.e. by synchronizing tumor cells in vivo. For instance, Klein et al. (1970, 1972, 1974, 1976) tried to arrest cells in mitosis by a small dose of VCR and assumed that after the release from the mitotic arrest the cells pass more or less synchronously to the next S phase. At that time an S phase specific drug (cyclophosphamide) is applied, in order to kill the increased number of tumor cells in S.

We were able to show that such an in vivo synchronization of cells cannot be achieved by VCR (Maidhof et al., 1975, Jellinghaus et al., 1975). In order to find out whether this is due to the mechanism of action of VCR the effect of VCR on cells in different cycle phases was studied. Such studies require synchronized cells and can only be carried out in vitro where synchronization can be achieved rather easily. Transferring those studies to in vivo conditions was not possible up to now, since there are no methods for synchronizing cells in vivo. However, with the double labeling method it is possible to study the effect of cytotoxic drugs on cells in different cycle phases without synchronization in the usual sense.

Fig. 12 shows the experimental design for studying the effect of VCR on cells in different cycle phases applied to the jejunal crypt cells of the mouse (Jellinghaus et al., 1977). Mice received first 14-C-TdR and 1 h later 3-H-TdR. This leads to a 1-h-wide subpopulation of purely 3-H-labeled crypt cells at the beginning of S. If the passage of these purely 3-H-labeled cells through the cycle is followed and VCR is injected when these cells are in different cycle phases, the effect of VCR on cells in different cycle phases can be studied. 1/2 h after double labeling, the animals received VCR and were killed after different time intervals. Thus, the purely 3-H-labeled cells were always situated at the beginning of S at the time of VCR injection. The purely 3-H-labeled mitoses and necrotic cells were counted as a percentage of all cells on two emulsion layer autoradiographs.

As shown in Fig. 12 no purely 3-H-labeled mitoses or necrotic cells are seen, before the purely 3-H-labeled cells are expected

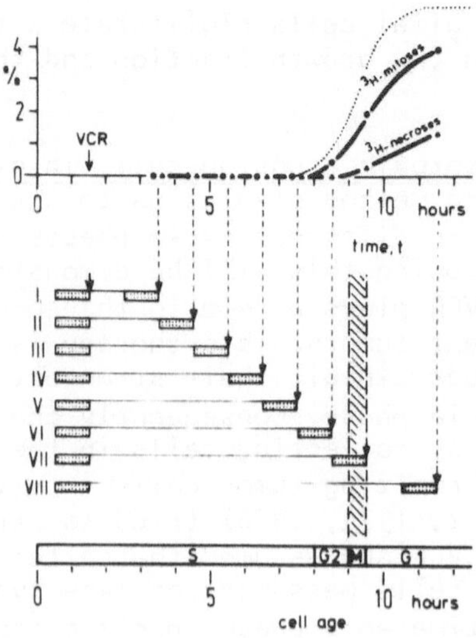

Fig. 12. Scheme and results for studying the effect of vincristin (VCR; 0.6 µg/g) on cells in different cycle phases. Jejunal crypt cells of the mouse. Abscissa: Time from beginning of double labeling. Ordinate: Purely 3-H-labeled arrested metaphases ( ) and necrotic cells ( ) as percentages of all cells (from Jellinghaus et al., 1977).

to have reached mitosis in the normal animal without VCR (see the scheme which represents the normal cycle parameters). As soon as this group of cells would normally be in mitosis, purely 3-H-labeled arrested mitoses appear. When the purely 3-H-labeled cells would normally have passed mitosis (VIII) the number of purely 3-H-labeled metaphases has further increased. That means those cells that are at the beinning of S at the time of VCR injection become arrested in metaphase during the next mitosis. With some delay purely 3-H-labeled necrotic cells appear.

The effect of VCR on cells in other cycle phases was studied in the same manner and it was found that all cells that are in S and G_2 at the time of VCR injection are arrested during the next mitosis, whereas all cells that are in G_1 pass normally through the next mitosis (Schultze et al., in prep.).

These experiments show that the mitotic poison VCR does not only affect cells in or shortly prior to mitosis, however, it damages cells which are in S and G_2 in such a way that they become arrested during the next mitosis. VCR does not influence the passage of the cells through the cycle, since the cells enter mitosis in due time. Furthermore, necrotic cells only arise from arrested metaphases but not from VCR affected interphase cells.

One question remained open in all these experiments, namely whether or not these arrested metaphases continue to proliferate at all. This is a decisive point, since this is the prerequisite for synchronizing cells in vivo with VCR. In order to clarify this point the following experiments were carried out (Camplejohn et al., 1980):

Mice with JB-1 ascites tumor received simultaneously 3-H-TdR and a small dose (0.03 μg/g)of VCR at time zero. 3 h later 14-C-TdR was injected. The scheme in Fig. 13 shows the position of those cells that will be differently labeled later on, at the time of VCR injection (t = 0). Those cells that later on will be purely 3-H-labeled are situated within the last 3 h of S at the time of VCR injection; the double labeled cells are in S minus the last 3 h and the purely 14-C-labeled cells are within the last 3 h of G_1. On two emulsion layer autoradiographs the differently labeled interphase cells, mitoses and necrotic cells were counted (as percent of all cells) as a function of time after VCR injection.

This double labeling method enables us to differentiate cells that are in different cycle phases at the time of VCR administration, and to follow their fate. Fig. 14 depicts the fate of the differently labeled and unlabeled cells.

The purely 3-H-labeled cells which were situated within the last 3 h of S at the time of VCR injection are first registered as

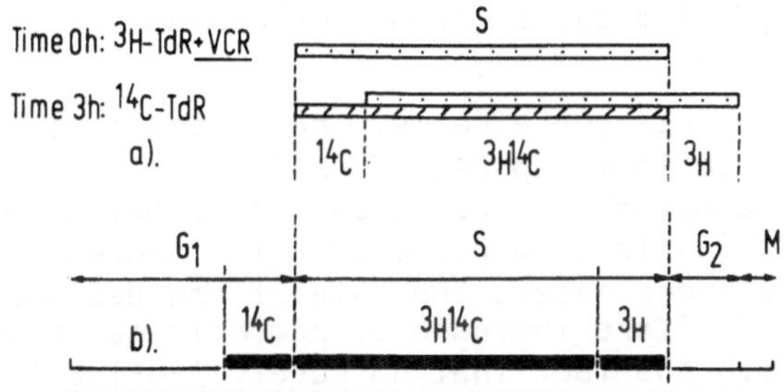

Fig. 13. Scheme of double labeling with 3-H- and 14-C-TdR. a: Time 0 h: injection of 3-H-TdR and VCR; time 3 h: injection of 14-C-TdR. b: Position at time 0 h of those cell populations that become differently labeled at 3 h. The scale at the bottom shows after how many hours the differently labeled cells reach mitosis (from Camplejohn et al., 1980).

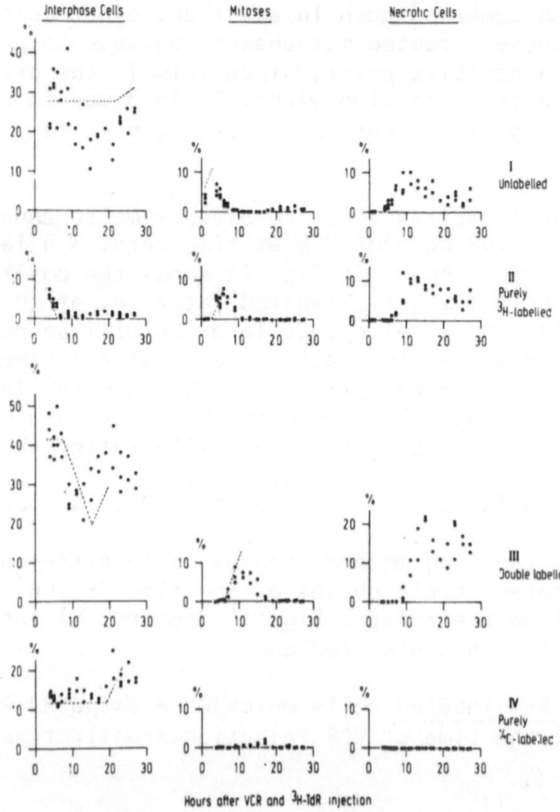

Fig. 14. Percentages of interphase cells, mitoses and necrotic cells for the various categories of differently labeled and unlabeled cells (from Camplejohn et al., 1980)

purely 3-H-labeled interphase cells. They enter mitosis 1 h later and are all arrested in metaphase. From 7 h on the percentage of purely 3-H-labeled metaphases decreases and simultaneously the percentage of purely 3-H-labeled necrotic cells increases to about 10% and then slowly decreases. The rate of the decrease of the non-necrotic cells and of the increase of necrotic cells is about the same. Together with the fact that no purely 3-H-labeled ana- or telophases are seen, this means that all arrested metaphases become necrotic. Thus, all cells that are within the last 3 h of S at the time of VCR injection are arrested in metaphase and subsequently become necrotic.

The purely 14-C-labeled cells which are situated within the last 3 h of G_1 at the time of VCR injection pass completely unaffected through mitosis in due time. No purely 14-C-labeled arrested metaphases or necrotic cells were seen. Thus, these cells are not damaged by VCR.

Things are more complicated with the double labeled cells. The double labeled cells which first enter mitosis are also arrested in metaphase. However, 13 h after VCR injection the first double labeled ana- and telophases were seen. This means from that time on normal mitoses occur.

Fig. 15 shows the percentages of total interphase cells, total mitoses and total necrotic cells, regardless of their type of labeling, as a function of time after VCR injection. 15 h after VCR administration a peak value of 35% necrotic cells is registered. From the age distribution of the JB-1 ascites tumor cells throughout the cycle at the bottom of Fig. 15 it can be seen that the amount of cells which enter mitosis within 13 h after VCR application corresponds to 35% of the entire cell population. This corresponds well to the 35% necrotic cells which are observed. According to the age distribution diagram these 35% of cells which are arrested in metaphase and subsequently become necrotic consist of all cells in G_2 and the cells that are in the last 2/3 of S.

These results clearly show that an in vivo synchronization of cells cannot be achieved by VCR, since all cells which are arrested

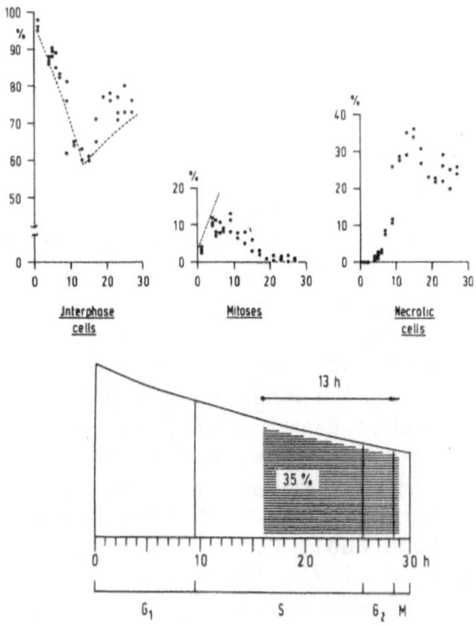

Fig. 15. Percentages of total interphase cells, mitoses and necrotic cells irrespective of type of labeling as a function of time after VCR injection. At the bottom: age diagram for JB-1 ascites tumor cells (from Camplejohn et al., 1980).

in mitosis are lethally damaged by VCR. Thus, there is no "synchronization therapy" by VCR.

The few examples of application of the various autoradiographic methods demonstrate that autoradiography can be a very valuable tool for studying metabolic processes as well as cell kinetics.

REFERENCES

Alpen, E. L., Cranmore, D., 1959, Observations on the regulation of erythropoiesis and on cellular dynamics by Fe-59 autoradiography, in: "The kinetics of cellular proliferation," F. Stohlman Jr.,ed., Grune & Stratton, New York-London, 290.

Amlacher, E., 1974, Autoradiographie in Histologie und Cytologie, Fischer, Stuttgart.

Antoni, F., Köteles, G. J., Hempel, K. and W. Maurer, 1965, Über die Eignung verschiedener Fixationen und perchlorsäurehaltiger Lösungen für autoradiographische Untersuchungen des RNS-, DNS- und Proteinstoffwechsels, Histochemie,5:210.

Baserga, R., Tyler, S. A. and W. E. Kisieleski, 1963, The kinetics of growth of the Ehrlich tumor. A comparative study of the kinetics of cellular proliferation with the use of tritiated thymidine, Arch.Pathol.Lab.Med., 76:9.

Burholt, D. R., Schultze, B. and W. Maurer, 1976, Mode of Growth of the Jejunal Crypt Cells of the Rat: An Autoradiographic Study Using Double Labelling with 3-H- and 14-C-Thymidine in Lower and Upper Parts of Crypts, Cell Tissue Kinet., 9:107.

Camplejohn, R. S., Schultze, B. and W. Maurer, 1980, An in vivo double labeling study of the subsequent fate of cells arrested in metaphase by vincristine in the JB-1 mouse ascites tumour, Cell Tissue Kinet., 13:239.

Carmona, Th., Kappeler, G. W., Schultze, B. and W. Maurer, Availability time of 3-H- and 14-C-thymidine in the jejunal crypt epithelia of the mouse, in prep.

Citoler, P.,Citoler, K., Hempel, K., Schultze, B. and W. Maurer, 1966, Autoradiographische Untersuchungen mit zwölf 3-H- and fünf 14-C-markierten Aminosäuren zur Größe des nucleären und cytoplasmatischen Eiweißstoffwechsels bei verschiedenen Zellarten von Maus und Ratte, Z.Zellforsch., 70:419.

Cleaver, J.E., 1967, Thymidine metabolism and cell kinetics, in: "Frontiers of biology", Vol.6, Neuberger, A. and E. L. Tatum, eds., North-Holland Publ. Comp., Amsterdam.

Cowan, W. M. and M. Cuénod, 1975, The use of axonal transport for studies of neuronal connectivity, Elsevier, Amsterdam, Oxford, New York.

Droz, B. and H. Warhawsky, 1963, Relability of the radioautographic technique for the detection of newly synthesized protein, J. Histochem. Cytochem., 11:426.

Fried, J., 1968, Estimating the median generation time of proliferating cell systems in steady state, Biophys. J.,8:710.

Fried, J., 1970, Mean, geometric mean, or median grain count in cell cycle studies, Exp.Cell Res., 59:447.

Fujita, S., 1962, Kinetics of cell proliferation, Exp.Cell Res., 28:52.

Garweg, G., 1969, Unterschiedliche Verteilung von L-Histidin und L-Prolin im Autoradiogramm des Mäuse-Kleinhirns, Naturwissenschaften,56:463.

Gerecke, D., 1970, An improved method for the evaluation of DNA synthesis time from the graph of labeled mitoses, Exptl. Cell Res.,62:487.

Jellinghaus, W., Maidhof, R., Schultze, B. and W. Maurer, 1975, Experimentelle Untersuchungen und zellkinetische Berechnungen zur Frage der Synchronisation mit Vincristin in vivo (Mäuseleukämie L 1210, Krypten-Epithelien der Maus), Z.Krebsforsch., 84:161.

Jellinghaus, W., Schultze, B. and W. Maurer, 1977, The effect of vincristine on mouse jejunal crypt cells of differing cell age: Double labelling autoradiographic studies using 3-H- and 14-C-TdR, Cell Tissue Kinet., 10:147.

Killmann, S. A., Cronkite, E. P., Fliedner, T. M. and V. P. Bond, 1962, Cell proliferation in multiple myeloma studied with tritiated thymidine in vivo, Lab.Invest., 11:845.

Klein, H. O., Lennartz, K. J., Habicht, W., Eder, M. and R. Gross, 1970, Synchronisation von Ehrlich-Ascites-Tumorzellen und ihre Bedeutung bei der Anwendung eines alkylierenden Cytostaticum, Klin.Wschr., 48:1001.

Klein, H. O., Lennartz, K. J., Gross, R., Eder, M. and R. Fischer, 1972, In-vivo- und In-vitro-Untersuchungen zur Zellkinetik und Synchronisation menschlicher Tumorzellen, Dtsch.med. Wschr., 97:1273.

Klein, H. O. and K. J. Lennartz, 1974, Chemotherapy after synchronization of tumor cells, Sem. in Haematol., 11:203.

Klein, H. O., Adler, D., Doering, M., Klein, P. J. and K. J. Lennartz, 1976, Investigations on pharmacologic induction of partial synchronization of tumor cell proliferation: Its relevance for cytostatic therapy, Cancer treatment Rep., 60:1959.

Korr, H., 1978, Autoradiographische Untersuchungen zur Proliferation verschiedener Zellelemente im Gehirn von Nagern während der prä- und postnatalen Ontogenese, Habilitationsschrift, University of Würzburg.

Korr, H., 1980, Proliferation of different cell types in the brain, Adv.Anat.Embryol. Cell Biol., Vol.61, Springer, Berlin,Heidelberg, New York.

Korr, H., Schultze, B. and W. Maurer, 1973, Autoradiographic investigations of glial proliferation in the brain of adult mice. I. The DNA synthesis phase of neuroglia and endothelial cells, J. Comp. Neurol., 150:169.

Korr, H., Schultze, B. and W. Maurer, 1975, Autoradiographic investigations of glial proliferation in the brain of adult mice. II. Cycle time and mode of proliferation of neuroglia and endothelial cells, J.Comp.Neurol., 160:477.

Lang, W., Müller, D. and W. Maurer, 1968, Prozentuale Beteiligung von exogenem Thymidin an der Synthese von DNS-Thymin in Geweben der Maus und in HeLa-Zellen, Exp.Cell Res.,49:558.

Maurer, W., 1972, Methodisches zur autoradiographischen Untersuchung des Eiweißstoffwechsels, Acta histochem., Suppl.-Bd. XII, 65.

Maurer, W. and E. Primbsch, 1964, Größe der ß-Selbstabsorption bei der 3-H-Autoradiographie, Exp.Cell Res., 33:8.

Maurer, W. and B. Schultze, 1968, Überblick über autoradiographische Methoden und Ergebnisse zur Bestimmung von Generationszeiten und Teilphasen von tierischen Zellen mit markiertem Thymidin, Acta histochem., Suppl. 8:73.

Maurer, W., B. Schultze and Th. Carmona, in prep., Tracer-Dosis für exogenes markiertes Thymidin und Beteiligung des exogenen Thymidin an der endogenen DNA-Synthese gemessen an den Krypten.

Maurer, W., B. Schultze, A. C. Schmeer and V. Haack, 1972, Autoradiographic studies on the mode of growth in jejunal crypt cells of the mouse, J. Microscopy,96:181.

Maidhof, R., Jellinghaus, W., B. Schultze and W. Maurer, 1975, Experimentelle und theoretische Untersuchungen zur Erzeugung einer teilsynchron proliferierenden Zellpopulation mit Vincristin in vivo, Dtsch.med.Wschr.,100:54.

Mendelsohn, M. L., 1960, The gr th fraction: a new concept applied to tumors, Science,132:1496.

Mendelsohn, M. L., 1962, Autoradiographic analysis of cell proliferation in spontaneous breast cancer of C3H mouse. III. The growth fraction, J. Natl. Cancer Inst.,28:1015.

Niklas, A., Quincke, E., Maurer, W. and H. Neyen, 1958, Messung der Neubildungsraten und biologischen Halbwertszeiten des Eiweißes einzelner Organe und Zellgruppen bei der Ratte, Biochem. Zeitschr., 330:1.

Oehlert, W., Schultze, B. and W. Maurer, 1958, Autoradiographische Untersuchung der Größe des Eiweißstoffwechsels der verschiedenen Zellen des Zentralnervensystems, Beitr.pathol.Anat.,119:343.

Quastler, H. and F. G. Sherman, 1959, Cell population kinetics in the intestinal epithelium of the mouse, Exp.Cell Res.,17:420.

Rogers, A. W., 1979, Techniques of autoradiography, 3rd ed., Elsevier/North-Holland Biomedical Press, Amsterdam,New York,Oxford.

Schneider, G. and W. Maurer, 1963, Autoradiographische Untersuchung über den Einbau von H-3-Cytidin in die Kerne einiger Zellarten der Maus und über den Einfluß des Fixationsmittels auf die H-3-Aktivität, Acta histochem.,15:171.

Schultze, B., 1968, Die Orthologie und Pathologie des Nucleinsäure-Eiweißstoffwechsels der Zelle im Autoradiogramm, in:"Handbuch der allgemeinen Pathologie", Bd. II/5, H. W. Altmann, F. Büchner, H. Cottier, G. Holle, W. Letterer, W. Masshoff, H. Meessen, F. Roulet, G. Seifert, G. Siebert and A. Studer, eds., Springer, Berlin-Heidelberg-New York, 466.

Schultze, B., 1969, Autoradiography at the cellular level, in: "Physical techniques in biological research" 2nd edn, Vol. III b, A. W. Pollister, ed., Academic Press, New York, London.

Schultze, B. and W. Maurer, 1963, Größe der RNS-Synthese in Nukleolus und Karyoplasma bei einigen Zellarten der Maus, Z. Zellforsch.,60:387.

Schultze, B. and W. Maurer, 1967, Nuclear and cytoplasmic protein synthesis in various cell types from rats and mice, in: "The control of nuclear activity",L. Goldstein, ed., Prentice-Hall, Inc., Englewood Cliffs, New Jersey, 319.

Schultze, B., W. Oehlert and W. Maurer, 1960, Vergleichende autoradiographische Untersuchung mit H-3, C-14 und S-35-markierten Aminosäuren zur Größe des Eiweißstoffwechsels einzelner Gewebe und Zellarten bei Maus, Ratte und Kaninchen, Beitr. pathol. Anat.,122:406.

Schultze, B., Citoler, P., Hempel, K., Citoler, K. and W. Maurer, 1965, Cytoplasmic protein synthesis in cells of various types and its relation to nuclear protein synthesis, in: "The use of radioautography in investigating protein synthesis", C. P. Leblond and K. B. Warren, eds., Symp.Int.Soc. Cell Biol., Vol.IV, 107, Academic Press, New York, London.

Schultze, B., Haack, V., Schmeer, A.C. and W. Maurer, 1972 a, Autoradiographic investigation on the cell kinetics of crypt epithelia in the jejunum of the mouse, Cell Tissue Kinet., 5:131.

Schultze, B., Rabbani, B. and W. Maurer, 1972 b, Blut-Hirn-Schranke und Placentar-Schranke für verschiedene Aminosäuren bei der Maus (Untersuchung mit Ganzkörper-Autoradiographie), Beitr. Path.,147:352.

Schultze, B., Nowak, B. and W. Maurer, 1974, Cycle times of the neural epithelial cells of various types of neuron in the rat. An Autoradiographic study, J.Comp.Neurol., 158:207.

Schultze, B., Maurer, W. and H. Hagenbusch, 1976, A two emulsion autoradiographic technique and the discrimination of the three different types of labelling after double labelling with 3-H- and 14-C-thymidine, Cell Tissue Kinet., 9:245.

Schultze, B., Kellerer, A. M. and W. Maurer, 1979, Transit times through the cycle phases of jejunal crypt cells of the mouse. Analysis in terms of the mean values and the variances, Cell Tissue Kinet., 12:347.

Schultze, B., Jellinghaus, W., Basler, U., Dettmer, J. and W. Maurer, in prep., Experimental studies concerning the so-called synchronization therapy with vincristine.

RECENT TRENDS IN ELECTRON MICROSCOPE AUTORADIOGRAPHY.

Nadir M. Maraldi

Institute of Histology and
General Embryology
University of Ancona (Italy)

INTRODUCTION

Since 1955 biological research utilizes some alpha and beta emitting radionuclides for monitoring the metabolic processes by determining the fate of radioisotope-labeled molecules (1). While biochemistry utilizes high energy radioactive tracers easy detectable by scintillation counting methods, the morphological investigations require tracers capable of being detected with a sufficient efficiency by photographic emulsions.

The application of the autoradiographic techniques to electron microscopy has been allowed by the production of fine grain nuclear emulsions and by the use of tritium as tracer (2-4). In fact this isotope is characterized by the emission of beta particles with a maximum energy of 18.6 KeV, which corresponds to a path in the emulsion of about 3 μm and which allows to obtain an high-sensitivity detection at low ionization, since the formation of the latent image depends on the amount of energy absorbed per unit length (5).

The aim of this chapter is to review the recent trends in EM ARG and the applications of these techniques to the problems of nuclear and chromatin functions.

IMPROVEMENTS IN EM ARG ON THIN SECTIONS

The limits of routine ultrastructural autoradiography are represented by the resolution, the efficiency and the correct quantitation of the results.

1. Resolution

Autoradiography has been currently applied to EM specimens in the early sixties. At that time the resolution power of the electron microscope for biological specimens reached 20 Å on average, while the resolution power of the EM autoradiography was in the range of 2000 Å. In twenty years the resolution power of the electron microscope for biological specimens has attained 5-7 Å for negatively stained isolated molecules and the resolution power for EM autoradiography has improved reaching about 600 Å. Therefore, despite the great efforts for ameliorating the technical procedures for EM autoradiography the gap between the two techniques still remains very wide (Fig. 1).

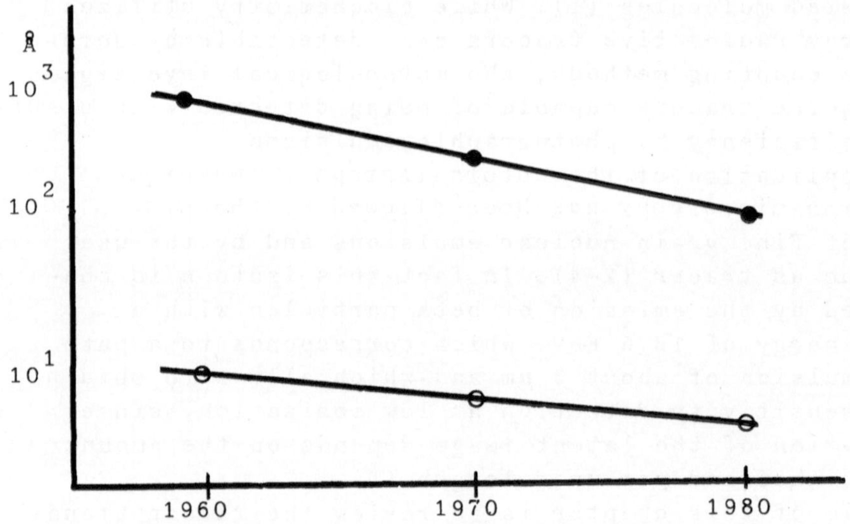

Fig. 1 Comparison between the improvement in resolution in trasmission electron microscope specimens from 1960 to 1980 (open circle) and that of electron microscope autoradiography (shut circles).

Numerous theoretical studies on the resolution problems have appeared in the last years (6-10). Resolution depends on the size of the silver halide crystals of the nuclear emulsion, on the distance between the radiation emitting source and the emulsion and on the developing procedures.

The most suitable method for evaluating the resolution is based on the use of very small sources which may be considered as ideal line-sources. The line-source or "hot line" consists in a 500 Å thick styrene-^{3}H film sandwiched between an Epon 812 block and a methacrylate film and then sectioned at right angles. By plotting the density distribution of developed grains around the hot line it is possible to deduce the half distance (HD), that is the distance from the hot line within which half of the grains fall, which represents a direct measure of the resolution (11).

The best mean values of the resolution, defined as "half distance" (HD), have been calculated to be, (using tritium as source), 1300 Å with the Ilford L4 and 800 Å with the Kodak NTE emulsion (11).

The efforts for improving the resolution have been mainly based on the preparation of thinner monolayer of nuclear emulsion and on the use of finer grain developers.

Monolayer preparation. By using the Ilford L4 emulsion monolayers 1500 Å thick can be obtained by the dipping method (12-14) and by the loop method (2,4). With the use of a particular expandable loop named "emulsion film drawer" a monolayer 1200 Å thick can be obtained having the same density of silver halide crystals but containing a lower content of gelatin as demonstrated by quantitative chemical evaluations. This type of monolayer allows a 20-30% improvement in resolution due to the lower value of the "source detector" geometry (15).

Significant improvements in the preparation of Kodak NTE monolayers have not been obtained in respect to the original dipping method (16) except for the use of a modified type of loop (17).

Fine grain developers. The fine grain development can be defined as a procedure by which the latent images will be developed into compact silver deposits with a mean diameter smaller than the silver halide crystals of the emulsion. Two methods are currently used for obtaining a fine grain development: the chemical development (direct method) and the "solution physical" development (18--24). However all the fine grain developing procedures and particularly the physical development reduce the efficiency of the autoradiographic procedure.

2. Efficiency

The efficiency or sensitivity of a nuclear emulsion is the ratio between the grain and the decays (25-27).

For evaluating the efficiency, biological point sources have been used, like lambda phages labeled with ^{3}H-thymidine with a final specific activity of about 0.05-0.1 count/phage/day (3,28). The efficiency is then evaluated by counting the number of grains for number of particles. The values of efficiency for Ilford L4 and Kodak NTE have been reported to vary from 20 to 40% depending on the experimental conditions (29,30).

In order to improve the efficiency two procedures have been proposed based either on the activation of the latent image or on the transformation of the electrons into photons.

Gold latensification. It has been hypothesized that the latent image consists partly of sub-image specks which are not developable without an "activation" by physical or chemical means. The activation causes a lowering of the electron levels in the sub-specks, allowing their reduction with the developer (31).

Remarkable improvement in efficiency have been obtained by means of specific gold solutions which accelerate the development kinetics without an appreciable increase of the chemical fog (30,32). This procedure, named gold latensification, is expected to improve the resolution since, in the case of β-emitting isotopes, the probability of a sub-speck formation is higher close to the radiation emitting source (31).

At the present time by using physical development and

gold latensification the maximum yeld in efficiency is of one silver grain for five emitted electrons (27).

Scintillation fluids. Some methods have been devised for enhancing the efficiency of the optical ARG based on the use of scintillation fluids (33,34). A similar procedure can be used in EM ARG by mixing the emulsion or the embedding plastic with scintillators (35) or by interposing a scintillator plastic film between the source and the emulsion (28). In this way an 80% increase of efficiency has been obtained, but a loss of resolution occurs both for diffusion of the photons and for the higher value of the source-detector geometry.

3. Statistical analysis of EM ARG

In order to obtain the maximum of the information from an EM autoradiogram it is necessary to use quantitative analyses, since the ARG techniques provide low resolution compared with the resolving power of the electron microscope. Moreover the β particle may travel some distance both in the section and in the emulsion and its scatter is proportional to the emission energy. Therefore the analysis of grain distribution must use statistical procedures which may increase the confidence of the obtained results.

HD curves. Resolution may be expressed in term of half-distance (HD) which is the distance from the source at which the possibility exists to find the 50% of the developed grains (8). The extrapolation of the experimental values of HD for various isotopes and different experimental conditions allows to calculate the expected distribution of grains for a variety of specimens. Therefore a circle of 1.7 x HD radius placed around a silver grain will have a 50% probability of including the actual source of radiation (11).

Random point gratings. Another method is based on the comparison between the actual grain distribution and that expected for a uniform distribution of radioactivity in the section. Practically HD circles are placed on the grains and the list of frequency of the circles over each cell component is compared with a similar list with circles placed at random. The statistical χ^2 test allows to identify the different degrees of labeling of the different compartments (36).

This procedure has been recently improved by using a point analysis within the circles (37) and by combining the advantages of HD method with that of random distribution (38-40).

NEW TRENDS IN EM ARG

EM ARG techniques have been applied mainly to ultrathin sections of embedded material examined by transmission electron microscopy.

In the last years many attempts have been made to extend autoradiography to other techniques employed in transmission and scanning electron microscopy.

1. Freeze-fracture ARG

Various types of artificial and biological membranes, labeled with ^{125}I or ^{3}H-precursors and freeze-fractured with the usual methods have been coated with nuclear emulsion and processed with the current developing procedures. The resolution of this technique has been estimated to vary between 1250 and 2500 Å, while the efficiency has been evaluated to range from 1 to 6% (41-43).

The obvious limitation of this technique is that it can be applied only to surface distribution of label. For this reason it seems particularly useful in the study of the localization of accumulated elements such as neurotransmitters or hormones.

2. Diffusible radioactive substances analysis

The problem of localizing the subcellular distribution of a large quantity of soluble metabolites such as peptides, ions, sugars and lipids by means of EM ARG is of great interest, owing to the difficulties of employing other techniques, except for X-ray microanalysis.

The technical procedures till now used are the freeze-drying (44) and the ultracryotomy (45,46). At present this technique shows some unresolved aspects such as the actual resolution and efficiency.

3. Scanning electron microscope ARG

Scanning electron microscope resolution is comparable with that of the transmission electron microscope autotadiography so that the gap between the two techniques is almost absent when applying autoradiography to the scanning electron microscopy.

The feasibility of this technique has been demonstrated though the biological applications are so far very scanty (47,48). This technique presents the advantage of allowing a quantitative evaluation of the grains by measuring the silver mass through the X-ray microanalysis (48).

4. Direct deposition autoradiography

Some papers have appeared claiming the possibility of localizing radioisotopes in thin sections by a direct deposition of metal from a solution of salt onto the disintegration sites, owing to the formation of latent sub-images through the trapping of metal ions by the negative holes formed in the specimens by β emission (49, 50). The results obtained with this technique were interpreted as an impressive increase of efficiency and resolution since short exposure times allowed the formation of a great quantity of grains. However strong evidences that the results obtained with direct deposition ARG may depend on artefacts like a reduction of the crystals by positive chemography by tissue components and by the light have been recently reported (51).

5. EM ARG of free specimens and spread molecules

A great part of the autoradiographic investigations deals with the localization inside the cell of the sites of synthesis and the pathways of transport through the cell compartments of a given metabolic compound.

However, some of the functions which take place in the intact cells can be also observed in cell-free systems and therefore in conditions which allow an easier identification of the metabolic events at the molecular level. Some nuclear functions like DNA replication and transcription can be visualized at high resolution by using the spreading of the isolated molecules (52,53). Therefore different attempts have been made in order to apply electron microscope autoradiography to isolated specimens or molecules.

The application of autoradiography to isolated specimens should provide some theoretical advantages like:
a) isolated specimens, stained with heavy metals, can be resolved at the highest resolving power allowed by the electron microscope;
b) the isolated molecules can be prepared at the wanted degree of purity and contaminants can be completely removed;
c) the autoradiographic resolution must improve since isolated molecules represent very thin sources in comparison to sections and since self-adsorption phenomena due to the staining with heavy metals reduce the radiation spread around the source (54).

The methods so far used for enhancing the contrast of isolated specimens for EM ARG, (shadow casting, positive and negative stainings), present both advantages and drowbacks so that they must be chosen according to the specimen type used.

The EM ARG studies based on shadowing have been principally used for identifying the sites of the DNA replication (55,56) and for analizing the incorporation of the precursors of preribosomal RNA in the RNP fibrils of isolated transcription units (57). The shadowing is therefore particularly indicated for spread long-chain molecules like DNA or RNP fibrils whose identification is allowed also by this low resolution method.

The use of positive staining with heavy metals seems to be more suitable for the autoradiographic study of isolated particles like phages (9) and especially for the analysis of macromolecules having morphologically identifiable substructures like the collagen fibers (58). On the contrary this technique seems to be not very useful for the study of spread molecules (59).

The negative staining methods are particularly useful when a high resolution power of the examined specimen is required. Therefore this technique has been used for

Table 1. The resolution power obtained with different procedures of EM ARG on isolated specimens is expressed in HD units.

Contrast Enhancing	Samples	Emulsion	Developing Procedure	Resolution (HD) in Å	References
Shadow casting	^{3}H DNA	Ilford L4	Physical Method	NR	(55)
	^{3}H DNA	"	Microdol X	NR	(56)
	^{3}H DNA genes	"	Physical Method	1300	(57)
Positive staining	^{3}H T2 phages	"	"	1000	(9)
	^{3}H SV 40 RNA	"	RN	NR	(59)
	^{125}I collagen	"	Microdol X	420	(58)
Negative staining	^{3}H ribosomes	Kodak NTE	Elon ascorbic acid	360	(60)
	^{3}H BK viruses	Sakura NR-H2	"	200	(61)

autoradiographic studies on isolated ribosome fractions, ribosome crystals and viruses (60,61).

While the shadowed specimens maintain their contrast also after the autoradiographic procedure (emulsion coating, developing and fixing baths) some cautions must be used in order to prevent the destaining of positively and negatively stained specimens. These cautions consist either in the deposition of the nuclear emulsion on the opposite site of the film supporting the specimen and in the staining after the photographic processes (58) or in the use of a fixative such as glutaraldehyde with the prestained specimens and in the reduction of the times of the photographic treatments (60).

Another problem affecting the EM ARG of isolated specimens is represented by the intensity of the labeling. Generally the labeling must be as high as possible because the isolation procedures necessarily lead to a dilution of the labeled specimen. However also a too much higher level of label must be avoided since, especially in the case of nucleic acids, a radiolysis of the molecule can occur (62). On the other hand the statistical analyses of the grain distribution are easier for isolated molecules since the background due to other sources is completely eliminated and since many of the isolated specimens should be considered as ideal point or line sources.

The resolution power obtained with EM ARG of isolated specimens obviously depends both on the nuclear emulsion and on the methods used for enhancing the contrast (Table I).

By using Ilford L4 emulsion on shadowed DNA molecules an HD value of 1300 Å has been reported (62), while an HD value of 420 Å has been obtained with positively stained collagen fibers (58). This difference may depend principally on the method used for enhancing the contrast. In fact the shadow casting of isolated molecules induces an uniform accumulation of metal particles increasing the source-detector geometry value, while the positive staining, consisting in a selective binding of the stain molecules to the specimen does not affect the source-detector geometry value.

The use of a fine grain emulsion like Kodak NTE and of negative staining as contrast method allows a further

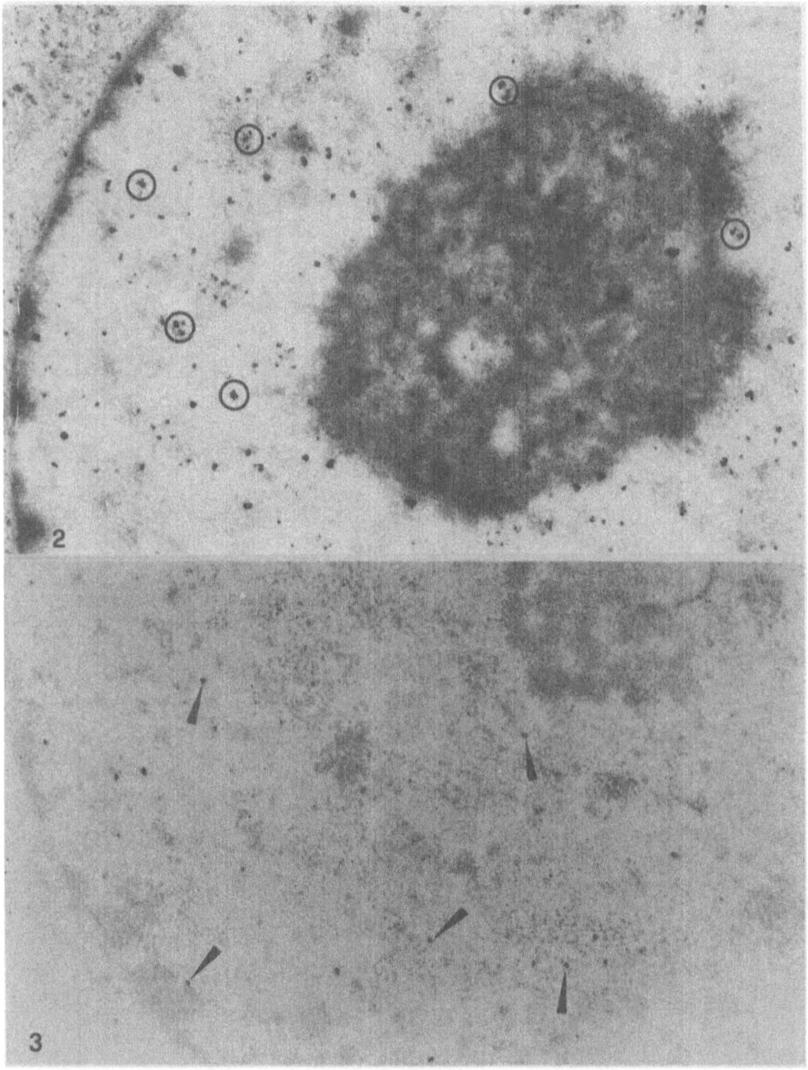

Figs. 2,3. ^{3}H-thymidine incorporation in regenerating rat liver thin sections, Sakura NR-H2 emulsion. In Fig. 2 the typical "cat paw pattern" of the multiple grains is evident (circles), while, after KOH digestion (Fig. 3), they are reduced to round shaped spots, 200 Å in diameter (arrows).

Note the decreased contrast of the specimen after the KOH digestion.

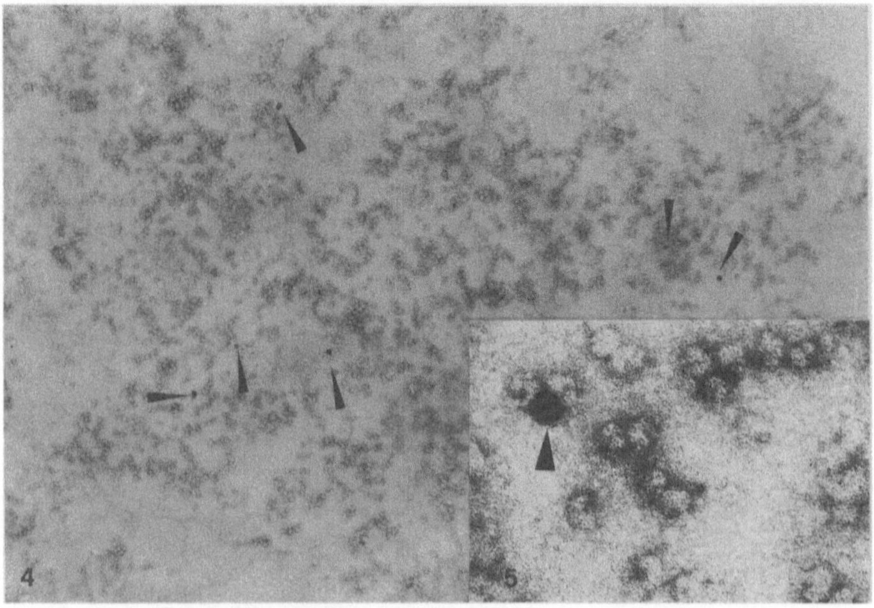

Figs. 4,5. Polysome fraction, uranyl acetate staining, Kodak NTE emulsion. The arrows indicate silver grains, 200 Å in diameter, on ribosome units.

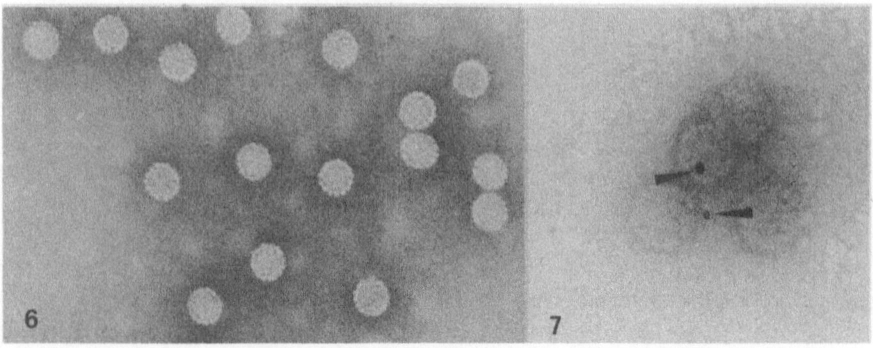

Fig. 6 Negatively stained BK viruses (control)

Fig. 7 EM ARG, Sakura NR-H2 emulsion. The arrows indicate the silver grains, 50-80 Å in diameter, on the capsides. Note the partial destaining of the virus particles with respect to control.

improvement of the resolution power (Table I). This may be due to the very reduced distance between the source and the detector, represented by a thin layer of evaporated carbon, and by the self-absorption phenomena due to the uniform distribution of an high molecular weight stain all around the emitting source (54). The resolution power obtainable with this technique (HD value of 360 Å) is very close to the best theoretically expected resolution value for a point source of zero thickness (61).

However a further gain in resolution can be obtained by using a particular developing procedure and the fine grain nuclear emulsion Sakura NR-H2. The rationals of this process (63) are the followings:

a) the radiations induce the formation of multiple latent images inside each silver halide crystal;

b) a suitable developing procedure may allow to obtain the reduction to metal of each latent image, so that the developed grains appear like "cat's paw patterns" formed by 50-100 Å subgrains;

c) the subsequent treatment with a staining solution containing KOH removes all the subgrains but one, that is the subgrain which is the nearest to the surface of the specimen, owing to its retainement by the surface tension forces. Since this grain is located in the plane closest to the radiation source the resolution is greatly improved owing to the reduced source-detector geometry value (Figs. 2,3).

The application of this method to negatively stained specimens has been demonstrated to be possible, though a partial destaining of the specimen occurs; however by this method an HD value of about 200 Å has been obtained which represents nowday better resolution power obtained with EM ARG (Figs. 4-7).

Despite to the great advantages both in resolution and in accuracy of the localization of the metabolic events at a molecular level the EM ARG of isolated specimens has been till now applied to the study of biological problems in few cases. In fact this method can by no means be considered a routine procedure, since it requires time-consuming sophisticated techniques and careful statistical evaluations of the results.

EM ARG STUDIES ON NUCLEAR FUNCTIONS

EM ARG represents the choice method for studies of topological character in cell biology. In particular the cell nucleus metabolism can be advantageously analyzed by EM ARG since nucleic acids are confined in identifiable structures, the synthesis of the nuclear components takes place at definite sites and differential migration and accumulation of the nuclear metabolites take place in short times through the intranuclear compartments.

EM ARG has played a key role in solving the problems of the distribution of DNA replication sites. The hypothesis that DNA replication should take place in association with the nuclear membrane has been suggested by autoradiographic data indicating a prevalent radioactivity concentration on the nuclear membrane-associated chromatin after very short pulses with DNA precursors. After the chase periods or with long labelings on the contrary the radioactivity appeared distributed in all the nucleus (64-69). The presence of an uniformly distributed label, suggesting for a random localization of the DNA replication sites has been, on the other hand, well documented by more recent studies (70-77). This suggests that the observed label concentration on the periphery of the nuleus may be due either to a higher concentration of the DNA in this region, or to the limits of the penetration rate of the precursors. In some cases however the localization of the label on the chromatin associated with the nuclear envelope has been considered as expression of a late DNA synthesis in the condensed chromatin (78).

The synthesis of the DNA, after the initiation and during the whole S phase, has been described to occur in the dispersed form of the chromatin, generally close to the periphery of the condensed chromatin regions (74, 79-81).

It has been therefore suggested that DNA replication takes place in dispersed chromatin regions but that, a continuous condensation and decondensation of chromatin occurs during S phase so that the replication sites are always very close to the condensed chromatin areas (81).

A large number of autoradiographic studies has been devoted to the identification of the transcription sites and to the kinetic of the transcript transport inside and outside the nucleus. The transcription sites of the nucleolar RNA appear prevalently distributed within the border of the intranucleolar chromatin threads. After few minutes the newly transcribed RNA molecules are localized in the fibrillar RNP region of the nucleolus (82-86)

The pathway of the rRNA transport outside the nucleolus is still a matter for speculations. In fact the hypothesis that interchromatin granules represent the sites of storage of the neosynthesized rRNA and therefore derive from the nucleolar granular component (87,88) has not been confirmed (89).

The extranucleolar RNA synthesis sites appear to be localized close to the border of the condensed chromatin areas without connection with the nuclear pore complex or with the nuclear membrane. The transcripts may be identified as perichromatin fibers with preferential staining or with spread preparation autoradiography (90-92).

The metabolic significance of the perichromatin granules also still remain not completely defined, though they may represent a storage form of stable RNA species, since it has been excluded that they contain HnRNA (64).

Finally it has been demonstrated that nuclear proteins accumulate both in perichromatin regions and in close relationship with the fibrillar nucleolar component, where they could serve for the assembling of neosynthesized nucleoproteins or for exerting a regulatory role on the transcription (93,94).

It can be expected that a larger application of EM ARG to isolated spread molecules may contribute to solve some still open questions on the mechanisms of nucleic acid metabolism and of the genome organization and functions at a molecular level.

ACKNOWLEDGMENTS

I am indebted to Drs. B. Bonora, P. Santi, E. Caramelli and S. Capitani for their help, support and friendship.

This work was partly supported by a grant of the National Research Council of Italy, N° CT 79.01927.

REFERENCES

1. G. Boyd, "Autoradiography in biology and medicine," Academic Press, New York (1955).
2. J. P. Revel, and E. D. Hay, Exp.Cell Res. 25:474 (1961).
3. L. G. Caro, J.Cell.Biol. 15:189 (1962).
4. L. G. Caro, and R. P. Van Tubergen, J.Cell.Biol. 15: 173 (1962).
5. R. Masse, J.Microsc.Biol.Cell. 27:83 (1976).
6. L. Bachmann, and M. M. Salpeter, Naturwissenschaften 51:237 (1974).
7. L. Bachmann, and M. M. Salpeter, Lab.Invest. 14:1041 (1965).
8. L. Bachmann, and M. M. Salpeter, J.Cell.Biol. 33:299 (1967).
9. L. G. Caro, Meth.Cell Physiol. 1:327 (1964).
10. M. Bouteille, J.Microsc.Biol.Cell. 27:121 (1976).
11. M. M. Salpeter, L. Bachmann, and E. E. Salpeter, J. Cell.Biol. 41:1 (1969).
12. P. Granboulan, in "The use of radioautography in investigating protein synthesis," C. P. Leblond, and K. B. Warren. eds., Academic Press, New York (1965) p.43.
13. M. B. Kopriwa, J.Histochem.Cytochem. 14:923 (1966).
14. G. F. J. M. Vrensen, J.Histochem.Cytochem. 18:278 (1970).
15. N. M. Maraldi, G. Biagini, P. Simoni, and R. Laschi, Meth.Cell Physiol. 5:289 (1972).
16. M. M. Salpeter, and L. Bachmann, J.Cell.Biol. 22:469 (1964).
17. A. Fantazzini, U. Serra, N. M. Maraldi, and R. Laschi, J.Submicr.Cytol. 4:114 (1972).
18. T. H. James, J. Phys.Chem. 66:2416 (1962).
19. C. E. K. Mees, and T. H. James, "The theory of the photographic process," The Macmillan Company, New York-London (1966).
20. M. M. Salpeter, and L. Bachmann, J.Cell.Biol. 22:469 (1964).
21. H. Lettré, and N. Paweletz, Naturwissenschaften 53: 268 (1966).
22. B. M. Kopriwa, J.Histochem.Cytochem. 14:923 (1967).

23. B. M. Kopriwa, J.Histochem.Cytochem. 15:501 (1967).
24. B. M. Kopriwa, Histochemistry 44:201 (1975).
25. L. Bachmann, and M. M. Salpeter, Lab.Invest. 14:1041 (1965).
26. E. Visse, and A. D. Tates, 4th European Regional Conference on Electron Microscopy, Rome, p.465 (1968).
27. G. F. J. M. Vrensen, J.Histochem.Cytochem. 18:278 (1970).
28. L. A. Buchel, E. Delain, and M. Bouteille, J.Microscopy 112:223 (1978).
29. H. C. Furtuck, and M. M. Salpeter, Proc.Nat.Acad.Sci. 71:1376 (1974).
30. M. M. Salpeter, and L. Bachmann, in: "Principles and Techniques of Electron Microscopy," M. A. Hayat. ed., Van Nostrand Reinhold Company, New York, vol. 2, p.220 (1972).
31. R. Rechenmann, and E. Wittendrop, J.Microsc.Biol.Cell. 27:91 (1976).
32. H. Heijnen, and H. Genze, Histochemistry 54:39 (1977).
33. G. S. Panayi, and W. A. Neill, J.Immunol.Meth. 2:115 (1972).
34. B. G. M. Durie, and S. E. Salmon, Science 190:1093 (1975).
35. H. A. Fischer, H. Korr, H. Thiele, and G. Werner, Naturwissenschaften 58:101 (1971).
36. M. A. Williams, in: "Advances in Optical and Electron Microscopy," R. Barer, and V. E. Cosslett. ed., Academic Press, London, vol.3, p.219 (1969).
37. C. Kent, and M. A. Williams, J.Cell.Biol. 60:554 (1974).
38. N. M. Blackett, and D. M. Parry, J.Cell.Biol. 57:9 (1973).
39. D. M. Parry, J.Microsc.Biol.Cell. 27:185 (1976).
40. N. M. Blackett, and D. M. Parry, J.Histochem.Cytochem. 25:185 (1977).
41. A. M. Downs, and M. A. Williams, J.Microscopy 114: 143 (1978).
42. K. Fisher, and D. Branton, J.Cell.Biol. 70:453 (1976).
43. M. W. Nermut, and L. D. Williams, J.Microscopy 118: 453 (1980).
44. J. R. J. Baker, and T. C. Appleton, J.Microscopy 108: 307 (1976).

45. T. Nagata, and F. Murata, Histochemistry 54:75 (1977).
46. J. Boyenva, and B. Droz, J.Microsc.Biol.Cell. 27:129 (1976).
47. G. M. Hodges, and M. D. Muir, Nature 247:383 (1974).
48. G. M. Hodges, and M. D. Muir, J.Microscopy 104:173 (1975).
49. D. K. Normandin, Trans.Am.Micros.Soc. 92:381 (1973).
50. W. A. Hemmings, and E. W. Williams, "Maternofoetal Transmission of Immunoglobulin," W. A. Hemmings, Cambridge University Press, London (1976).
51. E. W. Williams, J.Microscopy 112:319 (1978).
52. O. L. Miller, and B. R. Beatty, Science 164:955 (1969).
53. O. L. Miller, and A. H. Bakken, Acta Endocrin. 168: 155 (1972).
54. M. M. Salpeter, in: "Electron Microscopy and Cytochemistry," E. Wisse, W. Th. Deams, I. Molenaar, and P. Van Duijn. eds., North Holland, American Elsevier, Amsterdam, London, and New York, p.315 (1974).
55. A. Niveleau, J.Microscopy 11:175 (1971).
56. G. H. Weber, U. Heine, M. Cottler-Fox, and G. Beandreau, J.Cell.Biol. 61:257 (1974).
57. N. Angelier, D. Hemon, and M. Bouteille, Exp.Cell Res. 100:389 (1976).
58. T. Oda, S. Omura, and S. Watanabe, in: "Histochemistry and Cytochemistry," T. Takauchi, K. Ogawa, and S. Fujita, eds., Japan Society of Histochemistry and Cytochemistry, Kyoto, p.123 (1972).
59. R. A. Haworth, and J. A. Chapman, J.Microscopy 106: 125 (1976).
60. N. M. Maraldi, G. Biagini, P. Simoni, and R. Laschi, Histochemie 35:67 (1973).
61. N. M. Maraldi, in: "Principles and Techniques of Electron Microscopy," M. A. Hayat. ed., Van Nostrand Reinhold Company, New York, vol.6, p.271 (1975).
62. N. Angelier, M. Bouteille, J. J. Curgy, E. Delain, S. Fakan, S. Geuskens, M. Guelin, J. C. Lacroix, M. Laval, G. Steinert, and S. Van Assel, J.Microsc. Biol.Cell. 27:215 (1976).
63. G. Uchida, and V. Mizuhira, Arch.Histol.Jap. 31:291 (1970).
64. S. Fakan, in: "Cell Nucleus," H. Busch, ed., Academic

Press, New York, vol.5, p.3 (1978).

65. D. E. Comings, and T. Kakefuda, J.Mol.Biol. 33:225 (1968).

66. P. Hobart, R. Duncan, and A. A. Infante, Nature 267: 542 (1977).

67. R. L. O'Brien, A. B. Sanyal, and R. H. Stanton, Exp. Cell Res. 70:106 (1972).

68. G. G. Maul, and P. C. Cross, J.Ultrastruc.Res. 47: 115 (1974).

69. E. Sparvoli, M. G. Galli, A. Mosca, and G. Paris, Exp.Cell Res. 97:74 (1976).

70. B. Blondel, Exp.Cell Res. 53:348 (1968).

71. C. H. Ockey, Exp.Cell Res. 70:203 (1972).

72. J. A. Huberman, A. Tsai, and R. A. Deich, Nature 241:32 (1973).

73. G. E. Wise, and D. M. Prescott, Proc.Nat.Acad.Sci. 70:714 (1973).

74. S. Fakan, and R. Hancock, Exp.Cell Res. 83:95 (1974).

75. S. Fakan, L.R. Wallace, and N. Odartchenko, J.Microsc. Biol.Cell. 27:19 (1976).

76. W. W. Franke, E. Deumling, N. Zentgraf, H. Falk, and P. M. M. Rae, Exp.Cell Res. 81:365 (1973).

77. J. M. Vlak, P. H. Rozijn, and F. Spies, Virology 65: 535 (1975).

78. C. A. Williams, and C. H. Ockey, Exp.Cell Res. 63: 365 (1970).

79. G. R. Milner, J.Cell Sci. 4:569 (1969).

80. T. Kuroiwa, Chromosoma 44:291 (1973).

81. T. Kuroiwa, Exp.Cell Res. 83:387 (1974).

82. N. Granboulan, and P. Granboulan, Exp.Cell Res. 38: 604 (1965).

83. M. Genskens, and W. Bernhard, Exp.Cell Res. 44:579 (1966).

84. E. Puvion, G. Moyne, and W. Bernhard, J.Microsc.Biol. Cell. 25:17 (1976).

85. S. Fakan, and W. Bernhard, Exp.Cell Res. 67:129 (1971).

86. S. Fakan, E. Puvion, and G. Spohr, Exp.Cell Res. 99: 155 (1976).

87. K. Smetana, W. J. Steele, and H. Busch, Exp.Cell Res. 31:198 (1963).

88. I. I. Singer, Exp.Cell Res. 95:205 (1975).

89. S. Fakan, and W. Bernhard, Exp.Cell Res. 79:431 (1973).
90. D. Villard, and S. Fakan, C.R.Hebd.Seances Acad.Sci., Ser. D 286:777 (1978).
91. S. Fakan, J.Microscopy 106:159 (1976).
92. G. Moyne, Cytobiologie 15:126 (1977).
93. M. Bouteille, M. Laval, and A. M. Dupuy-Coin, in: "Cell Nucleus," H. Busch. ed., Academic Press, New York, vol.3, p.71 (1974).
94. S. Fakan, and N. Odartchenko, J.Microsc.Biol.Cell. 23:203 (1976).

GROWTH PARAMETERS IN NORMAL AND TUMOR CELLS: NON-CYCLING CELLS AND METASTATIC VARIANTS AS MONITORED BY FLOW CYTOMETRY

C. Nicolini, S. Lessin, S. Abraham, A. Chiabrera & S. Zietz

Temple University Health Sciences Center
Philadelphia, Pennsylvania USA and

National Research Council
ITALY

INTRODUCTION

Today in the biological sciences there is a growing interest and use in technologies capable of remote <u>in vivo</u> quantization of cellular constituents. The biophysical application of two such technologies, as scanning (1) and flow (2) cytometry, to elucidate the <u>in vivo</u> mechanisms of proliferation, differentiation and expression of metastatic potential in mammalian cell systems, has been recently introduced (for an updated review see also the chapters by Nicolini, Mendelsohn and Casperson in the previous book of this NATO-ASI series (3)). These technologies have made possible the detailed and quantitatively accurate characterization of "native interphase chromatin" (1,4,5) and membrane (2,6). This characterization has in turn led to a new model for normal and abnormal cell growth, that is related to and dependent upon the quinternary nuclear organization of the chromatin (1,4).

As shown in this chapter, computer data acquisition from laser flow systems permit rapid analysis (approx. 1000 cells/sec) of an entire population of cells from a single cell suspension in terms of multiple simultaneous physical-chemical parameters, i.e., any two chosen fluorescent emissions (related to chromatin-DNA, RNA or membrane), and light scatter (a measurement influenced by cell orientation or index of refraction, but grossly related under our operating conditions to cell size). Once obtained, these parameters can then be utilized to sort viable selected subpopulations of cells by means of operator selected windows within their frequency distributions (6,7).

The potential of this unique approach to the life sciences to characterize both membrane and chromatin in viable functionally intact cell preparations has yet to be fully exploited. In vivo cell populations are indeed quite heterogeneous in terms of functional properties and the physico-chemical analysis of such bulk preparations (as conducted so far) are bound to limited results. Only in functionally homogeneous cell populations (either in terms of cell proliferation and/or cell transformation) can structural studies yield useful and meaningful insights. By proper multiple staining of heterogeneous cell populations and subsequent viable cell sorting of functionally homogeneous subpopulations, further biological and physico-chemical characterization could then give new insights on the mechanisms controlling cell proliferation and transformation, at the level of various cellular constituents, such as chromatin and membranes.

Many new diagnostic tools for early cancer detection could then become available (8,9,10). Similarly, it could then be possible to monitor the perturbations in the growth kinetics of individual cellular subpopulations within complex in vivo systems either normal (11,7) or abnormal (12,13) as caused by drug treatment (14).

This chapter will focus on the use of automated multiparameter analysis of single flowing cells to search for new means to objectively discriminate and simultaneously monitor cell proliferation (i.e., cell cycle phase related changes) and cell differentiation (as related to the metastatic potential or to maturation and terminal differentiation) within three normal and abnormal animal cell systems.

Toward this end, the design and fabrication of an "in house" electronic modification to our flow system is described which allows an additional dimension of information to be obtained on selected homogeneous cellular subpopulations of heterogeneous cell suspensions, without increasing the complexity of data handling or storage. Particular attention will be addressed to non-cycling cells, either in tumor or normal cells, seen as necessary steps in any cell cycle progression (4), rather than separate states (15). Experimental evidences and theoretical hypothesis are intended here as a departure from the dominant nominalistic and semantic-empirical debates, and are within an analytical and rigorous physico-chemical framework.

THREE-PARAMETERS FLOW CYTOFLUOROMETRIC ANALYSIS

Cell fluorescence (green,red) and low angle forward light scatter were routinely measured on single flowing cells on our Cytofluorograf 4800A (Ortho Instrument Systems Inc., Mahopack,NY) with an argon laser at 488 nm. (See review in ref. 5) on line with a PDP11/40 computer (16).

In selected cases, to isolate and further characterize specific subpopulations, two parameter fluorescent activated cell sorter (FACS II), Becton Dickson, Mountain View, Calif.) was used to identify and sort selected (at a rate of 1000 - 4000 cells per second) cells by simultaneous measurement of fluorescence and low angle forward light scatter at an argon laser wavelength selected according to the dye utilized (6,7,16).

Normally, the 4800A Cytofluorograf measures low angle scattered light, red and green fluorescence, but in actuality only two of the signals are made externally available as 10 volt 100 ohm sources by a 3 x 3 switch matrix. When two signals are selected for output, a 5V Z-axis pulse is produced when either one exceed a preset threshold. This pulse is used to trigger simultaneous acquisition of data from both channels by the PDP 11/40. Thus, we have more information potentially available that can be used (with normal system configuration).

We have enhanced the original equipment configuration by constructing a circuit that enables us to set a selection window for any one parameter and to acquire data from one or more outputs that additionally satisfy their own threshold criteria. Circuit operation (Figure 1) is controlled by a 2-deck 4 position rotary switch. The first switch deck controls the path of the Cytofluorograf trigger signal. In the first position the trigger is routed from input to output. In all other positions the trigger is routed from input to output via the 7400 Quad 2-input NAND gate which performs a logical AND operation. The second switch deck controls the analog input to the circuit. In the first position there is no input selected. In the remaining three positions the switch selects the positive-going outputs of the green, red, and scatter detectors. The selected signal is passed in parallel to the inverting inputs of a pair of LM 311(National Semiconductor Inc.) voltage comparators whose non-inverting inputs receive reference voltages from a pair of 1458 unity gain operational amplifiers and 10-turm potentiometers. The outputs of the LM 311's are passed to a pair of 74121 one-shots that provide a (33 μsec width) Q output if the lower threshold is exceeded and a Q output if the upper threshold is exceeded. The outputs of the 74121's are logically ANDed in the 7400 with the Cytofluorograf trigger pulse. The 5 ns trigger

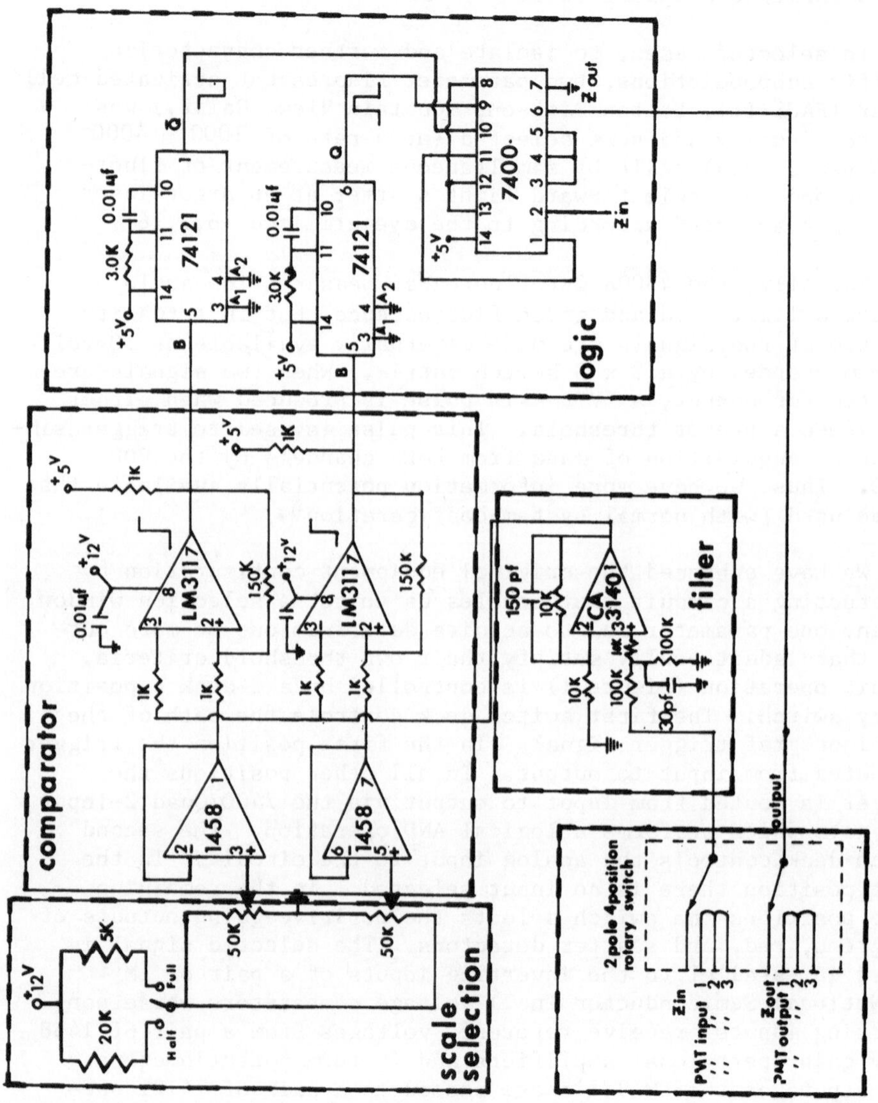

Figure 1. 3 Parameter Schematic

pulse from the Cytofluorograf itself is internally delayed in order to permit its 33μ sec. stretched output pulse to achieve their maximum values approximately 10 μs prior to its leading edge. This timing insures that no false triggers are produced by the small but finite signal rise-time dependent delay which exists between the outputs of the 74121's if both window voltages are exceeded.

In a typical application "red" and "green" fluorescence signals from a cell are acquired if either amplitude exceeds a preset threshold and if the scatter signal from the same cell falls within the independently selected amplitude window. Scattering windows are selected by means of a pulse-height analyzer and several threshold settings of both potentiometers are recorded according to the scattering amplitudes which they select. The Cytofluorograf is then set up for "red"-"green" operation and its triggering signal is passed through the logical portion of the window circuit. As a result the PDP11/40 acquires only data from events which exceeded "red" or "green" thresholds and which produced a scatter signal within the window.

MEMBRANE AND NUCLEIC ACID STAININGS

Acridine Orange (AO) Staining

The metachromatic, intercalating fluorescent dye AO at a molar ratio R(M AO/ M DNA-P) = 4 and at a final concentration of 2.5×10^{-5}M was added directly to cell suspensions of 5 to 8.5×10^5 cells/ml in 10 ml. At these conditions, following the mass action law as previously shown (12), AO provides a differential emission in the green (chromatin-DNA primary binding sites) and the red (cytoplasmic RNA) for intact live or fixed cells, without any (unnecessary) pretreatment by Triton X-100 or chelating agents at various pH.

Double Staining: Fluorescamine and Ethidium Bromide (EB)

Cells were first stained with fluorescamine followed by EB (6). The two methods of fluorescamine staining, adapted from previous procedures (17,18) varied in buffer and pH. The first method (17) involved dissolving fluorescamine and quickly mixing the solution in warm 0.2 borate buffer pH 9 and layering the acetone-borate buffer mixture on a monolayer of cells in a 75 cm^2 T-flask. Our alternative method (18) substituted dimethyl sulfoxide (DMSO) as a solvent for fluorescamine and a 0.15M NaCl 0.005M sodium phosphate buffer pH 8 for a staining environment. To stain with fluorescamine, medium was removed from a confluent monolayer (6×10^6 cells) grown in a plastic 75 cm2 T-flask and washed twice with Ca^{++} and Mg^{++} free Hanks' balanced salt solution (BSS) and once with staining buffer (pH 8 or pH 9).

Immediately after washing, 5 ml of staining buffer was added to the flask which was then kept in a vertical position. A solution of fluorescamine (Sigma, St. Louis, MO) 0.35 - 30.0 mg dissolved in 0.2 ml of solvent (DMSO or Acetone), was injected by pipet into the aqueous buffer giving a final solvent concentration of 4%. The flask was immediately shifted to a horizontal position and gently shaken so that the fluorescamine was evenly distributed over the monolayer. After 20 sec the cells were washed 3 times with Hanks' BBS and gently scraped off the plastic with the aid of a rubber policeman in 10 ml Hanks BBS. The cells were pipet mixed and passed through a 100 μ filter and counted and tested for viability (trypan blue exclusion). Ethidium bromide was then added directly to the resultant cell suspension so that the final (EB) = 10^{-4}M and a molar ratio R (μM total dye added/μM total DNA-P) = 10 (5,10). Such a high (EB) and R ensure a distinct nuclear emission (red) separated from the membrane emission (blue) but lose the capability to resolve fine structural differences in chromatin-DNA (2). Indeed even free EB, outside weekly bound EB and EB bound to RNA, do yield orange-red fluorescence even if with lower quantum yield, requiring perfect instrument calibrations, lengthy and complex analysis of broad multiband (20) peak to detect subtle differences in chromatin conformation (25, 26) as the one between "G0" and "G1" yielding ∿45% difference in primary binding sites (2).

Cytofluorometric Staining Standard

Chicken erythrocytes fixed in 80% ETOH were resuspended in PBS (pH 7.4) to a final concentration of 5 to 8.5 x 10^5 cell/ml. The suspension was then mixed with the cells to be analyzed in a volume ratio of 1 ml of CE to 2 ml of sample. Combined mixture was then directly stained with AO or EB, at conditions previously shown (7). Their utilization was extremely critical in assigning peak position and therefore cell cycle phases (7).

READILY-REVERSIBLE (G0) AND NON-READILY-REVERSIBLE (Q) OUT-OF-CYCLE CELLS: THEIR IDENTIFICATION, ISOLATION AND CHARACTERIZATION

Human Peripheral Blood Lymphocytes

Normal human lymphocytes stimulated with PHA were prepared as previously described (7). For live cytofluorometric analysis the above prepared samples were centrifuged at 900 rpm washed and resuspended in sterile PBS supplemented with fetal bovine serum, penicillin and streptomycin (7). Alternatively, samples were fixed by adding the concentrated and washed sample drop by drop into 20 ml of 80% ETOH (7).

The DNA synthesis that occurs during PHA induced proliferation of normal lymphocytes was monitored by C_{14} thymidine incorporation. DNA synthesis is not evident until 48 hours of PHA stimulation with the total population eventually going through approximately two cell cycle divisions (7,8). In terms of FMF analysis, during the first 48 hours the scatter profile remains invariant while there is a progressive broadening as early as 24 hours (8) of the green fluorescence emission profile which is associated with the G_0 (resting) - G_1 (cycling) transition; this becomes more clearly evident as bimodal distribution at 72 hour (Fig. 2). In Fig. 2, the 72 hour green emission profile shows a distribution characteristic of a population in log phase growth. To be noted also is the "Q" peak, a very low green fluorescence population, not evident at 0 hours. A comparison of this very low peak in Fig. 2, over all time points, reveals that its amplitude increases with time after PHA stimulation at the expense of all other fractions of the cell cycle (7,8). In lower panel of Fig. 2, the 144 hour scattergram reveals only two subphases of the cell cycle, namely Q and G_1. The G_1 assignment is based after normalization using chicken erythrocytes (7) on slightly higher green fluorescence (chromatin-DNA primary sites), quite larger red fluorescence (RNA amount), and similar scatter. Most notable is the large relative magnitude of the Q compartment at 144 hours.

To determine whether the low green fluorescent population contained viable cells or was a conglomeration of dead cells, broken cells or nuclei (which could be a possible explanation for this low fluorescence peak), Q phase lymphocytes from 72 hours stimulation were viably sorted (7). Our sorting criteria were centered on the peaks indicated by the designated letters in the two parameter scattergram (Fig. 2, middle panel), with a 5 to 10 channel null space between the selected windows in green fluorescence. The sorter fractions Q, G_0, G_1 and G_2+S, were then examined visually and tested for trypan blue exclusion. The Q cell (as well as G_0, G_1, and S) subpopulations contained whole cells with 75 - 90% viability after sorting depending on the experiments (7). In order to verify what kind of cells are contained in the Q subpopulation, viability sorted Q + G_0 compartments isolated from a 72 hour PHA stimulated culture were studied for lymphocyte membrane markers. Surface receptors were demonstrated by utilizing a double rosette technique (8) and at least 300 live cells were counted in the presence of trypan blue for each subfraction. Table I shows the percentage of live Q and G_0 cells with T or B markers and confirms that the Q cell are viable T-lymphocytes. A noticeable and interesting finding was the intensity of the SRBC rosette formation, yielding in Q an average of 20 to 30 SRBC around each cell, quite larger than for lymphocytes from routine peripheral blood (8).

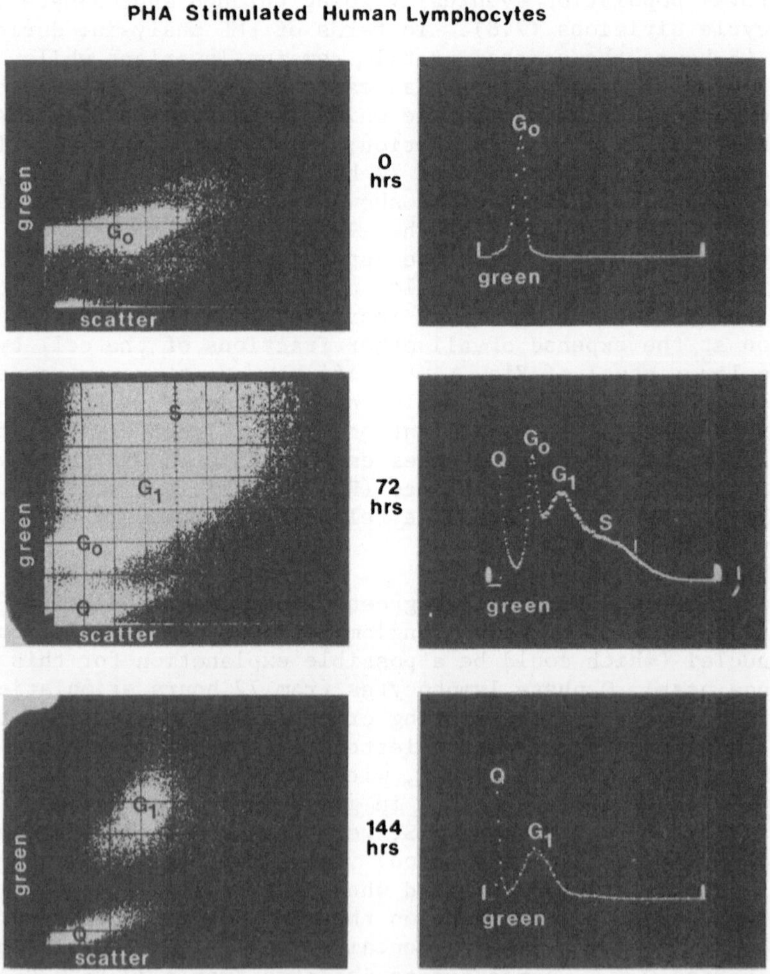

Fig. 2. Scattergrams of viable normal lymphocytes stimulated by PHA at 0, 72 and 144 hours by FACS II analysis. On right column are the associated one dimensional green fluorescent histograms (AO). Letters Q, G_0, G_1 and S are designated assignments of cell cycle subphases. Sorting criteria was centered about the designated letters with a 5 to 10 channel null space between selected windows in green fluorescence.

Table I. List of the Percentage of viably sorted Q and G_0 subpopulations with T or B Marker

LYMPHOCYTE SURFACE MARKERS		
	G_0 (%)	Q (%)
T-cell marker	96	89
B-cell marker	1	3
Double-cell	1	1
Null-cell	2	7

At least 300 live cells were counted for each fraction. Null cells are those without either membrane marker. Double cells are those with both markers.

To ascertain whether the Q cell is recruitable into the cell cycle, the same viable sorting procedure was again performed on 72 hour PHA stimulated lymphocytes. Each subphase was sorted into sterile sorting wells with RPMI medium, supplemented with 10% fetal calf serum, antibiotics and antimycotics (7). The samples were then sorted for viability via trypan blue exclusion and incubated in the presence of PHA for an additional 72 hours. Following restimulation each fraction was tested again for viability and stained with AO for cytofluorographic analysis. Table II summarizes the results in a typical experiment; which show equally high viability in all fractions (including Q) and high proliferation in all fractions (excluding Q).

The results of two parameter FMF analysis (8) of sorted fractions after secondary PHA stimulation confirms furthermore the above findings; the Q fraction exhibits indeed no apparent progression into the cell cycle while all other fractions do produce fluorescence histograms which indicate a proliferative response.

The evidence so far produced has proven that the Q cells are viable and not readily recruitable into the cycle, but what is the molecular mechanisms responsible for the extremely low dye uptake? The green emission, gated on log scale scatter of fixed samples at 72 and 144 hours as analyzed on the FMF (Fig. 3), yield histograms identical to the unfixed viable, including the Q compartment. Lack of acridine orange accessability to chromatin-DNA is then not apparently the cause of very low dye uptake in

Table II. Number of Cells x 10^3 After Restimulation of Separate Fractions

Fraction	TRIAL #1	
	0 hrs	72 hrs
Q	280	310
G_0	20	430
G_1	260	846
G_2+S	65	1217
Control	--	--

The viability immediately after sorting 72 hours is 95% and 90% respectively.

Table III. List of the Percentage of Cells in Q on fixed and Viable Samples of PHA Stimulated Lymphocytes at 0, 72 and 144 Hours as Determined by FMF Histograms Analysis.

% of Cells in Q	Fixed	Live
Control	2	2
72 hrs	33	33
144 hrs	55	62

PHA Stimulated Human Lymphocytes
Green Fluorescent Histograms

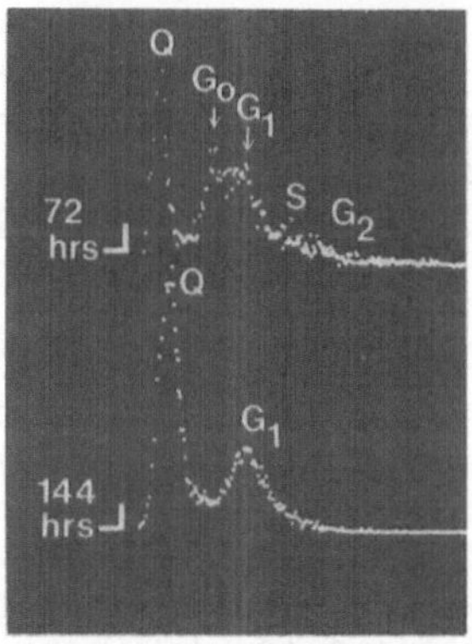

Fig. 3. Green Fluorescence distribution of fixed samples of PHA stimulated lymphocytes at 72 and 144 hours. R = 4; (A.O.) = 2.5×10^{-5}. Cells are gated in terms of scatter signal (logarithmic scale).

the Q cells, which indeed increase over time after stimulation also in fixed lymphocytes (Table III). The three fold increase of the non-cycling Q fraction from 0 to 144 hours may reflect the well known fact that PHA stimulated lymphocytes go only through two division cycles (15).

In order to ascertain whether lymphocytes enter the Q compartment from the G_0 or G_1 phase, or also from other phases, PHA stimulated lymphocytes were continuously labeled with C_{14} thymidine at 48 hours post PHA stimulation and harvested at 72 hours (Table IV). The thymidine incorporation of the Q fraction at 72 hours is too low to suggest a unique S or post-S phase entry into Q. Indeed, this combined with image analysis of Feulgen-stained sorted Q cells (7 and Fig. 4) strongly suggests that entry can be made into the Q compartment from all phases of the cell cycle. A detailed image analysis of Feulgen stained preparations of live sorted Q and G_0 fractions from a 72 hour sample (Fig. 4) confirms the multiploidy nature of the Q fraction, as compared to the G_0 fraction, shows 2C - 4C DNA content and supports an interpretation of multiphase entry. The same data shows that an enrichment in the G_0 cells causes a shifting in the 2C window towards larger nuclear DNA condensation with respect to the log-phase "unsorted" fraction (dominated by G_1 cells); that is, the higher chromatin condensation (G0) is accompanied by lower green fluorescence emission while lower chormatin condensation (G1) corresponds to higher green emission (Fig. 4). Statistical analysis of the image data revealed furthermore that the Q sorted fraction exhibits a significantly higher mean AVOD (nuclear chromatin condensation) and a significantly larger proportion of nuclei with 2C DNA content than the unsorted control samples. Statistical analysis of the events in both the 2C and 3-4C windows revealed that the increase in mean AVOD in the Q subphase was indeed significant in either case when compared to the unsorted control sample.

The red fluorescence distribution of AO stained cells at the proper R is indicative of RNA activity. The red fluorescence distribution of the various subphases of a 72 hour sample are shown in Fig. 5. On the right are the green emissions of selected narrow peaks of the Q, G_0 and G_1 fractions. On the left are the associated red fluorescence distributions, revealing a large increase in the peak position of the red fluorescence from Q and G_0 to a cycling G_1 state. Whereas the Q and G_0 distribution are monomodal and relatively homogeneous, the G_1 fraction exhibits a broad heterogeneous distribution which is characteristic of the increase in RNA activity of the G_1 cycling cell. The low green and red emissions of the Q cells suggests that these cells represent a population which is completely quiescent, i.e., with very low amount of RNA and very low primary binding sites.

Table IV. The ^{14}C thymidine incorporation (CPM) of unstimulated Q fraction, combined G_0-G1 fractions, and total population of stimulated Lymphocytes which were continuously exposed to a single dose of ^{14}C thymidine (0.02 μCi/200,000 cells) for 24 hours prior to collection. Normalized values correcting for Sorting Efficiency (70%) are given for sorted Q and G_0-G_1 fractions in parenthesis.

UNSTIMULATED	72 Hour STIMULATED LYMPHOCYTES		
G_0	Q	G_0-G_1	Total
78 ± 6	1546 ± 68	*2711 ± 283	14,998 ± 1952
(78 ± 6)	(1046)	(3942)	

*Some S phase contamination.

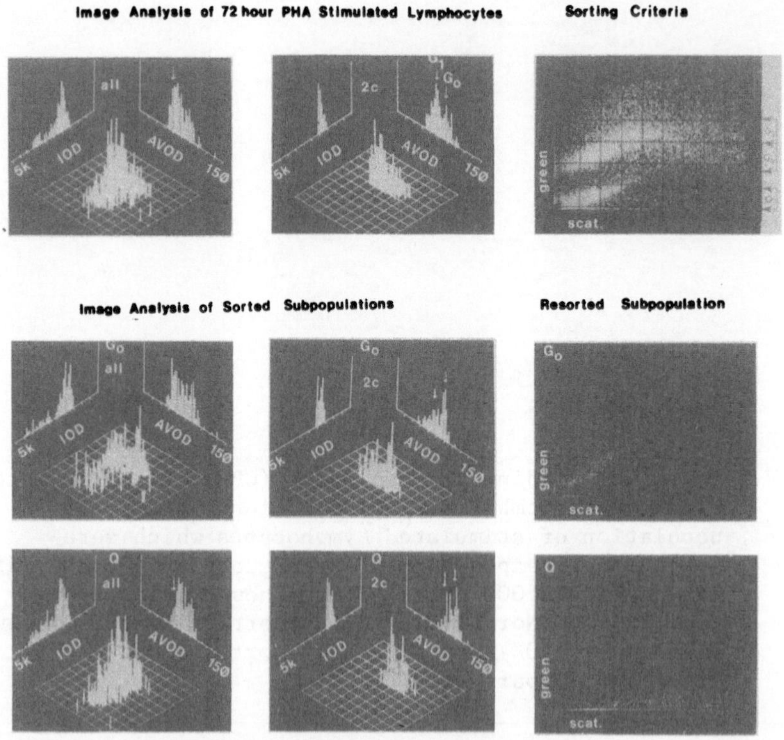

Fig. 4. Image Analysis of 72 Hour PHA Stimulated Lymphocytes and Their Sorted Subpopulation

Panel I - Left	IOD vs. AVOD at Feulgen-stained cells in unsorted control sample.
Panel I - Center	IOD vs. AVOD of 2C DNA cells in the control sample.
Panel I - Right	Sorting criteria for Q, G_0 and G_1 fractions of 72 hour sample from two parameter scattergram (green fluorescence vs. scatter)
Panel II - Left	IOD vs. AVOD of all sorted "G_0" cells
Panel II - Center	IOD vs. AVOD of 2C DNA sorted "G_0" cells.
Panel II - Right	Scattergram of the sorted "G_0 subpopulation"
Panel III - Left	IOD vs. AVOD of all "Q sorted" cells
Panel III - Center	IOD vs. AVOD of 2C DNA sorted Q cells
Panel III - Right	Scattergram of the sorted "Q fraction"

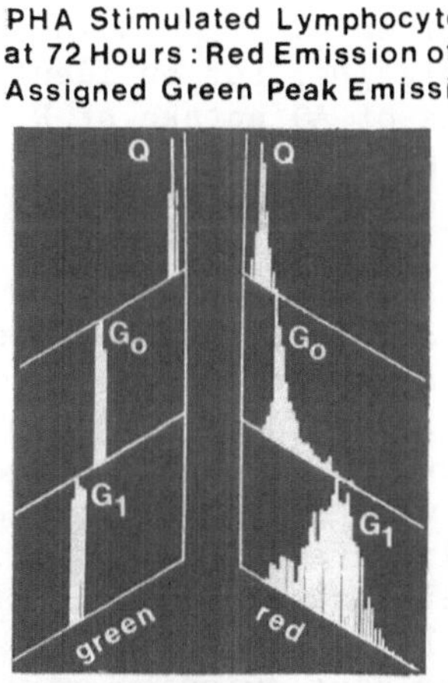

Fig. 5. Red fluorescence distribution of the narrowly selected peaks in green fluorescence of Q, G_0 and G_1 fractions of 72 hour PHA stimulated lymphocytes.

In order to be certain of peak assignments in all the histograms, fixed chicken erythrocytes were mixed and stained with the samples at all time points and subjected to FMF analysis (7).

In summary, the Q cells in PHA-stimulated lymphocytes are functionally viable T lymphocytes exhibiting multiploidy DNA content, low green and low red emission peaks under equilibrium AO staining and they cannot be induced to proliferation by PHA. Conversely G_0 cells (also quiescent) have only 40% less primary binding sites (respect to G_1), low RNA amount, intermediate between Q and G_1, and are completely recruitable into the cycle by PHA stimulation.

Human Fibroblasts

The existence of 3 levels of chromatin condensation (G_1, G_0, Q) also in 2C DNA WI-38 fibroblast at confluency has been recently demonstrated by automated image analysis of intact nuclei. Their existence is furthermore evident in Fig. 6 where WI-38 cells at confluency, 10 days after plating, for an identical cell size (scatter), exhibits 3 levels of AO uptake at R = 4 and 10^{-5}M AO final concentration, both in the linear scale (using the Beckton Dickenson) and logarithmic compressed scale (using the cytofluorograph).

Without two parameter acquisition and log-scale in the scatter, the cells with (Q) dye uptake could be easily confused with cell debris or PMT noise, by analyzing only the fluorescence frequency distribution (see Figs. 2 and 4) (14,15). Actually by double staining WI-38 fibroblasts with EB and fluorescamine it appears that these (G_0, G_1, Q) cells have not only similar size, but also the same number of bound fluorescamine, i.e., similar outer membrane properties (2) (also Fig. 6). Indeed the frequency distribution for cells with the same blue emission or with the same scatter, yield 3 levels of chromatin DNA uptake, followed by a few cells in S, G_2+M (Fig. 6).

The most striking example, which confirms both the existence of a differential emission for RNA versus DNA at the proper AO molar ratio, and our previous EB data on the G_0-G_1 transition of WI-38 fibroblast (20) have been recently obtained from WI-38 human diploid fibroblasts which have been grown up to 23 days into "deep" confluency with weekly changes of the medium (2, 16). Two days after plating, WI-38 cells show a log-phase distribution for the green fluorescence with a population of cells having the same scatter (respect to G_1) but more than twice lower chromatin primary sites (green) and lower amount of RNA (red): this subpopulation, likely relating to Q non-cycling cells (that is, after plating, even if adherent to the plastic, did not start growing), drastically reduces at 5 days (5%) and then progressively increases at 12 days (-30%) and 23 days (up to 95%).

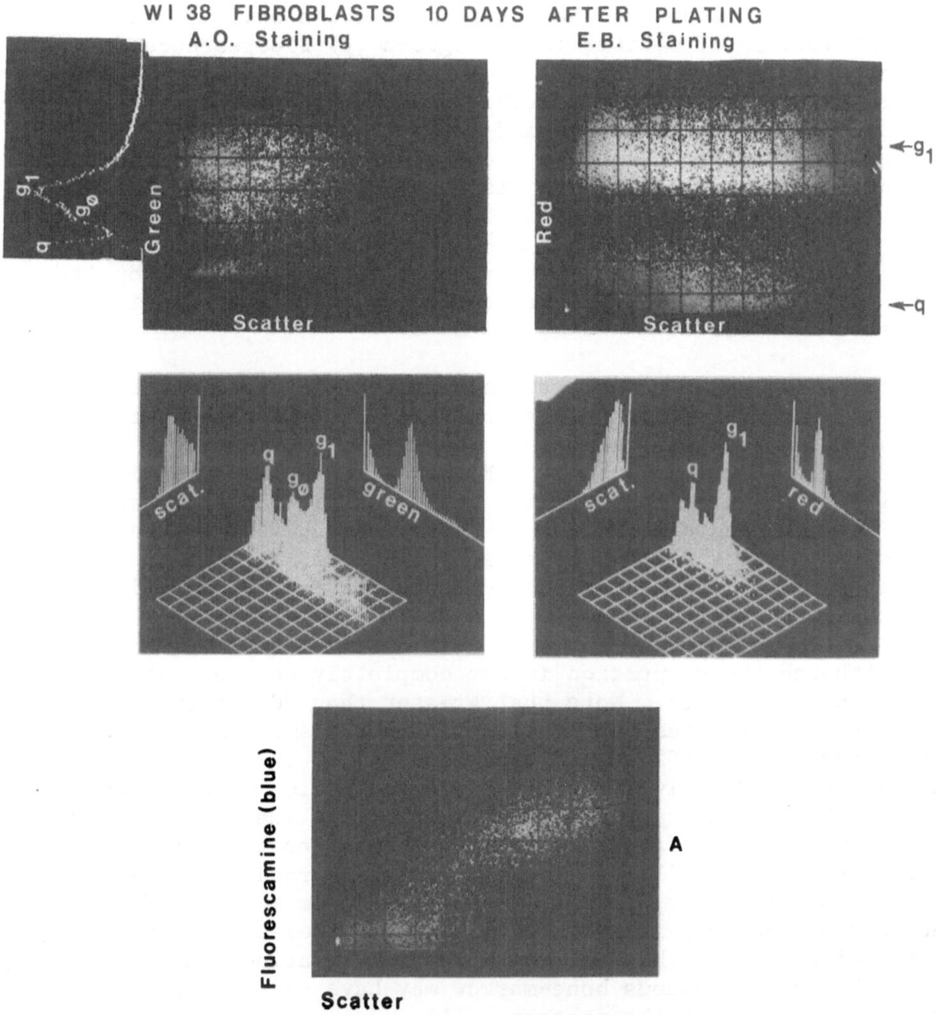

Fig. 6. (Left panels) scatter versus green fluorescence of WI-38 fibroblasts, stained with A.O.
(Right panels) as left, but stained with EB.
The lower panels are taken using in the abscissa a compressed logarithmic scatter (or fluorescamine uptake),
The upper panels are with a linear scatter signal.

Mouse Bone Marrow

The results obtained in bone marrow suspension (from mice) are presented in Fig. 7. Panel A shows the red (x-axis) vs. green (y-axis) distribution for the entire population while in Panel B the distribution of scatter (x) vs. green (y) fluorescence is presented for the entire population. In Panels C - E are presented the red (x) vs. green (y) distributions for the small scatter peak (C), medium scatter peak (D), and large scatter peak (E). In all the panels, the histograms above two dimensional distributions are obtained by projecting the distribution onto the corresponding axis.

Subsequent sorting and differential counting of the populations in Panel C - E, by Wright staining yielded the identification as listed in Table V.

It is important to observe that neither of the distributions which monitor the entire population (Panel A - B) reveals the wealth of informations obtainable when the circuit is used to gate on a third parameter (Panels C - E). Yet, only two dimensional distributions are collected which greatly facilitates the acquisition, storage, and handling of the data. Use of the 3-parameter acquisition with the heterogeneous populations of cells obtained from in vivo tissues as bone marrow shown in Fig. 7 can allow a selected subpopulation to be monitored.

Although the seapration is not completely unique a significant isolation is achieved. Note that greater than 90% of the mature cells and erythrocytes can be isolated in the small population, while greater than 85% of the blast and 85% of the lymphocytes can be found respectively in the large and medium scatter populations. The megakarocyte is the only cell type that shows no trend toward separation. Specifically, substantial enrichment of blast cells (from ∿8% to ∿71%) is apparent in the sorted "large scatter" population; similarly for the lymphocytes (from ∿4% to ∿50%) in the sorted "medium scatter" population. The selective monitoring of the blast cells, which is the most proliferating compartment in the highly heterogeneous bone-marrow may have useful applications in monitoring cancer chemotherapy (11).

B-16 Melanoma Tumor

Flow and static cytometry utilizing AO staining and Feulgen hydrolysis respectively of in vivo melanoma B16 tumor (13) reveals that 30% of the population is in the so-called Q compartment as characterized by lower AO uptake and higher chromatin condensation, with respect to cycling cells Sorting and subsequent subculturing and viability assay (with trypan blue exclusion) confirm that

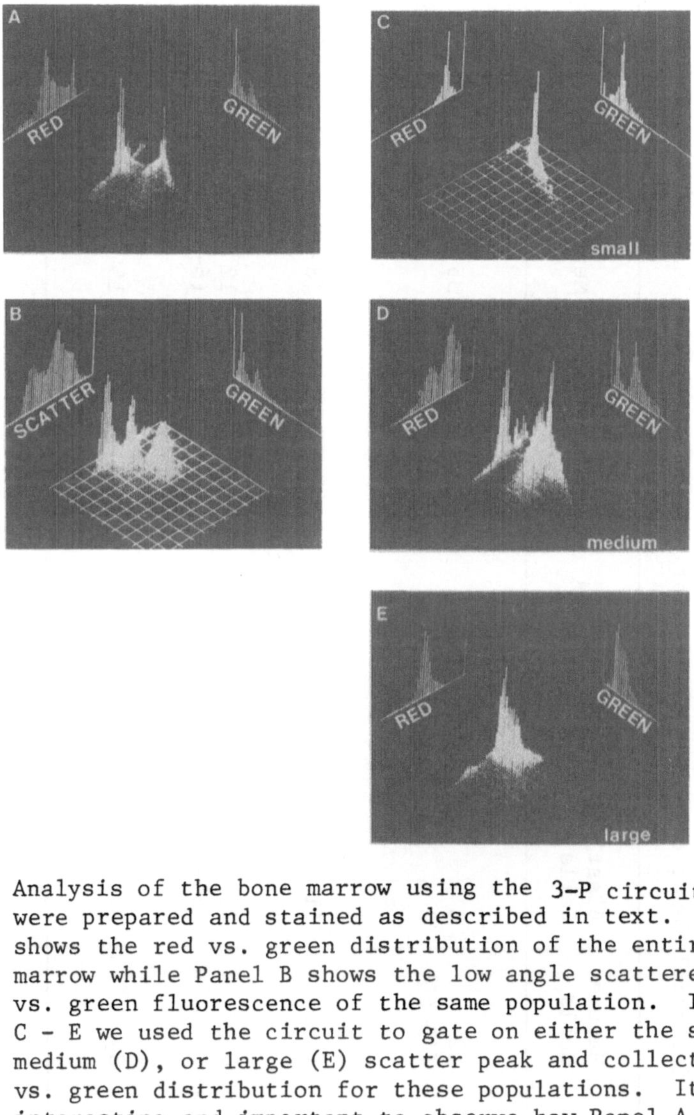

Fig. 7. Analysis of the bone marrow using the 3-P circuit. The cells were prepared and stained as described in text. Panel A shows the red vs. green distribution of the entire bone marrow while Panel B shows the low angle scattered light vs. green fluorescence of the same population. In Panels C - E we used the circuit to gate on either the small (C), medium (D), or large (E) scatter peak and collect the red vs. green distribution for these populations. It is interesting and important to observe how Panel A is made up of various subpopulations as evidenced in Panels C - E.

Table V. The Percentage (Mean and Range) of Cells Classified by Type Through a Wright Stain Differential Count of Sorted Subpopulations and Total Bone Marrow

	Mature Cells	Blast Cells	Intermediates	Erythrocytes	Megakarocytes	Lymphocytes
C - Small						
.Mean	52	0	4	5	35	6
.Range	(45-60)	0	(2-8)	(0-11)		(5-8)
D - Medium						
.Mean	3	9	25	0	20	50
.Range	(2-3)	(1-18)	(16-35)		(8-35)	(30-70)
E - Large						
.Mean	0	71	16	0	12	0
.Range		(65-78)	(10-22)		(6-17)	
Unsorted - Total Bone Marrow	34	8.3	33	37	15	3.7

even Q tumor cells are viable and non-cycling (without any 3H-thymidine incorporation when pulse-labeled). The rationale on the non-cycling properties of these Q cells in dynamic equilibrium with their cycling counterparts has been given previously (13). Cytofluorometric analysis of the same primary tumor after in vitro line establishment, revealed that the fraction of cells with low dye uptake, Q, decreases with the number of passages being 17% at the 15th passage and 2% at the 59th passage in vitro (Fig. 8). Compatible with the existence of a gradient in the amount of non-cycling cells in the primary tumor, a similar phenomena is evident when comparing F1 (low metastatic) and F10 (high metastatic) variants from the same B16 tumor (Fig. 8). F10 exhibits (2,6) a high growth fraction (low number of Q cells) while F1 exhibits a low growth fraction (higher number of Q cells). Numerically F10 has approximately 3% of the cells in Q while F1 has about 20%. Noticeably the G_0 fraction (Q < green fluorescence < G_1) is relatively non-existent in all B16 tumor lines, being present instead in both human fibroblasts (2, 4, 20) and human lymphocytes stimulated to proliferate (4, 7, 8).

A UNITARY MODEL OF CELL PROLIFERATION AND VARIOUS TYPES OF NON-CYCLING CELLS

Time-lapse cinematography on single cell culture (27) has shown that the lag time between the stimulus and the onset of DNA synthesis is similar to the minimum intermitotic time in subsequent generations. It then appears (see the chapter by Nicolini and Belmont in this book) that the duration of the lag is determined by the same process (first cycle (I) of chromatin condensation during G1) that determines the minimum intermitotic time (second cycle (II) of chromatin condensation between middle S and G2). It also suggests that normal quiescent cells have to undergo two random transitions (at the two restriction points) separated by a rather lengthy process of progressive chromatin condensation (of similar duration as during S+G_2 phases) before they can enter S phase, as indeed expected if normal cells would undergo (as apparently they do) changes similar to the one so far reported for synchronized transformed cells.

Experiments with 3T3 cells stimulated to proliferate by serum (27) suggest that cells are not committed to enter S phase more than 5 hours before they actually do so, even if their lag time is of 14 hours: this would be compatible with the existence of a second transition, as shown in Fig. 5. Assuming that the time required for chromatin condensation between early and late G1 is constant (about 4 hours for HeLa, as suggested by the preliminary time-lapse photography data shown in Fig. 7) the large variation in G_1 residence time is due to the variable time required for the very late (relaxed) chromatin to begin synthesizing DNA. Namely in a slowly growing population as WI-38 fibroblasts with an average

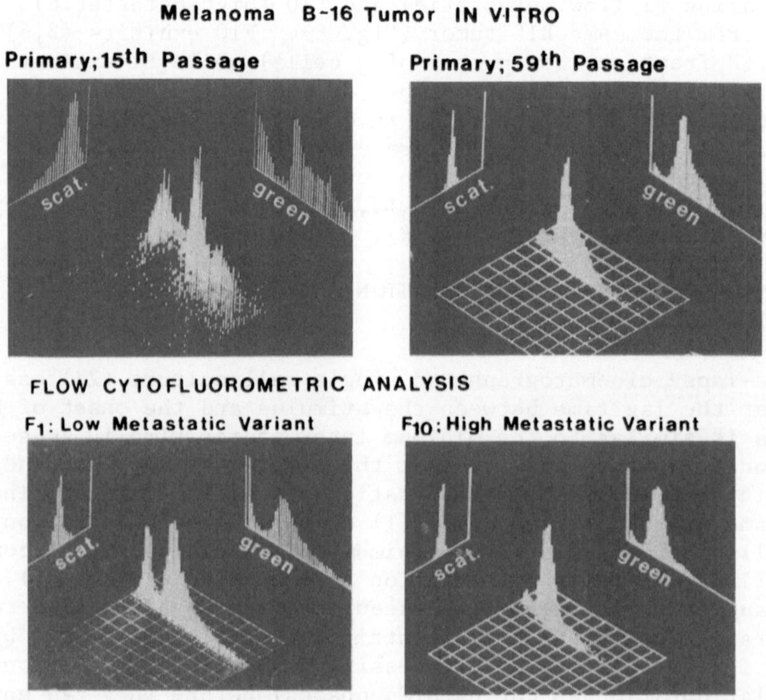

Fig. 8. Two parameter scattergrams of melanoma B16 in vitro. R = 4; A.O. concentration = 2.5×10^{-5}, respectively at 15th and 50th passages, or F1 and F10 metastatic variants (6).

12 - 24 hours G_1 period, most cells with 2C DNA will have in the log-phase a highly condensed chromatin (G_1), while at confluency ("G_0") the same fibroblasts, uncapable to undergo the abrupt relaxation, will be arrested at the I restriction point with a highly condensed "2C" chromatin; this indeed occurs (26).

All above findings combined with recent PCC and scanning cytometry (4) suggests that upon reaching a confluent phase or nutritional starvation, while normal cells are blocked in a "very early" condensed G_1 state (otherwise called "G_0"), most cells in transformed cell lines are accumulated in the "very late" decondensed G_1 or in a condensed S-G_2 state bypassing also the II restriction point. Non-cycling Q tumor cells and non-readily reversible Q normal cells, either lymphocytes or fibroblasts, do constitute the only exception to a well established rule (1), which has shown that a strict correlation exists between chromatin (or nuclear-DNA) condensation and biological parameters like template activity and between the average optical density of Feulgen-stained nuclei and the primary binding sites in intact cells differentially stained with acridine orange for chromatin-DNA. The latter correlation shown in the past for quiescent human fibroblast induced to proliferation by serum addition (26) or rat liver cells after partial hepatectomy (29), has been recently shown to exist in human lymphocytes induced to proliferation by PHA (7), even at the level of "sorted" homeogeneous subpopulations, as cycling (G_1) or resting (G_0) cells.

Since our early observations, originally in isolated chromatin by spectropolarimetry (25) and later in chromatin <u>in situ</u> by flow microfluorimetry (20), that the chromatin-DNA structure is changing so as to increase the number of primary binding sites for intercalating dyes in WI-38 quiescent cells (G_0) after the stimulation to proliferation (G_1), nuclear morphometry provided the conclusive evidence that a quantal decrease in chromatin is indeed occurring during the G_0-G_1 transition (26). These findings, which have been recently proved identical both in PHA-stimulated human lymphocytes (7, 10), and rat-liver after partial hepatectomy (29), show unequivocally that changes in native chromatin structures are indeed responsible for the differential dye uptake in intact cells which is presently a widely accepted criteria for G_0-G_1 discrimination (see reviews in ref. 2).

It is a pity that numerous people ignoring or misinterpreting this basic information for a long time in the literature (since 1974 when the FMF data were presented at an international meeting in Annapolis (31)), rather than directly staining "living" cells with supravital dye, as acridine orange, using the mass-action law (12) are still following "lethal" empirical staining recipes, based on a pretence to achieve differential denaturation of RNA and DNA.

It is also a pity and discomforting that ill-designed, ill-performed, and poorly interpreted experiments can still raise emotions and positive comments, indicative of the large extent of unawareness and lack of basic understanding. Fig. 6 in reference 4 should clarify the issue and doubts raised; incidentally, the original data on WI-38 fibroblasts (20, 31), reproduced with acridine orange, either on fixed or unfixed cells (see a review in ref. 4), were obtained by analyzing the unfixed sample immediately after staining.

The comparative study (4) was conducted on the same human lymphocytes, either fixed (F) or unfixed and maintained in ice (I) or at room temperature (T): cells are stained with acridine orange at R = 4 and 10^{-5}M to warrant differential green emission for chromatin-DNA primary binding sites (37). Identical results are obtained with ethidium bromide (34,39) with some complications in their physico-chemical interpretation, due to minor RNA contribution to the typical EB orange-red emission (20). In all experiments reported in ref. 4 cells are gated requiring a finite scatter signal in the logarithmically compressed scale: this warrants elimination of debris, broken cells, nuclei or dead cells and detection of Q non-cycling cells with very low dye uptake (7), otherwise impossible with the experimental configuration and the cell line utilized in ref. 30. As shown (7) for more complex populations, the frequency distribution of unstimulated lymphocytes, fixed (F) and immediately stained is identical at 0 hours and 1 hour later (with the dye constantly kept with the cell) to the same unfixed population stained and promptly analyzed. The fractions of cells with very low dye uptake (Q viable cells as clearly shown previously) and with dominant "G_0" uptake remain invariant also in unfixed cells when analyzed promptly after staining at 0 hours or 1 hour later, being kept either in ice or at room-temperature. Increase in cell death is not preceeded by an increase in the Q fraction: clearly changes in cell viability do not affect dye uptake, which confirms previous findings and cast doubts on a possible interpretation of Q cells as a simple "pre-death" state. The frequency distributions are instead dramatically affected by the permanence of the dye in suspension with the unfixed cells (30), either at room-temperature or in ice because the intercalation of each dye molecule (being E.B. or A.O.) in equilibrium stainings unwinds the DNA complex of several degrees (24), causing progressive changes in the higher order superhelical arrangement of native chromatin; in this instance the apparent decrease in AO-primary binding sites associated with the progressive change in DNA superpacking could then be reflective of changes in the configurational entropy and relative DNA-dye association constant (12) or constitute an indication of an increased supercoiling due to chromatin fibers attached by both extremity to the nuclear envelope (13). Incidentally, this critical time-dependence of fluorescence distribution is present only when cells are maintained for long times in suspension with the dye (prior to acquisition), but not when cells are analyzed within a few minutes after being stained regardless of cell viability (4).

The multiparameter acquisition and sorting of "viably stained" cells, followed by biological, immunological and image analysis of "selected" subpopulations, has recently permitted a previously unsuspected characterization of two non-cycling compartments (see Fig. 2) with two distinct levels of low dye uptake:

I. A resting G0 cell, readily-reversible into the cycle, characterized by a 2C DNA and by a more condensed and round nuclei with respect to G_1: interestingly this cell type which has about a 40 - 50% decrease in chromatin-DNA primary sites with respect to G_1 and 3-4 times less RNA (Fig. 5) seems present only in normal cells either in vitro or in vivo. Interestingly even PCC (3) shows that while normal cells are blocked by serum-deprivation in a condensed chromatin ("G_0 or very early G_1") transformed cells do not and in plateau phase they stop in a relaxed "very late G_1" state.

II. A non-cycling Q cell, still living, but recruitable with difficulty into the cycle (11) or irreversibly out-of-cycle (7), characterized in normal cells by condensed and round nuclei with mostly 2C but frequently also 3C - 4C DNA; this cell type, which has more than a two-fold decrease in AO chromatin-DNA primary binding sites with respect to G_1, and 4 - 5 fold decrease in RNA amount is instead present in both normal (7) and tumor tissue (13).

The fact that in tumors we are dealing with a different kind of non-cycling cell (nothing to do with the traditional "G_0" cells) is furthermore proven by recent PCC data showing a high correlation between the number of decondensed very late G1 cells and the lack of human tumor response to chemotherapy (28).

These recent findings, made possible by a combined utilization of fluorescence activated cell sorting, immunological and image analysis and by the data obtained by premature chromosome condensation induced by fusion with mitotic cells, shed new light on the physico-chemical and biological properties of non-cycling cells which play a critical role in normal and tumor growth. A new classification of various types of non-cycling cells (Fig. 9) can then emerge, based on the above experimental evidences that the non-cycling cells (regardless of their classification as Q, G_0, A or L states) are not separate states entered under particular conditions, as early empirical observations may have suggested, but represent necessary steps of any normal or abnormal cell cycle progression, where cells may be blocked for finite (G_0) or indefinite (Q) time.

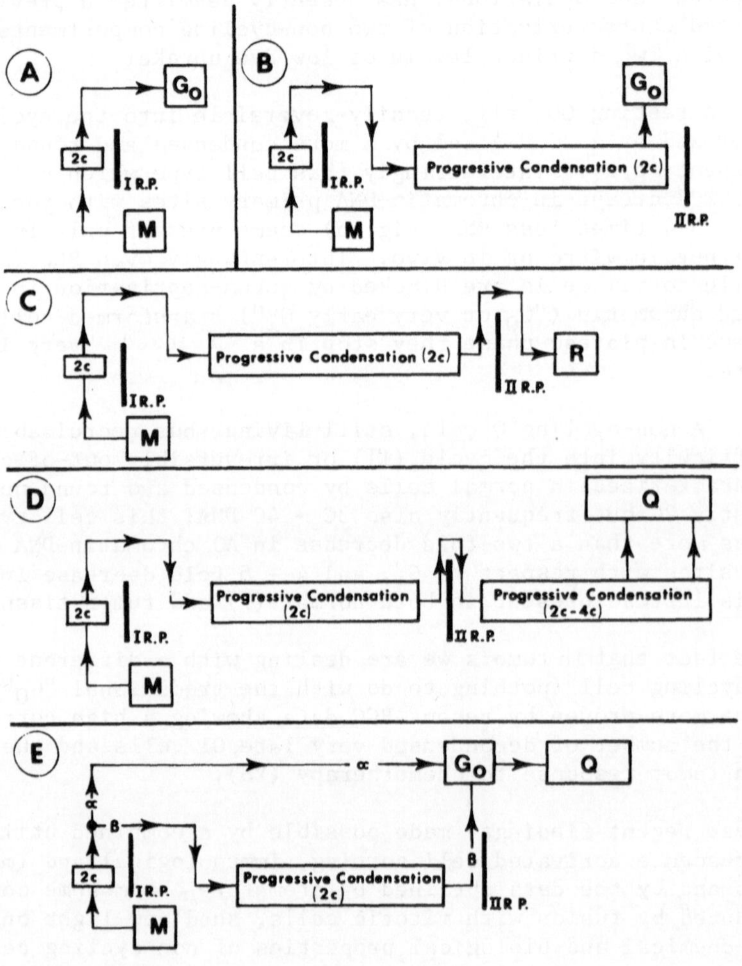
A
Go
2c
I R.P.
M
B
Go
2c
I R.P.
Progressive Condensation (2c)
M
II R.P.
C
2c
Progressive Condensation (2c)
R
II R.P.
I R.P.
M
D
Q
2c
Progressive Condensation (2c)
Progressive Condensation (2c-4c)
I R.P.
II R.P.
M
E
α
Go
Q
α
B
2c
Progressive Condensation (2c)
B
I R.P.
II R.P.
M

Fig. 9. Types of non-cycling cells, according to the experimental findings discussed in the text and within the analytical model described in Fig. 5.

A. Cells are blocked at the first restriction point (I R.P.) with a condensed chromatin typical of very early G_1 [i.e., confluent fibroblasts (22)].

B. Cells are blocked at the second restriction point (II R.P.) in the very condensed chromatin typical of late G_1 phase [i.e., theophylline-treated HeLa (12,16)].

C. Cells pass both restriction points undergoing the abrupt chromatin relaxation but are uncapable to initiate DNA synthesis (may because of the lack or reduced level of given chemical substrates and/or enzymes), and are blocked in a very relaxed chromatin typical of very late G_1 [i.e., ARA-C treated transformed cells (12, 25) or tumor cells in patients not responsive to chemotherapy treatment (3)].

D. Cells initiate DNA synthesis, but at the end (4C) or during middle-late S phase (3 - 4C), they exit the cycle cycle in a condensed chromatin state typical of "Q" phase [i.e., non-cycling lymphocytes after PHA-stimulation (6); non-cycling B16 mouse melanoma tumor cells (30, 51)].

E. Cells blocked at any of the two restriction points ("G_0") enter a deeper state of quiescency characterized by a very condensed chromatin typical of "2C DNA Q" non-cycling cells, either recruitable with difficulty [i.e., WI-38 fibroblasts in deeper confluency, 10-22 days after plating (6, 46, 52); B16 lung metastiasis after Hydroxiurea treatment (51)] or non-readily recruitable [i.e. differentiated lymphocytes induced by PHA-stimulation (6)].

IDENTIFICATION AND CHARACTERIZATION OF F1 AND F10 METASTATIC VARIANTS IN B16 MELANOMA

We utilized a model for blood borne metastates as developed by I.J. Fidler (19) which selects for cells of high metastatic potential by serial injection and collection of B16 melanoma in C57/BL/6 mice. The variants so isolated are respectively B16-F1 (one serial passage and collection) and B16-F10 (ten serial passage and collections), which represent a distinct subpopulation of the heterogeneous B16 primary tumor with a characteristics increase of metastatic potential in the lung of C57/BL/6 mice (IBID).

To further characterize the metastatic variants F1 and F10 we developed a double staining method utilizing fluorescamine, which stains the primary amines of all inner and outer membrane protein constituents, and ethidium bromide (EB), an intercalating dye for nucleic acids.

When double stained with fluorescamine (30 mg/0.2 ml acetone/ 5 ml borate buffer pH9) and EB (10^{-4}, R = 10), a confluent monolayer with 5 x 10^6 cells of B16-F1 and B16-F10 produced two dimensional histograms seen in Fig. 10. For both F1 and F10 the fluorescamine profile was bimodal, but presented a distinct difference in the distribution of dye uptake. In B16-F1 the increase in fluorescamine uptake appears to be a size dependent phenomena while B16-F10 clearly does not follow this trend since the fluorescamine uptake of its mid-scatter cells is considerably more than the B16-F1 mid-scatter cells (6). The EB profiles showed that the fluorescamine uptake did not correlate well with cell cycle parameters for in each of the scatter windows has one distinct level of fluorescamine uptake corresponded to two or more levels of EB uptake; i.e., Q (cells of middle and high scatter but very low EB uptake), G_1, S, G_2 and M. The amount of melanin, as determined by visual inspection after silver nitrate staining (6) was the same for both B16-F1 and B16-F10; having no apparent role in the differential fluorescamine uptake.

Apparent, from the scatter vs. EB histogram in Fig. 10, is the presence of a larger growth fraction in B16-F10 which corresponds to parallel A.O. FMF findings (Fig. 8). Evident in Fig. 10 is the heterogenity of the scatter signal (linear scale). This is not only due to a large heterogenity in cell size but also to flow system artifacts, for B16 is an adhesive cell type whose fibroblastic shape flow orientation artifacts. Also by scraping to achieve cell detachment the resultant suspension of cells contains cells with broken plasma membranes but intact nuclear DNA material (intermediate red, but low blue and low scatter) and

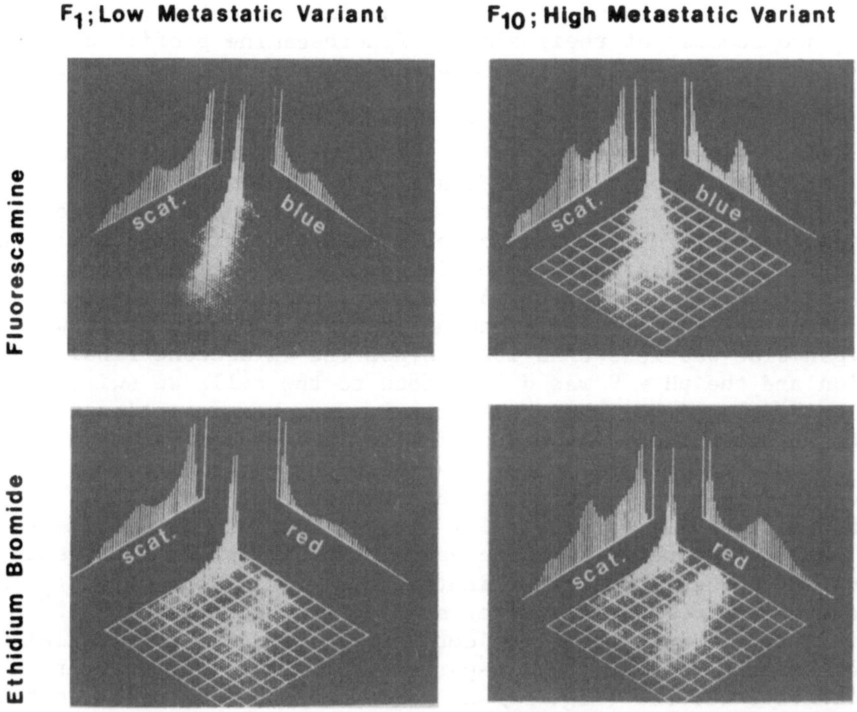

Fig. 10. Two dimensional histograms of double stained (fluorescamine: 30 mg/0.2 ml acetone/5 ml borate buffer; EB: 10^{-4}M, R = 10) B16-F1 and B16-F10. The B-D cell sorter was routinely calibrated using fluorescent microspheres (10μ) to yield a 4% coefficient of variation.

cellular debris which lie very close to the origin.

Cells with medium (II) and high (III) scatter (6) have discrete values of blue fluorescamine but exhibit a heterogenous distribution of EB uptake (red fluorescamine). Cells with medium scatter display a distinct level of very low EB-uptake (Q) followed by a typical log phase distribution (G_1, S, G_2) while the blue fluorescence for the same medium scatter cells in both F1 and F10 is distributed around a mean value, as a gaussian. Cells with very low dye uptake (Q) being present in both medium and high scatter are more dominant in B16-F1 than B16-F10 cells (Figs. 8 and 10) and because of their similar fluorescamine profile do therefore represent cells with membrane organization similar to the cells with G_1 dye uptake.

The increase in fluorescamine uptake of the mid-scatter cells of B16-F10 with respect to B16-F1 at 30 mg fluorescamine per 5 x 10^6 cells was exploited to see if fluorescamine was indeed a membrane marker for metastatic potential by sorting and replating the B16-F10 subpopulation and testing to see if they produced more pulmonary metastases in vivo (6). The viability of the B16-F10 cells after saturation fluorescamine staining (30 mg), as shown by trypan blue was less than 15%. Since the 4% acetone final solution and the pH = 9 was deleterious to the cells we switched to 4% DMSO (6) and for the same fluorescamine concentration similar results were obtained. By decreasing the fluorescamine concentration a dramatic increase in the viability was noted (i.e., 15% viable at 30 mg fluorescamine vs. 81% viable at 0.35 mg fluorescamine) (6). Paradoxically with an increase in viability there has to be concommitant decrease in the amount of fluorescamine uptake per cell resulting in the reduction of the differential uptake by the B16-F10 mid-scatter at 3 mg fluorescamine/5 x 10^6 cells. Further experimentation determined that the differential fluorescamine uptake in the two-paramater histogram of scatter vs. blue fluorescence (Fig. 10) was more pronounced in cells double stained with EB and fluorescamine, partly due to the presence of a limited resonant energy transfer between the two probes (22) since they both bind to cellular moieties that are within the acceptable transfer distance (200A^o), i.e., fluorescamine to inner and outer membrane surface primary amines and H1 and H2 nuclear histones while EB intercalates and slips into the interior of the double helix. Thus, the differential uptake we saw with saturation fluorescamine staining was due to both a complex interaction between EB and fluorescamine and to the saturation of primary amine sites which in B16-F10 cells occurred at high concentrations. Although lower concentrations of fluorescamine without EB produced a higher viability necessary for sorting and replating, it did not provide the distinct differential fluorescence distribution necessary for optimal identification. Furthermore, the addition of EB decreases cell viability even at low fluorescamine concentrations. To overcome the lack of non-differential emission at viable staining concentrations, a confluent monolayer of B16-F10 cells was stained with 3 mg fluorescamine/0.2 ml DMSO/5 ml buffer, pH 8 and utilizing the information gained from saturation double staining, cells with mid-scatter and mid-high blue fluorescence were sorted into sterilized wells (in reference 7) and designated F10M-b. To prove that the cells which represented 30% of the total population were in fact unique, another confluent monolayer of B16-F10 grown in parallel and stained, was sorted viably by triggering on the mid-scatter window but independent of fluorescence, i.e., using only cell size as criteria (6). This population was designated F10M-a. Because of the small amount of cells and volume of media, plating efficiencies were not measured, but by daily monitoring it was observed that growth resulted from a small percentage of the cells plated. Both F10M-a and F10M-b were passed once in vitro before being injected into the tail vein of C57BL/6 mice 14 days after sorting.

As assayed in our laboratory (6) the difference in median metastatic potential between B16-F1 and B16-F10 proved to be significant but not impressive in terms of absolute values: 1 vs. 5 respectively (6). The subpopulation F10M-a (unstained, mid-high scatter), assayed against the corresponding entire B16-F10 population produced a similar median number of metastases: 3 vs. 4 respectively (see Table VI). However, when we assayed the subpopulation F10M-b (stained mid-scatter and high fluorescence) against the corresponding entire B16-F10 population, a striking difference was seen as the respective median number, were 4 vs. 13 (6). In light of the results of F10M-a, F10M-b's ability to produce more pulmonary metastases than its parent B16-F10 is indicative of fluorescamine's capacity to be used as a membrane probe to identify highly metastatic subpopulations or populations exhibiting specific tissue, under proper dual staining conditions.

Densitometric and geometric image analysis of the Feulgen-stained sorted subpopulations (Fig. 11) shows that F10M-b constitutes a definite fraction of F10M-a cells, with a unique nuclear morphometry, characterized by a 2C-3C DNA content and a disperse chromatin. Indeed for an identical integrated optical density (2C DNA content), F10M-a cells are quite heterogenous in their nuclear area distribution while F10M-b appears relatively homogeneous around a large nuclear area value (disperse chromatin).

From Fig. 10 the highly metastatic cell variants appear to be confined in the mid-scatter population: to actually verify this conclusion, we sort on a latter passage in culture (35th) of the same F10-B16 tumor using scatter as the only criteria. At semi-confluency, the "sorted" mid-scatter cells (F10-M) were then assayed for metastatic potential in mice, against the remaining

Table VI. Metastasis Assay

A. B16-F10 vs. F10M-a and F10M-b

Cell Line	# of viable cells injected/0.2 ml	Pulmonary Metastases Median (Range)*
B16-F10	50,000	4 (0,0,1,1,1,2,2,2,3,3,4,6, 7,7,7,7,11,11,16,20,202)
F10M-a	50,000	3 (0,0,1,2,3,3,4,9,12,17)
F10M-b	50,000	13 (9,10,12,15,37,93)

1. 7-10 wk. old C57BL/6 male mice.
2. Cells were harvested from semi-confluent cultures with 0.2% trypsin - 0.02% EDTA soln.
3. Pulmonary metastases counted with the aid of a dissecting microscope 21 days after i.v. injections.

B. B16-F1 vs. B16-F10

Cell Line	# of viable cells injected/0.2 ml	Pulmonary Metastases Median (Range)*
B16-F1	25,000	1 (0,0,0,1,1,1,3)
B16-F10	25,000	5 (3,3,5,10,14)

1. 14 wk. old C57BL/6 male mice.
2. & 3. same as above.

C. F10 (mid scat) vs. F10-MW (Tot. pop. - mid scat) vs. B16-F10/35 (Tot. pop.)

Cell Line	# of viable cells injected/0.2 ml	Pulmonary Metastases Median (Range)
F10M	50,000	2 (1,1,1,2,2,3,4,11)
F10-MW	50,000	0 (0,0,0,0,1)
B16-F10/35	50,000	(1,5)

1. 7-9 wk. of C67BL/6 male mice.
2. Cells were harvested from semi-confluent cultures with 0.2% trypsin 0.02% EDTA solution.
3. Pulmonary metastases counted with the aid of a dissecting microscope 21 days after i.v. injections.
4. B16-F10/35 are at the 35th passage _in vitro_.

* Statistical analysis, showing significant differences, is now shown.

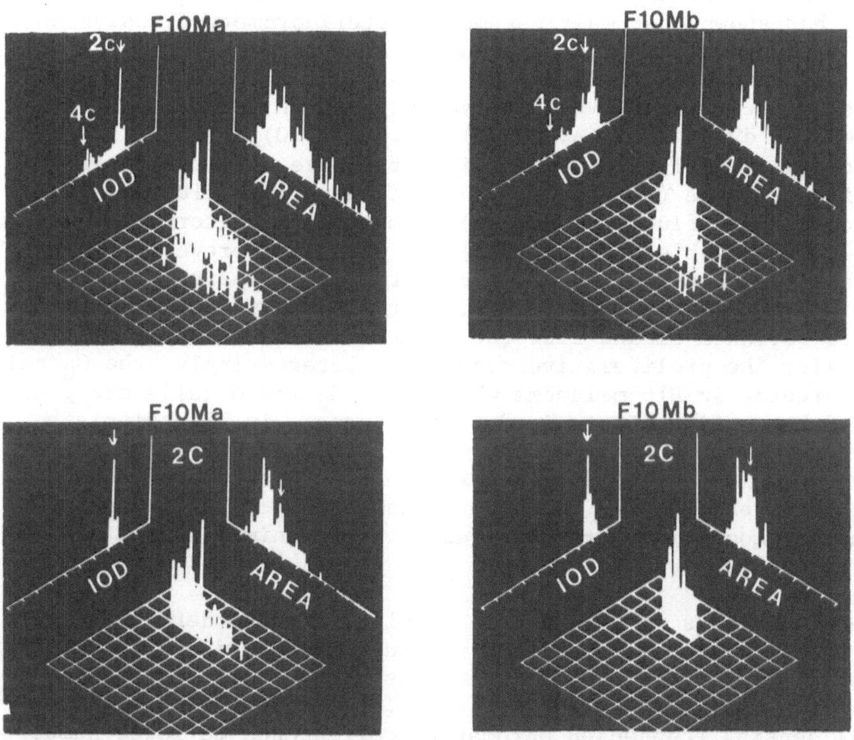

Fig. 11. Image analysis of "sorted" Feulgen-stained F10Ma (mid-high scatter) and F10Mb (mid-scatter and high blue fluorescence) subpopulation (see also chapter by Nicolini, Beltrame and Grattarola).

cells (F10MW, i.e., low and high scatter). The data (Table VI, panel C) shows that indeed the metastatic cells are mostly confined in the mid-scatter region: furthermore, the same data confirms previous observations that at such later passage in vitro (35th) B16-F10 variants decrease their metastatic potential significantly (experiments reported previously, as in panels A - B of Table II, were conducted with the 15-20th passages).

CONCLUSION

The data presented on the various cell systems, in vitro and in vivo has shown that multiparameter cytofluorometric analysis allows the separation of heterogeneous cell types and the characterization of cells with the same DNA content but different proliferative states (cycling G_1 vs. non-cycling G_0-0 in terms of their differential dye uptake (AO or EB) 2,4,6,7,8,20). This methodology of real time acquisition of scatter vs. chromatin-DNA-P fluorescence (AO staining) or cell membrane (fluorescamine) vs. nucleic acid fluorescence (EB) permits the monitoring of both G_0 cells (reversably out of cycle) and Q cells (irreversably or not readily reversible, out of cycle) the latter being present in normal and transformed cell lines. For B16 tumor in vivo and its in vitro variants F1 and F10, the greater the Q cell population the smaller the proliferative capacity. Interestingly, the G_0 cell is not present in B16 melanoma while both G_0 and Q cells are present in WI-38 human fibroblasts in deep confluency, in PHA stimulated lymphocytes at 72 and 144 hours, and in bone marrow. Thus, a more comprehensive objective determination of the growth fraction (number of proliferative cells/total number of cells) in human peripheral blood, in fibroblasts and even in animal normal and tumor tissue is possible.

Image analysis of sorted cell populations have determined that lower average optical density of the nuclear DNA (disperse chromatin) corresponds to higher number of chromatin DNA-P primary binding sites for AO staining in intact cells (G_1), while higher average optical density (condensed chromatin) corresponds to lower number of primary binding sites (G_0) in a linear fashion of 2C DNA content (1,2,4,6). Interestingly, the Q cell exhibits the highest average optical density and the lowest number of primary binding sites, but the ratio of binding sites (mean fluorescence peak) to AVOD for Q and G_1 cells is of a different order of magnitude than for the same binding site/AVOD ratio in G_1 to G_0 cells, suggesting that in Q cells other mechanisms than chromatin structural changes may also be responsible for the very low dye uptake. Yet, as characterized by fluorescamine uptake the Q cell membrane properties are identical to other phases of the cell cycle (being G_0 or G_1).

The dominant concern is the question on the nature of the Q cell, being apparently terminally out of cycle. The correct conclusion from the data is that the Q cell in lymphocytes does not respond to standard PHA stimulation as the FMF distributions of the sorted subpopulations confirm, while protein A (7) yield contrasting results, being occasionally capable to recruit these Q cells (work still in progress).

In lieu of the above, the application of other stimulatory techniques (as protein A) which could trigger recruitment in the Q cell are critical to a better understanding of its role in cellular proliferation and differentiation: indeed non-cycling Q tumor cells have been shown to be recruitable in vivo upon the administration of hydroxyurea (11). The various types of non-cycling cells, as summarized in Fig. 11, have been here shown (4; for a more complete justification see chapter by Nicolini and Belmont in this volume) to represent necessary steps in any mammalian cell cycle progression rather than separate states entered upon proper conditions.

Via the sorting of the F10M-b subpopulation stained with fluorescamine, we have also isolated and characterized through in situ analysis cells with enhanced metastatic potential (2,6) which is indicative of subtle biochemical and biophysical differences which are manifested in the tertiary or quarternary molecular structure of membrane moieties. While biochemical studies of surface differences have not been conclusive in their attempt to correlate membrane composition with metastatic potential, intact cell studies have shown a direct correlation between membrane properties, [i.e., resistance to wheat germ agglutinin (carbohydrate composition (21)], adhesion properties of detachment variants, suspected glycoprotein composition (21), resistance to lysis by syngenic lymphocytes [variation in antigenic determinants (22)], and pulmonary metastatic potential. Fusion studies of membrane vesicles' of B16-F10, which represent intact membrane configurations, have enhanced B16-F1's ability to be arrested in the lung and form pulmonary metastates (23).

Image analysis of F10M-b (Fig. 10) cells reveals that they are a definite fraction of F10M-a cells, with a unique nuclear morphometry, characterized by a 2C-3C DNA content, smaller AVOD, and for the 2C window a relatively dispersed chromatin (6). Such a chromatin organization with a 2C-3C DNA content within a log-phase population of cells, is typical of cells which are in early-middle S phase (5) which agrees with the finding that cells at the G_1-S boundary and middle S phase (2C-3C DNA) in murine fibrosarcoma Fsh1233 produced more pulmonary metastasis than those in G_1 and G_2 (24).

Due to the subtle nature of the variations within the intact membrane structure, metastatic potential appears to be a gradual phenomenon (6) in which different membrane configurations (even within a population of highly metastatic cells) may represent different capacities for metastasis. When correlated to chromatin organization (1; and chapter by Nicolini, et al., in this volume) by double staining (fluorescamine and EB), enhanced metastatic potential is related to active proliferation or smaller Q phase. Therefore, membrane organization appears to determine the arrest of metastatic cells and chromatin configuration appears to determine the rate of metastatic growth (as shown more clearly in refs. 1 and 6). In lieu of the above, the metastatic potential should be considered as a function of the interactions between these cellular structures, the membrane and the chromatin (6).

In summary, multiparameter laser flow microfluorometry and viable cell sorting analysis coupled with a knowledgeable understanding of the physico-chemical properties of higher order chromatin and membrane structure, represent a new powerful scientific methodology for the viable isolation and characterization of heterogeneous mammalian cell systems, in terms of cell proliferation, differentiation and expression of metastatic potential.

REFERENCES

1. C. Nicolini, J. Submicroscopy Cytology, 12:3 (1980).
2. C. Nicolini, in: "Advances in Neuroblastoma Research, Edited by A. Evans, 271-286 (1980).
3. "Chromatin Structure and Function", Edited by C. Nicolini, Plenum Publishing Co., NATO-ASI Series, Vol. 21a and 21b, pp. 1-917 (1979).
4. C. Nicolini, Cell Biophysics, 2:4 (1980).
5. C. Nicolini, W. Giaretti, and F. Kendall, Biophys. J., 19:163-176 (1977).
6. S. Lessin, S. Abraham, and C. Nicolini, Nature (1980).
7. S. Abraham, S. Zietz, E. Vonderheid and C. Nicolini, Cell Biophysics, Vol. 2, No. 4 (1980).
8. E.C. Vonderheid, S.M. Fang, S.R. Abraham, and C. Nicolini, J. Investigative Dermatology (In Press)
9. W.A. Linden, C. Nicolini, et al. J. Histochem. & Cytochem. 27:529-535 (1979).
10. M. Grattarola, C. DeSaive, S. Lessin, S. Zietz and C. Nicolini Cancer Biophysics & Biochemistry (1979).
11. S. Zietz, M. Grattarola, C. Desaive, and C. Nicolini Cell & Tissue Kinetics (In press)
12. C. Nicolini, S. Parodi, A. Belmont, S. Abraham and S. Lessin, J. Histochem. Cytochem. 27:102-113 (1979).
13. C. Nicolini, W.A. Linden, S. Zietz and C.T. Wu, Nature, 270:163-176 (1977).
14. S. Zietz and C. Nicolini, in: "Biomathematics and Cell Kinetics", Edited by A.J. Valleron, pp. 357-395, North Holland Publishers (1978).

15. R. Baserga, "Multiplication and Division in Mammalian Cells" Marcel Dekker, Inc., New York, N.Y., pp. 192-194 (1976)
16. C. Nicolini, S. Parodi, S. Lessin, A. Belmont, S. Abraham, S. Zietz and M. Grattarola, in: "Chromatin Structure and Function", Edited by C. Nicolini, pp. 293-323, Plenum Press, New York (1979).
17. S.P. Hawkes, J.C. Bartholomew, Proc. Natl. Acad. Sci., 74:1626-1630 (1977).
18. D.L. Poccia, B.A. Palevitz, et al., "Fluorescence Staining of Living Cells with Fluorescamine" Protoplasma, 98: 91-113 (1979).
19. I.J. Fidler, Nature New Biology, 242:248-250 (1973).
20. C. Nicolini, F.M. Kendall, C. Desaive, B. Clarkson and J. Fried, Exp. Cell Res., 106:111-117 (1977).
21. E.B. Briles and S. Kornfeld, J. Natl. Cancer Institutes, 60:1217-1221 (1977).
22. I.J. Fidler and C. Bucana, Cancer Res., 37:3945-3956 (1977).
23. G. Poste and G.C. Nicolson, PNAS-USA, 77:399-403 (1980).
24. T.W. Tao and M.M. Burger, Nature, 270:437-438 (1977).
25. C. Nicolini, S. Ng, R. Baserga, PNAS-USA (1975).
26. C. Nicolini, C. Desaive, F.M. Kendall and W. Giaretti, Exp. Cell Res., 106:118-127 (1977).
27. R. Brooks, D. Bennet, and A. Smith Cell, 19:493 (1980).
28. P. Rao and S. Hanks Cell Biophysics, Vol. 2, No. 4 (1980).
29. P. Miller, W. Linden, and C. Nicolini, Zietschirft fur Naturforschung, 390:442 (1979).
30. J. Bohmer, Exp. Cell Res. (1979).
31. C. Nicolini, F.M. Kendall, et al., Cancer Treatment Reports 60:1819-1827 (1976).

ACKNOWLEDGEMENT

This work was supported by NIH Grant CA 20034 (USA), Temple University Research and Study leave award to C.N. and by the National Research Council, Finalized Project on "Control of Neoplastic Growth", Italy.

CONDENSED CHROMATIN: SPECIES-SPECIFICITY, TISSUE-SPECIFICITY, AND CELL CYCLE-SPECIFICITY, AS MONITORED BY SCANNING CYTOMETRY

W. Nagl

Department of Biology
University of Kaiserslautern
6750 Kaiserslautern
Federal Republic of Germany

INTRODUCTION

The main point of this essay will be the question, how condensed chromatin can be interpreted in terms of its molecular determinants and its functional significance, in different taxa. I shall not discuss the lower levels of DNA packaging up to the quaternary structure (or solenoid) in details, as this was recently done by several authors (1-5). It will be shown that condensed chromatin at the quinternary level is species-specific in plants and probably insects, while it is tissue-specific in vertebrates. In all cases, the degree of condensation considerably varies during the cell cycle. The evolution of condensed chromatin including heterochromatin will be discussed. Its possible role in the control of gene activity, determination, differentiation, and morphogenesis is suggested and illustrated by hiterto unpublished quantitative data.

WHAT IS CONDENSED CHROMATIN?

In electron micrographs of interphase and working nuclei of both plants and animals, condensed chromatin can be seen in variable amounts. Although this portion of chromatin exhibits nearly the same electron density in all micrographs, it covers at least four different classes of chromatin (6):

(i) karyotypically condensed, constitutive heterochromatin (which can be visualized at the light microscope level by differential staining, and which often is enriched in highly repetitive DNA sequences, etc.);

(ii) sex-specifically condensed, facultative heterochromatin (female sex chromatin in mammals);

(iii) tissue-specifically condensed, inactivated euchromatin (in vertebrates), and species-specifically condensed euchromatin (in plants and perhaps insects and some other taxa).

In this terminology, heterochromatin is used in a structural sense, i.e., it is defined through its DNA sequences. The widely found practice to use the term heterochromatin in a functional sense, i.e., for any inactive, and hence condensed chromatin, may obscure fundamental differences between several portions of chromatin with respect to the mechanisms underlying condensation and its biological significance (for details see paragraph on heterochromatin). It seems, on the one hand, that species-specifically condensed euchromatin (i.e. chromomeres and chromonemata is plants, bands of polytene chromosomes in dipters) is just quantitatively different from constitutive heterochromatin. Both are karyotypical, evidently caused by the presence of binding sequences for a "packaging" protein (e.g. a modified H1 histone), and both are the result of the evolutionary history of the species. On the other hand, facultative heterochromatin and developmental- and tissue-specifically condensed chromatin are induced by proteins irrespectively of the DNA sequences present in this portion of the genome. These somatic condensation events are to be understood as part of ontogeny, although they depend on the phylogenetic state of the genome of a given species (7).

Table 1 gives an overview over the different levels of DNA packaging in chromatin, the terminology used by different authors, and the level of what is called "condensed euchromatin" in this article. Summing up, condensed chromatin in plants can be seen as the result of failure of decondensation after mitosis, while condensed chromatin in mammals is the result of active condensation after cessation of mitotic activity (in most cases the nuclei of proliferating cells exhibit a very diffuse structure).

DIGITALIZATION AND QUANTIFICATION OF CONDENSED CHROMATIN

Sophisticated automated microscope systems have been recently developed for texture or image analysis of chromatin organization in interphase nuclei (14-17). Moreover, biophysical methods allow to compare the in situ data with corresponding changes observable in isolated chromatin at lower levels of DNA packaging (4). Besides theoretical values of such studies with respect to the understanding of chromatin organization and its changes during the cell cycle, during replication and transcription, one should emphasize their enormous practical value for cytodiagnosis of tumorigenesis (18-21). In our lab we use a less sophisticated version, a scanning

Table I. Terminology, Factors, and Levels of Chromatin Organization

Structure level related to DNA	Over-all packaging ratio	Diameter of fiber (approx.)	Terminology according to:			Determined by	Comment
			Nagl (8)	Sedat and Manuelidis (9)	Nicolini (4)		
1	0	1 nm	p o l y n u c l e o t i d e c h a i n			-	-
2	1	3 nm	D N A d o u b l e h e x l i			-	-
3	1:7	10 nm	tertiary structure, super-helix	nucleosome fiber	tertiary helix, nucleosome fiber	histones	
4	1:40	30 nm	quaternary structure, super-super-helix	solenoid fiber	quaternary structure, superhelix	modified H_1, non-histone pr.	-
5	1:1000	200 nm	euchromatic elements	interphase tube	-	ions	condensed euchromatin in plants
6a	1:10,000	600 nm	chromatid, heterochro-matin	chromatid (except centromeres)	-	nonhistone proteins, modified histones	metaphase, constitutive heterochrom.
6b	?	variable	cross-linking of fibers	-	quinternary, drapery-like organization	ions	condensed euchromatin in animals

cytophotometer (Leitz MPV-2) interfaced to a pdp-8 computer (Digital Equipment).

Hydrolysis should be done with 5 N HCl at room temperature for 40 min., or with 0.1 N HCl at 37°C for 6-51 h, because "hot hydrolysis" at 60°C was found to yield less reproducible results, as in this case the liberation of aldehyde groups in the DNA follows different kinetics in large and small nuclei, and in condensed and diffuse chromatin (22, 23). The stained nuclei are then scanned, and the extinction of each measurement point is stored in the computer. The data can now be integrated to total extinction values of the measured nuclei, and statistically treated for estimation of the DNA content (using a standard of known DNA content). Also histograms may be printed to see the distribution of polyploidy in a tissue.

Certain programs allow to print the extinction values of each measurement point so that a digitalized picture of the nucleus arises. The input of thresholds and thus extinction ranges allows to separate euchromatin and heterochromatin, for instance (Fig. 1).

As structural components of chromatin may be beyond the level of resolution of the light microscope, and as electron micrographs may contain more information than is immediately apparent to the eye, electron microscopic morphometry of ultrathin sections of nuclei was established in our laboratory. Quantitative description of nuclear morphology is possible on electron micrographs by the use of an automatic or semi-automatic planimeter. However, morphometric electron microscopic studies have been yet seen only limited application in the establishment of quantitative parameters relating to chromatin structure and nuclear organization, particularly in plants.

In our laboratory we employ the semi-automatic equipment MOP/Digiplan (Kontron, Munich). First the circumference of the total nucleus is driven round with the connected stylo, then the nucleolus (whose area is eliminated), and finally the patches of electron-dense chromatin. The summarized areas of the latter is expressed as percentage of total area. As random thin sections are used, the method can only be used statistically.

The reliability of this quantitative electron microscopic morphometric method has been tested in a number of comparative light and electron microscopic studies (24-26, and unpublished results). Non-cycling nuclei of angiosperms show only minimal structural variation, except in the gametophyte and certain cells of theovule. The probability of evaluating an atypical nuclear section is reduced (a) by elimination of nuclei of very small area ("cup-sections"), and (b) by eliminating the nucleolar area (if present), which principally cannot be occupied by condensed chromatin; this leads

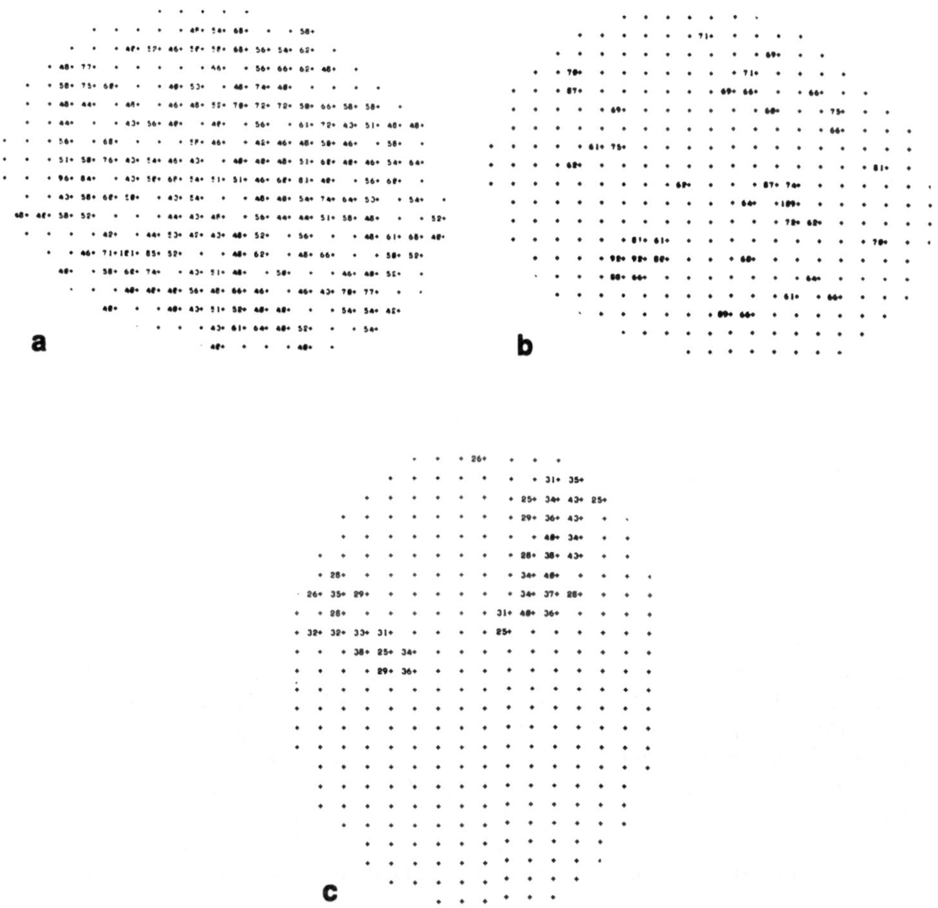

Fig. 1. Computer prints of Feulgen extinction values in individual nuclei at different thresholds. (a) and (b) Nucleus of Allium carinatum, shown is respectively total chromatin distribution and heterochromatin distribution. (c) Nucleus of Rhinanthis minor, shown is heterochromatin (euchromatin can not be visualized by the Feulgen technique).

to similar values for condensed chromatin in nuclei sectioned in the region of the nucleolus and nuclei sectioned in a different region.

In general, the probability of obtaining a real figure of the proportion of condensed chromatin is the higher, the more evenly the chromatin is distributed in the nuclear cavity. Therefore, the number of measurements has been the higher, the less condensed chromatin is present, and the more it is scattered over the nucleus in a given species. As a good criterion for the necessary number of measurements, a standard deviation of about 10% can be seen, which is similar to that obtained in scanning cytophotometric DNA measurements.

SPECIES-SPECIFICITY

Nuclear Types in Plants

At the light microscope level, species-, genus-, and family-specific nuclear structures are well known for a long time (27-30). At the ultrastructural level, 3 types of chromatin organization have been found: these are, arranged according to structural complexity (24, 27), the diffuse one, the chromomeric one, and the chromonematic (or reticulate) one (24; 31; 32; Figs. 2-4).

Nuclei with diffuse euchromatin always possesses heterochromatin in the form of chromocenters, so that this type is also known as "chromocenter type". Nuclei with chromomeric and chromonematic euchromatin organization may also possess chromocenters in addition to the patches or strands of condensed euchromatin. In general, the structural complexity is positively correlated with increasing 2C DNA content of the species (32). Moreover, the proportion of euchromatin which is species-specifically condensed (in the form of chromomeres or chromonemata) is postively correlated with the 2C DNA value, indicating that any excess DNA which may originate during evolution becomes condensed and thus "silenced" (24; Fig. 5). This could recently be proven by ^{3}H-uridine autoradiography (see below).

The high species-specificity of chromatin organization in plants allows, if the proportion of condensed chromatin is estimated quantitatively, to identify a species by this parameter, Nagl (33) used electron microscopic morphometry of thin sections of non-cycling nuclei and statistics for the diagnosis of species and hybrids.

Nuclear Types in Insects

Many insects such as Orthoptera, Phasmida, Hederopters, etc., and also some other avertebrates (among Crustacea, Gastropoda and

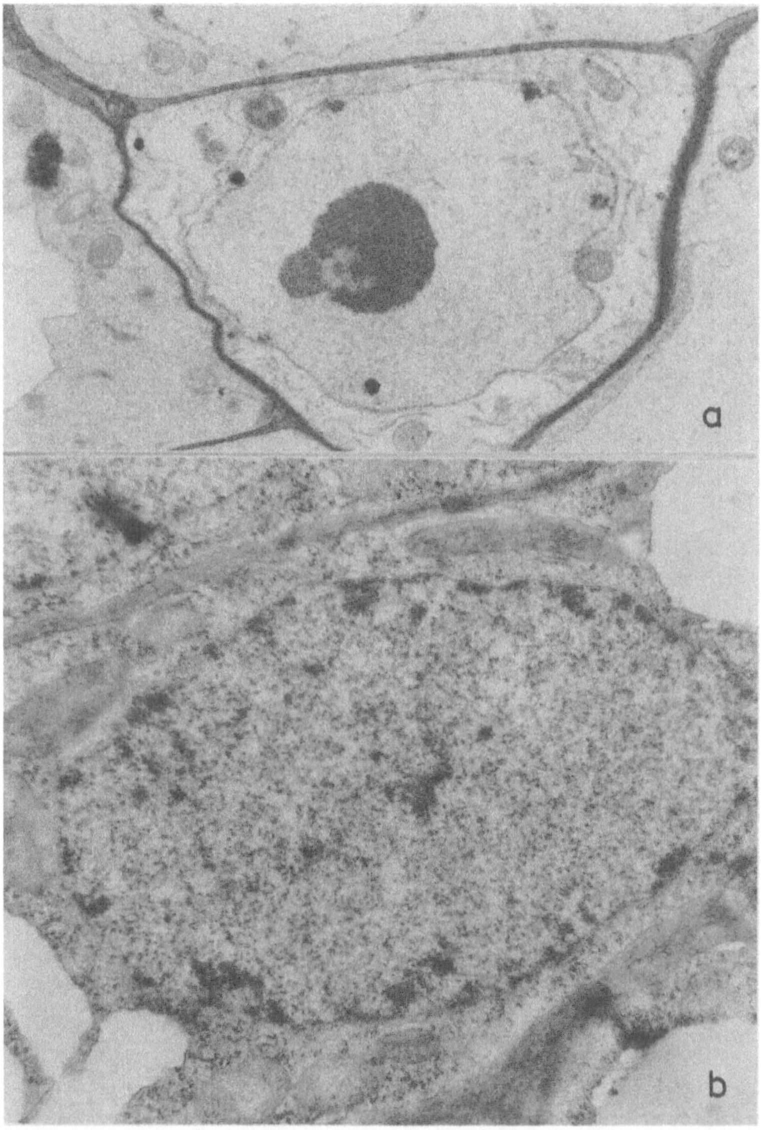

Fig. 2. Electron micrographs of plant nuclei of the "diffuse" or "chromocenter" texture type. (a) Arabidopsis thaliana (x 8,500), (b) Ruta graveolens (x 20,000). The little amount of condensed chromatin present represents constitutive heterochromatin (chromocenters), the euchromatin is completely diffuse.

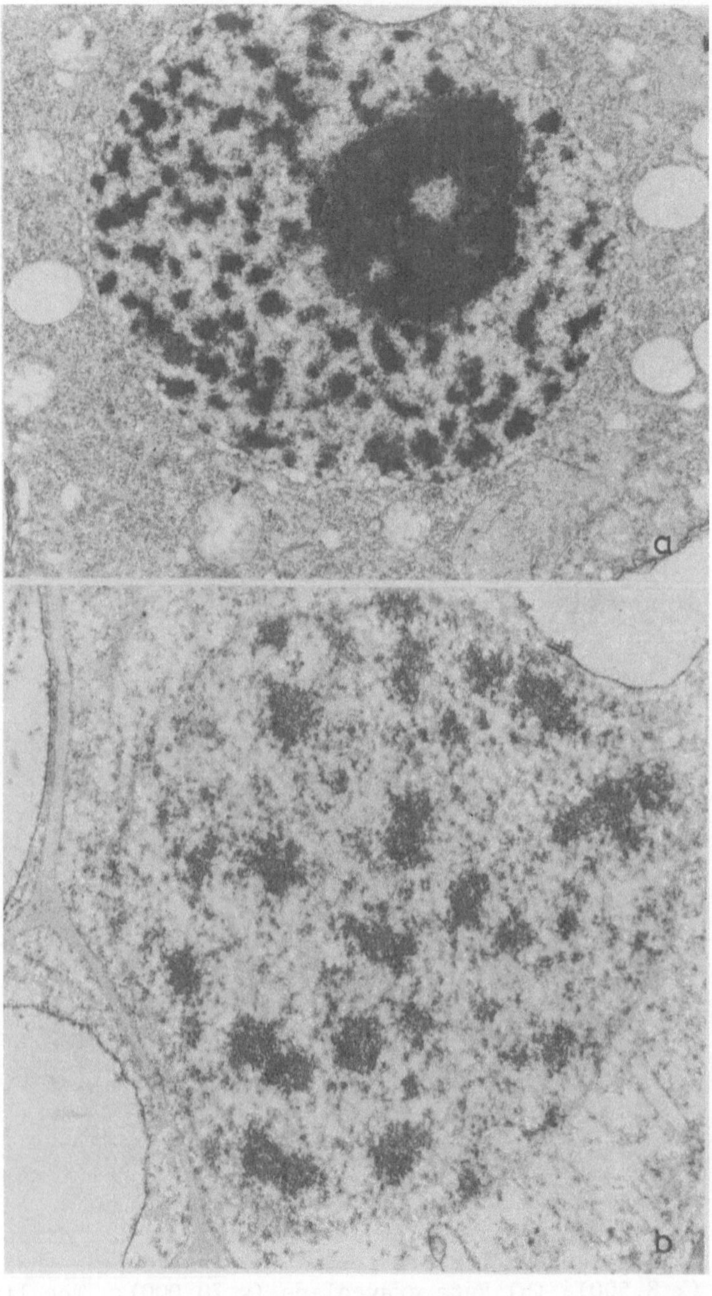

Fig. 3. Electron micrographs of plant nuclei of the "chromeric" structural type. (a) Lathyrus odoratus, (b) Oxalis acetosella (x 20,500). The spots of condensed chromatin represent euchromatin at the light microscopic level (chromomeres).

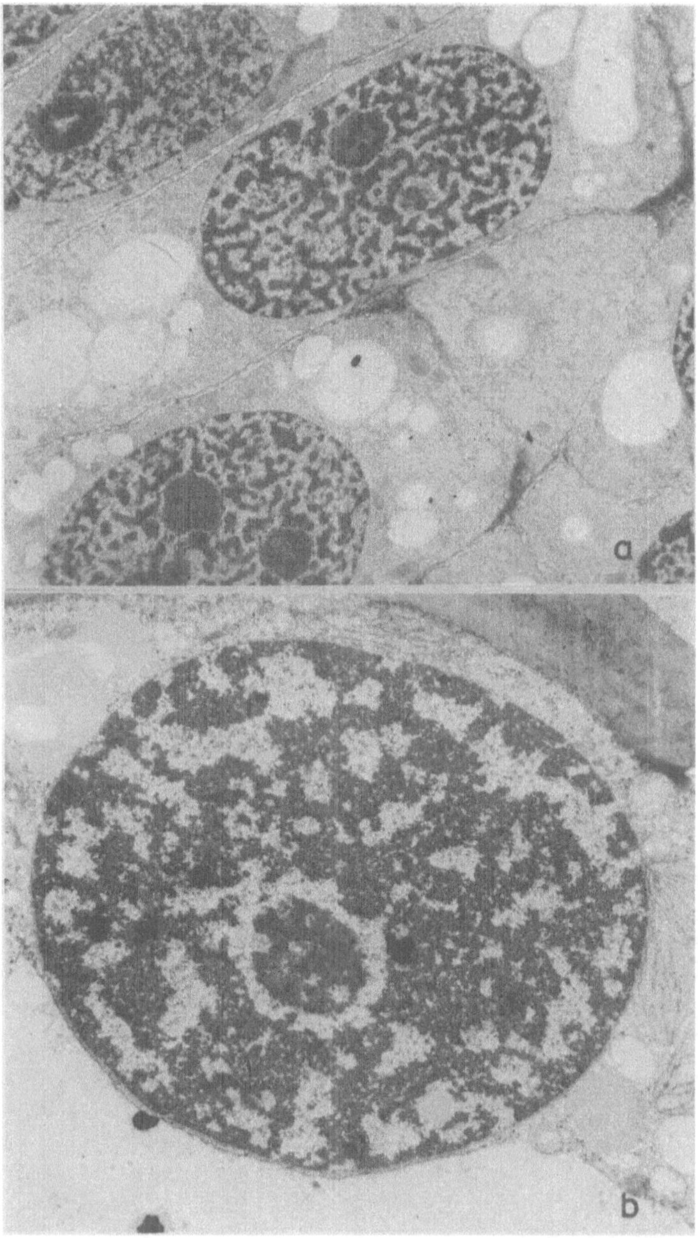

Fig. 4. Electron micrographs of plant nuclei of the "chromonematic" or "reticulate" type. (a) Allium cepa, (b) Viscum album (x 5,000). The condensed chromatin mainly represents euchromatin (chromonemata), heterochromatin can not be distinguished at the electron microscope level from euchromatin.

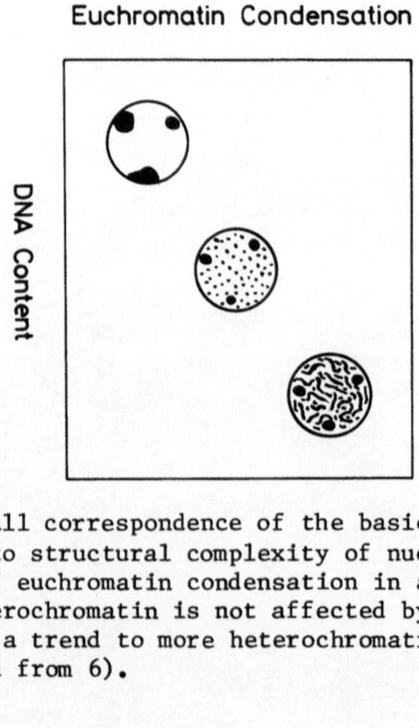

Fig. 5. The overall correspondence of the basic 2C nuclear DNA content to structural complexity of nuclei, i.e., the extent of euchromatin condensation in angiospermal plants (the heterochromatin is not affected by this rule, but there is a trend to more heterochromatin at low 2C values (modified from 6).

Collembola) are known to possess "chromosome nuclei", i.e., nuclei in which each chromosome can be recognized as an individual chromatin patch throughout interphase (28). Although heterochromatin can be discerned from the euchromatic rest of the chromosome at the light microscope level (Fig. 6), decondensation does not reach such a level that the nuclear cavity becomes homogeneously filled with chromatin. At the electron microscope level, the individual chromosomes can be clearly seen, but heterochromatin could not be detected (see chapter on heterochromatin at the end of this essay). This is probably due to the fact that the third dimension is lost in ultrathin sections. Fig. 7 shows electron micrographs of chromosome nuclei of the heteropteran, Gerris najas (pond skater), and of the phasmid, Extatosoma tiaratum (leaf insect). In Drosophila, the euchromatin is diffuse in diploid nuclei, as can be predicted on grounds of the low DNA content (2C) value of this species. The heterochromatin forms upon fusion one large collective chromocenter.

Variations of Species-Specific Structures

In spite of species-specificity, the nuclear structure and particularly the nuclear ultrastructure show some variation in an individual plant. Most of them are associated with the cell cycle (see the corresponding chapter), but some structural changes seem to be caused by extreme dehydration of nuclei (e.g. in resting seed, and dormant buds and roots (5, 33a,b) or by hydration and accumulation of acidic proteins in very active nuclei of the gametophyte and embryo. However, not one of these studies was quantitative, and inspection of the figures published is not convincing of the interpretation given by the authors. Moreover, cell cycle changes are apparently not remembered. The developmental changes, if there are any, may not represent real decondensation/condensation processes of the chromatin, but rather be the consequence of changes in the nuclear volume and hence alterations in the proportion of the volume occupied by chromatin. The only sitaution comparable to chromatin condensation in mammals apparently occurs in lower plants, where mobile sperm cells can be found (Fig. 8).

In insects, tissue-specific variation of the species-specific chromatin organization is known from ganglia, whose nuclei are more or less decondensed.

Gross changes of nuclear structure often take place during endopolyploidization (7), the most obvious being polyteny. Morphometric analysis of the polytene nuclei in the Phaseolus suspensor revealed that the pattern of chromatin distribution is drastically modified, but the species-specific fine texture and proportion of chromatin is evidently not altered (24), except perhaps in puff regions (which were, however, not yet studied). Also in

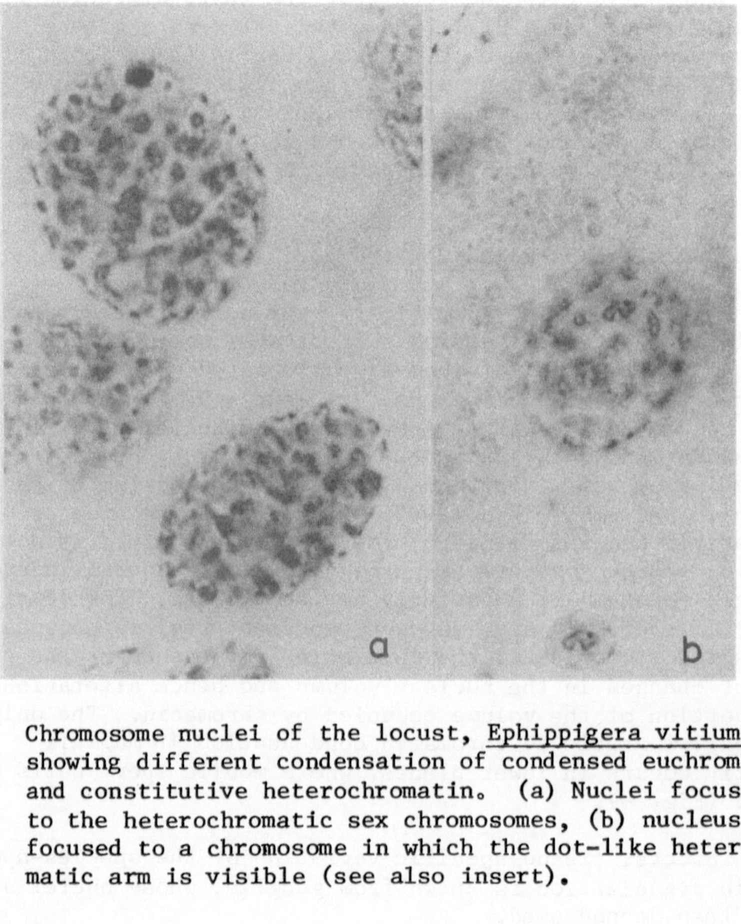

Fig. 6. Chromosome nuclei of the locust, Ephippigera vitium, showing different condensation of condensed euchromatin and constitutive heterochromatin. (a) Nuclei focused to the heterochromatic sex chromosomes, (b) nucleus focused to a chromosome in which the dot-like heterochromatic arm is visible (see also insert).

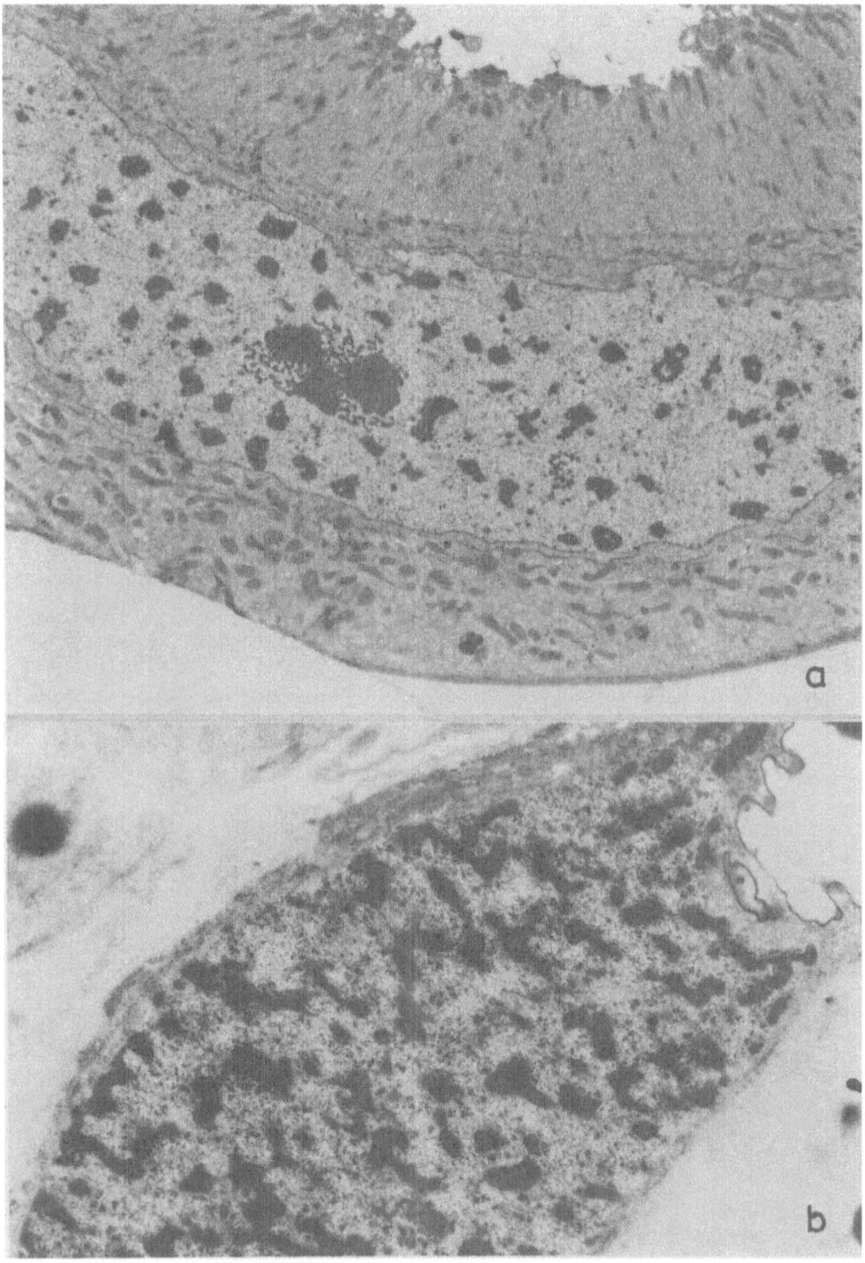

Fig. 7. Chromosome nuclei at the electron microscope level. (a) Gerris najas, (x 5,000), (b) Extatosoma tiaratum (x 8,000). The electron-dense patches represent the condensed parts of the chromosomes (each chromosome forms one patch, which is mainly composed of euchromatin).

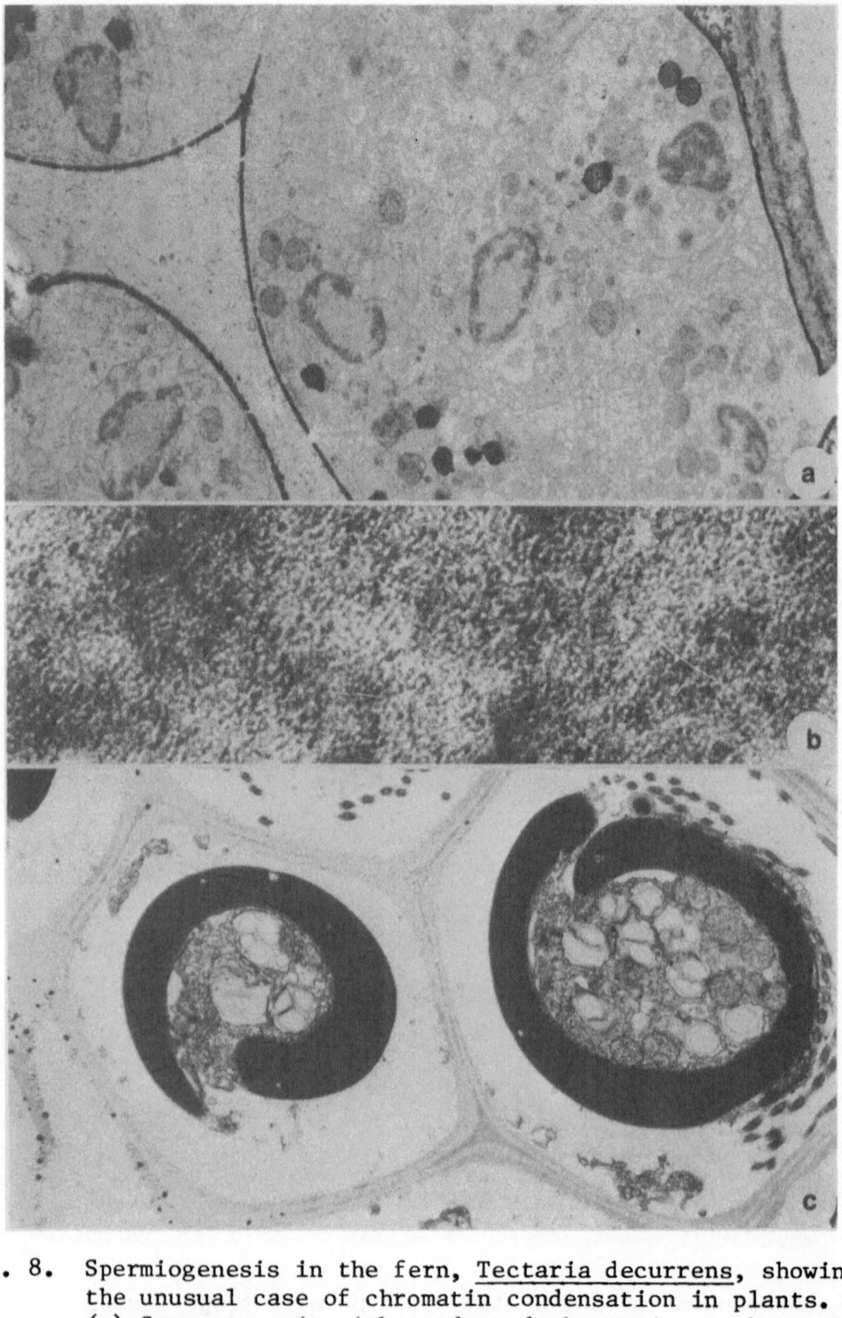

Fig. 8. Spermiogenesis in the fern, Tectaria decurrens, showing the unusual case of chromatin condensation in plants. (a) Spermatogonia with condensed chromatin at the nuclear envelope (x 12,000), (b) condensing chromatin exhibiting thick fibers (x 63,000), (c) maturing sperms with completely condensed chromatin (x 6,600).

polyploid nuclei of the rat liver, the proportion of condensed chromatin is stable (34).

Plant species with chromonematic nuclei, i.e., with much condensed chromatin, always exhibit higher 2C DNA values than mammals. Such high DNA contents may be necessary for the realization of the developmental program of certain plants as they affect cell cycle duration, cell size, minimum generation time, longevity, etc. (known as nucleotypic effects ofthe DNA, i.e. effects exerted by the DNA mass irrespectively of the information content) (35). The idea that the huge amount of DNA found in some species serves as resources in order to enable perhaps only one cell type, the egg cell, to enlarge upon chromatin decondensation (36), cannot be accepted for plants, as it is unlikely, and actually not evident, that the onion needs an egg cell nearly 30 times larger than, for instance, flax.

Although the variation in 2C DNA content among species actually seems to be in accordance with the concept of nucleotypic effects and may be understood as an evolutionary strategy (37-39), the amount of condensed chromatin in plant species is evidently nothing else than the consequence of the amount of DNA which is present, and which must be packed into a nucleus.

Generative Polyploidy, Allopolyploidy

DNA increase due to generative polyploidy does not exert any effect on the proportion of the condensed chromatin (25), as it does not show nucleotypic effects on the cell cycle duration (40). As the high DNA content due to polyploidy does not result in increased chromatin condensation (proportionally to the DNA content as in diploids), these findings support the hypothesis that polyploidy in general may have some advantage with respect to gene dosage effects, relation of nuclear surface to cell volume, and other functional parameters (7).

Hybrids (allopolyploids) do not exhibit an amount of condensed chromatin which is the sum of the parental values (as really is the DNA content), but the mean between the parental percentages (25, 33; Fig. 9). This behavior could also be proven in the intergeneric somatic hybrid Arabidobrassica (*Arabidopsis thaliana* x *Brassica campestris*) (26). These findings strongly suggest that not the DNA content *per se* determines chromatin condensation, but rather the quality of the DNA (i.e. the sequence organization), which, however, must be in some relation to the 2C value (see next chapter).

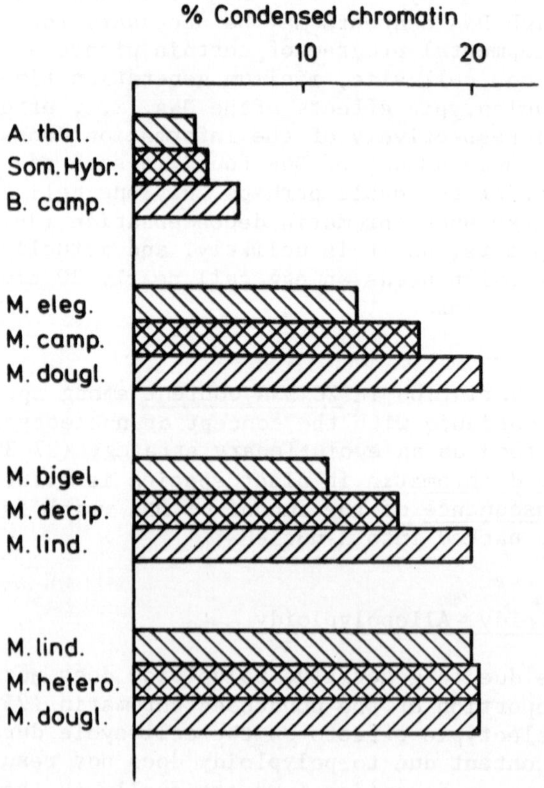

Fig. 9. Proportion of condensed chromatin in allopolyploid hybrids as estimated by electron microscopic morphometry of chromatin. Note that the allopolyploids show an intermediate percentage of condensed chromatin when compared to their parental species (A. thal. = Arabidopsis thaliana, Som. Hybr. = somatic hybrid, B. camp. = Brassica campestris, M. = Microseris, eleg. = elegans, camp. = campestris, dougl. = douglasii, bigel. = bigelovii, decip. = decipiens, lind. = lindleyi, hetero. = heterocarpa). Modified from (33).

MOLECULAR DETERMINATION AND SIGNIFICANCE OF CONDENSED CHROMATIN IN PLANTS

Determination

The correlation between 2C nuclear DNA content and the proportion of euchromatin in the condensed state suggests that the gross organization of chromatin in plant nuclei is directly controlled by the DNA (8, 24, 30, 33). The molecular determination could occur, for instance, via a nucleotypic effect. In this case the DNA amount itself, independent of the nucleotide sequence, would exert a condensation effect. This assumption is consistent with findings in dipters, where higher DNA amounts (2C) lead to higher band numbers as visible in polytene chromosomes (8) (the bands are a good model for condensed euchromatin in plants, as they correspond to interphase chromomeres). Moreover, in situ hybridization studies in DNA-rich anures revealed that some of the loops, which are developed during the lampbrush stage in oocytes, are entirely composed of repetitive DNA sequences (41). On the other hand, it is more likely that a specific repetitive DNA sequence, which is regularly interspersed with other coding and non-coding sequences, and which binds modified histone H1 or another "condensing protein", is involved in determining the degree of interphase chromatin condensation. Evidence for the latter suggestion comes, for instance, from the intermediate percentage of condensed chromatin in allopolyploid hybrids (25, 26), and from the preferential binding of a modified H1 to repetitive DNA sequences in *Drosophila* heterochromatin (42). If such binding sequences are regularly distributed over the genome, it should be expected that they also happen to fall into coding sequences and to interrupt genes. Actually, ideas have been proposed that *introns* are a consequence of the distribution of sequences solely involved in control of structural organization of chromatin (4, 43).

Figure 5 shows the probability with which increasing 2C nuclear DNA content leads to progressive euchromatin condensation. Although positive correlation between both quantitative parameters DNA, content and percentage of condensed chromatin, could be established if all species were plotted, which have been studied so far (33), Fig. 10 suggests that DNA-rich species and genera exhibit a regression line which differs from DNA-poor species: DNA-rich families and genera apparently show a flatter slope, i.e., less chromatin condensation per pg DNA increase. This is consistent on earlier findings by Barlow (30), who found that DNA-rich species display a flatter slope of the regression line between DNA content and chromosome and nuclear volumes.

Table II gives the statistical data for the correlation of 2C DNA content and proportion of condensed chromatin in certain angiosperm families and genera. It can be seen that the slope of the

Table II. Correlation Between 2C Nuclear DNA Content and the Percentage of Chromatin in the Electron-Dense State in Some Angiosperm Taxa (Diploids)*

Taxon	No. of Species Studies	Coefficient of Correlation	Slope of Regression Line	Cross With Y Axis	pg DNA/ 100% Condensed**
Microseris	9	0.89	12.66	-4.22	8.2
Anthemideae	5	0.87	3.50	44.72	15.8
Compositae	21	0.84	3.00	14.85	28.4
Cruciferae	6	1.00	4.36	2.61	22.3
Rutaceae	3	1.00	3.89	2.30	25.1
Luzula	4	0.98	9.36	0.69	10.6
Dicots, DNA-poor	15	1.00	4.23	0.40	23.55
Dicots, DNA-rich	6	1.00	0.53	3.67	182.12
Monocots	7	1.00	9.67	7.52	92.46

* 2C values were determined by scanning cytophotometry of Feulgen-stained telophase nuclei (100 - 4000 per species). The percentage of condensed chromatin was estimated by means of morphometry of electron micrographs of thin sectioned nuclei (25-60 per species).

** Prediction of 2C value at that (theoretically) all chromatin must be in the condensed state in a given taxon.

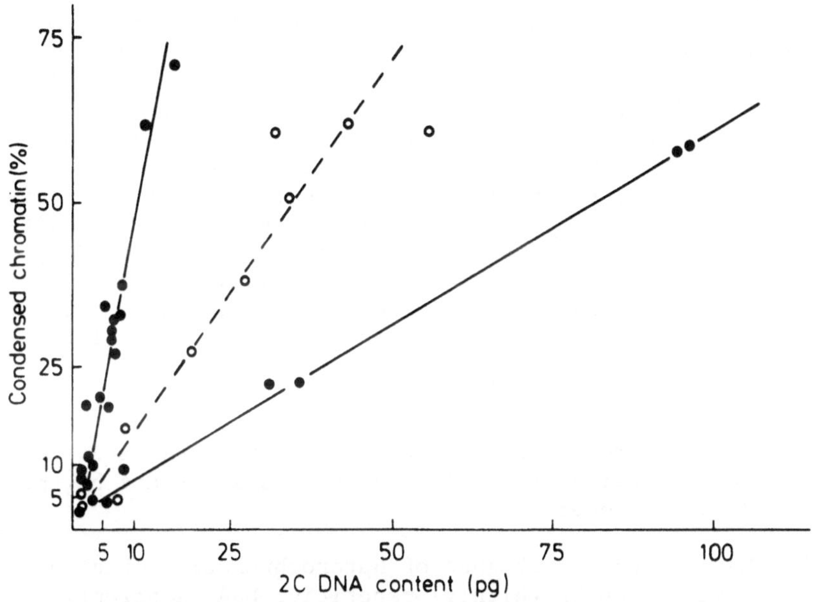

Fig. 10. Regression lines in DNA-poor and DNA-rich dicots (•) and monocots (o) between 2C nuclear DNA content and the percentage of condensed chromatin. Deviations from the regression may be primarily caused by variable amounts of heterochromatin.

regression line is specific for each genus and perhaps family (33). These established correlations allow to predict the chromatin organization for any species, if the genome size is known and vice versa, at least in a statistical manner. Moreover, one can calculate which genome size would lead to complete chromatin condensation and hence loss of viability. Thus, the different regressions between genome size and chromatin condensation occurring in different taxa could explain the differences in the nuclear DNA content which is maximally found in a given taxon.

The role of the proportion of repetitive and single-copy DNA sequences in species-specific chromatin condensation is not yet clear. Plotting the percentage of repetitive DNA against the percentage of condensed chromatin of all angiosperm species which have been studied so far, indicates a positive correlation (coefficient of correlation: 0.8, slope of regression line: 0.9% condensed chromatin per 1% repetitive DNA). Fig. 11 shows that the DNA of the DNA of conservation species with nuclei with diffuse euchromatin (such as members of the "primitive woody angiosperms") reassociates at rates indicating the presence of relatively little repetitive DNA (some 40%), while the DNA of progressive species with very complex chromatin organization (44) (such as weeds of the anthemidean subfamily of composites) is rich in repetitive sequences (60-90%). There are, however, three points which have to be taken into consideration:

(i) As the "condensing protein-binding sequence" is certainly only one among many different repetitive DNA sequences, the proportion of total repetitive DNA may not be the legal parameter.

(ii) The presence or absence of heterochromatin, which often is very rich in (highly) repetitive DNA, apparently influences the recognized percentage of total repetitive DNA, without any relation to euchromatin condensation. Actually, DNA-poor species with completely diffuse euchromatin are rich in satellite DNA and heterochromatin (45).

(iii) The determinations of repetitive DNA by means of reassociation kinetics (cot curves) differ between laboratories considerably, indicating that methodical errors may distort the true relationship. Moreover, the absolute amount of a certain fraction may be more important than its percentage.

Recent studies revealed that the structural complexity of a nucleus is somewhat related to the heterogeneity of its DNA with respect to the derivative melting profile (Fig. 12). The more the chromatin is condensed in the form of various structural elements,

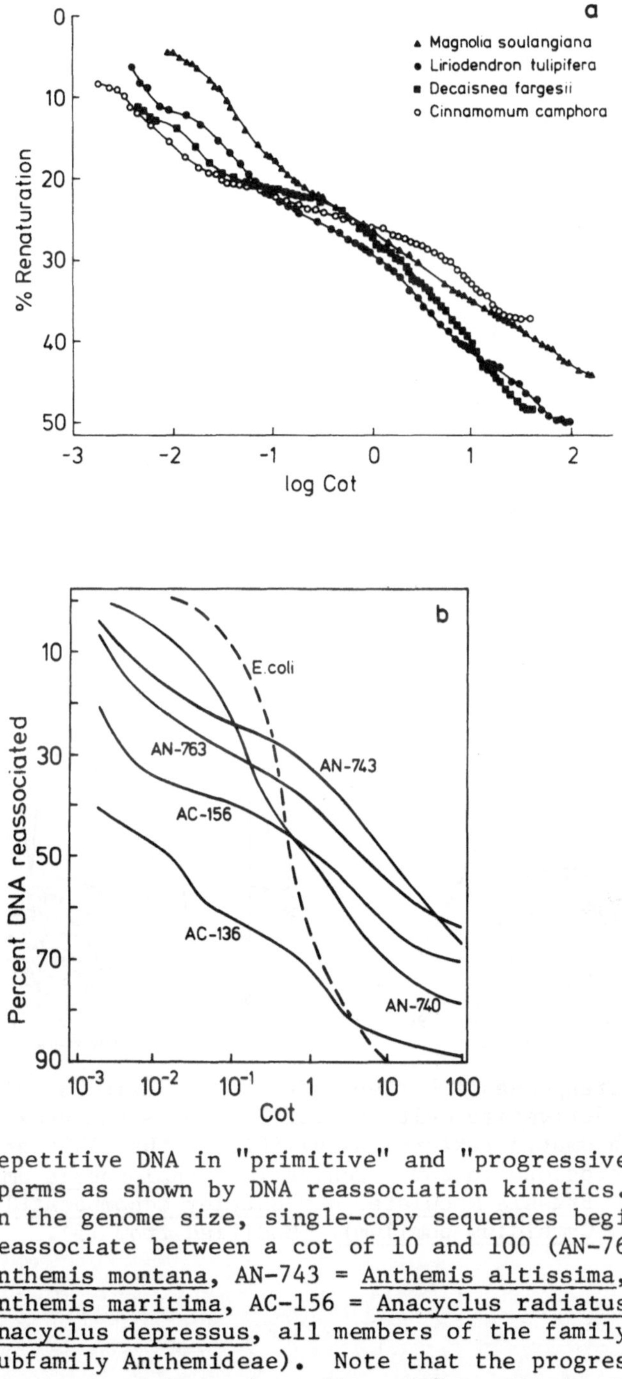

Fig. 11. Repetitive DNA in "primitive" and "progressive" angiosperms as shown by DNA reassociation kinetics. Depending on the genome size, single-copy sequences begin to reassociate between a cot of 10 and 100 (AN-763 = *Anthemis montana*, AN-743 = *Anthemis altissima*, AN-740 = *Anthemis maritima*, AC-156 = *Anacyclus radiatus*, AC-136 = *Anacyclus depressus*, all members of the family Compositae, subfamily Anthemideae). Note that the progressive species possesses about 60 to 90% repetitive DNA, while the primitive species possess only 35 to 45%.

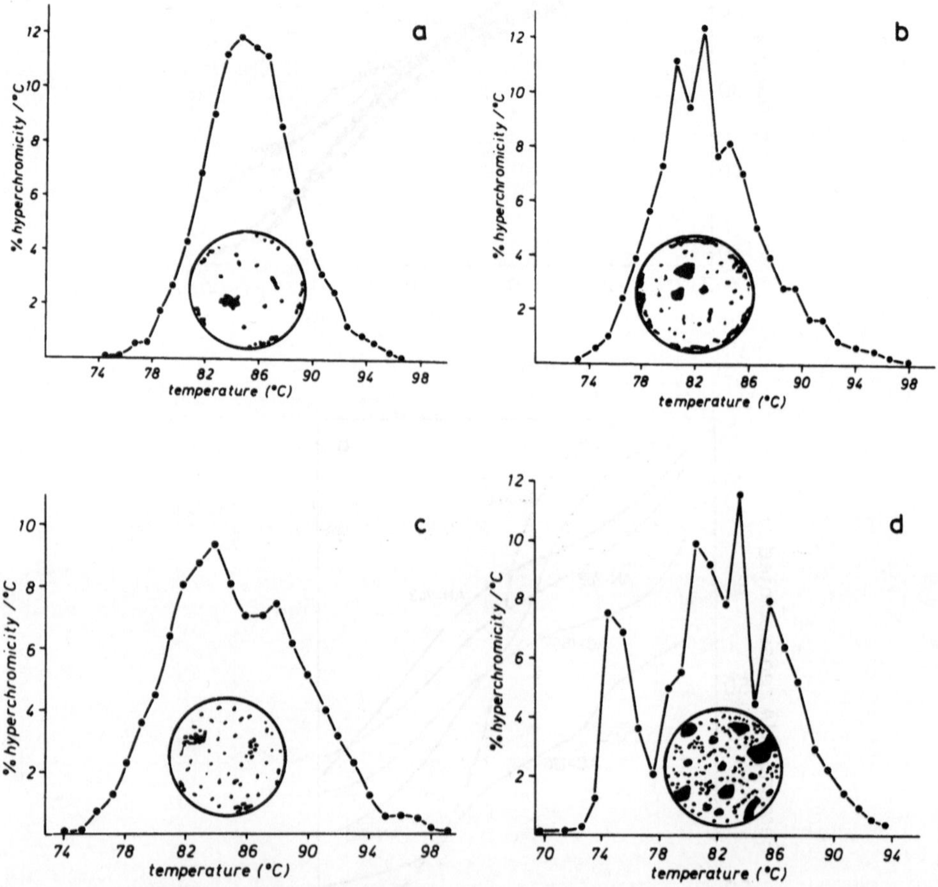

Fig. 12. Correspondence between DNA heterogeneity as visualized by derivative melting profiles and structural complexity (chromatin texture) as visible in the light and electron microscope in some orchids (a = Phalenopsis amabilis, b = Brassia maculata, c = Cattley schombocattleya, d = Cymbidium pumilum). Modified from (5).

the more peaks in the melting profile can be observed. Moreover, there is some evidence that DNA heterogeneity and structural complexity of nuclei are indications for an evolutionary progressive state of the species (29, 46).

Significance

As was suggested in the previous chapter, condensed chromatin may have no functional significance, but be only the necessary result of a high DNA amount, which must be packed into a nucleus. It seems that the significant factor is variation of 2C DNA values but not chromatin condensation. Variation of the basic nuclear DNA content can be understood as an evolutionary strategem involved in speciation processes (e.g. due to isolation effects) and in the control of growth (due to nucleotypic effects). However, as the relationship between genome organization and chromatin organization is no simple one, also the relationship between DNA content and growth parameters is rather complex. For instance, high DNA values may have selective advantage because they lead to large cells, particularly in the water transport system (36, 47), or in adaptation to colder climates which allow only slow growth (48). Annuals may increase their growth speed either by reduction of the 2C value (the common case) (35), or by reduction of the replicon number and shorter cell cycle times (49, 50), or by modeling the organism with a smaller number of larger cells, in comparison with the related perennials (51).

It has been estimated that an eukarotypic organism, including man, can be organized and managed by about 50,000 genes (52). It is illogical that a frog, or a lily, which have 2C values about one magnitude of size larger than humans, need about 10 times as much genes than humans. Actually, it was repeatedly shown that most of the variation of the nuclear DNA content, particularly among related species, is due to variation of the non-coding, repetitive sequences (7, 53; Fig. 13). When that variation in DNA content is accompanied by variation in the amount of condensed chromatin, this indicates that any excess DNA which may originate during evolution becomes condensed and thus "silenced" (24). This could recently be proved by ^{3}H-uridine autoradiography: the absolute amount of RNA synthesized is similar in different plants, because the transcription rate decreases with increasing 2C DNA content and proportion of condensed chromatin (Fig. 14). ^{3}H-uridine is mainly incorporated into decondensed chromatin, and the probability to find transcription complexes in spread chromatin is very rare in species with chromonematic, DNA-rich nuclei (5).

It cannot yet be excluded that also some "silenced genes" are located within species-specifically condensed chromatin (e.g. evolutionary old genes). If this is the case, then such genes should become active under certain abnormal circumstances. Probably,

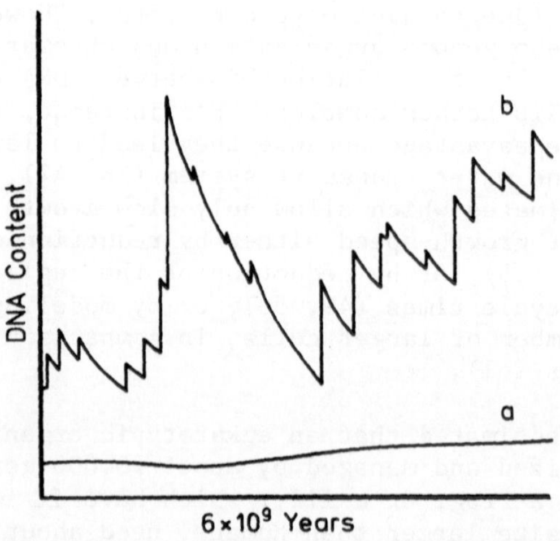

Fig. 13. Diagram to show the changes in 2C nuclear DNA content during evolution. (a) represents the "primary" DNA including genes, (b) the "secondary" DNA including non-coding and repetitive sequences. According to (53), redrawn from (7).

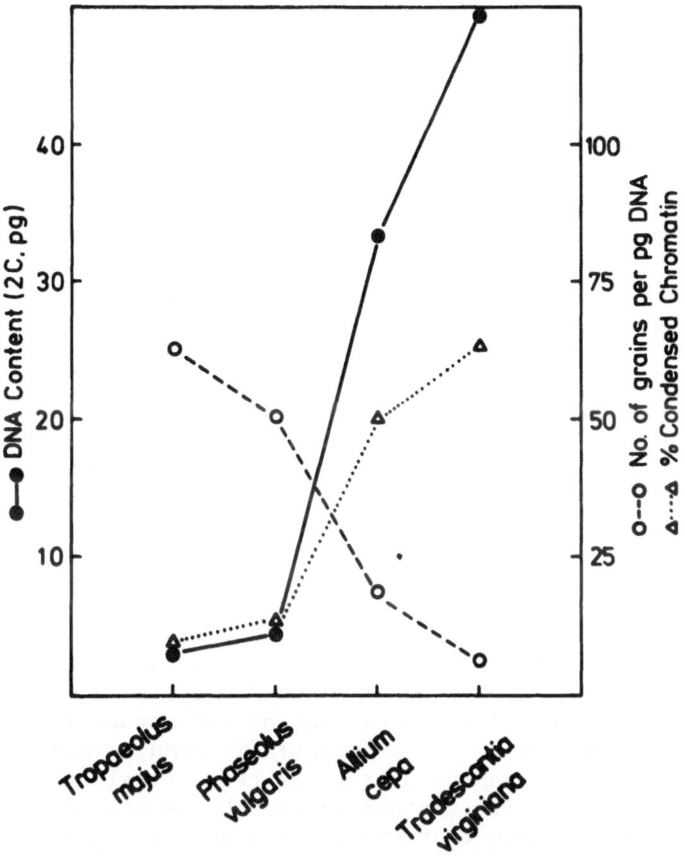

Fig. 14. Relationships between 2C nuclear DNA content, the proportion of condensed chromatin, and the overall rate of transcription in various higher plants. It can be clearly seen that the high amounts of DNA in some species are not reflected by an increase, but a decrease of the rate of RNA synthesis, as estimated by semi-quantitative autoradiography. Redrawn from (5).

atavisms can be taken as examples for the activation of evolutionary old, silenced genes.

TISSUE-SPECIFICITY

Patterns of Condensed Chromatin

In mammals and other vertebrates, mitotically cycling nuclei normally show a diffuse structure during interphase with some differences between G1 and G2 (see next chapter), while the chromatin undergoes tissue-specific condensation during differentiation (54-59). It seems that large portions of the genome become irreversibly silenced; i.e. cell determination occurs by inactivation of certain chromatin domains. Even if cell activity is enhanced by hormonal stimulation etc., the decondensed chromatin is the only further decondensing part of the nucleus, thus increasing the total nuclear volume, while the condensed chromatin is not affected (54, 60). The amount and pattern of condensed chromatin may even vary extremely between different cell types of the same tissue, as in the case in the Guinea pig retina (8, 59, 61) (such findings indicate that biochemical analysis of DNA and RNA isolated from whole tissues may be worthless, because they lead to results which represent a mixture of different data from different cells). Figure 15 shows electron micrographs of nuclei from different cell types of the Guinea pig eye.

The control of chromatin condensation in mammals is evidently exerted via modification of histones and nonhistone proteins and ions. Particularly the phosphorylation of H1 at a residue being different from that which is phosphorylated during mitotic chromosome condensation, causes chromatin decondensation in working nuclei (67-69). Condensation of chromatin during the maturation of avian red blood cells and during spermiogenesis is brought about by a specialized histone H5 and by protamines respectively, which are phosphorylated after their synthesis and subsequently dephosphorylated. The cross-linking proteins are known to give rise to the formation of macrocomplexes from DNA and core particles, respectively. From the viewpoint of electrostatic attraction, aggregation should be best promoted by those proteins which have a high specific positive charge (i.e. the net positive charge per amino acid residue, which indicates the degree of overall basicity of the polypeptide and hence its chromatin condensation ability; see table II in 70). Thus protein phosphorylation which reduced this value should also reduce the tendency of aggregate formation. However, in all instances investigated, including those of the recent literature, a protein phosphorylation promotes the formation of large complexes with a high internal order. Bode, et al. (71) suggested that the reduction of electrostatic affinities, which is achieved *in vivo* by protein phosphorylation, serves to enhance complex size, probably by permitting a thermodynamically controlled and ordered association

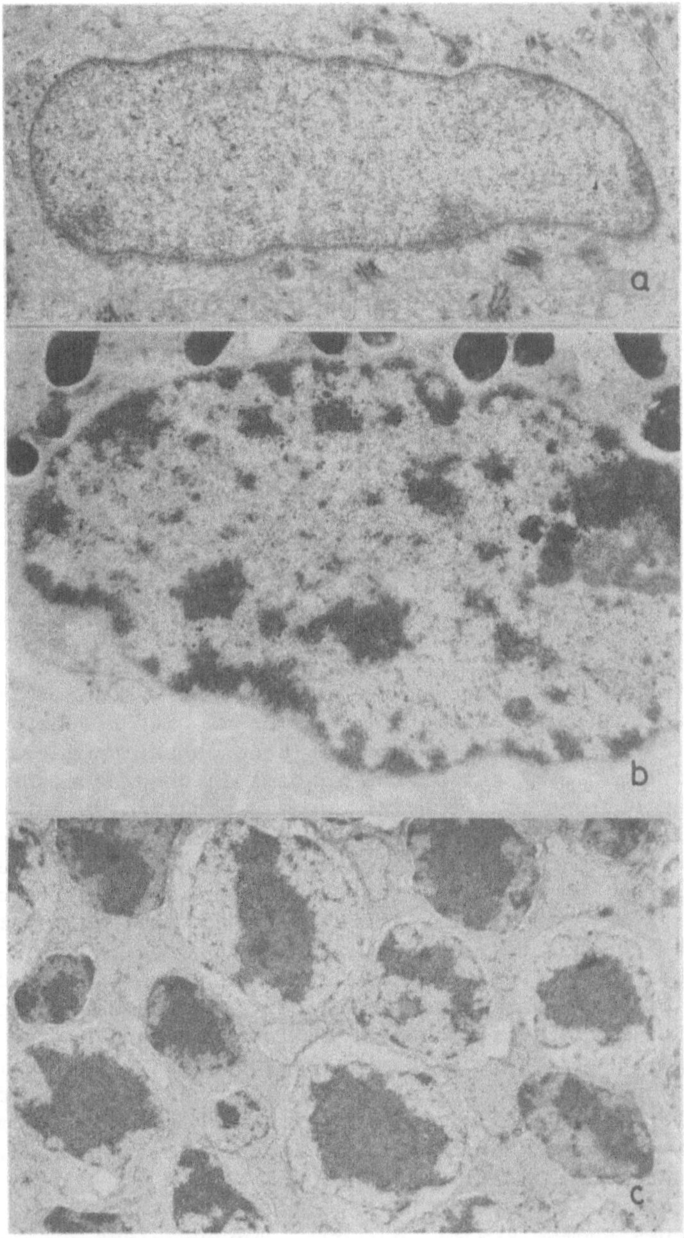

Fig. 15. Electron micrographs of nuclei from the Guinea pig eye showing different amounts and patterns of condensed chromatin. (a) Lens (x 10,000), (b) pigmented epithelium (x 18,000), (c) photoreceptor cells (x 2,000). Note the "drapery-like" pattern in (b) and the central patches in (c).

of DNA segments. Moreover, decondensation at the nucleosome/ solenoid level is apparently a prerequisite of quinternary structure formation (4). Actually, nucleosomes are sometimes visible in condensed chromosomes and chromatin (72), while they are normally not visible in decondensed chromatin.

Condensation is often the result of poly (ADP-ribose) polymerase activity. This chromatin-bound enzyme transfers the adenosine diphosphate ribose moiety of NAD (nicotineamide adenine dinucleotide) to the nuclear proteins (particularly H1), adds successively ADP-ribose to form a homopolymer, and liberates nicotin-amide. Both ends of the homopolymer become linked to chromosomal proteins, so that cross-linking takes place. Structurally, therefore, this kind of chromatin condensation is completely different from the chromosome condensation at mitosis, although it cannot be distinguished from the latter in electron microscopic sections. For effects of poly-(ADP-ribose) polymerase activity on chromatin condensation and differentiation see, for instance, ref. (73-75).

Models for DNA superpacking as visible in the pattern of quinternary chromatin condensation have been published by Nicolini (10) and Comings (76).

Chromatin organization in mammalian nuclei can also be understood as "functional structures" (60, 77). It was shown in many studies that condensed chromatin is inactive with respect to RNA synthesis (59-66). The tissue-specific condensation pattern (and thus inactivation of parts of the genome) is understandable in the light of suggestions that the chromosomes occupy distinct regions of the nuclear cavity during interphase (78-80). This position maybe fixed by attachment of telomeres, or centromeres, and/or other chromosomal regions to the nuclear membranes, especially to pore complexes (4, 81), or by interaction with the nuclear matrix (82-84). If the chromosomes are fixed in certain positions in the nucleus, then the diversity of condensation patterns should be the expression of the tissue-specific inactivation of certain chromatin domains. The pattern of condensation is predefined by the base sequence and induced by the chemical-electrostatic environment (Fig. 16). And this may be the critical point in the control of cell proliferation, cell differentiation and cell transformation (4), rather than the control of individual gene activities. This idea is in accordance with suggestions that the non-coding control sequences play the more important part in evolution and somatogenesis than the genes themselves (7, 85, 86). Nevertheless, it is not yet clear whether gene activity is the consequence, or the cause of chromatin condensation (66, 87, 88). This problem touches one of the fundamental questions of biology, which can be formulated: Is there a linear causality in living matter, or are there only feedback loops? "You're searching, Joe, for things that don't exist; I mean beginnings. Ends and beginnings - there are no such things.

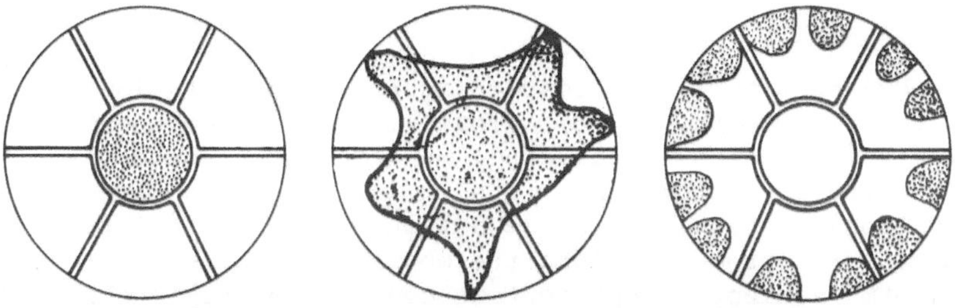

Fig. 16. Highly schematic diagram indicating the fixed location of chromatin domains (lines) in animal nuclei, and the tissue- and developmental-specific condensation of the domains (modified from 61).

There are only middles" (89). Moreover, the proposed understanding of chromatin condensation and gene activity allows us to see the problem of differentiation and morphogenesis in a new light. The ultimate question, how biological diversification may be controlled, can probably soon be traced back to the thermodynamic properties of DNA. The pattern of gene activity is marked out in the DNA sequence and hence DNA configuration by the pattern of melting and unfolding events according to their enthalpy changes. The program of ontogenetic realization of the Bauplan is, therefore, induced by ions, hormones, or environmental stimuli, but differentiation and morphogenesis themselves are the result of necessary step-by-step changes in base pairing and conformation of the DNA double helix and the DNA-protein complex, according to the optimal thermodynamic sequence of events. I guess that this way will be the only one which may some day explain the self-organizationand self-realization of living matter (see also the accompanying paper by Nagl in this volume).

CELL CYCLE - SPECIFICITY

The DNA of mitotic chromosomes is packed at a ratio of 1:5,000 to 1:10,000 (8, 9, 90). Thus, mitotic chromosomes exhibit the highest level of DNA packaging; meiotic chromosomes may even show one extra level more (Fig. 18a). In contrast to interphase chromatin condensation, the condensation of chromatin into the "transport form" proceeds along the whole chromosome complement in a uniform fashion, except certain conspicuous regions such as the centromeres and nucleolus organizing regions. As Table I shows the DNA double helix is compacted at several levels. While the arrangement of DNA and histones in the nucleosome and solenoid fiber is relatively well known, the arrangement and mechanisms involved in alteration of chromatin tracts in the scale of the entire genome still remains contradictory (see below).

Mitotic chromosome condensation is controlled independently of other mitotic events, such as spindle formation and breakdown of nuclear envelope, but normally takes place during a certain period of the mitotic cell cycle. The cell cycle is usually divided into interphase comprising G1, S (replication) and G2 periods, and mitosis (prophase, metaphase, anaphase, telophase). Several observations suggest a continuous chromosome cycle throughout the cell cycle (see the next chapter).

Because changes in the chromatin structure occur during the cell cycle have been recently reviewed (4, 91, 92), only the main points and some new results will be discussed here, in order to keep this essay on a complete and current level.

Mitotic Condensation

The critical question concerning mitotic chromosome condensation is, how the quaternary fiber, the solenoid, is further compacted. There exist three different approaches to explain this level of DNA packaging: folding, coiling, and looping of the solenoid fiber (Fig. 17).

The folded-fiber model as originally suggested by DuPraw (90) is based on the well-known "spaghetti-chromosomes" as seen in the electron microscope after whole-mount preparation (see also 93). As, however, this appearance is very likely an artifact (8), the representatives of the folded-fiber model reduced the folding concept to the level between the solenoidal fibril and the coiling of mitotis chromosomes as visible in the light microscope (94).

More elegantly, coiling models explain the condensation from the level of nucleosomes to the level of chromosomes (95, 96). However, there still exist considerable differences between the results of different authors. While Sedat and Manuelidis (9) suggest that the solenoid fiber is further coiled to a super-solenoidal tube of 200 nm diameter, and this tube again to a metaphase chromosome of about 0.6 μm, Bak and Zeuthen (12) found that the solenoid fiber is supercoiled to a supersolenoid, a tube some 400 nm in diameter with a wall made up of the solenoidal structure of 30 nm diameter. This tube was unfortunately termed "unit fiber". In that kind of "unit fiber" the DNA is contracted in the order of 1,300 - 1,500. The "unit fiber" is then further coiled to the chromosome by a factor of about 5 (97, 98).

Another attractive hypothesis is that of radial loops in chromosomes, held together by nonhistone proteins. Already in 1969, Sonnenbichler (99) observed subunits of chromosomes form which loops were projecting. More recently, spreading of chromosomes after extraction of most of the proteins revealed that there exists a chromosome scaffold built-up of residual proteins, from which many long loops of DNA protrude; if not all of the histones are removed, chromatin fibers of variable length and diameter are released (100-102). The residual (or scaffold) proteins are apparently firmly (perhaps covalently) bound to DNA and must polymerize to form the chromosome scaffold during mitotic chromosome condensation (103, 104).

Spreading of metaphase chromosomes, prepared by the acute angle metal deposition method (96), thin sections of treated (97, 98) and untreated (5) chromosomes and heterochromatin (see following chapter), and stereoscopic analysis of scanning electron micrographs indicate that chromatids are tubular (21, 22; Fig. 18b). Therefore these findings support coiling models of chromosome condensation.

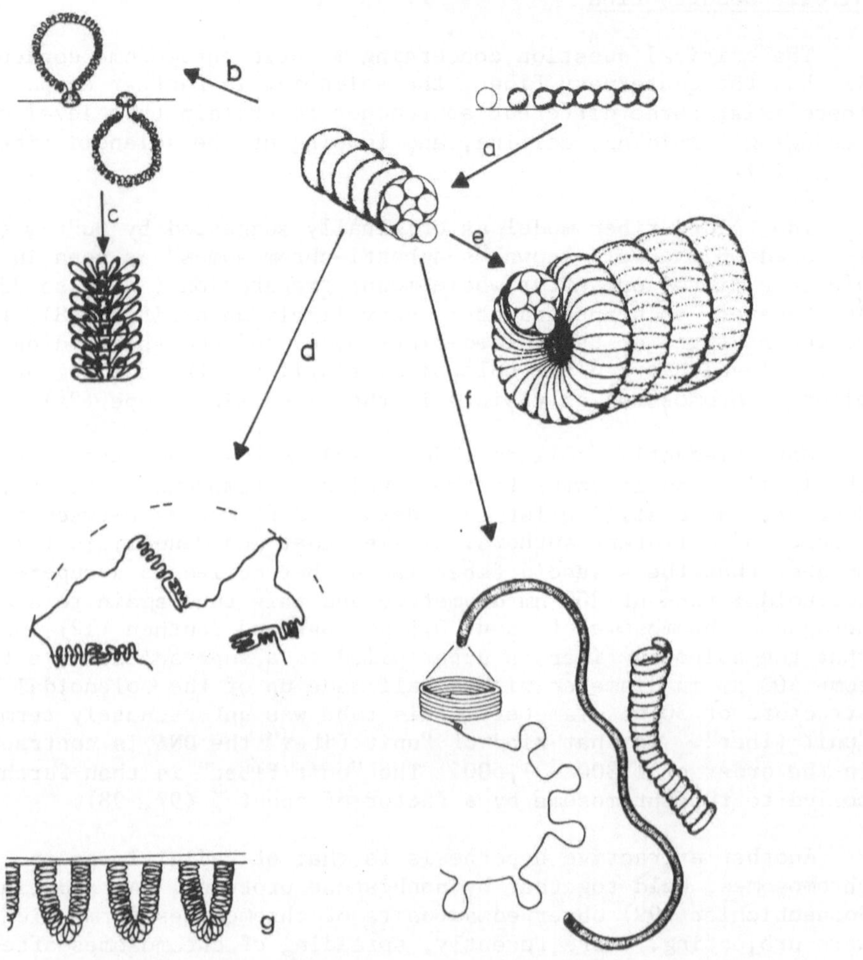

Fig. 17. Diagram to show various models of chromatin condensation from the nucleosome level to the mitotic chromosome and interphase chromatin, respectively. (a) Supercoiling of nucleosome fiber to the solenoid fiber (b, c) formation of mitotic chromosomes by polymerization of nonhistone chromosomal proteins (radial loop model), (d) formation of mitotic chromosomes by folding of the solenoid fiber (folded fiber model); note the interconnecting strands ("superchromosomal organization"), (e) coiling of the solenoid to the "tube", (f) coiling of the solenoid to the super-solenoid ("unit fiber") and further coiling to a metaphase chromatid (at right), (g) drapery-like condensation and fixation at the nuclear envelope of chromatin at interphase (according to 4, 5, 8, 9, 12, 13, 90, 97, 100).

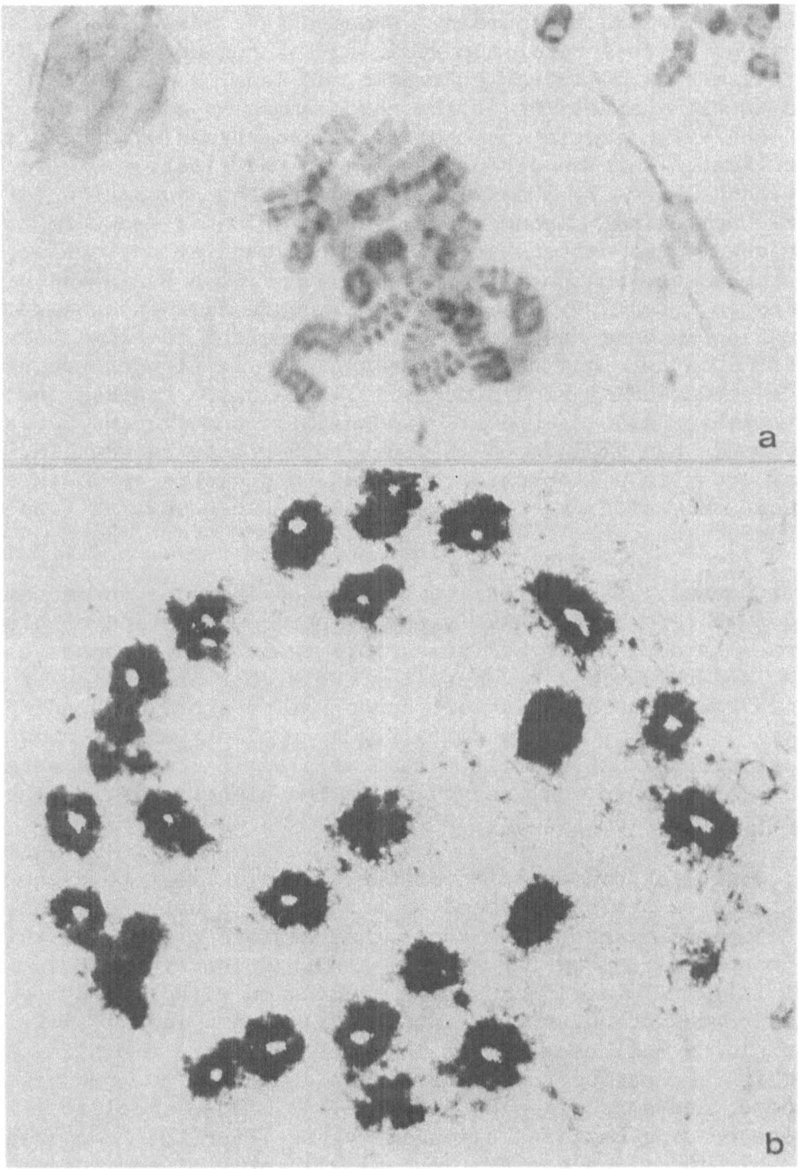

Fig. 18. Organization of condensed chromosomes. (a) Light micrograph of super-coiled metatic chromosomes in pollen mother cells of Tradescantia virginiana (aceto-carmine stained x 1,500), (b) electron micrograph of an ultrathin cross-section through anaphase chromosomes of Oxalis acetosella. Note the "holes" in the chromosomes (for better visualization the print was under-exposed, and some of the holes were retouched).

There is one aspect of chromatin condensation which is common to all three models: the domain. Domains have been detected in several ways, as regions of supercoiling, as regions defined by enzymatic excision of intact chromatin, as lengths of DNA seen attached to the scaffolding of the chromosome, as super-coiled discs of DNA after gentle lysis of cells and spreading of nucleids, and as chromatin domains directly shown in the electron microscope (for references see 5). One might speculate that domains are the basis of longitudinal chromosome differentiation as known to exist from the chromomere pattern at meiosis, the banding pattern of giant chromosomes, and the fluorescence and Giemsa bands which can be induced in metaphase chromosomes. Evidence is increasing now that the fluorescence and Giemsa-G bands, and the complementary pattern of R bands along chromosomes are a reflection of the predominance of GC and AT rich domains (5, 105-107). Besides these non-polymorphic, species-specific and highly conservative bands, Giemsa-C bands can be induced in constitutive heterochromatin, which may occur in polymorphic condition both within and between individuals and which are highly variable between species (see the corresponding chapter).

Unfortunately, G bands cannot be induced in plant chromosomes. The reason therefore probably is the higher condensation of plant chromosomes (about 3.5 to 6.5 times more than human chromosomes (8, 108), and this might be caused by the higher proportion of repetitive DNA (and thus also of "condensation-protein" binding sequences) in plants than in animals. A corollary of the above considerations is that one difference of plant chromosomes versus mammalian chromosomes must be sought in the higher order arrangement of the chromatin fiber (109).

The control of condensation of the solenoid fiber to higher order structures seems to depend on histone H1 phosphorylation. Since a couple of years it is known that certain residues of H1 become phosphorylated during mitotic condensation (references in 5, 8, 111). As was discussed in connection with interphase chromatin condensation, histone phosphorylation evidently unfolds the solenoid or nucleosome fiber in order to liberate binding sites, which are needed for higher order condensation (see also the evidence and data given by Nicolini, 4). Base specific chromosome banding patterns strongly suggest that the "condensation protein"-binding sites are AT-rich, as all forms of condensed chromatin such as bands, chromomeres, and heterochromatin can preferably be stained with AT-specific fluorochromes.

Interphase Changes

Mazia (112) suggested some 20 years ago that chromosomes

undergo a condensation-decondensation sequence extending throughout the entire cell cycle with the two extreme conformational states existing in mitosis (condensed) and S phase (decondensed). Actually, structural changes during mitotic interphase were observed in mouse liver (113) and wild onion species (114, 115) at the light microscope level and the electron microscope level (116, 117). By automated image analysis it has been possible to objectively characterize the nuclear structure during the cell cycle (14-16, 92, 118). Moreover, the geometric and densitometric descriptors introduced by Nicolini (16) prove that the chromatin alterations, detected in isolated chromatin at the tertiary-quaternary (nucleosome/solenoid) level, are reflected even at the quinternary level of chromatin superpacking in the intact cells. Specifically during the G1-S transition the increase in binding sites and molecular ellipticity and the decreased thermal stability of isolated chromatin are reflected in the increased dispersion and convolution of the overall chromatin-DNA *in situ* (4). These changes, on the other hand, directly mimic changes in transcriptional activity (4, 119) and parallel increasd synthesis of nonhistone chromosomal proteins (4). Similar changes in chromatin organization like those at the G1-S transition occur at the G0-G1 transition (4, 33b). Figures 19 and 20 show chromatin changes at interphase as recognized respectively at the light microscope level by Feulgen-DNA measurements and autoradiography and the electron microscope level by morphometry (24). After decrease of the proportion of condensed chromatin from the tightly packed chromatin in the small telophase nuclei, there is continuous increase of chromatin condensation from middle G1 to metaphase. Preliminary means obtained from electron micrographs of *Hyacinthus orientalis* are the following: telophase: 72%, G1: 23%, S: 33%, G2: 38%, prophase: 39%, condensed chromatin.

There is, however, some controversy, how S phase chromatin is organized. Some authors claim that there is maximum dispersion, and the heterochromatin (chromocenters) become decondensed during late S, while others did not find evidence for decondensation. It seems that various plant and animal species behave differently in this aspect (reviews: 7, 91, 116). Moreover, no large-scale, prolonged decondensation is essential for replication (117). There is also some confusion on whether "Z phase" (i.e. the dispersion stage of mitosis and endomitosis in plants; see article on "Cell growth and DNA increase..." by Nagl in this volume) is really an expression of early mitotis prophase and endomitosis, or an expression of late DNA replication in heterochromatin. In *Allium carinatum*, apparently both S phase and Z phase are passed under heterochromatin decondensation down to the level of condensed euchromatin (chromonema) (116).

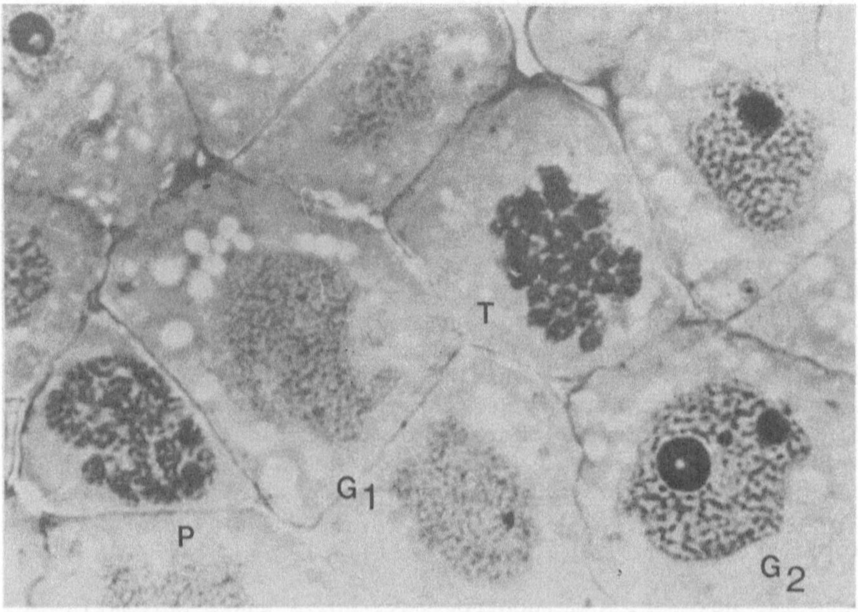

Fig. 19. Light microphoto of semi-thin sectioned nuclei in the root meristem of <u>Hyacinthus orientalis</u>. Note different structure of G1 and G2 nuclei, coiling of chromosomes in prophase (P) and "holes" in anaphase chromosomes (A) (phase contrast x 1,000).

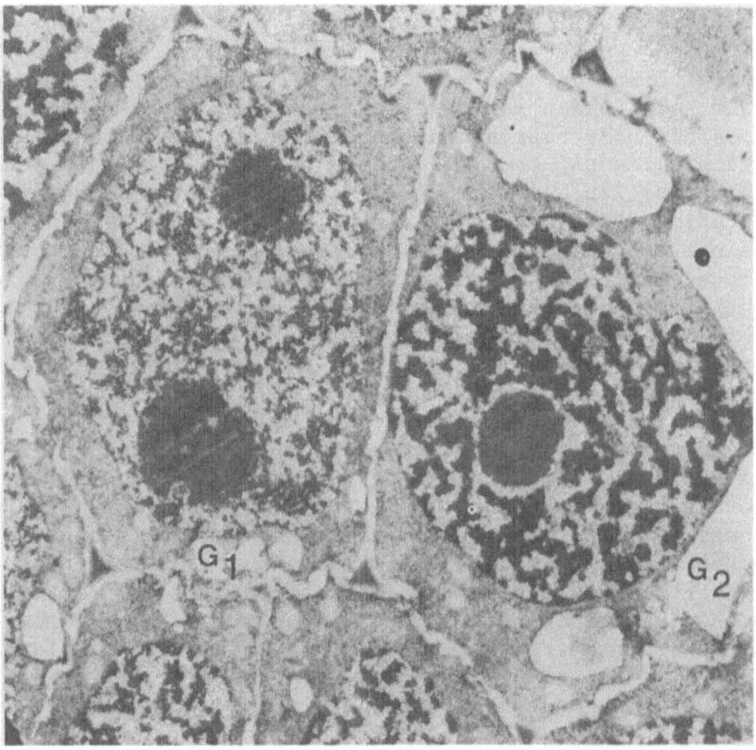

Fig. 20. Electron micrograph of Hyacinthus nuclei at G1 and G2 showing differences in chromatin organization (x 6,500).

HETEROCHROMATIN

The chromatin of interphase nuclei and non-cycling nuclei is normally divided into euchromatin and heterochromatin. These terms, however, are used in a different sense by different authors. While some understand them more in a functional sense (as template active and template inactive chromatin), heterochromatin is regarded by others in the original structural sense as those portions of the chromosomes, which remain condensed during interphase, which are rich in repetitive DNA, which do not carry genes, which replicate asynchronously when compared with euchromatin etc. (120-123). The latter definition applies to constitutive or karyotypical heterochromatin, which can be made visible in the mitotic karyotype by the C banding technique. In female mammals, the nuclei exhibits a sexchromatin body, which was designated as facultative heterochromatin (124). This body represents one of the two X chromosomes of females which evidently becomes inactivated due to a dosis compensation mechanism (125). Although the condensed X chromosome is late replicating, it possesses the same DNA nucleotide sequence as the active one, so that it cannot be regarded as true, or constitutive heterochromatin. In certain vertebrate nuclei also euchromatin can be in the condensed state, besides constitutive and facultative heterochromatin. This condensed euchromatin does not exhibit, of course, any of the characteristics of real heterochroamtin except that it appears condensed and inactive (the condensation mechanisms in euchromatin and heterochromatin may be quite different, e.g. cross-linking versus mitotic super-coiling. The increasing synonymous and indiscriminate use of these terms in the literature has led to a serious degree of conceptual and operational confusion. In this essay only the constitutive heterochromatin is considered as heterochromatin. This does not exclude that the term heterochromatin still represents a super-conception covering various types of chromatin. One can show this by the different staining behaviour of various heterochromatin blocks in diverse banding techniques (8, 107, 109). Mello (126) cites D.E. Comings with a fitting sentence: "Heterochromatin is somewhat like human society - it is a complex subject and simple slogans are inadequate to characterize it. For every apparent rule it is possible to cite an exception. The only solution is to become acquainted with its many and varied facets. The unsolved mysteries it presents remains the source of its fascination".

Structure

If heterochromatin actually remains in a mitotically coiled state at interphase, then there should be no difference between the organization of metaphase chromosomes and chromocenters (i.e. a piece of interphase heterochromatin). Actually, there often is a close structural similarity between mitotic chromosomes and chromocenters at both the light and ultrastructural level. As

Figs. 21 and 22a show, chromocenters more often than not exhibit a "whole" in thin sections (5). Reconstruction of serial sections strongly suggest that most chromocenters are super-solenoidal tubes (Fig. 22b). This indicates that heterochromatin may actually be in the mitotic state, and that this state is achieved by super-coiling. An exceptional structure was recently found in chromocenters of Lupinus polyphyllus, which resemble somewhat meiotic poly-complexes (127, 128). Their origin and functional significance are not known, but they show that interphase heterochromatin may represent various classes or types of constitutively condensed chromatin.

As already indicated, the permanently condensed state of heterochromatin ultimately depends on the primary sequence of the involved DNA, but other chromosomal constituents must mediate this condensation. Heterochromatin is known to be rich in highly repetitive DNA, although this DNA is not necessarily discernible as satellite DNA. Moreover, it is mostly AT-rich, but there are exceptions (these do not exclude, however, that the "condensing-protein" binding sites are AT-rich). Candidates for such "condensing-proteins" have actually been found (42, 129).

Function

Understanding the function of heterochromatin is as difficult as to understand its structure. Brutlag et al. (130) pointed out that the misconception that satellite DNA has no essential function is based on the lack of genes in heterochromatin and on the viability of individuals with large heterochromatin deletions. The authors reviewed the literature which indicates that heterochromatin does have a function in the germ-line, and that variations in satellite DNA could lead to speciation (see also 131). Nagl (7, 86) reviewed indications for a role of quantitative and spatial variation of heterochromatin (and chromatin in general) in differentiation and speciation, i.e. in ontogenesis and phylogenesis. Lima-de-Faria (121) attributed to heterochromatin and satellite DNA, the highest importance in the control hierarchy of the karyotype and the chromosomes. John and Miklos (132) emphasized that the predominant feature of satellite DNA (and heterochromatin) is its variation in several respects. At least some types of satellite DNA appear to have no somatic functions but do play a role in the germ line, particularly in the regulation of recombination between homologous chromosomes (see also 133).

A problem: Where does heterochromatin begin, and where does species-specifically condensed euchromatin end?

Some of the most frustrating aspects of cell biology are the difficulties to distinguish between euchromatin and heterochromatin in electron micrographs, when the former builds chromomeres or chromonemata (unfortunately many workers are not aware of this!). Where is the delimitation between chromomeres and chromocenters,

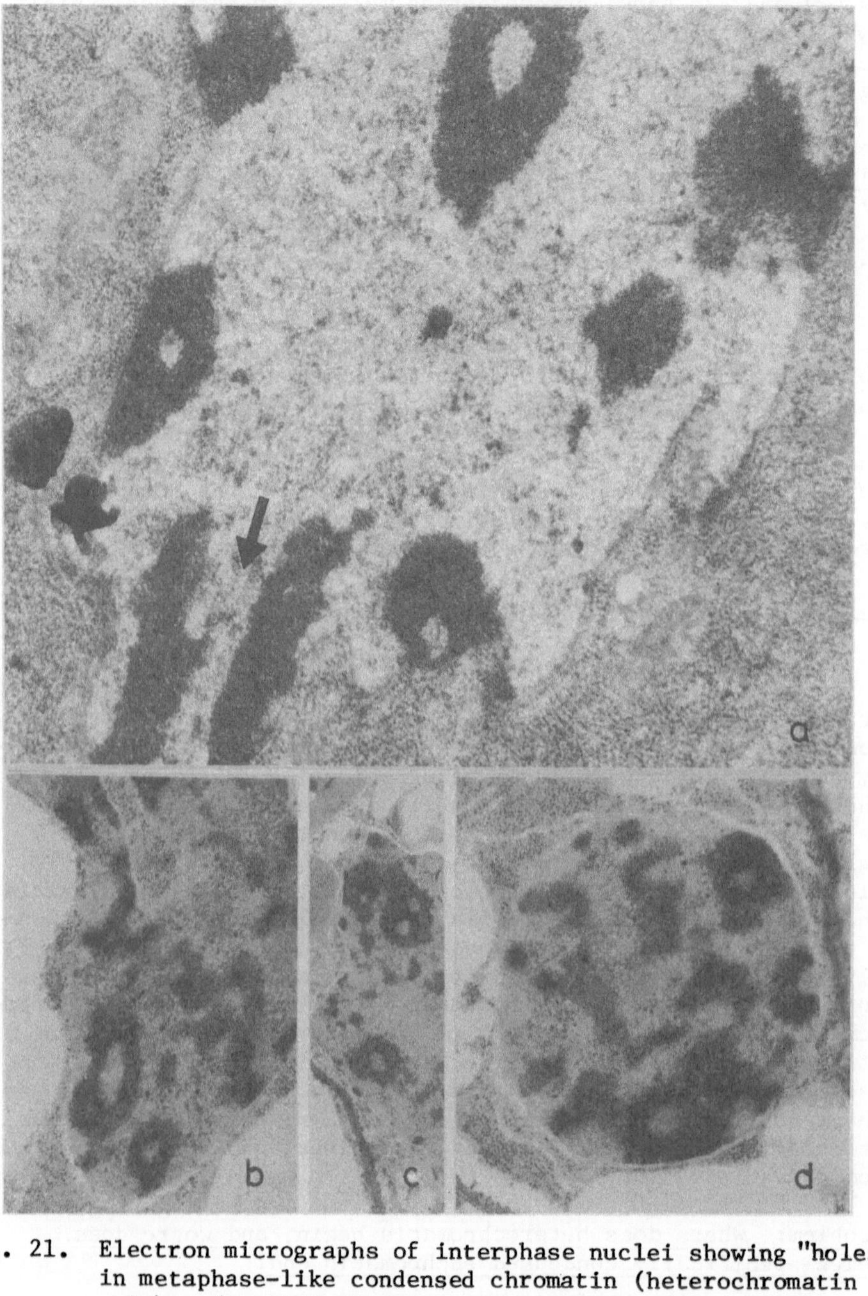

Fig. 21. Electron micrographs of interphase nuclei showing "holes" in metaphase-like condensed chromatin (heterochromatin s.l.). (a) Diploid nucleus of Gerris najas, showing cross-sectioned and longitudinally sectioned (arrow) chromocenters (x 32,000). (b-d) Nuclei of buds (juvenile phase) of the ivy, Hedera helix showing "holes" in chromocenters (x 25,000, except c = x 10,000).

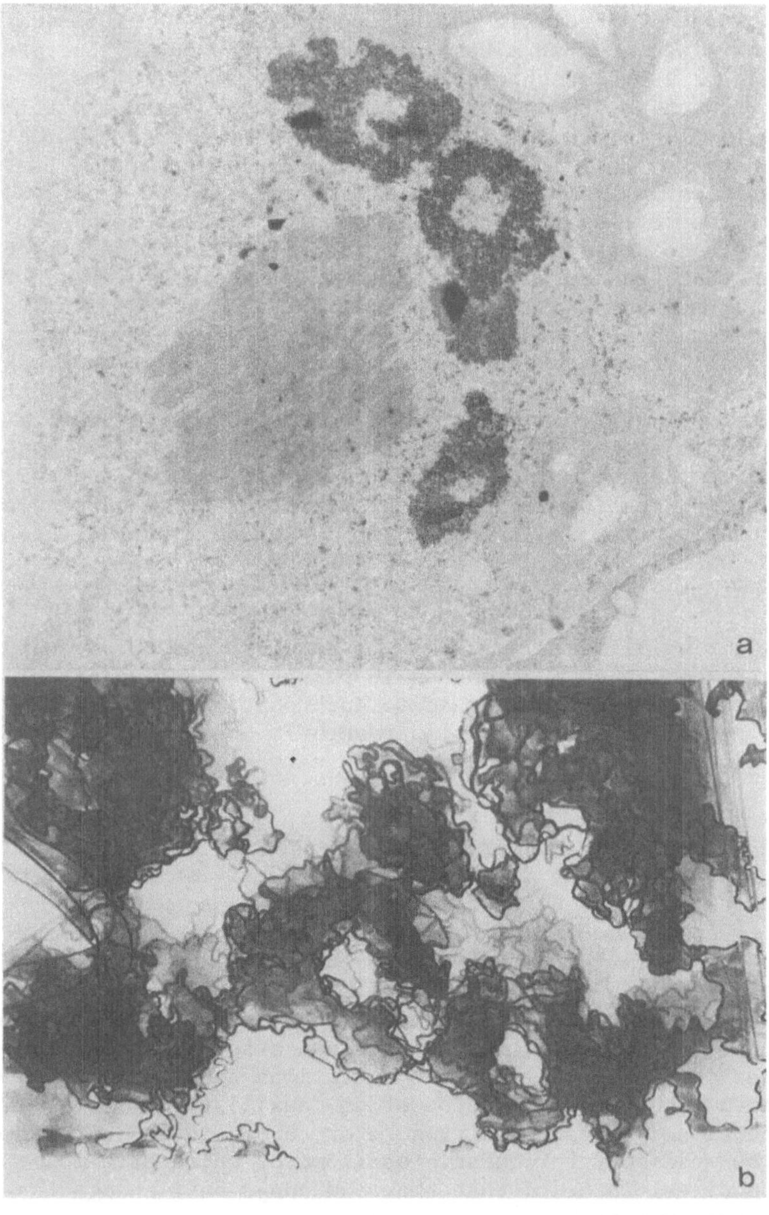

Fig. 22. Evidence for super-solenoidal organization of heterochromatin. (a) Electron micrograph of a nucleus of Rhinanthus minor, the euchromatin is completely diffuse, the heterochromatic chromocenters exhibit "holes" (note also the species-specific protein crystal in the nucleus; x 10,000). (b) Reconstruction of chromocenters from serial thin sections of a hyacinth nucleus; the "holes" of single sections are continuous and form a channel.

that is, between euchromatin and heterochromatin, if it is just a question of size, whether a band of a giant chromosome reacts as euchromatin or as heterochromatin after differential Giemsa staining? (see 134). What is the difference between euchromatin and heterochromatin in plants, when both exhibit the same electron-density and when both may fail to hybridize with highly repetitive DNA as in many species (e.g. 135). If we are aware of these problems, we might have the feeling that there is something wrong with our terminology. As stated at the beginning of this article, it can be assumed that the degree of euchromatin condensation is controlled by a similar mechanism as heterochromatin condensation, i.e. by a nucleotide sequence. Both types of chromatin are species-specific (karyotypical), and both depend on the remaining of some mitotic coiling during interphase. The coincidence between high amounts of repetitive DNA in plants and high degrees of overall chromatin condensation is suggestive of a government of chromatin fiber coiling by periodically interspersed repetitive DNA seuqences. Another way of creating condensed chromatin during evolution, namely heterochromatin, lies in the sudden formation of blocks of highly repetitive satellite DNA, but also heterochromatin can be created by different ways (136).

Considering all data available, I suggest that species-specifically condensed euchromatin in plants is more related to constitutive heterochromatin than to euchromatin in mammals. The biological significance of both, permanently condensed "euchromatin" and constitutive heterochromatin may be seen as formulated by Bostock (136) for heterochromatin.

"The selection would not be for any particular satellite sequence, but simply for a DNA structure that is able to maintain the condensed state of heterochromatin. The apparent rapid variability in the amount of satellite DNAs (and heterochromatin) and their chromosomal locations may provide a means of altering the overall genotype of an organism that is faster than, and additional to, that which can be achieved by mutation of structural genes alone. Undoubtedly, many things can affect the behaviour of chromosomes and thus the genotype - mutations to structural genes amongst them - but the possession of an "auxiliary" system such as that mentioned above, involving heterochromatin and satellite DNA, may be necessary for eukaryotes (some of which have long generation times) in order that they can adapt fast enough to changing environments."

Similar emphasis on the dominant role of chromatin organization for cell proliferation and differentiation - in contrast to traditional ideas on the dominant role of structural genes - have been put forward by Lima-de-Faria (121), Nagl (7), Ayala (85), Nicolini (4) and others.

CONCLUDING REMARKS

Cell growth, proliferation and differentiation are more likely a consequence of changes in the three-dimensional chromatin organization and of thenuclear DNA content (see accompanying paper by Nagl) than of differential gene activity. The pattern of both condensation and DNA variation is encoded by the DNA nucleotide sequence and its realization during ontogenesis follows according to the laws of thermodynamics due to the different enthalpy of nucleotide clusters and DNA configurations in chromatin.

Angiosperms, which have a lower complexity than mammals (but higher DNA contents), exhibit species-specifically condensed euchromatin in addition to constitutive heterochromatin. The proportion of condensed euchromatin is primarily correlated to the 2C DNA content. The latter, and not the chromatin condensation may be the biologically significant variable parameter, exerting nucleotypic effects on cell and organ size, cell cycle and minimum generation time, etc., i.e. many growth parameters. Variation on the amount of non-coding DNA seems to be an important tool in the evolution and adaptation of plants, while the variation of chromatin organization is just a consequence of DNA variation, in order to package the huge amounts of DNA into a nucleus. Chromatin condensation in plants is an expression of the evolutionary state of a species, not the expression of gene activity in a given tissue. The close correlation between 2C value and percentage of condensed chromatin allows to predict the nuclear structure from the DNA content and vice versa.

In vertebrates, a new control system evolved and found its highest efficiency: variation of chromatin condensation during development and differentiation. Although the induction of a tissue-specific pattern of condensed chromatin is brought about by enzymes (e.g. histone phosphorylases, poly (ADP-ribose) polymerase, etc.) and ions (e.g. Mg^{++}, Ca^{++}), the patterns themselves are predefined in the nucleotide sequence and DNA conformation according to thermodynamic laws ("thermodynamic code of the DNA"). The pattern of gene activity is not the cause of differentation, but just the consequence of the thermodynamically determined pattern of condensed versus decondensed chromatin, causing determination, and limiting the sensitivity to hormones, ions, etc. Any disturbance in the nucleotide sequence must result in a change of the thermodynamic code and may easily lead to carcinogenesis, i.e. to a change to the evolutionary oldest and preferred proliferation program.

REFERENCES

1. D.M.J. Lilley and J.F. Pardon, Ann. Rev. Genet., 13:197-234 (1979).

2. C.A. Nicolini (Editor) "Chromatin Structure and Function" Plenum Press, New York (1979).
3. M. Balabam (Editor) "Molecular Mechanisms of Biological Recognition" Elsevier/North-Holland, Amsterdam (1979).
4. C.A. Nicolini, in: "Chromatin Structure and Function" C. Nicolini, Editor, Plenum Press, New York, pp. 613-666 (1979).
5. W. Nagl, in: "Encyclop. Plant Physiol." New Ser.,Vol. 16 Springer, Berlin-Heidelberg-New York, B. Parthier and D. Boulter, Editors (in press).
6. W. Nagl, in: "Genome and Chromatin: Organization, Evolution, Function", W. Nagl, V. Hemleben and F. Ehrendorfer, editors, p. 247-260, Springer, Vienna-New York (1979).
7. W. Nagl, in: "Endopolyploidy and Polyteny in Differentiation and Evolution" North-Holland, Amsterdam (1978).
8. W. Nagl, Zellkern und Zellzyklen. Ulmer, Stuttgart (1976).
9. J. Sedat and L. Manuelidis, Cold Spring Harb. Symp. Quant. Biol., 42: 331-350 (1977).
10. W. Nagl, Chromosomen. Paul Parey, Berlin (1980).
11. A.L. Bak and J. Zeuthen, Cold Spring Harb. Symp. Quant. Biol. 42:367-377 (1977).
12. J.R. Paulson and U.K. Laemmli, Cell, 12:817-828 (1977).
13. J.T. Finch and A. Klug, Proc. Natl. Acad. Sci. US, 73:1897-1901 (1976).
14. W. Sawicki, J. Rowinski and J. Abramczuk, J. Cell Biol., 63:227-233 (1974).
15. F. Kendall, R. Swenson, T. Borun, J. Rowinski and C. Nicolini Science, 196:1106-1109 (1977).
16. C.A. Nicolini, F. Kendall and W. Giaretti, Biophys. J., 19:163-176 (1977).
17. J.S. Ploem, N. Verwoerd, J. Bonnet and G. Koper, J. Histochem. Cytochem. 27:136-143 (1979).
18. I. Al, C.J. Cornelisse, P.L. Pearson and J.S. Ploem, Acta Histochem. Suppl., 20:211-215 (1978).
19. K.D. Kunze, W.R. Herrmann, and W. Meyer, Arch. Geschwulstforsch. 48:131-139 (1978).
20. A.W.M. Smeulders, L. Leyte-Veldstra, J.S. Ploem and C.J. Cornelisse, J. Histochem. Cytochem. 27:199-209 (1979).
21. J.H. Tucker, J. Histochem. Cytochem., 27:613-620 (1979).
22. W. Vahs, Histochemie, 33:341-348 (1972).
23. L. Dennhofer, Plant Syst. Evol., Suppl. 2:91-84 (1979).
24. W. Nagl, Protoplasma, 100:53-71 (1979).
25. W. Nagl and K. Bachmann, Theor. Appl. Genet. xx,xx-xx (1980).
26. W. Nagl, and F. Hoffmann, Europ. J. Cell Biol. xx,xx-xx (1980).
27. J. Doutreligne, Cellule, 48:191-214 (1939).
28. E. Tschermak-Woess, "Strukturtypen der Ruhekerne von Pflanzen und Tieren.", Springer, Vienna-New York (1963).
29. R. Tanaka, Bot. Mag. Tokyo, 84:118-122 (1971).

30. P. Barlow, Ann. Sci. Nat., Bot. Biol. Veget., Ser. 12, 18:193-206 (1977).
31. J.G. Lafontaine, in: "The Cell Nucleus", Vol. 1, p. 149-185, Academic Press, New York (1974).
32. W. Nagl and H.P. Fusenig, in: "Genome and Chromatin: Organization, Evolution, Function", W. Nagl, V. Hemleben, F. Ehrendorfer, editors, p. 221-233, Springer, Vienna-New York (1979).
33. W. Nagl, Microscop. Acta xx,xx-xx (1980).
33a. Risueno and S. Moreno Diaz de la Espina, J. Submicr. Cytol., 11:85-98 (1979).
33b. J. Sans and C. de la Torre, Europ. J. Cell Biol., 19:294-298 (1979).
34. K. Brasch, Ninth Int. Congr. Electr. Micr., Toronto, Vol. 2: 208-209 (1978).
35. M.D. Bennett, Brookhaven Symp. Biol., 25:344-366 (1973).
36. T. Cavalier-Smith, J. Cell Sci., 34:247-278 (1978).
37. W. Nagl, The Nucleus, 20:10-27 (1977).
38. W. Nagl, Nature, 261:614-615 (1976).
39. F. Ehrendorfer, in: "Origin and Early Evolution of Angiosperms" C.B. Beck, Editor, p. 220-240, Columbia Univ. Press, New York (1976).
40. G.M. Evans, H. Rees, C.L. Snell and S. Sun, Chromosomes Today, 3:24-31 (1972).
41. H.C. Macgregor, S. Mizuno and M. Vlad, Chromosomes Today, 5:331-339 (1976).
42. M. Blumenfeld, J.W. Orf, B.J. Sina, R.A. Kreber, M.A. Callahan, and L.A. Synder, Cold Spring Harb. Symp. Quant. Biol., 42:273-276 (1977).
43. M.I. Lerman and S.V. Degtyarev, Mol. Biol. Rep., 4:117-120 (1978).
44. B. Fuhrmann and F. Nagl, in: "Genome and Chromatin: Organization, Evolution, Function", W. Nagl, V. Hemleben and F. Ehrendorfer, editors, p. 235-245, Springer, Vienna-New York (1979).
45. J. Ingle, J.N. Timmis, and J. Sinclair, Plant Physiol., 55:496-501 (1975).
46. I. Capesius and W. Nagl, Plant Syst. Evol., 129:143-166 (1978).
47. J.P. Miksche, Can. J. Genet. Cytol., 9:717-722 (1967).
48. M.D. Bennett, Environm. Exptl. Bot., 16:93-108 (1976).
49. W. Nagl, Nature, 249:53-54 (1974).
50. W. Nagl, Devel. Biol., 39:342-345 (1974).
51. H.J. Price and K. Bachmann, Plant Syst. Evol., 126:323-330 (1976).
52. T. Ohta and M. Kimura, Nature, 233:118-119 (1971).
53. R. Hinegardner, in: "Molecular Evolution" F.J. Ayala, editor, p. 179-199, Sinauer Ass., Sunderland, MA (1976).
54. K. Brasch, Cell Biol. Int. Rep., 4:217-226 (1980).

55. J.C. Jeanny and M. Gontcharoff, Wilhelm Roux's Arch., 184: 195-211 (1978).
56. M.M. Al-Hamdani, M.E. Atkinson and T.M. Mayhew, Cell Tiss. Res., 200:495-509 (1979).
57. J.M. Teich, I.T. Young, S.E. Sher and J.S. Lee, J. Histochem. Cytochem., 27:193-198 (1979).
58. M.D. Kendall, Cell Tiss. Res., 199:63-74 (1979).
59. B. Schmalenberger and W. Nagl, in: "Genome and Chromatin: Organization, Evolution, Function", W. Nagl, V. Hemleben, and F. Ehrendorfer, editors, p. 119-125, Springer, Vienna-New York (1979).
60. G. Kiefer and W. Sandritter, Acta Histochem. Suppl., 20: 193-201 (1978).
61. B. Schmalenberger, Protoplasma, xx,xx-xx (1980).
62. T.C. Hsu, Exptl. Cell Res., 27:332-334 (1962).
63. V.G. Allfrey, V.C. Littau and A.E. Mirsky, Proc. Natl. Acad. Sci. US, 49:414-421 (1963).
64. K. Brasch and G.D. Sinclair, Virchow's Arch. B Cell Pathol. 27:193-204 (1978).
65. M. Derenzini, E. Lorenzoni, V. Marinozzi and P. Barsotti, J. Ultrastr. Res., 59:250-262 (1977).
66. M. Derenzini, F. Novello and A. Pession-Brizzi, Exptl. Cell Res., 112:443-454 (1978).
67. A.J. Louie and G.M. Dixon, Nature New Biol., 243:164-168 (1973).
68. M.G. Ord and L.A. Stocken, Biochem. J., 178:173-185 (1979).
69. H. Busch, Editor "The Cell Nucleus", Vol. 4 and 5, Chromatin Academic Press, New York (1978).
70. K.G. Wagner, J. Bode, K.W. Hoffmann and L. Willmitzer, Studia Biophys. 67:85-86 (1978).
71. J. Bode, L. Willmitzer, H.A. Arfmann and K. Wehling, Intern. J. Biol. Macrimol., 1:73-78 (1979).
72. H. Fuge, Europ. J. Cell Biol., xx,xx-xx (1980).
73. N.A. Berger, S.J. Petzold and S.J. Berger, Biochim. Biophys. Acta, 564:90-104 (1979).
74. A.I. Caplan, C. Niedergang, H. Okazaki and P. Mandel, Devel. Biol., 72:102-109 (1979).
75. B. Farina, M.R.F. Menella and E. Leone, in: "Macromolecules in the Functioning Cell", F. Salvatore, G. Marino and P. Volpe, editors, p. 283-300, Plenum Press, New York (1979).
76. D.E. Comings, Human Genet., 53:131-144 (1980).
77. W. Sandritter, Funktionsstrukturen des Zellkerns. Schattauer, Stuttgart (1970).
78. D.B. Brown, S.M. Stack, J.B. Mitchell and J.S. Bedford, Cytobiologie, 10:398-412 (1979).
79. C. Cremer, T. Cremer, C. Zorn and V. Cioreanu, Hoppe-Seyler's Z. Physiol. Chem., 360:244-245 (1979).
80. A.B. Murray and H.G. Davies, J. Cell Sci., 35:59-66 (1979).
81. K. Brasch and G. Setterfield, Exptl. Cell Res., 83:175-185 (1974).

82. D.E. Comings and A.S. Wallack, J. Cell Sci., 34:233-246 (1978).
83. F. Wunderlich, Naturwiss. Rdschau, 31:282-288 (1978).
84. R. Berezney, in: "The Cell Nucleus", H. Busch, Editor, vol. 7, p. 413-456, Academic Press, New York (1979).
85. F.J. Ayala, Editor, "Molecular Evolution" Sinauer Ass., Sunderland, MA (1976).
86. W. Nagl, in: "Genome and Chromatin: Organization, Evolution, Function" W. Nagl, V. Hemleben, F. Ehrendorfer, editors, p. 3-25, Springer, Vienna-New York (1979).
87. W. Nagl, Chromosoma, 28:85-91 (1969).
88. M. Derenzini, A. Pession-Brizzi, E. Bonetti and F. Novello, J. Ultrastr. Res., 67:161-179 (1979).
89. B. Wright, Trends Biochem. Res., 4:N110-N111 (1979).
90. E.J. DuPraw, in: "DNA and Chromosomes", Holt, Rinehart and Winston, Inc., New York (1970).
91. W. Nagl, in: "Mechanisms and Control of Cell Division", T.L. Rost and E.M. Gifford, Jr., editors, p. 147-193, Dowden, Hutchinson and Ross, Inc., Stroudsburg, PA (1977).
92. W. Sawicki, in: "Chromatin Structure and Function" C. Nicolini, editor, p. 667-681, Plenum Press, New York (1979).
93. D.E. Comings, in: "The Cell Nucleus", H. Busch, Editor, vol. 1., p. 537-563, Academic Press, New York (1974).
94. B. Lewin "Gene Expression-2. Eucaryotic Chromosomes", John Wiley and Sons, London, (1974).
95. M.A. Person, and D.T. Suzuki, Can. J. Genet. Cytol., 10: 627-647 (1968).
96. D.A. Filip, C. Gilly and C. Mouriquand, Humangenetik, 30: 155-165 (1975).
97. A.L. Bak, B. Bak and J. Zeuthen, J. Theor. Biol., 76:205-217 (1979).
98. P. Bak, A.L. Bak and J. Zeuthen, Chromosoma, 73:301-315 (1979).
99. J. Sonnenbichler, Nature, 223:205-206 (1969).
100. K.W. Adolph, S.M. Cheng and U.K. Laemmli, Cell, 12:806-816 (1977).
101. J.R. Paulson and U.K. Laemmli, Cell, 12:817-828 (1977).
102. M.P.F. Marsden and U.K. Laemmli, Cell, 17:849-858 (1979).
103. P.G.N. Jeppersen and A.T. Bankier, Nucleic Acids Res., 7: 49-67 (1979).
104. W. Kraut and D. Werner, Biochim. Biophys. Acta, 564:390-401 (1979).
105. D.E. Comings, Ann. Rev. Genet., 12:25-46 (1978).
106. D. Schweizer, Chromosoma, 58:307-325 (1976).
107. W. Schnedl, Intern. Rev. Cytol., Suppl. 4:237-272 (1974).
108. J. Greilhuber, Theor. Appl. Genet., 50:121-124 (1977).
109. D. Schweizer, in: "The Plant Genome", R. Davies and R. Markham, editors, p. 61-72, Academic Press, London (1979).

110. T.R. Mellem and M.M. Laane, Mikrokosmos, 68:111-113 (1979).
111. L.R. Gurley, J.A. D'Anna, S.S. Barham, L.L. Deaven and R.A. Tobey, Eur. J. Biochm., 84:1-15 (1978).
112. D. Mazia, J. Cell. Comp. Physiol., Suppl. 1:123-140 (1963).
113. H.W. Altmann, Verhdl. Deutsch. Ges. Pathol., 50:15-51 (1966).
114. W. Nagl, Oesterr. Bot. Z., 115:322-353 (1968).
115. W. Nagl, Caryologia, 23:71-78 (1970).
116. W. Nagl, Protoplasma, 91:389-407 (1977).
117. G. Setterfield, R. Sheinin, I. Dardick, G. Kiss and M. Dubsky, J. Cell Biol., 77:246-264 (1978).
118. C. Nicolini, K. Ajiro, T.W. Borun and R. Baserga, J. Biol. Chem., 250:3381-3385 (1975).
119. W. Nagl, Chromosoma, 44:203-212 (1973).
120. E. Heitz, Jahrb. Wiss. Bot., 69:762-818 (1928).
121. A. Lima-de-Faria, Hereditas, 81:249-284 (1975).
122. A. Lima-de-Faria, in: "Specific Eukaryotic Genes" J. Engberg, H. Klenow and V. Leick, editors, p. 25-38 Munksgaard, Copenhagen (1979).
123. J.J. Yunis and W.G. Yasmineh, Science, 174:1200-1209 (1971).
124. S.G. Smith, J. Hered., 36:194-196 (1945).
125. M.F. Lyon, Ann. Rev. Genet., 2:31-52 (1968).
126. M.L.S. Mello, Ciencia e Cultura, 30:290-303 (1978).
127. K. Carniel, Plant Syst. Evol., 131:235-242 (1979).
128. K. Carniel, Plant Syst. Evol., 133:301-307 (1980).
129. T.S. Hsieh and D.L. Brutlag, Proc. Natl. Acad. Sci. US, 76:726-730 (1979).
130. D. Brutlag, M. Carlson, K. Fry and T.S. Hsieh, Cold Spring Harb- Symp. Quant. Biol., 42:1137-1146 (1977).
131. W.J. Peacock, A.R. Lohe, W.L. Gerlach, P. Dunsmuir, E.S. Dennis and R. Appels, Cold Spring Harb. Symp. Quant. Biol., 42:1121-1135 (1977).
132. B. John and G.L.G. Miklos, Intern. Rev. Cytol., 58:1-114 (1979).
133. G.L.G. Miklos, and B. John, Amer. J. Human Genet., 31:264-280 (1979).
134. K. Hagele, Chromosoma, 59:207-216 (1977).
135. B. Deumling and W. Nagl, Cytobiologie, 16:412-420 (1978).
136. C. Bostock, Ternds Biochem. Res., 5:117-119 (1980).

DISCUSSION (SECTION II)

PRESCOTT: Have you found evidence of turnover of DNA labelled with 3H-TdR independent of cell reproduction?

MAURER-SCHULTZ: Positively not. We can find no evidence of turnover of DNA independent of cell reproduction although we looked very carefully for it. Although Pelc claimed that there might be turnover of metabolic DNA and some incorporation of 3H-TdR into neurons we did experiments prelabelling precursors of rats with a lot of hot thymidine during the embryonic development and then killed our animals between 1 and 90 days after birth and then counted the mean grain number per nucleus and did not find any decrease in grain number. So there is no metabolic turnover of DNA. Again contrary to Pelc, we showed by double labeling that in the mouse small intestinal crypt, each labeled cell divides. Also although we examined thousands of neurons after 3H-TdR we could find no labeled cells.

PRESCOTT: Are there diurnal variations which affect measurements of cell turnover?

MAURER-SCHULTZ: Of course we have diurnal variations of mitotic and labeling index, but with the usual cell kenetic methods, the diurnal variations do not disturb us too much. With the PLM curve, we only measure the relative percentage of the labeled mitosis. So if the M.I. is somewhat down because of diurnal variation we don't measure that in the PLM curve.

ZIETZ: One comment to your last remarks, the PLM measurement strongly depends on the fact that the cells are in asynchronous growth. If we have diurnal variations, we are no longer in this situation and thus the PLM can yield marked differences from the true parameters.

MAURER-SCHULTZ: I totally agree. However, the cell types we measure did not have such variation.

ALABASTER: In a paper in Nature in December 1975, it was printed out that thymidine disturbed the natural progression of cells through the cycle. Also, clinical studies are being done using thymidine as a chemotherapeutic agent for leukemia. Therefore, thymidine may affect cell cycle progression. Would you care to comment?

MAURER-SCHULTZ: Yes, it maybe, although if you give thymidine continuously you have to give it in a big amount because it is catabolized very quickly.

SCHREIL: Does grain density as measured in your experiments on amino acid incorporation require standardization of exposure time of the emulsion on the tissue or is the activity so high that the emulsion exposure time is neglectable?

MAURER-SCHULTZ: Of course if we are to compare the incorporation into the different cell types we must use the same exposure time. We might try to correct for different exposure times, however, this require standardization of the emulsion. However, we clunge on the same exposure time for the tissue we compare.

NICOLINI: *Since radioactivity is emited isotropically, how do you know that twin labeled cells within proximity of one another are not due to cross contamination?*

MAURER-SCHULTZ: It is for that reason that we use squashed preparations and only consider cells at 2 diameters apart.

NICOLINI: I wonder if 2 diameters is enough, especially for your thick emulsion used for double labeling.

ALABASTER: I would like to know if you would agree that the labeling index is subject to many errors, like emulsion exposure, pool size, grain correcting, etc. Often the labeling index is presented as some sacrosant number, while in fact you can change it by altering your experimental conditions.

MAURER-SCHULTZ: I don't agree. If you use the method correctly, the results are correct. We have no problems with the back-ground. At least for a labeling index measurement, we can give enough thymidine so that unambiguous results can be obtain after an expose time of 3 days.

BARLATTI: *Is there any difference in labeling depending on the method of injection of the radioisotope?*

MAURER-SCHULTZ: No. The same results are obtained whether the dose is given IP of IV. Even for aminoacids the same results are obtained for IP or IV injection, or even if you feed the animals the material.

ALABASTER: You seemed to indicate this morning that you may measure the growth fraction from labeling studies.

MAURER-SCHULTZ: Yes, you must get the quotient of the labeling index divided by the proliferating fraction. The latter you get by dividing the S phase cells/cycle time. There is no other method to measure the growth fraction. Mendelson measured the growth fraction from continuously labeling experiments, but I don't like that method since non-proliferating cells must eventually be labeled to keep a constant growth fraction.

ZIETZ: I must remark that many including Fried and myself have shown that estimation of the growth fraction by any labeling technique is a extremely tricky and ill conditioned problem and thus any such estimate is probably unreliable. As you are probably aware, using techniques from automated cytology, as shown in this Institute by Nicolini, we may now objectively and accurately measure the growth fraction even in perturbed or drug treated populations.

SECTION III:
THE CELL CYCLE

GROWTH AND DIVISION OF *ESCHERICHIA COLI*

N. Nanninga, C.L. Woldringh and L.J.H. Koppes

Department of Electron Microscopy and Molecular Cytology
University of Amsterdam
Plantage Muidergracht 14, 1018 TV AMSTERDAM
The Netherlands

INTRODUCTION

Though from a structural point of view, the prokaryotic cell is quite different from the eukaryotic one, both types do not differ with respect to the most essential characteristics of the living cell. These are genome duplication and cell division. It seems reasonable to expect that the regulation of both processes will underly the same principles in the two groups of organisms. Of course numerous differences occur on the chemical and structural level and such details will count to reach a general understanding of genome duplication and cell division.

Mitchison (1973) made a useful more or less explicit distinction between cytodifferentiation and chemodifferentiation. Whereas in the first process a given type of cell changes into another one, the last process is confined to one cell type. Chemodifferentiation comprises the chemical and macromolecular arrangements during cell growth, i.e. from the newborn cell to the dividing cell and thus during the division cycle. (Strictly speaking cytodifferentiation is also based on chemodifferentiation).

Examples of bacterial cytodifferentiation are *Caulobacter crescentus* which comprises three cell types (review: Dow and Whittenbury, 1979; Dworkin, 1979), sporulation in Grampositives like *Bacillus subtilis* (Chambliss, 1979) and fruiting body formation in *Myxococcus xanthus* (Dworkin, 1979; Parish, 1979). The latter system has striking similarities with respect to *Dictyostelium discoidum:* upon starvation cells aggregate to form a fruiting body.

In this paper we will limit ourselves to chemodifferentiation as defined above. Chemodifferentiation is being studied in bacteria like *Escherichia coli*, *Bacillus subtilis* (Mendelson, 1977; Burdett, 1979; Sargent, 1979; Nanninga et al., 1979), *Streptococcus faecalis* (Daneo-Moore and Shockman, 1977; Higgins and Daneo-Moore, 1980) and *Staphylococcus aureus* (Giesbrecht et al. 1976). The references apply to some very recent papers on the respective organisms.

This paper will be particularly concerned with the rod-shaped *E. coli*. Its structure is schematically depicted in Fig. 1. Note the absence of a nuclear envelope and the presence of a cell wall. The latter maintains the rod-shape

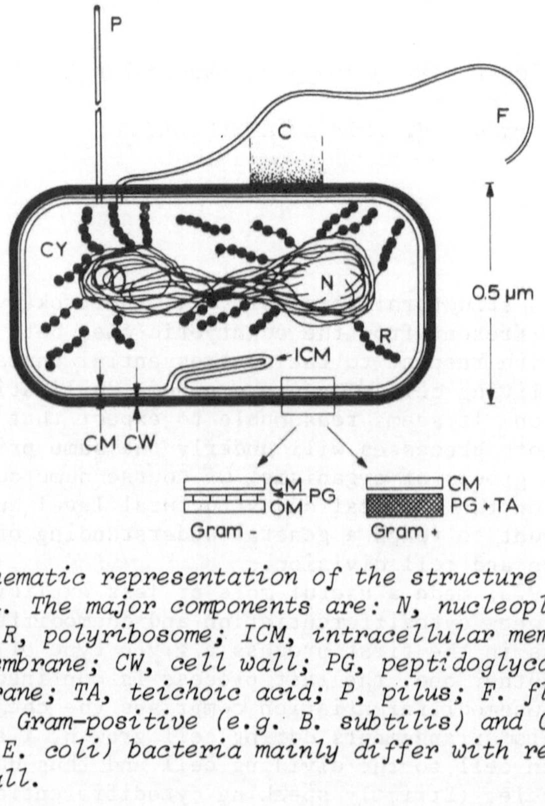

Fig. 1 Schematic representation of the structure of a bacterial cell. The major components are: N, nucleoplasm; CY, cytoplasm; R, polyribosome; ICM, intracellular membrane; CM, cell membrane; CW, cell wall; PG, peptidoglycan; OM, outer membrane; TA, teichoic acid; P, pilus; F. flagellum; C, capsule. Gram-positive (e.g. B. subtilis) and Gram-negative (e.g. E. coli) bacteria mainly differ with respect to the cell wall.

which is of a considerable experimental advantage. Because changes in diameter are small during the division cycle (Trueba and Woldringh, 1980) measurement of length stands for size. We will start with some general aspects of the

E. coli division cycle (Fig. 2) and proceed then more specifically with envelope growth and DNA replication. This is followed by a discussion of the coordination of the two events. Finally some of the outstanding problems will be presented.

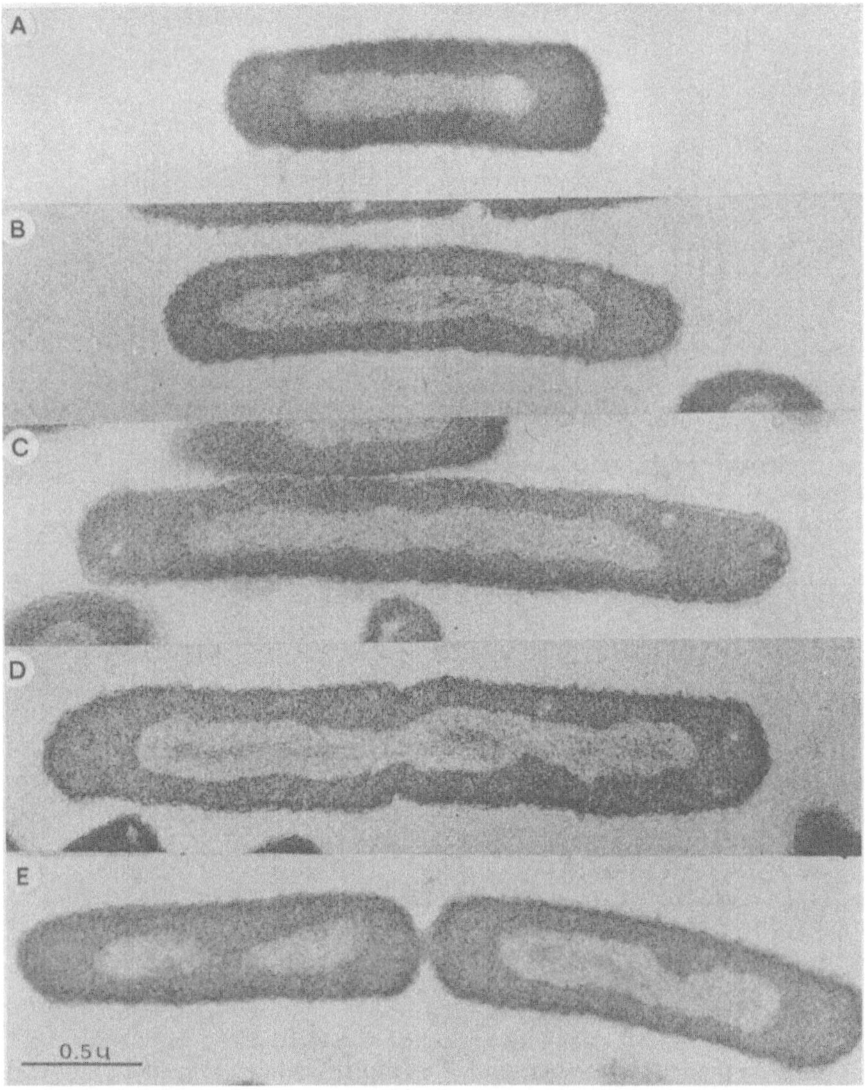

Fig. 2 Simulation of the division cycle of E. coli (T_D = 60 min) by sectioned cells. Note initiation of cell constriction before nucleoplasmic separation in D.

GENERAL

For the division cycle periods of *E. coli* a different terminology is used as compared to the eukaryotic cell (Fig. 3). B is defined as the period between birth and the onset of DNA replication and equals G1. C represents the DNA replication period and equals S. The D period lies between

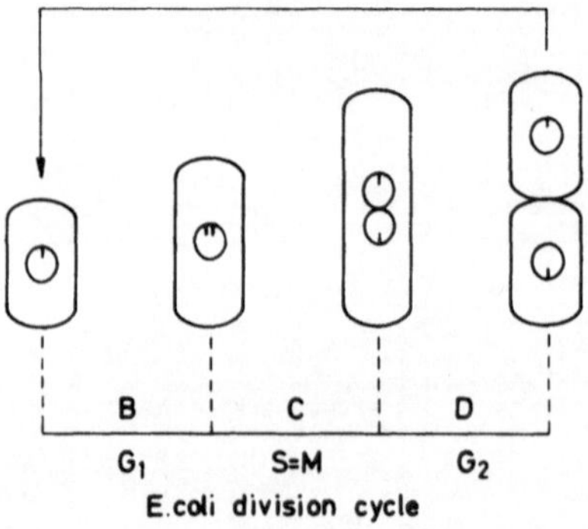

Fig. 3. Terminology of E. coli division cycle. B, period between onset of DNA replication; C, DNA replication period; D, period between termination of DNA replication and cell separation. Compare with G_1, S = M and G_2, respectively.

termination of DNA replication and cell division (more properly: cell separation) and is largely the equivalence of G2. Note that there is no separate M period. In bacteria DNA replication and DNA segregation go hand in hand. Thus the genomes appear spatially separated when DNA replication has terminated (Woldringh, 1976).

Genome segregation

Centrioles or spindle plaques as they occur in *Saccharomyces cerevisae*, have not been demonstrated in bacteria. This is not surprising since the above structures have such a size that, if present, they would occupy a too predominant part of the bacterial cell. Microtubules would fit, but they have likewise not been observed. The early concept (Jacob, Brenner and Cuzin, 1963) that the cell surface might function as a mitotic equivalent is still relevant. It implies that replicating (and segregating) DNA is attached to the

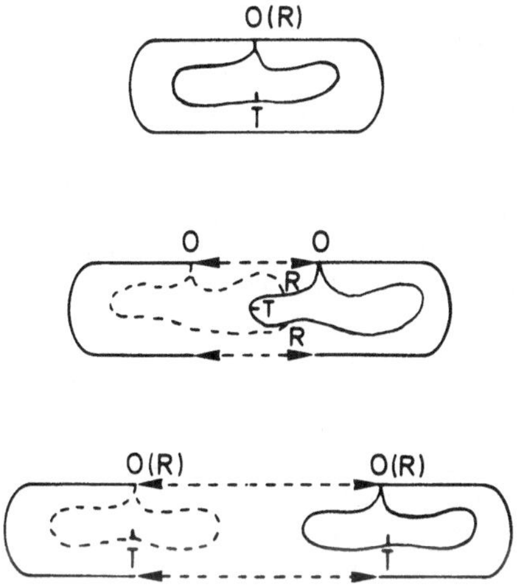

Fig. 4 Coordination of DNA replication and DNA segregation. Envelope growth between DNA envelope attachment points results in segregation. O, origin of DNA replication; R, replication fork.

cell envelope. Growth of the cell would separate the newly replicated parts of the DNA molecule (Fig. 4). This notion has led to numerous attempts to demonstrate a chemical and/ or physical link between DNA and the cell membrane. For a comprehensive review see Leibowitz and Schaechter (1975). It has to be admitted that though progress has been made, this problem is not yet conclusively resolved (Meijer et al. 1976a, 1976b; Seiki et al. 1979; Winston and Sueoka, 1980; Nagai et al., 1980). A recent approach is to study the relationship of sites near the origin of replication of *E. coli* (cloned in plasmids as *ori C* Yasuda and Hirota, 1977; von Meyenburgh et al., 1978; Messer et al. 1978) to outer membrane proteins (Wolf-Watz and Masters, 1979; Jacq et al., 1980). However, these minichromosomes are defective in segregation, because they are easily lost when the selective pressure is removed (Messer et al., 1978). Removal of the defect will contribute to understanding minichromosome partitioning (Meacock and Cohen, 1980; Ogura et al., 1980).

We have digressed somewhat to bacterial genome segregation in order to facilitate comparison with the eukaryotic system.

The Helmstetter-Cooper model

An important and stimulating framework within which to place the *E. coli* division cycle is the Helmstetter-Cooper model (Cooper and Helmstetter, 1968; Helmstetter et al., 1968). This model relates growth rate, cell size, initiation of DNA replication and cell division. It contains three constants I, C and D. I represents the time (min) to accumulate a so-called "initiator complement" of constant size per chromosomal origin (Helmstetter et al., 1968). I is growth rate dependent and equals the doubling time (T_D). C represents the duration of the DNA replication period and is about 40 min at 37°C and independent of growth rate for T_D's between 20 and 60 min. C is growth rate dependent for $T_D > 60$ min and increases proportionally (Helmstetter and Pierucci, 1976). D, the period between termination of DNA replication and cell division is about 20 min for T_D's of 20 to 60 min. It should be noted that D does not stand for a defined biochemical process like C, though it is clear that during D at least part of the preparatory processes for cell division take place. From the foregoing it will be clear that cell division ensures after I + C + D minutes. A complication arises when T_D is smaller than 60 min, then DNA replication should start in the previous cycle. A further complication arises when T_D is smaller than 40 min. This problem is solved by multifork replication i.e. a new DNA replication cycle starts before the previous one has ended. These aspects are schematically shown in Fig. 5 for T_D's of 60, 40 and 20 min, respectively. For simplicity we have omitted I in this figure. The model has been worked out for *E. coli* B/r A (ATCC 12407). Up to a T_D of 60 min it also applies to other B/r strains (Helmstetter and Pierucci, 1976). Differences occur at slower growth which amongst others led to the introduction of the B period (Fig. 3 and Fig. 22). A derivate of the Helmstetter-Cooper model and the growth rate dependence of size is that initiation of DNA replication starts at a constant mass (M_i; Donachie, 1968) or when a unit of initiator has accumulated (Helmstetter et al., 1968). Thus, when a cell has two genomes (Fig. 5), the simultaneous start of two DNA replication cycles would require a cell size equal to two times M_i.

Before discussing some more specific points we would like to draw attention to some valuable recent reviews on the bacterial division cycle: Daneo-Moore and Shockman 1977;

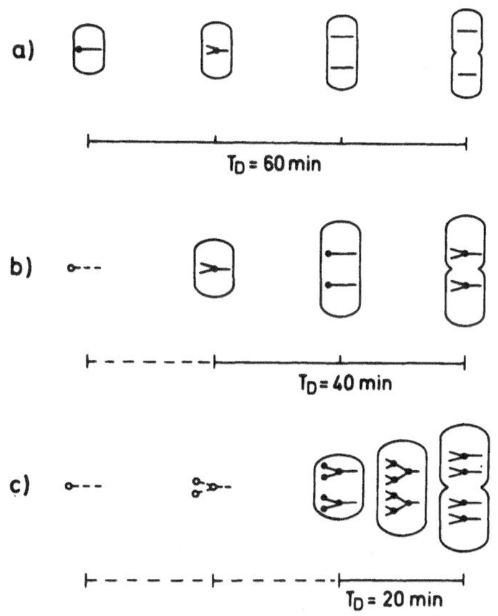

Fig. 5 Relation between DNA replication pattern and doubling time in E. coli B/r A (Helmstetter and Cooper, 1968). DNA replication takes 40 min and cell division follows 20 min later at the three respective doubling times. Initiation starts after the elaps of one doubling time, Note that cells become bigger and that the number replication forks increases.

Mendelson, 1977; Koch, 1977; Donachie, 1979; Helmstetter et al., 1979; Sargent, 1979.

FAMILY RELATIONS

A consequence of the I + C + D-concept (Helmstetter et al., 1968; Helmstetter et al., 1979) is that for a given division cycle preparations have been made in previous ones. For instance for a doubling time of 20 min the I period started two generations ago which implies the involvement of a heredity element. Thus, a newly born cell does not have a fresh start and the term cell cycle seems, therefore, misleading (Helmstetter et al., 1968, 1979; Bleecken 1971; Cooper, 1979; Koppes et al., 1980). Basically there is a linear sequence of events $I_1 + C_1 + D_1$ after the onset of I_1; I_2 starts T_D minutes later, etc. These partly parallel-running sequences are interrupted by divisions (Newman and

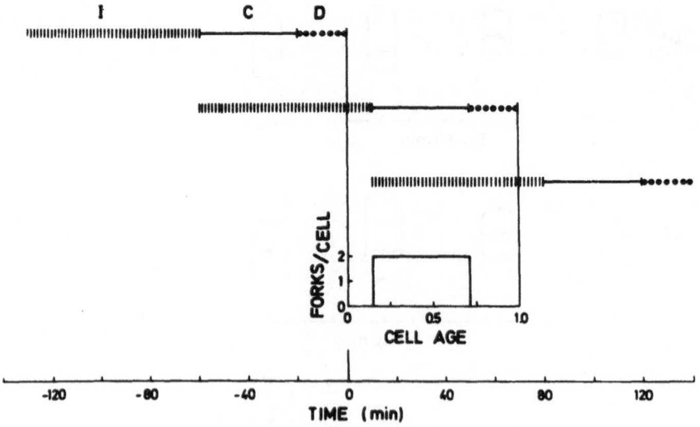

Fig. 6 Replication and cell division patterns for E. coli with I = 70 min, C = 40 min, D = 20 min. The construction is based on the I + C + D sequence beginning at -130 min with a hypothetical cell containing a single chromosome (cf. Fig. 5). Preparation for replication requires I minutes (hatched line), replication occupies C minutes (solid line), and the cell divides D minutes later (dotted line). At the time replication begins (-60 min), preparation for the next (I + C + D) sequence commences. The vertical lines indicate divisions. The rate of replication is shown in terms of the number of forks per cell. (cf. Fig. 1 in Helmstetter et al., 1979)

Kubitschek, 1978; Koppes et al., 1980). Schematically this is depicted in Fig. 6.

We feel that the above considerations should be kept in mind with respect to generation time correlations. For instance, between sisters and sisters and between mothers and daughters. Note that the doubling time (T_D) refers to the culture as a whole. T_D can be determined by measuring optical density or cell number in a steady state culture. Generation time (τ) refers to the length of the division cycle of individual cells. This information can be obtained by growing cells in a microculture under the light microscope. The term interdivision time is used for individual cells (τ) as well as growing cells ($\bar{\tau}$), the average generation time.

Direct measurements on individual cells have been carried out in the late fifties and early sixties (Powell, 1956; Powell, 1958; Schaechter et al. 1962; Kubitschek, 1962). The correlation coefficient of the generation times of sister-pairs was found to be positive, in the order of 0.5 (Schaechter et al., 1962; Kubitschek, 1962; Powell, 1964).

With respect to mother-daughter relationships the results are variable (same references). Kubitschek (1962) pointed out that in the case of extreme sizes negative correlations apply, whereas no correlation exists between the generation times of medium sized mothers and daughters. Unfortunately, these pioneering studies have not been continued within the framework of the Helmstetter-Cooper model (1968).

It is reasonable to assume that the above correlations are modulated by cell size (growth rate dependent) and cell shape. Systematic studies with well characterized strains and steady state growth conditions are urgently needed. We will continue this subject in the section on cell division (see below). At the moment we wish to emphasize that a cell should not be looked upon as a separate entity (cell cycle) but rather as a dependent element in a family tree, each cell having its own history.

ENVELOPE GROWTH

The cell envelope of *E. coli* comprises three layers (Fig. 1): the cell membrane, the murein (peptidoglycan) and the outer membrane (for a recent collection of reviews see Inouye, 1979). The murein layer and outer membrane together constitute the cell wall. The murein layer is composed of

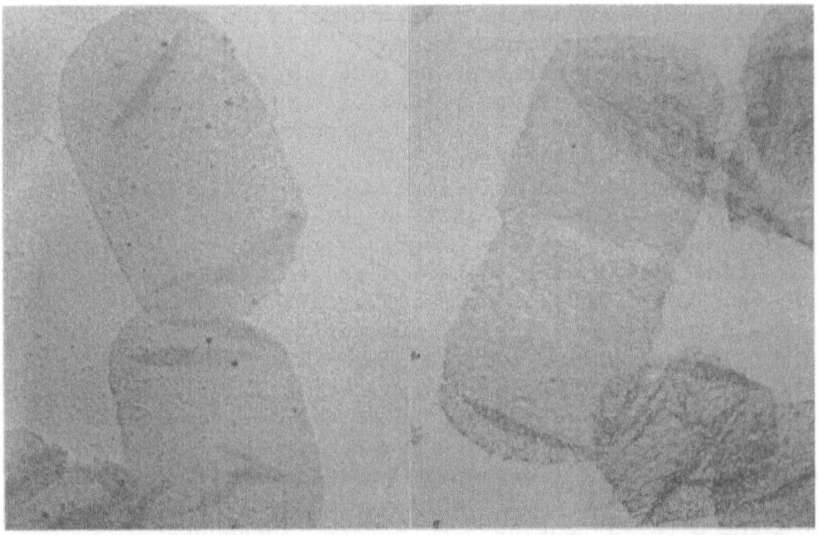

Fig. 7 (a) Isolated sacculus; (b) sacculus treated with endopeptidase. This enzyme disrupts the peptide bridges between the flycan chains (cf. Verwer et al., 1978). Magnification: 30.000 x

glycan chains which are interconnected with short peptides. The whole layer is a covalently closed structure which in purified form still maintains the bacterial rod-shape (*sacculus*; Weidel and Pelzer, 1964; Fig. 7a). In the *sacculus* the glycan chains run more or less parallel to the length axis of the cell in a tangential way (Fig. 7b, Verwer et al., 1978; Verwer et al., 1980). Cytoplasmic and outer membrane possess a lipid bilayer structure in which various components are integrated. The integrity of the membrane layers is maintained by non-covalent bonds, whereas lipoprotein links the *sacculus* with the outer membrane. The association of cytoplasmic membrane and *sacculus* is less well understood.

For surface extension of the *sacculus* hydrolytic and synthesizing enzymes will be needed to disrupt and to reconnect covalent bonds, respectively. By contrast, surface extension of the two membranes will be essentially different. Presumably the *sacculus* acts as a kind of scaffold for the outer membrane (Yamada and Mizushima, 1978). Growth of the envelope thus requires coordination between surface extension of the three respective layers. So far this coordination is a little explored field (cf. Boyd and Holland, 1979).

Surface extension can be studied on two different levels. One may pose the question, irrespective of envelope structure, how the cell increases its size, thereby determining and changing its shape. Alternatively, there is the question where and when building stones are incorporated into one of the respective envelope layers. Is incorporation everywhere or are there discrete growth zones?.

Doubling in size can be achieved in numerous ways, which may be different for either mass, volume or surface. Mass and volume have been assumed to increase exponentially (Ecker and Kokaisl, 1969; Pritchard, 1974), whereas linear models have been proposed for both surface growth (Rosenberger et al., 1978a) and volume growth (Kubitschek, 1968). In the exponential model for surface growth (Koch and Schaechter, 1962; Koppes et al., 1978) the number of growth sites increases with cell size. This is comparable to the exponential increase in the number of ribosomes during growth. In the linear models the rate of surface synthesis doubles at a particular time once during the division cycle. The activity of such a site may be constant or growth rate dependent.

The various growth models have to explain two basic properties which characterize rod-shaped Gram-negative cells growing under steady-state conditions: *average* cell size ($\bar{M}$, $\bar{V}$) increases with increasing growth rate (Schaechter, Maaløe and Kjeldgaard, 1958) and the surface to volume ratio ($\bar{A}/\bar{V}$) decreases (Schaechter et al., 1958; Grover et al., 1977). These two penomena are illustrated here for *E. coli* B/r H266 (Fig. 8), but have been reported for other strains as well (Woldringh et al., 1977; Donachie et al., 1976).

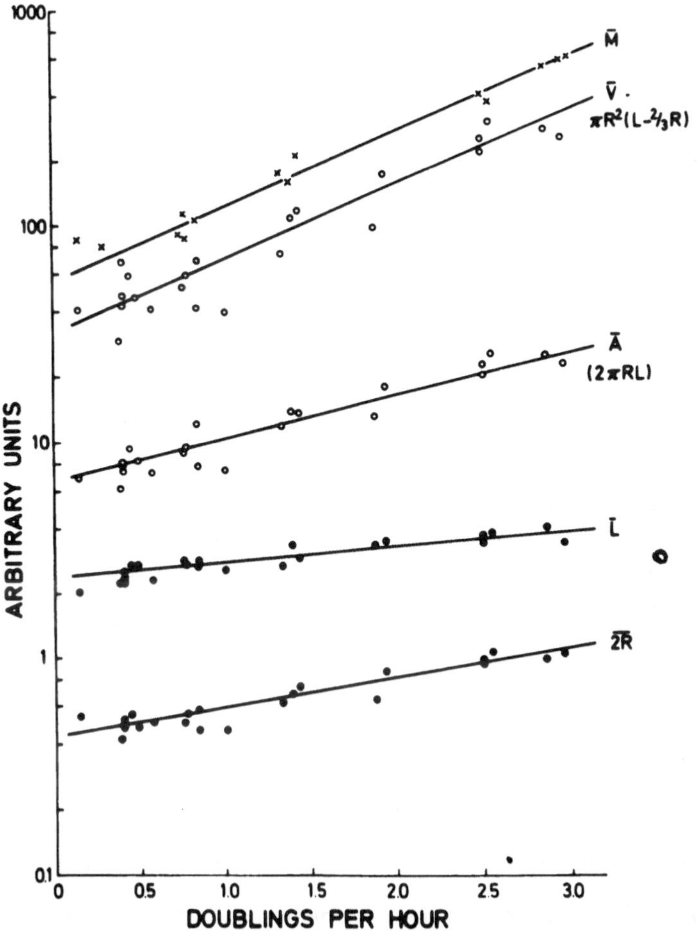

Fig. 8 Average cell mass ($\bar{M}$, 450 nm absorbance units per 10^9cells), volume ($\bar{V}$ m^3), surface ($\bar{A}$ m^2), length ($\bar{L}$ m) and diameter ($\overline{2R}$ m) of E. coli B/r H266 as a function of growth rate. Lines drawn by eye.

In the early studies of Collins and Richmond (1962) and of Schaechter et al., (1962) length increase of individual cells in a microculture was followed with a light microscope. Later synchronuously growing cells were used to measure cell volume electronically with a Coulter Counter (Kubitschek, 1968; Ward and Glaser, 1971) and total cell length with a light microscope (Donachie et al., 1976). Another approach has been to determine the growth kinetics from the shape of the size distribution of a population growing in steady state (Collins and Richmond, 1962; Harvey et al., 1967; Koch, 1977; Koppes et al., 1978a, 1978b). More indirectly, one may study the variation of cellular dimensions with growth rate (Fig. 8) or the mode of adjustment when cells are shifted from one growth rate to another (Woldringh et al., 1980). We will start our discussion with the two latter approaches.

Surface growth and growth rate

The rationale of this approach to overall size increase is to compare cell size as a function of growth rate (Donachie et al., 1976; Grover et al., 1977; Pierucci, 1978). It is well known that bacteria increase in size when they grow faster (Schaechter et al., 1958; Donachie et al., 1976; Grover et al., 1977; Rosenberger et al., 1978a). An extension of this is the study of the transition when cells are shifted to a richer medium (Pierucci, 1978; Woldringh et al., 1980). For earlier work and references on *B. subtilis* the reader should consult Sargent (1979).

It is clear that one should distinguish between (i) the size of an individual cell as it changes from birth to division at a given doubling time, and (ii) the average cell size of a population as a whole as it changes with growth rate. The kinetics of the first change determine the second and not *vice versa*. The approach is therefore indirect and is only able to exclude certain models.

The various linear models which have been proposed (Sargent, 1975; Donachie et al., 1976; Pierucci, 1978) differ with respect to the time or event in the division cycle at which the number of growth sites doubles, and whether the activity of the growth site is dependent on the growth rate of the culture.

Two analogous, linear models can be considered. Both assume surface synthesis to take place at a discrete number of growth sites which double at a fixed time (d) before division, the rate per site being proportional to the growth rate. In the first model (Grover et al., 1977) diameter is adjusted passively to accomodate the increase in volume (Fig. 9a) whereas length is actively determined (Fig. 9b). The second model considers cell surface to be actively determined (Fig. 9d), whereas both length (Fig. 9e) and diameter (Fig. 9f) adjust passively.

The growth kinetics of the individual cell do not only define the variation in average dimensions with growth rate, but determine also the dimensional changes following transfer of the culture to a richer medium. After deriving expressions (Grover et al., 1980) for average cell dimensions as a function of time following such a nutritional shift-up, it was found that the two above models for elongation and sur-

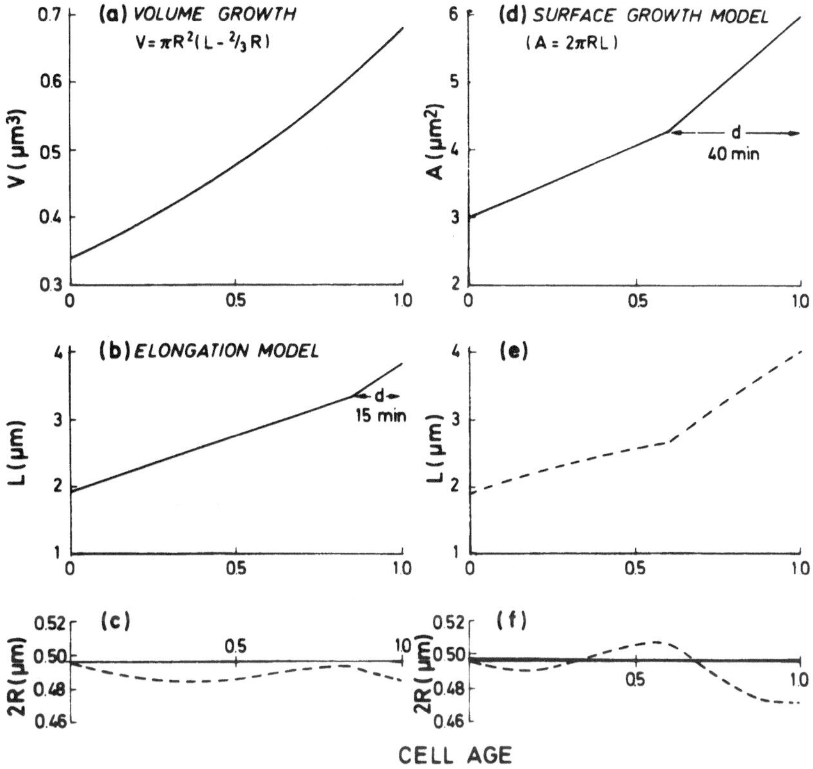

Fig. 9 Changes in dimensions (c, e, f) during the cell cycle (T_D = 100 min) as predicted by the elongation (b) and surface growth model (d). Dashed lines were computed assuming exponential volume growth (a) and cell shape to be a cylinder with hemispherical polar caps.

face synthesis predict different transient behaviour.

Fig. 10 shows graphically what happens to the volume (Fig. 10a), length (Fig. 10b) and diameter (Fig. 10c) of a cell during nutritional shift-up, the behaviour of one individual cell growing according to the elongation model (Fig. 9b). One phenomenon, evident in this figure is that length overshoots its final value. Such an overshoot is not predicted by several other models. For instance, with the assumption (Donachie et al., 1976) that after shift-up the elongation rate is maintained at its preshift level and only accelerates at the time of the first doubling after transition, the overshoot does not occur (dashed line in Fig. 10b).

The results of the shift-up analysis (Woldringh et al., 1980) are shown in Fig. 11. Observations of two independent

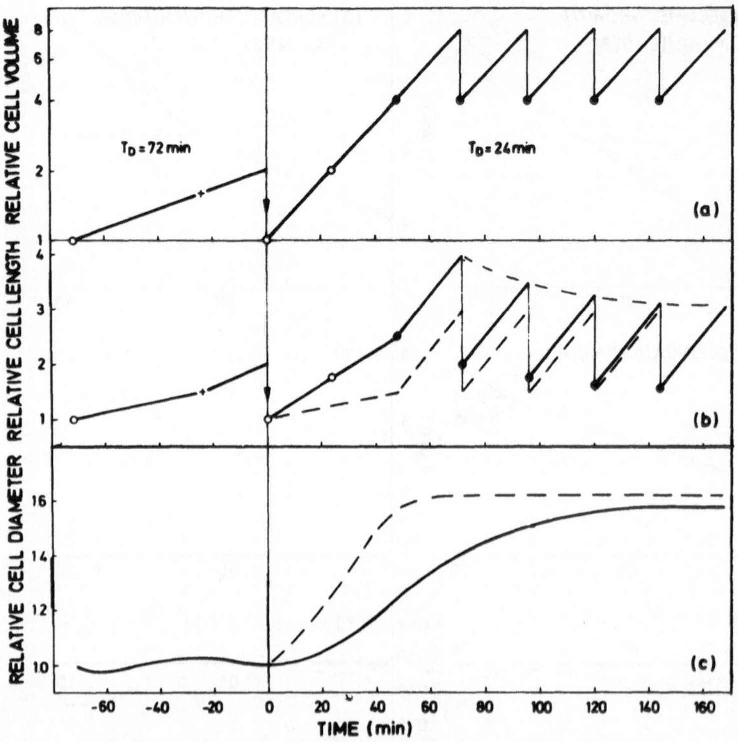

Fig. 10 Growth of an individual cell before and after shift-up at the time of cell division (arrow). Volume (a) increases exponentially and length (b) linearly with rate doubling (crosses) 24 min before division. Circles: initiation of DNA replication; solid lines: elongation rate growth rate dependent; dashed lines: preshift elongation rate is maintained until first doubling after shift-up.

experiments indicate that mean length ($\bar{L}$) rises rapidly after the shift-up and indeed overshoots its final steady-state value in about 75 min. Mean cell diameter ($2\bar{R}$) increases monotonically, but both dimensions approach their asymptotic levels only after 4 hours. The predictions of the dimensional changes during shift-up by various models appeared to differ sufficiently to allow their discrimination (Zaritsky et al., 1980) and to choose conclusively the linear surface-growth model defined previously (Rosenberger et al., 1978a).

Whether cells do indeed increase linearly in surface as predicted by this model has to await the results of more direct approaches. Another aspect of the model which remains to be confirmed is the growth rate dependency of surface synthesis.

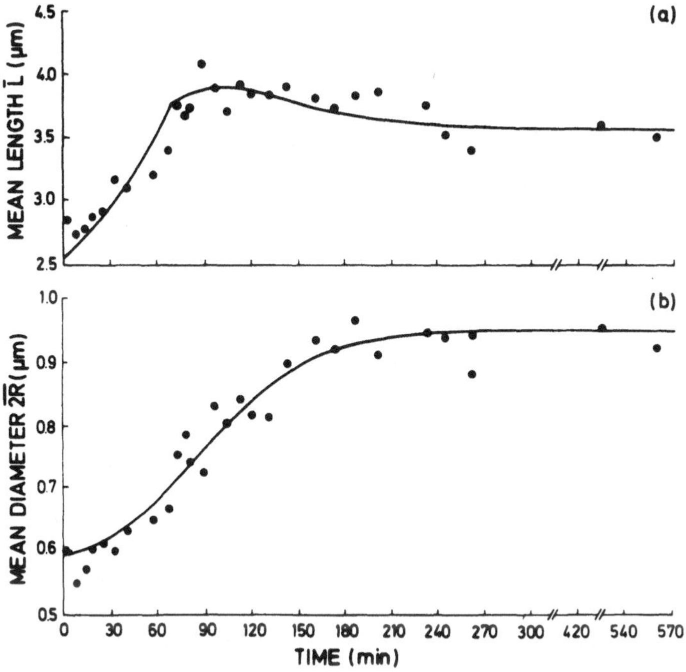

Fig. 11 Cell dimensions as a function of time following shift-up from T_D = 72 min to T_D = 24 min. Dots: experimental data; lines: computed from average cell volume and average cell surface as predicted by the linear surface growth model (cf. Fig. 9d).

Changes in cell density and cell diameter

As discussed above, doubling in cell size may involve different patterns of increase in mass, volume and surface. Of these parameters the first two ($\bar{M}$, $\bar{V}$) define cell density, whereas cell dimensions or cell shape can be derived from the latter two ($\bar{V}$, $\bar{A}$) if the geometry of the cell is given.

If the cell's dry mass is assumed to increase exponentially and volume is taken to increase linearly (Kubitschek, 1970), cell density will oscillate during the cycle with a difference of about 2 %, if the organism is assumed to be made up of 27 % dry solids (Koch & Blumberg, 1976).

Determination of cell density by equilibrium centrifugation of living *E. coli* cells in Percoll gradients revealed that the individual cells in a steady-state population differ less than 1 % in density (Woldringh et al., manuscript in pre-

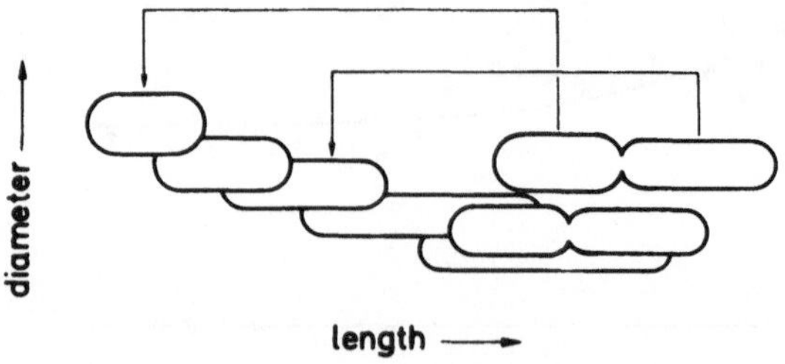

Fig. 12 Diagrammatic representation of the change in cell dimensions during the division cycle of E. coli.

E. coli cells and in *B. subtilis*, cell diameter was found to remain virtually constant. This raises the possibility that the cell wall is more rigid and that density may fluctuate during the cycle (Rosenberger et al., 1978b).

During cell constriction the cell may behave between the two extremes indicated in Fig. 13. In the first case (Fig. 13, A → C) cell density is kept constant and both surface and volume remain the same. As a result cell shape has to change, the cells becoming 18% shorter and 22% thicker in this particular case. Alternatively (Fig. 13, A → B), cell diameter is kept constant and cell constriction proceeds by a decrease in cell volume, resulting in a 10 % increase in density. Evidently, such large fluctuations in either parameter have not been observed.
paration). Within this range small density differences seem still to be present, because, especially in rapidly growing cells, the highest density fractions contained increased amounts of constricting cells.

Nevertheless, as an approximation, cell density can be assumed to remain constant during the division cycle. This implies that in all linear-log models for surface-mass (volume) growth, cell diameter has to vary with cell age (Fig. 9c and f).

For a long time cell diameter has been assumed to be constant in accordance with the observations of Marr et al. (1966). However, we recently have observed that in slowly growing *E. coli* populations cell diameter shows oscillations of 8 % during the division cycle. Extensive measurements of both living cells with the light microscope and of fixed and air-dried cells with the electron microscope have indicated that cell diameter decreases with increasing cell length (Fig. 12). At the end of the cycle diameter seemed to increase again during the constriction process. In rapidly growing

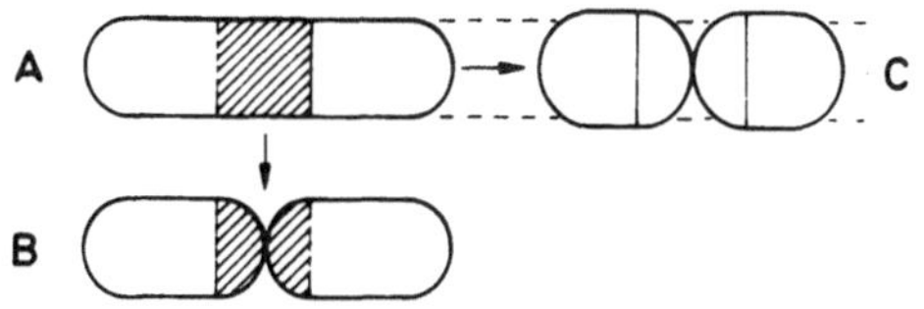

Fig. 13 Two possibilities for cell constriction. A → B: Mass and surface remain constant, volume decreases. As a result density increases A → C: Mass, volume and surface remain constant. As a result density remains constant but cell shape changes. During elongation the diameter of the cell decreases again gradually(cf. Fig. 12)

It seems probable that changes in the rigidity of the cell wall (expressed, for instance, as the degree of peptidoglycan cross-linkage), in cellular density, hydrostatic pressure and diameter are interdependent and that the decision of the cell to divide depends on their relative proportions. Accurate qualifications of the above properties may therefore help to understand the mechanism for the control of cell division.

Surface growth and size distribution

Size (= length) distributions can readily be obtained by electron microscopy of spread whole cells (agar filtration; Woldringh, 1976). A typical example is shown in Fig. 14. The hatched area represents the dividing cells. $\bar{L}_o$ is the mean cell length of the newborn cell and $\bar{L}_s$ is the mean length of the separating cell. From the percentage of constricted cells the T period can be determined (Woldringh, 1976). This is the period from initiation of constriction till cell separation, i.e. the division period.

The length distribution can also be used to derive the distributions of the extant $\lambda(x)$, separating $\phi(x)$ and newborn cells $\psi(x)$. The distribution of the extant cells is measured (Fig. 14). The distribution of the separating cells is derived from the observed distribution of the dividing cells by shifting the last one somewhat to the right (see below). The distribution of the newborn cells is derived from the distribution of the prospective daughters as they occur in the dividing cells. The accuracy of the distributions is thus in the order extant, separating and newborn cells. By means of the Collins and Richmond equation (1962) a relationship can be calculated between length and age.

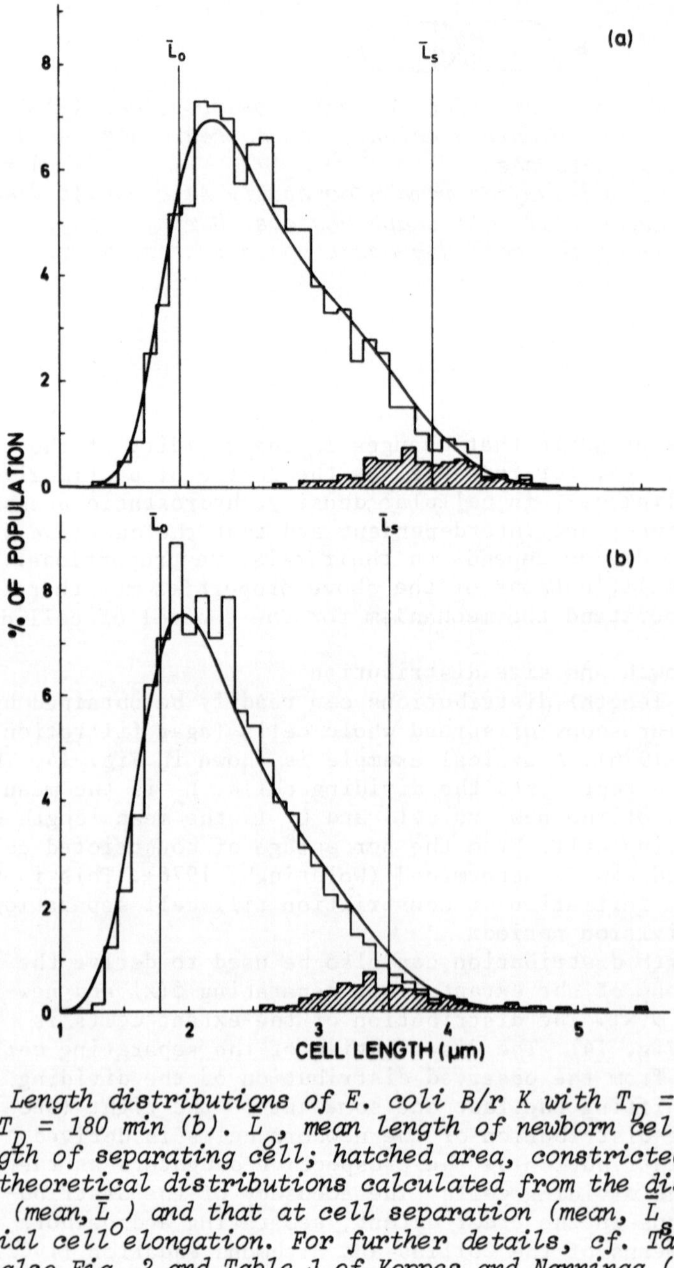

Fig. 14 Length distributions of E. coli B/r K with T_D = 90 min (a) and T_D = 180 min (b). $\bar{L}_o$, mean length of newborn cell; $\bar{L}_s$, mean length of separating cell; hatched area, constricted cells; curves, theoretical distributions calculated from the distribution at birth (mean, $\bar{L}_o$) and that at cell separation (mean, $\bar{L}_s$) assuming exponential cell elongation. For further details, cf. Table 2 and see also Fig. 2 and Table 1 of Koppes and Nanninga (1980).

Alternatively, the theoretical length distribution $\lambda(x)$ can be calculated from the $\psi(x)$ and $\phi(x)$ distributions and can then be compared with the measured length distribution.

For a better estimation of the mean length at cell separation ($\bar{L}_s$) and at birth ($\bar{L}_o$), in the present example the constricted cells (hatched area in Fig. 14) were classified by eye as slightly, medium and very constricted. The average length was found to increase during constriction with a constant coefficient of variation. Therefore, the lengths of the medium constricted cells, when multiplied with the factor $f = \bar{L}$ very constricted cells/ $\bar{L}$ medium constricted cells ($1.04 < f < 1.09$), gave a distribution which did not differ significantly from that of the very constricted cells. The corrected distribution of the medium constricted cells and the distribution of the very constricted cells were pooled. The resulting distribution was used to estimate the distribution at cell separation $\{\phi(x)\}$; the distribution of the prospective daughters of the separating cells was used to estimate the distribution at birth $\{\psi(x)\}$. The parameters of the $\psi(x)$ and $\phi(x)$ distributions are shown in the legend of Fig. 14. Log normal fits to the estimated $\psi(x)$ and $\phi(x)$ distributions could not be rejected ($\alpha = 0.10$) and were routinely used in the further analysis for the sake of computational convenience. Both a model in which the cell elongates with a constant rate that doubles at length L_d with a coefficient of variation of 15%, and a model in which length increases exponentially were tested. The latter model gave the best fit and this is shown as a continuous curve in Fig. 14 (Koppes and Nanninga, 1980). This method would favour an exponential relationship between size and age.

Surface growth during synchronous growth

A different appraoch is to follow cell length during synchronous growth. The example applies to *E. coli* B/r F 26 (Meijer et al. 1979) synchronized with the membrane elution technique (Helmstetter, 1969). In this technique cells are attached to a Millipore membrane filter. Elution with prewarmed growth medium releases newborn cells which are collected to start synchronous growth. In the example subsequent samples have been taken every 10 minutes to measure cell number with the Coulter counter and to make whole mount preparations for the electron microscope. The obtained length distributions are shown in Fig. 15. One observes changes in the length distributions as well as the emergence and disappearance of the constricted cells with time. The corresponding change of total cell length is shown in Fig. 16. Total cell

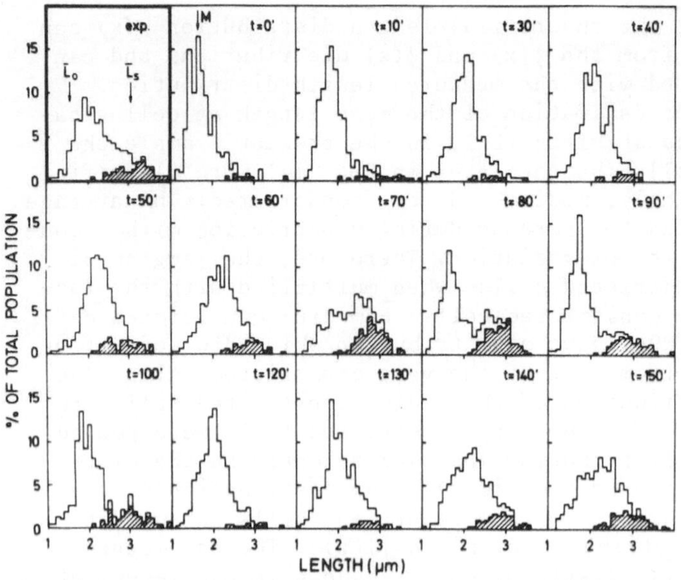

Fig. 15 E. coli B/r F (T_D = 82 min). Length distributions in an exponential culture (exp.) and at various times during synchronous growth. Constricting cells are indicated by the hatched areas. $\bar{L}_o$, average length of the newly born cell; $\bar{L}_s$, average length of the separating cell. L_o and L_s were determined by the method of Harvey et al. (1967) from the length distribution of the exponential culture (Meijer et al., 1979).

length represents relative cell number times the arithmatic mean of the cell length at the given times. Note the increase in growth during the later half of the division cycle in the first as well as in the second generation. This result does not agree with exponential length extension.

The same data (Figs. 15 and 16) were analyzed again in more detail (Koppes et al., 1980) and total cell length was plotted as shown in Fig. 17. The straight line was calculated by linear regression (ρ = 0.99). The data points may indicate alterations in the rate of length growth. However, after application of a runs test it had to be concluded that the scatter of the points along the fitted lines was random (α = 0.05; Koppes et al., 1980). This would indicate exponential lenth extension in accordance with a Collins and Richmond analysis of the length distribution of the steady state culture (Fig. 15; Meijer et al., 1979).

Recently, we have again synchronized *E. coli* B/r F 26 (T_D = 43 min) with membrane elution technique (Olijhoek and Nanninga, manuscript in preparation). Samples have been taken more frequently, i.e. every 3 or 5 minutes. Length was

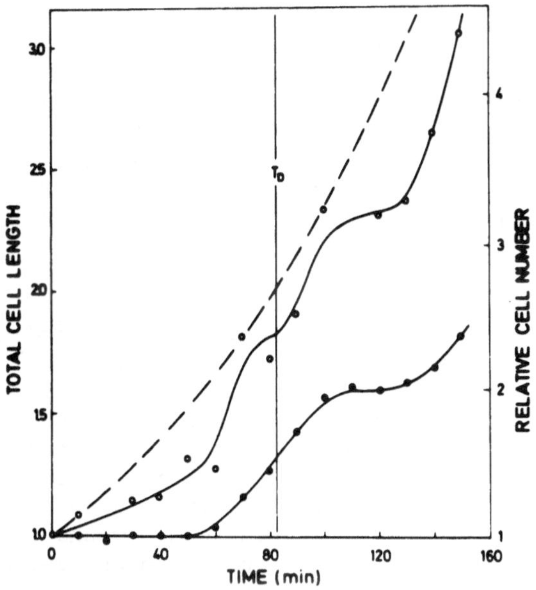

Fi.g 16 Total cell lenth as a function of time in synchronized E. coli B/r F (cf. Fig. 15). o, relationship between length and time of the first and second generation; ●, relative cell number as determined with a Coulter counter. The broken line denotes exponential cell elongation (Meijer et al., 1979).

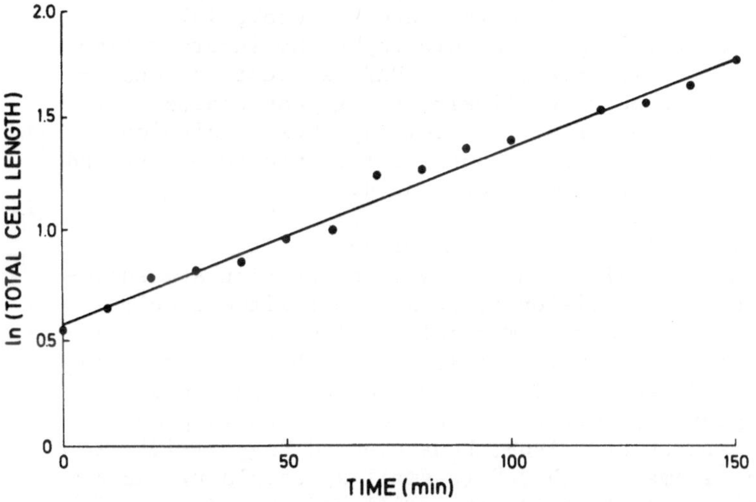

Fig. 17 Total cell length of E. coli B/r F (T_D = 84 min) as a function of time. The straight line was calculated by linear regression (cf. Koppes et al., 1980).

checked with an electron microscope and volume with a Coulter counter. In both cases a periodic change in cell length and in cell volume was found during the division cycle. A runs test indicated that the deviation from exponential length or volume increase was significant. The objection could be made that the observed growth pattern is influenced by the previous attachment of cells to the membrane filter. The above serves to underline the difficulties in obtaining conclusive data.

Sites of length growth

As mentioned before length growth requires the coordinated biogenesis of cell membrane, *sacculus* (murein layer) and outer membrane. In view of the fluidity of the cell membrane it is difficult to envisage the presence of conserved areas in this layer. With respect to the outer membrane attention has been focussed on the induction and location of phage receptors (Begg and Donachie, 1973, 1977; Ryter et al. 1975). In these studies phages have been found around the cell center. However, the response to induction and deinduction were not as clear-cut. This area of research would need more attention (cf. Sargent, 1979).

We will now concentrate on the murein layer which is responsible for rod-shape maintenance. To assess the location of incorporation sites with respect to cell length two ways have been followed. Firstly, one can interfere with cell elongation and cell division on the level of the *sacculus*. This has been done by adding β-lactam antibiotics like penicillin (Donachie and Begg, 1970), ampicillin (Staugaard et al., 1976) or cephaloridin (Mathys and Van Gool, 1979). Secondly, one can follow by autoradiography the incorporation of radioactive diaminopimelic acid (DAP; a specific constituant of the short peptides linking the glycan chains). It is assumed that sites of active murein synthesis coincide with those where the β-lactam antibiotics interfere most and where DAP is predominantly incorporated.

Localized action of β-lactam antibiotics

Penicillin and ampicillin at low concentration are supposed to prevent cell division of *E. coli* B/r without cell elongation being inhibited. Morphologically, the site of action can be discerned by way of so-called "bulges" occurring at the center (Schwarz et al., 1969; Burdett and Murray, 1974). At higher concentrations cell elongation is prevented as well (Schwarz et al., 1969; Donachie and Begg, 1970). In this instance a small "sphere" (a small spheroplast) can generally be detected alongside the still rod-shaped bacterium

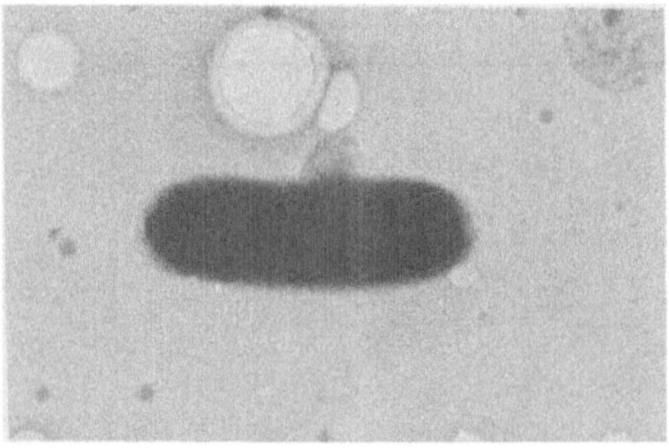

Fig. 18 E. coli B/r ATCC 12407 treated with ampicillin. After agar filtration the site of sphere formation is easily located.

(Donachie and Begg, 1970; Staugaard et al., 1976; Fig. 18). Ultimately cells lyse completely. On the other hand, cephaloridine is supposed to affect specifically cell elongation and not cell division (Spratt, 1975).

The distribution of spheres along the length axis of the cell has been used as a basis for the so-called unit cell model of growth for *E. coli* (Donachie and Begg, 1970). This model implies that during growth the sphere should remain at a fixed distance (the length of the unit cell) from the oldest cell pole. For cells with a T_D of 60 min there would be one asymmetrically located growth site which would end up gradually at the division site. We have repeated these experiments with the same bacterial strain (*E. coli* B/r ATCC 12407) and the same T_D (65 to 70 min). The differences were the following: (i) Ampicillin was used instead of penicillin. This should not affect the results. (ii) An electron microscope was used instead of a light microscope, and (iii) We did not pool cells of the same length which came from cultures of different growth rates. We found some prominent sites at which spheres occurred (Fig. 19; Staugaard et al., 1976). In the shortest cells, which have a length of about 1.5 μm they were found at the presumed new cell pole. In slightly older cells (about 1.8 μm in length) the position of the sphere was not well defined. In longer cells, whether dividing or not spheres occurred predominantly at the cell center. The position of the spheres bore a striking resemblance to sites were pulse-labeled DAP is incorporated (see below). Our observations on ampicillin-treated cells have been confirmed by Mathys and Van Gool (1979). They neither observed a gradual displacement of the sphere from cell pole to cell

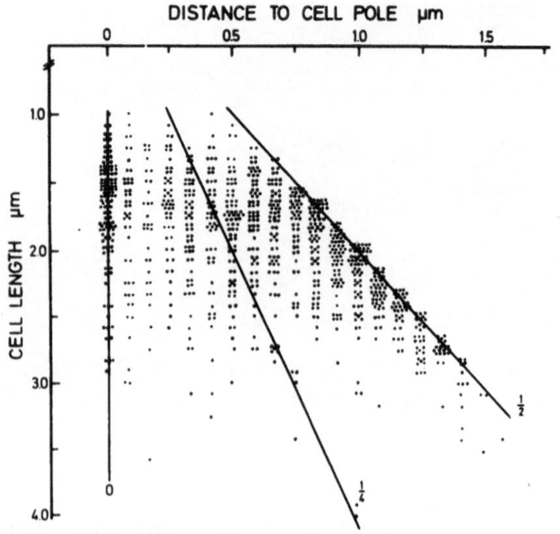

Fig. 19 Position of spheres with respect to the cell pole in 2,250 cells of different length. Lines indicate cell pole (0), cell centre (½) and one-quarter of cell length (¼). Average length of newborn cell is 1.4 μm.

center with increasing cell length. This particular aspect of the unit-cell concept can therefore not be confirmed.

Localized incorporation of diaminopimelic acid

Earlier autoradiographic investigations on [^{3}H] DAP pulse-labelled *E. coli* cells (Ryter et al., 1973; Schwarz et al., 1975) revealed a central area of incorporation and an indication for additional lateral zones in the longest cells. These studies were carried out on a limited number of length classes (three). However, a more detailed analysis of [^{3}H] DAP incorporation in *E. coli* PAT 84 confirmed the earlier results (Koppes et al., 1978a).

In the above mentioned studies the silver grain location has been measured with respect to the geometric center (in non-dividing cells) or to the visible cell constriction. Moreover, it was implicitly assumed that the two cell halves in non-dividing and in dividing cells are interchangeable. A possible asymmetry in grain location would not have detected in this way. For this reason, we have studied the question whether the two cell halves are equivalent with respect to location and quantity of DAP incorporation. Each cell was divided in a left and a right half; the left half being defined as the compartment with the largest number of grains. Grain counting and grain localization has been carried out separately in the various length classes. The positioning of cells for this type of analysis is shown in Fig. 20. The mean of the grain numbers of the left compartment are then

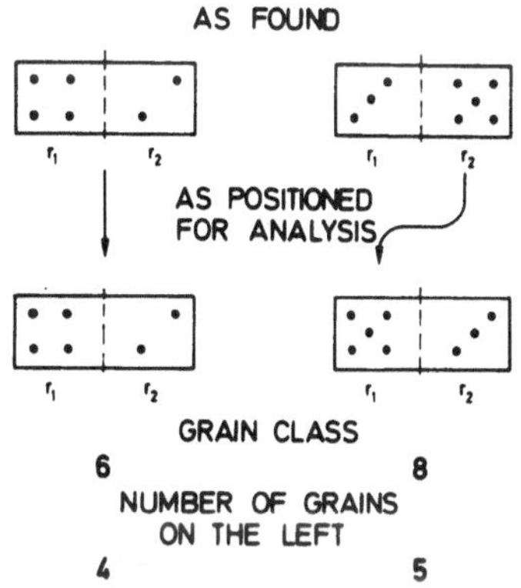

Fig. 20 Orientation of cells before analysis of both cell halves. r_1, number of grains on the left side; r_2, number of grains on the right side (cf. Verwer and Nanninga, 1980).

compared to the expected number of grains on the basis of random placement of grains over the two cell halves (Table 1). It can be seen that the observed mean number of grains in the left cell half is generally close to the expected mean. However, some significant deviations occur (Verwer and Nanninga, 1980). This might be explained by the consideration that the geometric cell center does not exactly coincide with the prospective division site (F.J. Trueba, personal communication). Taken this into account it will be fair to assume that grains are randomly placed over the two cell halves. An example of grain localizations in the two cell halves of non-dividing cells is shown in Fig. 21. The highest probability of finding grams is in the central area in all length classes. In longer cells (Fig. 21) and in dividing cells (not shown) lateral incorporation sites emerge. It is not possible to distinguish sharply between diffuse growth and zonal growth. Perhaps this distinction does not apply, since grains can be found everywhere over the cells. This leads one to the concept of an incorporation gradient with the highest probability of incorporation in the actual or prospective division site (Verwer and Nanninga, 1980).

TABLE 1. *Comparison of the theoretical and observed distribution of grains over the left halves of non-dividing cells*

Number of grains per cell	Number of cells	Mean number of grains over the left cell half			D max[b]	P[D≥dcrit][c]
		theoretical	observed	SD[a]		
2	21	1.50	1.62	0.50	0.12	0.35-0.38
3	22	2.25	2.41	0.50	0.16	0.04
4	34	2.75	2.85	0.74	0.08	0.34-0.37
5	31	3.44	3.68	0.74	0.14	0.13-0.14
6	34	3.94	4.32	1.09	0.17	0.04
7	32	4.59	5.09	0.96	0.23	0.01
8	34	5.09	5.33	1.12	0.15	0.08
9	39	5.73	5.87	0.95	0.06	0.57-0.68
10	34	6.23	6.79	1.07	0.22	0.02
11	24	6.85	7.42	1.44	0.16	0.17-0.18
12	35	7.35	7.74	1.58	0.17	0.06
13	22	7.97	8.68	1.46	0.19	0.10
14	21	8.47	9.24	1.73	0.16	0.18-0.19
15	15	9.07	9.93	1.62	0.30	0.03
16	13	9.57	10.16	2.08	0.25	0.10
17	13	10.17	11.62	2.33	0.31	0.04
18	10	10.67	12.00	2.40	0.40	0.02
19	7	11.26	12.29	1.98	0.27	0.19-0.20
20	3	11.76	13.00	1.00	0.50	0.25-0.27

a Since the theoretical as well as the observed distribution possess necessarily exactly the same but limited range, it is appropriate to compare the shape of the respective distributions and inappropriate to consider the difference of the means (see Appendix in Verwer and Nanninga, 1980).
b See Appendix in Verwer and Nanninga (1980).
c Level of significance calculated according to Conover (cited in Verwer and Nanninga, 1980).

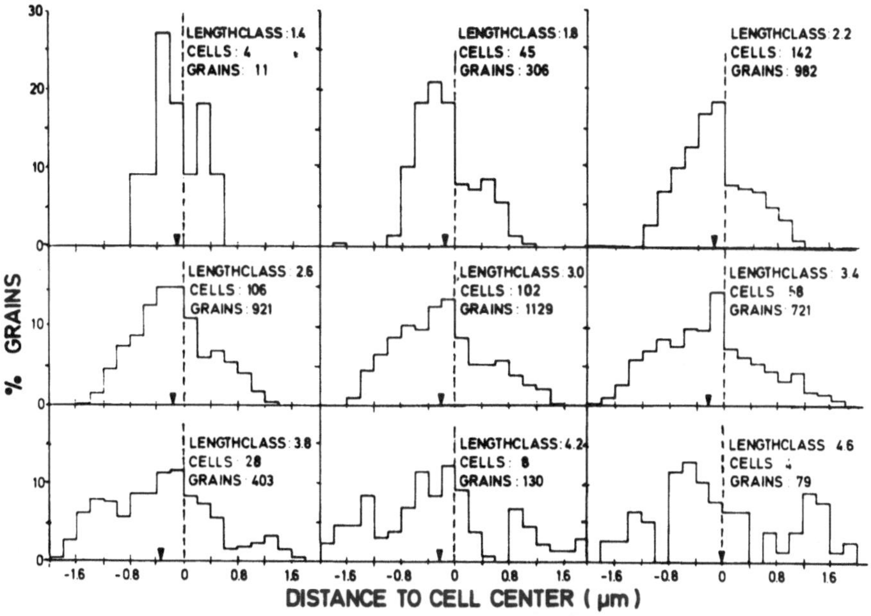

Fig. 21 Localization of grains over the two cell halves of non-dividing cells (cf. Fig. 20). Broken line: geometric cell center. Arrow: mean distance of grains to the cell center (cf. Verwer & Nanninga, 1980)

The central location of ampicillin or cephaloridin spheres and of DAP incorporation areas would favour a growth model in which new cell envelope material would be continuously inserted in the cell center. From there it would be displaced in the direction of the cell pole. This idea is attractive because it would support the notion that the growing bacterial envelope serves as the equivalent of the mitotic apparatus (Fig. 4). The crucial experiment would be to show that newly incorporated material is shifted away from the cell center with time. This simple thought is in practice fraught with complications (Begg and Donachie, 1973, 1977; Schwarz et al., 1975; Ryter et al., 1975). However, it should be stressed that wall synthesis in *B. megaterium* (de Chastellier et al., 1976) and in *B. subtilis* (Archibald and Coapes, 1976) seem to occur all over the cell surface except for the poles. This would preclude a DNA-segregating mechanism as has been depicted in Fig. 4. Segregation might be determined by the physical properties of replicating DNA itself. This idea is strengthened by the observation that segregation can occur in the absence of cell growth (Grossman, Ron and Woldringh, manuscript in preparation.

DNA REPLICATION

The Helmstetter-Cooper model accounts for the location of the DNA replication cycle within the division cycle of *E. coli* (Fig. 5). The model in particular applies to *E. coli* B/r A and for T_D's between about 20 and 60 minutes. At slow growth C and D increase and for some strains a gap appears at the beginning of the division cycle: B period (Fig. 3). In slow growing *E. coli* PAT 84 (T_D = 170 min) replication starts in one cycle and finishes in the next one (Koppes et al., 1978a). This unexpected finding resembles the DNA replication pattern of *E. coli* B/r A with T_D = 50 min (Helmstetter et al., 1968). The respective patterns at slow growth as detected by autoradiography is shown in Fig. 22. Despite

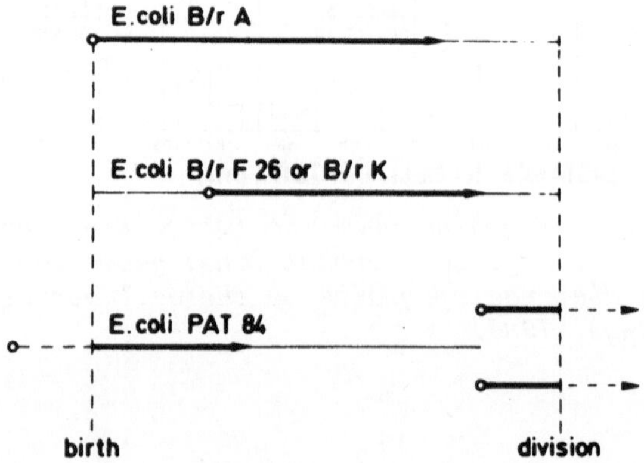

Fig. 22 Replication pattern of slowly growing E. coli as determined by autoradiography (cf. Koppes et al, 1878 a,b).

the large doubling time (170 min) PAT 84 cells are fairly big. This is not unexpected, because the presence of two initiating chromosomes at the end of the division cycle suggests a size two times M_i (initiation mass). A size of one M_i will not occur unlike the situation for slowly growing *E. coli* B/r A and B/r K (Fig. 22).

It should be recognized that M_i may be (sub)strain (Koppes et al., 1978b) and to some degree growth rate dependent, because increasing C + D periods have been found with increasing doubling time (Kubitschek, 1974; Churchward and Bremer, 1977). Clear deviations from the dogma that M_i is a constant have been found in two studies in which the steady-state conditions of *E. coli* populations were altered: In the first it was found that after amino acid starvation of synchronized

E. coli B the cells initiated DNA replication upon restoration of the amino acid, long before they could reach their steady-state M_i (Grossman et al., manuscript in preparation). A complementary observation in the other direction was made after nutritional shift-up, which resulted in an increase of M_i (Zaritsky and Zabrovitz, 1980).

Slowly growing strains like B/r F or B/r K (T_D about 100 to 150 min) are of experimental advantage since divison cycle events are well separated in time. In particular the occurrence of a B period enables the measurement of the size and age distribution at which DNA replication starts. Slowly growing cells are also more comparable to eukaryotic ones (Fig. 3).

Size at initiation of DNA replication

Up till now the concept of initiation mass has not been translated into chemical and/or physical terms. One has considered the accumulation of an initiator (positive regulation) or the dilution of an inhibitor (negative regulation) to a critical concentration upon attainment of M_i (for a review see Pritchard, 1978).

As a quite reasonable approximation mass can be equated to cell length in whole mount electron microscopic preparations, whereas DNA replication can be followed by autoradiography. For the estimation of length at initiation of DNA replication (L_i), the population of non-dividing cells that had been fixed subsequently to pulse labeling and washing (0 min chase), was divided into twenty length classes containing about an equal number of cells. For each class the frequency distribution of grains per cell was assessed. The theoretical probability P(n) of observing a cell with n grains is given by the Poisson law (cf. Koch, 1977; see the Appendix of Koppes et al., 1978a):

$$P(n) = (1-S).e^{-G}.\frac{(G)^n}{n!} + S.e^{-C}.\frac{(C)^n}{n!}.\text{ in which}$$

S is the fraction of DNA synthesizing cells which have on the average C grains per cell and G is the average number of background grains per cell. This equation was fitted to the observed grain distribution by the method of maximum likelihood (Rao, 1973). The parameters G and C that gave the best fit approximately doubled with increasing length from L_o to L_s. By linear regression of G and C on length an improved estimate of the parameter S could be obtained which is shown as a function of the average length of each cell class in Fig. 23 (filled circles).

In *E. coli* B/r K average lengths at termination of DNA replication ($\bar{L}_f$) and at initiation of cell constriction ($\bar{L}_c$) coincide (Woldringh et al., 1977; Koppes et al., 1978b) and almost all unlabeled constricted cells occur before initiation of DNA replication i.e. the B period. Therefore, the theoretical fraction P(x) of labeled unconstricted cells is approximated by (see Appendix of Koppes et al., 1978b):

$$P(x) = \frac{\Omega(x) - \Xi(x)}{1 - \Xi(x)} \quad \text{in which } \Omega(x) = \int_o^x \omega(y).dy,$$

i.e. the cumulative length distribution at initiation of DNA replication and $\Xi(x) = \int_o^x \xi(y).dy$, i.e. the cumulative length distribution at cell constriction. The above equation was fitted to the estimated fractions of unlabeled cells S by the method of maximum likelihood (Finney, 1973). The parameters of the distribution at initiation of DNA replication ($\omega(x)$) that gave the best fit are shown in Fig. 23 and in Table 2 (Koppes and Nanninga, 1980).

The coefficient of variation of length at initiation of

TABLE 2. *Estimated average length and CV at different events during the cell cycle*

	cell birth		initiation of DNA replication		initiation of cell constriction		cell separation	
T_D	L_o	CV_o	L_i[a]	CV_i[a]	L_c	CV_c	L_s	CV_s
(min)	(µm)	(%)	(µm)	(%)	(µm)	(%)	(µm)	(%)
90	1.93	12.8	2.39	22.3	3.63	10.0	3.87	9.9
180	1.78	14.3	2.42	14.8	3.30	10.2	3.55	11.4

[a] Corrected for the duration of the pulse (Δt) as follows: $L_i = L_i' - \frac{1}{2}\Delta L$ and $CV_i = 100\ [(L_i' . CV_i'/100)^2 - (\Delta L)^2/12]\ \frac{1}{2}L_i$ in which L_i and CV_i are the uncorrected parameters and $\Delta L = \ln 2.\ L_i'.\ t/T_D$ (see appendix (iii) of Koppes et al., 1978b).

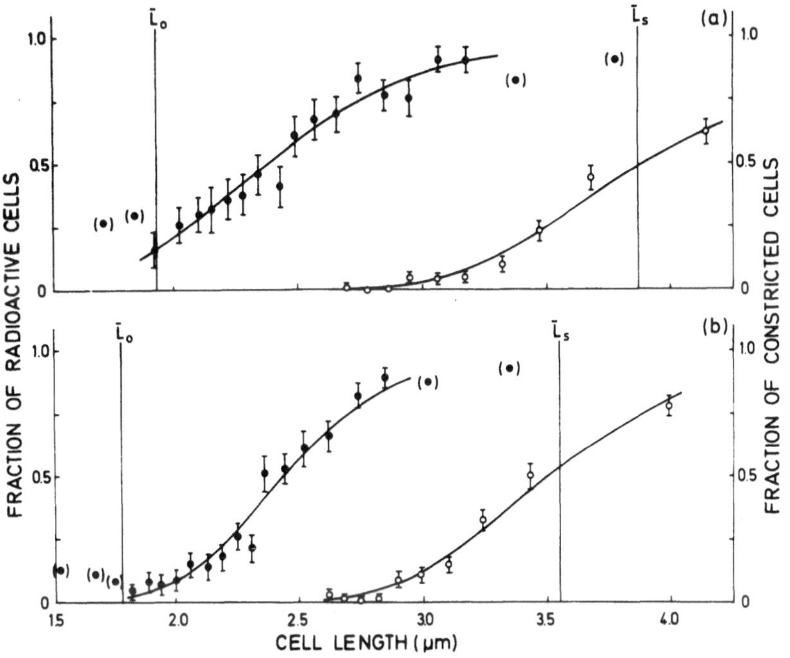

Fig. 23 Fractions of radioactive and constricted cells per length class. (a) T_D = 90 min. (b) T_D = 180 min. (●) Fractions of radioactive cells in unconstricted cells (0 min chase sample). The curves were calculated assuming: L_i = 2.44 μm, CV = 21.9% (a); L_i = 2.46 μm, CV = 14.6% (b). Fractions within brackets were neglected in the calculation. (o) Fractions of constricted cells of the overall length distribution. The curves were calculated assuming: L_c = 3.63 μm, CV = 10.0% (a); L_c = 3.30 μm, CV = 10.2% (b). The SE is given by the flags.

DNA replication (CV_i) is within the range of 15 to 25% reported by Koch (1977). The large difference of CV_i of size at T_D = 90 min and T_D = 180 min (Table 2) is presumably due to the different growth rates. Table 2 also contains data on the distribution of cell lengths at birth ($\bar{L}_o$), initiation of cell constriction ($\bar{L}_c$) and cell separation ($\bar{L}_s$). These will be referred to further on.

The above data apply to exponentially growing cells which had been classified according to length. The alternative is to use a synchronized culture and to follow L_i as a function of time. For *E. coli* B/r F (which can be better synchronized by membrane elution than B/r K (Helmstetter and Pierucci, 1976)

Koppes et al., 1978b) CV_i was 22.9 ± 5.2% for T_D's in the order of 150 min (Koppes et al., 1980).

Age at initiation of DNA replication

DNA starts at the end of the B period in slowly growing cells (cf. Fig. 3). The age distribution at initiation of DNA replication is reflected in the distribution of the B period. The experimental set up was as follows (Koppes et al. 1980). Part of membrane eluted cells of B/r F (T_D = 165 min) were incubated at t = 0 in the presence of [^{3}H] thymine at 37°C. At successive times samples were fixed and the fraction of radioactive cells (i.e. cells that had initiated

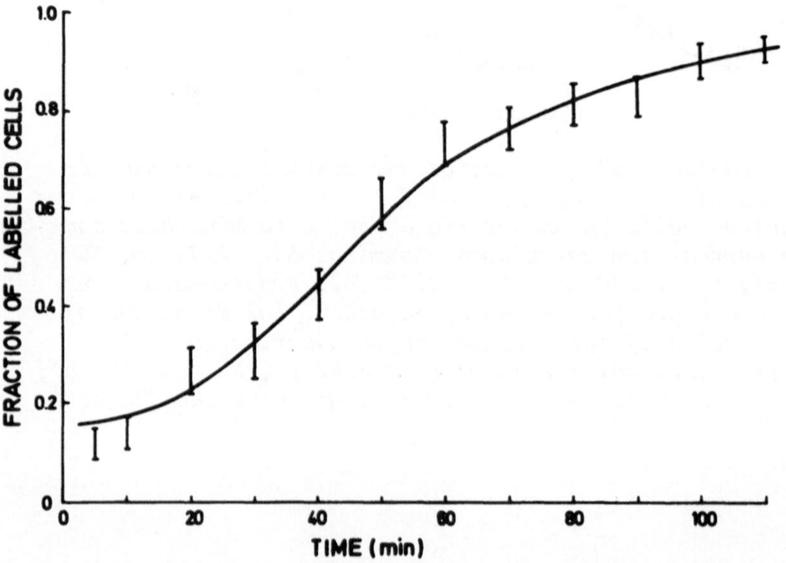

Fig. 24 Fraction of labeled cells as a function of time after start of synchronized growth of E. coli B/r F. Membrane eluted cells (T_D = 165 min) were incubated at t = 0 at 37°C in the presence of (^{3}H) thymine. At successive times afterwards samples were fixed and the fraction of labeled cells estimated by autoradiography. The 95% confidence intervals are given by the flags. The curve was calculated assuming a lognormal distribution. The kinetics of labeling of the asynchronous fraction (0.32) was assumed to be that of a steady state culture and is outlined in Appendix (ii) of Koppes et al., 1980.

DNA replication) estimated by electronmicroscopic radioautography. In Fig. 24 95 % confidence intervals for the labelled fractions are plotted as a function of time after start of synchronized growth. Such a figure represents the kinetics of entrance of cells in the C period (period of DNA synthesis) as a function of their age. It can be seen, that from the beginning more than 10 % of the cells is already labelled probably due to some contaminating asynchronous cells (see below). The curve is a best fit calculated on the assumption of a lognormally distributed B period. In the calculation a different kinetics of labelling of the contaminating asynchronous fraction was taken into account (see Appendix (ii) in Koppes et al., 1980). The fit is rather poor as indicated by the chi-square test ($P > 0.005$) and was not much improved by assuming an exponentially distributed B period (Smith and Martin, 1973).

We conclude that though our data are not accurate enough to establish the precise kinetics of initiation of chromosome replication, they do indicate, however, a large spread in the age at which this event occurs: $CV_B = 60$ %. G_1 is the most variable period of the mitotic cycle (review: Prescott, 1976) and it seems that this is also the case for B. We may also conclude on the basis of the respective coefficients of variation that initiation of DNA replication is controlled by size (CV_i around 20 %) rather than by age (CV_B around 60 %). This makes it unlikely that a transition probability element (Smith and Martin, 1973) is associated with the start of DNA replication.

Distribution of the U period

The U period is defined as the time interval between initiation of DNA replication and initiation of visible constriction (Table 3).

The variability of this period has been estimated by application of the fraction of unlabelled constrictions method. (Koppes and Nanninga, 1980; Koppes et al., 1980). This is an analogue of the fractions of labelled mitoses method (Howard and Pelc, 1953). When a [^{3}H] thymidine pulse is given to a steady-state culture, cells that are in the U period will be labelled whereas cells in the B period will not. With time the labelled cells will appear as labelled constricted cells. Later on unlabelled constricted cells emerge i.e. cells not labelled in the B-period. The kinetics of the appearance of the latter cohort of cells in the "constriction window" of the division cycle has been used to estimate the variability of the U period (Fig. 25). This was done as follows. Theo-

Table 3. *Symbols used for the parameters of the cell cycle*

Event	Length	Period
Birth	L_o	B (Birth to initiation of DNA replication)
Initiation of DNA replication	L_i	C (initiation to termination of DNA replication); U (initiation of DNA replication to initiation of cell constriction)
Termination of DNA replication	L_f	D (termination of DNA replication to cell separation)
Initiation cell constriction	L_c	T (initiation of cell constriction to cell separation)
Cell separation	L_s	

The interdivision time τ = B + C + D = B + U + T

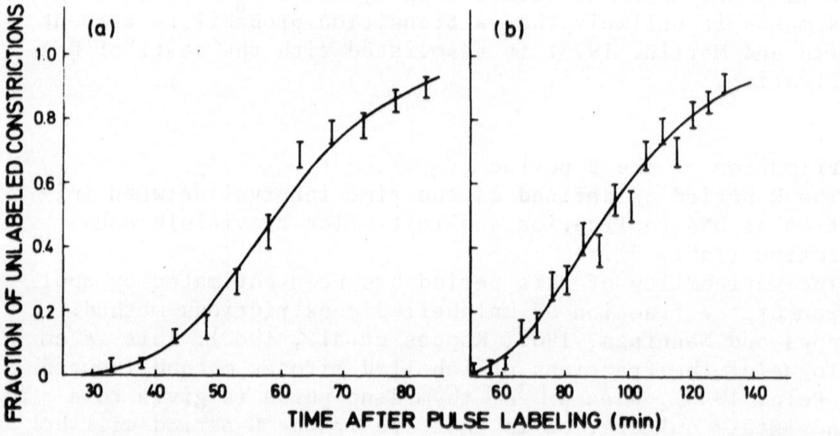

Fig. 25 Fraction of unlabeled constrictions as a function of time after pulse labeling of E. coli B/r K. (a) T_D = 90 min. (b) T_D = 180 min. The 68% confidence interval is given by the flags. The curves were calculated assuming a lognormal distribution of the U period with parameters: $\bar{U}$ = 55.1 min, CV = 25.4%, T = 7.8 min (a); $\bar{U}$ = 84.5 min, CV = 27.9%, T = 13.9 min (b).

retical fractions of unlabelled constricted cells were calculated according to the method of Nachtwey and Cameron (1968). This method corrects for the average duration of the T-period and assumes a log normal distribution of the U period for computational convenience. In addition, exponential increase of the population during the chase period was taken into account. The theoretical fractions were fitted to the observed ones by the method of maximum likelihood (Finney, 1973). For *E. coli* B/r K variation coefficients of 25.4 % and 27.9 % were estimated for T_D = 90 min and T_D = 180 min, respectively (Fig. 25).

The fraction of unlabelled constrictions method can be applied to synchronized cells (Koppes et al., 1980). For *E. coli* B/r F (T_D = 150 min) a variation coefficient of 29 ± 2 % was found for the U period. The U period is thus less variable than the B period but more variable than the interdivision period (see below).

Summarizing this section on DNA replication we may state that cells of different ages initiate DNA replication, on the average, at the same length (size). Cells that are born small have to accrue more and they are older at initiation of DNA replication than cells that are born large.

CELL DIVISION

Cell division implies physical separation of two new compartments by their cell membranes. Cell division should be distinguished from cell separation. In *E. coli* cell division is followed by cell separation almost instantaneously. By contrast, in *B. subtilis* cell separation as such may take 20 min (Nanninga et al., 1979).

Physiological division might occur before completion of the physical barrier (cell membrane). It is defined by the agent whose action is measured (Clark, 1968). For instance, physiological division as measured by T_4 phage killing was found to occur well before cell division (Clark, 1968; Onken and Messer, 1973). However, analysis of recombination and complementation between different phage genomes led to the conclusion that physiological and physical division more or less coincide. In other words, cell envelope and cytoplasm behave differently.

In *E. coli* cell division is always started by invagination of the cell envelope (Burdett and Murray, 1974a, 1974b). It is likely that preparations for cell division, including chemical alterations at the prospective division site, take place before the T period, i.e. before a constriction becomes visible. An indication for this can be found in the early central location of the ampicillin-spheres (Figs. 18 and 19).

Cell division can be studied from a variety of view points. One may analyze envelope composition, be it cell wall or cell membrane as cells approach division or activities of enzymes involved in the biogenesis of the *sacculus* (murein layer). For a discussion and references the reader should consult the reviews of Sargent (1979) and Helmstetter et al., (1979). One can also study defects in division, which may become manifested as filamentation. This reminds one of the rather cynical remark of Slater and Schaechter (1974) in their incisive review: "In fact, it seems possible that any chemical at some concentration, whether attainable in the laboratory or not, could cause filament formation". A more recent line is to characterize so-called cell division genes and their products (review: Donachie, 1979).

Apart from biochemical measurements one should consider physical alterations pertaining to osmotic pressure (Pritchard, 1974) and cell density (Koch and Blumberg, 1976; Poole, (1977). Unfortunately, this field of research received little attention so far. Presumably in part because of technical limitations. However, what ever facet one wishes to study, we believe that this should be done within the context of size and age variability as they occur at initiation of cell division.

Size at initiation of cell constriction

Our example pertains to the length distributions of constricted cells of *E. coli* B/r K shown in Fig. 23. For the estimation of length at initiation of cell constriction (L_c), the overall length distribution $\lambda(x)$ (Fig. 14) was divided into 10 length classes such that each class contained about an equal number of cells. The observed fractions of constricted cells in each class are shown in Fig. 23 (open circles) as a function of the average length of that class. Theoretical fractions P(x) are approximated by (see section of surface growth and size distribution):

$$P(x) = \frac{\Xi(x) - \Phi(x)}{1 - \Phi(x)}$$

in which: $\Phi(x) = \int_0^x \phi(y).dy$ i.e. the cumulative length distribution at cell separation. The curves shown in Fig. 23 were obtained by fitting the observed equation to the observed fractions by the method of maximum likelihood (Finney, 1973). For $\bar{L}_c$ the estimated coefficients were 10.0 % and 10.2 % for T_D's of 90 min and 180 min, respectively. Similar values have been found for *E. coli* B/r A and B/r F26 (Koppes et al., 1978b).

Correlation between length and age at division

The mean interdivision time ($\bar{\tau}$) can be derived from synchronized cultures. The range observed (CV_{τ}) was from 12 - 19 % (Koppes et al., 1980). There is thus a greater variation in age at division than in size (see above). This confirms the earlier light microscopic observations of Koch and Schaechter (1962) and their conclusion that division is governed by size rather than by age. Note that the same conclusion has been drawn concerning the initiation of DNA replication. The correlation between length and age at cell constriction can be calculated when the mean length of constricted cells ($\bar{L}_c$) is plotted against time. In Fig. 26 the $\bar{L}_c$ of slightly constricted cells (filled circles) is plotted as a function of time after start of synchronized growth (T_D = 165 min; Fig. 24). The straight lines were calculated by linear regression. The coefficients of correlation (ρ = -0.23 and -0.38 for slightly and far constricted cells, respectively) did not differ significantly from zero ($P>0.10$). The coefficient of variation of $\bar{L}_c$ for slightly and far constricted cells were 8.7 and 9.1 %, respectively. It can be concluded that cells of very different ages divide, on the average, at the same length (Fig. 26). This implies a negative correlation between size at birth and interdivision time (τ).

COORDINATION BETWEEN DNA REPLICATION AND CELL GROWTH

DNA replication and growth towards cell division can be looked upon as two separate processes (Mitchison, 1971; Donachie et al., 1973). If division is inhibited filaments arise in which DNA replication continues. Blocking of growth allows completion of ongoing rounds of DNA replication, whereas initiation of new ones is prevented. In some instances defects in DNA replication produce DNA-less cells. For details the reader is referred to the reviews of Helmstetter et al. (1979) and Donachie et al. (1979). These observations emphasize the independence of the two processes. Nevertheless it is clear that normally coordination should occur.

Above we have outlined that cells require a certain size rather than a certain age to start DNA replication. We have referred to the term initiation mass (M_i) without being able to give it a concrete meaning. Several workers have suggested that doubling in the rate of length or surface growth during the division cycle might be associated with (i) the initiation of DNA replication (Pierucci, 1978) (ii) explicitly not with the initiation of DNA replication (Ward and Glaser, 1971) (iii) with no particular phase of the DNA replication period (Grover et al., 1977; Zaritsky and Woldringh, 1978; Woldringh et al., 1980), and (iv) as (iii), but explicitly no relation with termination of DNA replication (Donachie et al., 1976). This enumeration reflects the present state of the art and we should add the possibility of exponential length extension (Fig. 14).

Correlation between *length* at initiation of DNA replication (L_i) and length at initiation of cell constriction (L_c).

In Table 2 we have presented the parameters of the size distributions of L_i and L_c (cf. Fig. 23). The variation in duration between L_i and L_c (the U period) has been specified in Fig. 25. The correlation coefficient (Feller, 1976) can be calculated according to:

$$\rho = \frac{(L_i.CV_i)^2 + (L_c.CV_c)^2 - CV_U{}^2.(L_c - L_i)^2}{2\ L_i.L_c.CV_i.CV_c}$$

For both growth conditions of *E. coli* B/r K (T_D = 90 min and T_D = 180 min) a positive correlation was found: ρ = 0.76 - 1.00 and ρ = 0.69 - 0.98, respectively. It was assumed that cells elongate exponentially during the U period. In the linear case the ρ's are only slightly smaller (Koppes and Nanninga, 1980). For *E. coli* B/r F (T_D = 150 min) a similar ρ (L_i, L_c) was calculated (0.8) assuming exponential cell elongation (Koppes et al., 1980).

The positive correlation between L_i and L_c indicates that cells that start DNA replication at a big size proceed faster to division than cells that initiate at a small size, thus reducing the variation coefficient of cell size (homeostasis; cf. Fantes, 1977). The physiological basis of the highly positive correlation between L_i and L_c is unclear. It might be that cell growth leading to cell division and DNA replication are triggered by a common event before the start of the U period.

Correlation between *age* at initiation of DNA replication and age at initiation of constriction.

From the average duration and the respective coefficients of variation of B, U and (cf. Table 3 and previous sections) the correlation was calculated between B (DNA replication starts at the end of the B period) and B + U (cell constriction starts after the U period): ρ = 0.66. By constrast a negative correlation was found between the B and U period: ρ (B, U) = -0.85. Initiation of DNA replication is thus more positively correlated with cell division than with cell birth. From the same material the coefficient of correlation between length at birth (L_0) and length at cell separation (L_S) was calculated $\rho(L_o, L_s) = 0.55$ (Koppes et al., 1980), in other words size at division also depends on size at birth.

Our data (Fig. 24 and 25) indicate that the B period is the most variable part of the cell division cycle (CV = 60%). The B period can be compared to the G1 period of the eukaryotic cell cycle. It is well known that growth conditions affect the length of G1 rather than the other periods (S, G2 and M) of the division cycle. At one particular growth condition, it is the variation in G1 which would primarily be responsible for the distribution of interdivision times (Prescott, 1976). We found that the variation of age at initiation of chromosome replication is considerable larger than the variation of size. (This suggests a size control over chromosome replication rather than an age control). Two hypotheses have been put forward to explain this variation: i. According to Smith and Martin (1973), cells enter an indeterminate state in G_1 from which they leave at random with a

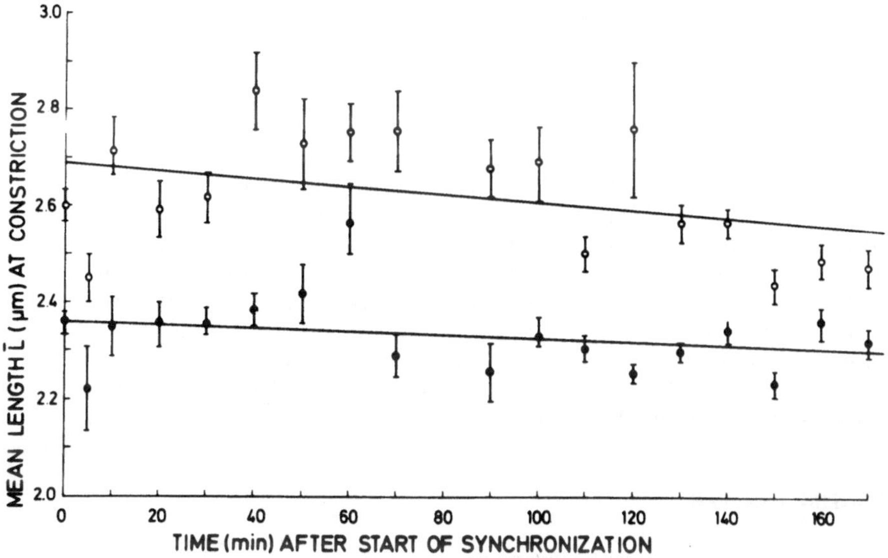

Fig. 26 Average length at cell constriction as a function of time. A culture of E. coli B/r F (T_D = 165 min) was synchronized by membrane elution. At successive times samples were fixed and analyzed by electron microscopy. (●) $\bar{L}_c \pm$ SD of slightly constricted cells; $\bar{L}_c$ = 2.33 µm; CV = 8.7%. (o) $\bar{L}_s \pm$ SD of far constricted cells; $\bar{L}_s$ = 2.58 µm; CV = 9.1%. The straight lines were calculated by linear regression (cf. Koppes et al., 1980).

constant probability per unit of time. The variable age at initiation of chromosome replication is the consequence of the variable time spent in the hypothetical indeterminate state. Therefore, this hypothesis predicts positive correlation between size and age at initiation of chromsome replication. ii. According to Donachie (1968) and Pritchard (1968), cells initiate chromosome replication when they achieve a particular size independent of the size at birth. The variable age at initiation is the consequence of the independent size and age at chromosome initiation. This finding is not consistent with the model of Smith and Martin in which initiation is primarily controlled by time. (However, the introduction of a size element in the model (L. Martin, personal communication) would reduce the expected correlation between size and age). We favor the view that initiation occur at a particular size independent of birth. A similar size control over initiation of chromosome replication has been described for fission yeast (Nasmyth et al., 1979), budding yeast (Tyson et al., 1979) and animal cells (Killander and Zetterberg, 1965) Size as such is a vague concept and it is not clear how it should be understood in terms of, for instance, the critical concentration of a chemical compound and/or a characterisitic cytological feature.

Initiation of DNA replication is preceded by the I period (Fig. 6). The B period is thus the difference between the I period and C + D period of the preceding cycle (Fig. 27): $B_2 = I_2 - (C_1 + D_1)$. The age at which DNA replication starts can therefore be considered in two ways. Firstly, the age after cell birth (B period), secondly, the age since the previous initiation (I period). It should be remembered that the variation in interinitiation time (I) of DNA replication is smaller than the variation in interdivision time (τ) (Newman and Kubitschek, 1978). The additional finding of the extreme variability of the B period (Koppes et al., 1980) suggest a new age definition. That is, age is measured starting from the previous initiation of DNA replication. In this way both size and age could be equally important to describe the control of chromosome initiation.

CONCLUDING REMARKS

In this section we will summarize a number of problems of which we feel that they should be solved for a further understanding of the *E. coli* division cycle.

First of all one can witness a generation gap with respect to the study of family relations of individual cells. The

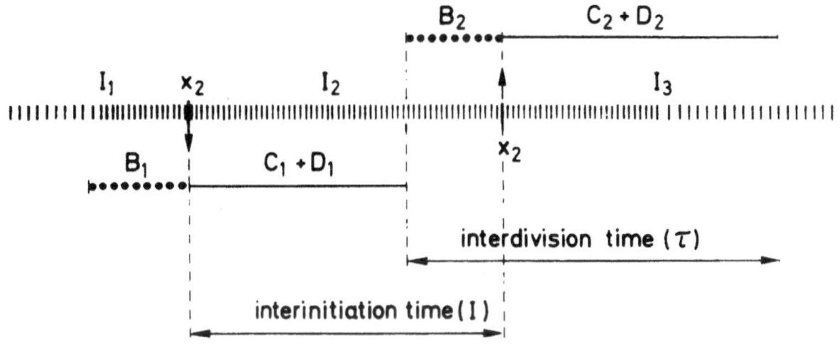

Fig. 27 Diagram to illustrate the relationship between the chromosome replication cycle (I) and the cell division cycle ($\tau = B + C + D$). At the end of the I period replication is initiated resulting in the doubling of the number of origins. x2 indicates that two new I periods are initiated. Arrow indicates that initiation is followed by cell division after C + D min. B is the difference between I and C + D of the preceding cycle: $B_2 = I_2 - (C + D)_1$. The duration of I, B, C + D and τ should not be conceived as constant, as suggested in the diagram, but distributed.

relevant studies at the late fifties and early sixties have not been taken up again. Systematic studies in this area, with our present knowledge as a background, should be extended. Such investigations are perhaps not spectacular and are very time consuming, but would be very informative.

A second problem concerns the mode of overall surface growth. We have pointed out the different approaches and we stressed the absence of a general accepted view. A solution can be sought in a thorough analysis of cell synchrony.

Insufficient knowledge exists also with respect to growth of the three surface layers (Fig. 1). It has not yet convincingly been demonstrated that incorporation in a central growth site is followed by lateral spreading. Though it is likely that the murein layer serves as a scaffold for outer and inner membrane, the coordinated growth of the three has still to be resolved.

Hardly investigated is the change (if at all) of physical parameters during the division cycle. This applies for instance to cell density, osmotic pressure, flexibility of the cell envelope. It might be naive to think that the signal leading to cell division will only be found by application of recombinant DNA techniques for the isolation of gene products. Cell division might well be triggered by structural properties and processes that follow simple physical laws.

Concerning cell division and also DNA replication we have mentioned the importance of size rather than age for these processes to start. Nevertheless a concept like initiation mass has so far not obtained a concrete physiological meaning. It is expected that more knowledge about regulation of initiation of DNA replication will remove some of the uncertainties. In particular *ori C* containing minichromosomes seem promising tools.

In this paper we have stressed the variability of cell division cycle events with respect to size and age. We could also deduce correlations between them. Especially between the initiation of DNA replication and the onset of division. However, the task remains to give these correlations a chemical and/or physical meaning.

We have limited our discussion to *E. coli* and for the other prokaryotic organisms we have referred to some recent reviews.

Finally, we hope that those who work with eukaryotes will recognize their own research problems in our presentation.

ACKNOWLEDGEMENTS

We are grateful to A.C. Gräper and C.E.A. van Wijngaarden for help in preparing this manuscript and to J.H.D. Leutscher and J. Woons for making the illustrations.

We also wish to acknowledge many fruitful discussions we had in the past with Drs. W.D. Donachie, N.B. Grover, A.L. Koch, H.E. Kubitschek, L. Martin, M. Meijer, W. Messer, R.H. Pritchard, R.F. Rosenberger, U. Schwarz, F.J. Trueba, R.W.H. Verwer and A. Zaritsky.

REFERENCES

Begg, K.J. & Donachie, W.D. (1973) Nature (London) 245, 38-39.

Begg, K.J. & Donachie, W.D. (1977) J. Bacteriol. 129, 1524-1536.

Bleecken, S (1971) J. Theor. Biol. 32, 81-92.

Boyd, A. & Holland, I.B. (1979) Cell 18, 287-296.

Burdett, I.D.J. & Murray, R.G.E. (1974) J. Bacteriol. 119, 303-324.

Burdett, I.D.J., E. Murray, R.G.E. (1974) J. Bacteriol. 119, 1039-1056.

Burdett, I.D.J. (1979) J. Bacteriol. 137, 1395-1405.

Chambliss, G.H. (1979)in Developmental Biology of Prokaryotes, ed. Parish, J.H. (Blackwell, Oxford), pp. 57-71

Churchward, G. & Bremer, H. (1977) J. Bacteriol. 130, 1206-1213.

Clark, D.J. (1968) Cold Spring Harb. Symp. Quant. Biol. 33, 823-838.

Collins, J.F. & Richmond, M.H. (1962) J. Gen. Microbiol. 28, 15-33.

Cooper, S. & Helmstetter, C.E. (1968) J. Mol. Biol. 31, 519-540
Cooper, S. (1979) Nature (London) 280, 17-19.
Daneo-Moore, L. & Shockman, G.D. (1977) in Cell Surface Reviews eds. Poste, G. & Nicolson, G.L. (Elsevier/North Holland, New York), vol. 4, pp. 597-715.
De Chastellier, C., Hellio, R. & Ryter, A. (1975) J. Bacteriology 123, 1184-1196.
Donachie, W.D. (1968) Nature (London) 219, 1077-1079.
Donachie, W.D. & Begg, K.J. (1970) Nature (London) 227, 1220-1224.
Donachie, W.D., Begg, K.J. & Vicente, M. (1976) Nature (London) 264, 328-333.
Donachie, W.D. (1979) in Developmental Biology of Prokaryotes, ed. Parish, J.H. (Blackwell, Oxford), pp. 11-35.
Dow, C.S. & Whittenbury, R. (1979) In Developmental Biology of Prokaryotes, ed. Parish, J.H. (Blackwell, Oxford), pp. 139-165.
Dworkin, M. (1979) in The Bacteria, eds. Sokatch, J.R. & Ornston, L.N. (Academic Press, New York) vol. 7, pp. 2-84.
Ecker, R.E. & Kokaisl, G. (1969) J. Bacteriol. 98, 1219-1226.
Fantes, P.A. (1977) J. Cell Sci. 24, 51-67.
Feller, W. (1976) An introduction to probability theory and its applications. vol. I, p. 236.
Finney, D.J. (1973) Probit Analysis (3rd ed.) Cambridge Univ. Press.
Giesbrecht, P., Wecke, J. & Reinicke, B. (1976) Int. Rev. Cytol. 44, 225-318
Grover, N.B., Woldringh, C.L., Zaritsky, A. & Rosenberger, R.F. (1977) J. Theor. Biol. 67, 181-193.
Grover, N.B., Zaritsky, A., Woldringh, C.L. & Rosenberger, R.F. (1980) J. theor. Biol. 85, in press.
Harvey, R.J., Marr, A.G. & Painter, P.R. (1967) J. Bacteriol. 93, 605-617.
Helmstetter, C.E., Cooper, S., Pierucci, O. & Revelas, E. (1968) Cold Spring Harb. Symp. Quant. Biol. 33, 809-822.
Helmstetter, C.E. (1969) In: Methods in Microbiol., vol. 1, Norris, J.R. & Ribbons, D.W. (eds.) Acad. Press, New York, p. 327-363.
Helmstetter, C.E. & Pierucci, O. (1976) J. Mol. Biol. 102, 477-486.
Helmstetter, C.E., Pierucci, O., Weinberger, M., Holmes, M. & Tang, M.S. (1979) In: The Bacteria, vol. 7, p. 517-579, Ornston, L.N. & Sokatch J.R. (eds.), Acad. Press, New York.
Higgins, M.L. & Daneo-Moore, L. (1980) J. Bacteriol. 141, 938-945.
Howard, A & Pelc, S.R. (1953) Heredity, Suppl. 6, 261-273.
Inouye, M. (1979) ed. Bacterial Outer Membranes (John Wiley & Sons, New York).

Jacob, F., Brenner, S. & Cuzin, F. (1963) Cold Spring Harb. Symp. Quant. Biol. 28, 329-347.
Jacq, A., Lother, H., Messer, W. & Kohiyama, M. (1980) ICN-UCLA Symp. on Molec. and Cell Biol. vol. 19 (in press).
Killander, D. & Zetterberg, A. (1965) Exp. Cell Res. 40, 12-20.
Koch, A.L. & Schaechter, M. (1962) J. Gen. Microbiol. 29, 435-454.
Koch, A.L. & Blumberg, G. (1976) Biophys. J. 16, 389-405.
Koch, A.L. (1977) Adv. in Microbial Phys. 16, 49-98.
Koppes, L.J.H., Overbeeke, N. & Nanninga, N. (1978a) J. Bacteriol. 133, 1053-1061.
Koppes, L.J.H., Woldringh, C.L. & Nanninga, N. (1978b) J. Bacteriol. 134, 423-433.
Koppes, L.J.H., Meijer, M., Oonk, H.B., De Jong, M.A. & Nanninga, N. (1980) J. Bacteriol. in press.
Koppes, L.J.H. & Nanninga, N. (1980) J. Bacteriol.143, 89-99.
Kubitschek, H.E. (1962) Exptl. Cell Res. 26, 439-450.
Kubitschek, H.E. (1968) Biophys.J. 8, 792-804.
Kubitschek, H.E. (1970) J. theor. Biol. 28, 15-29.
Kubitschek, H.E. (1974) Biophys. J. 14, 119-123.
Leibowitz, P.J. & Schaechter, M. (1975) Int. Rev. Cytol. 41, 1-28.
Marr, A.G., Harvey, R.J. & Trentini, W.C. (1966) J. Bacteriol. 91, 2388-2389
Mathys, E & van Gool, A.P. (1979) J. Bacteriol. 138, 642-646.
Meacock, P.A. & Cohen, S.N. (1980) Cell 20, 529-542.
Meijer, M., de Jong, M.A., Woldringh, C.L. & Nanninga, N. (1976a) Eur. J. Biochem. 63, 469-475.
Meijer, M., de Jong, M.A., Woldringh, C.L. & Nanninga, N. (1976b) Eur. J. Biochem. 65, 409-414.
Meijer, M., de Jong, M.A., Demets, R. & Nanninga, N. (1979) J. Bacteriol. 138, 17-23.
Mendelson, N. (1977) In: Microbiology p. 5-24, Schlessinger, D. (ed.), American Soc. Microbiol., Washington, D.C.
Messer, W., Bergmans, H.E.N, Meijer, M. & Womack, J.E. (1978) Mol. Gen. Genet. 162, 269-275.
Mitchison, J.M. (1973) Symp. Soc. Gen. Microbiol. 23, 1-11.
Nachtwey, D.S. & Cameron, I.L. (1968) In: Methods in Cell Physiol., vol. III, p. 213-258. Prescott, D.M. (ed.) Acad. Press, New York.
Nagai, K., Hendrickson, W., Balakrishan, R., Yamaki, H., Boyd, D. E Schaechter, M. (1980) Proc. Natl. Acad. Sci. USA 77, 262-266.
Nanninga, N., Koppes, L.J.H & de Vries-Tijssen, F.C. (1979) Arch. Microbiol. 123, 173-181.
Nasmyth, K., Nurse, P. & Fraser, R.S.S. (1979) J. Cell Sci. 39, 215-233.
Newman, C.N. & Kubitschek, H.E. (1978) J. Mol. Biol. 121, 461-471.

Ogura, T., Miki, T. & Hiraga, S. (1980) Proc. Natl. Acad. Sci. USA 77, 3993-3997.
Onken, A. & Messer, W. (1973) Mol. Gen. Genet. 127, 349-358.
Parish, J.H. (1979) in Developmental Biology of Prokaryotes, ed. Parish, J.H. (Blackwell, Oxford), pp. 227-253.
Pierucci, O. (1978) J. Bacteriol . 135, 559-574.
Poole, R.K. (1977) J. Gen. Microbiol. 98, 177-186.
Powell, E.O. (1956) J. Gen. Microbiol. 15, 492-511.
Powell, E.O. (1958) J. Gen. Microbiol. 18, 382-417.
Powell, E.O. (1964) J. Gen. Microbiol. 37, 231-249.
Prescott, D.M. (1976) Repr. of Euk. Cells (Ac. Press, N.Y.)
Pritchard, R.H. (1968) Heredity 23, 472 (Abstract).
Pritchard, R.H. (1974) Phil. Trans. Royal. Soc. London 267, 303-336.
Pritchard, R.H. (1978) In: DNA synthesis: Present and Future, p. 1-26. Molineux, I. & Kohiyama, M. (eds.) Plenum.
Rao, C.R. (1973) In: Linear statistical inference and its applications p. 366-373. John Wiley & Sons.
Rosenberger, R.F., Grover, N.B., Zaritsky, A. & Woldringh, C.L. (1978a) J. theor. Biol. 73, 711-721.
Rosenberger, R.F., Grover, N.B., Zaritsky, A. & Woldringh, C.L. (1978b) Nature (London) 271, 244-245.
Ryter, A, Hirota, Y. & Schwarz, U. (1973) J. Mol. Biol. 78, 185-195.
Ryter, A., Shuman, H. & Schwartz, M. (1975) J. Bacteriol. 122, 295-301.
Sargent, M.G. (1975) J. Bacteriol. 123, 1218-1234.
Sargent, M.G. (1979) Adv. in Microbiol. 18, 105-176.
Schaechter, M., Maaløe, O. & Kjeldgaard, N.O. (1958) J. Gen. Microbiol. 19, 592-606.
Schaechter, M., Williamson, J.P., Hood, J.R. & Koch, A.L. (1962) J. Gen. Microbiol. 29, 421-434.
Schwarz, U., Asmus, A. & Frank, H. (1969) J. Mol. Biol. 41, 419-429.
Schwarz, U., Ryter, A., Rambach, A., Helio, R. & Hirota, Y. (1975) J. Mol. Biol. 93, 749-760.
Seiki, M., Ogasawara, N. & Yoshikawa, H. (1979) Nature (London) 281, 699-701.
Slater, M. & Schaechter, M. (1974) Bacteriol. Rev. 38, 199-221.
Smith, J.A. & Martin, L. (1973) Proc. Natl. Acad. Sci USA 68, 2627-2630.
Spratt, B.G. (1975) Proc. Natl. Acad. Sci USA 72, 2999-3003.
Staugaard,P., van den Berg, F.M., Woldringh, C.L. & Nanninga N. (1976) J. Bacteriol. 127, 1376-1381.
Trueba, F.J. & Woldringh, C.L. (1980) J. Bacteriol. 142, 869-878.
Tyson, C.B., Lord, P.G. & Wheals, A.E. (1979) J. Bacteriol. 138, 92-98.

Verwer, R.W.H., & Nanninga, N. (1980) J. Bacteriol 144, (in press).
Verwer, R.W.H., Nanninga, N.. Keck, W. & Schwarz, U. (1978) J. Bacteriol. 136, 723-729.
Verwer, R.W.H., Beachey, E.H., Keck, W., Stoub, A.M. & Poldermans, J.E. (1980) J. Bacteriol. 141, 327-332.
Von Meyenburg, K., Hansen, F.G., Nielsen, L.D. & Riise, E. (1978) Mol. Gen. Genet. 160, 287-295.
Ward, C.B. & Glaser, D.A. (1979) Proc. Natl. Acad. Sci USA 68, 1061-1064.
Weidel, W. & Pelzer, H. (1964) Adv. Enzymol. 26, 193-232.
Winston, S. & Sueoka, N. (1980) Proc. Natl. Acad. Sci. USA 77, 2834-2838.
Woldringh, C.L. (1976) J. Bacteriol. 125, 248-257.
Woldringh, C.L., de Jong, M.A., van den Berg, W. & Koppes, L.J.H. (1977) J. Bacteriol. 131, 270-279.
Woldringh, C.L., Grover, N.B., Rosenberger, R.F. & Zaritsky, A. (1980) J. theor. Biol. 85, in press.
Wolf-Watz, H. & Masters, M. (1979) J. Bacteriol. 140, 50-58.
Yamada, H. & Mizushima, S. (1978) J. Bacteriol. 135, 1024-1031
Yasuda, A. & Hirota, Y. (1977) Proc. Natl. Acad. Sci. USA 74, 5458-5462.
Zaritsky, A. & Woldringh, C.L. (1978) J. Bacteriol. 135, 581-587.
Zaritsky, A. & Zabrovitz, S. (1980) Mol. Gen. Genet., in press.
Zaritsky, A, Grover, N.B., Naaman, J., Woldringh, C.L. & Rosenberger, R.F. (1980) manuscript in preparation.

DEVELOPMENTAL REGULATION OF ENZYME SYNTHESIS IN *SACCHAROMYCES CEREVISIAE*

James G. Yarger, Keith A. Bostian and Harlyn O. Halvorson

Rosenstiel Basic Medical Sciences Research Center
Brandeis University
Waltham, Massachusetts 02254

The yeast *Saccharomyces cerevisiae* has long been a model organism for studying developmental processes in an eukaryote. The life cycle involves both haploid and diploid cells. These cells undergo spherical vegetative growth or polar growth leading to the formation of a discreet morphological bud early in the mitotic cell cycle. During sporulation in diploid cells, meiosis and ascospore formation leads to haploid cells which may then conjugate to complete the life cycle.

The mitotic cell cycle, as diagrammed in Fig. 1, begins with a single unbudded cell. This cell enlarges progressively throughout the cell cycle. Bud initiation precedes DNA synthesis (S period) which occurs in the first quarter of the cell cycle. After a further growth during the G2 period, nuclear division and nuclear migration occur. The cells then enter the G1 period and continue until cell division. Throughout these specific periods numerous cellular events have been identified (1,2). Many of these events are periodic and spacially related.

In this chapter we will present current information on the basic molecular controls which regulate macromolecular synthesis during growth in *S. cerevisiae*. Our attention will be drawn to those levels of control (transcription, post-transcription, trans-lation or post-translation) at which regulation of the synthesis of individual proteins occurs.

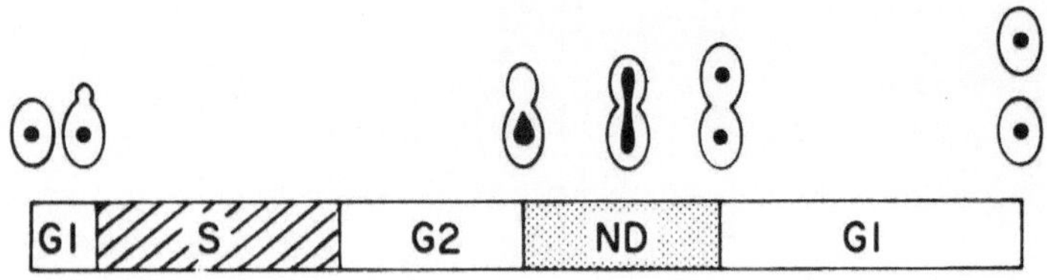

Fig. 1. The Cell Cycle of Saccharomyces cerevisiae

THE YEAST CELL CYCLE

Considerable data has accumulated in the past decade on the timing of macromolecular synthesis in the cell cycle. Numerous alternative models have been generated (3-8), which disagree over the pattern and underlying controls of individual macromolecular synthesis. There is agreement that total RNA, tRNA and total protein concentrations increase continuously throughout the cell cycle, (9-14), whereas DNA replication is confined to a brief S period. Observations with budding yeasts from our laboratory and others have shown that with the exception of RNA polymerase I (6,15) and ribosomal proteins (17), over 30 enzymes show step-wise increases in activity which are not confined to a single period of the cell cycle (18,19). Three methods have been employed to study macromolecular synthesis during the cell cycle: synchronous cultures made by induction or selection from gradients, age fractionation on a zonal rotor, or single cell enzyme assays on exponential cultures. In contrast to the step functions mentioned above, Elliott and McLaughlin (14,20), using pulse labelling and O'Farrell two dimensional gel electrophoresis, reported that the rates of synthesis of 108 out of 110 proteins examined in S. cerevisiae were exponential throughout the cell cycle. Cells were age-fractionated by centrifugal elutriation. A lack of periodicity has also been observed in 1D gels from S. pombe (D. Dickerson, personal communication) and in 2D gels from other organisms (chlorella, physarium, mammalian cells and in E. coli). In one case, Howell (personal communication) has shown evidence for periodic synthesis in Chlamydomonas induced by light/dark cycles. All of these experiments, however, have been subject to limitations similar to those in yeast. Studies by Mitchison on the fission yeast S. pombe, which showed changes in the rate of synthesis of individual enzymes (4), have been recently reinterpreted as perturbations induced by pre-growth conditions. These perturbations appear in both asynchronous and synchronous cultures (21). Immediate questions therefore arise as to whether the observed experimental differences in budding yeasts are due to perturbations induced by pregrowth conditions, methods of fractionation, degree of fractionation of cell types, variation in the mixing of precursor pools, or analysis of the electrophoretic data. Moreover, do the step increases observed in budding yeast reflect intrinsic differences between budding and fission yeast? If step-wise patterns represent a normal cell cycle event, as the assay of β-galactosidase in single cells (22) leads one to believe, a critical question becomes whether they represent *de novo* protein synthesis or some post-translational regulation.

GENERAL REGULATORY CONTROLS

Normal proliferation of yeast cells by mitotic or meiotic cell division clearly encompasses controls on metabolism. Numerous examples in yeast of regulatory control systems directing inducible or repressible metabolic pathways have been documented. Although some eucaryotic genetic loci have been mapped and identified as structural genes, little is known about their regulatory mechanisms. A number of yeast structural genes have recently been cloned. Among these are the genes for: (a) alcohol dehydrogenase (23); (b) orotidine-5'-phosphate decarboxylase (ura3) (24); (c) imidazol-glycerolphosphate dehydratase (*his3*) (66); (d) argininosuccinate lyase (*argH*) (25); (e) cytochrome c (*cyc1*) (26); (f) acid phosphatase (27; Bostian, Lemire, Rogers and Halvorson, unpublished observation); (g) transferase (*GAL7*), epimerase (*GAL10*), and galactokinase (*GAL1*) (28). Among these genes, *cyctochrome c* (26) and *ura3* (24) have been shown to be regulated at the level of transcription.

In bacterial systems, gene expression is highly efficient and multiple forms of regulation occur. In the *E. coli* arabinose operon, there are frequently multiple forms of transcriptional regulation within the same pathway. The complex, multiple controls regulating bacterial metabolism have probably arisen from selective pressures on bacteria to utilize available nutrients as efficiently as possible and to achieve optimal growth rates. Multiple controls regulating protein synthesis in yeast and other lower eucaryotes may also have arisen from similar selective pressures. Mechanisms for transcriptional control of enzyme synthesis can be considered at three general levels that need not be mutually exclusive.

Coarse control is generally a position-specific effect rather than gene-specific regulation. In many eucaryotic systems, coarse control is exerted through the action of heterochromatin regions of chromosomes. For example, varigated position effects in Drosophila are seen whenever a gene normally lying in euchromatin is brought next to or into a heterochromatin region (29,30). This effect on gene activity is a phenomenon brought about by the influence of specific portions of heterochromatin on genes abnormally set next to it. Thus, genes normally located in euchromatin frequently have their activity suppressed when they are translocated to heterochromatin. This situation is exaggerated in the case of mammalian X-chromosome inactivation (31-34). During early embryological development some interaction between the X-chromosomes and autosomes apparently occurs with the result that one entire X-chromosome becomes condensed into a heterochromatin state. Currently there is no data to suggest that yeast contain heterochromatin-like regions. However, yeast do exhibit other forms of position-specific effects, notably in the expression of mating type genes. Mating type in S. cerevisisae is controlled by transposable alleles of the mating type (MAT) locus. Mating type information (either a or α) can only be expressed when it is present at the MAT locus (35,36). The conversion of mating type depends upon the alleles of two other loci, HML and HMR that are located distant from the MAT locus on chromosome III (35). The switching of mating-type involves the transposition of a copy of a or α information from the unexpressed genes at HML or HMR to replace the sequence at the mating-type locus, MAT (37). In homothallic strains which carry the dominant allele HO, a MATa allele can switch to MATα or vice versa as often as once every cell division. However, in normal heterothallic (ho) strains a particular MAT allele is very stable and changes in mating-type information occur only rarely (37).

The second level of control is fine control. This level includes control of specific genes through the action of inducers and repressor upon small regions of euchromatin. Examples of these include the acid phosphatase (38-40), and galactose systems (41,42) in S. cerevisiae, the regulation of phosphorus metabolism in Neurospora (43) and proline metabolism (44) and acetamidase synthesis (45) in Aspergillus nidulans. All of these metabolic pathways contain both positive and negative regulatory genes which control structural gene activities. Fine control also includes induction of transcriptional activity through the action of an earlier intermediate in a pathway. For example, transcription from the yeast ura3 gene is inducible by earlier intermediates in the pyrimidine pathway, i.e., ural (46).

Many structural and regulatory genes fall under the control of catabolite repression. Organic compounds which can be metabolized efficiently and support rapid growth tend to inhibit the utilization of carbon sources that are less readily metabolized. In general, catabolite repression upon the inhibition of inducible enzyme synthesis is not due to the carbon source itself, but rather to intermediary metabolites which act as repressors (47,48). In bacteria, and possibly yeast, the intermediary metabolite responsible for catabolic repression in most cases is cyclic 3'-5'-AMP (cAMP) (49). In *S. cerevisiae*, synthesis of iso-1-cytochrome c is regulated by catabolite repression. Glucose represses both the rate of protein synthesis and cellular levels of translatable *cyc1* RNA (50,51). Zitomer et al. (52) used the cloned yeast *cyc1* to assay the rate of *in vivo cyc1* mRNA transcription by DNA excess hybridization of pulse-labelled RNA. They determined that the effects of catabolite repression on the *cyc1* gene is at the level of transcription. Thus it appears that the mechanism of catabolite repression for at least one eucaryotic gene is similar to that observed in procaryotic systems.

The third level of control concerns *fine tuning* of gene expression; i.e., mechanisms operating which control the amount of RNA transcription occuring from either constitutively expressed genes or from genes under the influence of inducers and repressors. Such fine tuning can operate through several different mechanisms. Fine tuning controls are not limited to transcriptional regulation but include all mechanisms effecting modulations in the levels of the transcription products. There is evidence that yeast mRNA concentrations can be controlled at the level of mRNA stability (Osley and Hereford, personal communication; 53). In gene dosage studies, a 2X amount of histone genes will produce a 2X amount of histone mRNA at any point in time (i.e., during pulse labelling). However, the "extra" histone mRNA becomes quickly degraded to produce a constant 1X amount of mRNA (Osley and Hereford, personal communication) suggesting effects of *specific* mRNA species can operate. It is known that in both bacteria and yeast individual mRNA molecules can have widely differing half-lives. The half-life of total yeast mRNA has been determined to be approximately 20 min for *S. cerevisiae* growing with a three hr generation time (54-56). When the half-lives of individual yeast mRNAs are determined by their ability to synthesize detectable proteins, a very wide range of values is obtained (53,57; Perlman and Halvorson, unpublished results). This is most dramatically shown in a recent study of 80 separate mRNAs by Chia and McLaughlin (personal communication). The analysis of protein products upon O'Farrell two-dimensional gels showed half-lives ranging from 4.5 to 41 min with an average value of 22 min (Chia and McLaughlin, personal communication). Thus, individual mRNA molecules in yeast appear to have specific half-lives although what specifies a particular mRNA half-life is unknown.

Poly(A) sequences at the 3' termini of mRNA molecules have been implicated in the control of mRNA stability (58,60) and mRNA turnover (61). mRNAs assayed by in vitro translation contain a wide range of poly(A) lengths (62). There are at least two explanations: (a) mRNAs are polymerized with varying poly(A) lengths initially, or (b) poly(A) lengths of a given mRNA are uniform initially but through degradation and/or repolymerization of poly(A) in the cytoplasm, become heterogeneous.

Saunders and Bostian (62) showed that the length of poly(A) sequences on yeast uridyl-transferase mRNA do indeed decrease with age of the mRNA although they did not demonstrate that this decrease was correlated with changes in mRNA stability or half-life. By comparing the ratio of translation products for galactose-uridyl transferase mRNAs and for glyceraldehyde 3-phosphate dehydrogenase in each poly(A) size fraction at both steady-state for galactose induction and 15 min post-induction, a shift in the poly(A) size distribution from a fairly narrow one soon after induction to a more disperse one (virtually identical to that of the constitutive mRNA) in the case of steady state galactose grown cells were observed. This supports model (b) above.

Alterations in mRNA stability could possibly come about through mechanisms recently discovered for eucaryotic mRNAs in E. coli (63). When the expression of two cloned eucaryotic genes (Neurospora crassa qA-2^{+} gene (catabolite dehydroquinase) and the S. cerevisiae HIS3 gene (imidazol-glycerolphosphate dehydratase)) are compared in a wild type E. coli strain and an E. coli strain deficient in polynucleotide phosphorylase, expression is dramatically increased in the E. coli strain deficient in polynucleotide phosphorylase (63). The qA-2 and HIS3 mRNAs are significantly stabilized in polynucleotide phosphorylase deficient strains such that 20-100 fold increases in the specific enzyme activities are seen. This stabilization is unique for eucaryotic mRNAs since mRNAs for the pBR322-encoded Ap^{r} and Tc^{r} proteins were not stabilized nor was the level of expression of the E. coli biosynthetic dehydroquinase increased. This suggests that eucaryotic mRNAs may possess some unique secondary or tertiary structures not present in procaryotic mRNAs which can be recognized and processed through the intervention of polynucleotide phosphorylase. In the absence of polynucleotide phosphorylase, the eucaryotic mRNAs may be partially protected from E. coli degradative enzymes. It is possible that this same type of destabilization of eucaryotic mRNAs may occur in some form or another in eucaryotic cells.

Fine tuning also includes autogenous regulation which can show either transcriptional or post-transcriptional controls as well as protein turnover (64). The latter is an important mechanism for controlling cellular enzyme accumulation (23,65,66).

REGULATION OF THE GALACTOSE METABOLIC PATHWAY

1. The Galactose System

The galactose pathway in *Saccharomyces cerevisiae* contains four inducible structural genes and one constitutive structural gene (see Fig. 2 and 3). The inducible structural gene activities are under the control of three regulatory genes.

Externally supplied galactose is actively transported into the cell through the action of the structural protein, galactose permease which is coded for by *GAL2* (67). Although *GAL2* is required to accumulate galactose, its activity is inducible.

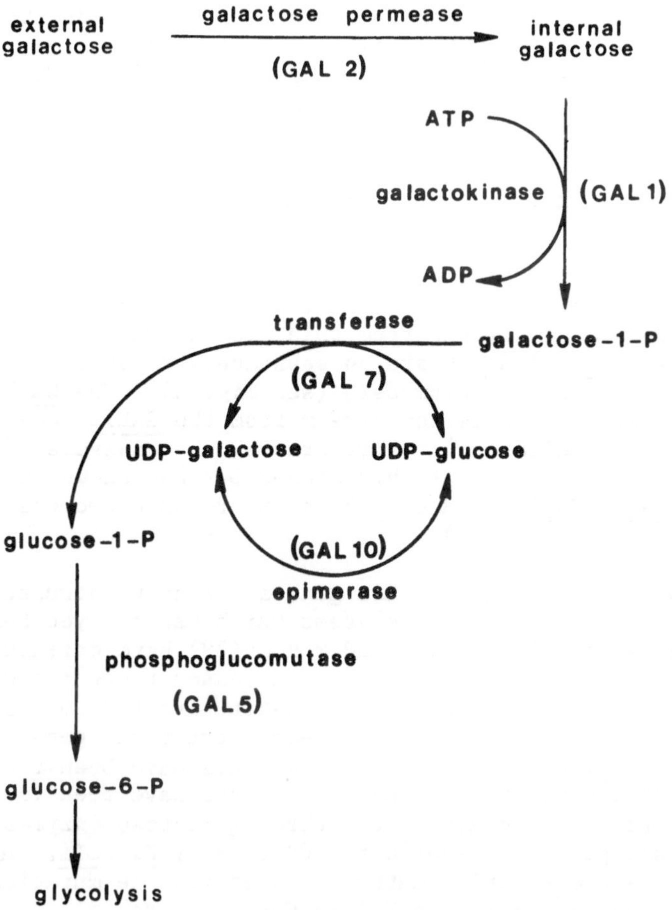

Fig. 2. The Galactose Metabolic Pathway

Internal galactose is converted to glucose-1-phosphate, a substrate for the glycolytic pathway, through the action of the "Leloir" enzymes (68-71). GAL2 maps to chromosome XII (72). All three of the Leloir enzymes are unlinked to GAL2 and map as a tightly linked trio near the centromere on chromosome II (71,73,74). The first enzyme of the pathway, galactokinase (GAL1 or "kinase", EC 2.7.1.6), catalyzes the phosphorylation of galactose to galactose-1-phosphate. Galactose-1-phosphate is toxic to cells. In wild-type cells it is quickly converted to glucose-1-phosphate by the next inducible enzyme, uridyl transferase (GAL7 or "transferase", EC 2.7.7.10). The third inducible enzyme in the pathway, uridine diphosphogalactose-4-epimerase (GAL10 or "epimerase", EC 5.1.3.2) recycles UDP-galactose, a byproduct in the formation of glucose-1-phosphate, to UDP-glucose. The end-product produced by these Leloir enzymes, i.e., glucose-1-phosphate, is converted to the substrate for the glycolytic pathway, glucose-6-phosphate, by the action of the constitutive enzyme phosphoglycomutase (GAL5 or "mutase", EC 2.7.5.1) (71,75,76). Nonsense mutations in any of the three inducible structural genes (GAL7, GAL10, GAL1) result in a loss of enzyme activity only for the enzyme specified by the mutant gene (77). There are no observed polarity effects for these genes.

2. Transcriptional Control of the Structural Genes

St. John and Davis (28) have recently isolated a cloned yeast DNA fragment containing all three of the structural gene trio, GAL7, GAL10, and GAL1 by differential hybridization. Hybridization studies with separated strands from this cloned DNA has revealed that the genes GAL1 and GAL10 are on separate strands and are transcribed from divergent promoters (see Fig. 3). The GAL7 gene is on the same strand as and is downstream from the GAL10 gene (St. John, personal communication). Furthermore, the analysis of *in vitro* synthesized deletions of this cloned DNA has shown that all three genes, GAL7, GAL10, and GAL1 have their own promoters (St. John, personal communication).

The association of an operator gene adjacent to structural genes is a feature of bacterial systems which has not yet been identified in yeast (71). Jacob and Monod (79) have defined the *E. coli Lac* operator locus as the site of repressor recognition. This locus was identified by *cis* operator-constitutive (O^c) mutations. The data obtained by St. John showing each structural gene has its own promoter explains why no simple O^c mutants have been isolated for the galactose gene trio in yeast. Mutants have been isolated in yeast that fail to synthesize the three galactose enzymes and therefore phenotypically resemble the O^c type of *E. coli*. However, none of these were *cis*-acting mutants. Instead, the mutations mapped to a locus on chromosome XVI (McCusker, personal

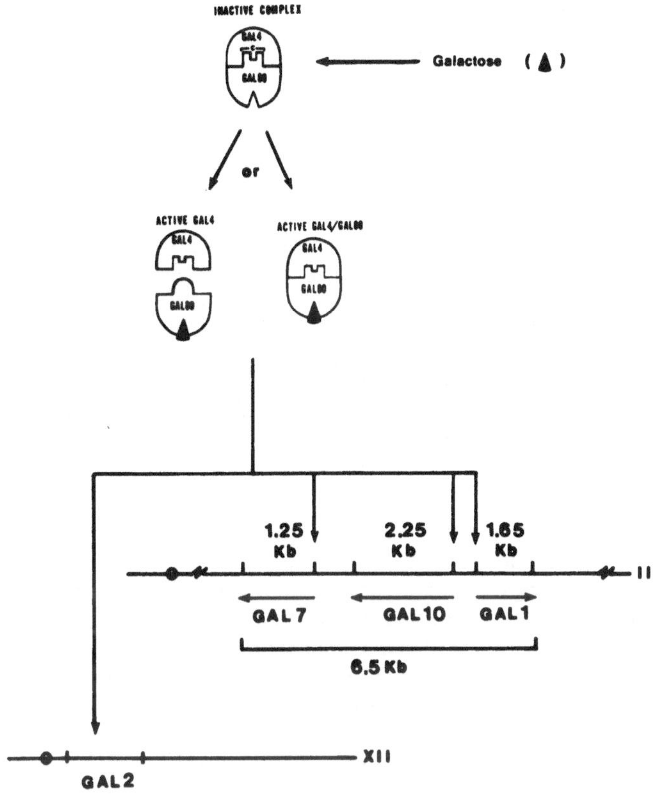

Fig. 3. Model for Regulation of the Galactose Metabolic Pathway

Model of Douglas and Hawthorne (41) as modified by Perlman and Hopper (42) and Matsumoto et al. (78) for the GAL4/GAL80 interaction. The transcriptional organization of the GAL7, GAL10 and GAL1 cluster is from St. John (personal communication).

communication) which segregated independently of the galactose structural genes. This corresponds to the positive regulatory locus, GAL4 (80-83). GAL4 codes for a polypeptide since some GAL4 mutations are suppressible by external nonsense suppressors (D. Hawthorne, personal communication). Temperature sensitive mutations in GAL4 do not produce thermolabile structural enzymes (80). Thus GAL4 does not code for a polypeptide common to the structural enzymes. Regulation of the galactose structural genes is at the level of mRNA synthesis since the appearance of translatable galactokinase and transferase mRNA depends upon wild-type GAL4 function (82,83). Using plasmids containing DNA sequences homologous to GAL7, GAL10, and GAL1, it has been shown that the GAL4 gene is required for the induction of the structural gene RNA transcripts (28). Since GAL4 is constitutively expressed, activity of the GAL4 protein itself must be regulated (42,78).

Another regulatory gene, the GAL80 locus, was identified through mutations that result in a constitutive phenotype (41,71) and by mutations that result in an uninducible phenotype (84). The $GAL80^s$ uninducable phenotype is dominant to the wild-type allele (GAL80) and to the recessive constitutive (gal80) allele of the gene. These mutants are termed supersuppressible ($GAL80^s$). GAL80 gene product has been found to regulate the activity of GAL4 protein by interacting directly with GAL4 at a site on the GAL4 enzyme identified as c or GAL81 (85,86). gal81 mutations result in a constitutive phenotype and are epistatic to the $GAL80^s$ mutations (84). Presumably, the GAL80 gene product is also constitutively synthesized.

Mutations in the last regulatory gene, GAL3, map close to the centromere on chromosome IV (87). These mutations are described in terms of long-term adaption and are characterized by a delay of 24 to 36 hrs in the onset of normal galactose fermentation in contrast to 8 to 10 min delay as found in wild-type cells. Strains that are both respiratory-deficient (petites) and $gal3^-$ are non-inducible for the galactose structural genes (88). However, the galactose-negative phenotype can be rescued by either of two constitutive mutations, i.e., either GAL4, gal81 mutants or gal80 mutants (41,88). It has been postulated by Adams (89,90) that the GAL3 gene might specify a function required for the establishment, but not the maintenance, of the induced state.

Broach (77) has shown that in a $gal3^-$ strain any mutation in the GAL7, GAL10, GAL1 cluster prevents even long-term induction of the galactose metabolic enzymes. Possibly GAL3 encodes a protein which converts galactose into a product which functions as the real inducer of the galactose enzymes (77,89,90). However, the real inducer of the galactose enzymes may be an intermediate in normal galactose metabolism (77). This seems unlikely based on the analysis by Hopper (personal communication) of the MELI gene. The product of the MELI gene (α-galactosidase) converts melibiose to glucose and galactose (91). Thus either melibiose or galactose can induce the galactose structural enzymes. In addition, melibiose or galactose also induced the MELI gene (91). Strains carrying a deletion for GAL7, GAL10, and GAL1 still induce α-galactosidase activity (Hopper, personal communication). Therefore, the presence of GAL7, GAL10, or GAL1 gene products are apparently not required for the induction of the MELI gene and therefore presumably not required for galactose enzyme induction.

3. Response to Catabolite Repression

The synthesis of galactose pathway enzymes respond to catabolite repression. The addition of galactose to a culture of cells previously grown in glucose exhibits a lag of several hrs before galactose induction occurs in contrast to the 8 min required for induction of cells previously grown on lactate (92). When glucose is added to yeast cultures growing on galactose, a severe, transient repression of enzyme synthesis occurs followed by a resumed lower rate of galactose enzyme synthesis (92).

In the absence of galactose (or melibiose) there is no detectable galactose enzyme activity (77,92). Ron Davis (personal communication) has estimated galactose-specific mRNAs are present in concentrations of approximately one copy per 20-50 cells in the absence of galactose. With the addition of galactose to the growth media, all four galactose structural enzyme activities are rapidly induced although the time of appearance of these enzymes is dependent upon the nature of the carbon source present in the growth media prior to the addition of galactose (92). When cells are pregrown on lactate or acetate, galactose induces detectable structural enzyme activity within 8 min. Once galactose is added to cells pregrown on acetate or lactate, full expression of the galactose enzymes is attained in approximately one hr (77).

Upon full induction, galactokinase (*GAL1*) and transferase (*GAL7*) each constitute 1.5% of the soluble protein in galactose-induced yeast cells (77).

4. Constitutivity of the Regulatory Proteins

Douglas and Hawthorne (41) originally proposed a model for the regulation of the galactose pathway genes which was based primarily on the Jacob-Monod model (93) of bacterial gene regulation. This model has now been extensively modified to accommodate the constitutive nature of the *GAL4* gene (see Fig. 3). *GAL4* protein is constitutively synthesized and is required for the *de novo* transcription of the galactose structural genes (28,42,78). The activity of the *GAL4* enzyme is regulated through the action of the *GAL80* protein (41,84).

Regulation of the *GAL4* enzyme activity by the *GAL80* repressor can most easily be explained by the existence of a *GAL4-GAL80* protein complex (42,78,86). This interaction occurs through the region of the *GAL4* protein encoded by the DNA sequence defined as *GAL81* (85,86). In the absence of the inducer galactose, the enzyme complex is normally inactive. The presence of galactose could dissociate the complex allowing GAL4 protein to effect structural gene transcription or galactose could activate the complex itself which could now activate structural gene transcription (42). Thus the galactose system is known to contain both negative and positive regulatory control elements.

REGULATION OF PHOSPHOROUS METABOLISM

1. The Phosphatase System

The phosphatase genetic control system in S. cerevisiae consists of a complex array of structural and regulatory genes cooperatively involved in the expression of both alkaline and acid phosphatases (94-97). These nonspecific phosphatases exist both as constitutive enzymes and enzymes repressible by inorganic phosphate. Alkaline phosphatase (pH optimum 8.0) is located cytoplasmically, whereas acid phosphatase (pH optimum 4.3) is a highly glycosylated exocellular enzyme. Acid phosphatase has been cytochemically localized (98) in the endoplasmic reticulum (involved in yeast cell wall synthesis), in the flat vesicles underneath the plasma membrane (considered subsurface cisternae and related to protein synthesis (99)), and in Golgi-like structures as well as other small vesicles. These observations support the view that the exocellular acid phosphatase is synthesized following a pathway similar to that of glycoproteins in higher eucaryotes (100). The structural genes for the constitutive and repressible acid phosphatases, PHOC and PHOE, respectively, are tightly linked (94,101). The PHOH gene is thought to be either the structural gene for repressible alkaline phosphatase, or a gene for some function in its expression (95).

There are seven unlinked regulatory genes: PHOS, PHOR, PHOU, PHOB, PHOD, PHOF and PHOG (38,94,101,102). Another gene, PHOT, is involved in active transport of inorganic phosphate (103). Mutations in PHOD and PHOS (38,95,102) repress both acid and alkaline phosphatase, whereas mutations in PHOR and PHOU are recessive and result in constitutive acid and alkaline enzyme production. Mutations in PHOB do not show the same pleiotrophic effect but rather give a phenotype lacking only the repressible acid phosphatase (94). PHOF and PHOG are involved in constitutive acid phosphatase expression.

Based on the genetic analysis of various double mutants, Oshima and colleagues (97,102) originally proposed a regulatory model wherein a repressor, encoded by PHOR and PHOU, is under negative control by PHOS. Repressor blocks transcription of the PHOD gene by binding to an adjacent controlling site PHOO. The PHOD gene product is essential for structural gene expression and acts as a positive regulatory element for both the repressible acid and alkaline phosphatase structural genes. For acid phosphatase this element acts in concert with the PHOB gene product. Several recent observations, however, are inconsistent with this model: (i) fine structure meiotic mapping situate the PHOO site far within the PHOD gene (39,97) (ii) the phoo/PHOD genotype is always semi-dominant in heterozygous diploids over the PHOO/PHOD (wild-type) and PHOO/phoD (PHOD mutant) genotypes (39,94,104), and (iii) coupled experiments of temperature and pH shifts with ts mutants of the regulatory genes suggest that acid pH affects the function of the cytoplasmic products of those regulatory genes in the expression of acid phosphatase, but not with alkaline phosphatase, even though they share a common regulatory mechanism (104). These findings have led Toh-e et al. (104) to propose a new regulatory model as outlined in Fig. 4. Accordingly, the PHOO region codes for a site within the positive factor which has an affinity for a negative factor. This

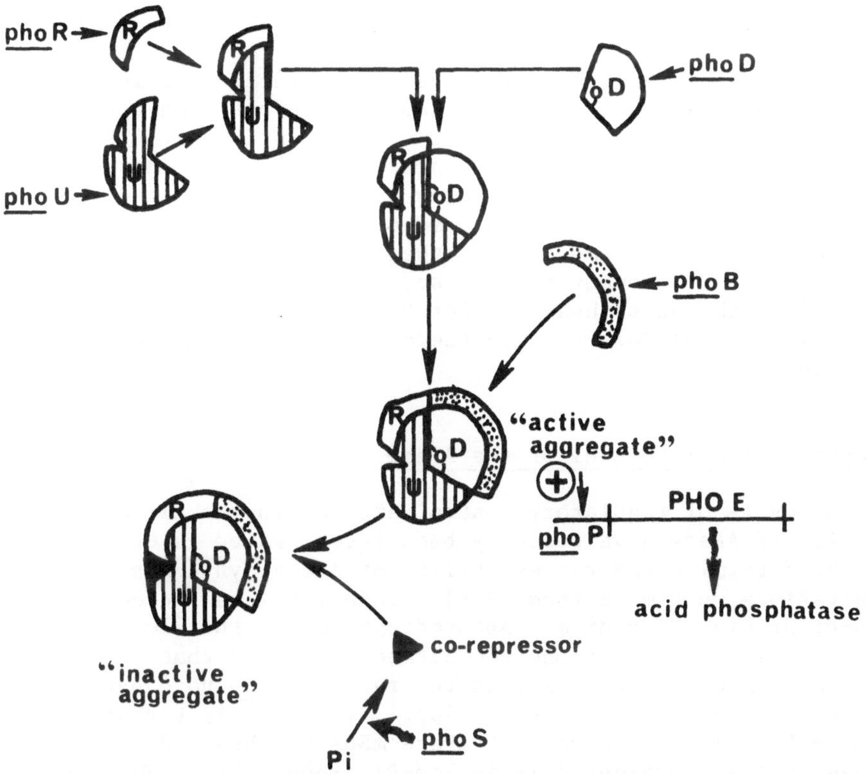

Fig. 4. Current Regulatory Model for Repressible Acid Phosphatase in S. cerevisiae Proposed by Toh-e, Kobayshi and Oshima (104)

positive regulatory factor is made constitutively from the PHOD gene. A product for negative control (apo-repressor), cooperatively produced by the PHOR and PHOU genes is aggregated to form an active regulatory complex which functions by binding to a control site (PHOP) adjacent to PHOE. The co-repressor (inorganic phosphate) has an affinity for the aggregate through the apo-repressor and when bound represses expression of PHOE. The cellular level or activity of the co-repressor might then be controlled by PHOS. At low pH the function of the aggregate is active in alkaline phosphatase structural gene expression but not PHOE.

Bostian et al. (40,105) recently have detected by cell-free translation three mRNAs (p60, p58 and p56) that code for distinct acid phosphatase polypeptides. These mRNAs are present only in cells grown in low-Pi medium. Utilizing hybridization selection procedures two structural genes among a number of isolated yeast genes controlled by the level of inorganic phosphate (27) have been shown to code for p60 and p56 mRNA (Bostian, Lemire and Rogers, unpublished results). This confirms the transcriptional control of APase expression.

Understanding the events involved in secretion and integration of acid phosphatase into the periplasmic space has been approached by Schekman and coworkers (106-108). Incorporation into the cell wall of APase and other cell wall material is restricted to the bud, in which membrane bound vesicles have been implicated. The isolation of conditional secretory mutants into a large number of complementation groups has enabled them to identify structural inermediates in the movement of APase, invertase and sulfate permease through the pathway of secretion. These morphogenic events have obvious relevance to the cell cycle control of APase expression.

2. Transcriptional Control of the APase Genes

The levels of regulatory control exerted in the differential expression of APase have recently been investigated in our laboratory. To determine the repressibility of APase synthesis, immunoprecipitations were made from [^{35}S]-methionine labelled cells grown in low-Pi or high-Pi medium. Autoradiography of the electrophoretically fractionated immunoprecipitates revealed that derepression of APase occurs primarily as the result of *de novo* synthesis of repressible APase polypeptide (40). Immunoprecipitations were also used to analyze products of APase mRNA synthesized *in vitro*. Messenger RNA was present only in low-Pi grown cells. This is consistent with the *in vivo* observation, and indicates a *de novo* synthesis of functional repressible APase mRNA. These data support the transcriptional control model of Toh-e et al. (104) rather than a model of post-translational control (109). That the regulation of translatable APase mRNA is primarily transcriptional comes from the identification of APase structural genes as induced by low-Pi.

By detection of immunogenic polypeptides synthesized in cell-free translation system programmed with RNA from wild type strain P28-24C (phoc, PHOE) we have characterized three mRNAs coding for three antigenically related APase polypeptides (p60, p58, p56) that have different but substantially homologous sequences. As shown in Fig. 5, the deglycosylated enzyme from P28-24C consists of several polypeptides (D1-D4). By analysis of

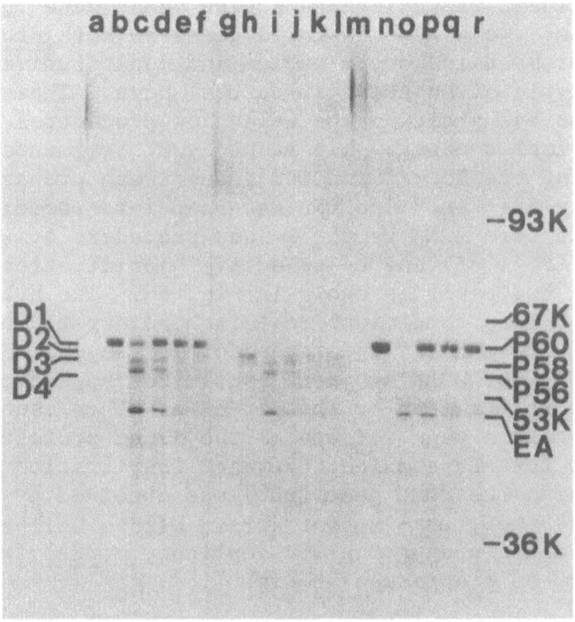

Fig. 5. Composite Gel Autoradiograms of Deglycosylated Enzyme and Cell-free Synthesized Proteins

Lanes (a)-(f) are strain P28-24C: (a) native and (b) deglycosylated enzyme; (c) immunoprecipitated cell-free synthesized proteins; (d) 2:1 mixture; (e) 4:1 mixture and (f) 8:1 mixture of (b) and (c). Lanes (g)-(l) are strain P142-4A: (g) native and (h) deglycosylated enzymel (i) immunoprecipitated cell-free synthesized proteins; (j) 1:1 mixture, (k) 2:1 mixture, and (l) 4:1 mixture of (h) and (i). Lanes (m)-(r) are strain A430: (m) native and (n) deglycosylated enzyme; (o) immunoprecipitated cell-free synthesized proteins; (p) 1:2 mixture; (q) 1:1 mixture and (r) 2:1 mixture of (n) and (o).

enzymes from mutant strains lacking p60 (P142-4A) or p58 (A430) mRNA activity, we have confirmed that at least 3 of the enzyme polypeptides are coded by APase mRNA. The existence of three mRNAs coding for three different proteins *in vitro*, and the appearance of multiple polypeptides in the native enzyme preparation suggest that the primary translation products of these mRNAs are modified and glycosylated *in vivo* in the formation of one or more secreted glycoprotein molecules.

3. Identification of the APase Structural Genes

A number of yeast genes which are controlled by inorganic phosphate have been isolated by Kramer and Anderson (27). These were identified by differential colony hybridization to a bank containing recombinant bacteriophage λ with yeast gene inserts. The transcription of two of the isolated genes in both *phoB* and *phoD* mutants is repressed in low-Pi medium and constitutive in a *phoR* mutant, phenotypic of an APase structural gene. These genes have been identified by hybridization selection procedures utilizing an *in vitro* translation assay. Two yeast EcoRI fragments of 8 kb and 5 kb, containing the low-Pi induced genes, were prepared from the λ clones shown in Fig. 6A. The DNA was used for hybridization selection of translatable mRNA by the R-loop procedure of Woolford and Rosbash (110) (Fig. 6B) and by selective hybridization to the DNA immobilized on DMB paper as shown in Fig. 6C. The hybridization under relaxed R-loop conditions reveals homology for both fragments to all three APase mRNAs. Under the more stringent conditions with immobilized DNA, the 8 kb fragment hybridizes specifically with p60 mRNA and the 5 kb fragment to the p56 mRNA. Thus each fragment appears to contain a gene for one of the APase proteins indicating multiple genes for APase mRNAs. Further confirmation that the 8 kb fragment contains the *PHOE* gene (p60) was obtained by transformation of a yeast *phoE*, *ura3* mutant strain with a self-replicating yeast plasmid containing a 2 μ DNA replicon, a *URA3* fragment and the 8 kb insert as diagrammed in Fig. 7.

A possible explanation exists for the failure to detect a third gene for the p58 polypeptide. Strain S288C, which was used to construct the genomic bank for initial screening, upon examination, was found to lack translatable p58 mRNA. This suggests that S288C lacks a copy of this gene. We therefore surveyed a number of yeast strains from several stock collections and found a large number of strains lacking translatable p58 mRNA. Relying on the expected cross-hybridization of the putative gene with the 8 kb and 5 kb fragments, we have been able to correlate the presence of a weakly hybridizing 4.5 kb Eco RI fragment found in genomic blots from strains that possess p58 mRNA (Fig. 8) (Thill, Bostian and Kramer, unpublished observations). This 4.5 kb fragment thus becomes a candidate for a p58 gene.

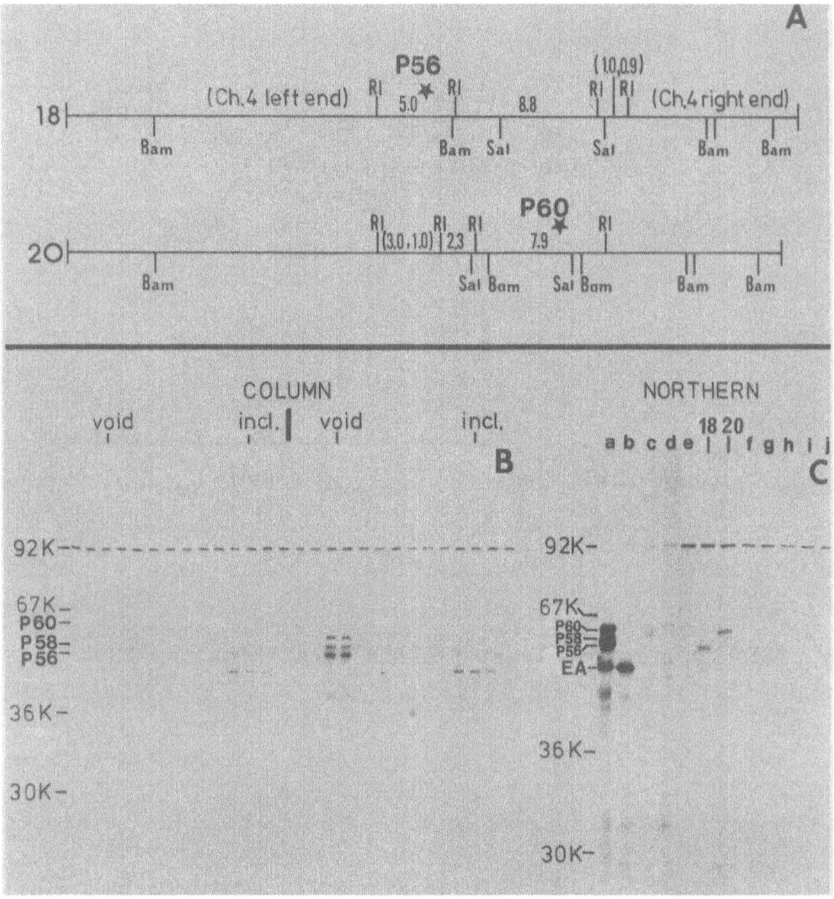

Fig. 6. Identification of Low-Pi Induced Genes in Yeast

(A) Restriction of maps of bacteriophage λ 18 and 20, containing yeast DNA fragments (5 kb and 8 kb) possessing a low-Pi induced gene (Kramer, personal communication).

(B) R-loop hybridization selection RNA from low-Pi grown cells were hybridized to purified 8 kb EcoRI fragment under conditions favoring R-loop formation (110) and passed through a Sepharose A150 column. Fractions were heat treated (right side) or not heated (left side). Each lane shows the translation products following immunoprecipitation with APase and enolase antibodies and gel electrophoresis. Similar results were obtained with the 5 kb fragment.

(C) Hybridization-selection using immobilized DNA EcoRI fragments 18 and 20 were immobilized onto DBM paper and used for hybridization to polyA containing RNA from low-Pi grown cells of immunoprecipitated translation products corresponding to each fragment.

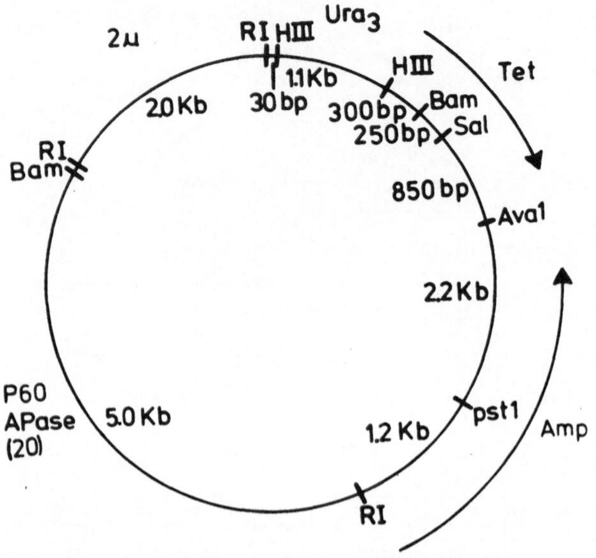

Fig. 7. Self-replicating Yeast Plasmid Containing the PHOE Gene

4. Constitutivity of the APase Regulatory Gene Products

The kinetics of repression and derepression of APase mRNA were compared to the accumulation and disappearance of translatable mRNA in temperature-shift experiments using conditional mutants in the phosphatase regulatory genes. From the results summarized in Table 1, expression of translatable APase mRNAs require PHOB and PHOD for derepression and are constitutive in phoR and phoU mutants. From these results we conclude that the interplay of these positive and negative regulatory genes occurs prior to the production of translatable mRNA. Utilizing the ability to rapidly inhibit protein synthesis with levels of cycloheximide that do not immediately inhibit RNA synthesis, we have assessed the constitutivity of the regulatory gene products as accomplished for the galactose system by Perlman and Hopper (42). The appearance and disappearance of APase mRNAs in derepression and repression experiments were examined in the presence of absence of cycloheximide. Whereas protein synthesis was immediately inhibited by cycloheximide, no effect was

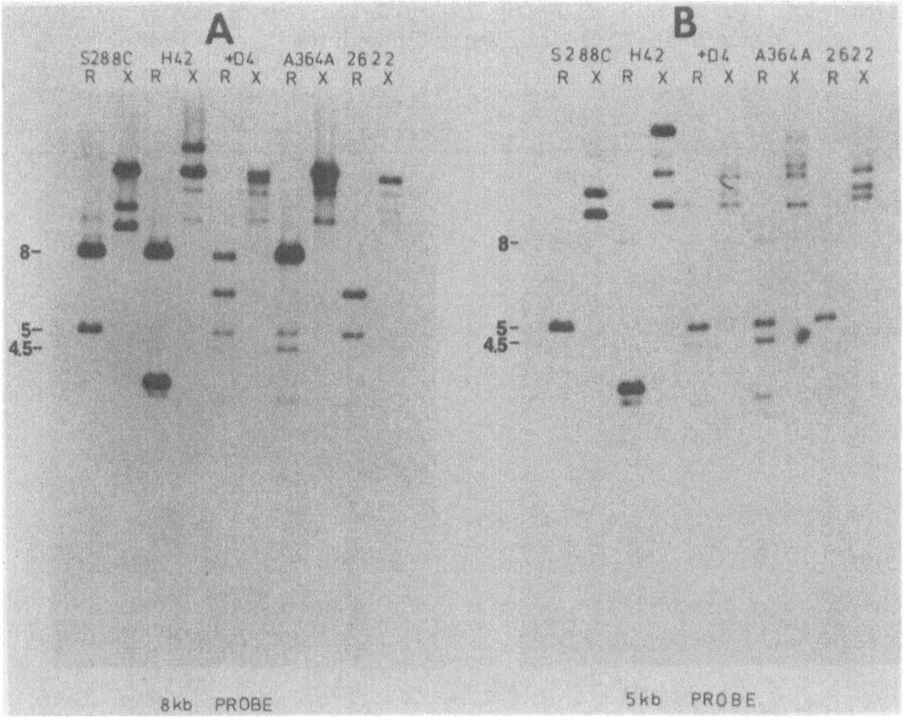

Fig. 8. Southern Genomic Blots of DNA Isolated from Various Yeast Strains Hybridized to the 8 kb or 5 kb Probe

Strains H42, +D4, and A364A possess translatable p58 mRNA when cells are grown on low-Pi medium. Strains S288C and 2622 show no detectable p58 mRNA. Data from Thill, Bostian and Kramer (unpublished observations).

observed on either the repression or derepression of translatable APase mRNAs.

If any part of the regulatory mechanism is inducible then one would predict that cycloheximide would inhibit APase mRNA synthesis when a ts phoR mutants is shifted in high-Pi to the non-permissive temperature. As shown in Fig. 9, the accumulation of translatable APase mRNA upon shift in temperature is insensitive to cyclohexmide inhibition of protein synthesis. Thus in high-Pi the entire machinery for the transcriptional expression of the APase structural gene is present. This supports the revised Toh-e et al. (104) model for the direct interaction of these regulatory gene products in the control of APase structural gene expression.

Table 1. Enzyme and mRNA Levels of Cells Grown in Low (lo) and High (hi) Inorganic Phosphate

Genotype	Const. APase lo	Const. APase hi	Rep. APase lo	Rep. APase hi	Rep. Alk. Pase lo	Rep. Alk. Pase hi	*in vitro* Rep. Apase mRNA activity lo P60	lo P58	lo P56	hi P60	hi P58	hi P56
H42 (w.t.)	+	+	+	-	+	-	+	+	+	-	-	-
phoC	-	-	+	-	+	-	+	+	+	-	-	-
phoE (phoC)	-	-	l	-	+	-	-	+	+	-	-	-
phoK (tent.)	-	-	l	-			+	-	+	-	-	-
phoP (phoC)	-	-	+	w	+	-	+	+	+	w	w	w
phoD (phoC)	-	-	-	-	-	-	-	-	-	-	-	-
phoB (phoC)	-	-	-	-	+	-	-	-	-	-	-	-
phoR (phoC)	-	-	++	w	++	w	++	++	++	w	w	w
phoU (phoC)	-	-	++	w	++	w	++	++	++	w	w	w
phoS (phoC)	-	-	-	-	-	-	-	-	-	-	-	-

l = leaky; w = weak

5. Evidence for Post-translational Protein Modification

Despite the presence of common regulatory genes (PHOD, PHOS) for APase and alkaline phosphatase, shift of cells to low pH medium allows for the derepression of alkaline phosphatase but not APase. The inability to derepress APase when both the pH and temperature are elevated in ts phoB and ts phoS mutants led Toh-e et al. (104) to revise their original model to one involving the simultaneous functioning of cytoplasmic factors produced by the regulatory genes. These results lead to the prediction that expression of functional APase mRNA would be inhibited at low pH even at low concentrations of Pi. Although accumulation of enzyme activity ceases at the low pH in such experiments, all three APase mRNAs continue to be synthesized and are functional in the *in vitro* synthesis of immunoprecipitable polypeptide. Under these conditions, we have detected APase polypeptide synthesis by an indirect radioimmune as-

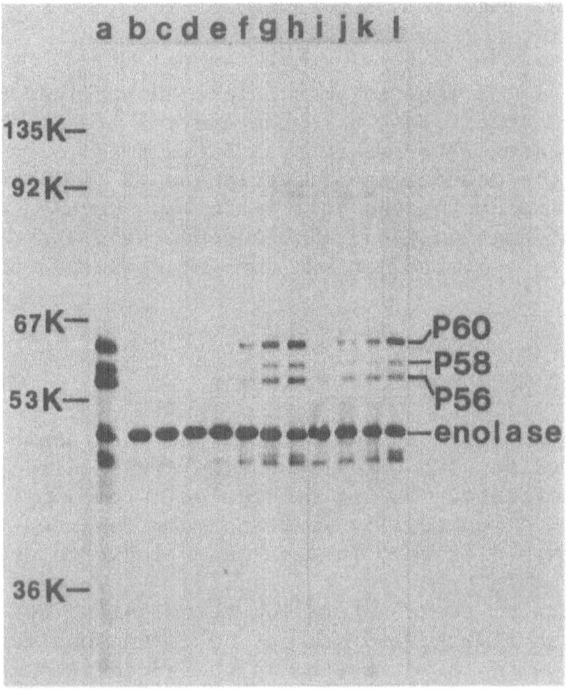

Fig. 9. Autoradiograph of a 10% SDS/polyacrylamide Gel of the Immunoprecipitates from Cell-free Translation of mRNA from a ts phoR Strain

Content of each lane is: (a) wild type (P28-24C); (b)-(d) from cell at 25°C at zero time, 20 min and 40 min after temperature shift; (e)-(h) from cell temperature-shifted to 35°C for 10, 20, 30 and 40 min.; (i)-(l) from cells to which cycloheximide was added to 100 μg/ml 5 min before temperature shift to 35°C, at 10, 20, 30 and 40 min.

say. It therefore appears most likely that the regulatory step influenced by growth at low pH is a post-translational modification, a processing step of a primary translation product into an active enzyme molecule, or enzyme inactivation.

REGULATION DURING THE CELL CYCLE

The synthesis of many enzymes have been examined during the cell cycle under conditions where the enzymes are neither fully induced or repressed, but are autoregulated (112). In relatively few cases have the mechanisms of regulation or the integrated controls operating during the cell cycle been documented. Considerable information is available concerning macromolecular synthesis, and to a limited extent the galactose and phosphatase system enzymes.

1. Regulation of Macromolecule Synthesis

The linear increases in the rate of rRNA synthesis throughout the cell cycle raises the question whether it is directly subject to gene dosage effects. During studies on chromosomal localization of the 5S and rRNA genes (113) we found that repeatedly subculturing of a strain deficient in rRNA genes leads to magnification of DNA genes to wild type levels (114,115). The added rRNA genes in magnified strains are clustered on chromosomes, not closely linked to any centromere, and are not on chromosome I. In experiments with otherwise isogenic strains the added rRNA genes in magnified strains, neither increases the growth rate nor the rate of RNA synthesis. These results, and previous studies on RNA polymerase levels (116), show that the regulation of mRNA synthesis differs from that of periodic enzyme synthesis.

There are fundamental differences between the kinetics of RNA and protein synthesis, and the pattern of step enzymes during the cell cycle. Yeast has two forms of poly(A) polymerase (117): one is nuclear and the other cytoplasmic. The two forms are distinguished by primer requirement, pH and salt optima, cation preference, and molecular weight. Absence of the cytoplasmic form in stationary phase cells implies that the two poly(A) polymerase activities are independently regulated.

Stationary phase cells of S. cerevisiae were found to have a reduced poly(A) content compared to exponential phase cells (118). A sizing procedure for poly(A) was devised to distinguish between alternative hypotheses to explain this reduction. Two major size classes of poly(A) were found. The decreased representation of the larger of the two classes accounted for the majority of the poly(A) loss. The remainder of the loss was accounted for by fewer poly(A) containing sequences. The smaller of the two poly(A) classes was apparently not of mitochondrial origin and may be added transcriptionally.

If poly(A) function is responsible for increased mRNA stability in the cytoplasm, the most stable mRNA species would be enriched for poly(A) over the total mRNA population. One can select for the "stable" mRNA fraction by using the rnal mutant, ts 136, which fails to transport mRNA from the nucleus to the cytoplasm at the restrictive temperature. Polysomes decayed at the restrictive temperature with the expected half life of yeast mRNA (17.5 min). The RNA isolated from polysomes after varying times at the restrictive temperature showed an increasing percentage of poly(A). A 3-fold enrichment was found after 60 min. The remaining material was found to sediment as messenger sized molecules, indicating that poly(A) fragments from degraded messengers were not artificially increasing the observed poly(A) content.

2. Position Effects on Enzyme Synthesis

Galactose enzyme synthesis can be induced at any point of the cell cycle (112). However, once induction has occurred, further increases in the galactose enzyme activities appear only as step-increases during a specific point of each cell cycle (112). The mechanisms of cell-cycle control of enzyme synthesis is unknown. However, the chromosomal location of the galactose structural genes has been reported to effect the time of appearance of the specific step-increase in enzyme activity. Cox and Gilbert (119) analyzed the timing of the step-increase in enzyme activity in two strains which contained different chromosomal locations for both GAL1 (galactokinase) and LYS2 (α-aminoadipic acid reductase (Fig. 10). The synthetic period for galatokinase began at 0.45-0.55 fraction of cell generation and the reductase synthetic period at 0.25-0.35 fraction of a generation in the wild-type strain S288C. However, in the mutant strain 61009, where the two genes now map in entirely different locations, the increased galactokinase enzyme activity began earlier in the cell generation (at 0.15-0.25 fraction of a cell generation). Thus, increases in induced enzyme activities (GAL1 and Lys2) may occur periodically throughout the yeast cell cycle with the period of synthetic activity being influenced by the genes' position upon the chromosome.

3. β-galactosidase Activity in Single Cells

Cell cycle analysis on large populations of yeast cells, by separating random populations of cells using gradient centrifugation, has provided an approximation of events which occur at the level of single cells (120). The measurement of enzyme levels in single cells of a random population provide an unambiguous test for

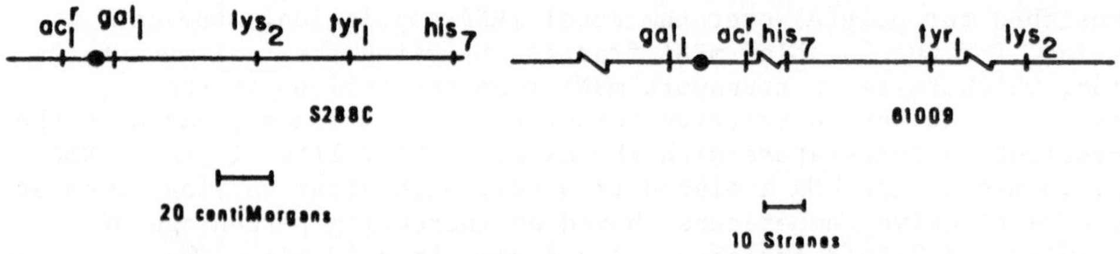

Fig. 10. Genetic map of Chromosome II in S. cerevisiae S288C and 61009

the pattern of enzyme synthesis. We therefore measured the β-galactosidase activity in single cells of Kluyveromyces lactis (22) by a modification of the procedure previously reported by Rotman and Papermaster (121). The periodic doubling of the level of β-galactosidase in single cells during the cell cycle agrees with data previously obtained by gradient fractionation and by the use of synchronous cultures.

4. Turnover of Functional APase mRNA During the Cell Cycle

If the rate of synthesis of any two enzymes are continuous throughout the cell cycle, then one would anticipate a continuous increase and constant ratio of messenger RNA for those two enzymes determined by the rates of mRNA synthesis and turnover. The inhibition of APase mRNA synthesis by high-Pi provides an experimental procedure to approximate messenger half-life. Derepressed cells of P28-24C were grown to mid-logarithmic phase and repressed by the addition of high-Pi. At intervals, aliquots were removed and the levels of translatable APase mRNAs determined by measurement of their cell-free products by densitometry of the electrophoretically fractionated immunoprecipitates. Linearity of the assay under standard translation conditions was determined by showing that product yield was proportional to the amount of low-Pi mRNA used (see insert, Fig. 11). From the data in Fig. 11 the half-life of APase functional mRNA in a culture dividing every 90 min is approximately 8 min.

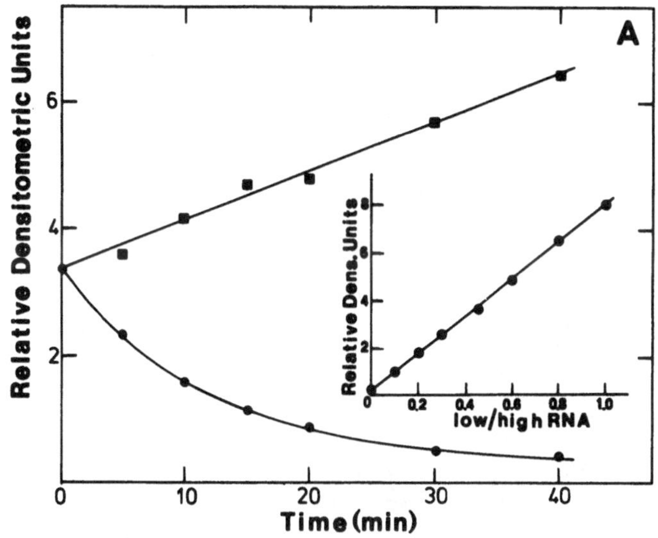

Fig. 11. Estimation of APase mRNA Half-life

The first step toward clarification of mRNA turnover during the cell cycle is to generate populations of yeast cells which accurately represent different cell cycle stages. A critical examination of cell fractionation and synchronization methodologies is underway in a number of laboratories because of the controversy over patterns of enzyme synthesis. Our approach has been to age fractionate yeast cells by zonal fractionation (120). An intrinsic problem is masking of potential periodic synthetic events by poor cell separation and delays in fractionation. Cell fractionation was assessed by a number of control assays defining cell cycle events (DNA/cell, % budded cells, mean cell volume), as shown in Fig. 12A, that are incorporated within an experimental run. As seen in Fig. 12B, a step in in acid phosphatase activity occurs concommitant with bud formation. In contrast, the rate of synthesis of enolase, a constitutive, cytoplasmic enzyme rises continuously across the cell cycle, implying a continuous increase in translatable enolase mRNA (Fig. 12B). Possibly, the step in APase activity reflects periodic synthesis and turnover in mRNA. Functional mRNA levels were followed directly after fractionating expontentially growing derepressed cells on a zonal rotor. mRNA activity for APase essentially decreases relative to enolase as the cell ages (Fig. 12B). These changes in the ratio are consistent with a continuous rise in enolase mRNA throughout the

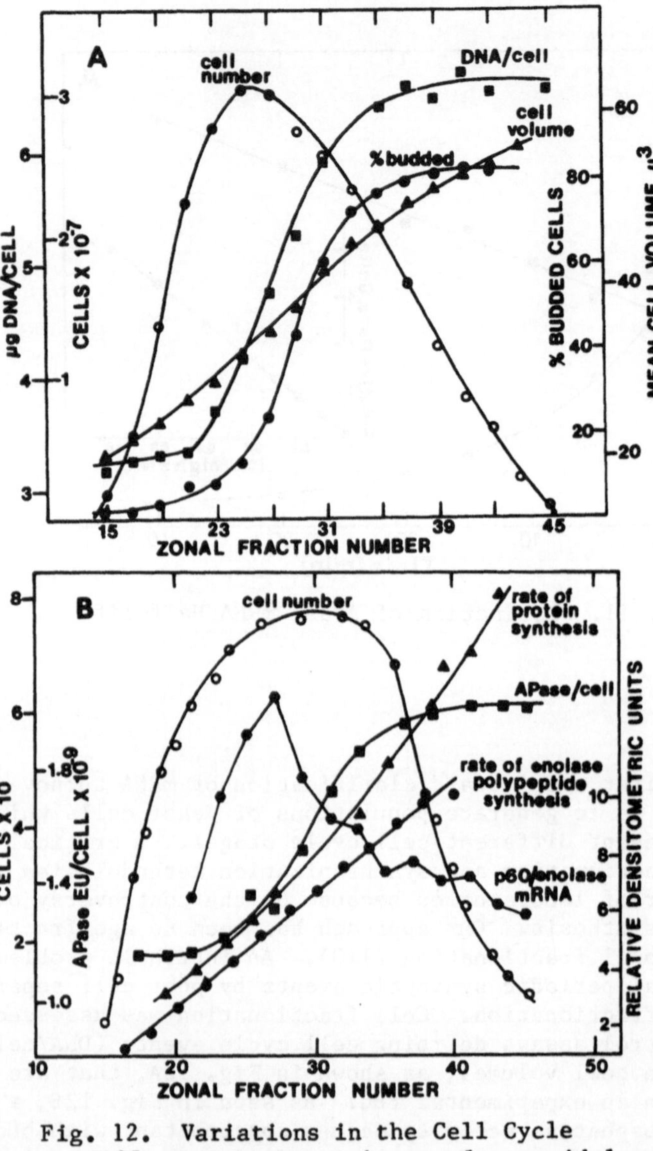

Fig. 12. Variations in the Cell Cycle
Strain Y185 was grown asynchronously to mid-log phase in low-Pi medium and the harvested cells fractionated on a zonal rotor. Zonal rotor fractions, representing different stages of the cell cycle were analyzed for: (A)○, cell number; ▲, cell volume; ●, % budded cells; ■, DNA/cell. (B)○, cell number; ▲, rate of protein synthesis; ●, rate of enolase polypeptide syntheaia; ■, APase activity/cell; ⬢, the ratio of translatable p60/enolase mRNA.

cell cycle and a rapid rise and decline in APase mRNA during the step in APase activity. These findings suggest that under these conditions a periodic transcription, processing, or transport of functional APase mRNA occurs during the cell cycle.

This work was supported in part by a U.S. Public Health Service Grant AI 1060 (H.O.H.), by an American Cancer Society Postdoctoral Fellowship (K.A.B.) and by a Brandeis University Graduate fellowship (J.G.Y.). We thank Drs. Akio Toh-e, Yasuji Oshima, Richard Kramer, James Hopper and David Rogers for their generous provision of information, strains and materials, and for unpublished information used in this manuscript.

REFERENCES

1. Mitchison, J.M. (1971) In The Biology of the Cell Cycle, Cambridge University Press, London.

2. Lewin, B. (1974) In Gene Expression Vol. 2, John Wiley & Sons N.Y.

3. Halvorson, H.O. (1977) Leopoldina Symposium. VEB Gustav Fisher Verlag Jena, p. 361-376.

4. Mitchison, J.M. (1977) Leopoldina Symposium on Cell Differentiation in Microorganisms, Plants and Animals. Nover Mothes eds. VEB Gustav Fisher Verlag Jena p. 377-401.

5. Hartwell, L.H. (1974) Bacteriol. Rev. 38:164-198.

6. Masters, M. and Donachie, W.D. (1966) Nature 209:476-479.

7. Edmunds, L.N. (1978) Aging and Biological Rhythms. Plenum Pub. Co. H.V. Samis and S. Capobianco, eds. p. 125-184.

8. Fraser, R.S.S. and Nurse, P. (1978) Nature 271:726-730.

9. Wain, W.H. and Staatz, W.D. (1973) Exp. Cell Research 81:269-278.

10. Fraser, R.S.S. and Moreno, F. (1976) J. Cell Sci. 21:497-521.

11. Williamson, D.H. and Scopes, A.W. (1960) Exp. Cell Research 20:338-349.

12. Halvorson, H.O. Gorman, J., Tauro, P., Epstein, R. and LeBerge, M. (1964) Fed. Proc. 23:1002-1008.

13. Sogin, S.J., Carter, B.L.A. and Halvorson, H.O. (1974) Exp. Cell Research 89:127-138.

14. Elliott, S.G. and McLaughlin, C.S. (1978) P.N.A.S. USA 75:4384-4388.

15. Sebastian, J., Takano, I. and Halvorson, H.O. (1974) P.N.A.S. USA 71:769-773.

16. Carter, B.L. and Dawes, I.W. (1975) Exp. Cell Res. 92:253-258.

17. Shulman, R.W., Hartwell, L.H. and Warner, J.R. (1975) J. Mol. Biol. 73:513-525.

18. Saunders, C.A., Sogin, S.J., Kaback, D.B. and Halvorson, H.O. (1977) Control Mechanisms in Development. Plenum Press, N.Y. R.H. Meints and E. Davies eds. p. 21-34.

19. Matur, A. and Berry, P. (1978) J. Gen. Microbiol. 109:205.

20. Elliott, S.G. and McLaughlin, C.S. (1979) J. Bacteriol. 137:1185-1190.

21. Mitchison, J.M. and Carter, B.L.A. (1975) Methods in Cell Biology, 9:201-219.

22. Yashphe, J. and Halvorson, H.O. (1976) Science 191:1283-1284.

23. Williamson, V.N., Bennetsen, J., Young, E.T., Nasmyth, K. and Hall, B.D. (1980) Nature 283:214-216.

24. Bach, M.L., Lacroate, F. and Botstein, D. (1979) P.N.A.S. USA 76:386-390.

25. Clarke, L. and Carbon, J. (1978) J. Mol. Biol. 120:517-523.

26. Montgomery, D.L., Hall, B.D., Gillam, S. and Smith, M. (1978) Cell 14, 673-680.

27. Kramer, R.A. and Andersen, N. (1980) P.N.A.S. USA 77, 6541-6545.

28. St. John, T.P. and Davis, R.W. (1979) Cell 16:443-452.

29. Baker, W.K. (1968) Adv. Genetic 14:133-169.

30. Lewis, E.B. (1950) Adv. Genetics 3:73-115.

31. Lyon, M.F. (1961) Nature (London) 190:372-373.

32. Lyon, M.F. (1972) Biol. Rev. 47:1-36.

33. Salzmann, J., DeMars, R. and Benke, P. (1968) Proc. Natl. Acad. Sci. USA 60:545-552.

34. Gartler, S.M. and Andina, R.J. (1976) Adv. Human Genetics 7:99-140.

35. Hicks, J., Strathern, J. and Herskowitz, I. (1977). In DNA Insertion Elements, Plasmids and Episomes, Bukhari, A., Shapiro, J., and Adhya, S., (eds.), Cold Spring Harbor, NY. p. 457-469.

36. Hicks, J., Strathern, J. and Klar, A.J.S. (1979) Nature 282, 478-483.

37. Haber, J.E., Mascioli, D.W. and Rogers, D.T. (1980) Cell 20:519-528.

38. Toh-e, A., and Oshima, Y. (1974) J. Bacteriol. 120:608-617.

39. Oshima, Y. (1978) IX Int. Cong. on Yeast Genet. and Mol. Biol. Rochester, N.Y. p. 15.

40. Bostian, K.A., Lemire, J.M., Cannon, L.E. and Halvorson, H.O. (1980) P.N.A.S. USA 77 , 4504-4508.

41. Douglas, H.C. and Hawthorne, D.C. (1966) Genetics 54:911-916.

42. Perlman, D. and Hopper, J.E. (1979) Cell 16:89-95.

43. Metzenberg. R.L. and Nelson, R.E. (1977). In Eucaryotic Genetic Systems, ICN-UCLA Symposia on Molecular and Cellular Biology, VIII, p. 253.

44. Arst, H.N., Jr. and MacDonald, D.W. (1975) Nature 254:26-31.

45. Hynes, M.J. (1978) Molec. Gen. Genetics 161:59-65.

46. LaCroute, F. (1968) J. Bacteriol. 95:824-832.

47. Neidhardt, F.C. and Magasanik, B. (1956) Nature 178:801-802.

48. Magasanik, B., Neidhardt, F.C. and Levin, A.P. (1958). In Physiological Adaptation, Prosser, C.L. (ed.), Lord Baltimore Press, MD, p. 159.

49. Mandelstam, J. and McGuillen, K. (1973). In Biochemistry of Bacterial Growth, Blackwell Scientific Publications, London, pp. 453.

50. Zitomer, R.S. and Hall, B.D. (1976) J. Biol. Chem. 251:6320-6326.

51. Zitomer, R.S. and Nichols, D.L. (1978) J. Bacteriol. 135:39-44.

52. Zitomer, R.S., Montgomery, D.L., Nichols, D.L. and Hall, B.D. (1980) Proc. Natl. Acad. Sci. USA 76:3627-3631.

53. Bossinger, J. and Cooper, T.G. (1977) J. Bacteriol. 131:163-173.

54. Hartwell, L.H., Hutchison, H.T., Holland, T.M. and McLaughlin, C.S. (1970) Molec. Gen. Genetics 106:347-361.

55. Hutchison, H.T., Hartwell, L.H. and McLaughlin, C.S. (1969) J. Bacteriol. 99:807-815.

56. Tonnesen, T. and Freisen, J.D. (1973) J. Bacteriol. 115:889-896.

57. Lawther, R.P. and Cooper, T.G. (1975) J. Bacteriol. 121:1064-1073.

58. Hereford, L. and Rosbash, M. (1977) Cell 10:453-462.

59. Huez, G., Marbaix, G., Hubert, E., LeClerq, M., Nudel, V., Soreq, H., Solomon, R., LeBleu, B., Revel, M. and Littauer, N. (1974) Proc. Natl. Acad. Sci. USA 71:3143-3146.

60. Wilson, M., Sawicki, S.G., White, P.A. and Darnell, J.E. (1978) J. Mol. Biol. 126:23-36.

61. Sussman, M. and Newell, R. (1972). In Molecular Genetics and Developmental Biology, Prentice-Hall, NJ, p. 245.

62. Saunders, C., Bostian, K. and Halvorson, H.O. (1980) Nucl. Acid Res. 8:3841-3849.

63. Hautala, J.A., Bassett, C.L., Giles, N.H. and Kushner, S.R. (1979) Proc. Natl. Acad. Sci. USA 76: 5774-5778.

64. Goldberger, R.F. (1974) Science 183:810-816.

65. Dickson, R.C., Abelson, J., Barnes, W.M. and Reznikoff, W.S. (1975) Science 187:27-35.

66. Struhl, K. and Davis, R.W. (1977) Proc. Natl. Acad. Sci. USA 74:5255-5259.

67. Douglas, H.C. and Condie, F. (1954) J. Bacteriol. 68:662-670.

68. Leloir, L.F. (1951) Archiv. Biochem. 33:186-190.

69. Kosterlitz, H.W. (1943) Biochem. J. 37:322-325.

70. Kalckar, H.M., Braganca, B. and Munch-Peterson, A. (1953) Nature 172:1038.

71. Douglas, H.C. and Hawthorne, D.C. (1964) Genetics 49, 837-844.

72. Mortimer, R.K. and Hawthorne, D.C. (1966) Genetics 53, 165-173.

73. Bassel, J. and Mortimer, R.K. (1971) J. Bacteriol. 108, 179-183.

74. Mortimer, R.K. and Hawthorne, D.C. (1969). In The Yeasts, Vol. I., Rose, A.H. and Harrison, J.E. (eds.), Academic Press, NY, p. 385.

75. Bevan, P. and Douglas, H.C. (1969) J. Bacteriol. 98: 532-535.

76. Douglas, H.C. (1961) Biochim Biophys. Acta 52:209-211.

77. Broach, J.R. (1979) J. Mol. Biol. 131:41-53.

78. Matsumoto, K., Toh-e, A. and Oshima, Y. (1978) J. Bacteriol. 134:446-457.

79. Jacob, F. and Monod, J. (1965) Biochem. Biophys. Res. Commun. 18: 693-701.

80. Klar, A.J.S. and Halvorson, H.O. (1974) Mol. Gen. Genet. 135:203-212.

81. Klar, A.J.S. and Halvorson, H.O. (1976) J. Bacteriol. 125:379-381.

82. Hopper, J.E. and Rowe, L.B. (1978) J. Biol. Chem. 253:7566-7569.

83. Hopper, J.E., Broach, J.R. and Rowe, L.B. (1978) P.N.A.S. USA 75:2878-2882.

84. Douglas, H.C. and Hawthorne, D.C. (1972) J. Bacteriol. 109:1139-1143.

85. Nagi, Y., Matsumoto, K., Toh-e, A. and Oshima, Y. (1977) Molec. Gen. Genetics 152:137-144.

86. Matsumoto, K., Adachi, Y., Toh-e, A. and Oshima, Y. (1980) J. Bacteriol. 141:508-527.

87. Hawthorne, D.C. and Mortimer, R.J. (1960) Genetics 45:1085-1110.

88. Douglas, H.C. and Pelroy, G. (1963) Biochem. Biophys. Acta 68:155-156.

89. Tsuyuma, S. and Adams, B.G. (1973) Proc. Natl. Acad. Sci. USA 70:919-923.

90. Tsuyuma, S. and Adams, B.G. (1974) Genetics 77:491-505.

91. Kew, O. and Douglas, H.C. (1976) J. Bacteriol. 125:33-41.

92. Adams, B.G. (1972) J. Bacteriol. 111:308-313.

93. Jacob, F. and Monod, J. (1961) J. Mol. Biol. 3:318-356.

94. Toh-e, A., Veda, Y., Kakimoto, S., Oshima, Y. (1973) J. Bacteriol. 113:727-738.

95. Toh-e, A., Nakamura, H., Oshima, Y. (1976) Bioch. Biophys. Acta 428:182-192.

96. Schurr, A. and Yagil, E. (1971) J. Gen. Microbiol. 65:291-303.

97. Toh-e, A., Oshima, Y. (1977) CSH Symp. Mol. Biol. Yeasts, Cold Spring Harbor, N.Y. p. 129.

98. van Rijn, H.J., Linnemans, L.A. and Boer, P. (1975) J. Bacteriol. 123:1144-1149.

99. Hereword, F.U. (1974) Exp. Cell Res. 87:213-218.

100. Sentandreu, R. and Elorza, M.U. (1973) In Yeasts, Mold and Plant Protoplasts, Villanueve, et al. eds. Academic Press, N.Y. p. 187.

101. Toh-e, A., Kakimoto, S., Oshima, Y. (1975) Molec. Gen. Genet. 141:81-83.

102. Ueda, Y., Toh-e, A. and Oshima, Y. (1975) J. Bacteriol. 122:911-259.

103. Ueda, Y. and Oshima Y. (1975) Molec. Gen. Genet. 136:255-922.

104. Toh-e, A., Kobayashi, S., Oshima, Y. (1978) Molec. Gen. Genet. 162:139-149.

105. Bostian, K.A., Lee, R.C. and Halvorson, H.O. (1979) CSH Symp. Mol. Biol. Yeasts, Cold Spring Harbor, N.Y. p. 136.

106. Novick, P. and Schekman, R. (1979) P.N.A.S. USA 76:1858-1862.

107. Field, C. and Schekman, R. (1980) J. Cell. Biol. 86:123-128.

108. Novick, P., Field, C. and Schekman, R. (1980) Cell 21:205-215.

109. Schweingruber, M.E. and Schweingruber, A.M. (1979) Molec. gen. Genet. 173:349-351.

110. Woolford, J.L. and Rosbash, M. (1979) Nucl. Acid. Res. 6: 2483-2497.

111. Alwine, J.C., Kemp, D.J., Parker, B.A., Reiser, J., Renart, J., Stark, G.R. and Wahl, G.M. (1979) In Methods in Enzymology 68:220-242.

112. Halvorson, H.O., Carter, B.L.A. and Tauro, P. (1971) Adv. Microbial. Phys. 6:47-106.

113. Kaback, D.B. and Halvorson, H.O. (1976) J. Mol. Biol. 107:385-390.

114. Kaback, D.B. and Halvorson, H.O. (1977) P.N.A.S. USA 73:1177-1180.

115. Kaback, D.B. and Halvorson, H.O. (1978) J. Bacteriol. 134:237-245.

116. Sebastian, J., Mian, F. and Halvorson, H.O. (1973) FEBS letters 34:159-162.

117. Saunders, C.A., Sogin, S.J. and Halvorson, H.O. (1977) Cold Spring Harbor Symp. Mol. Biol. Yeasts, p. 54.

118. Sogin, S.J. and Saunders, C.A. (1980) J. Bacteriol. 144:74-81.

119. Cox, C.G. and Gilbert, J.B. (1970) Biochem. Biophys. Res. Commun. 38:750-757.

120. Sebastian, J., Carter, B.L.A. and Halvorson, H.O. (1971) J. Bacteriol. 108:1045-1050.

121. Rotman, B. and Papermaster, B.W. (1966) Proc. Natl. Acad. Sci. USA 55:134-141.

THE CELL LIFE CYCLE AND THE G1 PERIOD

David M. Prescott, R. Michael Liskay and George M. Stancel

Department of Molecular, Cellular and Developmental Biology
University of Colorado
Boulder, Colorado 80309

The classical experiment of Howard and Pelc (1) showed that the reproductive cycle of plant root cells was divisible into four periods of time: G1, S, G2, and M. This organization of cell reproductive events has come to serve as a conceptual foundation for the study of cell reproduction in a wide variety of unicellular and multicellular organisms. It is generally accepted that the cell cycle is a sequence of events that is based in turn on a set of specific genes termed the cell cycle genes. Presumably, the sequential occurrence of these cell cycle events is established by causal relationships between successively functioning products of cell cycle genes, e.g. in the simplest sense the occurrence of one event triggers the next, and so on. Thus, the orderliness of the events by which a cell progresses from division to division resides in underlying causal relationships among these events based on the cell cycle genes.

This concept has been firmly established in yeast cells by the identification of more than 30 cell cycle genes (2, 3), the function of each gene product being essential for progression through particular, specific events of the cell cycle, for example, initiation of DNA synthesis or entry into and transit through cell division. A few such cell cycle genes have also been identified by mutation in other cell types, including mammalian cells (for reviews, see 4, 5).

The question we raise here is whether the concept of cell cycle specific events applies to the G1 period. It is clear that initiation of DNA synthesis and progression through the S, G2, and M period represent a sequence of specific events most likely based on gene expressions in all cell types. However it is by no means clear that movement of the cell through G1 consists of any specific G1 events. Indeed, in yeast, where the search has been the most thorough, only one gene (the so-called start function) has been identified whose expression can be considered specifically or uniquely associated with G1. Mutations in the start function arrest yeast cells in G1, but the arrest point (called "Start") might be at the G1-S border and not within the body of the G1 period itself (work of C.J. Rivin quoted in 6). It is our judgment that no G1 specific events, that is, events that occur only between M and S, have been identified in any cell type. It is commonly assumed that preparations for DNA synthesis occur in G1 and that these are specific G1 functions. No convincing evidence for such specific G1 functions has to our knowledge been published, and the original definition of G1 remains; it is a period of time measure between cell division and initiation of DNA synthesis. Several mutant lines of mammalian cells that arrest in G1 have been described (4, 7), implying G1 functions; but the functions of these several genes remain unknown as does the question do these functions only occur in G1 and never elsewhere in the cycle. Indeed, the arrest points caused by these mutations may well be at the G1-S border or might define functions that can under certain conditions take place in other phases of the cycle.

Despite this failure to demonstrate specific G1 events, special importance can be ascribed to G1 because unquestionably the rate of cell reproduction in multicellular organisms, and in some unicellular organisms is in general ultimately controlled by regulating the rate at which the cell transits from mitosis to DNA synthesis. This strongly implies that G1 does contain at least one unique event, whose execution may be delayed or blocked, resulting in delay or block in entry of the cell into DNA synthesis. Pardee (8) has called this point in G1 in mammalian cells the R point (R for restriction). The R point then may contain the "switch" by which a variety of signals, intracellular conditions, or extracellular conditions transiently or permanently arrest the cell cycle. In this regard the R point is equivalent to the "Start" point in yeast cells.

We present experiments in this paper that are consistent with the idea that G1 is not based on specific events. Rather, the experiments suggest that G1, when present, is simply a period for cell growth. We further propose that the G1 period is not an essential part of the cell cycle but instead is the result of a transient delay in transition from mitosis to DNA synthesis. This general view has recently been proposed by Cooper (9).

THE ABSENCE OF A G1 PERIOD IN SOME MAMMALIAN CELLS

Several types of mammalian cells have cycles with no G1 period (see 10 for review). Our experience has been that many researchers are reluctant to accept the existence of a mammalian cell cycle that lacks G1. We attribute this reluctance to an incorrect definition of G1. It is essential to remember that the definition of G1 is a period of time between mitosis and DNA synthesis. It has not been possible to translate this time period into multiple unique G1 events. We assume that one or more events forms a specific causal link between mitosis and DNA synthesis (for example, synthesis of an initiator of DNA synthesis), but this event(s) occurs too quickly to occupy measurable time and is not the basis for a G1 period.

Thus, one of the strongest pieces of evidence that the search for G1 specific events as the underlying cause of the G1 period is illusory is the observation that certain mammalian cells transit immediately from mitosis to DNA synthesis without a G1. The most important example of the G1 less ($G1^-$) cell cycle occurs in the blastomeres of embryos in species ranging from sea urchins to mice (see 10 for review). Indeed, the blastomere may reasonably be considered to express the "arch-type" of cell cycle, proceeding without impedments imposed by growth limitations or regulatory signals associated with differentiation. There are other cell types known to be $G1^-$ for part of their life histories, e.g. neuroblasts of grasshopper embryos and normoblasts in mammalian bone marrow, but the cycles of relatively few cell types of early embryos have been analyzed. The $G1^-$ cell types make it quite clear that the G1 interval is not an essential time element of the cell cycle. In the subsequent history of these cells G1 periods do appear in their cycles by imposition of delays between mitosis and DNA synthesis as the cells progress further along their developmental pathways.

Almost all cell types belonging to renewing tissues have G1 periods of various length in their cell cycle, reflecting different rates of reproduction of the different cell types. When cells from such tissues are placed in cell culture, they continue to express G1 periods although in general the G1 periods are probably much shorter in culture than in the tissues of origin. These cells have been the primary focus for the analysis of G1, particularly efforts to identify events associated with release of cells from quiescence (release from G_0) to reproduction, and events associated with transit from mitosis to DNA synthesis in the cycling situation.

There are at least three lines of cultured cells whose cycles are G1 less. These are the Chinese hamster lines V79-8 (11,12) and DON (13) and the mouse teratocarcinoma cell F-9 (14). The $G1^-$ condition of these cells has been shown by conventional means of cell cycle analysis. For example, 90% or more of the cells in a logarythmic asynchronous population are labeled with a pulse of

^{3}H-thymidine, showing that the S period occupies about 90% of the cell cycle. The G2 and M periods account for the remaining 10%. Because of the lack of a G1 period, the generation times of these cells are unusually short, on the order of 9 to 10 hours.

We have used the $G1^-$ V79-8 cell line to obtain information about the G1 period in ways that are now described.

DOMINANCE OF THE $G1^-$ PHENOTYPE IN CELL HYBRIDS

Cells of the V79-8 line reproduce with an average cell cycle time of 9.2 to 9.5 hours in which the S period occupies ~9.0 hours and G2 + M are together equal to ~0.5 hours. V79-8 cells have been fused with cells of several other Chinese hamster cell lines all of which have G1 periods of various lengths ($G1^+$ cells). Careful analysis of the resulting hybrids have shown that the cycles of all hybrids are $G1^-$ (15). Similarly, fusion of V79-8 cells with primary fibroblast $G1^+$ cells also produces hybrids that are all $G1^-$ (16). We interpret this dominance of the $G1^-$ phenotype to indicate the presence of some condition or activity in $G1^-$ cells that is deficient in $G1^+$ cells.

CONVERSION FROM THE $G1^-$ TO A $G1^+$ PHENOTYPE BY MUTATION

We treated V79-8 cells with the mutagen N-methyl-N'-nitro-N-nitrosoguanidine and using ^{3}H-thymidine suicide selected for cells that had acquired G1 periods. The procedure yielded seven $G1^+$ mutant cell lines with G1 periods varying from two to 4.9 hours in length (Table 1) (12, 17). The lengths of S, G2 and M in these $G1^+$ mutants remained as they were in the V79-8 parent cell.

Two of the mutants ($G1^+$-4 and $G1^+$-5, which are most likely <u>sibs</u>) proved to be temperature sensitive for expression of the G1 phenotype. At 37° and 39° these lines were $G1^+$ but at 33° these cells were $G1^-$. The other mutants were $G1^+$ at all three temperatures. Cell cycle times for the $G1^+$ mutants are given in Table 1.

<u>$G1^+$ Mutants of Line V79-8 have G1 Periods for Different Reasons</u>

When $G1^+$-1 cells were fused to $G1^-$ cells (V79-8), the resulting hybrids were all $G1^-$, indicating recessiveness of the $G1^+$-1 phenotype. We assume that the $G1^+$-1 mutant expresses G1 because it is partially deficient in some function compared to the $G1^-$ parent (V79-8). Complementation tests were done for mutants $G1^+$-1 through $G1^+$-5 by crossing (fusing) mutant cells in various combinations to determine whether mutants were $G1^+$ for the same or for different reasons. For example when mutant 1 was fused to mutant 2, the resulting hybrids were all $G1^-$, that is, mutants 1 and 2 were complementary and we assume therefore have G1 periods because they have different deficiencies. The complete set of tests for mutants

1 through 5 showed the presence of 4 complementation groups. Thus, 4 of 5 mutants have G1 periods for different reasons, i.e. different deficiencies. Expressed in a reciprocal manner the mutants do not have a primary common basis for G1. Further, it is improbable that a given mutant line has suffered more than one G1-inducing mutation. Thus, conversion of a G1$^-$ cell to a G1$^+$ cell is presumably achieved by mutation in a single gene. Therefore, the entire G1 period in any given mutant is based upon a limiting function of one genetically determined event-- not a group of events. Therefore, the bulk of the G1 period in these cells cannot be based upon sequence of events.

Non-Mutant or Natural G1$^+$ Lines of Chinese Hamster Cells can Complement Each Other to Form G1$^-$ Hybrids

Many of the experiments described to this point deal with mutant cells. We have begun to extend the analyses to standard cell types. Thus, complementation tests of the type done with mutant cells have been done with 4 standard lines of Chinese hamster cells: CHO, DeDe, V79-743, and CH III (6). These cells have G1 periods of 3.0, 2.0, 3.5, and 2.3 hours respectively. Crossing (fusing) these four cell types in all possible combinations revealed three complementation groups. Lines CHO and DeDe do not form G1$^-$ hybrids, i.e. do not complement each other. Lines CH III and V79-743 complement each other and both lines CHO and DeDe, i.e. form only G1$^-$ hybrids. These crosses therefore define three complementation groups. Thus, these standard cell lines give results similar to those obtained with mutant G1$^+$ cells. We conclude that among these four lines there are three different bases for the presence of the G1 period. It is possible that cells in each complementation group have growth cycles that are longer than their chromosome cycles, but the primary reason

Table 1. Cell Cycle Characteristics of the G1$^-$ Cell Line V79-8 and Seven G1$^+$ Mutants Derived From It

Cell Line or Mutant	G1	S	G2 + M	Generation Time
V79-8	0 (-0.2)	9.0	0.5	9.5
G1$^+$-1	2.9	9.3	1.0	13.2
G1$^+$-2	2.5	9.0	0.5	12.0
G1$^+$-3	2.0	9.5	0.5	12.0
G1$^+$-4	3.5	9.0	0.5	13.0
G1$^+$-5	3.1	9.2	0.7	13.0
G1$^+$-6	4.9	9.4	0.7	15.0
G1$^+$-7	4.0	9.3	0.7	14.0

or mechanism for the longer growth cycle (i.e. G1) is different in each complementation group. We should stress that the primary basis of G1 in the G1 mutants was mutation whereas in the natural G1$^+$ lines we feel other explanations for their G1 periods apply.

Conversion of G1$^-$ V79-8 to a G1$^+$ Phenotype by Partial Inhibition of "Growth"

Normally, V79-8 cells reproduce with an average generation time of about 9.5 hours with no G1. When protein synthesis is inhibited by about 40% with addition of 0.05 µg/ml of cycloheximide, the average generation time increases to 14.2 hours (17). The increase from 9.5 to 14.2 is accounted for by the appearance of a G1 period of 4.2 hours with no significant increase in S + G2 + M. One interpretation of this is that the cell must achieve some particular rate of protein synthesis or reach a particular size to initiate DNA synthesis. When it is delayed in reaching the necessary state by partial inhibition of protein synthesis, a G1 period is then created for completion of growth. The idea that cell size and initiation of DNA synthesis are related is not new, and is in accord with the large body of data for prokaryotes and some for eukaryotes. The cell probably does not monitor size or growth rate, but rather senses the amount or rate of synthesis of some specific factor whose level is normal proportional to cell size.

The induction of G1 by partial inhibition of protein synthesis led to the idea that induction of G1 by mutation might be due to impairment of protein synthesis. The relative rates of protein synthesis were therefore measured in mutants 1, 5, 6, and 7 and compared to the rate in the parental G1$^-$ cell line (17) (Table 2).

The rates of protein synthesis in mutants 1, 6, and 7 are decreased to 68%, 59%, and 63% of the parental G1$^-$ line. Therefore,

Table 2. Relative Rates of Protein Synthesis in the Parental G1$^-$ Line V79-8 and G1$^+$ Mutant Lines

Cell	G1	GT	Relative Rate of Protein Synthesis
V79-8	0	9	100%
G1$^+$-1	2.9	13.2	68%
G1$^+$-5	3.1	13.0	108%
G1$^+$-6	4.9	15.0	59%
G1$^+$-7	4.0	14.0	63%

it may reasonably be concluded that the appearance of G1 periods in these cells is the result of mutation that ultimately impairs protein synthesis. Thus, this conforms to the demonstration that cycloheximide induces a G1 period in $G1^-$ V79-8 cells. However, we admit that cause and effect has not been demonstrated.

Mutant 5, however, has the same rate of protein synthesis as V79-8, and the G1 might not be due to impairment of overall protein synthesis. Mutant 5 also has the same rate of protein degradation as V79-8 (17). Hence, we tentatively conclude that the mutation in $G1^+$-5 does not affect the overall rate of protein accumulation and therefore induces G1 through a different route (for example, through an effect on growth not reflected in the rate of protein synthesis or in an event specific to initiation of the S period) than in mutants 1, 6, and 7.

The induction of G1 in $G1^-$ cells either by slowing protein synthesis with cycloheximide or by mutation offers a consistent explanation for different bases of the G1 period in different $G1^+$ mutant cells. Presumably, any mutation that decreases the rate of protein synthesis will induce a G1 period in a $G1^-$ cell (or lengthen the G1 period in a cell that already has a G1). Literally hundreds of different G1-inducing mutations are therefore possible.

<u>The Cell Life Cycle is Composed of a Chromosome Cycle and a Cell Growth Cycle</u>

The experiments described to this point support the idea that the G1 period is not composed of specific or unique events but is rather a period of general "growth". The G1 period ends when the cell achieves a size or certain amount of a crucial factor required to trigger DNA synthesis. Viewed in another way the cell life cycle is made up of two interacting cycles; (1) a chromosome cycle consisting of replication of DNA (S period), preparation for mitosis (G2), and distribution of chromosomes to daughter nuclei at mitosis; and (2) a growth cycle in which the cell doubles its mass in each generation time.

The interaction of these two cycles is observed by inhibiting progress of one and observing the other. If DNA synthesis is totally blocked by one or another inhibitor, cell growth continues for a limited period and then stops. Alternatively, if growth is blocked by deprivation of one or another factor, e.g. an essential amino acid, serum growth factor, etc., initiation of DNA replication is blocked. Thus, one cycle can modulate the progress of the other. Within this frame, if the length of the growth cycle is equal to the length of the chromosome cycle (S + G2 + M), then the cell will proceed through its life cycle without a G1 period, i.e. be $G1^-$. This is so because the cells grow fast enough so that the new daughter cells already possess the critical size or growth rate to initiate

DNA synthesis immediately after mitosis.

On the other hand if the growth cycle is longer than the chromosome cycle, then in each generation the cell must wait for the growth cycle to be completed before it can initiate the next chromosome cycle. The period of waiting is G1-- the slower the growth cycle relative to the chromosome cycle, the longer the G1 period. In this model the wide differences in the average length of G1 for different cell types (while maintaining relatively constant length of S + G2 + M) are readily accounted for.

If this model of the cell cycle is correct, it should be possible to convert a $G1^-$ cell into a $G1^+$ cell by reducing the rate of cell growth. This expectation is borne out as already described by reducing the rate of protein synthesis in the $G1^-$ V79-8 through mutation or cycloheximide. Conversely, the model also leads to the predicition that a $G1^+$ cell may be converted to a $G1^-$ cell by any means that makes the chromosome cycle equal in length to the growth cycle. We do not have the means to shorten the growth cycle so that it equals the chromosome cycle, but we can attempt to equalize the two by lengthening the chromosome cycle. This attempt is the basis of the experiment in the next section.

The G1 Period in Mutant $G1^+$-1 can be Markedly Shortened by Increasing the Length of the S period

Hydroxyurea (HU) inhibits DNA synthesis in mammalian cells by inhibiting the enzyme pyrimidine nucleoside diphosphate reductase thereby preventing the synthesis of deoxycytidine and thymidine triphosphates. The degree of inhibition of DNA synthesis can be varied over a wide range with different concentrations of HU. For mutant $G1^+$-1 a concentration of 2×10^{-5}M HU inhibits DNA synthesis without increasing the average generation time (18). The S period is thereby lengthened from 9.0 hours to 11 hours with a concommitant shortening of G1 from 2.9 to 0.9 hours. Whether more careful adjustment of the HU concentration will reduce G1 to "zero" remains to be seen. In any case the experiment, even at this point, bears out the prediction that extension of the chromosome cycle to a length approaching that of the growth cycle reduces the length of G1 accordingly. The experiment therefore supports the thesis that G1 is a period of growth, i.e. part of the growth cycle, but is not part of the chromosome cycle.

Analysis of G1 in Lung and Liver Cells in Culture

So far we have dealt only with cell lines that have been in culture for long periods. We next ask whether normal cells from different tissues might also have different bases for their G1 periods when put into culture (16). We have examined the G1 basis of "liver" and "lung" cells from a single hamster. Briefly, the results show that lung cells have at least two bases for their G1 period, only one of which it shares with "liver" cells. Furthermore, long term cultivation of the cells did not detectably change the cause(s) for their G1 periods.

CONCLUSIONS

The observations reported in this paper, taken together with the large body of published studies, including especially the arguments of Cooper (9), lead us to the following tentative model of the mammalian cell cycle.

The "arch-type" of cell cycle lacks a G1 period because cell growth keeps pace with the chromosome cycle (equal to S + G2 + M). Interruptions in the "arch-type" of cell cycle can occur between mitosis and DNA replication, creating a G1 period. Such interruption is achieved by conditions that impinge on a switch located immediately before the start of S (R point or "Start"). In cells in which the growth cycle is longer than the chromosome cycle (most cultured mammalian cells), the switch can't be turned on until the growth cycle is completed. Thus, the G1 period in cultured mammalian cells is a period of growth and contains no uniquely defining events.

In contrast to cultured cells, the G1 period in cells in tissues is not created by a limited growth rate, but is the result of specific signals that impinge on the switch at the start of S. Cells in different tissues are regulated by different signals. Cells blocked at the initiation switch for S progress into other states, including so called quiescent or G_0 states.

How G1 length is controlled in tissues, whether it be by limiting growth and/or by specific signals, remains unknown.

REFERENCES

1. Howard, A. & Pelc, S.R. (1953) Heredity, Suppl. 6, 261-273.
2. Hartwell, L.H. (1978) J. Cell Biol. 77, 627-637.
3. Nurse, P., Thuriaux, P. & Nasmyth, K. (1976) Molec. gen Genet. 146, 167-178.
4. Basilico, C. (1977) Cancer Res. 24, 223-266.
5. Siminovitch, L. (1976) Cell 7, 1-11.
6. Liskay, R.M., Leonard, K.E. & Prescott, D.M. (1979) Somatic Cell Genet. 4, 615-623.
7. Liskay, R.M. (1974) J. Cell. Physiol. 84, 49-55.
8. Pardee, A.B. (1974) Proc. Natl. Acad. Sci. USA 71, 1286-1290.
9. Cooper, S. (1979) Nature 280, 17-19.
10. Prescott, D.M. (1976) Reproduction of Eukaryotic Cells, Academic Press, New York.
11. Robbins, E. & Scharff, M.D. (1967) J. Cell Biol. 34, 684-686.

12. Liskay, R.M. & Prescott, D.M. (1978) Proc. Natl. Acad. Sci. USA 75, 2873-2877.
13. Liskay, R.M. Unpublished results.
14. Liskay, R.M. & Rosenstraus, M. Unpublished results.
15. Liskay, R.M. & Prescott, D.M. (1978) Cell Reproduction, ICN-UCLA Symposium on Molecular & Cellular Biology, eds. Dirksen, E.R., Prescott, D.M. & Fox, C.F. (Academic Press, New York), Vol. 12, pp. 115-125.
16. Liskay, R.M., Fullerton, P. & Kornfeld, B. (1980) Exptl. Cell Res. 128, 191-197.
17. Liskay, R.M., Kornfeld, B., Fullerton, P. & Evans, R. (1980) J. Cell. Physiol. 104, 461-467.
18. Stancel, G., Liskay, R.M. & Prescott, D.M. Unpublished results.

THE CONTINUUM MODEL: APPLICATION TO G1-ARREST AND G(0)

Stephen Cooper

Department of Microbiology and Immunology
University of Michigan Medical School
Ann Arbor, MI 48109

The G1 period, the period between mitosis and the start of DNA synthesis, is generally believed to contain the crucial events which are involved in the regulation of cell growth. A number of papers have appeared supporting the idea that there are functions occurring in, and specific for, the G1 period which are required for the regulation of the initiation of DNA synthesis (1-11).

Two types of experimental findings have supported the existence of G1-specific events. The first is the observation that variability in the length of time between cell divisions is primarily due to variations in the G1 period, with the S and G2 erpiods remaining fairly constant (11-14). This G1 variability is observed between cells grown in different media (15-17), among cells in a culture (12-14), between cells of different ages (18,19), and between variants or mutants and the parental cells (20-22). It was assumed that changes in the rate of passage through G1 was an indication of the rate at which G1-specific events were occurring. This led us to the current model which proposes that the major events controlling growth occur in the G1 period.

The second type of experimental support for the existence of regulatory functions in G1 is the finding that cells appear to be "arrested in the G1 period" under a variety of conditions inhibiting growth. These results have been interpreted as indicating that there are various "restriction points" or specific G1-arrest points at which cells can come to rest (23-25). It is assumed that these arrest points imply the existence of G1 specific functions. The G(0) state, a special case of G1-arrest, is postulated to exist in normal tissues (26) or when cells are placed in adverse conditions (2, 27).

I have recently described an alternative and very different model for the regulation of cell division in eukaryotic cells (28). I proposed:

1. that there are no specific G1 functions;

2. that the G1 period is found when the time between initiations of the S phase is greater than the sum of the S, G2, and M (mitotic) periods; and

3. that the events involved in the initiation of DNA synthesis are occurring continuously during the division cycle (28).

The model described a unified view of the division cycle and the G1 period for both prokaryotes and eukaryotes.

In the original presentation of this model, I dealt with the relationship of the variability in the G1 period to the variation in the length of the division cycle and the proposal that there are G1 specific events. The model accounted for the absence of a G1 period in some cells, for the variation of the G1 period among different cells, and G1 periods in cells grown under different conditions (28). In this paper I will apply this model to the observation of G1-arrest and the specific case of G(0). I propose that these quiescent cells are not "arrested in G1" but are in a state unlike any period of the normal division cycle. I will present a formalism describing such cells, and show how a number of different experimental observations can fit the proposed model.

THE CURRENT VIEW OF G1-ARREST

The current model for the nature of G1-arrest (reviewed in 2, 4, 7, 29) is illustrated in Figure 1. All cells, independent of their position in the division cycle, enter a resting state from the G1 period (Figure 1, upper left). Therefore, all resting cells have a G1 content of DNA. Cells which have initiated DNA synthesis (are in S) or have twice the G1 content of DNA (are in G2) will complete mitosis prior to entering G1-arrest. The current model is illustrated by such statements as, "the mutation ... causes cells to stop traversing the cell cycle and to become arrested in the G1 phase of the cycle with respect to DNA synthesis" (30), "cells prepare for DNA synthesis during G1" (20), and "control of growth in mammalian cells seems to be exerted exclusively during G1 phase of the cell cycle" (10). It has been suggested that there is a "single point in G1 called the restriction point" (23) at which cells starved for various substances are arrested, and other (31) have proposed that "the G1 phase of

	ENTRANCE TO G1–ARREST ("G1")	EXIT FROM G1–ARREST ("G1")
CURRENT MODEL	G1 → "G1"	"G1" → G1 → S → G2 → M
	M → G1 → "G1"	"G1" → G1 → S → G2 → M
	G2 → M → G1 → "G1"	"G1" → G1 → S → G2 → M
	S → G2 → M → G1 → "G1"	"G1" → G1 → S → G2 → M
CONTINUUM MODEL	$G1_1 \equiv$ "$G1_1$"	"$G1_1$" $\equiv G1_1 \rightarrow S_S$
	$M_M \rightarrow G1_M \equiv$ "$G1_M$"	"$G1_M$" $\equiv G1_M \rightarrow G1_1 \rightarrow S_S$
	$G2_2 \rightarrow M_2 \rightarrow G1_2 \equiv$ "$G1_2$"	"$G1_2$" $\equiv G1_2 \rightarrow G1_M \rightarrow G1_1 \rightarrow S_S$
	$S_S \rightarrow G2_S \rightarrow M_S \rightarrow G1_S \equiv$ "$G1_S$"	"$G1_S$" $\equiv G1_S \rightarrow G1_2 \rightarrow G1_M \rightarrow G1_1 \rightarrow S_S$

Fig. 1. Comparison of the current model for G1-arrest with the Continuum Model. At the left the two models are compared for entry into a state of G1-arrest (designated "G1") and at the right the exit of the cells from the resting state is compared. Exit from the resting state could be due to a large number of causes from stimulation by addition of serum to shifting to the permissive temperature for a temperature-sensitive mutant. The subscripts in the Continuum Model denote the initiator per origin for those cells that enter arrest. When a cell enters a G1-arrest state, the value for initiator per origin is fixed, and the cells progress through the replication-segregation sequence as shown by the large letters.

the mammalian cell cycle has special significance because the rate of cell reproduction is regulated by modulation of transit through G1." A recent article summarized the general feeling of the workers in this field as follows: "It is the general consensus among investigators that the critical period in the cell cycle for the control of proliferation in eukaryotic cells is in the G1 phase" (32). The current model proposes that cells are arrested in a particular state which is entered from the normal G1 phase of the division cycle. It envisions cells leaving G1-arrest by reentering the division cycle at some point in G1 (2, 4, 7) and progressing through S, G2, and M (Figure 1, upper right).

THE CONTINUUM MODEL

I propose the Continuum Model, illustrated in the lower half of Figure 1. There are two parameters which characterize the state of a cell at various times during the division cycle. One parameter is the position in the replication-segregation sequence specified by the content and state of DNA in the cell. Cells with a 2n DNA content are in the G1 period. Cells that are replicating DNA and have a DNA content between 2n and 4n are in the S period. Cells that have a 4n DNA content are in the G2 period, and dividing cells which have condensed chromosomes are in the short M period. I now propose that cells are also characterized by a second parameter, the amount of "hypothetical initiation" (33). Initiator is being synthesized continuously during the division cycle. Thus, the amount of initiator per origin in a cell varies continuously. This is diagrammatically illustrated in Figure 2. (The formalism of Figure 2 is only one of many different ways of envisioning the accumulation of initiator according to the Continuum Model. For example, it is possible that the initiator is "used up" or "destroyed" at initiation and rather than halving the initiator concentration goes to zero. In this case the next round of initiator synthesis would have to proceed at twice the rate in order to have the initiator a constant fraction of the cell mass.) When the amount of initiator per origin present is equal to a threshold value, DNA synthesis will begin. The amount of initiator per origin is greatest just prior to the start of the S period. Due to the sudden doubling in the number of origins of replication, the amount of initiator per origin decreases and is lowest just after the start of the S period.

The production of cells "arrested in G1" according to the Continuum Model is illustrated in Figure 1 (lower left). The subscripts denote the amount of initiator per origin in cells at different times during the division cycle or during a period of growth arrest. The Continuum Model proposes that cells found to be arrested in growth and having a G1 content of DNA result from an inhibition of initiator synthesis. Therefore the amount of initiator per origin is fixed in all cells of the population. The cells in the S period finish replicating DNA and, along with the cells of the G2 and M periods, divide to produce cells with a G1 content of DNA. The cells produced all have the same G1 content of DNA, yet remain different in their amounts of initiator per origin (Figure 1, lower left). When G1-arrested cells leave their arrested state they do not leave synchronously. This prediction of the Continuum Model can be understood by observing that resting cells differ in the amounts of initiator per origin present in each cell (Figure 1, lower half). Becuase the cells have different amounts of initiator per origin (the original G1 cells the most, the S phase cells the least (see Figure 2, lower graph) the different cells require different times to synthesize the threshold

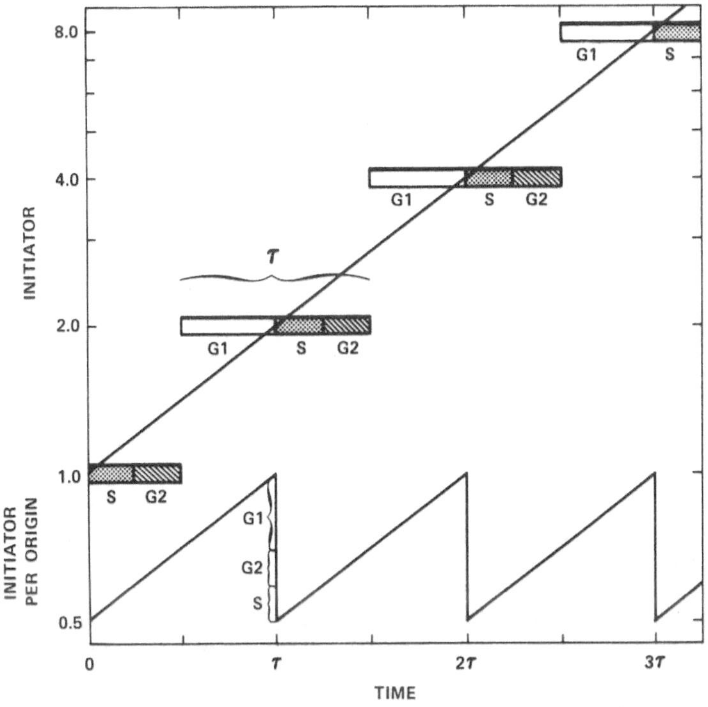

Fig. 2. A formal description of the synthesis of "hypothetical initiator" in relation to the phases of the division cycle. The lower saw-tooth graph is the amount of initiator per origin, varying from 0.5 to 1.0, the critical amount required for the initiation of DNA synthesis. At the start the cells have just entered an S period and the cells are progressing through S and G2. The amount of initiator per origin is increasing from 0.5 to 1.0 over one doubling time ( ). When it reaches 1.0 another S period is initiated but since at initiation there is a sudden doubling in the number of available origins of replication the amount of initiator per origin decreases immediately to 0.5. Note that the Gl period is the time between the end of G2 and the start of S (neglecting the short M period). If the rate of initiator synthesis was decreased and S and G2 remained constant then Gl would increase. There is no Gl-specific synthesis of initiator but the synthesis is occurring continuously. As shown in the upper graph, this formal statement of the continuum model does not visualize any sudden appearance or disappearance of any particular molecules. It is not postulated that any particular time during the division cycle there should be sudden alterations in the amount of any particular molecule.

amount of initiator required for initiation of DNA synthesis. Thus, the first cells to initiate DNA synthesis were in G1 at the time of inhibition of initiator synthesis. These cells are followed into S phase by the original M, G2, and S phase cells and in that order. This behavior is not predicted by the existing, and generally accepted, model.

The Continuum Model may be summarized as follows:

1. Initiator is synthesized continuously. DNA synthesis starts when a threshold amount of initiator per origin is achieved.

2. Cells arrested with a G1 content of DNA are produced by conditions that cause an impairment in the synthesis of initiator.

3. Cells in S, G2, and M are allowed to complete DNA synthesis and/or division even though initiator synthesis is impaired.

4. The pattern of DNA synthesis during the division cycle of any cell is merely a function of the rate of synthesis of initiator relative to the rate of passage of the cells through the S and G2 periods (28).

One way of contrasting the Continuum Model with the current model is to note that the current model visualizes cells entering the G1-arrest state from a distinct period of the division cycle, the G1 period, and that cells are arrested in a unique "state". In contrast, the Continuum Model proposes that the cells in the resting state do not enter the resting state from a distinct phase of the division cycle. The population of cells with a G1 content of DNA are not in a single state determined by their DNA content. They are different from one another due to differences in the amount of initiator per origin which are retained during passage through DNA replication and cell division. Since initiation of DNA synthesis stops when synthesis of initiator ceases, the amount of initiator per origin will remain constant.

The Continuum Model proposes that existence of cells "arrested in G1" is no reason to conclude that there exists a specific step in G1 or a specific function in G1 at which the cells are arrested. A function or process which occurs in all phases of the division cycle may be inhibited and still produce cells which are apparently arrested in a specific part of the division cycle. All that is required is that cells continue to finish replication of DNA and divide, even when inhibition of initiator synthesis blocks new initiation of DNA synthesis. The resulting cells will all have a G1 content of DNA, and will superficially appear to be arrested in

G1, but in fact the arrested cells will be unlike normal G1 cells. The Continuum Model emphasizes the idea that there is no fundamental difference between cells with a G1 period (primarily eukaryotes) and cells devoid of a G1 period (primarily prokaryotes). Both types of cells are part of a larger <u>continuum</u> (28) in which it is merely the doubling time of the cell, in relation to the S, G2 and M periods, which determines whether or not a G1 period is observable.

THE NATURE OF THE INITIATOR

At this time the biochemical events involved in the initiation of DNA replication, in either prokaryotes or eukaryotes, are unknown, and the chemical nature of the "hypothetical initiator" is still a mystery. Rubin and Sanui (34) have summarized their experiments on the requirements for the intiation of DNA synthesis and concluded that "It seems most likely, in agreement with Castor, that the overall rate of protein synthesis (or more correctly, protein accumulation) is the major determinant of the onset of DNA synthesis." A large number of other experiments have shown that protein synthesis is required for the initiation of DNA synthesis in eukaryotes (6, 35-41). I propose that the "hypothetical initiator" is, or is related to, the protein of the cell. When enough protein, or specific protein, is produced, DNA synthesis begins.

Whether the "initiator" substance is ultimately shown to be protein in general (23), or a specific protein (5, 41), or cell mass (42), or cell size (43, 44) or some related substance, does not affect the formal aspects of the Continuum Model. However, the assumption that the "initiator" is protein does allow one to analyze a number of experiments on G1-arrest in a less formal and artificial manner. I will show that a number of experiments are consistent with the notion that protein synthesis (or some property related to protein synthesis) is involved in the initiation of DNA synthesis, and that inhibiting protein synthesis is quite likely the cause of cells arrested in G1.

Let me stress, however, that the relationship of protein synthesis to the initiation of DNA synthesis does not distinguish between the current and the Continuum Models. Both models may postulate that initiation is related to some accumulation of protein. Rossow, Riddle and Pardee (41) have stated that "cells must accumulate a specific protein in a critical amount before they can proceed beyond the restriction point." These authors, however, predict "the existence of a serum-sensitive, rapidly turning over protein unique to early G1" (41), while the Continuum Model does not. I note that their data are consistent with the idea, proposed here, that there is a continuous synthesis throughout the cell cycle, of some protein or proteins required for the initiation of DNA synthesis.

The same relationship of protein synthesis to the initiation of DNA synthesis in eukaryotes has also been proposed for prokaryotes. Inhibition of protein synthesis in prokaryotes (45-47) leads to a cessation of initiation of DNA synthesis but does not affect the completion of rounds of replication in progress. Further, measurements of cell size during the division cycle of prokaryotic cells growing at different rates have provided evidence that the cell mass (similar to cell protein) per origin at initiation of DNA synthesis is constant (45, 48). Although the mechanisms involved are probably quite different in the two types of cells, the essential logic of the regulation of cell division and the cell cycle is present in both types of cells. This logic requires that a certain amount of growth and cell mass, represented by initiator, must be present before the events of DNA synthesis and cell division are allowed to occur.

It is important to distinguish between protein synthesis and "net protein accumulated". Although cells may be metabolically active and may be synthesizing large amounts of protein, it may be incorrect, in terms of the Continuum Model, to count excreted protein as protein which can lead to initiation. It is not merely protein synthesis alone, but protein syntehsis which leads to cell growth which initiates DNA synthesis. Turnover, excretion, and other metabolic events involving protein syntehsis may not produce a proliferating cell.

APPLICATION OF THE CONTINUUM MODEL

I will now proceed to apply the Continuum Model to various reports of G1-arrest, to the evidence that there are specific G1 functions, and to the proposed G(0) period. Rather than an exhaustive treatment of the experimental reports, I have chosen to describe representative results and to show how the Continuum Model reformulates the previous conclusions.

Temperature-Sensitive Mutants

A number of mammalian temperature-sensitive mutants with ostensible cell cycle specific lesions have been described. Most of these mutants are arrested in the G1 phase of the cell cycle at non-permissive temperatures (49-59). Cells arrested with a G1 content of DNA are common; cells arrested in S or G2 are relatively rare (57, 58). Analysis of these mutants has led to the conclusion that the temperature-sensitive lesions affect G1 functions (60, 61).

A temperature-sensitive mutant of Syrian hamster BHK/21 cells, tsAF8 (30, 49), which grows normally at 32-34 C but becomes "blocked in the G1 phase of the cell cycle at the non-permissive temperature" provides an instructive example (62). Rossini and Baserga (62) summarized the studies of this mutant and stated that the block was located in mid-G1, 6-8 hours before the start of the S period. Biochemical studies revealed that the mutation affected the activity of RNA polymerase II (62) and at elevated temperatures net protein synthesis ceased (30). The first paper (30) on this mutant noted that amino acid incorporation was unimpaired at elevated temperatures while thymidine and uridine incorporation was inhibited, implying a specific defect in DNA synthesis. This obscured the observation in the same paper (30), that there was no net protein synthesis at the elevated temperatures. It is probable that the incorporation of amino acids is due to alterations, with temperature, in the mechanisms of uptake, turnover, or pool size of the mutant cells. The arrest of this mutant in G1 can now be explained by the Continuum Model. The cells arrive in what appears to be G1 because the initiation of DNA synthesis is dependent upon protein synthesis while passage through the S, G2, and M periods (i.e., DNA synthesis and cell division) is not dependent upon protein synthesis. Protein synthesis was not considered a candidate for a "G1-specific function" because protein synthesis occurs in all phases of the division cycle (63). The inhibition of a biosynthetic function, which occurs in all parts of the division cycle, may produce cells arrested in what appears to be a particular phase of the division cycle.

Complementation studies of these eukaryotic temperature-sensitive mutants, arrested in the G1 period, have revealed that these mutations are recessive, and that there are a large number of complementing groups (20-22, 64-68). This has been interpreted as indicating that there are a number of different functions which are involved in the G1 period. For example, this is clearly enunciated by the statements of Liskay and Meiss (64): "We interpret these results to mean that AF8 and B54 represent defects in different functions specific to the G1 phase and that furthermore, each mutant line is capable of supplying the 'missing' function of the other line, thus enabling the hybrid cells to grow at 39 C. If we accept this view then AF8 and B54 define two distinct steps specific to the G1 phase of the cell cycle.". A number of similar studies, with essentially identical observations and conclusions, have been reported (65-68).

The alternative interpretation of the complementation data states that the mutants affect general biosynthetic mechanisms that are not cell cycle specific. The Continuum Model proposes that when the temperature is raised, cells in S, G2, and M are able to complete the cycle and produce the two cells with a G1 content of DNA. The large number of complementation groups merely defines different steps in the synthesis of initiator and does not define different cell cycle specific functions.

It is interesting to note that Smith and Wigglesworth (69, 70) have reported a "G1-arrested" temperature sensitive mutant that resumes DNA synthesis after the cell is returned to the permissive temperature if protein synthesis is allowed. It cannot resume DNA synthesis when protein synthesis is inhibited. This result is consistent with the hypothesis that the initiation of DNA synthesis is dependent upon the synthesis of cell protein.

The physiology of different temperature sensitive mutants has also been used to support the notion that there are different steps in G1 affected by the different mutations. For example, the lag prior to resumption of DNA synthesis after cells are replaced at the permissive temperature is different in different mutants (56). It was proposed that the longer the lag, the "earlier" the affected function occurs in G1. Also supporting the idea that mutants affect functions occurring at different times in the G1 period is the finding that different mutants have different execution points (59, 60). If a cell is at a point prior to the execution point (60), then raising the temperature will prevent DNA synthesis. If the cell is at a point after the execution point then raising the temperature will allow that cell to initiate DNA synthesis. By comparing different execution points, it is thought that different points in G1 may be identified (30, 60). I would propose that the different lags in DNA synthesis may be explained by different delays prior to resumption of protein synthesis in the different mutants. Interestingly, this does not ever appear to have been studied in the various mutants. Since this aspect of the physiology of the mutants could be allele specific, different rates of resumption of protein synthesis need not have any relationship to "points" in G1. In a similar way, the different execution points may be explained by varying degrees of "leakiness" of the different mutants. I have previously discussed the problems associated with the use of execution points in a different, but very appropriate, context (71).

An interesting example of mutants affected in the G1 period is reported by Liskay and Prescott (21, 22). They worked with V79-8 cells (derived from the lung tissue of the Chinese hamster) which have no G1 period (G1-less). Liskay and Prescott (21, 22) used tritiated thymidine suicide to enrich for and isolate cells that contained a G1 period. Complementation studies revealed that the G1-less cells were dominant in hybrids with the G1-containing variants. I suggest that what Liskay and Prescott actually selected were slow growing cells, slightly impaired in some aspect of protein synthesis. These slower growing cells were enriched in the suicide experiments because they synthesized DNA over a smaller fraction of the division cycle than the rapidly growing parental, G1-less, cells.

This slight impairment in protein synthesis led to a slowed synthesis of initiator. The increased interdivision time led, in a very trivial way, to the appearance of a G1 period. It is trivial because the reason the G1 period was produced in these slower cells is that the S and G2 periods remained constant. It was not because the selected cells had a rate limiting G1-function which the parental cells did not have. This case has been analyzed previously (28).

G1-Arrest By Chemicals

It is possible to arrest cells in specific stages of the division cycle by addition of specific inhibitors (4, 72-75). For example, Actinomycin D, which inhibits RNA synthesis by binding to DNA, inhibits cell growth and division and leads to the accumulation of cells in the G1 phase of the division cycle (72). This has been interpreted as the demonstrating of a requirement for a specific RNA synthetic step in the G1 period (2). According to the Continuum Model, the arrest in G1 is not due to inhibition of G1-specific RNA synthesis, but rather, it is due to the general inhibition of initiator synthesis as illustrated in Figure 1. The Continuum Model I have proposed explains how specific cycle arrest can be produced without having any specific events occurring in that part of the division cycle.

Another instructive example are the studies of the regrowth of cells inhibited by puromycin (73), streptovitacin A (73) and indomethacin (74, 75). It has been reported that after removing these drugs the cells grew and divided "synchronously" as would be expected from the current view of the arrest of cells in the G1 period (Figure 1, upper right). After examining the experimental data in support of such synchronous growth, I conclude that synchrony was not demonstrated. The "synchrony" observed was primarily due to a lag in the start of cell division upon removal of the inhibitors. Also, the rise time of the cells was much too long to support synchronous division. I suggest that if protein synthesis was measured in these cases there would probably have been a lag until protein synthesis, and hence initiator synthesis, resumed. The prime importance of these examples is that they clearly demonstrate the nature of the current model which predicts, as shown in Figure 1, synchronous growth upon release from G1-arrest.

"Restriction Points" and G1-Arrest in Depleted Media

The most popular method of producing cells "arrested in G1" is to alter the medium. There are a number of different depleted media that produce G1-arrested cells (76-80), but the most common media used is one with the serum reduced or removed (23, 27, 81).

Cells in serum-depleted media are found to be arrested in the G1 phase (Pardee, 1974) as determined by the DNA content. Addition of fresh serum leads to the eventual initiation of DNA synthesis and cell proliferation (27, 78-86). There is a delay until DNA synthesis resumes and many earlier biosynthetic events are initiated quite soon after serum addition (82-84). I suggest that all of these varied events following serum addition lead primarily to an increase in the rate of protein synthesis, and it is the protein synthesis that leads to the initiation of DNA synthesis.

Pardee (23) analyzed three different starvation regimens for obtaining G1-arrested cells. These were the removal of serum, or glutamine, or isoleucine. He measured the times required after addition of complete media for the initiation of DNA synthesis, and found that all three starvation regimens produced cells arrested at a single point, which he termed the restriction (R) point. This was indicated by the constant lag until resumption of DNA synthesis after readdition of serum, glutamine, or isoleucine. Switching cells from one starvation regimen to a different starvation regimen in all combinations (e.g. serum-deprived to isoleucine starvation and vice versa) failed to bring about resumption of DNA synthesis thereby supporting the conclusion that all of the cells were arrested at the same "point". If one state of arrest had occurred before any other, it would have been expected that DNA synthesis would have resumed in one of the switching experiments. If a cell inhibited at a later point was placed in medium which arrested cells only at an earlier point the DNA synthesis would be expected to resume.

I propose that these observations of Pardee (23) can be explained simply on the basis of the inhibition of protein synthesis (82-86). If protein synthesis is needed for initiation of DNA synthesis, and all three starvation media cause a cessation of protein synthesis, changing the means by which protein synthesis is inhibited would not be expected to alter the final result. This G1-arrested state, R, is the result of the inhibition of initiation of DNA synthesis and not the inhibition of the synthesis of DNA or the process of cell division. The inhibition of initiation was due to the cessation of protein synthesis in serum and amino acid deficient media. The uniqueness and constancy of the restriction point observed by Pardee was probably fortuitous. Other workers (87), using similar methods, found that whereas the lag prior to resumption of DNA synthesis may be different for different modes of starvation, changing starvation regimens did not lead to the resumption of DNA synthesis. Also, whereas Pardee (23) reported that changing starvation times did not lead to alterations in the lag until resumption of DNA synthesis, other workers found that the lag until DNA synthesis resumed was dependent upon the length of time of starvation (88). The conclusion to be made from this brief summary is that Pardee's restriction point need not signify any G1-specific step. G1-arrest is the consequence of the inhibition of a process (protein synthesis) which occurs continuously throughout the division cycle.

Other workers have studied the effect of serum starvation and have postulated the existence of other arrest points in the G1 period (89). These points, called the "V" and "W" points, were determined by allowing progression of cells through G1 and removing serum at particular times and then measuring the time for resumption of DNA synthesis when serum was restored. These measurements of the "execution points" (60) for initiation of DNA synthesis can be explained, in terms of the Continuum Model proposed here, by the moving of cells from the various states in which they were originally arrested (Figure 1, lower left) into states "closer" to initiation than the originally starved cells (89). In these experiments, because continuous labelling protocols were used, only the movement of cells closer to initiation was observed). Different execution points for different growth requirements (platelet derived growth factor, plasma, serum, and isoleucine) have also been determined to occur at different times in the G1 period (25). As discussed above in the section on temperature-sensitive mutants, these different execution points are explained by different levels of inhibition of protein synthesis in media depleted of the different substances. As I have noted in a different, but similar context (71), such experiments would be expected to lead to an infinite number of such "G1-arrest points" and unless strict criteria are used to study such points the data are subject to misinterpretation.

There have been reports that particular molecules are made specifically in the G1 period (41). I have previously argued that these are not necessarily synthesized in the G1 period (28) and are therefore not G1-specific proteins. In a similar way, one may also analyze the proposal that there are G1-specific proteins made in the G1 period which follows release from a G1-arrested state. For example, following serum stimulation of serum-deprived G1-arrested cells, a 50,000 molecular weight polypeptide was synthesized (90). Synthesis was observed to decrease when cells entered S phase. This "characteristic peaking profile of mid-G1 protein synthesis", it was suggested, "can serve as a useful marker for the progression of events in G1 prior to S" (90). The Continuum Model explains this particular example of a "G1-event" by proposing that the increased synthesis of the polypeptide is a result of readdition of serum to serum-starved cells. It is merely a fortuitous consequence of the apparent arrest of cells in G1 by serum starvation that this synthesis is called a G1-event. Until it is shown that only cells in G1 make the peptide, it is not clear that its synthesis marks a G1 event.

Nature of "G(0)"

It has been postulated that cells choose between proliferation or quiescence by making a choice in the G1 period at which time cells can withdraw from the cell cycle and enter a qualitatively distinct G(0) state (2, 4, 27, 81, 91-95). The G(0) phase has been postulated to be distinct from the G1 phase (2). G(0) cells are characterized by a lack of cell proliferation and a G1 content of DNA. Such G(0) cells are able to proliferate when properly stimulated and they synthesize DNA prior to cell division. This confirms the suggestion that they are in the "G1 phase" of the division cycle rather than any other part of the division cycle. The existence of the G(0) phase was originally proposed by Lajtha (94). He observed that after a partial hepatectomy there is a lag period after which there is a large increase in the rate of DNA synthesis which is followed in turn by a large increase in mitosis. The nonproliferating cells were able to proliferate if they were given the proper stimulus. The liver cells which were not proliferating until stimulated were postulated to be outside of the normal cycle and in the G(0) state.

It was soon suggested, without rigorous proof, that normal cells in culture respond to suboptimal growth conditions by entering the G(0) state (2). It was not thought possible to produce a G(0) resting state in cell culture by various growth conditions. These resting or quiescent cells were able to proliferate when properly stimulated. Although they contained a G1 content of DNA, they were shown to be kinetically and chemically different (see discussion below) from G1 cells found in proliferating cultures. Thus it was postulated that the cells are in a different and out-of-cycle state, the G(0) state (2). The choice of calling such resting cells G(0) cells, or R cells, or quiescent cells change with the different workers, and there are no definite criteria for differentiating G(0) cells from any cells which are arrested at a point in G1. One indication that the definition of G(0) is not as pellucid as it might be, is attested by the statement of Brooks (35) that "normal fibroblast lines such as 3T3 cells arrest in the G(0)/G1 compartment of the cell cycle when starved of serum." Because there is no clear definition of G(0) it is possible for Brooks to postulate the existence of a cell in what should be two mutually exclusive compartments. I should note that there have been some excellent criticisms of the G(0) concept (96-98) but they have not had the recognition they deserve.

There are two major types of experimental support for the existence of the distinct G(0) state in resting cells. One type of experimental support is the observation that upon stimulation of such resting cells the time prior to initiation of DNA synthesis is longer than the G1 period alone (87, 99). This increased length has been explained as the time required for G(0) cells to leave G(0) and to enter the G1 period (2, 87). If the cells were merely resting in G1, it was thought, then the time until resumption of DNA synthesis would never be greater than the total G1 period. The model proposed here, as exhibited in Figure 1 (lower right) may explain this observation. Since the arrested cells are only superficially in G1, but have amounts of initiator which correspond to the S, G2 and M periods, it is clear that the average time until initiation of DNA synthesis can be longer than the G1 period. By contrasting the current model and the Continuum Model with regard to resumption of DNA synthesis (Figure 1, upper and lower right) it can be seen that one might indeed expect that the time for resumption of DNA synthesis can be longer than the G1 period. The exact kinetics would depend upon how the measurements are made (generally the time for half of the cells to initiate DNA synthesis is measured), and upon the cell cycle parameters (i.e., the relative lengths of G1, S, G2, and M) of the particular cell being investigated. I also note that in none of the experiments which have been performed has mention been made of the possibility that "starvation" or any other deleterious condition could lead to some lag in the resumption of protein synthesis. If there was such an effect, it would lead to a lag in the resumption of DNA synthesis. When the predictions of the Continuum Model and such deleterious effects of starvation on resumption of protein synthesis are considered, it is not at all unexpected that the alleged G(0) cells take a relatively long time to initiate DNA synthesis. A different explanation for the lag before resumption of DNA synthesis is the possibility that the amount of initiator can decrease during different starvation or resting conditions. If total protein synthesis stops and there is turnover of protein, it is possible to consider that there can be a decrease in the amount of initiator present in the cell. Such a decrease would lead to an increase in the time until the next initiation of DNA synthesis.

A second type of experimental support for the notion of a distinct "G(0) phase" is the observation that cells which are supposedly arrested in that state are chemically different than the normal G1 arrested cells found in exponentially growing, cycling populations (2, 100-102). This result is not strong evidence because it is also predicted by the Continuum Model proposed here. For example, it has been suggested that G(0) cells have fewer ribosomes per cell than regular G1 cells (2, 103), but reference to Figure 1 (lower left) shows that if inhibition of protein synthesis leads to continued cell division in a portion of the population without a concomitant increase in ribosome number then the ribosomes per cell should be observed to decrease.

Other measurements have shown that the chromatin of a cell is altered in any of a number of physical (104) or chemical (105, 106) or biochemical (107-111) properties when the cells are presumably in a G(0) state. I suggest that such alterations in the chromatin may very well be expected because the ratio of DNA per mass increases during the various experiments which produce (G0) cells. If the DNA must be associated with a particular amount of protein to be in a particular physical condition, and if the DNA increases more than the protein, it may be expected that physical and the chemical nature of the chromatin can be different from the normal G1 cells.

One major problem with the notion of the "G(0) state" is illustrated by the proposal that further starvation of cells in serum-depleted media leads to the entry of the cells into "deeper" and "deeper" G(0) states (88). One simple explanation for this abundance of "states" is to consider that prolonged starvation causes the cells to become altered in such a way that they require longer and longer times until resumption of protein synthesis and hence delays in DNA synthesis are observed. For example, a normal G1 period of 8 hours in WI-38 human diploid fibroblasts was extended, after a starvation period of 18 days, to a prereplicative period of 20 hours (88). It is not difficult to imagine that starvation of any cells for 18 days can lead to some deleterious effects which would extend or delay the time until resumption of protein synthesis. The problem of the "deeper G(0)" states is primarily a problem of lack of a precise and rigorous definition of G(0). Further, as noted above, if turnover of protein was occurring, and there was an inhibition of synthesis of some specific initiator molecule then it would be expected that there could be a decrease in the amount of initiator present. If this was so, then one could at least expect delays on the order of the interdivision times of the culture. Thus by postulating that it is possible for initiator to decrease under some conditions, an explanation of the "deeper G(0)" state can be made in terms of the Continuum Model.

The G(0) state can be understood in terms of the Continuum Model as cells in which mass synthesis is so slow that cells appear to have an extremely long G1 period. As Rubin and Steiner (98) have noted, up to 96% of quiescent chick embryo cells become labeled after 120 hours of continuous exposure to tritiated thymidine. This experiment is clearly in support of the notion of a slow growing cell with a long G1 rather than a non-cycling G(0) state. Because of the problems inherent in the G(0) concept I suggest that the use of this term be avoided.

RESTATEMENT OF THE CONTINUUM MODEL

I propose that the model for the regulation of cell growth and division, described in this paper, be called the Continuum Model. I note two reasons for this name. One reason is to emphasize the fact that the synthesis of any sort of "initiator" is a continuous process, which does not occur during just a portion of the cell cycle. There are no periods of the cell cycle devoid of initiator synthesis. Secondly, the name stresses the concept that cells of diverse and superfically different division cycles are related in a continuum of cell cycle types. Those cells with a G1 (initiation of DNA synthesis after cell division, primarily eukaryotes), those without a G1 (initiation at division, both prokaryotes and eukaryotes), and those with initiation occurring prior to cell division (primarily prokaryotes but perhaps some eukaryotes (112)) are related over a continuum by the ratio of interdivision times to the S and G2 periods. When the interdivision time is greater than the S and G2 periods a G1 period is found. When it is equal no G1 is found. And when the interdivision time is less than the time for S and G2 initiation can occur prior to cell division. This has been explicitly illustrated and analyzed in a prior discussion (28).

A book by Prescott (7) summarizes the current and widely held model: "Cell proliferation is regulated primarily by interruption of the G1 period several hours before the initiation of the S period, ... Interruption of G1 traverse is probably achieved by a cell cycle regulatory gene acting in early G1. Understanding regulation of cell proliferation and loss of regulation will depend upon discovery of the molecular events that underlie the G1 period, particularly those G1 events concerned with the G1 arrest of the cell cycle." I suggest that the Continuum Model is a simpler description of the experimental observations, and that there is no evidence at this time for obligatory G1 events. If any events do occur in G1 in a particular situation they are neither necessarily in G1 nor responsible for the arrest of a cell in G1 (28). I expect the Continuum Model proposed here will serve as a guide for experiments which can decisively test the two models for the regulation of the division cycle.

ACKNOWLEDGEMENTS

This work was supported by grants from the National Science Foundation, the Rackham School of Graduate Studies of the University of Michigan, and the Vice President for Research of the University of Michigan. I extend my sincere thanks to Drs. William Murphey, David Jackson, Rolf Freter, William Brockman, Ernest Chu, Donald Chambers, Michael Savageu, Frederick Neidhardt, and Elliot Juni, and Mr. Robert Mames for their helpful discussions. I also thank Alexandra Cooper, whose special editing skills, patience, and support were priceless.

REFERENCES

1. R. Baserga, Cell Tissue Kin. 1:167-191 (1968).

2. R. Baserga, "Multiplication and Division in Mammalian Cells", Dekker, New York (1976).
3. R. Baserga, J. Cell Physiol. 95:377-386 (1978).
4. A.B. Pardee, R. Dubrow, J.L. Hamlin and R.F. Kletzien, Ann. Rev. Biochem. 47:715-750 (1978).
5. D.M. Prescott, Cancer Res. 28:1815-1820 (1968).
6. D.M. Prescott, in: "Adv. Cell Biol." D.M. Prescott, Lester Goldstein, Edwin McConkey, eds., Appleton-Century-Crofts, New York, 57-117 (1970).
7. D.M. Prescott, "Reproduction of Eukaryotic Cells", Academic Press, New York (1976).
8. L. Jimenez de Asua, M.K. O'Farrell, D. Clingan and P.S. Rudland, Biochem. Soc. Trans. 5:937-939 (1977).
9. L. Jimenez de Asua, K.M.V. Richmond, A.M. Otton, A.M. Kubler, M.K. O'Farrell, and P.S. Rudland, in: "Hormones and Cell Culture" G. Sato & R. Ross, eds., Cold Spring Harbor Biological Laboratories, 403-424 (1979).
10. H.A. Armelin, M.C.S. Armelin, S.E. Faras, A.G. Gambarini and E. Kimura, in: "Hormones and Cell Culture" G. Sato & R. Ross, eds., Cold Spring Harbor Biological Laboratories, 269-279 (1979).
11. I. Cameron, in: "Cellular and Molecular Renewal in the Mammalian Body" I. Cameron & J.D. Thrasher, eds., Academic Press, New York, 45-85 (1971).
12. T.O. Fox, and A.B. Pardee, Science, 167:80-82 (1970).
13. A.B. Pardee, B.Z. Shilo, and A.L. Koch, in: "Hormones and Cell Culture" G. Sato & R. Ross, eds., Cold Spring Harbor Biological Laboratory, 373-392 (1979).
14. J.A. Smith, and L. Martin, Proc. Natl. Acad. Sci. USA, 70:1263-1267 (1973).
15. M.D. Enger, R.A. Tobey, and A.G. Saponara, J. Cell. Biol., 36:583-593 (1968).
16. V. Defendi and L.A. Manson, Nature 198:359-361 (1963).
17. I.L. Cameron and R.C. Greulich, J. Cell. Biol. 18:31-40 (1963).
18. G. Grove and V.J. Cristofalo, J. Cell Physiol. 90:415-422 (1977).
19. V.J. Cristofalo, J.M. Ryan and G.L. Grove, in: "Cell Culture and its Applications" R.T. Alton & J.D. Lynn, eds., Academic Press, New York, 223-245 (1977).
20. R.M. Liskay, Exp. Cell. Res. 114:69-77 (1978).
21. R.M. Liskay and D.M. Prescott, Proc. Natl. Acad. Sci. USA 75:2873-2877 (1978a).
22. R.M. Liskay and D.M. Prescott, in: "Cell Reproduction, ICN-UCLA Symposium on Molecular and Cellular Biology", Vol. 12. E.R. Dirksen, D.M. Prescott, C.F. Fox, eds., Academic Press, New York, 115-125 (1978b).
23. A.B. Pardee, Proc. Natl. Acad. Sci. USA 71:1286-1290 (1974).

24. R. Dubrow, V.G. Riddle and A.B. Pardee, Cancer Res. 39:2718-2726 (1979).
25. A. Yen and V.G.H. Riddle, Exp. Cell Res. 120:349-357 (1979).
26. L.G. Lajtha, J. Cell. Comp. Physiol. 62:143-145 (1963).
27. C.D. Stiles, W.J. Pledger, J.J. Van Wyk, H. Antoniades and C.D. Scher, in: "Hormones and Cell Culture" G. Sato & R. Ross, eds., Cold Spring Harbor Biological Laboratories, 435-439 (1979).
28. S. Cooper, Nature 280:17-19 (1979).
29. D.M. Prescott, Adv. Genet. 18:100-177 (1976).
30. S.J. Burstin, H.K. Meiss, and C. Basilico, J. Cell. Physiol. 84:397-408 (1974).
31. R.M. Liskay, K.E. Leonard, and D.M. Prescott, Som. Cell Genet. 5:615-623 (1979).
32. M. Rossini, J. Floros, R. Baserga, and R. Weinmann, in: "Hormones and Cell Culture" G. Sato & R. Ross, Eds., Cold Spring Harbor Biological Laboratories, 393-402 (1979).
33. S.G. Margolis, and S. Cooper, Comp. Biomed. Res. 4:427-443 (1971).
34. A.H. Rubin and H. Sanui, in: "Hormones and Cell Culture" G.H. Sato & R. Ross, Eds., Cold Spring Harbor Biological Laboratories, 741-750 (1979).
35. R.F. Brooks, Cell 12:311-318 (1975).
36. A.H. Rubin, M. Terasaki and H. Sanui, Proc. Natl. Acad. Sci. USA 76:3917-3921 (1979).
37. I. Lieberman, R. Abrams, and P. Ove, J. Biol. Chem. 238: 2141-2149 (1963).
38. L. Castor, J. Cell Physiol. 92:457-468 (1977).
39. H. Rubin and D. Fodge, Cold Spr. Conf. Cell Prolif. 1:801-816 (1974).
40. M.W. Unger and L.H. Hartwell, Proc. Natl. Acad. Sci. USA 73:1664-1668 (1976).
41. P.W. Rossow, V.G.H. Riddle and A.B. Pardee, Proc. Natl. Acad. Sci. USA 76:4446-4450.
42. D. Killander and A. Zetterberg, Exp. Cell Res. 38:272-284 (1965).
43. R.F. Kimball, S.W. Perdue, E.H.Y. Chu, and J.R. Ortiz, Exp. Cell Res. 66:17-32 (1971).
44. J.M. Mitchison, in: "Cell Reproduction" E.R. Dirksen, D.M. Prescott, C.F. Fox, eds., Academic Press, New York, 93-102 (1978).
45. C.E. Helmstetter, O. Pierucci, M. Weinberger, M. Holmes, and M.S. Tang, in: "The Bacteria" J.R. Sokatch & L.N. Ornsten, eds., Academic Press, New York, 517-579 (1979).
46. O. Maaloe and P.C. Hanawalt, J. Mol. Biol. 3:144-155 (1961).
47. P.C. Hanawalt, O. Maaloe, D.J. Cummings, and M. Schaechter, J. Mol. Biol. 3:156-165 (1961).

48. W.D. Donachie, N.C. Jones and R. Teather, In: "Microbial Differentiation", J.M. Ashworth & J.E. Smith, eds., Cambridge University Press, Cambridge, pp. 9-44, 1973.
49. A. Kan, C. Basilico, R. Baserga, Exp. Cell Res. 99:165-173, 1976.
50. C. Basilico, J. Cell Physiol. 95: 367-376, 1978.
51. R.M. Liskay, J. Cell Physiol. 84: 49-55, 1974.
52. D.H. Roscoe, H. Robinson, and A.W. Carbonell, J. Cell Physiol. 32: 333-338, 1973.
53. I.E. Scheffler and G. Buttin, J. Cell. Physiol. 81: 199-216, 1973.
54. T. Ashihara, S.D. Chang and R. Baserga, J. Cell Physiol. 96: 15-21, 1978.
55. J. Hatzfield and G. Buttin, Cell 5: 123-129, 1975.
56. P.M. Naha, A.L. Meyer, K. Hewitt, Nature 258: 49-53, 1975.
57. J. Melero, J. Cell Physiol. 98: 17-30, 1979.
58. H.L. Ozer, J. Cell Physiol. 95: 373-375, 1978.
59. J. Floros, T. Ashihara, and R. Baserga, Cell Biol. Int'l Rep. 2: 259-269, 1978.
60. J.R. Pringle, J. Cell Physiol. 95: 393-406, 1978.
61. A.J. Levine, J. Cell Physiol. 95: 387-392, 1978.
62. M. Rossini and R. Baserga, Biochemistry 17:858-863, 1978.
63. L.D. Hodge, T.W. Borun, E.W. Robbins and M.D. Scharff, in: "Biochemistry of Cell Division" R. Baserga, Ed., C.C. Thomas, Springfield, Illinois, pp. 353-374, 1969.
64. R.M. Liskay and H.K. Meiss, Som. Cell Genet. 3: 343-347, 1977.
65. P.L. Ming, H.L. Chang, and R. Baserga, Proc. Natl. Acad. Sci. USA 73: 2052-2055, 1976.
66. P.M. Naha, J. Cell Sci. 35: 53-58, 1979.
67. G.J. Jonak, and R. Baserga, Cell 18: 117-123, 1979.
68. A. Talavera and C. Basilico, J. Cell Physiol. 92: 425-436, 1977.
69. B.J. Smith and N.M. Wigglesworth, J. Cell Physiol. 82: 339-349, 1973.
70. B.J. Smith and N.W. Wigglesworth, J. Cell Physiol. 84: 127-134, 1974.
71. S. Cooper, J. Theo. Biol. 46: 117-127, 1974.
72. R. Baserga, R. Estensen and R. A. Peterson, Proc. Natl. Acad. Sci. USA 54: 1141-1148, 1965.
73. A.B. Pardee and L. James, Proc. Natl. Acad. Sci. USA 72: 4994-4998, 1975.
74. B.M. Bayer and M.A. Beaven, Biochem. Pharmacol. 28: 441-443, 1979.
75. B.M. Bayer, H.S. Kruth, M. Vaughan and M.A. Beaven, J. Pharm. Exp. Therapeut. 210: 106-111, 1979.
76. L.P. Everhart and D.M. Prescott, Exp. Cell Res. 75: 170-174, 1972.
77. K.D. Ley and D.A. Tobey, J. Cell Biol. 47: 453-459, 1970.

78. R.W. Holley and J.A. Kiernan, Proc. Natl. Acad. Sci. USA 71: 2942-2945, 1974.

79. R.R. Burk, Exp. Cell Res. 63: 309-316, 1970.

80. J.C. Mauck and H. Green, Proc. Natl. Acad. Sci. USA 70: 2819-2922, 1973.

81. E. Rozengurt, In: "Hormones and Cell Culture" G. Sato and R. Ross, eds., Cold Spring Harbor Biological Laboratories, pp. 773-788, 1979.

82. A. Herschko, P. Mamont, R. Shields, and G. Tomkins, Nature 232: 206-211, 1971.

83. P.S. Rudland, Proc. Natl. Acad. Sci. USA 71: 750-754, 1974.

84. L.F. Johnson, H.T. Abelson, H. Green and S. Penman, Cell 1: 95-100, 1974.

85. D.D. Cunningham, and A.B. Pardee, Proc. Natl. Acad. Sci. USA 64: 1049-1056, 1969.

86. D. Greenberg, G. Barsh, T.S. Ho, and D. Cunningham, J. Cell Physiol. 90: 193-210, 1977.

87. R. Martin and S. Stein, Proc. Natl. Acad. Sci. USA 73: 1655-1659, 1976.

88. L. Augenlicht and R. Baserga, Exp. Cell Res. 89: 255-262, 1974.

89. W.J. Pledger, C.D. Stiles, H.N. Antoniades, and C. Sc-fr.

90. B.J. Gates and M. Friedkin, Proc. Natl. Acad. Sci. USA 75: 4959-4969, 1978.

91. O.I. Epifanova and V.V. Terskikh, Cell Tissue Kinet. 2: 75-93, 1969.

92. M. Costlow and R. Baserga, J. Cell Physiol. 82: 411-419, 1973.

93. M. St. J. Crane and D.B. Thomas, Nature 261: 205-208, 1976.

94. L.G. Lajtha, J. Cell Comp. Physiol. 67: 133-148, 1966.

95. O.I. Epifanova, In: "Int. Rev. Cytol." Suppl 5, G.H. Bourne, J.F. Danielli and K.W. Jeon, eds., pp. 305-335.

96. L.M. Van Putten, Biomedicine 20: 5-8, 1974.

97. J.M. Brown, Exp. Cell Res. 52: 565-570, 1978.

98. H. Rubin and R. Steiner, J. Cell. Physiol. 85: 261-270, 1975.

99. G. Sander and A.B. Pardee, J. Cell. Physiol. 80: 267-272, 1972.

100. R. Baserga, C.H. Huang, M. Rossini, H. Chang, and P.M.L. Ming, Cancer Res. 36: 4297-4300, 1976.

101. O.I. Epifanova, M.K. Abuladze, and A.I. Zosimovskaya, Exp. Cell Res. 92: 23-30, 1975.

102. G. Rovera, and R. Baserga, Exp. Cell Res. 78: 118-126, 1973.

103. S.R. Farmer, A. Ben-Ze'ev, B. Benecke, and S. Penman, Cell 15: 627-637, 1978.

104. C. Nicolini, F. Kendall, R. Baserga, C. Dessaive, B. Clarkson and J. Fried, Exp. Cell Res. 106: 111-118, 1977.

105. Z. Darzynkiewicz, F. Traganos, T. Sharpless, and M.R. Melamed Proc. Natl. Acad. Sci. USA, 73: 2881-2884, 1976.

106. V.G.H. Riddle, R. Dubrow, and A.B. Pardee, Procl. Natl. Acad. Sci. USA 76: 1298-1302, 1979.

107. T. Ashihara, F. Traganos, R. Baserga, and Z. Dorzynkiewica, Cancer Res. 38: 2514-2518, 1978.
108. M. Rossini, J.C. Lin, and R. Baserga, J. Cell. Physiol. 88: 1-11, 1976.
109. J. Hochstadt, D.C. Quinlan, A.J. Owen, and K.O. Cooper, in: "Hormones and Cell Culture" G. Sato and R. Ross, eds., Cold Spring Harbor Biological Laboratories, pp. 751-777, 1979.
110. G. Rovera, J. Farber, and R. Baserga, Proc. Natl. Acad. Sci. USA 68: 1725-1729, 1971.
111. A. Novi and R. Baserga, J. Cell Biol. 55: 554-562, 1974.
112. C.J. Bostock, Exp. Cell Res. 60: 16-26, 1970.

PROTEIN AND RNA SYNTHESIS

Renato Baserga

Fels Research Institute and Department of Pathology
Temple University School of Medicine
Philadelphia, Pennsylvania 19140

In studying cell proliferation we often forget that in ordinary circumstances a cell, to divide, must not only replicate DNA, but must also double its size. We have been so taken by the simplicity of measuring the incorporation of Tdr-{^{3}H}- into nuclei of cells that we have forgotten that cell DNA replication and cell division can sometimes be dissociated. Yet when cells go from one mitosis to the next one, they must double their size. This is almost intuitive since, if dividing cells were not to double their size from G_1 to M, they would become progressively smaller and eventually vanish. In fact, cellular size progressively increases from G_1 to M and with size there is also a doubling of proteins and nucleic acids which are good indicators of the size of the cells. Proteins and nucleic acids constitute about 50% of the dry weight of a cell and among the nucleic acids ribosomal RNA constitute about 85% of the total RNA and roughly 70% of the total nucleic acids. The ribosomal RNA genes are, therefore, a reasonable target for growth signals. By increasing ribosomal RNA synthesis the cell provides most of its nucleic acids, as well as the framework where proteins are synthesized. It is, therefore, not surprising that when cells are stimulated to proliferate an increase in ribosomal RNA synthesis is almost invariably noticed.

Before proceeding to a genetic analysis of the role of RNA genes in cell proliferation I would like to withdraw from semantic arguments, by simply stating that the terms of G_0, G_1, S etc., as I use them, are purely operational conveniences.

No person in his right mind believes that G_1, G_0, or S actually exist, just as no biochemist in his right mind will claim that K_m and V_{max} actually exist. Or that when we divide the bacterial chromosomes in map units, that the chromosome is actually divided in units. K_m and V_{max} are convenient notations that describe an enzymatic reaction, and map units facilitate the localization of genes on the chromosome. Similarly, G_1, G_0, etc., are simple notations that help us in describing a cell population. Anyone who has worked with proliferating cells knows that 99% of the cell lines have an interval of several hours between completion of mitosis and cellular DNA replication. It is simply convenient shorthand to say that cells in this interval are in G_1, just as I find it convenient to call serum-deprived cells G_0 cells, because they have different characteristics from G_1 cells. But one should not give too much importance to these names, and I, personally, do not find it very useful to argue about G_1, G_0, transition probabilities, Q state, A state, etc. It seems to me that these discussions give too much importance to simple notations that we have adopted for convenience. The problem at hand is not the cell cycle, but cell proliferation, which is controlled externally by growth factors and internally by genes and their products. All subdivisions are a matter of semantics, are useful as descriptive methods, but add only limited information to our quest for the biochemical and molecular basis of cell proliferation.

ROLE OF RIBOSOMAL RNA SYNTHESIS ON THE CONTROL OF CELL PROLIFERATION

The notion that increased transcriptional activity occurs in G_0 cells stimulated to proliferate goes back to the experiments of Lieberman and coworkers (1). Since then a large body of evidence has accumulated in support of Lieberman's findings, evidence that has been analyzed in detail in reviews and books (2). This increased transcriptional activity in nuclei or chromatin of cells stimulated to proliferate has been confirmed repeatedly in several systems and by many laboratories. Recently, it has been shown that most of the quantitative changes in transcriptional activity described in cells stimulated to proliferate can be attributed to an increased synthesis of nucleolar RNA, i.e. ribosomal RNA. The suggestion that the synthesis of ribosomal RNA (rRNA) may be increased in growing mammalian cells again goes back several years and again to Lieberman and collaborators. In work from several laboratories the evidence for an increased synthesis of rRNA in stimulated cells varied from an increased incorporation of radioactive precursors into nucleolar RNA or cytoplasmic RNA to an increased activity of RNA polymerase I and to the effect of a variety of drugs which include Actinomycin D, Lucanthone and finally α-amantin (for a review see the paper by Baserga, et al. (3)). However, it is only with a measurement of endogenous RNA polymerase activity in isolated nucleoli that it has become possible to accurately determine the magnitude of the increase. A number of reports have now appeared indicating that the doubling in RNA synthesis is nuclei isolated from cells stimulated to proliferate can largely be attributed to a fourfold increase, or even more, in RNA synthesis in nucleoli. Since the synthesis of preribosomal RNA ordinarily accounts for about 40% of the total nuclear RNA synthesis, a fourfold increase in pre-rRNA synthesis can explain most of the quantitative changes in transcriptional activity described in cells stimulated to proliferate. There is, of course, a precedent to the conclusion that nucleolar rRNA synthesis is important in cell proliferation. The pathologists of the 19th century had observed that tumor cells had particularly large nucleoli. It is, therefore, particularly interesting that recent results have indicated that the nucleolus may be involved in the stimulation of cell proliferation that occurs in virally-transformed cells.

ROLE OF T ANTIGEN IN THE STIMULATION OF RIBOSOMAL RNA SYNTHESIS

It has been known for a number of years that infection with certain oncogenic DNA viruses, like SV40 and Polyoma, can induce cellular DNA synthesis in the host cell (for a review see Weil (4)). In some respects these viruses, which Weil calls mitogenic viruses, behave in a similar way as serum in density-inhibited cultures. The stimulation of cellular DNA synthesis by oncogenic DNA viruses also occurs after a lag period as in resting cells stimulated by serum, but it is preceded by the appearance of a nuclear antigen which, in the case of SV40 has been called the T antigen. It has been clearly demonstrated that the T antigen is a virally-coded protein and that the large T antigen has all the necessary information to induce host DNA synthesis in resting cells. This evidence stems partly from genetic studies with ts A and deletion mutants of SV40 and more formally, on the demonstration that purified T antigen, manually microinjected into resting kidney cells, causes these cells to enter S phase (5). With this background, it seemed reasonable to us to ask whether SV40 T antigen may stimulate cellular proliferation by acting on nucleolar genes, i.e., by increasing rRNA synthesis. Previous reports in the literature have already indicated that infection with either SV40 or Polyoma can increase RNA synthesis and/or accumulation in the cytoplasm of infected cells. In 1977 we reported that addition of T antigen preparation to the incubation mixtures stimulates RNA synthesis in nuclei and in nucleoli isolated from resting cells. The evidence that rRNA is the product of this direct in vitro stimulation of nuclear and nucleolar RNA synthesis and that stimulation is due to T antigen and not to contaminating proteins has already been given in previous reports. However, although these findings were highly reproducible, in vitro systems are notoriously riddled with artifacts and for this reason we have been seeking evidence that T antigen actually stimulates rRNA synthesis IN VIVO, i.e. in intact cells.

IN VIVO STIMULATION OF rRNA SYNTHESIS BY SV40 T ANTIGEN (See Soprano et al., 6,7).

{^{3}H}-uridine incorporation into RNA of cells is not a reliable measurement of RNA synthesis, because of variations in the uptake and pool size of the RNA precursors with growth rates. We, therefore, had to find other ways to investigate the relationship of T antigen to increased synthesis of rRNA in vivo. Croce has developed a hybrid cell line between fibrosarcoma HT1080 cells and mouse Balb/c macrophage in which the hybrid cells retain all human chromosomes, and 18 mouse chromosomes, including 12, 15 and 18 where genes for mouse rRNA are located. Although the mouse rRNA genes are present, these hybrid cells synthesize only human rRNA and not mouse rRNA. Human and mouse 28S RNA can be distinguished from each other, because the human 28S rRNA has in polyacrylamide gels a slightly lower electrophoretic mobility than 28S rRNA of rodents. In addition, the activity of ribosomal RNA genes can also be evidenced by silver staining of nucleolus organizers and by restriction endonuclease analysis of the products. When these hybrid cells, called 55-54, are infected with SV40, mouse 28S rRNA appears in the cytoplasm with a lag period of 16-18 hr. Simultaneously, the nucleolus organizers of some mouse chromosomes become stainable by silver. We further investigated the role of SV40 T antigen in the derepression of silent mouse rRNA genes by using different viruses.

Adenovirus 2, at variance with SV40 and Polyoma, does not stimulate rRNA synthesis. In agreement with these findings, infection of G55-54 cells with Adenovirus 2 does not reactivate the silent mouse rRNA genes. With this information a decisive experiment could be carried out using the hybrid virus, D1. This virus is an Adeno-2 virus containing an insertion in the SV40 genome. More precisely, D1 lacks the Adeno-2 genome sequences that map between 64 and 74 units of the conventional Adenovirus map and has in its place that part of the SV40 genome that maps between .10 and .71 of the SV40 map. These coordinates include all of the SV40 A gene that codes for T antigen. When cells are infected with D1 virus, they become strongly positive for the SV40 T antigen. Because Adenovirus 2 does not stimulate rRNA synthesis, a reactivation of silent mouse rRNA genes by D1 could conclusively demonstrate that it is the SV40 T antigen that reactivates the repressed rRNA genes. When we infected G55-54 cells with D1 virus, mouse 28S rRNA appeared in the cytoplasm of the infected cells (Figure 1).

The A genes of SV40 codes for two proteins, large T and small t. We used the deletion mutant of SV40, dl2005 that does not produce small t, although the large T is normal. This mutant reactivated the silent mouse rRNA genes in G55-54 cells, thus indicating that

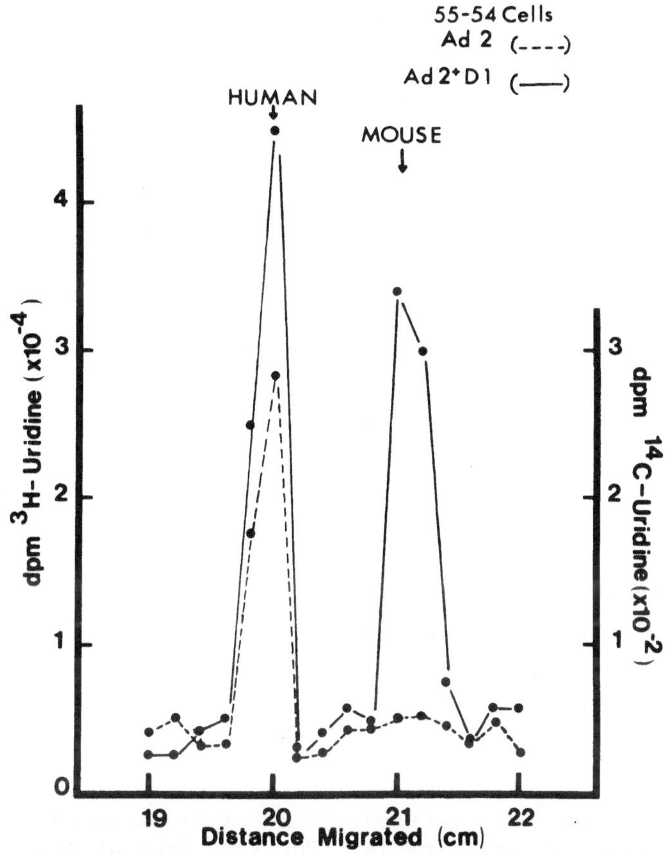

Figure 1. Reactivation of Mouse rRNA Genes by the A Gene of SV40.

Appearance of 28S or rRNA in the cytoplasm of human>mouse hybrid cells. In cells infested with Adenovirus 2 (as in control cells), only human 28S rRNA is present (---). Mouse 28S rRNA is induced in cells infected with $Ad2^{+}D_1$(——), a hybrid Adenovirus that contains the entire early region of SV40.

the small t antigen is not needed for the reactivation of repressed rRNA genes.

There are several other hybrid viruses between Adenovirus and SV40 that contain smaller fragments of the SV40 A gene. By using these viruses, one can determine what are the sequences in the SV40 A gene that are not required for the reactivation of silent mouse rRNA genes. Table 1 shows a list of these viruses, the map coordinates of the SV40 sequences that they contain and their ability or not to reactivate silent rRNA genes. In addition, fragments of SV40 DNA were microinjected into 55-54 cells and their

Table 1. A Map of SV40 DNA Sequences That Reactivate Silent rRNA Genes

Virus	SV40 Map Coordinates	Reactivation of rRNA genes
Adenovirus 2	-	-
D1	.10 - .70	+ + +
$Ad2^{+}ND_{1}$	.11 - .28	-
$Ad2^{+}ND_{2}$	.11 - .44	+ + +
$Ad2^{+}ND_{3}$	.11 - .18	-
$Ad2^{+}ND_{4}$	.11 - .59	+ + +
$Ad2^{+}ND_{5}$	.11 - .39	+ + +
SV40 HpaII-PstI fragment	.27 - .73	+ + +
SV40 Hpa I B fragment	.37 - .75	-
SV40 PvaII A fragment	.32 - .70	+
SV40 94G2 clone	.17 - .99	+ + +

With hybrid Adeno-SV40 viruses, rRNA reactivation was determined after infection of 55-54 cells. The same type of cells were used for the microinjection of SV40 DNA fragments. The assay for rRNA reactivation has been described by Soprano et al., (6,7).

ability to reactivate rRNA genes determined as usual. From that Table it is clear that the sequences of the A gene, mapping between 39 and 27 on the conventional SV40 map, are necessary and sufficient for the reactivation of mouse rRNA genes. Interestingly enough, these same sequences are those where most of the tsA mutants map but overlap only slightly with the sequences of the SV40 A gene that are responsible for the stimulation of cellular DNA syntehsis (.51 - 37 map units).

THE SIGNAL TO GROW IN SIZE CAN BE INDEPENDENT FROM THE SIGNAL TO REPLICATE DNA

It is clear from the studies of Ross and collaborators (8) and Pledger et al. (9,10), that there are at least two signals for the entry of cells into S. The first signal comes from the PDGF that "primes" the cell (10), stimulates RNA synthesis (11), but does not, by itself, cause cells to enter into S. A second signal comes from a factor in platelet-poor plasma that makes cells, made competent by PDGF, enter into S (for review see Scher et al. (12)). The simplest explanation is that a plasma factor (somatomedin) interacts with a PDGF-induced cellular product to initiate DNA synthesis. Our hypothesis is that PDGF gives the signal for doubling the size of the cell, while the signal for DNA replication comes from the plasma factor plus a PDGF-induced cellular product.

Available data support this hypothesis.

a. It is possible to dissociate an increase in rRNA synthesis (presumably a marker of increase in size) from DNA replication. Thus, nucleolar RNA synthesis increases in tsAF8 serum-stimulated at the nonpermissive temperature, i.e. blocked in mid-G_1 (13), and PDGF, by itself, can stimulate RNA but not DNA synthesis (11).

b. 422E cells, a mutant of BHK, that fail to accumulate ribosomes at the nonpermissive temperature (14), will enter DNA synthesis even under nonpermissive conditions, but they fail to divide (15).

c. Adenovirus 2 stimulates synthesis in infected cells (16), but does not cause an increase in rRNA synthesis (6). Infection by Adenovirus also fails to increase the accumulation of cellular RNA (17) (Figures 2 and 3). This is at variance with SV40, which stimulates rRNA synthesis (6) and accumulation of cellular RNA (18).

d. The SV40 T-antigen stimulates cellular DNA synthesis (5) even in ts13 cells at the nonpermissive temperature (see above). ts13 cells are a ts mutant of BHK cells that, like tsAF8, arrest in G_1 at the nonpermissive temperature (19). Adenovirus 2 can also bypass the ts block of these G_1 mutants (16), but the gene or genes responsible for the stimulation of cellular DNA synthesis have not yet been identified.

Our interpretation of these results is the following: Two signals exist for cell growth, one to double the cellular size, one to replicate DNA. PDGF gives the first signal only, Adenovirus gives only the second signal. SV40 T antigen gives both signals, while platelet poor plasma plus a PDGF-induced cellular factor can give the signal for DNA replication.

In this view, certain virally-coded proteins can be seen as replacing different growth factors, a sort, so to speak, of internal growth factors. In addition, the stimulation of ribosomal RNA synthesis and of cellular DNA replication can be separated but both are necessary for the ordinary growth and division of cells.

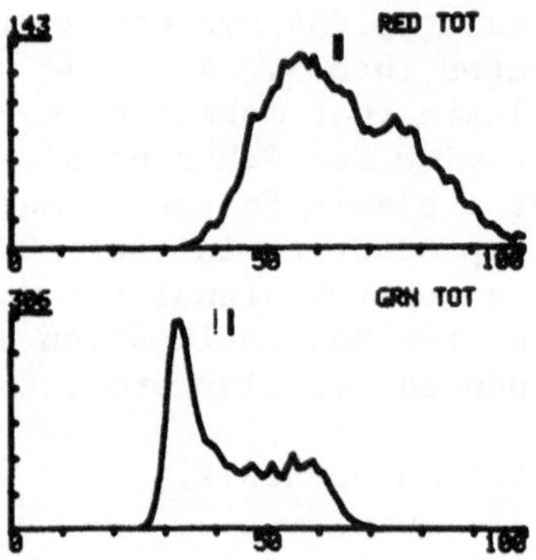

Figure 2. Accumulation of RNA in Serum-stimulated AF8 Cells.

The figure represents the distribution of green (lower panel) and red (upper panel) fluorescence of individual cells, stained with acridine orange, as computed from data obtained with a flow microfluorimeter. Green fluorescence measures DNA amounts: the peak at $\sim$40 arbitrary units represents G_1 cells. S and G_2 cells are present ($>$40 arbitrary units). Red fluorescence measures RNA amount per cell. In controls, it would average 30 arbitrary units. Here it is shifted to the right, indicating accumulation of RNA in serum-stimulated cells.

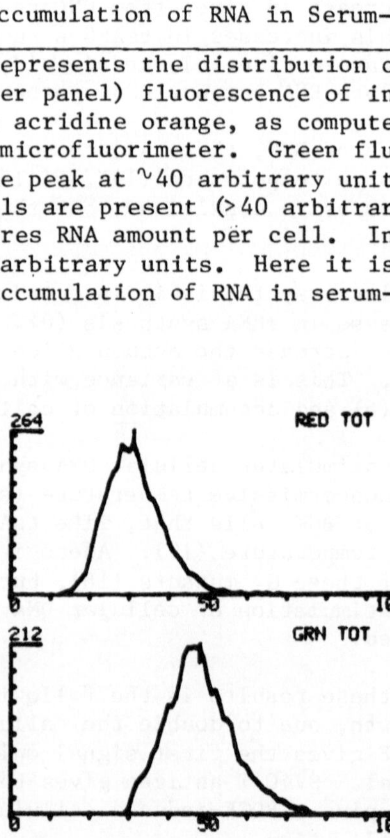

Figure 3. Failure of Accumulation of RNA in AF8 Cells Infected by Adenovirus.

Same parameters as in Figure 2. There are quite a few cells in DNA synthesis (40 arbitrary units), but the amount of RNA (upper panel) is that of G_0 cells, i.e. 20-40 arbitrary units.

REFERENCES

1. I. Lieberman, R. Abrams and P. Ove, J. Biol. Chem. 238:2141-2149 (1963).
2. R. Baserga, in: "Multiplication and Division in Mammalian Cells" Marcel Dekker, New York, pp. 239 (1976).
3. R. Baserga, C.H. Huang, M. Rossini, H. Chang and P.M.L. Ming, Cancer Res. 36:4297-4300 (1976).
4. R. Weil, Biochim. Biophys. Acta 516:301-388 (1978).
5. R. Tjian, G. Fey and A. Graessmann, Proc. Natl. Acad. Sci. 75:1279-1283 (1978).
6. K.J. Soprano, V.G. Dev, C.M. Croce and R. Baserga, Proc. Nat. Acad. Sci. 76:3885-3889 (1979).
7. K.J. Soprano, M. Rossini, C. Croce and R. Baserga, Virology 102:317-326 (1980).
8. R. Ross, C. Nist, B. Kariya, M.J. Rivest, E. Raines and J. Callis, J. Cell Physiol. 97:497-508 (1978).
9. C.D. Stiles, J.T. Capone, C.D. Scher, H.N. Antoniades, J.J. Van Wyk and W.J. Pledger, Proc. Nat. Acad. Sci. 76:1279-1283 (1979).
10. W.J. Pledger, C.D. Stiles, H.N. Antoniades and C.D. Scher, Proc. Nat. Acad. Sci. 74:4481-4485 (1977).
11. H.T. Abelson, H.N. Antoniades and C.D. Scher, Biochim. Biophys. Acta 561:269-275 (1979).
12. C.D. Scher, R.C. Shepherd, H.N. Antoniades and C.D. Stiles, Biochim. Biophys. Acta 560:217-241 (1979).
13. M. Rossini and R. Baserga, Biochemistry 17:858-863 (1978).
14. D. Toniolo, H.K. Meiss and C. Basilico, Proc. Natl. Acad. Sci. 70:1273-1277 (1973).
15. M. Mora, Z. Darzynkiewica and R. Baserga, Exp. Cell Res. 125:241-249 (1980).
16. M. Rossini, R. Weinmann and R. Baserga, Proc. Nat. Acad. Sci. 76:4441-4445 (1979).
17. S. Pochron, M. Rossini, Z. Darzynkiewica, F. Traganos and R. Baserga, J. Biol. Chem. (in press).
18. R. Weil, C. Soloman, E. May and P. May, Cold Spring Harbor Symposium 39:381-395 (1975).
19. J. Floros, T. Ashihara and R. Baserga, Cell Biol. Int'l Rpts. 2:259-269 (1978).

CELL CYCLE DEPENDENCE OF ERYTHROID MATURATION

J. Paul, D. Conkie and P. R. Harrison

The Beatson Institute for Cancer Research
Garscube Estate, Switchback Road
Bearsden, Glasgow, G61 1BD, Scotland

Although the term was only introduced by Holtzer in 1964 (Holtzer, 1964), the concept of the quantal cell cycle in differentiation has been in existence for a long time. It is used to describe a cell division which is necessary for a phenotypic change to occur in a cell lineage and has been particularly studied by Holtzer's group and others in myogenesis. It does not necessarily imply that the two daughter cells are different from each other, but in some models one daughter cell is identical with the parent and the other is not. It has, however, become clear that, while quantal cell division may be implicated in the determination of some phenotypic characteristics, they are not mandatory for all, for there are some well defined exceptions. Among the most striking are experiments in which nuclei have been injected into oocytes (De Robertis et al, 1977; Gurdon et al, 1978). These have shown quite conclusively that inactive genes in the injected nuclei can be activated without any nuclear division or DNA synthesis. It is possible that oocytes are special but the inescapable implication is that cell division is not essential for all kinds of gene activation. In this paper we discuss evidence that a round of DNA synthesis is mandatory for the final stages of erythroid maturation (in many respects analogous to the final stages of muscle differentiation).

In erythroid differentiation the first identified precursor is a pluripotent haemopoietic stem cell (CFU-S) which can be identified by colony formation in the spleens of irradiated mice. It gives rise by a process which is not well understood to a series of committed precursor cells which in the erythroid line are, first, the BFU-E and then the CFU-E. The distinction between the BFU-E and the CFU-E is that both of these are identified by the formation of erythroid colonies in a tissue culture culture assay in response to erythropoietin. Whereas both kinds of cell respond in this way, the BFU-E cells are less sensitive to erythropoietin and give rise to larger colonies than CFU-E cells. BFU-E are probably precursors of CFU-E and CFU-E may be identical with cells morphologically identified as proerythroblasts which, following stimulation with erythropoietin, go through a series of well-recognized maturation stages, respectively basophilic erythroblasts, polychromatic erythroblasts, normochromic erythroblasts, reticulocytes and erythrocytes. The determination step, that is, the step which determines that a lineage will give rise to erythrocytes, appears to occur during the transition from the CFU-S to the BFU-E. The stages from proerythroblast onwards during which the characteristic mature phenotype emerges, involve maturation events according to an already predetermined programme and it is during this maturation stage that we have some evidence that a quantal cell cycle occurs.

The first evidence we had for this was an experiment (Paul and Hunter, 1969) which showed that in primary cultures of mouse fetal liver the maturation response to erythropoietin was prevented by inhibitors of DNA synthesis. Due to technical limitations it was not possible to take these studies with primary clutures further at the time but with the discovery of the Friend cell the question could be asked quite precisely. The Friend cell line (Friend et al, 1971) is derived from a mouse erythroleukaemia. The cells have many of the characteristics of proerythroblasts and in many respects behave like transformed CFU-E except that they are completely refractory to erythropoietin. However, an effect very similar to that of erythropoietin on CFU-E, can be obtained by treating these cells with a number of different substances, notably dimethylsulfoxide (DMSO), hexamethylene bisacetamide (HMBA), butyric acid and haem. In response to these substances the cells accumulate globin messenger RNA and haemoglobin; erythrocyte-specific membrane proteins, such as spectrim and glycophorin, make their appearance and the nuclei of the cells become condense (Harrison, 1976). Friend cells in suitable medium normally replicate continuously but, following treatment with an inducer, cell division stops after two or three divisions.

Globin messenger RNA and haemoglobin do not appear immediately after treatment with inducer, but only following a lag period of some 12-18 hours. McLintock and Papaconstantinou (1974) found a correlation between the cell cycle time and the time of onset of haemoglobin production which was maintained when the cell cycle time was artificially manipulated. Further experiments by Levy et al (1975) indicated that cells needed to be in contact with inducer during the S-phase after release of a metabolic block in order to accumulate haemoglobin. We also (Harrison et al, 1978) found that induction of haemoglobin synthesis could be prevented by inhibitors of cell division of DNA synthesis. An apparently contradictory result of Leder et al (1975) was possibly due to the fact that in their experiment, while cell division was inhibited, nuclear division was not, with the result that binucleate cells accumulated which were capable of haemoglobin synthesis.

There are always serious doubts about the interpretation of results derived from experiments with metabolic inhibitors of nucleic acid or protein synthesis when the parameter being measured is synthesis of a specific polypeptide. For this reason we sought a more satisfactory experimental system by isolating conditional mutants of the Friend cell which were temperature sensitive for DNA synthesis and/or cell division. In particular we developed a mutant which at the premissive temperature of 33.5^{o} replicated at exactly the same rate as wild type cells (generation time of 20 hours) and which on treatment with 5mM HMBA at the permissive temperature synthesised haemoglobin at the same rate as wild type cells. However, at non-permissive temperatures of 37.5^{o} or above, these cells accumulated in G1 phase (Conkie et al, 1980). The particular subclone employed in these studies, t_sA54S3, was selected to maintain high viability during prolonged maintenance in non-permissive conditions (45% plating efficiency after 4.5 days' maintenance at 39°). Moreover, by careful experimentation it was determined that these cells exhibited permissive behaviour at 34.25^{o} but non-permissive behaviour at 37.25^{o}.

Efficient induction of haemoglobin synthesis could be achieved in both the wild type cell and the temperature sensitive variant by exposing them to HMBA according to the following regime. First, both kinds of cells were incubated for 12 hours at the non-permissive temperature to permit the temperature sensitive cells to accumulate in G1. Then both cells were exposed to 5mM HMBA for 30 hours, the temperature sensitive cells at 34.25^{o} and the wild type cells at 37.25^{o}. Subsequently both sets of cells were cultured without inducer for 114 hours at 34.25^{o}. The wild type cells yielded 53% benzidine-positive cells and had a mean cellular haemoglobin of 11 pg per cell, whereas the temperature sensitive mutants yielded 50% benzidine-positive cells with a mean cellular haemoglobin of 8 pg per cell (Table 1). The crucial experiment was to follow the same regime with the following modifications. (Only temperature-sensitive cells were used.) During 30 hours' exposure to inducer they were incubated at either the permissive or the non-permissive temperature, and subsequently the cells were plated out at 10^{4} per ml in methocel medium at 33.5^{o} and grown for 8 days without inducer. When the cells had been exposed to inducer at the permissive temperature, 45% of the resulting colonies were benzidine-positive and every cell in these colonies gave a positive reaction. By contrast, those cells which had been exposed to the non-permissive

Table 1. Induction of Hb synthesis at permissive and non-permissive temperatures

Cell type	Preincubation 37.25°(h)	Exposure to HMBA (h) 37.25°	Exposure to HMBA (h) 34.25°	Culture without HMBA at 34.25° (h)	Benzidine positive cells	Mean cellular Hb (pg/cell)
707	12	144	0	0	84	13
(wild type)	12	30	0	114	53	11
t_s	12	0	114	0	82	13
	12	0	30	114	50	8
	12	30	0	114	1	0

Table 2. Accumulation of globin mRNA in t_s Friend cells treated with inducer at permissive and non-permissive temperatures

preincubation h at 37.25°	Exposure to HMBA (h) 37.25°	Exposure to HMBA (h) 34.25°	Culture without HMBA (h) 37.25°	Culture without HMBA (h) 34.25°	Globin mRNA (ppm cytoplasmic RNA)
12	0	0	30	72	22
12	0	30	0	72	77
12	30	0	0	72	23

temperature during the induction period gave rise to colonies among which only 4% gave a positive benzidine reaction, and within the colonies only 5% of the cells stained. Globin mRNA levels in these cultures increased only when the exposure to inducer was at the permissive temperature (Table 2).

DISCUSSION

The overall conclusion from these results is:

1. Exposure of wild type cells or temperature-sensitive mutants to inducer at the permissive temperature for 30 hours leads to extensive differentiation and haemoglobin synthesis during a subsequent 114 hours' incubation without inducer.

2. However, if cell division is prevented during the induction period of 30 hours by exposure of the temperture sensitive mutant to non-permissive conditions, then insignificant differentiation occurs during subsequent incubation at the permissive temperature.

Hence, cell division and/or DNA synthesis during exposure to inducer seems to be necessary for subsequent maturation.

There could be several technical objections to this experiment. First, exposure to inducer at the non-permissive temperature might cause irreversible damage to the cells. Two observations suggested that this was not so. First, the plating efficiency of the cells treated in this way was still high (70%). Secondly, in all these experiments it was shown that if the growth arrest in the presence of inducer was reversed by shifting the cells to a permissive temperature, the cells replicated and accumulated haemoglobin, showing that they were competent to do so.

Another possible objection is that the temperature-sensitive lesion might be in protein or RNA synthesis. This was checked directly and it was found that the G1 cells arrested for 30 hours in inducer synthesised RNA and protein at about 45% to 50% of the rate found for randomly proliferating induced cells, whereas DNA synthesis was reduced to 2% of that of proliferating cells. Rudland et at (1975) have shown that total RNA and protein synthesis are regulated co-ordinately with changes in the cellular growth rate. They reported that the relative protein content in synohronised G1 cells was about 45% of that in G2 cells. A third possibility is that the non-permissive conditions interfere with haemoglobin synthesis directly. However, when cells were treated with inducer for 30 hours at the permissive temperature and then shifted to the non-permissive temperature, they accumulated haemoglobin at the same rate as cells maintained at the permissive temperature. Therefore, post-commitment expression is apparently independent of cell division. Hence, these findings demonstrate that a cell-cycle-dependent event is involved in the induction of maturation.

The nature of this event is not know, but experiments by Levenson and Housman (1979) and Housman et al (1978) indicated that, during the lag phase which probably corresponds to the period during which the cell-cycle related event operates, a slow steady accumulation of same stable molecular species may occur. These authors obtained evidence that a rate-limiting protein synthetic step and a non-rate-limiting RNA synthesis were involved. Similar conclusions were reached by Paul and Hunter (1968) in relation to erythropoietin dependent haemoglobin synthesis in fetal liver. These events seem to be associated with a prolongation of G1 following exposure to inducer (Terada et al, 1977; Adolph and Swetly, 1978; Geller et al, 1978) although whether this is a mandatory component of the response is unclear. Recently, some interesting results have been reported concerning the role of methylation of DNA in gene expression which may have a bearing on it. The results of Van Der Ploeg and Flavell (1980) indicate that heavily methylated genes are not expressed, whereas expressed genes are probably always undermethylated. The implication is that undermethylation may be a necessary (but not sufficient) condition for gene expression. Similar observations have been made by a number of other workers. A mechanism such as this may be implicated in a cell-division dependent step in erythroid maturation. However, this hypothesis should be considered with caution because it has been demonstrated that Friend cells, even when uninduced, still maintain a low rate of globin gene transcription. Moreover, the globin genes in both induced and uninduced Friend cells appeared to be sensitive to digestion of nuclie by DNAse 1, a phenomenon which may be correlated with undermethylation of DNA.

This work was supported by grants from MRC and CRC.

REFERENCES

1. Adolph, G.R. and Swetly, P., 1978, Poly (A) polymerase activity during cell cycle and erythropoietic differentiation in erythroleukemic mouse spleen cells, Biochim. Biophys. Act., 581, 334-344.
2. Conkie, D., Young, B.D. and Paul, J., 1980, Friend cell variants temperature-sensitive for growth, Exp. Cell Res., 126, 439-444.
3. De Robertis, E.M., Partington, G.A., Longthorne, R.F. and Gurdon, J., 1977, Somatic nuclei in amphibian oocytes: Evidence for selective gene expression, J. Embryol. Exp. Morphol., 40, 199-214.

4. Friend, C., Scher, W., Holland, J.G. and Sato, T., 1971, Haemoglobin synthesis in murine virus induced cells in vitro: Stimulation of erythroid differentiation by dimethyl sulphoxide, Proc. Natl. Acad. Sci. USA, 68, 378.
5. Geller, R., Levenson, R. and Housman, D., 1978, Significance of the cell cycle in commitment of murine erythroleukaemia cells to erythroid differentiation, J. Cell Physiol., 95, 213-222.
6. Gurdon, J.B., De Robertis, E.M. Laskey, R.A., Mertz, J.E., Partington, G.A. and Wyllie, A.D., 1978, Cytoplasmic Control of Gene Expression in Oogenesis, "Cell Differentiation and Neoplasia". Grady Saunders, Ed., Raven Press, New York.
7. Harrison, P.R., 1976, Analysis of erythropoiesis at the molecular level, Nature, 262, 353-356.
8. Harrison, P.R., Conkie, D., Rutherford, T. and Yeoh, G., 1978, Molecular Regulation of Erythropoiesis in "Stem Cells and Tissue Homeostasis", B. Lord, C.S. Potten and R. Cole, Eds., Cambridge University Press.
9. Holtzer, H., 1964, Control of chondriogenesis in the embryo, Biophys. J., 4, 239-251.
10. Housman, D., Gusella, J., Geller, R., Levenson, R. and Weil, S., 1978, Differentiation of murine erythroleukemia cells: the central role of the commitment event, in "Differentiation of Normal and Neoplastic Hematopoietic Cells", Cold Spring Harbor Conferences on Cell Proliferation, 5, 193-208.
11. Leder, A., Orkin, S. and Leder, P., 1975, Differentiation of erythroleukemic cells in the presence of inhibitors of DNA synthesis, Science, 190, 893-894.
12. Levenson, R. and Housman, D., 1979, Developmental program of murine erythroleukemia cells, J. Cell Biol., 82, 715-725.
13. Levy, J., Terada, M., Rifkind, R.A. and Marks, P.A., 1975, Induction of erythroid differentiation by dimethyl sulphoxide in cells infected with Friend virus: Relationship to the cell cycle, Proc. Natl. Acad. Sci. USA, 72, 28-32.
14. McClintock, P.R. and Papaconstantinou, J., 1974, Regulation of haemoglobin synthesis in a murine erythroleukaemic cell: The requirement for replication to induce haemoglobin synthesis, Proc. Natl. Acade. Sci. USA, 71, 4551-4555.
15. Rudland, P.S., Weil, S. and Hunter, A.R., 1975, Changes in RNA metabolism and accumulation of presumptive messenger RNA during the transition from the growing to the quiescent state of cultured mouse fibroblasts, J. Mol. Biol., 96, 745-766.
16. Terada, M., Fried, J., Nudel, V., Rifkind, R.A. and Marks, P.A., 1977, Transient inhibition of initiation of S-phase associated with dimethyl sulphoxide induction of murine

erythroleukemia cells to erythroid differentiation, Proc. Natl. Acad. Sci. USA, 74, 248-252.
17. Van Der Ploeg, L.H.T. and Flavell, R.A., 1980, DNA methylation in the human $\gamma\delta$-globin locus in erythroid and non-erythroid tissues, Cell, 19, 947-958.

INITIATION OF DNA SYNTHESIS AND PROGRESSION THROUGH THE S PERIOD

David M. Prescott

Department of Molecular, Cellular and Developmental Biology
University of Colorado
Boulder, Colorado 80309

Cell reproduction has three major components: growth, replication of the genetic apparatus, and cell division. These components are closely integrated with one another, but the molecular interactions that are responsible for integration are still poorly understood. Integration is obviously essential to the smooth progression of a cell cycle during which, on the average, the cell alternately doubles in size and is halved by cell division, and the chromosomes replicate and distribute to daughter nuclei coordinately with division of the cytoplasm.

THE CELL CYCLE IS COMPOSED OF TWO INTERACTING CYCLES

A convenient convention is to arrange events in a cycle of the familiar kind in the diagram in Figure 1. The cell cycle is really made up of two cycles, a growth cycle and a chromosome cycle, an idea first articulated by Mitchison (1). The chromosome cycle is depicted in Fig. 1 by three periods, namely the S period, the G2 period, and the M period (mitosis). The growth cycle on the other hand occurs without subdivision; growth is almost continuous and proceeds in parallel with the chromosome cycle. The single discontinuity in growth occurs transiently during mitosis, when RNA synthesis virtually stops, and the rate of protein synthesis falls by 75 per cent. By convention both cycles are considered to end and begin with the completion of mitosis.

Figure 1 shows the "arch-type" of chromosome cycle-- one that lacks the interruption between mitosis and DNA synthesis known as the G1 period. The "arch-type" of cycle occurs generally in cleavage stage embryos of animals ranging from marine invertebrates to mammals

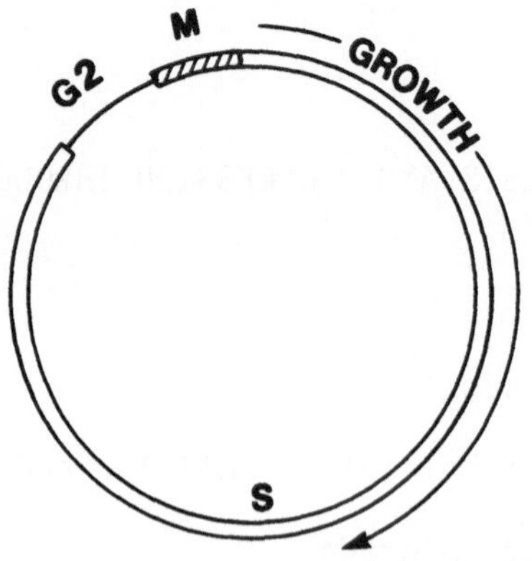

Fig. 1. The "arch-type" of cell cycle in which there is no interruption between mitosis and DNA synthesis.

(2). Such G1-minus cycles occur in some unicellular organisms, but in this paper I will deal exclusively with animal cells. In animal embryos a G1 period is introduced into the cell cycle after the cleavage stage and in conjunction with very early sorting out of cells into the very first stage of cell differentiation, e.g. into embryonic germ layers. In addition, however, the arch-type chromosome cycle occurs in at least one rapidly reproducing cell type of an adult mammal, i.e. bone marrow cells in the differentiation pathway leading to red blood cell formation. Several cell lines maintained in culture also reproduce without a G1 period (3).

RELATIONSHIP BETWEEN CELL GROWTH AND DNA REPLICATION

An essential point, illustrated in Fig. 1, is that the G1 period probably should not be considered as a part of the chromosome cycle, but rather as an interruption in the chromosome cycle. There are probably two general mechanisms by which the cycle may be interrupted between mitosis and DNA replication, creating a G1 period. In the first instance, if the rate of cell growth is slowed, increasing the length of the growth cycle so that it is now longer than the chromosome cycle, the initiation of the chromosome cycle (initiation of S) is delayed until the growth cycle is completed. In this situation G1 is the final part of the growth cycle by which the cell reaches a size needed to initiate DNA synthesis.

Clearly, the fact that cells maintain the same average size during proliferation proves that cell growth and initiation of S are interrelated. The presumptive relationship between cell size or growth and start of S are discussed by Prescott et al. (3), this volume. Experimentally the relationship is demonstrable by slowing cell growth through partial deprivation of an essential nutrient, for example, an amino acid. The lower the level of the amino acid in the medium, the slower cell growth and the longer the G1 period. At nearly complete deprivation of the essential amino acid cell growth stops, and the cells arrest in the G1 period. Thus, the inhibition of growth is sensed by the cell in such a way that a new chromosome cycle cannot commence.

In contrast, progression of the chromosome cycle is far less sensitive to growth inhibition than its initiation. The rate of growth can be severely reduced by partial amino acid deprivation or a partial inhibition of protein synthesis with a low level of cycloheximide (3) without interfering with completion at the normal rate of S, G2, and mitosis, but the cell cannot reenter S under such conditions. This is the basis of G1 arrest that occurs following deprivation of serum, growth factors, amino acids, etc. Although cells in S or G2 are less sensitive to arrest by growth inhibition than are cells in G1, they are not totally insensitive. If the growth of cells in S or G2 is completely inhibited, for example, by total deprivation of an amino acid or a high concentration of cycloheximide, such cells do not progress but arrest in S and G2. Complete inhibition of growth apparently does not cause arrest of cells in M, at least those cells that have reached late prophase. But the growth rate of cells from late prophase to early telophase is normally sharply reduced anyway.

The particularly high sensitivity of cells at a point between mitosis and DNA synthesis to a slowdown in growth is observed for many animal cell types in culture, but not for at least some transformed cell lines. For example, the Chinese hamster lung line known as V79-8 cannot be efficiently arrested in G1 by partial deprivation of isoleucine. These cells arrest instead more or less randomly in the cycle. Apparently, transformation can disrupt the interaction that normally integrates cell growth with initiation of DNA synthesis.

How the cell senses its own growth state and translates the information to govern the initiation of DNA synthesis is not known, and there are few clues. One clue derives from cells of different chromosome ploidies. For example, tetraploid cells are approximately twice as large as diploid cells. To initiate DNA replication a tetraploid cell must reach a much larger size than a diploid cell. Therefore, the crucial factor in initiation of DNA replication is not absolute cell size but rather a ratio of nuclear size (ploidy) to cytoplasmic mass. The importance of such a ratio in regulating cell reproduction was clearly pointed out by R. Hertwig (4) more than 70 years ago. Part of the difficulty in analyzing the relationship of cell growth to initiation of DNA replication resides in the fact that very little is known about cell growth. We know essentially nothing about the mechanisms that control and coordinate all of the components of growth such as the production of ribosomes, endoplasmic reticulum, mitochondria, enzymes, plasma membrane, etc. One or another component of growth, however, holds the key to the interaction of the cell growth cycle with the chromosome cycle. Similarly, we do not know how initiation of DNA synthesis might be modulated by growth because we do not understand the molecular events by which DNA replication is initiated.

SPECIFIC ARREST OF THE CELL CYCLE

The second case of interruption of the chromosome cycle between mitosis and DNA synthesis is not growth related, but is accomplished by specific regulatory signals in the cellular environment. This is the mechanism by which cell reproduction is differentially controlled in tissues, but very little is known about it. Most speculations about such specific arrest in G1 center around the idea of a group of regulatory hormone-like substances called chalones.

The clearest model currently available to describe specific control of the cell cycle derives from observations on yeast cells; the yeast model incorporates a chalone-like substance and is in overall form analogous to what is believed to occur in animal cells (5, 6, 7). Mating type in yeast is determined by a single gene locus. Yeast of mating type "α" produce and secrete a polypeptide (mating pheromone) designated "$\underline{\alpha}$". This chalone-like substance inhibits the initiation of the chromosome cycle in cells of the opposite mating type "a", thereby causing G1 arrest. In short, yeast cells secrete and respond to specific signal molecules by undergoing G1 arrest. Indeed, yeast cells arrested by mating pheromone appear to stop at the same point in G1 (at the G1-S border) as cells arrested by limiting growth (8). This suggests that a single switch controlling entry into S may be operated by different signal inputs, i.e. both mating pheromone or a signal generated by a growth-sensing mechanism.

LOCATION OF ARREST POINTS IN THE CELL CYCLE

Less is known about control of entry into S in animal cells. The simplest hypothesis is to assume a single switch that can be operated by a variety of signals. Certainly, at least one switch exists but we know nothing of its nature. A first step in studying such a switch is to define its location in the cell cycle, but even this has been difficult. A simplistic approach is to arrest cells in G1 by withholding an essential amino acid, serum, or other growth requirement, release the cells by reprovision of the missing element, and use the time interval from release to entry into S as a measure of the location of the arrest point. For example, if the time from release of the growth restriction to entry into S (recovery time) is four hours, then the arrest point is presumably located in G1 four hours before the start of S. Determination of the arrest point in this way, however, gives erroneous results because the observed recovery time is greater the longer cells are kept in G1 arrest. The shorter the arrest time, the closer to S is the apparent arrest point. For technical reasons it is not possible to shorten the arrest time sufficiently to obtain an accurate determination of the arrest point.

Another approach to the problem is to impose an arrest condition on a cell population and observe how long cells continue to enter S. For example, in an experiment in which CHO cells were switched to a medium lacking leucine, cells continued to enter S for no more than 15 minutes suggesting that the arrest point is located 15 minutes before S. Indeed, it may be closer to S than 15 minutes depending on how long it takes the cells to react to the change to leucine deficient medium. In any event, this is a many-fold shorter interval than observed when leucine-deprived cells are resupplied with leucine, further pointing up the inappropriateness of using release of arrested cells to locate an arrest point.

The leucine arrest experiment also reemphasizes the fact that arrested cells change their temporal relation to initiation of the S period. Cells positioned earlier than 15 minutes before the start of S undergo a change that delays their entry into S when leucine is subsequently resupplied. For example, a cell located at a point 30 minutes before the start of S at the time of leucine deprivation fails to reach S without leucine; when reprovided with leucine such a cell does not reach S in 30 minutes but requires several hours to start DNA synthesis. The nature of the delaying changes that occur in arrested cells is not known, but clearly they are profound and invalidate estimates of the location of arrest points obtained by measuring the time needed to reach S following reversal of a block.

A number of temperature sensitive mutant cell lines have been obtained that arrest in G1 at a restrictive temperature (9). Assuming that the mutations directly affect the switch mechanism in G1, one may estimate the location of the switch by measuring how long cells continue to enter S after transfer to a restrictive temperature. For several mutant cells these estimates place the arrest point (switch point) one to two hours before DNA synthesis. In one case the arrest point is one hour or less before S (10). All of these estimates probably represent the maximum possible time between the arrest point and S since some delay may occur between transfer of the cells to a restrictive temperature and the expression of the mutant phenotype.

Determining the position of an arrest switch is important because it would probably provide clues about its nature. For example, arrest immediately at the G1-S border might imply that the switch is part of the initiation mechanism for DNA synthesis. If the switch is located some time before S (9), we would need to assume that one or more events must bridge the interval. What these events might be is a matter of speculation (11).

The observation that the arch-type of cell cycle completely lacks a G1 period suggests that the regulatory switch is located immediately before the start of S. It may therefore be useful to study the events that surround the G1-S border. Expression of histone genes (transcription and translation) is initiated very near the G1-S border (12). A number of genes whose products are needed for synthesis of deoxynucleoside triphosphates also begin expression near the G1-S border. These events are temporally coordinated with the initiation of S but they are probably not causally involved in initiation of DNA synthesis. The elucidation of the molecular events that start DNA replication would likely reveal clues about regulation of S initiation.

CYTOPLASMIC INDUCER OF DNA SYNTHESIS

The well known cell fusion experiments of Rao and Johnson (13) provide another kind of perspective on the initiation of S. When a HeLa cell in the S period was fused with a cell still in G1 to make a binucleated cell, the G1 nucleus was induced to begin DNA synthesis precociously. Fusion of an S period cell with a G2 cell did not induce DNA synthesis in the G2 nucleus. These experiments indicate the presence of some factor or condition in the cytoplasm of an S phase cell that induces DNA synthesis. A nucleus in G1 can respond to the inducing stimulus but a G2 nucleus cannot. One interpretation is that the appearance of the cytoplasmic inducing stimulus is triggered by the completion of the growth cycle. Alternatively, the inducing stimulus may accumulate continuously throughout G1 and earlier parts of the cycle (14); in this view S is initiated when a critical level of the inducer is reached. It is reasonable to assume that the cytoplasmic inducer of DNA replication is responsible for the extremely high synchrony with which DNA replication is induced in multiple nuclei in the same cytoplasm. It is well known that the two nuclei in a binucleated mammalian cell almost always begin S simultaneously.

Further proof that the factor or condition that induces DNA synthesis is cytoplasmic was provided by fusing only the cytoplasm derived from an S period CHO cell to a cell in G1. S phase cytoplasm was obtained by enucleating S phase cells by the cytochalasin-centrifugation technique (15). Enucleated cells (called cytoplasts) were fused to intact G1 cells with the result that the G1 nuclei were induced to initiate DNA synthesis precociously.

A study of the cytoplasmic inducer in G1-minus cells has provided some further insight. V79-8 cells, which lack a G1 period (3), were collected at mitosis, and mitotic cells were fused to G1 cells of another type (16). As expected in the fusion products, the chromosomes of the G1 nucleus were induced to undergo premature mitotic condensation. Remarkably, however, the DNA of the condensing chromosome also initiated DNA synthesis. Thus, the cytoplasmic inducing activity is present in the mitotic cells of the G1-minus line. The DNA of the original mitotic chromosomes however, were not capable of responding to the inducing signal. With completion of mitosis, however, these chromosomes (of $G1^-$ cell line V79-8) gain responsiveness and immediately enter the S period of the next chromosome cycle.

The presence of the cytoplasmic inducer of DNA synthesis in mitotic cells of V79-8 suggests that it is constitutively present, as opposed to intermittantly present in cells with a G1 period. The constitutive presence of the inducer may perhaps account for the absence of a G1 period in these cells.

From these experiments we may tentatively conclude that initiation of S requires not only the cytoplasmic inducer but also a responsive state of the DNA. What accounts for responsiveness of G1 chromosomes and non-responsiveness of G2 and mitotic chromosomes is completely unknown.

The S period of animal cells proceeds with an orderly series of events (see below). For example, it is well documented that DNA present in euchromatic chromatin replicates in the first part of the S period, and DNA in heterochromatic chromatin replicates late. Since heterochromatin tends strongly to be condensed against the inner surface of the nuclear envelope and around the nucleolus, autoradiography of cells labeled late in S shows a distinct pattern of labeling, with autoradiographic grains almost exclusively over the region near the nuclear envelope and nucleolus. The autoradiographic pattern obtained in early S is the opposite, with grains only over the central part of the nucleus.

We asked whether the progression from replication in euchromatin to replication in heterochromatin might be brought about by a corresponding switch in the kind of cytoplasmic inducer (17). We fused a late S period cell with a G1 cell (CHO line), pulse labeled with ^{3}H-thymidine, and observed the autoradiographic patterns for the two nuclei in the same cell. The late S period nucleus showed the usual pattern of heterochromatin-associated autoradiographic grains. The G1 nucleus was induced to initiate DNA synthesis by the inducing activity present in the late S period cell. The autoradiographic pattern for this nucleus was, however, the opposite of the late S nucleus, with grains only over the central, euchromatic chromatin. Thus, the cytoplasmic inducer of DNA synthesis is non-specific in the sense that it does not influence the pattern of DNA replication.

Apparently, DNA replication, once initiated, proceeds according to a pattern established by an intranuclear program.

DNA REPLICATION AND THE NUCLEAR ENVELOPE

In bacteria some evidence suggests that DNA replication is initiated at an attachment site of the chromosome to the plasma membrane. Further, it has been suggested that the replication machinery is attached to the plasma membrane, and all replication is therefore localized at the membrane. By analogy it has been suggested that DNA replication in eukaryotes is initiated at the nuclear envelope, and replication remains localized there as the S period continues. Following fractionation of nulei pulse labeled with ^{3}H-thymidine, newly replicated DNA was found preferentially associated with the fraction containing the nuclear envelope material. Interpretation of this experiment is made uncertain by possible artefacts encountered in nuclear fractionation. A more secure way of examining the relationship of replication to the nuclear envelope is by autoradiography, which leaves the cells and nuclei intact. When cells (CHO) labeled at the start of S with a very short pulse of ^{3}H-thymidine were subsequently sectioned and examined by autoradiography in the electron microscope, virtually all grains were over the central region of the nucleus (18; see also 19,20). From this it appears that neither the initiation nor the continuation of DNA replication are associated with the nuclear envelope.

THE ORDERLY PROGRESSION OF THE S PERIOD

A diploid mammalian cell contains about 150 cm of DNA, all of which replicates in an S period of 8 to 9 hours (in cultured cells). The original autoradiographic studies of Huberman and Riggs (21) and those of others that followed (see 22) showed that replication begins at many points along each DNA molecule and proceeds bidirectionally at a rate of about 3,000 base pairs (3 KBP) per minute at each replication fork. On the average the origins of replication are about 30 μm (90 KBP) apart. This average interorigin distance is used to define the average size of the replicating units or replicons. It may be calculated, therefore, that the mammalian diploid nucleus contains about 30,000 replicons.

At the rate of replication fork travel of 1 μm per minute the average replicon requires about 15 minutes to replicate. Because the total S period lasts at least 480 minutes, it must be assumed that the initiation of the replication of different replicons is staggered throughout the S period. What mechanism produces the staggered initiations is not known, but some observations indicate that replicons replicate in a fixed sequence. For example, by using a double labeling technique in which DNA of synchronized cells is labeled at a known point in the S period with ^{14}C thymidine and then labeled in the next S period with BUDR, it was demonstrated that given segments of DNA replicate at approximately the same position within the S period from one cycle to the next (23, 24).

An ordered sequence of replication of replicons is also implied by autoradiographic studies showing that different parts of different chromosomes replicate according to a distinct pattern. Also, in several kinds of mammalian cells the S period is characterized by replication of relatively GC rich DNA early in the period with a gradual shift to replication of relatively AT rich DNA late in the S period (see 2, for review). Similarly, the different density components of DNA known as density satellites, which are composed of a high number of repeats of a short nucleotide sequence, replicate preferentially in particular parts of the S period [reviewed by Prescott (2) and by Bostock and Sumner (25)].

More refined information about the order of DNA replication, for example, replication of particular genes is not available, with one possible exception. DNA coding for rRNA has been reported to replicate in the early part of the S period in Chinese hamster cells (26, 27). However, no such restricted period of replication of rDNA was found in HeLa cells (28).

Related to the problem of how order is imposed on DNA replication, presumably in a replicon-by-replicon fashion, is the problem of assurance that <u>every</u> replicon replicates and that <u>no</u> replicon replicates more than once during a given S period. Perhaps the mechanism by which mistakes such as rereplication are prevented is related to the inability of chromosomes in G2 to respond to cytoplasmic initiation of DNA replication.

CONCLUDING COMMENT

The information brought together in this brief review shows that at least the principal questions about the initiation and progression of DNA replication can now be defined. Questions about the control of initiation of DNA replication are especially important because they are the most direct questions we can now ask about regulation of the cell cycle and cell proliferation. Like so many problems in cell biology the questions must finally be formulated in terms of molecules and the interactions of molecules that create the regulatory mechanisms that are reflected in every aspect of cell reproduction.

REFERENCES

1. Mitchison, J.M. (1971) "The Biology of the Cell Cycle," Cambridge University Press, London and New York.
2. Prescott, D.M. (1976) "Reproduction of Eukaryotic Cells," Academic Press, New York.

3. Prescott, D.M., Liskay, R.M. & Stancel, G.M., this volume.
4. Hertwig, R. (1908) Arch. Zellforsch. 1, 1-32.
5. Bücking-Throm, E., Duntze, W., Hartwell, L.H. & Manney, T.R. (1973) Exptl. Cell Res. 76, 99-110.
6. Reid, B.J. & Hartwell, L.H. (1977) J. Cell Biol. 75, 355-365.
7. Duntze, W., Stötzler, D., Bücking-Throm, E. & Kalbitzer, S. (1973) Eur. J. Biochem. 35, 357-365.
8. Reed, S.I. (1980) Genetics 95, 561-577.
9. Baserga, R., Potten, C. & Ming, P.M.L., this volume.
10. Liskay, R.M. (1974) J. Cell. Physiol. 84, 49-56.
11. Pardee, A.B., this volume.
12. Stein, G.S. & Stein, J.L., this volume.
13. Rao, P.N. & Johnson, R.T. (1970) Nature 225, 159-164.
14. Rao, P.N., Sunkara, P.S. & Wilson, B.A. (1977) Proc. Natl. Acad. Sci. USA 74, 2869-2873.
15. Prescott, D.M. & Kirkpatrick, J.B. (1973) Methods in Cell Biology (D.M. Prescott, ed.), Vol. VII, 189-201. Academic Press, New York.
16. Hanks, S.K. & Rao, P.N. (1980) J. Cell Biol. 87, 285-291.
17. Yanishevsky, R.M. & Prescott, D.M. (1978) Proc. Natl. Acad. Sci. USA 75, 3307-3311.
18. Wise, G.E. & Prescott, D.M. (1973) Proc. Natl. Acad. Sci. USA 70, 714-717.
19. Huberman, J.A., Tsai, A. & Deich, R.A. (1973) Nature 241, 32-36.
20. Fakan, S., Turner, G.N., Pagano, J.S. & Hancock, R. (1972) Proc. Natl. Acad. Sci. USA 69, 2300-2305.
21. Huberman, J.A. & Riggs, A.D. (1968) J. Mol. Biol. 32, 327-341.
22. Hand, R. (1979) Cell Biology: A Comprehensive Treatise (D.M. Prescott & L. Goldstein, eds.), Vol. 2, 389-430. Academic Press, New York.
23. Mueller, G.C. & Kajiwara, K. (1966) Biochim. Biophys. Acta 114, 108-115.
24. Taylor, J.H., Myers, T.L. & Cunningham, H.L. (1971) In vitro 6, 309-321.
25. Bostock, C.J. & Sumner, A.T. (1978) "The Eukaryotic Chromosome," North Holland Publishing Co., Amsterdam, New York, Oxford.
26. Amaldi, F., Giacomoni, D. & Zito-Bignami, R. (1969) Eur. J. Biochem. 11, 419-423.
27. Stambrook, P.J. (1974) J. Mol. Biol. 82, 303-313.
28. Balazs, I. & Schildkraut, C.L. (1971) J. Mol. Biol. 57, 153-158.

PRIMARY CILIA AND THEIR ROLE IN THE REGULATION OF DNA REPLICATION AND MITOSIS

R.W. Tucker
Cell Proliferation Laboratory
The Johns Hopkins Oncology Center
Baltimore, Maryland 21205

A.B. Pardee
Cell Growth and Regulation
Sidney Farber Cancer Institute
Boston, Massachusetts 20114

CENTRIOLES AND PRIMARY CILIA

Motile cilia have a "9 and 2" organization of microtubules, with 9 doublets of microtubules on the periphery and 2 doublets in the center. Primary cilia are non-motile "9 and 0" without dynein arms and a central doublet of microtubules. The name primary was used for this structure because they were thought to be a precursor to the complete "9 and 2" cilium (47). However, it is now known that the primary cilium can be found during some part of the cell cycle in many, if not most cells in a multicellular organism (3, 4, 6, 13, 14, 17, 23, 25, 30, 36, 37, 44, 45, 50, 51, 58, 61, 62). In the majority of these cases, the "9 and 0" structure never becomes a "9 and 2" complete cilium. There are also some interesting differences in the formation of the "9 and 2" and "9 and 0" cilia. The classic "9 and 2" cilia are formed by basal bodies which are arranged next to the plasma membrane at some distance from the cyto-center. In a single cell there are often multiple basal bodies, each forming a cilium. The primary "9 and 0" cilium seems to be exclusively formed in vertebrates by a centriole usually near the nucleus away from the cell surface. By some mechanism the ciliary membrane forms around the primary cilium, even in its internal position next to the nucleus (48). In some cells the ciliary membrane is exposed to the external medium (31), but in most vertebrate cells it is surrounded by a vacuole lined by plasma membrane. This space around the primary cilium is not open to the external medium, as judged by serial thin section EM (2) and by the addition of exogenous Horseradish Peroxidase (R.W. Tucker, unpublished observation). The plasma membrane and the ciliary membrane join at the base of the centriole in a structure called the ciliary necklace. This structure is a specialized membrane with a distinctive array of intramembranous particles (15). The ciliary necklace is also found in a "9 and 2" cilium in which the two membranes (plasma and ciliary) have been shown to have different histochemical characteristics. Near the base of the centriole are more specialized structures (rootlets, basal feet, and transitional fibers) of unknown origin or function. Presumably they function at least in part to anchor the centriole and cilium in the cell.

The centriole itself remains a poorly understood structure. In the electron microscope the centriole is seen as a cylinder, .25 μm by .5 μm, with 9 triplets of microtubules arranged around the periphery of the cylinder (49). Out of each triplet arises a doublet of microtubules in the primary cilium. A daughter centriole, or procentriole, is formed at the proximal end of the centriole at right angles to the parent centriole. The mechanism of centriole duplication is not clear, but morphological considerations suggest that a template requiring protein synthesis initially forms with further elongation of the centriole by the addition of tubulin and other proteins. In fact, only a restricted time of protein synthesis sometime prior to DNA synthesis is necessary for procentriole formation (33). RNA is present in the centriole and has been suggested, but not proven, to be a direct part of the generative process (21, 32, 39). Recently karyoplasts, which lack centrioles, have been shown to be capable of generating completely intact centrioles (64). Thus, the nucleus must monitor the number of centrioles and is capable of producing exactly one centriole pair with a minimum of cytoplasm. Two general mechanisms for the formation of the centriole have been proposed. One model suggests that there are small molecular weight RNA molecules which leave the nucleus and form the seed for the procentriole (18). Another model suggests that reverse transcriptase is used to translate the centriole RNA into a DNA template, which then forms another RNA molecule (60). Neither model is excluded at present by the available evidence. In addition to the procentriole, the proximal end of the centriole is also surrounded by electron-dense material (pericentriolar material) with which the cytoplasmic microtubules seem to be associated (see Figure 1). Recent *in vitro* evidence suggests that this pericentriolar material is indeed responsible for stimulating the polymerization of cytoplasmic and mitotic microtubules (16). The centriole itself actually can directly organize only a primary cilium at its distal end, much in the same way that basal bodies form motile flagella and cilia.

CENTRIOLE CYCLE

Centriole pairs duplicate once and only once each cell generation. Thus, the events associated with centriole duplication form

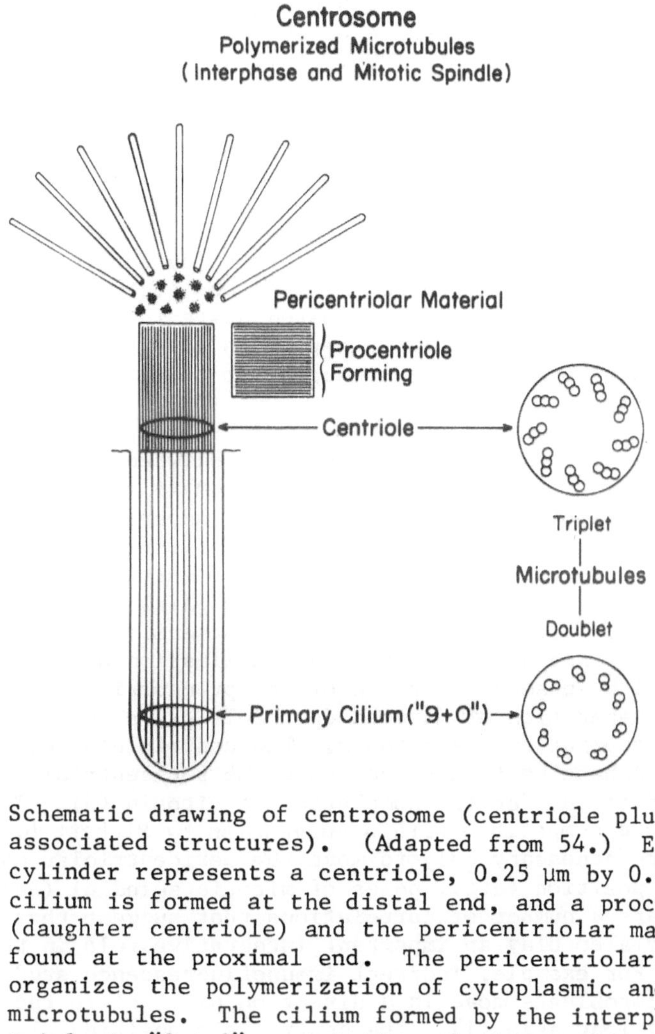

Fig. 1. Schematic drawing of centrosome (centriole plus associated structures). (Adapted from 54.) Each shaded cylinder represents a centriole, 0.25 μm by 0.5 μm. A cilium is formed at the distal end, and a procentriole (daughter centriole) and the pericentriolar material are found at the proximal end. The pericentriolar material organizes the polymerization of cytoplasmic and mitotic microtubules. The cilium formed by the interphase centriole is "9 + 0", consisting of nine peripheral microtubular doublets without a central doublet.

a cycle just as events of DNA synthesis do. On the other hand, unlike DNA, the centriole pairs change position dramatically during the cell cycle.

The relationship between centriole events and DNA synthesis cycles has been most extensively studied in two systems, yeast and sea urchin. In fertilized sea urchin eggs with no G_1 period centriole duplication occurs near telophase of the previous division, and thus, before the next DNA synthesis (29). Initial attempts (29) to test whether centriole duplication is necessary before DNA synthesis are difficult to interpret (46) and are probably inconclusive. In yeast, by contrast, the relationships are much clearer. A series of genetic studies in yeast has shown that a mutation in function or structure of the spindle plaque (centriole analogue) stops the cell at a growth arrest point and prevents DNA synthesis (19, 20). The growth arrest point is followed in yeast by duplication and separation of the spindle plaque and then DNA synthesis. Moreover, mating hormone also produces an arrested state in which the spindle plaque remains unduplicated and unseparated (11). Thus, centriole or spindle plaque events in yeast appear to be necessary before DNA synthesis can occur.

In mammalian cells the relationships between DNA and centriole cycles are unclear (Figure 2). Studies by EM in synchronized mammalian cells have shown that an initial small separation (a few microns) of two mature centrioles is followed by the formation of two procentrioles and further separation of the centriole pairs and procentriole elongation until two centriole pairs are situated at opposite poles at the start of mitosis. Presently we only know that the centriole's initial separation and duplication occur near the G_1/S border with subsequent growth of the procentriole during S (41, 47) and separation of matured centriole pairs during G_2 or prophase in preparation for mitosis. The centriole's role in mitosis ends after prophase so that by anaphase the pericentriolar material itself is sufficient for the completion of mitosis (5). These experiments are compatible with a suggestion by Pickett-Heaps that centrioles are necessary only to move the pericentriolar material into correct position (e.g., poles of mitotic spindle) (34). However, there are a number of correlations that suggest that the centriole may also play an important integrative role in the interphase cell. For example, indirect immunofluorescence studies have shown that fibroblasts move in a direction parallel to the centriole's (or cilium's) long axis (1). The centriole's position also seems to determine the axis of microfilament constriction during cytokinesis and to correlate with the direction of movement of polys in a chemotactic field (28). These observations suggest that the centriole may play a role in organizing the contractile machinery in both the interphase and mitotic cell.

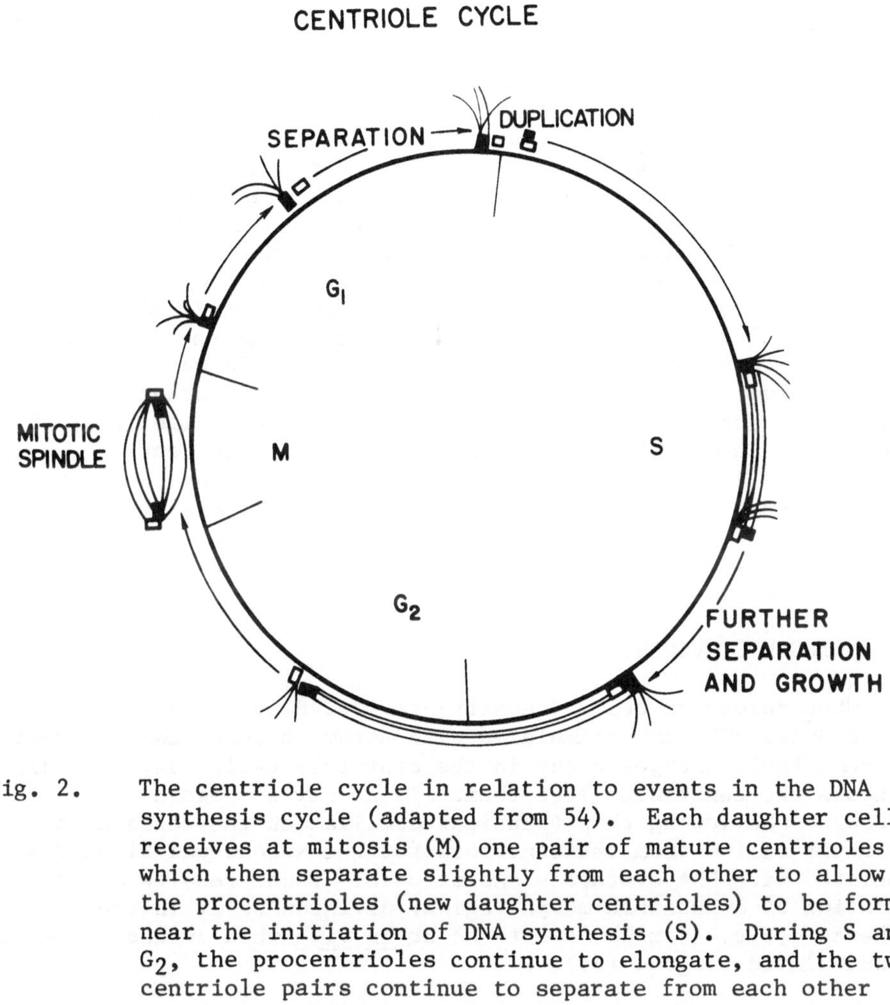

Fig. 2. The centriole cycle in relation to events in the DNA synthesis cycle (adapted from 54). Each daughter cell receives at mitosis (M) one pair of mature centrioles which then separate slightly from each other to allow the procentrioles (new daughter centrioles) to be formed near the initiation of DNA synthesis (S). During S and G_2, the procentrioles continue to elongate, and the two centriole pairs continue to separate from each other until the two mature pairs of centrioles are positioned at opposite ends of the mitotic spindle.

QUIESCENCE

The role of ciliated centrioles in quiescent cells has figured prominently in theories about growth control. In 1898 early workers (22, 24) hypothesized that ciliation of centrioles would switch growing cells into a quiescent, non-growing state. Indeed, many workers have found that quiescent or differentiated cells have ciliated centrioles (13, 37, 50). However, it was later found that rapidly dividing cells also had ciliated centrioles (13). Since growing and quiescent cells could both have ciliated centrioles, ciliation itself was not a growth control signal. We recently found an explanation for these findings in our studies on Balb/c 3T3 cells in culture. Quiescent 3T3 cells all have a primary cilium which disappears sometime after the cells are stimulated to enter DNA synthesis (54, 55). The kinetics indicate that the cells produce a "9 and 0" cilium from their centriole sometime in late G_1 during which time the cells can also leave the cell cycle if conditions are not optimal for growth. Thus, both growing and quiescent cells have ciliated centrioles because at the time the cell commits itself to either quiescence or growth the centriole has formed a primary cilium. Thus, the primary cilium is not a marker of quiescent cells, but its formation is associated with a time in G_1 which is important in growth control. Exactly when in the centriole cycle the centriole is ciliated is not completely clear, but the cilium must form sometime in G_1 and resorb sometime in S or G_2. It is also possible that some remnant of the cilium remains at one pole when the mitotic spindle is formed as has recently been shown for PTK_1 cells (40).

EARLY MITOGENIC EVENTS

When quiescent cells (density-arrested Balb/c 3T3 or serum starved Swiss 3T3) are stimulated with serum to enter DNA synthesis, some remarkable changes occur in the centriole cycle (54). Within 1-2 hours the centrioles lose their cilia only to regain them by 6-8 hours. Following this transient deciliation the cells next lose their cilium from the centriole for the second and final time associated with DNA synthesis and mitosis. These results are an indication of a possible morphological division of G_1 in the DNA synthesis cycle. Similar events occur <u>in vivo</u> when quiescent cells are stimulated to grow (50).

We next studied the effect of different components of serum on the centriole cycle (55). Scher et, al. (44) and Pledger et al. (35) have recently shown that serum can be divided into two functional components, platelet-derived growth factor (PDGF) and platelet-poor plasma (PPP). Neither PDGF nor PPP alone will induce DNA synthesis and PPP followed by PDGF also will not stimulate DNA synthesis. However, a pulse of PDGF (3 ng/ml) as short as 2 hours can be sufficent to prepare the cells to respond to the subsequent addition of PPP and progress into DNA synthesis 12 hours later. Furthermore, the state of "competence" induced by the platelet factor lasts at least 13 hours, since the addition of PPP within this time period stimulates the cells to enter S with a constant lag time (12-14 hours) after the addition of PPP. Therefore, the transition from quiescence to growth in Balb/c 3T3 cells involves at least two functionally different and temporally separated steps (competence and progression) which are under the control of separate serum components. We found that these two different steps are associated with two different steps in the centriole cycle. "Competence" produced by PDGF is associated with the initial, rapid, reversible deciliation of the centriole, while "progression" is associated with the preparation for centriole duplication and deciliation associated with subsequent DNA synthesis and mitosis. PDGF produces only the initial deciliation of the centriole while PPP alone produces no change in the centriole. However, PPP produces the second and final deciliation of the centriole if the cells are first made competent with PDGF. Other factors (fibroblast growth factor, calcium) produce "competence" only at doses which also produce deciliation (25). Similarly, other progression factors (multiplication stimulatory activity, epidermal growth factor) do not by themselves produce centriole deciliation. Thus, this transient deciliation of the centriole may be a morphological marker for the early mitogenic events ("competence") necessary for growth of non-neoplastic cells.

We next studied these early deciliation events in virally infected cells. This was of particular interest since these early mitogenic steps ("competence") are only necessary for non-neoplastic cells, and in fact, virally transformed cells can grow in plasma alone (44). We approached this problem by adding SV40 to quiescent Balb/c 3T3 cells. Such infected cells become autonomous, override density inhibition of growth and initiate DNA synthesis. In such cells there is no initial reversible deciliation of the centriole (55). It appears then that the SV40-induced cell cycle differs from that induced by serum and that abortive transformation by SV40 does not simply increase the effect of exogenous growth factors. Thus, the early mitogenic events which distinguish virally transformed and perhaps neoplastic cells from non-neoplastic cells are closely associated with a reversible deciliation of the centriole.

DNA SYNTHESIS

As mentioned before, the timing of centriole duplication is near the G_1/S border in mammalian cells (41). There is also a second and final deciliation of the centriole associated with DNA synthesis and subsequent mitosis produced by all of the growth factors (serum, milk growth factor, cartilage growth factor) and even SV40 virus. Thus, factors which do not produce an initial deciliation of the centriole still must produce a second final deciliation. These results emphasize the close association between centriole deciliation and events associated with DNA synthesis and mitosis. In preliminary experiments with EM autoradiography we have been able to detect deciliated centrioles in cells which have initiated DNA synthesis (R.W. Tucker, unpublished observation). This result is compatible with the original suggestion by Archer and Wheatley (3), and with the recent work on PTK_1 cells by Reider and Jensen (40) and on 3T3 cells by Lockwood (27). It is possible that the final deciliation of the centriole is a gradual process, with removal of tubulin and other proteins, leaving a matrix which is more slowly resorbed. It is unknown whether this resorption process is different from the reversible deciliation associated with the early mitogenic events. The detailed morphological changes in the loss of the cilium from the centriole during early and late periods in the cell cycle remain to be clarified.

MITOSIS

Many unicellular organisms lose their flagella before forming a mitotic spindle (7, 26). However, the cilium and spindle formation do occur at opposite ends of the centriole, and can occur together in meiotic divisions in, for example, insect spermatocytes (42). There are also rare instances of a centriole forming a flagellum while it is also forming a mitotic spindle. The fact that formation of an intact "9 and 0" or "9 and 2" cilium probably does not occur throughout most mitotic divisions, but can occur in some mitotic and meiotic spindles suggests that the cell imposes this particular order. The interrelationship between centriole deciliation, duplication and formation of the mitotic spindle remain questions for further investigation. At present, we only know that the centriole must deciliate or resorb its cilia before completion of mitosis so that each daughter cell receives a pair of centrioles without an attached cilium.

REGULATORY ASPECTS

The role of a primary cilium in regulation of the cell cycle is still an active area of research. Our studies have indicated the strong association or correlation between transient loss of a cilium and early mitogenic events, and between a final deciliation (or cilium resorption) and eventual mitosis. This final loss of cilium is reminiscent of primative flagellates which also lose their "9 and 2" microtubular appendages associated with growth and reproduction (26). Interestingly, there are other microtubular structures which must be lost before growth and reproduction can occur in ciliates (7). Nassula must resorb their cytopharyngeal basket before mitosis (53); stentor (12) and tetrahymena (64) must resorb an oral apparatus and form another before division occurs. These structures contain centrioles with microtubules and are examples of the importance of centriolar structures in coordination of growth even when the centriole structures do not directly participate in the partitioning of DNA at mitosis. Tetrahymena can also be synchronized by conditions which affect these centriolar structures. Thus, even in the ciliates the centriole and its associated cilium may be closely correlated with growth regulatory steps. Tetrahymena can also be made quiescent by increased levels of cyclic AMP which also inhibits ciliary regeneration (63). So, in these unicellular organisms ciliation and cell growth are coupled to the extent that they respond to the same chemical stimuli. The change in the primary cilium associated with mitosis in mammalian cells may be a vestige of these primary controls or associations. On the other hand, the first transient change in cilium may reflect changes in intracellular messengers which have become important in the evolution of multicellular organisms.

Calcium is a candidate for such a second messenger. Paramecium can be deciliated by calcium fluxes (43, 52) and calcium can depolymerize microtubules in vivo and in vitro (59). Thus, the deciliation of the centriole in mammalian cells could result from an increase in intracellular calcium. This increase in calcium need not occur via an increased flux of calcium through the plasma membrane. Rather, a release of calcium from internal stores may trigger the initial reversible deciliation of the centriole associated with early mitogenic events (competence) in mammalian cells. This is simply a hypothesis at the present time, but it is consistent with our observation that calcium ionophore can produce both competence and centriole deciliation (56).

The initial deciliation and postulated associated intracellular changes in second messenger are not absolutely necessary for growth since SV40 infected cells bypass this step. In this regard, it is interesting that SV40 transformed cells also bypass the requirement for extracellular calcium (8, 56). We have also found that deciliation of the centriole by itself does not necessarily cause competence but instead may reflect other changes which themselves produce the desired growth changes. Ciliation of the centriole may be related to the organization of the pericentriolar material or to the formation of the procentriole, either of which may be fundamental to the coordination of DNA and centriole cycles. At present there are only hints of what are the changes in the cell which ciliation of the centrioles actually reflect but already cell kinetic models involving aspects of of centriole cycle in the control of cell growth have been postulated (9).

Thus, it is premature at the present time to talk about how the primary cilium regulates the cell cycle. We have certain associations and correlations which imply that the formation and disappearance of the primary cilium may reflect important changes in the cell which are related to growth control. We hope that these correlations indicate that studies of centriole duplication and ciliation will teach us more about growth regulatory differences between neoplastic and non-neoplastic cells in the future.

REFERENCES

1. G. Albrecht-Buehler, Cell, 12:333-339 (1977).
2. G. Albrecht-Buehler and A. Bushnell, Exp. Cell Res., 126: 427-437 (1980).
3. F.L. Archer and D.N. Wheatley, J. Anat., 109:277-292 (1971).
4. B.G. Barnes, J. Ultrastructure Res., 5:453-467 (1961).
5. M.W. Berns, J.B. Rattner, S. Brenner and S. Meredith, J. Cell Biol., 72:351-367 (1977).
6. T.J. Biscoe and W.E. Stehbens, J. Cell Biol., 30:563-578 (1966).
7. R.A. Bloodgood, Cytobiol., 9:143-161 (1974).
8. A.L. Boynton, J.F. Whitfield, R.J. Issacs and R. Tremblay, J. Cell Physiol., 92:241-248 (1977).
9. R.F. Brooks, D.C. Bennett and J.A. Smith, Cell, 19:493-502 (1980).
10. N.L.R. Bucher and D. Mazia, J. Biophys. Biochem. Cytol., 7:651-655 (1960).
11. B. Byers and L. Goetsch, J. Bacteriol., 124:511-523 (1975).
12. N. deTerra, in: "Cell Reproduction: In Honor of Daniel Mazia", E.R. Dirksen, D. Prescott and C.F. Fox (Editors), Academic Press, New York, pp. 525-537 (1978).
13. V.G. Fonte, R.L. Searles and R.S. Hilfer, J. Cell Biol., 49:226-229 (1971).
14. H. Fritz-Niggli and T. Suda, Cytobiologie, 5:12-41 (1972).
15. N.C. Gilula and P. Satir, J. Cell Biol., 53:494-509 (1972).
16. R.R. Gould and C.C. Borisy, J. Cell Biol., 73:601-615 (1977).
17. M.A. Grillo and S.L. Palay, J. Cell Biol., 52:430-436 (1962).
18. H. Hartman, J. Theor. Biol., 51:501-509 (1975).
19. L.H. Hartwell, J. Cell Biol., 77:627-637 (1978).
20. L.H. Hartwell, Fourth European Cell Cycle Workshop, April 17-19, 1978, Bern. Nature, 273:594
21. S.R. Heidemann, G. Sander and M.W. Kirschner, Cell, 10:337-350 (1977).
22. L.F. Henneguy, Arch. Anat. Microscop. Morphol. Exp., 1:481-496 (1898).
23. H. Latta, A.B. Maunsbach, and S.C. Madden, J. Biophys. Biochem. Cytol., 11:248-251 (1961).
24. M. Von Lenhossek, Iber flimmerzellen. Verh. Anat. Ges. Kiel., 12:106-128 (1898).
25. H.S. Lin and I. Chen, Z. Zellforsch, 96:186-205 (1969).
26. C. Lloyd, Nature, 280:631-632 (1979).
27. A. Lockwood, Personal Communication (1980).
28. H.L. Malech, R.K. Root and J.I. Gallin, J. Cell Biol., 75: 666-693 (1977).
29. D. Mazia, P.J. Harris and T. Bebring, J. Biophys. and Biochem. Cytol., 7:1-21 (1960).

30. E. Meier-Vismara, N. Walker and A. Vogel, Expl. Cell Biol., 47:161-171 (1979).
31. Y. Mori, H. Akedo, K. Tanigaki and M. Okada, Exp. Cell Res., 120:435-436 (1979).
32. S.P. Peterson and M.W. Berns, J. Cell Sci., 34:289-301 (1978).
33. S.C. Phillips and J.B. Rattner, J. Cell Biol., 70:9-19 (1976).
34. J.D. Pickett-Heaps, Annal N.Y. Acad. Sci., 238:352-361 (1975).
35. W.J. Pledger, C.D. Stiles, H.N. Antoniades and C.D. Sher, Proc. Nat. Acad. Sci. USA, 74:4481-4484 (1977).
36. R.J. Przybylski, J. Cell Biol., 48:214-221 (1971).
37. J.E. Rash, J.W. Shay, J.J. Biesele, J. Ultrastructure Res., 29:470-484 (1969).
38. J.B. Rattner and S.G. Phillips, J. Cell Biol., 57:359-372 (1973).
39. C.L. Rieder, J. Cell Biol., 80:1-9 (1979).
40. C.L. Rieder, C.G. Jensen and L.C.W. Jensen, J. Ultrastructure Res., 68:173-188 (1979).
41. E. Robbins, G. Jentzsch, A. Micali, J. Cell Biol., 36:329-339 (1968).
42. L.E. Roth, H.J. Wilson and J. Chakraborty, J. Ultrastructure Res., 16:460-483 (1966).
43. B. Satir, W.S. Sale, and P. Satir, Exp. Cell Res., 97:83-91 (1976).
44. C.D. Scher, W.J. Pledger, P. Martin, H.N. Antoniades, and C.D. Stiles, J. Cell Physiol., 97:371-380 (1978).
45. J.P. Scherft and W.T. Daems, J. Ultrastructure Res., 19: 546-555 (1967).
46. G. Sluder, in: Cell Reproduction: In Honor of Daniel Mazia" E.R. Dirksen, D. Prescott and C.F. Fox, (Editors), Academic Press, N.Y., pp. 563-569 (1978).
47. J.A. Synder and R.M. Liskay, J. Cell Biol., 79:13 (1978).
48. S.P. Sorokin, J. Cell Sci., 3:207-230 (1978).
49. E. Stubbelfield and B.R. Brinkley, J. Cell Biol., 30:645-652 (1966).
50. S. Tachi, C. Tachi and H.R. Lindner, Biol. Reprod., 10:391-403 (1974).
51. Y. Tanuma, and M. Ohata, Arch. Histol. Jap., 41:367-376 (1978).
52. C.A. Thompson, L.C. Baugh, L.C. Walker, J. Cell Biol., 61: 253-257 (1974).
53. J.B. Tucker, J. Cell Sci., 6:385-429 (1970).
54. R.W. Tucker, A.B. Pardee and K. Fujiwara, Cell, 18:527-535 (1979).
55. R.W. Tucker, C.D. Scher and C.D. Stiles, Cell, 18:1065-1072 (1979).
56. R.W. Tucker, C.D. Stiles, C.D. Sher, and A.B. Pardee, J. Cell Biol., 83:12a (1979).
57. J.T. Tupper and Sorgniotti, J. Cell Biol., 75:12-22 (1977).
58. N.J. Wilsman, J. Ultrastructure Res., 64:270-281 (1978).
59. R.C. Weisenberg, Science, 177:1104-1105 (1972).

60. H.A. Went, J. Theor. Biol., 68:95-100 (1977).
61. D.N. Wheatley, J. Anat., 105:351-362 (1969).
62. D.N. Wheatley, J. Anat., 110:367-382 (1971).
63. J. Wolfe, J. Cell Physiol., 82:39-48 (1973).
64. E. Zeuthen and N.E. Williams, in: "Nucleic Acid Metabolism Cell Differentiation and Cancer Growth" E.V. Cowdry and S. Seno, editors, Oxford Pergamon Press, pp. 203-217 (1967).
65. C.A. Zorn, J.J. Lucas and J.R. Kates, Cell, 18:659-672 (1979).

REGULATION OF HISTONE GENE EXPRESSION DURING THE CELL CYCLE AND COUPLING OF HISTONE GENE EXPRESSION WITH READOUT OF OTHER GENETIC SEQUENCES

Gary S. Stein and Janet L. Stein

Department of Biochemistry and Molecular Biology and
Department of Immunology and Medical Microbiology
University of Florida, College of Medicine
Gainesville, Florida 32610

INTRODUCTION

In this chapter the regulation of histone gene expression during what has functionally been defined as the cell cycle will be addressed. In doing so three very different biological situations will be considered: continuously dividing cells, nondividing cells which are stimulated to proliferate, and early stages of development where the cell cycle is significantly abbreviated. Because it would be unrealistic to review extensively all the information currently available regarding histone genes and the regulation of histone gene expression, this chapter will be limited to a consideration of the following points: 1) level(s) at which regulation of histone gene expression resides; 2) the relationship between expression of histone genes and other genes expressed when DNA replication occurs; 3) approaches which have been taken to clone human histone sequences; and 4) although it would be presumptuous to propose a comprehensive model for control of histone gene expression, throughout the chapter an attempt will be made to consider the requirements for explaining the regulation of histone genes within the context of what is known about their structural, functional and biological properties.

Understanding the regulation of histone gene expression can provide useful information about control of cell proliferation from several standpoints. The gene products, histones, play an important role in chromatin structure and function. There is apparently a stringent requirement for histones to package newly replicated DNA. Histones are involved in determining the availability or lack of availability of newly replicated DNA sequences to be expressed. It is also becoming increasingly evident that knowledge of histone gene control under various biological circumstances can provide information regarding regulation of gene expression during the cell cycle at several levels.

HISTONE PROTEINS AND HISTONE GENES

Before considering the regulation of histone gene expression it would be appropriate to consider briefly some of the salient features of histone proteins and the histone genes. No attempt will be made to review extensively the available literature since an in-depth presentation of these topics can be found in several reviews (1-13) and monographs (14-22). The chapter by Professor Bradbury in this volume contains a lucid description of the involvement of histones in chromatin structure.

Histone Proteins

General Features. Over the past two decades a considerable amount of attention has been focused on the histones and it has now been generally acknowledged that these basic chromosomal proteins, enriched in arginine, lysine and histidine residues, play an important role in the structural as well as in the functional properties of the eukaryotic genome. There are five principal species of histones (H1, H2A, H2B, H3 and H4) represented ubiquitously throughout the plant and animal kingdoms. While there are some cell, tissue, and species variations in several of the histones (particularly H1, and to some extent H2A and H2B), these proteins are amongst the most highly conserved.

Post-translational Modifications. Some of the heterogeneity that exists in individual histone fractions can be largely but not entirely attributed to modifications of the proteins in the nucleus and/or in the cytoplasm of cells after completion of synthesis. Without changes in the amino acid sequences of histones, acetate (summarized in 23), phosphate (23-26), and methyl groups (27-30) may be covalently added to and removed from the amino acids of certain histones. In addition, the sulfur in cysteine can be modified and poly(ADP- ribose) chains can be added (31-36). In the case of phosphorylation, while it has been known for some time that phosphate groups may be post-translationally added to serine and threonine, it has also been observed that acid-labile aminophosphate is associated with other amino acids (37,38). The latter mode of phosphorylation is often undetected due to the acid extraction procedures commonly used for the preparation of histones.

Acetyl-CoA is the principal donor of acetate groups (7,23) and enzymes which are involved with acetylation and deacetylation of histones have been isolated (39,40). S-adenosyl methionine is the primary methyl donor and several methylases have been identified (29). The addition of phosphate to histones is catalyzed by a series of histone kinases (24), and phosphatases which are involved in the removal of phosphate from histones have been observed. One arginine-rich histone (H3) contains cysteine residues which chemically, and perhaps enzymatically, undergo oxidation of sulfhydryl groups (SH$\rightarrow$ 2SS) and reduction of disulfide linkages (SS$\rightarrow$ SH + SH).

Post-translational modifications of histone fractions in cells have been correlated with changes in transcriptional, replicative and structural properties of the genome. These modifications in histones are thought to alter histone-DNA binding and histone-histone relationships as well as the modes of interaction of histones with other genome-associated proteins. Such a role for histone modifications in influencing chromatin structure is suggested by results of Vidali and coworkers (41). Acetylation and phosphorylation of histones precede or occur concomitantly with the activation of transcription associated with numerous biological processes; e.g., stimulation of cellular proliferation, stimulation of cells and tissues by hormones, development, and transformation by oncogenic viruses or chemical carcinogens. Changes in the phosphorylation of specific histones also occur at defined points during the cell cylcle (see chapter by Bradbury in this volume). Removal of acetate and phosphate groups from histones, which occurs as a function of time, may provide a mechanism for repression of activated genetic sequences.

Evidence has been presented which indicates that the metabolism of phosphate and acetate groups associated with defined amino acid residues of histone fractions is regulated by specific enzymes. Langan and Hohmann (6) have demonstrated that in nonproliferating cells the serine 37 residue of the lysine-rich histone H1 is phosphorylated by a cyclic AMP-dependent protein kinase. These investigators have also shown that in proliferating cells phosphorylation of the lysine-rich histone involves sites other than serine 37 and occurs to a large extent on threonine as well as serine residues. The enzyme responsible for the growth-associated phosphorylation, in contrast to the enzyme present in nongrowing cells, is unaffected by cyclic nucleotides. Sequence analysis of peptides containing the phosphorylated sites of the lysine-rich histone carried out by Langan and coworkers may provide insight into the mechanism by which histone kinases recognize the appropriate amino acid residues to be phosphorylated. Because enzymes capable of dephosphorylating histones have also been identified, understanding phosphorylation at this level should enhance our knowledge of the biological relevance of the phenomenon. The elegant studies carried out by Bradbury and his collaborators (see chapter by Bradbury in this volume) in which histone phosphorylation and the relevant histone kinases were critically examined during the growth and division cycle of Physarum are very instructive within this context. Allfrey and co-workers have studied histone acetylases and have also identified a deacetylating enzyme which specifically utilizes acetylated lysine residues of arginine-rich histones as substrate (39,40). The enzyme is ineffective in removing acetate groups from proteins other than histones.

Methylation and changes in the oxidation state of the sulfhydryl groups of arginine-rich histones have been purported to alter the structural properties of the genome (4,42). These latter two post-translational modifications of histones have been correlated with chromosomal condensation during mitosis. However, the evidence for a role in mitosis for disulfide linkages between H3 histone molecules is very weak, particularly when one recognizes that one of the H3 cysteine residues has not been conserved during evolution while mitosis obviously has been. In contrast to histone acetylation and phosphorylation which are reversible processes, histone methylation appears to be irreversible.

Because it is necessary to explain the turning on and turning off of genes as well as modifications in chromatin structure in response to changes in biological activity, the well-documented observation that certain post-translational modifications of histones are reversible makes such modifications particularly attractive from a functional standpoint. Consistent with the purported structural and regulatory implications for post-translational modifications of histones is the presence in the nucleus and associated with chromatin of many of the enzymes responsible for addition and/or removal of acetate, phosphate, methyl and poly ADP-ribose groups.

Turnover. Results from experiments in which the specific activity of radioactively labeled histones is followed for several consecutive cycles of cell proliferation show that the five principal histones do not exhibit any significant extent of turnover (43,48). In fact, histones appear to be about as stable as DNA, which places them among the most stable molecules in eukaryotic cells (43-48). Some studies suggest the possibiltiy that the histones exhibit turnover; however, it must be carefully considered whether the molecules exhibiting turnover are histones or histone-like nonhistone chromosomal proteins. Additionally, losses of histones during extraction and/or incomplete extraction of histones must be evaluated. Recovery of histones may vary as a function of extraction conditions, proteolytic activity, or post-translational modifications of histones--resulting in altered histone-histone interactions, histone-DNA interactions and histone-nonhistone interactions. Differential extractability of histones should not be taken lightly since this may reflect variations in the functional state of the macromolecules.

Chromatin Structure. Biochemical, biophysical and electron-microscopic evidence has shown that the histones H2A, H2B, H3 and H4 are intimately involved in organizing DNA into nucleosomes and that H1 histone is probably involved in the next higher order of genome packaging (19,49-54). For some time it has been apparent that histones may be responsible for the nonspecific repression of genetic sequences for transcription (3,5,9,14,15). Although there generally is a tendency for macromolecular structure and function to be viewed as independent problems, understanding the biological properties of chromatin has effectively employed an amalgamation of both approaches.

Histone Genes

The hybridization of histone mRNA or histone-specific cDNA to vast excesses of genomic DNA has revealed that the histone genes belong to the moderately repetitive class of DNA sequences; they are the only repetitive genes known to code for proteins. The reiteration frequency of histone genes per haploid genome has been estimated to be twenty-fifty for humans (55,56) and mice (55,57), 20-50 for Xenopus laevis (57), 2 for yeast, approximately 110 for Drosophila, 10 for chicken (58) and 400-1200 for various species of sea urchins (13,59-63). One possible reason for multiple copies of histone genes is the requirement for rapid synthesis of histones when DNA replication occurs and particularly during embryonic development in some organisms, especially sea urchins (reviewed in Davidson (64)). Another possiblity, that the histone genes are arranged in tissue-and/or developmental-stage-specific operons will be discussed in more detail below.

The amino acid content of the histones, combined with properties of the genetic code, led to the prediction that the DNA that coded for the histones should have a high G + C content (50-56%) and potentially be separable from the bulk DNA on this basis (59). The availability of purified histone mRNAs as specific hybridization probes has allowed this prediction to be confirmed, resulting in the purification of histone genes from sea urchins (13,59-66).

The presence of sea urchin genes coding for all the histones in the same density gradient fractions first suggested that the histone genes might be clustered and interspersed. This has been definitively shown for several sea urchin species by using recombinant DNA technology to clone the genes in prokaryotic vectors (for review see Kedes, 13,67,68). Analysis of these cloned DNA fragments and of total histone DNA by restriction enzyme analysis, histone mRNA-DNA hybridization, and partial denaturation mapping has revealed that the G + C-rich genes for all five histones, interspersed with A + T-rich nontranscribed spacers, occur in repeating units of H1-H4-H2B-H3-H2A (3' to 5' direction) in the sea urchins Psammechimus miliaris, S. purpuratus, and L. pictus (13,67,68). The sizes of the histone gene repeat units in the sea urchin species thus far examined range from 5.6 - 7.3 kb. Furthermore, digestion with the 5'-specific lambda exonuclease has demonstrated that in sea urchins all five genes have the same polarity and are therefore transcribed from the same strand (69). A molecular map of the histone gene repeat units from these sea urchin species is shown in Figure 1. Of particular importance is the ability to directly read the genetic code for complete histone proteins from contiguous genomic histone DNA sequences. It therefore follows that in contrast to genes such as globin, ovalbumin and immunoglobulin, where in genomic DNA protein-coating sequences are

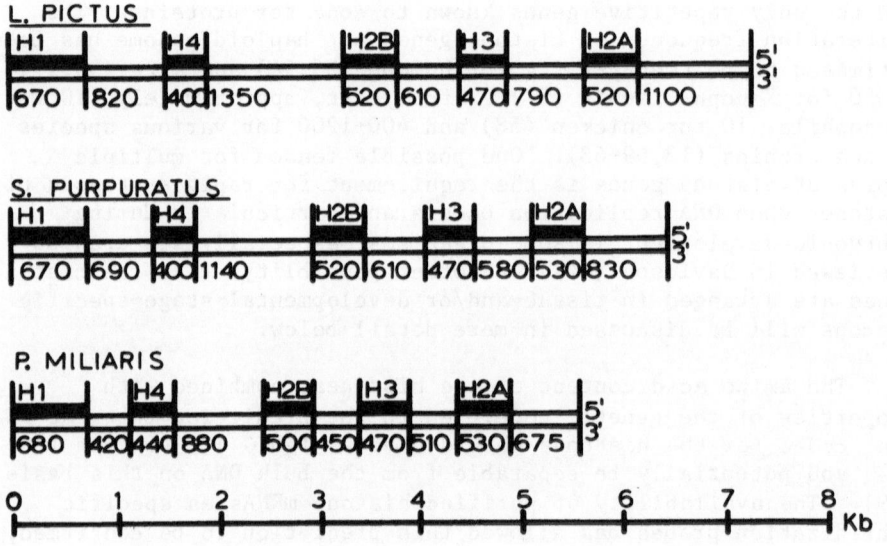

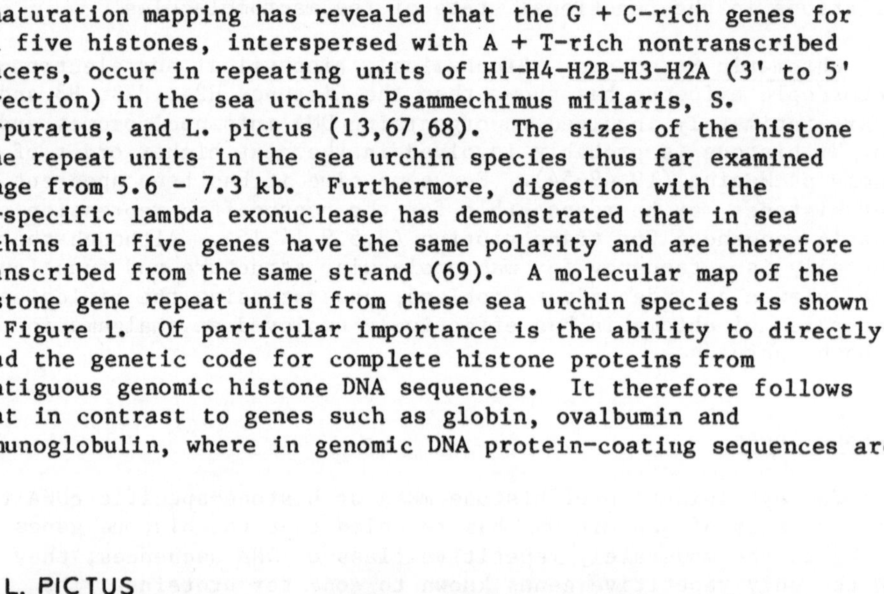

Fig. 1. Molecular maps of histone genes from three sea urchin species, Lytechinus pictus, strongylocentrotus purpuratus and Psammechinus miliaris. The mRNA coding regions are indicated by heavy lines on the DNA strand from which they are transcribed. The sizes of the various segments of the histone gene repeats are indicted.

interspersed with nonprotein-coating sequences which must be spliced out during post-transcriptional processing, histone gene organization is far less complex permitting a reduction in the number of obligatory post-transcriptional processing steps. Several important implications for regulation of histone gene expression are therefore raised which will be subsequently discussed in this chapter. However, to comprehend fully the biological significance of these observations and the total extent of their applicability, one must bear in mind that histone genes are reiterated and that generalizations regarding histone gene structure and expression must be reserved until a number of histone gene repeats from a single species have been cloned and characterized. Then one can address questions such as the types of regulation and regulatory molecules which are compatible with such a "primitive" pattern of genetic organization.

In situ hybridization studies with cloned sea urchin histone DNA have shown that the histone DNA sequences of Drosophila melanogaster are clustered in a single region of the genome (70,71) while those of the newt Trituris cristatus carnifex are localized in one major and a few minor regions (72). In situ hybridization studies also suggest that in humans the histone genes may be clustered on the distal end of the long arm of chromosome 7 in the G-negative band, q34 (73,74).

Additional information regarding the structure and organization of invertebrate histone genes can be gleaned from studies in which restriction endonuclease-digested total genomic DNAs were electrophoretically fractionated, transferred to nitrocellulose paper, and hybridized with radiolabeled, cloned sea urchin DNA (75). Using this approach the sizes of the histone gene repeats for several marine invertebrates were shown to be as follows: Limulus polyphemus (horseshoe crab) 4.1KB, Echinarachinus parma (sand dollar) 6.1KB, Spisula solidissima (clam) 4.5KB, Crassostrea virginica (oyster) 6.3KB and Chaetopterus pergamentaceous (worm) 5.2KB. In the case of avian histone genes where restriction endonuclease-digested genomic DNA was similarly analyzed by hybridization to ^{32}P-labeled (nick-translated), cloned genomic sea urchin histone DNA it appears that the size of the chicken histone gene repeat is approximately 15KB (58). It is most striking and perhaps extremely important from a functional standpoint that for proteins which have been as highly conserved phylogenetically as the histones there are dramatic divergences with respect to the organization of the genomic sequences. For example, in Drosophila the order of the histone genes differs from that of sea urchins and within a single repeat H1, H3 and H2A mRNA-coding sequences are transcribed from one DNA strand while H4 and H2B sequences are transcribed in the opposite direction from the complementary DNA strand (78, Figure 2). Rather obviously, such a pattern of

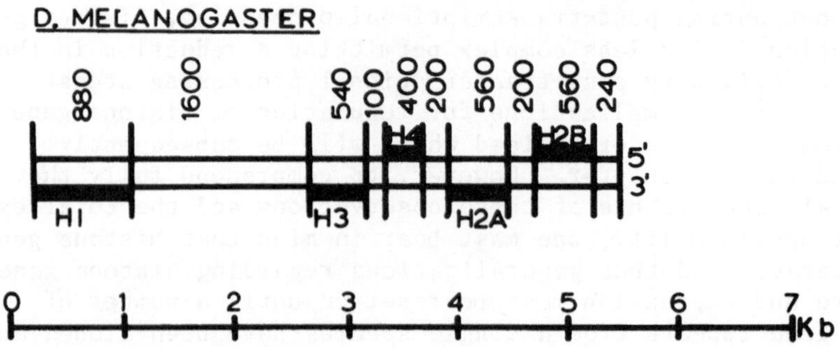

Fig. 2. Molecular map of histone gene repeat from Drosophila melanogaster. The mRNA coding regions are indicated by heavy lines on the DNA strand from which they are transcribed. The sizes of the various segments of the histone gene repeats are indicated.

genetic organization is incompatible with transcription of a polycistronic precursor containing the genetic information of an entire histone gene repeat. Within this context it is also interesting that for the yeast Saccharomyces cerviciae evidence has been presented indicating that the DNA sequences which code for H2A and H2B histones are located in different regions of the genome and are found on opposite DNA strands. From this discussion it should be apparent that at present our understanding of histone genes is limited and that an analysis of the structure and organization of histone genes in a variety of organisms should be undertaken. At the same time, attention should be focused on genetic polymorphism with respect to histone genes in various tissues and cells of specific organisms. With the high resolution techniques now available for construction and analysis of histone sequences such an approach is realistic and should provide valuable insight into the control of histone gene expression.

LEVEL AT WHICH REGULATION OF HISTONE GENE EXPRESSION OCCURS

Gene Regulation During the Cell Cycle

When approaching the problem of differential gene expression during the cell cycle, that is, the expression of specific genetic sequences which are required for progression of the proliferative process, the question of how to approach the problem of gene regulation arises. To appreciate the complexity of the problem, it is necessary to bear in mind that there are numerous levels at which regulation of histone gene expression may be mediated and that there are variations in the level(s) at which control of cell cycle stage specific gene expression resides under different biological circumstances.

Whether or not a gene is expressed may be determined at a multitude of steps beginning in the nucleus with the availability of a genetic sequence for transcription. Other points of control include the transcription process itself, with the many required enzymes and factors, and the various post-transcriptional chemical modifications and rearrangements undergone by a genomic transcript prior to export to the cytoplasm. The association of RNA molecules with proteins may play important roles in such processes. Once in the cytoplasm further chemical modifications of transcripts occur, protein-RNA interactions may be altered, and the decision whether or not to associate with ribosomes and enter into a functional translation complex must be made. In many instances, completion of translation is not the final step in expression of a genetic sequence because post-translational changes such as chemical modifications of polypeptides, cleavage of terminal amino acid sequences, or association of two or more polypeptide subunits are required before a functional gene product is available.

A striking example of differences in the levels at which cell cycle stage specific gene expression occurs as a function of the biological circumstances can be seen by comparing the requirement of RNA synthesis for DNA replication and mitosis during early stages of development with such a requirement for RNA synthesis in more differentiated cells. The ability of post-fertilization cleavages (DNA replication and cell division) to progress in an apparently normal manner in the absence of RNA synthesis is consistent with activation and/or recruitment of pre-existing mRNAs; i.e., post-transcriptional control. In contrast, the dependence of DNA replication and mitosis on newly synthesized RNA in more differentiated, continuously dividing cells and in nondividing cells after stimulation to proliferate can best be interpreted as reflecting a requirement for transcription.

Despite the complexity of the regulation problem, one must select a specific approach to studying the control of gene expression during the cell cycle. One viable possibility is to pursue a biochemical and molecular analysis of specific genes and/or their gene products. From the complex and interdependent biochemical events occurring during the cell cycle there are numerous candidates. Two such genetic sequences that are being actively examined are dihydrofolate reductase genes (77-79) and histone genes. Another very powerful approach is the construction and analysis of cell cycle mutants. Once characterized, mutants that differ with respect to the presence or absence of G1 periods or that appear to have biochemical defects that map at various points during the cell cycle can be used to address the regulation of specific genes. The latter approach has been used successfully in several laboratories to study cell cycle stage specific gene expression--most notably those of Prescott (see chapter in this volume) and Baserga (see chapter in this volume).

Histone Gene Regulation

From the information available about the structure and organization of histone genes, and from our understanding of the properties of histone mRNAs, it is apparent that many of the complexities in genes such as globin, ovalbumin and immunoglobulin may not be encountered. Thus, examining the regulation of histone gene expression presents a situation that may be somewhat simplified. For example, newly transcribed histone mRNAs, which can be directly read from contiguous DNA sequences, are rapidly transferred to the cytoplasm with what appears to be a minimal number of processing steps--cleavages and chemical modifications. Despite early claims to the contrary, most data available are inconsistent with a large molecular weight (polycistronic) histone mRNA precursor. Splicing and rearrangements of histone gene transcripts do not seem to occur. Although histone mRNAs contain capped 5'-termini (80,81), the 3'-ends of histone mRNAs (with the exception of stored maternal histone mRNAs found in oocytes of some species) do not generally contain tracts of poly A.

In addressing the regulation of histone gene expression, there are two basic questions that must be considered: 1) When are the histone proteins synthesized during the cell cycle? 2) When are histone mRNA sequences transcribed during the cell cycle? Do these events occur throughout the cell cycle or during a restricted period of the cell cycle?

Synthesis of histone proteins during the cell cycle. The first indication that the synthesis of histone proteins occurs during a restricted period of the cell cycle came from a series of studies carried out in the mid 1960's by Mueller and his collaborators (82) and simultaneously by Robbins and Borun (83). These investigators pulse-labeled synchronized, continuously dividing HeLa S_3 cells with radio-labeled amino acids and demonstrated that newly synthesized histones were associated with chromatin only during the S phase. Consistent with these observations Stein and Borun (84) showed that the nuclear histone content doubled during each cell generation and that doubling occurred during S phase. Subsequently, the synthesis of histone proteins at the time of DNA replication has been confirmed for numerous populations of continuously dividing cells and of nondividing cells stimulated to proliferate, for cells in culture as well as cells in intact animals. A possible exception is a recent report suggesting that in two mammalian cell lines histones are synthesized at equivalent rates during G1 and S phase (85), although the observation is incompatible with the well-documented absence of histone mRNAs on the polysomes of cells not undergoing DNA replication. There is also general agreement that, perhaps with the exception of H_1 histones which under some circumstances may exhibit a limited amount of turnover during G_1, the histones are extremely stable macromolecules, probably as stable as the DNA with which they are associated (86).

Early studies on the metabolism of histone proteins also indicated that inhibition of DNA synthesis by drugs such as cytosine arabinoside, hydroxyurea, or aminopterin resulted in a rapid and complete shutdown of histone synthesis (48,82-84, 87-92). While these findings raise the possibility of a functional relationship between DNA replication and histone biosynthesis, the level at which such a relationship may be mediated remains to be resolved.

Caution must be exercised in invoking teleological arguments to explain biological and biochemical phenomena. However, the concomitant synthesis of DNA and histones during the S phase of the cell cycle can be explained in terms of a requirement for histones to package newly replicated DNA into nucleosones and to facilitate inactivation of those regions of the newly replicated genome that are not transcribed in particular cells, tissues, or organs. Such an argument is strengthened by the apparent absence of a nuclear or cytoplasmic pool of histones (histones not complexed with DNA) during G1, G2 or mitosis in most cells which have been examined. Yet one must not lose sight of the well-documented presence of "stored" histones in oocytes, which points to the possibility of regulation of histone gene expression at different levels depending on the biological circumstances.

The observation that synthesis of histone proteins occurs primarily or exclusively during S phase is indicative of cell cycle stage specific histone gene expression but does not by itself provide sufficient information to assess the level or levels at which regulation of histone gene expression resides. Amongst the critical points to be determined is whether histone proteins are translated from mRNAs that are transcribed and associated with polysomes during S phase or whether histone protein synthesis is, at least in part, accounted for by mRNA templates synthesized during other periods of the cell cycle and recruited into the polysomes when DNA replication occurs.

<u>Synthesis of mRNAs during the cell cycle</u>. To assess further the level(s) at which regulation of histone gene expression resides, in vivo synthesized RNAs can be isolated from the nucleus and cytoplasm of G1 and S phase cells and assayed for the representation of histone mRNA sequences by hybridization with

histone cDNA. RNAs transcribed in vitro from nuclei or chromatin isolated from G1 and S phase cells can be assayed similarly for the presence of histone mRNA sequences. The rationale for pursuing these approaches is that the presence of histone mRNA sequences only in S phase would imply nuclear or transcriptional control, whereas the presence of histone mRNA sequences during G1 would be consistent with post-transcriptional control.

Continuously dividing cells. Results from a typical experiment carried out in our laboratory in which nuclear, polysomal, and post-polysomal RNAs from G1 and S phase Hela cells were assayed for histone mRNA sequences are shown in Fig. 3. G1 cells were obtained by mitotic selective detachment, a synchronization procedure we routinely use that yields a G1 population containing less than 0.1% S phase cells. Using RNA excess hybridization to a homologous ^{3}H-labeled cDNA probe complementary to one of the Hela cell H4 histone mRNAs, we

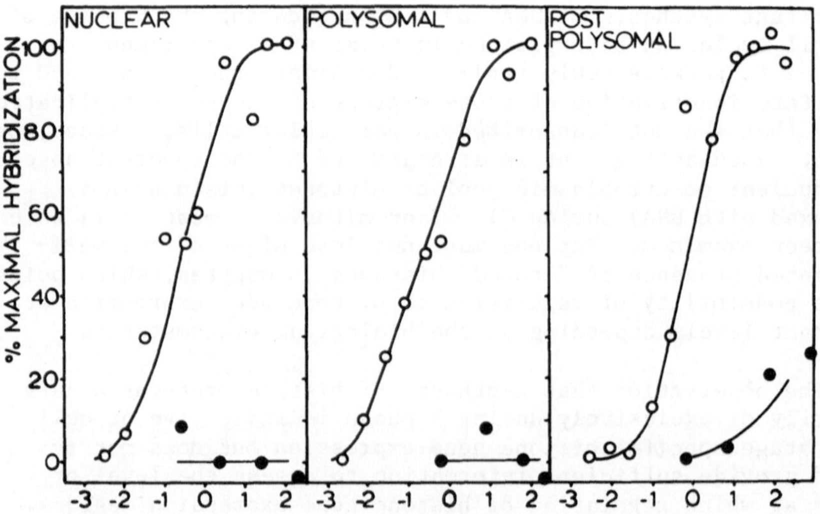

Fig. 3. Hybridization of RNAs from S phase cells (O) and G1 cells synchronized by mitotic selective detachment (●). HeLa cells were fractionated and the RNA from each fraction was isolated. S phase RNA (55 ng or 1.1 μg) was hybridized with 15 pg of H4 histone ^{3}H-cDNA. For G1 RNA fractions 0.86 μg of nuclear, 3 μg of polysomal or 2.2 μg of post-polysomal RNA was hybridized with 15 pg of cDNA. Average maximal hybridization was 91%. Cr_0t = moles of ribonucleotides x time. Conditions for hybridization have been reported (93).

observed histone mRNA sequences in the nucleus and cytoplasm of S phase but not of G1 cells. Similar results were obtained using a homologous ^{3}H-labeled probe complementary to Hela cell H2A, H2B, H3 and H4 histone mRNAs (93-96). These results are consistent with control of H4, H2A, H2B, and H3 gene expression during the cell cycle of continuously dividing Hela cells being mediated, at least in part, at the transcriptional level. The data are also in agreement with earlier studies in which nascent histone polypeptides were detected on the light polyribosomes of S phase

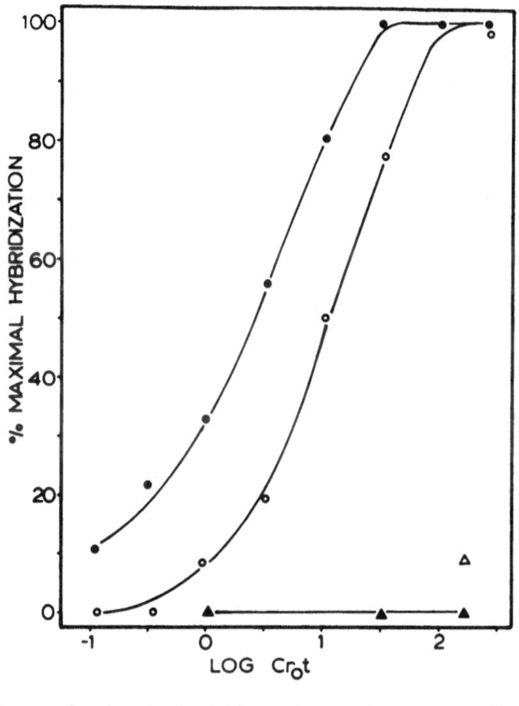

Fig. 4. Kinetics of the hybridization of transcripts of S phase and G1 nuclei to histone ^{3}H-cDNA. S phase nuclei at a concentration of 3.8×10^{7} nuclei/ml and G1 nuclei at 4.6×10^{7} nuclei/ml in individual reactions were incubated as described (95). RNA was isolated and 3.5 μg of S phase RNA or 1.2 μg of G1 RNA were hybridized for various periods of time. S phase nuclei transcribed in the presence of nucleotides (●). S phase nuclei transcribed in the absence of nucleotides (O). The percent maximal hybridization for G1 nuclei transcribed in the presence (▲) or absence (Δ) of nucleotides was calculated by dividing the percent of the hybridization of the G1 RNA by the percent of maximal hybridization for RNAs of S phase nuclei.

but not G1 cells (97) and similar results were observed with respect to in vitro translation of histone polypeptides by RNAs isolated from these polyribosomes (87,89-91, 97,98). While our experimental results do not preclude the existence in G1 cells of a rapidly turning over histone transcript that would not be detected by RNA excess hybridization, data from in vitro nuclear and chromatin transcription studies (Figs. 4 and 5), despite their limitations, support the possibility that histone mRNA transcription is α-amanitin sensitive and S phase specific (93). With the recent development of in vitro transcription systems that exhibit a requirement for a single class of homologous RNA polymerase, execution of in vitro transcription experiments and interpretation of data thereof should be very instructive for further analysis of histone gene regulation. More direct evidence for or against transcriptional control can be obtained by DNA excess hybridization experiments which are currently in progress.

Transcriptional level control of histone gene expression during the cell cycle appears to be at odds with reports of Melli and co-workers (99,100) that high molecular weight precursors of histone mRNAs are present in G1 as well as S phase HeLa cells. Specifically, by hybridizing pulse labelled RNA to cloned sea urchin histone DNA, Melli et al. detected newly synthesized

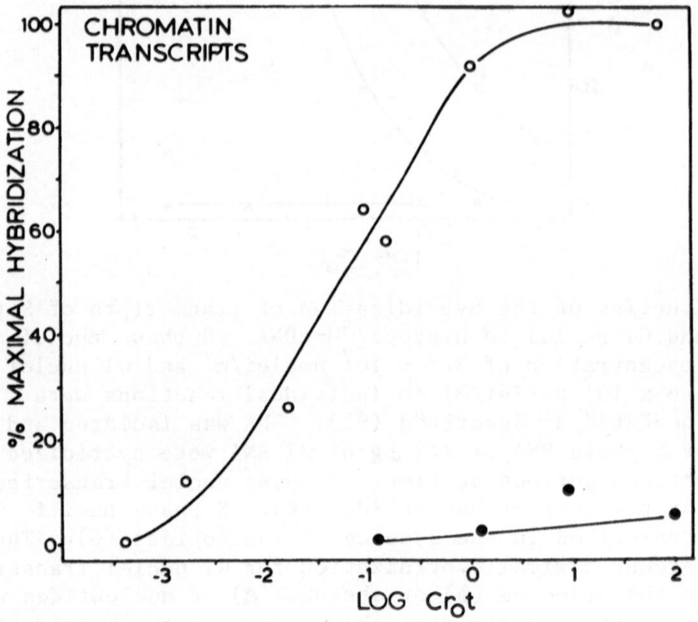

Fig. 5. Hybridization of histone cDNA to in vitro transcripts from G1 (●) and S phase (O) chromatin.

histone mRNA sequences in HeLa cells twelve hours after release from a double thymidine block ("G1"). These investigators similarly reported hybridization of pulse labeled HeLa cell RNA to sea urchin histone DNA two to three hours after release from thymidine block (S). However, when cells are synchronized by double thymidine block, the "G1" population obtained contains 20-30% S phase cells. We, therefore, interpret detection of histone mRNA sequences in "G1" cells produced by double thymidine block to reflect hybridization with RNA derived from the S phase cells in the G1 population--an artifact of the synchronization procedure. Such an interpretation is supported by a comparison of our results in Figs. 3 and 6 where we demonstrate by homologous cDNA hybridization analysis that histone mRNA sequences can be detected in "G1" cells obtained by double thymidine block (30% S phase cells) but not in those obtained by mitotic selective detachment (less than 0.1% S phase cells) (96,101). But it seems inescapable that to resolve in an unambiguous manner when during the cell cycle histone sequences are transcribed, it is an absolute necessity to carry out DNA excess hybridization analysis of RNAs pulse labeled in vivo, using unlabeled homologous histone DNA as a probe.

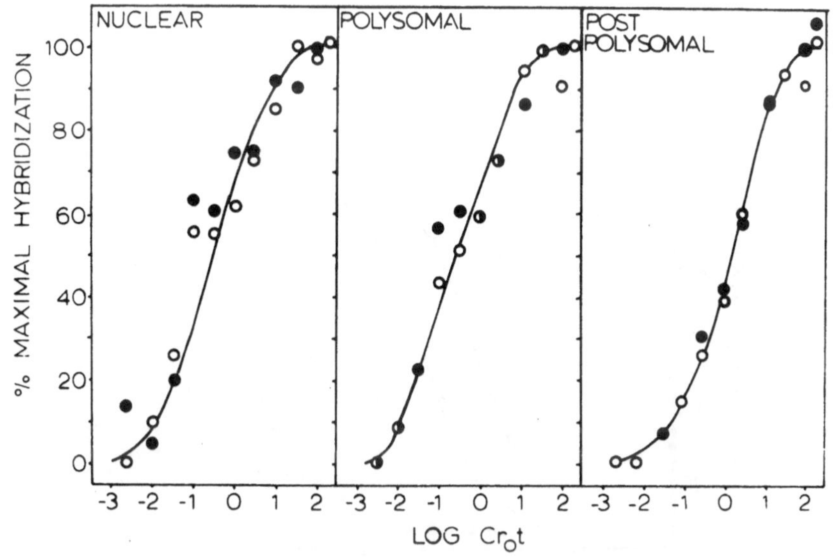

Fig. 6. Hybridization of RNAs from S phase cells (O) and from G1 cells isolated 12 hours after release from the second of two 2 mM thymidine blocks (●). HeLa cells were fractionated and the RNAs from each fraction were isolated. RNA (44 ng or 1.1 μg) was hybridized with 15 pg of cDNA. Nuclear (A). Polysomal (B). Post-polysomal (C).

Stimulation of nondividing cells to proliferate. In other studies carred out in our laboratory, the representation of histone mRNA sequences in human diploid fibroblasts prior to and after stimulation to proliferate was examined (102). Hybridization analysis of in vivo as well as in vitro transcribed RNA, using a ^{3}H-labeled human histone cDNA probe, indicates that stimulation of DNA synthesis is accompanied by a simultaneous elevation in the representation of histone mRNA sequences (Fig. 7). It should be noted that Marzluff and co-workers (103) recently assayed pulse-labelled RNAs isolated from G1 and S phase mouse myeloma cells, stimulated to proliferate following growth arrest by the isoleucine deprivation method, for the presence of histone mRNA sequences by hybridization to cloned sea urchin histone DNA. Consistent with transcriptional level control of histone gene expression during the cell cycle of continuously dividing cells, they found that synthesis of H3 and H4 histone mRNAs was prominent only during S phase. Parker and Fitschen (104) have also observed a difference (greater that 500 fold) in the representation of histone mRNA sequences in G1 and S phase mouse 3T6 cells. In the latter experiments, histone mRNA sequences were identified by hybridization of nuclear and cytoplasmic RNAs to homologous histone cDNA.

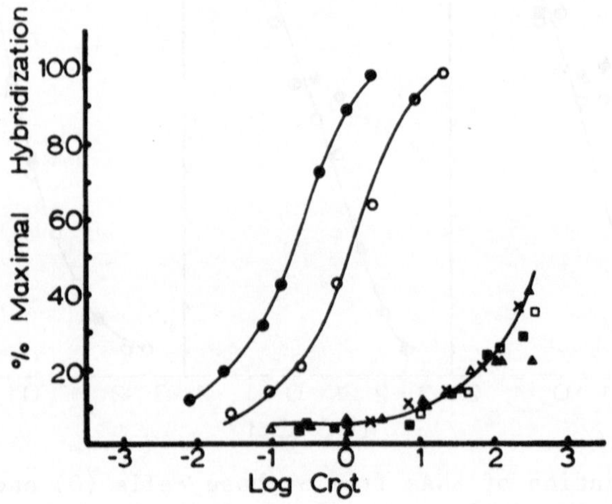

Fig. 7. Hybridization of RNAs from human diploid fibroblasts prior to and following stimulation to proliferate to human histone cDNA. Non-proliferating WI-38 cells (x) and WI-38 cells at 1 hour (□), 4 hours (■), 7 hours (▲), 10 hours (○) and 12 hours (●) after serum stimulation.

Histone gene expression during the S to G2 transition. Our considerations regarding the regulation of histone gene expression during the cell cycle have been confined primarily to the G1 and S phase transition. Therefore, if transcriptional control of histone gene expression is operative, we have been focusing on the "turn on" phase of the regulatory process while ignoring the regulatory mechanism(s) operative when DNA replication is completed and the cell enters G2. While there are a number of models which can be invoked to address the S to G2 transition, the two most obvious are: 1) transcription level control in which histone mRNA transcription terminates when DNA replication is completed, and 2) post-transcriptional control in which the synthesis of histone mRNA sequences continues but a block is imposed at some step of mRNA processing, protein synthesis, protein modification or association of histones with the genome. Unfortunately, to date, information regarding the regulation of histone gene control during G2 is not well documented, primarily because of the difficulty in obtaining pure populations of G2 cells. However, a model of histone gene control in which expression would be initiated at the transcriptional level when the histone proteins are required for packaging the newly replicated DNA (i.e., at the onset of S phase) and expression would be blocked at some post-transciptional step at the end of S phase, is an intriguing possibility. Viewed within the context of gene expression during the cell cycle in general, one can envision the following sequence of events for some genes, beginning at the onset of G1: 1) transcriptional activation of genetic sequences when the gene products are required; 2) a post-transcriptional block in expression when the gene product (enzyme or structural protein) is no longer required or when an accumulation of the gene product would be detrimental to cell function, 3) terminaton of transcription when RNA synthesis in general is blocked during metaphase; and then 4) initiation of transcription when the protein is required during the next cell cycle.

An overview of histone gene expression during the cell cycle of continuously dividing cells and nondividing cells stimulated to proliferate. Refinements in our ability to analyze in vivo and in vitro synthesized histone proteins are required. Also, there is a need for refinements in procedures employed for analysis of histone gene transcripts--high resolution probes for detection of in vivo synthesized histone mRNAs and precursors thereof, as well as in vitro transcription systems that retain in vivo fidelity. Yet, from the evidence presently available, it is reasonable to postulate that regulation of histone gene expression during the cell cycle of continuously dividing cells and of nondividing cells stimulated to proliferate is regulated at least in part at the transcriptional level. Whether or not such a proposal can withstand the rigors of an unequivocal proof that histone proteins are synthesized exclusively during S phase and that histone mRNA sequences are transcribed only when DNA is replicated remains an open-ended question. None-the-less, transcriptional control of histone gene expression during the cell cycle is a valid working model since it is one that can be and is being tested experimentally.

Histone gene expression during early development. As cells progress through the cell cycle, a complex and highly interdependent series of biochemical events occurs which by itself suggests the necessity for differential gene expression. Such cell cycle stage specific biochemical events have been extensively discussed in several reviews and monographs (1,8-12,16,105-116). Differential gene expression during the cell cycle is also reflected by quantitative differences in RNA and protein synthesis as well by the cell cycle stage specific appearance of defined mRNAs and proteins. But perhaps most important is the suggestion of a functional relationship between the changes in gene expression and progression through the cell cycle, as indicated by what appears to be a universal requirement for protein synthesis for cells to commence DNA replication and to divide mitotically.

In a multitude of cells and tissues (continuously dividing cells and non-dividing cells stimulated to proliferate) that have been examined, RNA synthesis is required for DNA replication and mitosis to occur, suggesting that transcriptional level control is at least in part operative. However, the inital rounds of cell division that follow fertilization can take place in the presence of inhibitors of RNA synthesis. Because of the necessity for protein synthesis for post-fertilization proliferation to proceed, it appears that expression of genes prerequisite for the rapid series of "cleavage divisions" occurring during early development can be post-transcriptionally mediated. Several lines of evidence indicate that in the absence of transcription those proteins needed for cleavage divisions are templated by stored maternal mRNAs or precursors thereof (117,118).

Specifically, with regard to regulation of histone gene expression during early development, a departure from the pattern that appears to be operative during the cell cycle of more differentiated cells is observed. "Stored" histone proteins, synthesized in oocytes, can become associated with the genome and histone proteins can be translated from oocyte-derived histone mRNAs. Both of the latter processes occur in the absence of RNA transcription. Differences observed in the expression of histone genes during cleavage stages of embryos and during the cell cycle in more differentiated cells are reflected by variations in the properties of the histone mRNAs. For example, oocyte histone mRNAs are stable and in some situations have poly (A) tails; this is quite different from the general situation of rapidly turning-over histone mRNAs which lack poly(A) at their 3'-termini. In summary, the conclusion is inescapable that during early development regulation of histone gene expression involves post-transcriptional control mechanisms including some which operate at the cytoplasmic level. Taken together with the evidence discussed previously in this article for nuclear level control of histone gene expression, it is reasonable to postulate that histone genes can be regulated at different levels depending upon the biological circumstances.

RELATIONSHIP OF HISTONE GENE EXPRESSION TO EXPRESSION OF OTHER S PHASE-SPECIFIC GENES

Although there is convincing evidence to support the concomitant synthesis of DNA and histones, to date little is known regarding the molecular nature of this relationship of DNA replication to regulation of histone gene expression. A problem which is even more complex, yet potentially very important with regard to understanding cell cycle stage specific gene expression, is the relationship of DNA replication to "other" genetic sequences that may be exclusively or preferentially expressed during the S phase of the cell cycle. Resolution of such issues permits one to determine whether it is reasonable to think of "regulatory molecules" that act exclusively on histone genes or generally on genes that are expressed during S phase; i.e., is it valid to think in terms of a model which incorporates "coordinate type control"?

Histone Gene Expression and DNA Replication

Are all histone genes expressed concomitantly? Because the genes which code for histone proteins are reiterated, a very basic question which arises is whether all copies of the histone genes are "operative" when DNA replication is occurring or whether there are variations in those histone genes which are expressed depending upon the specific biological circumstances. In sea urchins, there are several lines of evidence suggesting differences in those histone genes which are expressed during various stages of early development. High resolution electrophoretic analysis of histones in acrylamide gels containing non-ionic detergents, coupled with tryptic peptide analysis, indicates that there are subtle variations in those histones synthesized at different points during early development. This has been clearly shown for H2A, H2B, and H1 histones. The corresponding mRNAs which code for these "histone variants" have also been shown to be developmental stage specific (119-121). Coupled with sequence analysis data demonstrating that all copies of the histone genes of a specific sea urchin species are not identical (which suggests that developmental stage specific variations in histone mRNAs and proteins do not arise during processing of gene transcripts), it seems inescapable that there is developmental stage specific expresssion of subsets of the histone gene complement of an organism. However, it is important to bear in mind that amongst the basic questions to be answered are: whether mechanisms operative in control of the expression of the variant histone genes are similar or different, and whether or not all of the genes coding for a particular histone variant are expressed concomitantly.

Variant forms of histone proteins are not restricted to lower eukaryotes. Variant forms of various histone species have been observed in rodents. But perhaps more important tissue-specific differences in the representation of the variant histone species have been noted. The biological implications of variant forms of histones and mechanisms operative in regulating their expression remain to be resolved. But, the utilization of defined subsets of the reiterated histone genes by specific tissues and the ability to modify those histone genes which are expressed depending upon cellular requirements appears to be a reality. Within this context it should be noted that in human tissue culture cells variant forms of H4 histone mRNAs have also been observed (101). Although these H4 mRNAs code for identical H4 histone polypeptides (101) they appear to represent the products of different genes.

What is the effect of inhibiting DNA replication on expression of histone genes? As previously indicated, it is well documented that in many cells and tissues inhibition of DNA replication by drugs such as cytosine arabinoside, hydroxyurea, or aminopterin is followed by a rapid and complete shutdown of histone synthesis (82-84,87-91,97). Therefore, one way to assess further the relatioship of DNA replication and histone gene expression is to compare the representation and functional properties of histone mRNA sequences in various intracellular compartments prior to and following DNA synthesis inhibition. Consistent with the inhibition of in vivo histone protein synthesis, results from in vitro translation studies carried out in several laboratories suggest that after cytosine arabinoside or hyroxyurea treatment significant amounts of translatable histone mRNAs are not present in S phase cells (87,89-91,97). Nucleic acid hybridization analysis supports the absence of histone mRNA sequences associated with polysomes of S phase cells treated with DNA synthesis inhibitors(122). However, during the initial period after inhibition of DNA synthesis, histone mRNA sequences can be detected by nucleic acid hybridization procedures in the nucleus as well as in the post-polysomal cytoplasmic cellular fraction (122). Inihibition of DNA replication does not appear to block in vitro transcription of histone mRNA sequences from isolated nuclei or chromatin although such histone mRNA sequences are presumably non-translatable(122).

One interpretation of these results is that histone gene expression and DNA replication are functionally related but that inhibition of histone gene expression, as brought about by drugs such as cytosine arabinoside or hydroxyurea, is not mediated at a nuclear or transcriptional level. Whether or not the cessation of histone biosynthesis which follows such drug treatment is analogous to that which occurs during the S phase to G2 transition, as cells progress through the cell cycle in an uninterrupted manner, is an open-ended question. An alternative interpretation is that histone mRNA transcription is programmed to coincide with DNA replication but that the two events occur in a parallel manner, i.e., are not functionally coupled. Rather, the synthesis of histone proteins is the component of histone gene expression which is functionally coupled with DNA replication. These models are indeed testable experimentally using cell cycle mutants, with those having S phase restriction points being especially instructive in this regard.

Nonhistone Gene Expression and DNA Replication

There are a number of approaches that one can pursue to identify genes that are preferentially or exclusively expressd during the S phase of the cell cycle, that is, genes which can be potentially useful "markers" for studying the relationship of histone gene expression to that of other genes expressed when DNA is replicating. Enzymes involved with DNA synthesis and deoxynucleotide metabolism are reasonable candidates. One such enzyme that falls into this category is dihydrofolate reductase which is responsible for the NADPH-dependent reduction of folic acid and dihydrofolic acid to tetrahydrofolic acid, and which is also required for many facets of single carbon metabolism including biosynthesis of purines and thymidylic acid. Johnson and coworkers (78,79) have elegantly shown that in mouse 3T3 and 3T6 cells stimulated to proliferate there are dramatic increases in dihydrofolate reductase and in dihydrofolate reductase mRNAs at the onset of S phase. Similar observations have been made for thymidylate synthetase and thymidine kinase. It is particularly interesting to note that the increase in dihydrofolate reductase mRNA does not appear to be sensitive to inhibition of DNA synthesis by hydroxyurea, adding additional complexity to elucidating the regulation of S phase specific gene expression. Since recombinant DNA plasmids containing histone sequences have been constructed and the mRNA sequences of the above-mentioned enzymes have also been cloned in bacteral plasmids (77), examining their mode of regulation with emphasis on whether control is independent or coordinate should be a realistic challenge.

CLONING OF HUMAN HISTONE SEQUENCES

From the preceding discussion, it should be evident that cloned human histone sequences can significantly facilitate the study of the control of histone gene expression. Such cloned histone sequences can be used as hybridization probes for detection of histone mRNAs and/or DNAs as well as for isolation and characterization of chromosomal proteins and/or other macromolecules that interact with histone sequences. We will now briefly outline the approaches we have taken to: a) prepare recombinant plasmids containing cloned human histone mRNA sequences, and b) isolate and characterize genomic human sequences.

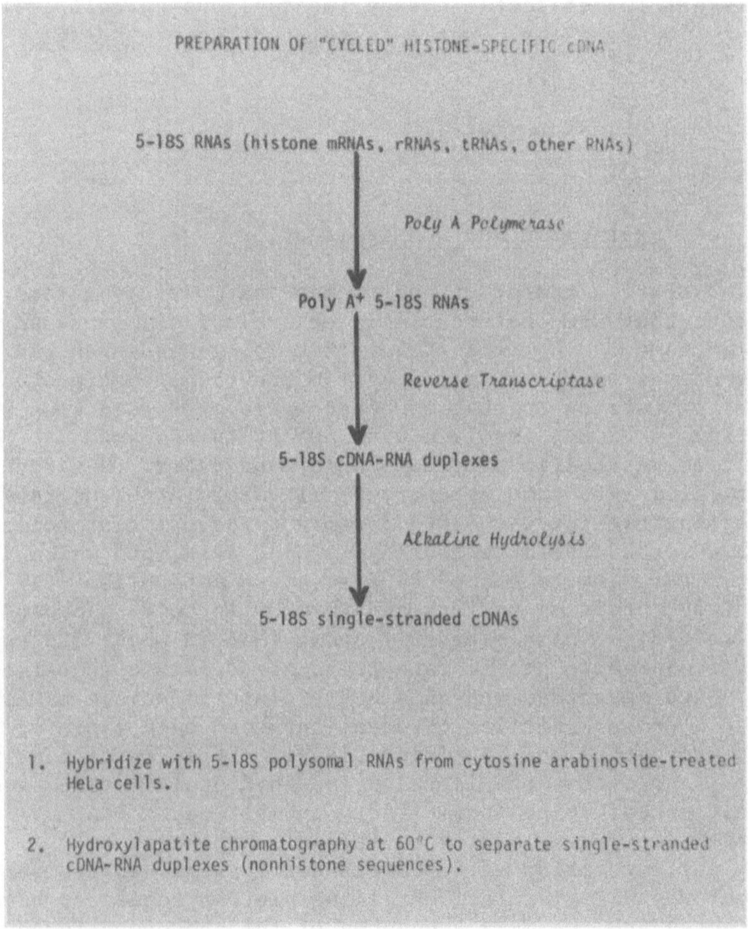

Fig. 8. A. Flow diagram for preparation of histone-specific cDNA.

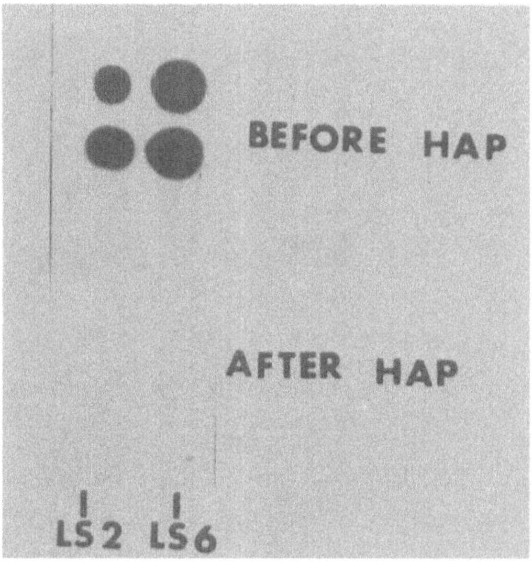

Fig. 8. B. Hybridization of ^{33}P-labeled histone cDNA to cloned human ribosomal DNA immobilized on nitrocellulose. Radiolabeled cDNA was assayed prior to and following purification by hybridization to polysomal RNA from cytosine arabinoside-treated cells and hydroxylapatite chromatography. The inability of the hisonte-specific DNA probe to hybridize with ribosomal DNA following hydroxylapatite chromatography reflects the effectiveness of the purification procedure.

Cloned Human Histone mRNA Sequences

Plasmids containing double-stranded cDNA to 7-11S RNA from the polysomes of S phase HeLa S_3 cells, which contains sequences coding for H2A, H2B, H3, H4 and H1 histone polypeptides, were prepared by ligating cDNA to Hind III linkers and inserting the cDNA into the Hind III site of the pBR322 plasmid. E. coli HB101 cells were transformed with these recombinant pBR322 plasmids. Ampicillin-resistant, tetracycline-sensitive colonies were selected and screened with ^{32}P-labeled histone-specific cDNA. Histone-specific cDNA was prepared as described in Figure 8. The critical step in preparation of the probe is hybridization with 5-18S RNAs from cytosine arabinoside-treated S phase HeLa cells. Because cytosine arabinoside treatment results in a selective loss of histone mRNAs from polysomes with no apparent effect on the representation of other mRNA sequences, hybridization with the RNA from cytosine arabinoside-treated cells eliminates nonhistone mRNAs, ribosomal RNAs, tRNAs, etc. In Figure 9, a typical primary and secondary screening of colonies by the in situ hybridization procedure of Grunstein and Hogness (123), using ^{32}P-labeled histone-specific cDNA, is shown. The histone cDNA inserts in

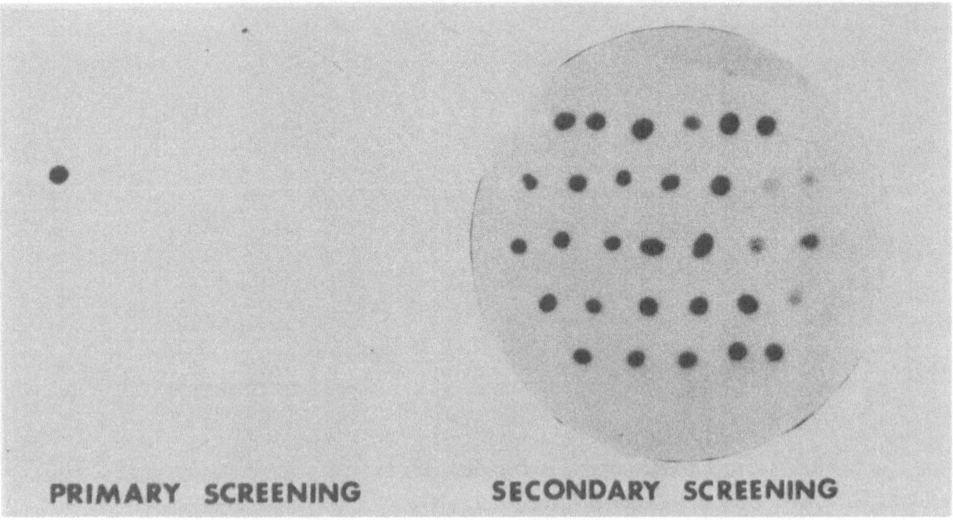

Fig. 9. Screening of cDNA plasmids with histone-specific cDNA. Plasmids containing double-stranded cDNA to total 7-11S polysomal RNA from S phase HeLa cells were prepared by ligating cDNA to Hind III linkers and inserting the cDNA into the Hind III site of pBR322 plasmids. E. coli HB101 cells were transformed with recombinant pBR322 plasmids. Ampicillin-resistant, tetracycline-sensitive colonies were selected. Histone-specific cDNA (^{32}P-labeled) was used to screen the colonies.

"positive" plasmids were visualized by restriction endonuclease digestion, transfered to nitrocellulose paper and hybridized with ^{32}P-labeled histone-specific DNA as shown Figure 10. The histone sequences contained in various plasmids can then be identified by hybridizing a preparation of each plasmid with 5-18S polysomal RNA from S phase HeLa cells and in vitro translation of the hybridized mRNA sequences.

Cloned Human Genomic Histone Sequences

The cDNA clones discussed above reflect at best only mRNA-coding segments of the human histone genes. Because of the numerous questions to be addressed regarding the structure and organization of human histone genes and the regulation of histone gene expression, we have used the cloned cDNA sequences for identification, isolation, and characterization of human genomic histone sequences. A genomic library in λ Charon 4A phages containing random 15-20 Kb fragments of human DNA (kindly provided to us by Dr. T. Maniatis) was screened using ^{32}P-labeled DNA from a cDNA clone containing HeLa cell H4 histone mRNA sequences.

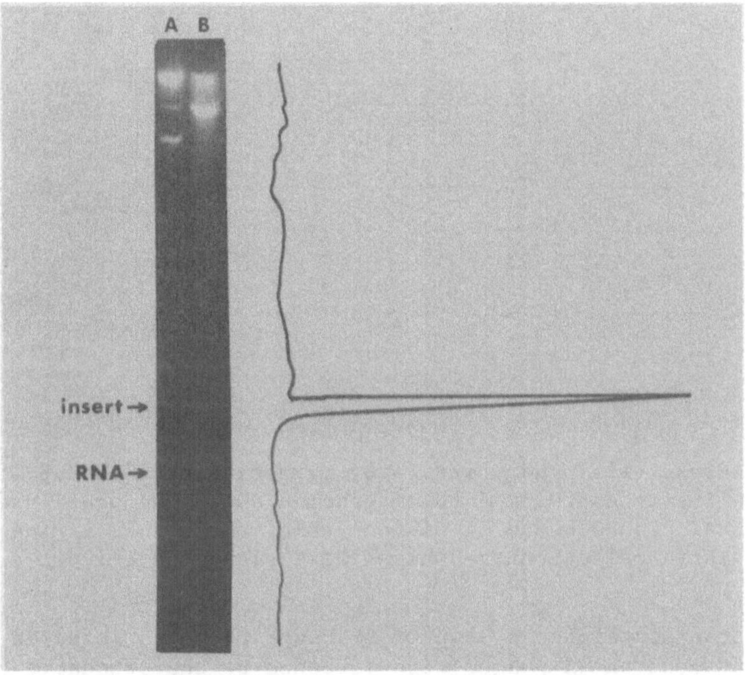

Fig. 10. Identification of a histone cDNA plasmid. Plasmid DNAs were electrophoresed on 3% agarose gels before (A) and after Hind III digestion (B) and stained with ethidium bromide. The tracing on the right is an optical density scan of an autoradiogram of well B after hybridization to ^{32}P-labeled cDNA. Note that hybridization was observed only with the histone cDNA insert. A partial restriction map of the plasmid is shown. The numbers on the restriction map refer to base pairs.

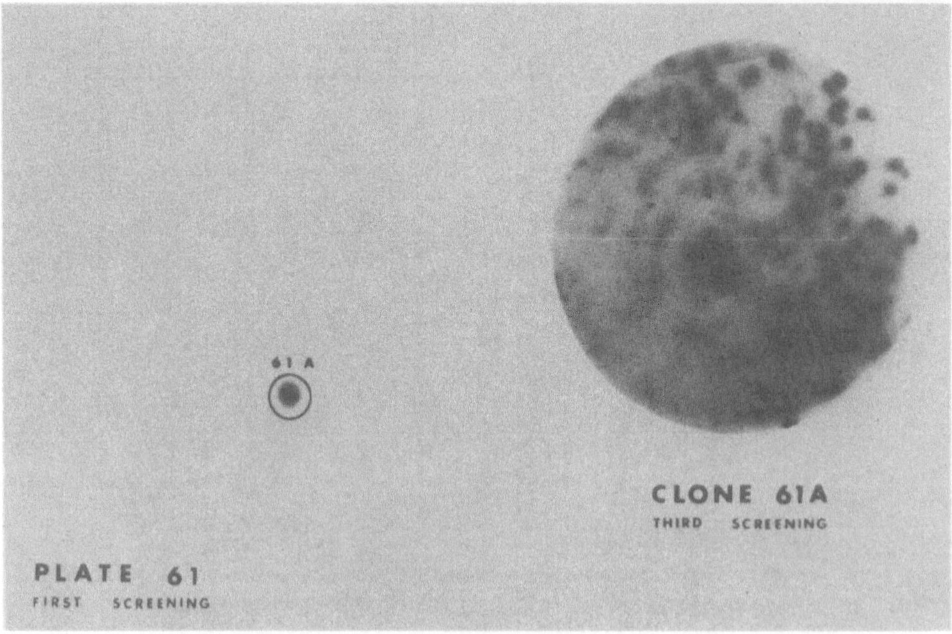

Fig. 11. First (11A) and third (11B) screening of a cloned library containing human genomic DNA. The probe used was cloned human H4 histone cDNA. Note that by the third selection the phages were plaque purified.

Eighty plates containing a lawn of DP50supF bacteria infected with recombinant λ Charon 4A phages were screened by the procedure of Benton and Davis (124). Figure 11a shows a "positive plaque" indicating hybridization of recombinant phage to cloned human H4 histone cDNA and Figure 11b shows that by the third screening the recombinant phage was plaque-purified. Eleven such clones were isolated and grown in liquid culture. Phage DNAs were extracted and restricted with EcoRI. The restriction fragments were separated on a 0.8% agarose gel, transferred to nitrocellulose and hybridized to histone-specific cDNA. As expected none of the vector (λ) bands and some of the insert-derived (human genomic) bands do not hybridize with histone-specific cDNA (Figure 12).

To characterize further the genomic clones, DNA was restricted with EcoRI, transferred to nitrocellulose paper and hybridized first to ^{32}P-labeled RNA from S phase HeLa cells and subsequently to ^{32}P-labeled RNA from S phase HeLa cells treated with cytosine arabinoside (Figure 13). Because histone mRNA sequences are selectively lost from polysomes of cytosine arabinoside-treated cells, hybridization with "control RNA" but

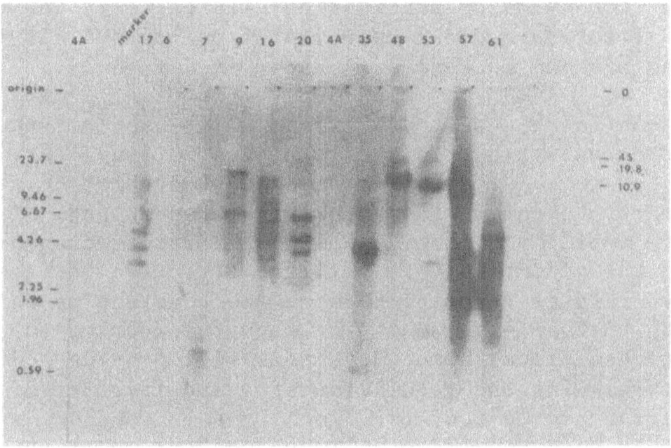

Fig. 12. Hybridization of ^{32}P-labeled histone-specific cDNA to 11 EcoRI-digested λ Charon 4A clones containing human histone sequences. Eleven different clones were isolated and grown in liquid culture. Their DNAs were extracted and restricted with EcoRI. The fragments were electrophoretically fractionated on an 0.8% agarose gel, transferred to nitrocellulose and hybridized to ^{32}P-labeled histone-specific cDNA. The molecular sizes of the λ Hind III markers (right side) or vector fragments obtained from EcoRI digestion (left side) are indicated.

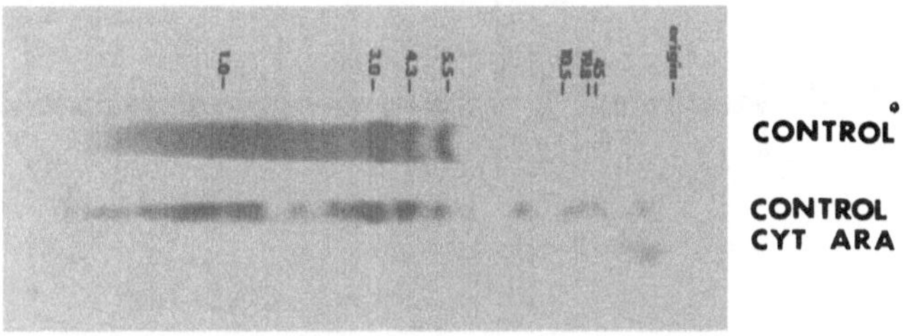

Fig. 13. Hybridization of EcoRI-digested clone 17 DNA with ^{32}P-labeled polysomal RNA from S phase control and cytosine arabinoside-treated HeLa cells. DNA from clone 17 was restricted with EcoRI, transferred to nitrocellulose, and hybridized with ^{32}P-labeled S phase polysomal RNA (control). Following autoradiography the hybridized RNA was eluted by heating the filters in boiling water and the filters were cut vertically. Half of the filter was rehybridized with ^{32}P-labeled S phase polysomal RNA (control) and half was hybridized with ^{32}P-labeled polysomal RNA from cytosine arabinoside-treated cells.

not with RNA from cytosine arabinoside-treated cells was observed. A partial restriction endonuclease map of one of the clone α genomic human histone sequences is shown in Figure 14.

We are confident that these cloned human histone sequences will provide effective hybridization probes to facilitate further analysis of the level(s) at which regulation of histone gene expression occurs under various biological conditions. It is also reasonable to anticipate that the cloned histone sequences, particularly the genomic clones, will be effective for the isolation by affinity chromatography of macromolecules which interact with histone sequences. We hope that during the next several years the cloned human sequences will provide additional information regarding the regulation of a specific set of genetic sequences during the cell cycle. In a broader sense we are optimistic that such knowledge will enhance our understanding of control of the proliferative process.

ACKNOWLEDGEMENT

These studies were supported by grants from the National Science Foundation (PCM 77-15947) and the National Birth Defects Foundation (5-217).

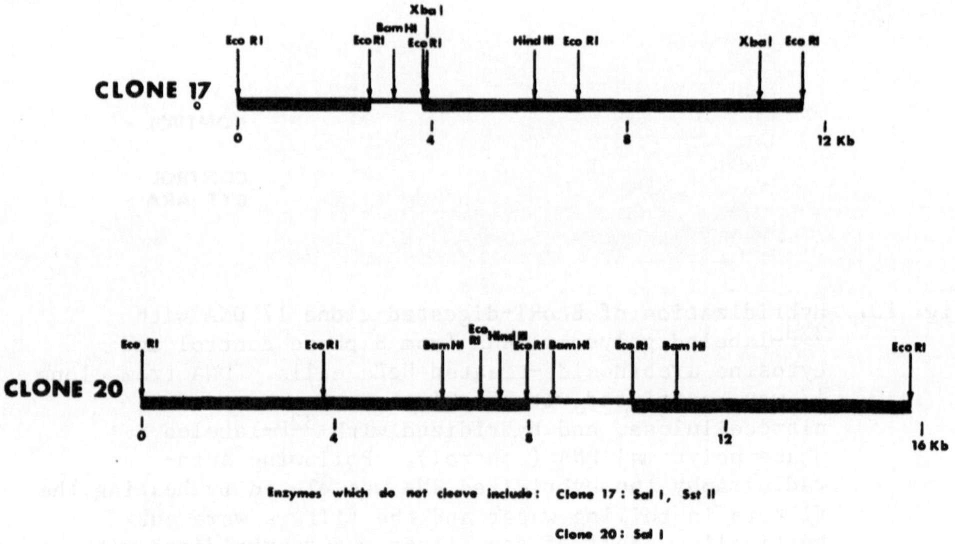

Fig. 14. Partial restriction endonuclease map of two cloned genomic human histone sequences.

REFERENCES

1. G.S. Stein, S. Hochhauser, and J.L. Stein, Histone Genes: Their Structure and Control, in: "The Cell Nucleus," H. Busch, ed., vol. VII, pp. 259-307. Academic Press, New York, NY (1979).
2. R. Baserga, Life Sciences 15:1057 (1974).
3. J. Bonner, M. Dahmus, D. Fambrough, R.-C. Huang,, K. Marushige and D. Tuan, Science 159:47 (1968).
4. P. Byvoet and C. S. Baxter, in: "Chromosomal Proteins and Their Role in the Regulation of Gene Expression," G.S. Stein and L.J. Kleinsmith, eds., p. 127. Academic Press, New York.
5. S.C.R. Elgin and H. Weintraub, Annu. Rev. Biochem. 44:725 (1975).
6. T.A. Langan, and P. Hohmann, in: "Chromsomal Proteins and Their Role in the Regulation of Gene Expression," G.S. Stein and L.J. Kleinsmith, eds., p. 113. Academic Press, New York.
7. V.G. Allfrey, Cancer Res. 26:2026 (1966).
8. G.S. Stein and R. Baserga, Adv. Cancer Res. 15:287 (1972).
9. G.S. Stein, T.C. Spelsberg, and L.J. Kleinsmith, Science 183:817 (1974).
10. S.J. Hochhauser, G.S. Stein and J.L. Stein, Intl. Rev. Cytol. in press (1981).
11. G.S. Stein, J.L. Stein and J.A. Thomsom, Cancer Research 38:1101 (1978).
12. I.R. Phillips, E.A. Shephard, J.L. Stein and G.S. Stein, Role of Nonhistone Chromosomal Proteins in Selective Gene Expression, in: "Eukaryotic Gene Expression," G. Kolodny, ed., CRC Press, West Palm Beach, Florida, in press.
13. L.H. Kedes, Ann. Rev. Biochem. 48:837 (1979).
14. H. Busch, "Histones and Other Nuclear Proteins." Academic Press, New York (1965).
15. L.S. Hnilica, "The Structure and Biological Function of Histones." CRC Press, Cleveland, Ohio (1972).
16. I. Cameron and J.R. Jeter eds., "Acidic Proteins of the Nucleus," Academic Press, New York, NY (1974).
17. H. Busch, ed, "The Cell Nucleus" vols. I-VII, Academic Press, New York, NY (1975-1979).
18. G.S. Stein and L.J. Kleinsmith, eds, "Chromosomal Proteins and Their Role in the Regulation of Gene Expression," Academic Press, New York, NY (1975).
19. C. Nicolini, ed., "Chromatin Structure and Function", Plenum Publishing Co., New York, New York (1979).
20. I.R. Phillips, E.A. Shephard, J.L. Stein, and G.S. Stein, in: "Eukaryotic Gene Expression," G. Kolodny, ed., CRC Press, West Palm Beach, Florida.

21. G.S. Stein, J.L. Stein, and L.J. Kleinsmith, ed., "Methods in Cell Biology", Vols. 16-19, Academic Press, New York.
22. G.L. Whitson, ed., "Nuclear-Cytoplasmic Interactions in the Cell Cycle", Academic Press, New York (1980).
23. A. Ruiz-Carrillo, J.L. Waugh, and V.G. Allfrey, Science 190:117 (1975).
24. T.A. Langan, Science 162:579 (1968).
25. R. Balhorn, M. Balhorn, and R. Chalkley, Dev. Biol. 29:199 (1972).
26. D. Marks, W.K. Paik, and T.W. Borun, J. Biol. Chem. 248:5660 (1973).
27. V.G. Allfrey, R. Faulkner, and A.E. Mirsky, Proc. Natl. Acad. Sci. U.S.A. 51:786 (1964).
28. P. Byvoet, Biochim. Biophys. Acta 238:375a (1971).
29. W.K. Paik, and S. Kim, Science 174:114 (1971).
30. T.W. Borun, D. Pearson, and W.K. Paik, J. Biol. Chem. 247:4288 (1972).
31. P. Chambon, J.D. Weill, J. Doly, M.T. Strosser, and P. Mandel, Biochem. Biophys. Res. Commun. 25:638 (1966).
32. M. Futai, D. Mizuno, and T. Sugimura, J. Biol. Chem. 243:6325 (1968).
33. L.A. Stocken, J.S. Smith, and M.G. Ord., in: "Poly(ADP-Ribose)-An International Symposium," M. Harris, ed., National Institutes of Health, Bethesda (1974).
34. M. Miwa, M. Tanaka, T. Matsushima, and T. Sugimura, J. Biol. Chem. 249:3475 (1974).
35. J.H. Roberts, P. Stark, C.P. Giri, and M. Smulson, Arch Biochem. Biophys. 171:305 (1975).
36. D.W. Mullins, C.P. Giri, and M. Smulson, Biochemistry 16:506 (1977).
37. C. Chen, D.L. Smith, B. Bruegger, R.M. Halpern, and R.A. Smith, Biochemistry 13:3785 (1974).
38. D.L. Smith, C.C. Chen, B.B. Bruegger, S.L. Holtz, R.M. Halpern, and R.A. Smith, Biochemistry 13:3780 (1974).
39. G. Vidali, L.C. Boffa, and V.G. Allfrey, J. Biol. Chem. 247:7365 (1972).
40. D.E. Krieger, R. Levine, R.B. Merrifield, G. Vidali, and V.G. Allfrey, J. Biol. Chem. 249:332 (1974).
41. G. Vidali, L.C. Boffa, and V.G. Allfrey, Cell 12:409 (1977).
42. T. Tidwell, V.G. Allfrey, and A.E. Mirsky, J. Biol. Chem. 243:707 (1968).
43. R. Hancock, J. Mol. Biol. 40:457 (1969).
44. L.R. Gurley and J.M. Hardin, Arch. Biochem. Biophys. 130:1 (1969).
45. W.J. Garrard and J. Bonner, J. Biol. Chem. 249:5570 (1974).
46. J.A. Duerre and C.T. Lee, J. Neurochem. 23:541 (1974).
47. Y. Ohba, K. Hayashi, Y. Nakagawa, and Z. Yamaguchi, Eur. J. Bichem. 56:243 (1975).
48. T.W. Borun and G.S. Stein, J. Cell Biol. 52:308 (1972).
49. G. Felsenfeld, Nature (London) 271:115 (1978).

50. R. Kornberg, Annu. Rev. Biochem. 46:931 (1977).
51. J.L. Roti Roti, G.S. Stein, and P. Cerutti, Biochemistry 13:2900 (1974).
52. I. Isenberg, Ann. Rev. Biochem. 48:159 (1979).
53. M. Noll, Cell 8:357 (1976).
54. J.-C. Lin, C. Nicolini, and R. Baserga, Biochemistry 13:4127 (1974).
55. M.C. Wilson, M. Melli, and M.L. Birnstiel, Biochem. Biophys. Res. Commun. 61:404 (1974).
56. S. Detke, A. Lichtler, I.R. Phillips, J.L. Stein, and G.S. Stein, Proc. Natl. Acad. Sci. 76:4995 (1979).
57. E. Jacob, G. Malacinski, and M.L. Birnstiel, Eur. J. Biochem. 69:45 (1976).
58. A. Scott and J. Wells, Nature 259:635 (1976).
59. L.H. Kedes and M.L. Birnstiel. Nature (London) New Biol. 230:165 (1971).
60. E.S. Weinberg, M.L. Birnstiel, I.R. Purdom, and R. Williamson Nature (London) 240:225 (1972).
61. E.S. Weinberg, G.C. Overton, R.H. Shutt, and R.H. Reeder, Proc. Natl. Acad. Sci. U.S.A. 72:4815 (1975).
62. M.N. Farquhar and B.J. McCarthy, Biochemistry 12:4113 (1973).
63. M. Grunstein, and P. Schedl, J. Mol. Biol. 104:323 (1976).
64. M. Birnstiel, J. Telford, E. Weinberg, and D. Stafford, Proc. Natl. Acad. Sci. U.S.A. 71:2900 (1974).
65. W.J. Garrard and J. Bonner, J. Biol. Chem. 249:5570 (1974).
66. W. Schaffner, K. Gross, J. Telford, and M. Birnstiel, Cell 8:471 (1976).
67. L.H. Kedes, Cell 8:321 (1976).
68. L.H. Kedes, in: "Recombinant Molecules: Impact on Science and Society", R.F. Beers, Jr. and E.G. Basset, eds., p. 399, Tenth Miles International Symposium, Raven, New York.
69. K. Gross, W. Schaffner, J. Telford, and M. Birnstiel, Cell 8:479 (1976).
70. M.L. Pardue, Genetics 79:Supplement 159 (1975).
71. M.L. Pardue, E. Weinberg, L.H. Kedes, and M.L. Birnstiel, J. Cell Biol. 55:199a (1972).
72. R.W. Old, H.G. Callan, and K.W. Gross, J. Cell Sci. 27:57 (1977).
73. L.C. Yu, P. Szabo, T.W. Borun, and W. Prensky, Cold Springs Harbor Symp. Quant. Biol. 42:1101 (1978).
74. M. Chandler, L.H. Kedes, and J. Yunis, Am. Soc. Hum. Genet. p. 76A-(Abstract) (1978).
75. N. Fregien, M. Marchionni, and L.H. Kedes, Bio. Bull. 151:407(Abstract) (1976).
76. R.P. Lifton, M.L. Goldberg, R.W. Karp, and D.S. Hogness, Cold Spring harbor Symp. Quant. Biol. 42:1047 (1977).
77. F.W. Alt, R.E. Kellems, and R.T. Schimke, J. Biol. Chem. 251:3063 (1976).

78. L.F. Johnson, C.L. Fuhrman, and L.M. Wiedemann, J. Cell. Physiol. 97:397 (1978).
79. L.M. Weidemann and L.F. Johnson, Proc. Natl. Acad. Sci. 76:2818 (1979).
80. J.L. Stein, G.S. Stein, and P.M. McGuire, Biochemistry 16:2207 (1977).
81. B. Moss, Gershowitz, A., L. Weber, and C. Baglioni, Cell 10:113 (1977).
82. J. Spalding, K. Kajiwara, and G. Mueller, Proc. Natl. Acad. Sci. 56:1535 (1966).
83. E. Robbins, and T.W. Borun, Proc. Natl. Acad. Sci. 57:409 (1967).
84. G.S. Stein and T.W. Borun, J. Cell Biol. 52:292 (1972).
85. V.E. Groppi and P. Coffino, Cell 21:195 (1980).
86. R. Hancock, J. Mol. Biol. 40:457 (1969).
87. D. Gallwitz and G.C. Mueller, J. Biol. Chem. 244:5947 (1969).
88. G.S. Stein and C.L. Thrall, FEBS Lett. 34:35 (1973).
89. T.W. Borun, F. Gabrielli, K. Ajiro, A. Zweidler, and C. Baglioni, Cell 4:59 (1975).
90. M. Breindl and D. Gallwitz, Eur. J. Biochem. 45:91 (1974).
91. D. Gallwitz and M. Breindl, Bichem. Biophys. Res. Comm. 47:1106 (1972).
92. G.S. Stein and C.L. Thrall, FEBS Letters 32:41 (1973).
93. G.S. Stein, W. Park, C. Thrall, R. Mans, and J.L. Stein, Nature 257:764 (1975).
94. J.L. Stein, C.L. Thrall, W.D. Park, R.J. Mans, and G.S. Stein, Science 189:557 (1975).
95. S. Detke, J.L. Stein and G.S. Stein, Nucl. Acids Res. 5:1515 (1978).
96. S. Detke, A. Lichtler, I. Phillips, J. Stein and G. Stein, Proc. Natl. Acad. Sci. 76:4995 (1979).
97. T.W. Borun, M.D. Scharff, and E. Robbins, Proc. Natl. Acad. Sci. 58:1977 (1967).
98. M. Jacobs-Lorena, C. Baglioni, and T.W. Borun, Proc. Natl. Acad. Sci. 69:2095 (1972).
99. M. Melli, G. Spinelli, H. Wyssling, and E. Arnold, Cell 11:651 (1977).
100. M. Melli, G. Spinelli, and E. Arnold, Cell 12:167 (1977).
101. A.C. Lichtler, S. Detke, I.R. Phillips, G.S. Stein, and J.L. Stein, Proc. Natl. Acad. Sci. 77:1942 (1980).
102. R.L. Jansing, J.L. Stein, and G.S. Stein, Proc. Natl. Acad. Sci. 74:173 (1977).
103. I.-M. Chiu, D. Cooper, and W.F. Marzluff, "Abstracts of the Second Annual Cancer Research Seminar", American Cancer Society, Florida Division, Inc., Tampa, Florida, abstract #38 (1979).
104. I. Parker and W. Fitschen, Cell Diff. 9:23 (1980).
105. G.S. Stein and J.L. Stein, BioScience 26:488 (1976).
106. R. Baserga, Cell Tissue Kinet. 1:167 (1968).

107. A.B. Pardee, R. Dubrow, J.L. Hamlin, and R.F. Kletzien, Ann. Rev. Biochem. 47:715 (1978).
108. S. Gelfant, Cancer Res. 37:3845 (1977).
109. B. Clarkson and R. Baserga, eds., "Control of Proliferation in Animal Cells", Cold Spring Harbor Laboratory, Cold Spring Harbor, New York (1974).
110. R. Baserga, Life Sci. 15:1057 (1974).
111. I. Cameron and G. Padilla, eds., Cell Synchrony, Academic Press, New York, New York (1966).
112. R. Baserga, "Multiplication and Division of Mammalian Cells" Marcel Dekker, New York, New York (1976).
113. D.M. Prescott, "Reproduction of Eukaryotic Cells", Academic Press, New York, New York (1976).
114. J.M. Mitchison, "The Biology of the Cell Cycle", Cambridge Univesity Press, London, UK (1971).
115. R. Baserga, ed., "The Cell Cycle and Cancer", Marcel Dekker, New York, New York (1971).
116. L.F. Lamerton and R.J.M. Fry, eds., "Cell Proliferation", Blackwell, Oxford, UK (1962).
117. K.W. Gross, M. Jacobs-Lorena, C. Baglioni, and P. Gross, Proc. Natl. Acad. Sci. 70:2614 (1973).
118. Skoultchi, A. and Gross, P.R., Proc. Natl. Acad. Sci. 70:2840 (1973).
119. N.S. Kunkel and E.S. Weinberg, Cell 14:313 (1978).
120. K.M. Newrock and L.H. Cohen, Cell 14:327 (1978).
121. G.C. Overton and E.S. Weinberg, Cell 14:247 (1978).
122. I.R. Phillips, E.A. Shephard, J.L. Stein, L.J. Kleinsmith, and G.S. Stein, Biochem. Biophys. Acta 565:326 (1979).
123. M. Grunstein and D. Hogness, Proc. Natl. Acad. Sci. 72:3961 (1975).
124. W.D. Benton and R.W. Davis, Science 196:180 (1977).

CHROMATIN STRUCTURE, HISTONE MODIFICATIONS AND THE CELL CYCLE

E. M. Bradbury and H.R. Matthews

Department of Biological Chemistry
School of Medicine
University of California
Davis, California 95616

INTRODUCTION

Credit for the recent major advances in our understanding of chromatin structure must be attributed largely to the finding by Hewish and Burgoyne (1) that Ca^{++} activated endogenous nucleases preferentially cleaved the DNA in rat liver chromatin at sites separated by about 200 base pairs (bp)(2). This led to their proposal that chromatin was a simple repeating subunit structure. It also led to the research industry that has built up on the uses of different nucleases as biochemical probes of chromatin structure and on the additional use of staphylococcal(s.) nuclease in biochemical procedures for the preparation of large amounts of the chromatin subunit, the nucleosome, oligomers of nucleosomes and large pieces of chromatin.

The structure of the nucleosome is now understood in outline. It consists of the histone octamer 2(H2A, H2B, H3 & H4), one histone H1 and a variable length of DNA depending on tissue and organism though for most tissues in higher organisms it is 195 $\pm$ 5 base pairs (bp) DNA (3,4). On nuclease digestion of the DNA within the nucleosome, it was found that histone H1 was released from the nucleosome when the length of the DNA was digested from 168 to about 140 bp (5) and this left a well-defined subnucleosomal particle, the core particle (6), containing 146 $\pm$ 2 bp of DNA (7) and the histone octamer 2(H2A, H2B, H3, H4). Because of the regularity of the structure of the core particle, it has been studied intensively by a wide range of physical techniques.

Chromatin is made up of a string of nucleosomes joined by the continuity of the DNA molecule. The different orders of chromatin structure are generated by interactions involving histones and subsets of non-histone chromosomal proteins. In this chapter we shall describe the current state of our understanding of chromatin structure and discuss the proposed roles for the reversible chemical modifications of histones and the involvement of non-histone proteins (NHP) in the control of chromosome structure and function.

STRUCTURE OF CHROMATIN

The chromatin core particle.

The structure of the chromatin core particle is now understood at low resolution from both neutron scatter studies of core particles in solution (8-13) and from X-ray diffraction and electron microscopy studies of crystals (14,15). Both approaches converge on essentially the same structure showing that the solution and crystal structures of the core particle are very similar.

By using neutrons the scatter from the DNA component can be largely separated from that of the histones (see 16,17) and it was shown unambigously that the DNA was coiled around a core of the histone octamer. Analysis of a series of neutron scatter curves from core particles in a range of mixtures of light (^1H_2O) and heavy (^2H_2O) water gave the fundamental scatter functions related to the particle's shape and internal structure (9,10). From these the maximum dimension of the core particle was found to be 11.0 nm (9) and the overall shape which best fitted the data was an oblate spheroid of axial ratio 0.5 (8,9). Thus the dimensions of the particle could be estimated to by 11.0 x 11.0 x 5.5 nm. Systematic calculations of the fundamental scatter functions from proposed models for the core particle have been compared with the observed functions and show that the neutron scatter data severely restricts the number of models which can be valid for the structure in solution. The best model (10) has the overall shape given above with the histone octamer forming a core of approximate dimensions 4 nm x 6.4 to 7.5 nm around which is coiled 1.7 ± 0.1 turns of DNA of pitch between 3.0 and 3.5 nm. A hole probably penetrates the histone core but the data does not allow a diameter greater than 1 nm. This model is given in Figure 1.

A very similar low resolution model to that given in Figure 1 has also been proposed from X-ray diffraction and electron microscopy studies of small crystals of core particles. In this model the pitch of the DNA is given as 2.7 nm. It is probable that within 3 years single crystal X-ray diffraction will give the core particle structure at atomic resolution.

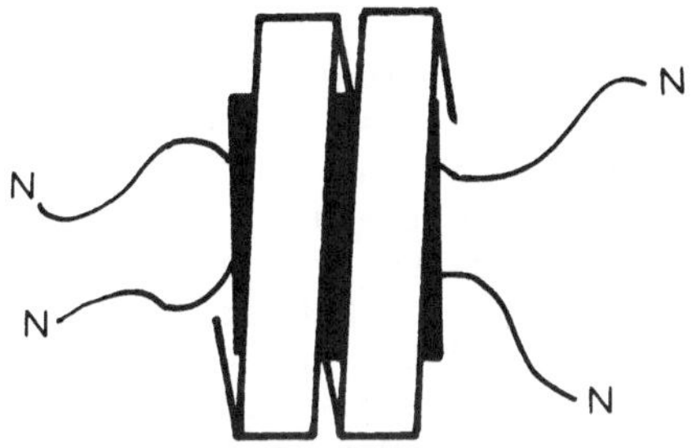

Figure 1. Model of a chromatin core particle. The black core represents the histone octamer 2(H2A, H2B, H3 & H4) with free N-terminal regions of histones H2A and H2B. The encircling white band represents 1.7 ± 0.1 turns of DNA of pitch 3.0 - 3.5 nm. The disc shaped core particle has an outer diameter of 11.0 nm and a thickness of 5.5 - 6.0 nm. A hole of ca 1.0 nm probably penetrates the protein core.

Structure of the nucleosome.

Because of the variability of the DNA content of nucleosomes, even within a single cell type (18,19), it has not proved possible to obtain nucleosomes with sufficient regularity of structure to grow crystals for diffraction studies. Solution studies of nucleosomes with the full complement of histones have also proved difficult because of their strong tendency to aggregate when histone H1 is present. Chicken erythrocyte nucleosomes with a reduced DNA content from 212 bp to 195 ± 20 bp and depleted of the very lysine rich histones H1 and H5 have been studied by neutron scatter techniques (17). The shape of the model which best fitted the low angle scatter data was an oblate spheroid of axial ratio 0.5. The fundamental scatter functions were very similar to those found later for the core particles (9,10) showing that the structure of the H1 and H5 depleted nucleosome was very similar to the core particle structure of Figure 1. Additional information on the structure of the nucleosome has been obtained from studies of the structure of extended chromatin.

Structure of extended chromatin.

The structure of chromatin is very sensitive to the ionic strength of the aqueous medium. At low ionic strength chromatin is in an extended form which, in many electron microscopy studies, has been called the 10 nm diameter filament. Increase in ionic strength, particularly with divalent cations, results in the transition to the '30 nm' diameter filament (20). The finding that the chromatin core particle was disc-shaped raised questions as to how nucleosomes were arranged in the 10 nm filament, i.e., were the discs of Figure 1 arranged face-to-face, edge-to-edge or with some intermediate arrangement. From neutron scatter studies of extended chromatin in mixtures of ^1H_2O and ^2H_2O at low ionic strength the tranverse radius of gyration of the DNA component could be obtained largely independently of the histone component and vice versa (21). The mass per unit length of the 10 nm filament correponded to 1 nucleosome per 10 $\pm$ 2 nm, i.e., a DNA compaction ratio on folding into the nucleosomes in the extended structure of between 6 and 9. Models which were consistent with the above parameters had the disc-shaped nucleosomes arranged edge-to-edge or with the faces of the discs inclined to within 20° to the axis of the 10 nm filament as shown in Figure 2. Similar arrangements have been suggested from electric dichroism studies of dimers of nucleosomes (22) and of extended chromatin (23).

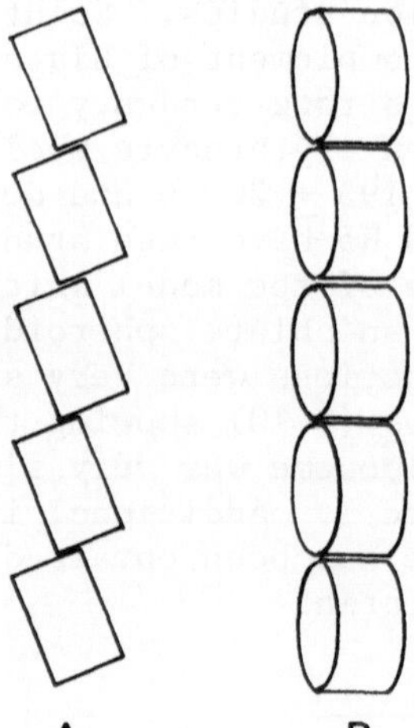

Figure 2. Model for the extended 10 nm filament of chromatin. The nucleosomal discs are arranged edge-to-edge within the limits of 20° (A) or parallel to the filament axis (B). (From Suau et al. (21)).

A different edge-to-edge arrangement of nucleosomes has been proposed from electron microscopy studies of short lengths of chromatin (24). Electron micrographs showed that the nucleosomes were arranged in a regular zig-zag with their faces down on the grid. Such arrangements have been observed over small regions of high molecular weight chromatin in many electron micrographs and this raises the question as to whether it is representative of the basic structure of extended chromatin or whether it is an intermediate structure induced by the preparative procedures for electron microscopy. It is relatively easy to make the transition from the linear form of the edge-to-edge arrangement of nucleosome discs to the zig-zag form by performing a one half of a rotation per nculeosome about the filament axis. The neutron scatter parameters given above are in much better agreement with the linear form than with the zig-zag form. It should be stressed that the neutron scatter studies are of hydrated chromatin in aqueous solution and thus correspond more closely to the in vivo form than dehydrated chromatin prepared for electron microscopy.

Irrespective of whether the zig-zag form is the basic form of extended chromatin or some intermediate structure an important finding of this study was the dependence of the regularity of the zig-zag form on the presence of histone H1. With H1-depleted chromatin this regularity was completely lost and it was suggested that one function of histone H1 was to 'seal-off' two turns of DNA coiled around the core of the histone octamer. This is in accord with our understanding of the structure of histone H1 which it has been shown consists of three structural domains: i) a variable, basic non-structured N-terminal region 1-40, ii) a constant apolar, globular central region 41-120 and, iii) a very basic variable, non-structured C-terminal half 121-216 (25-28). It has been suggested (26,27) that the constant globular region had a specific binding site on the subunit structure. Very recent nuclease digestion studies of the role of the three domains of H1 in chromatin structure have supported and extended this suggestion (29). As described earlier in the nuclease digestion kinetics of chromatin a pause is observed at 168 pb DNA; H1 is released from the nucleosome as the digestion proceeds to 146 bp DNA and H1 was thought to be associated with the linker DNA between core particles (5). H1 can be dissociated from chroamtin without losing the 200 bp DNA ladder. The pause at 168 pb in the nuclease digestion kinetics is, however, not observed with H1 depleted chromatin and the digestion goes direct to 146 bp DNA. If H1 is added back to the H1 depleted chromatin, the 168 bp particle, the chromatosome, is observed. On adding back each of the H1 peptides corresponding to the three structural domains i, ii and iii described above the pause in the digestion kinetics was restored only by the conserved globular region of H1. A model for the nucleosome which incorporates these findings and the properties of H1 is given in Figure 3. Also

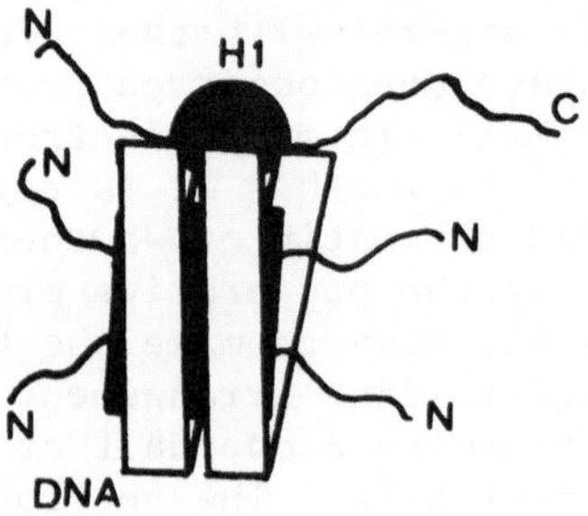

Figure 3. Model of a nucleosome which is the same as the core particle (Figure 1) but with 2 turns of DNA of 80 bp per turn. The globular region of H1 seals off the two turns of DNA.

included in this model are the findings (30-32) that the basic variable regions of histones H2A and H2B and the basic constant regions of histones H3 and H4 are not essential for the core particle structure. The possible functions of the basic N- and C-terminal regions of histone H1 and the basic N-terminal regions of the core histones will be discussed later.

Higher order chromatin structure.

Neutron scatter studies have been made of the ionic strength induced transition from the 10 to the 30 nm filament (21). Above 20 mM NaCl or 0.4 mM $MgCl_2$ there was a sudden transition to a higher structure with a transverse radius of gyration of about 9.0 nm and packing ratio of 0.2 nucleosomes/nm. With further increase of ionic strength the radius of gyration increased to 9.5 nm and the packing ratio to 0.6 nucleosomes/nm. This behavior suggests a family of supercoils of nucleosomes which contract with increasing ionic strength (Figure 4). The diameter of the most compact form was found to be 34 nm and this structure, Figure 4C, clearly corresponds to the supercoil of nucleosomes proposed from neutron fibre diffraction (33) or the solenoid structure proposed from an electron microscopy study of the 10 nm to 30 nm filament (see 24). The pitch of the most compact supercoil is 10-11 nm and contains 6-7 nucleosomes/turn. The arrangement of the nucleosome discs in

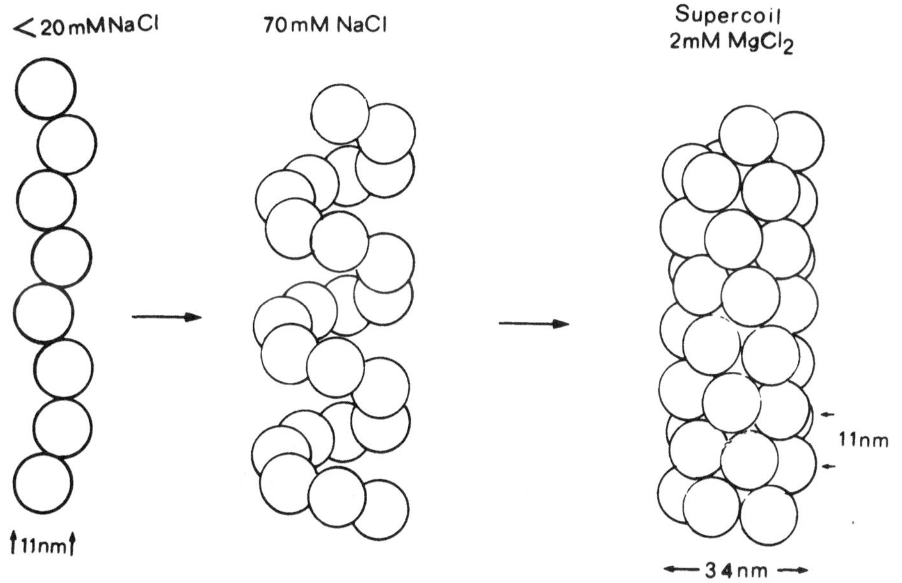

Figure 4. Ionic strength induced transition from the extended form of chromatin at low ionic strength to the most compact supercoil of nucleosome at high salt.

the supercoil is not known with certainty. Of the three possible orthogonal arrangements of nucleosome discs (Figure 5) one can be eliminated by the neutron scatter data (21). This is model 5A with the faces of the discs pointing out radially from the supercoil axis. The transverse radius of gyration R_G for this model would be about 14.2 nm which is much higher than the measured value of 9.5 nm for chicken erythrocyte chromatin. In more recent measurements (34) R_G values of 10.5 - 11 nm have been measured for calf thymus supercoiled form. Again this is too low to be consistent with model 5A. Both models 5B and 5C would give a transverse R_G of 11.9 nm which is closer to the measured values. On the basis of neutron experiments, it is not possible to distinguish between mdoels 5B and 5C or any inclination of the nucleosome discs between the extremes represented by these models. Model 5C, however, has been proposed on the basis of electron microscopy studies (24) and also electric dichroism studies (23). A feature of the former model is the location of the H1 molecule on the inside of the superciol. It should be possible by neutron scatter techniques to determine whether histone H1 is on the inside or on the outside of the supercoil and these experiments are in progress. A somewhat different proposal concerning the structure of the '30 nm' filament has been made by Renz and colleagues (35) that the nucleosomes instead of forming a regular coil are somehow clustered into

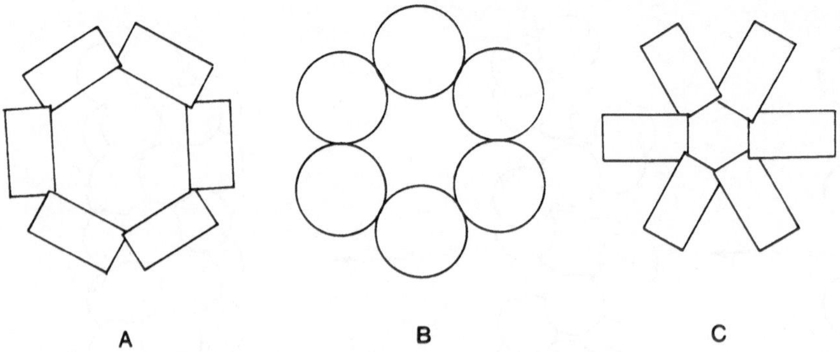

Figure 5. Possible arrangement of nucleosomes in the 34 nm diameter supercoil. A) nucleosome discs facing out from the helix axis; B) nucleosome discs facing along the helix axis; and C) nucleosome discs arranged radially.

discrete superbeads 25-30 nm in diameter. A range of '30 nm' filament structures from regular uniform arrays to clustered arrays have been described by Olins (36). It is not known whether this variability is functional and due to an irregular distribution of histone variants on modified histones along the chromatin fibre or whether it results from the E.M. preparative procedures. Further studies are required to resolve these issues.

CONTROL OF CHROMATIN STRUCTURE

Properties of histones.

The sequence properties of histones have recently been the subject of a review by Von Holt and his colleagues (37) which includes many new sequences from a range of organisms and tissues. This extensive survey confirms and extends our earlier understanding of the properties of histones.

All histone sequences so far studied (37) are highly asymmetrical with respect to the distribution of basic, acidic and apolar residues. The N-terminal quarters of the histones H2A, H2B, H3 and H4 contain about 70% of the lysine residues of each molecule and are very basic. The central regions of H2A and H2B and the central and C-terminal regions of H3 and H4 contain most of the apolar and acidic residues and a high proportion of arginine residues; there are short basic tails at the C-terminal ends of H2A and H2B.

As regards variability in histone sequences, H3 and H4 remain the most highly conserved of all protein sequences and this implies an essential function for each residue along the polypeptide chain. Histones H2A and H2B are known to be more variable than H3 and H4 and recent comparison by Von Holt (37) show that this variability is restricted to the N- and C-terminal regions of H2A and to the N-terminal region of H2B. The apolar central region of H2A and the central and C-terminal regions of H2B appear to be as highly conserved as histones but again most of the sequence changes are located in the basic N- and C-terminal regions leaving the central apolar region highly conserved.

The sequence properties of all the histones are summarized in Figure 6. We now attempt to relate these properties to the interaction and functions of histones.

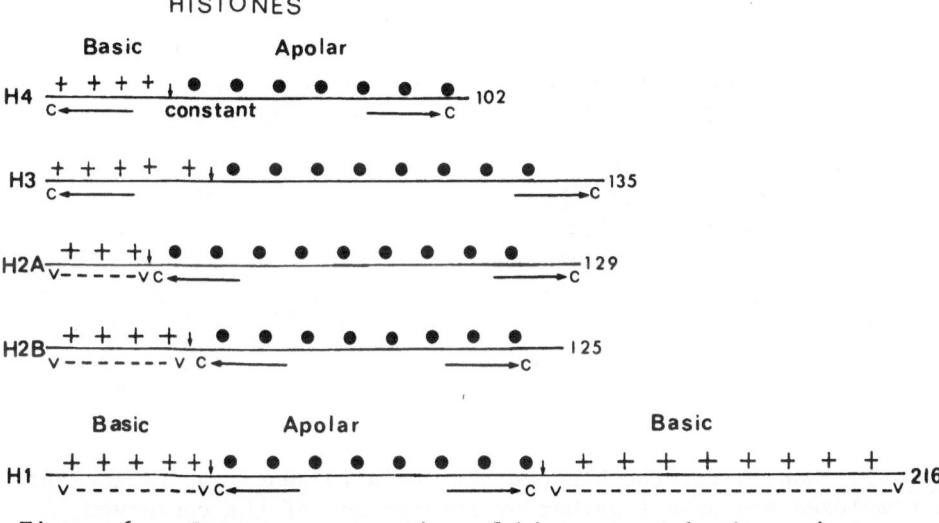

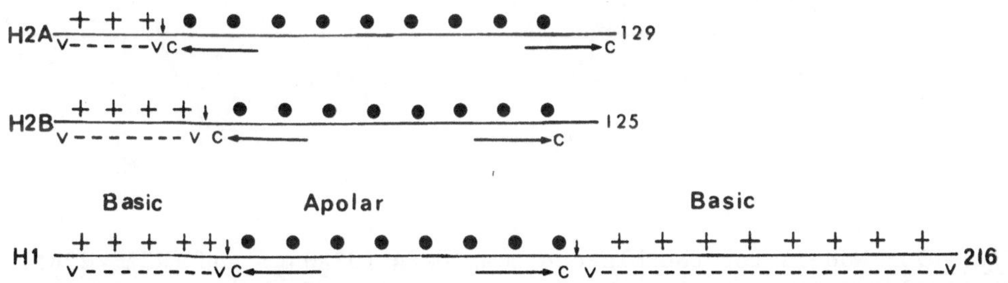

Figure 6. Sequence properties of histones; + basic regions, ● apolar regions, C - highly conserved sequences, V - variable sequence regions.

Histone complexes.

In Figure 6 it can be seen that histones H3 and H4 and histones H2A and H2B form pairs of similar histones. It has been known for some time now that these pairs of histones also form the specific complexes ($H3_2H4_2$) (38) and (H2A, H2B) (39) which are important structural elements of chromatin. Weaker interactions have also been found between other pairs of histones (40).

Histone interactions.

NMR studies of the ($H3_2H4_2$) tetramer (41) and of interactions between H3 and H4 peptides (42) show unambigously that the H3H4 complexes are held together by interactions between the central and C-terminal apolar regions of these molecules. Similarly the histone dimer (H2A, H2B) is held together by interactions between their conserved central apolar regions of H2A and H2B (43). In both of these complexes and in the histone complex released from core particles at high salt (30) the basic N-terminal regions are free and mobile. Physical models for the tetramer ($H3_2H4_2$) and dimer (H2A and H2B) are given in Figure 7. Histone H1 does not form specific complexes with other histones and a model for its structure in solution is given in Figure 8 (26-28).

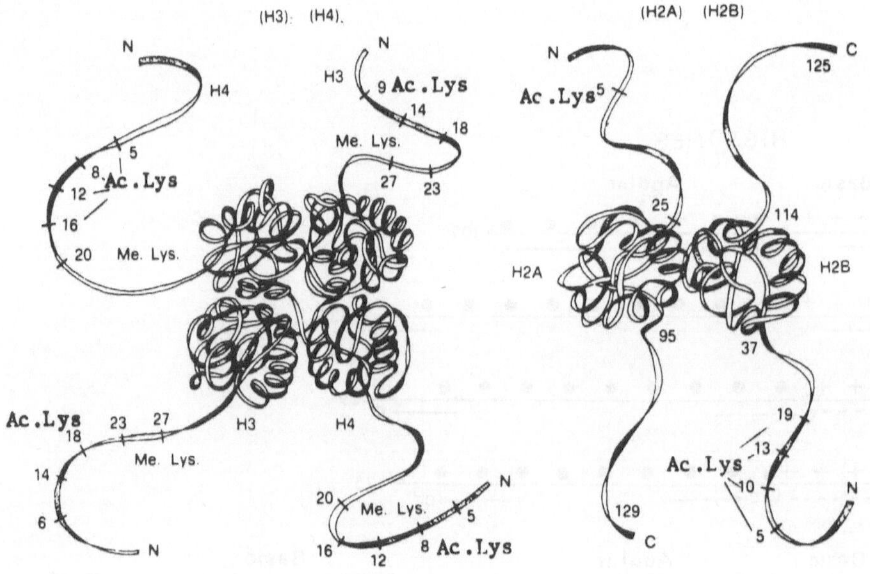

Figure 7. Models for the histone complexes ($H3_2H4_2$) and (H2A, H2B). The complexes are held together by interaction of the conserved apolar regions. The sites of acetylations of lysines in the basic N-terminal regions are indicated. (Taken from E.M. Bradbury (1978) La Recherche 9, 644-658).

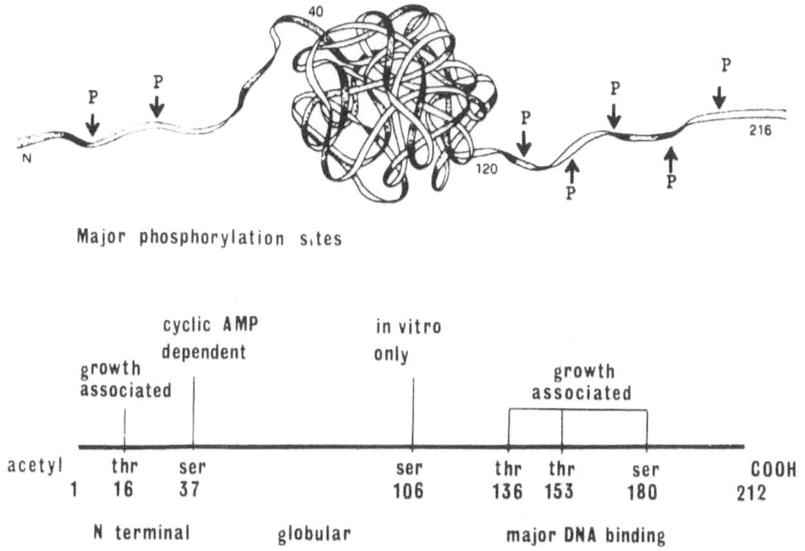

Figure 8. Model for histone H1 and sites of in vivo post synthetic phosphorylation of serines and threonines in the basic N- and C-terminal regions.

Included in the above models are the sites of the major reversible chemical modifications of histone; these are acetylation of lysines in histones H2A, H2B, H3 and H4 and phosphorylation of serines and threonines in histone H1. As can be seen all of the in vivo sites of modifications are located in the basic N-terminal regions of the four core histones and in the basic N- and C-terminal regions of H1. These modifications will be discussed later in more detail.

Histone domains.

These seems little doubt that by several criteria histones have distinct domains with quite different properties and it is presumed different roles in chromatin structure and function. In figures 6 and 7 it can be seen for histones H2A, H2B, H3 and H4 that compared to their central and C-terminal regions the N-terminal regions i) have different sequence properties; ii) for H2A and H2B different mutation rates; iii) have different conformational behaviors, and iv) contain all of the sites of chemical modification. By the same criteria histone H1 contains three well-defined domains.

Following the above findings for histones, it has become increasingly apparent that nucleic acid binding proteins can have more than one structural domain (44) usually a nucleic acid binding domain and a functional domain. For histones we shall outline possible roles for their different domains.

Structural role for histones.

It has been shown that histones H3 and H4 are essential for the structure of chromatin core particles (45-48). Core particle-like structures can be reconstituted from DNA and histones H3 and H4 but not from histones H2A and H2B. This result does not imply that H2A and H2B are not important for nucleosome structure because the histone octamer $(H2A, H2B, H3, H4)_2$ is found in all nucleosomes. It suggests, however, that the histone tetramer $(H3_2H4_2)$ interacts with DNA and the resulting complex provides the structural framework for the subsequent binding of H2A and H2B which then completes the core particle structure. In the absence of H2A and H2B it would appear that an additional $(H3_2H4_2)$ tetramer can complete a core-particle-like structure.

NMR studies of core particles in solution (30) show that the variable basic N-terminal regions of H2A and H2B are not firmly bound within the core particle but give NMR signals corresponding to a mobile, random coil polypeptide chain. The conserved apolar central domains of H2A and H2B are, however, bound in the particle. It was also found that, whereas histones H3 and H4 were bound within the particle at low salt on increasing the ionic strength to 0.6 M CaCl, the N-terminal regions of these histones were released although the particle was not unfolded at this ionic strength.

These results are fully consistent with the properties of histones. The core particle, which is the conserved structural unit of all chromatins so far studied, is generated by the highly conserved histones H3 and H4 together with the conserved apolar central regions of H2A and H2B. The variable basic N-terminal domains of histones H2A and H2B probably act outside of the core particle.

At first sight it appears paradoxical that the basic N-terminal region of histones H3 and H4 should be less tightly bound within the core particle than the central and C-terminal regions. This observation is consistent, however, with the idea that the different domains of histones may have different functions. The N-terminal domains of histones H3 and H4 contain a much higher proportion of lysines compared to arginines; in the N-terminal 30% of histone H3 there are 8 lysines and 5 arginines whereas in the remaining central and C-terminal domain there are 5 lysines and 13 arginines; similarly with histone H3, in the N-terminal 30% there are 6 lysines and 4 arginines compared to 5 lysines and 10 arginines in the central and C-terminal regions. The higher basicity of the N-terminal domains of these histones is conferred by a higher proportion of lysine residues of which four in both histones H3 and H4 are subjected to acetylation.

A question which is now being addressed concerns the relative strengths of binding of the lysine and arginine side-chains to the DNA phosphates. Williams (44) has recently discussed the chemistry of lysines and arginine side-chains in a review of the conformational properties of proteins in solution. A comparison of the solubilities of ammonium salts with salts of other cations such as Na^+, K^+ and Rb^+ shows that the "NH_4+ ion has a special interaction with some weak acid anions, e.g., with some phosphates". This specificity involves the formation of hydrogen bonds which introduces a directional character into the interactions of the tetrahedral ion NH_4+ with, e.g., phosphates and carboxylates. This behavior is enhanced when the ammonium ion is substituted in the lysine side-chain $-(CH_2)_4-NH_3+$. The interaction of a lysine side-chain with a DNA phosphate contains components of both electrostatic and hydrogen bond formation. Acetylation converts the charged lysine side-chain to the neutral acetyl lysine side-chain-$(CH_2)_4-NH_2-CO-CH_3$ and would be expected to suppress the interactions of the lysine side-chains with the DNA phosphates. Methylations of lysines, however, do not discharge the side-chain but result in a series of methyl lysines depending on the degree of substitution, i.e., $-(CH_2)_2NH_2(CH_3)+$, $-(CH_2)_4NH(CH_3)_2+$ and $-(CH_2)_4N(-CH_3)_3+$. These unusual cations, both charged and hydrophobic, are capable of specific charge-charge interactions within the hydrophobic interiors of proteins (44).

The arginine side-chain contains the guanidinium group

NH_2+

-NH-CH

NH_2

which is a large cation with a relatively low charge density. However, the phosphate groups in DNA are good hydrogen bond acceptors and the geometry of the quanidinium group allows a strong hydrogen bonded charge-charge complex to form according to the following scheme

H NH_2....O
|
$\alpha CH-(CH_2)_3-N-CH$ +   -   P

NH_2....O

A similar complex is possible also between guanidinium groups and the carboxyl groups of aspartic and glutamic acids.

In comparing the properties of arginine and lysine residues, studies have been made of the random coil → helix transition of the homopolymers of these residues (49-51). It was found that poly-L-arginine but not poly-L-lysine could be induced to undergo a random coil → helix transition by titrating with tetrahedral anions. Such behaviors show that arginines can form stable complexes in solution with these anions resulting largely in charge neutralization but that lysines form weaker more labile complexes. From these studies, it is expected therefore that the guanidinium groups of arginine side-chains would form stronger complexes with the DNA phosphate groups than the amino groups of the lysine side-chains. This behavior accords with NMR studies of histone interaction with DNA which show that arginine residues remain bound to DNA at ionic strengths at which the lysine side-chains are free and mobile. This differential bindings of lysines and arginines are clearly of importance in considerations of the functions of the different domains of histones and the effects of post-synthetic chemical modifications on these domains.

Cell-cycle studies of histone modifications.

In our studies of the control of chromosome structure through the cell-cycle we have used the naturally synchronous growth cycle of the multinuclear plasmodia of the true slime mould *Physarum polycephalum*. *Physarum* plasmodia can be grown easily to 15 cm diameter and this size of plasmodium contains of the order of 10^9 nuclei which provides about 1 mg DNA and 2 mg of chromosomal protein. The nuclear division cycle in plasmodia is about 9-11 hours; there is no G1 phase, S-phase occupies 4-5 hours and G2 phase 5-6 hours. Within a single plasmodium the nuclei enter metaphase within 3-5 minutes in the 9-11 hour cycle. The nuclear division cycle is therefore extremely precise and is achieved naturally because all the nuclei within a plasmodium are in the same cytoplasm and come under the same biochemical controls. *Physarum* is being increasingly used as a model system for eukaryotes. Two of the histones, H3 and H4, are very similar to the same histones from higher eukaryotes. This can be seen in Figure 9A where *Physarum polycephalum* histones (P) are compared with calf thymus histones (C) by three different types of polyacrylamide gel electrophoresis. The bands labelled have been identified by electrophoresis of purified histones. In all three gel systems *Physarum* H4 comigrates with calf thymus H4; H3 from both organisms migrates in a very similar, though not always identical, manner. Also histones H2B and H2A from both organisms are not very different in their migratory behaviors. Histone H1 shows the greatest differences and although the amino acid composition of *Physarum* H1 is not very different from calf thymus H1, there are differences in size in the C-terminal region. Figure 9B shows the 2-dimensional gel electrophoresis pattern of *Physarum* histones.

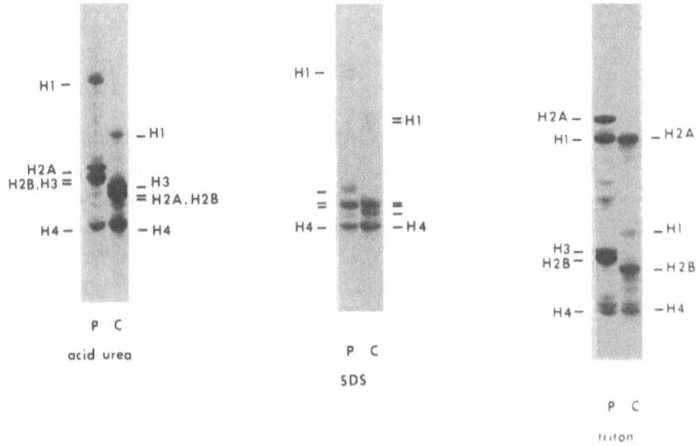

Figure 9A. Comparison of histones from Physarum plasmodia and calf thymus by acrylamide gel electrophoresis. C = calf thymus; P = Physarum. Gel electrophoresis in the presence of i) acetic acid and urea acid/urea; ii) sodium dodecyl sulphate (SDS); and iii) triton X-100. The labelled bands were identified by electrophoresis of purified histone fractions. (From Chahal et al. (106)).

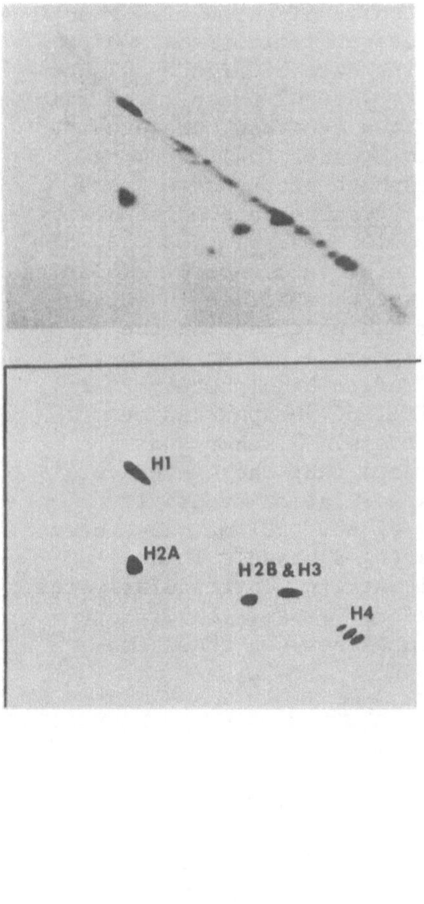

Figure 9B. 2-D gels of Physarum histones as indicated; both dimensions have acetic acid/urea also. Figure provided by Dr. Reinhold Mueller.

PHOSPHORYLATION OF HISTONE H1

Cell cycle dependence of H1 phosphate (^{32}P).

Physarum incorporates ^{32}P into histones if plasmodia are grown in the presence of labelled inorganic phosphate. The potassium phosphate normally present in the growth medium can be omitted. Most of the incorporation, over a long period, is into H1 histone. Looking back in our earlier work (53) we can now see that there is also substantial incorporation into H2A, at a fairly constant level in the cell-cycle, and mitosis-specific incorporation into H3 or H2B. There is also substantial ^{32}P incorporation into a minor component running slightly faster than H1 and except in mitosis, into non-histone material near the top of the gel (53).

The incorporation of ^{32}P into H1 was measured through the cell-cycle using ^{3}H-lysine and ^{32}P to label plasmodia continuously from fusion to harvesting. Total histone was isolated and fractionated by acrylamide gel electrophoresis and the $^{32}P/^{3}H$ ratio in the H1 band measured. Figure 10 shows the results (53). These data were used to propose a correlation between H1 phosphate content and the "initial mechanism" of chromosome condensation in mitosis. Later studies of histone kinase supported the association of H1 phosphorylation with the initial phase of chromosome condensation and it is now widely accepted that H1 phosphorylation is associated with chromosome condensation in mitosis. The exact time-course of phosphorylation and the stage of mitosis at which it is important remain controversial. The data of Figure 10 show an increase in H1 phosphate during mid to late G2 phase. The main differences, to be outlined in the next two sections, concern the drop in H1 phsophate. Gurley and his colleagues (54), working with CHO cells, find the drop in H1 phosphate occurs after metaphase. They suggested that the earlier Physarum data might be affected by phosphatase activity during histone isolation (53, 55). Inglis, in our laboratory, showed this to be incorrect by isolating histone from metaphase plasmodia using the phosphatase inhibitor 50 mM sodium bisulphite. The inhibitor had no effect on the results although we showed that the phosphatase present in metaphase nuclei is inhibited 50% by 50 mM sodium bisulphite. It is possible that Physarum avoids the problems of phosphatase at metaphase by having an intra-nuclear mitosis. Fischer and Laemmli (56), using SDS gels, obtained data that they interpret as showing the absence of any H1 dephosphorylation at all in Physarum and they suggest that Bradbury et al. (53) may have been selectively losing phosphorylated H1 during S phase. This suggestion is inconsistent with the gel patterns, particularly the relative amounts of H1, reported by Bradbury et al. (53). More work is required to resolve the differences between these two experimental approaches in Physarum.

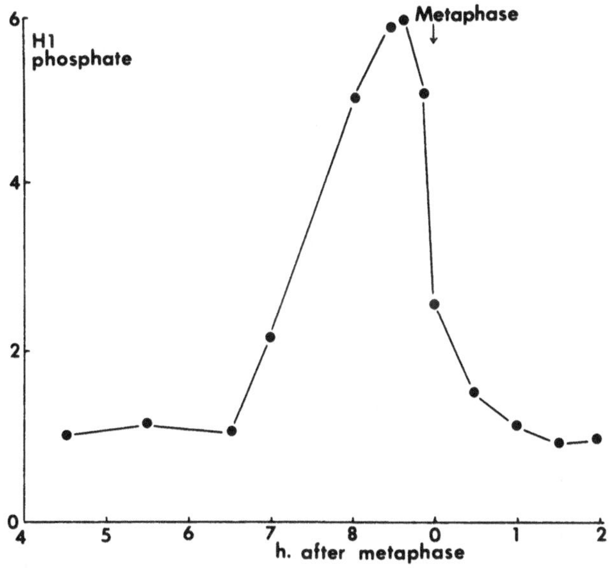

Figure 10. Phosphate content of Physarum H1 histone in the cell-cycle (53). The phosphate content of Physarum H1 was determined from long-term labelling with ^{3}H-lysine and ^{32}P-phosphate and normalized to 1.0 in late S phase. (Mammalian cells have approximately 1 phosphate per H1 molecule in S phase.)

Cell-cycle dependence of H1 phosphate (SDS gel mobility).

Mohberg and Rusch (57) and Bradbury et al. (53) observed differences in the electrophoretic mobility of H1 isolated from different stages of the mitotic cycle, using gels containing acetic acid and urea. Fischer and Laemmli (56) studied Physarum H1 using gel electrophoresis in SDS. In acid-urea gels the changes were assumed to be due to changes in net charge, probably due to phosphorylation. In SDS gels the changes must be assumed to be due to changes in SDS-binding capacity. Treatment of slowly migrating H1 with alkaline phosphatase increased its mobility on SDS gels but not to the maximum mobility observed for newly synthesized H1. The basis of the alkaline-phosphatase-resistant decrease in mobility is unknown, as is the mechanism whereby phosphorylation of H1 has a major effect on the SDS-binding capacity. It is interesting to speculate that this effect may reflect the in vivo function of H1 phosphorylation.

Fischer and Laemmli (56) found that H1 labelled with ^{14}C-lysine in S phase and extracted immediately migrated rapidly although the total H1 showed both a slow and a fast band. At later times the mobility of H1 labelled in the previous S phase was lower, reaching a minimum early in mitosis. In the next S phase this H1 remained in the low mobility position. On the basis of these and other experiments it was concluded that newly synthesized H1 was phosphorylated during G2 phase and then remained phosphorylated although some turn-over of phosphate apparently occurred (56). This interpretation is inconsistent with the conclusions of Bradbury et al. (53) that H1 phosphate falls dramatically in late mitosis. It also differs from the situation in CHO cells where H1 phosphate also falls dramatically after metaphase (54, 55, 58). The reason for these inconsistencies is not yet clear but it should be resolved by parallel studies on acid-urea and SDS gels of H1s with differing phosphate contents and by extractions, in parallel, by different methods. This is particularly necessary because the more recent paper (56) contains no detailed, quantitative cell-cycle studies.

Comparison with mammalian cells.

The most detailed cell-cycle studies in mammalian cells have been carried out by Gurley et al. (54, 55, 58), although Lake (59-61), Chalkley's group (62-68), and others (69-73) have also made important observations. Gurley et al. (54, 55) synchronized CHO cells by extended treatment with hydroxyurea and isoleucine deprivation and then used Colcemid to block the cells in metaphase. They harvested cells at various times after release from hydroxyurea and analyzed them by electron microscopy which allowed them to measure the proportion of cells in a number of stages of G2 phase and mitosis. H1 was isolated from the cells and analyzed by gel electrophoresis. The phosphate content of the H1 can be deduced, approximately, from these gel electrophoresis profiles and plotted as a function of time after hydroxyurea block (Figure 11). This data is not directly comparable with the Physarum data because Physarum is much more synchronous and not blocked at metaphase. However, the Physarum data, Figure 10 (53) can be used to calculate the phosphate content that would be observed in a mixture of cell-cycle stages equivalent to that obtained by Gurley et al. (54, 55) at each time point. Figure 11 also shows the result of this calculation and allows a direct comparison of the CHO cell data and the Physarum data. Clearly, the two cell types have a very similar time-course of H1 phosphorylation. It is likely, however, that the maximum phosphate content in CHO cells occurs at the point of Colcemid arrest which, if this is equated with metaphase, is slightly later in mitosis than the maximum phosphate content in Physarum. This apparent minor difference may be an artifact due to the use of Colcemid or may reflect the fact that Physarum has no G1 phase in this stage of its life cycle and so must be ready for S phase immediately after metaphase. In CHO cells, the H1 phosphate content falls very rapidly after release of the cells from Colcemid arrest.

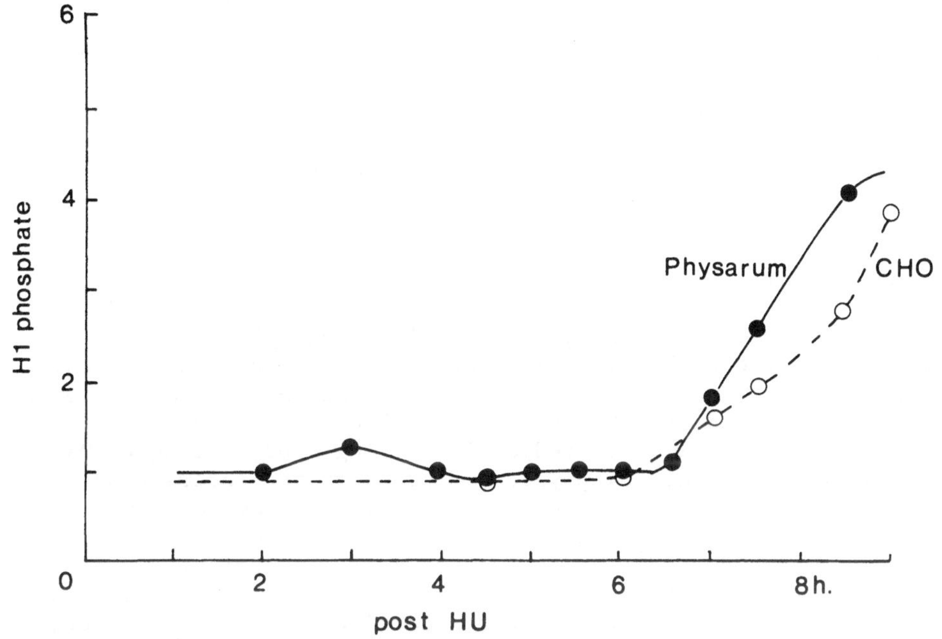

Figure 11. Comparison of histone H1 phosphate content in Physarum and a CHO cell line (53, 54). CHO cells were synchronized by isoleucine deprivation and prolonged treatment with hydroxyurea (HU). Histone H1 was isolated and analyzed by gel electrophoresis at various times after release from hydroxyurea. The cells were not very synchronous by Physarum standards and the cell-cycle was further confused by adding Colcemid so that the cells accumulated in metaphase (54). However, at each time point the fraction of cells at each of several stages recognized by electron microscopy was determined. This allowed us to construct an imaginary Physarum culture with the same proportion of nuclei at each stage and to calculate its average H1-phosphate content. The results shown in the figure indicate a remarkable similarity between Physarum H1 and CHO cell H1. The small difference in early prophase may be real or it may reflect experimental uncertainties such as the effect of Colcemid.

Tanphaichitr et al. used $ZnCl_2$ to inhibit histone phosphatase in growing cells and showed that chromosome de-condensation was not inhibited. This result supports the original suggestion that H1 phosphorylation is involved in the initiation of chromosome condensation. Much more significantly, Matsumoto et al. (74) have isolated a mutant mammalian cell line that is blocked in late G2 phase and fails to phosphorylate H1. Finally, it should be mentioned that the reports of the absence of H1 from the condensed chromosomes of Tetrahymena micronuclei are incorrect and H1 (like H3) has now been found in this system (M. Gorovsky, unpublished communication).

Interaction with DNA.

The foregoing data suggest that H1 phosphorylation in G2 phase is involved with initiating chromosome condensation and this is supported by studies of histone kinase described below. However, it has been pointed out that the simple electrostatic effect of phosphorylation of H1 is to reduce its strength of binding to DNA. This effect can be seen by physical studies of the interaction of phosphorylated H1 with DNA (56, 73, 75-78). Some authors have gone on to postulate that the reduced strength of binding would cause chromosome decondensation, contrary to the indications given by in vivo studies (79, 80). There is no detailed basis for such arguments (81) but neither is there a detailed experimental picture of the effect of phosphorylation of H1 on chromatin structure, although rapid progress can be expected here.

An indication of the effect of phosphorylation on the H1-induced aggregation of DNA has been obtained from turbidity studies (82). Studies with H1 isolated from Physarum plasmodia in either S phase (low phosphate) or early prophase (high phosphate) have been reported (83). Figure 12 shows that both H1s show the typical pattern of aggregation of H1+DNA at 0.1 to 0.4 M NaCl. However, the phosphorylated H1 shows much greater ability to aggregate DNA at moderate salt concentration. This shows that the interaction of H1 with DNA is more complex than a simple electrostatic interaction and that phosphorylation at the sites associated with early prophase enables H1 to aggregate DNA more effectively, thus supporting the idea that early prophase H1 phosphorylation is involved with the initiation of chromosome condensation. At high salt concentration, highly phosphorylated H1 is released from DNA and so the turbidity falls before the turbidity of low-phosphorylated H1+DNA.

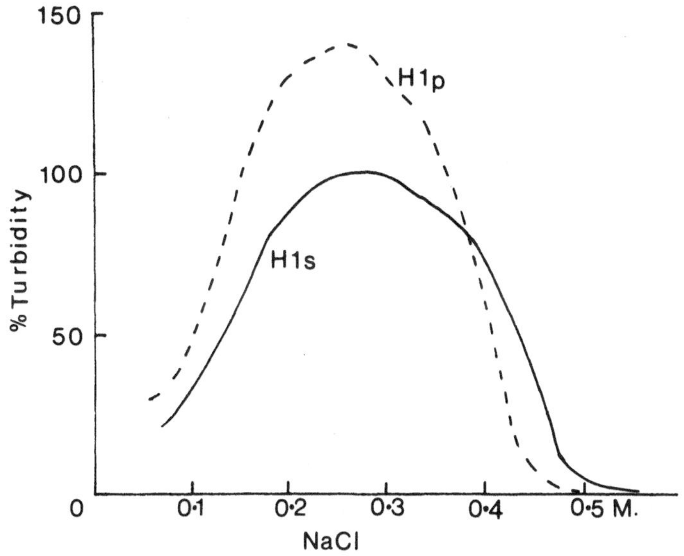

Figure 12. Interaction of S phase- and prophase-histone H1 with DNA. Histone H1 was isolated from Physarum plasmodia in S phase or prophase. The turbidity of H1.DNA solutions was studied (cf. Figure 12) and the figure shows that prophase H1 aggregated or condensed DNA much better than S phase H1, at NaCl concentrations below 0.35 M (83). Experiments with enzymically phosphorylated H1 and alkaline-phosphatase treated H1 suggest the difference between S phase H1 and prophase H1 can be accounted for by the high phosphate content of prophase H1 (82).

NUCLEAR HISTONE KINASE

Cell-cycle dependence on activity.

The cell-cycle dependence of Physarum nuclear histone kinase was determined by assaying 0.4 M NaCl extracts of nuclei and very large changes were observed. Figure 13 shows that the kinase activity was low (but not zero) during S phase and increased 15-fold during early and mid G2 phase. In late G2 phase it fell very rapidly to reach its low level again just before the next S phase. The data suggested that this increase in kinase activity was the immediate cause of the rise in H1 histone phosphate described above (Figure 10) (84). There was also a correlation between the histone kinase activity and the results of plasmodial fusion experiments and heat shock experiments in Physarum.

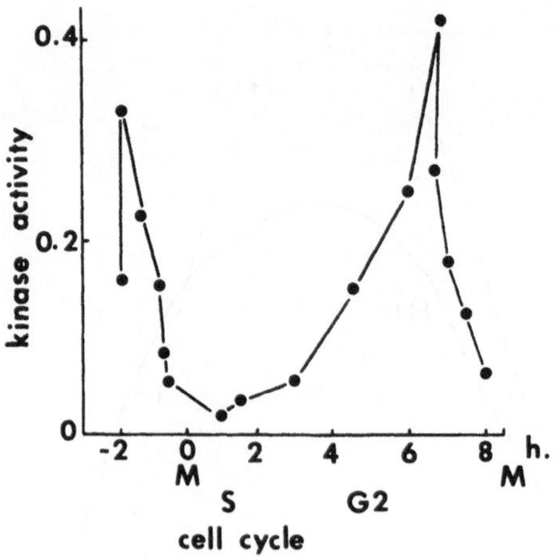

Figure 13. Cell-cycle changes in nuclear histone kinase.

Activation of histone kinase.

The original data on enzyme activity did not address the question: "Is the change in activity due to enzyme synthesis or to another mechanism such as activation or transport?" This question was approached using the technique of deuterium labelling and fractionation on Metrizamide density gradients. The enzyme was located in the gradient from its activity and the resolution was sufficient to separate normal enzyme from enzyme grown in the presence of deuterated amino acids, which are more dense. Plasmodia grown from fusion to mitosis-2 in normal medium, then transferred to deuterium containing medium and harvested in G2 phase showed 2 major peaks of histone kinase activity on a Metrizamide gradient. One coincided with that of normal enzyme, the other was taken to be deuterium labelled enzyme. The two peaks were of comparable size showing that the main cause of the 15-fold increase in activity was not enzyme synthesis but was an activation process (85).

It has been suggested that the equalibration of amino acid pools might be too slow for the deuterium labelling experiment but this is probably not the case since the peak of labelled enzyme was as sharp as the peak of normal enzyme. Slowly equilibrating pools would give a range of densities for labelled enzyme, which was not observed. There were also two smaller peaks on the Metrizamide gradients, probably due to different levels of hydration of the kinase but they both behaved like the major peak as far as their labelling went.

The search for an activation mechanism has so far been unsuccessful. Physarum has cyclic AMP dependent kinase activity in the cytoplasm (A.M. Campbell, unpublished) but the histone kinase activity extracted from nuclei is not affected by cyclic AMP (86). The total activity also appeared not to be affected by the heat stable inhibitor of cyclic AMP dependent protein kinases. However, as described below, the kinase activity can be resolved into 3 components. One of these components remains unaffected by cyclic AMP and the inhibitor; the other two components are unaffected by cyclic AMP but do respond to the inhibitor and evidence is accumulating that one of these components is analogous to the catalytic subunit of mammalian protein kinase (87). Chambers (87) also looked for an effect of alkaline phosphatase on Physarum histone kinase using alkaline phosphatase bound to polyacrylamide beads (suggested to us by Dr. S. Shall). He was unable to demonstrate a specific effect due to the phosphatase suggesting that the kinase is not activated by phosphorylation but this cannot be regarded as a rigorous conclusion.

Three nuclear histone kinases.

Hardie *et al*. (88) showed that Physarum nuclear histone kinase could be separated into 3 components by ion exchange chromatography but were unsuccessful in purifying any of the components further. Chambers (87) confirmed the fractionation of DEAE-cellulose and obtained a small degree of purification by prior gel filtration on Sepharose 6B. However, the enzyme has remained unstable except in crude extracts and has consequently resisted efforts to purify it. Three peaks are obtained for histone kinase on Metrizamide gradients but these three peaks do not correspond to the different enzymes (85), rather each of the separated enzyme runs on Metrizamide like the major peak from unfractionated enzyme. Small quantities of a nuclear histone kinase have been purified from Ehrlich ascites cells (89).

The three Physarum kinases were assayed through the separation procedures using H1 histone substrate. In the case of one of the enzymes, the most strongly bound to DEAE-cellulose, H1 histone is a better substrate than any of the other histones and much better than protamine. In contrast, the other enzyme that binds to DEAE-cellulose, called kinase-A, phosphorylates histone H2B twice as well as histone H1 and protamine 3 times as well as H1. H2A is about half as good as H1 while H3 and H4 alone are very poor substrates. There is evidence of H2A phosphorylation in Physarum but the substrate specificity measured with isolated histones may not be a good guide to their availability as substrates in nucleosomes (88). The differential behavior of the enzymes with protamine as substrate can be used to assay for both enzymes in a mixture of the two enzymes or to readily distinguish the enzymes (85).

The cell-cycle dependence of the two enzymes that bound to DEAE-cellulose was measured (88) and both showed a large peak of activity in G2 phase, similar to that observed with the total activity. However, the precise timing was different with kinase-A peaking one hour later than the strongly bound kinase. Hardie et al. (88) speculated that the enzymes might phosphorylate different sites sequentially on H1. The substrate specificity and cell-cycle dependence of the enzyme that did not bind to DEAE-cellulose have not been measured.

SITES OF PHOSPHORYLATION *IN VITRO*

Location on calf histone H1

Chambers (87), partly in our laboratory and partly in collaboration with T.A. Langan, phosphorylated calf H1 with *Physarum* kinase-A. The phosphorylated H1 was subjected to peptide analysis and a major phosphorylation site, serine-37, was found. This is the same site specificity as is shown by the catalytic subunit of cyclic AMP dependent protein kinase (69,90) and since *Physarum* kinase-A is also inhibited by the inhibitor of cyclic AMP dependent protein kinase, *Physarum* kinase-A is tentatively regarded as analogous to the catalytic subunit of cyclic AMP dependent protein kinase (87).

In a later study the three *Physarum* kinases and Ehrlich ascites kinase (70) were used to phosphorylate calf thymus H1 histone. The phosphorylated histone was partially digested with chymotrypsin as described above and the fragments sepatated by gel electrophoresis. The gels were autoradiographed and scans of the autoradiographs are shown in Figure 14. *Physarum* kinase-A phosphorylated only the N-terminal fragment, residues 1 to 106, consistent with serine-37 being the major phosphorylation site. *Physarum* kinase-B (the one which binds strongly to DEAE-cellulose and is specific for H1) phosphorylated mainly the N-terminal peptide but showed some phosphorylation in the C-terminal region. *Physarum* kinase-G (which does not bind to DEAE-cellulose) and Ehrlich ascites kinase-G phosphorylated H1 at sites in both peptides in a very similar manner (87).

Location on Physarum histone H1

Physarum histone H1 was phosphorylated in the same way with the separated *Physarum* enzymes. Partial chymotryptic digestion was carried out and the N- and C-terminal peptides separated. The site specificity showed some differences from the case of calf thymus H1, with kinase-A showing substantial phosphorylation

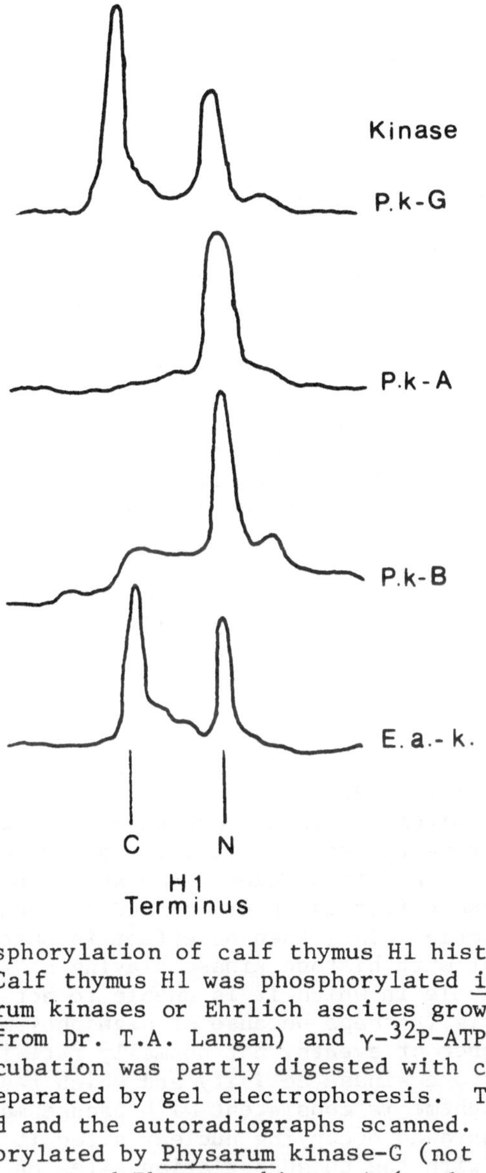

Figure 14. Phosphorylation of calf thymus H1 histone by various kinases (87). Calf thymus H1 was phosphorylated *in vitro* by separated *Physarum* kinases or Ehrlich ascites growth associated kinase (a gift from Dr. T.A. Langan) and γ-^{32}P-ATP. The labelled H1 from each incubation was partly digested with chymotrypsin and the fragments separated by gel electrophoresis. The gels were autoradiographed and the autoradiographs scanned. The top scan shows H1 phosphorylated by *Physarum* kinase-G (not bound to DEAE-cellulose; the next used *Physarum* kinase-A (analogous to the catalytic subunit of cyclic AMP-dependent protein kinase); the next used *Physarum* kinase-B (strongly bound to DEAE-cellulose) and the bottom scan used Ehrlich ascites growth associated kinase. Electrophoresis was from right to left and the two major peaks are the N- and C-terminal halves of H1.

in the C-terminal peptide as well as in the N-terminal peptide and kinase-B actually showing more phosphorylation in the C-terminal peptide than in the N-terminal peptide. These results imply that the C-terminal region of Physarum contains a phosphorylation site or sites that is absent in calf thymus H. Physarum kinase-G showed phosphorylation of Physarum H1 very similar to its phosphorylation of calf thymus H1 (87).

Cell-cycle role of histone kinase.

The cell-cycle behavior of histone kinase activity correlates with plasmodial fusion experiments and heat shock experiments (84, 91-94). The data suggested that histone kinase activity was a controlling or rate-limiting factor in the timing of mitosis and attempts were made to affect the timing of mitosis directly by adding exogenous histone kinase. Prior et al. (95) have subsequently shown that Physarum will take up and use exogenous protein in an intact form, in the case of histone H3. The kinase preparations available were from Ehrlich ascites cells but Figure 15 shows that even this heterologous kinase was very effective at changing the time of mitosis bringing it forward by up to 60 min. Various control substances including inactivated kinase preparations were without effect (96-98). Antibody to the histone kinase preparation specifically delayed mitosis when added to plasmodia (99). There is, however, a major difficulty with these experiments namely that the histone kinase was only partly purified (97) and so the possibility that the effect was due to another component of the preparation cannot be ruled out. Nevertheless, an important indication that histone kinase is the active ingredient is provided by the cell-cycle dependence of the advancement of mitosis. If histone kinase is added during the normal rise of histone kinase activity, it is very effective at advancing mitosis; if addition takes place after the normal rise of activity, then no effect is observed; and if addition takes place before the normal rise, then no effect is observed. The data are consistent with histone kinase playing a "triggering" role in the cell-cycle in which it is unable to act before the end of DNA synthesis (perhaps because of high phosphatase activity) and in which a number of events, not normally rate-limiting, is "triggered" by histone kinase activity and which result in mitosis (91, 100). This scheme is consistent with Sachsenmaier's proposal of a "trigger" substance occupying nuclear sites if the nuclear sites are equated with the phosphorylation sites on H1 histone (101).

It is possible, for example with cycloheximide, to block mitosis much later in the cell-cycle than the maximum of histone kinase activity. This does not mean that histone kinase is not normally the rate-limiting step; rather it confirms that there are other steps before mitosis that are "triggered" by histone kinase (84).

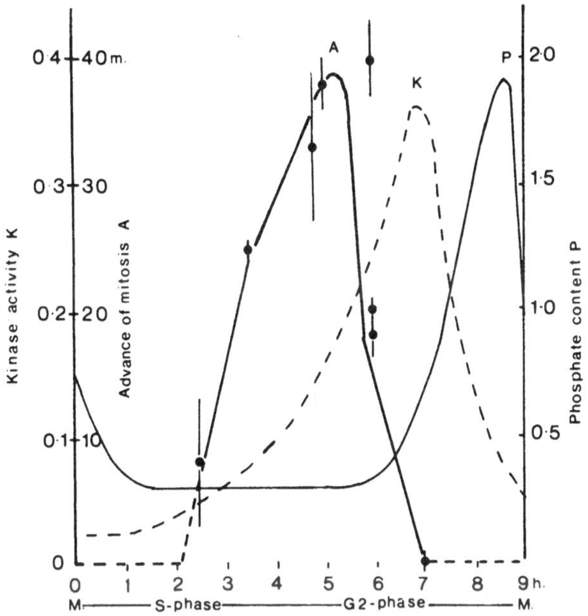

Figure 15. Control of mitosis by histone kinase (96). The thin solid line shows the change in histone H1 phosphate content through the Physarum cell-cycle. The broken line shows the change in histone kinase through the cycle. The heavy solid line shows the advance of mitosis produced by adding histone kinase at various times in the cycle.

ACETYLATION OF HISTONE H4

Incorporation of labelled acetate.

Histone H4 is modified after synthesis by acetylation of the N-terminal lysines in all eukaryotes studied. Allfrey and co-workers (102, 103) has suggested this is associated with transcription and Dixon (104) has suggested this is associated with DNA replication. Physarum plasmodia incorporate ^{14}C from 1-^{14}C-acetate into H4. After short pulses, less than 10 min, the label is found only in the acetylated species although after longer pulses it is found in all species of H4. Amino acid analysis showed that incorporation after longer pulses included incorporation into several unmodified amino acids (105). The different acetylated forms of H4 can be separated by gel electrophoresis. Figure 16 shows incorporation of ^{3}H-lysine and ^{14}C-acetate into the different forms of Physarum H4 as separated by gel electrophoresis. There is a number of minor bands preventing accurate measurements on gels containing total histone so subsequent experiments have been carried out with H4 purified by chromatography. The gel electro-

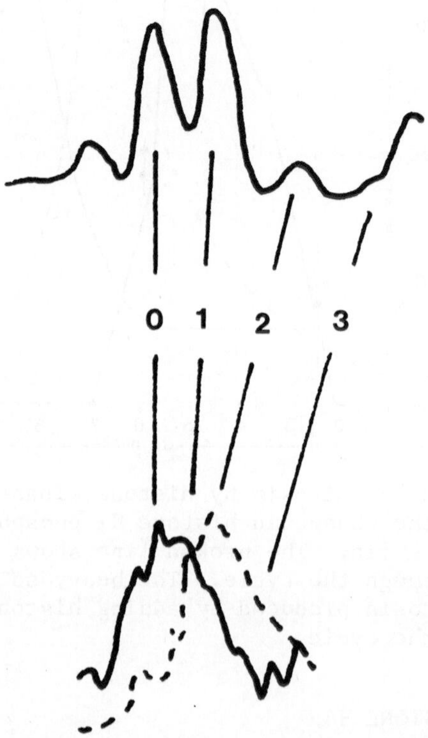

Figure 16. Incorporation of ^{14}C-acetate into Physarum H4 histone (105). Total Physarum histone was fractionated by gel electrophoresis. The part of the gel containing H4 was stained and scanned, and a duplicate gel was slices and radioactivity determined. The upper trace shows the scan of the stained gel, electrophoresis is from right to left. The lower traces show radioactivity incorporated, solid line for ^{3}H-lysine, dotted line for ^{14}C-acetate. After some small corrections (Robinson and Matthews, unpublished) the relative net incorporation of ^{14}C-acetate into bands 0 to 3 was 0, 1.0, 2.2, 2.8, ($^{14}C/^{3}H$). Other bands on the gel overlap the H4 region so subsequent work was carried out with pre-purified H4 (106).

phoresis conditions can then be optimized for H4 fractionation and good separation obtained (106). Preliminary results from pulse labelling with ^{3}H-acetate show a high turn-over of acetate in the diacetyl form, ac_2H4, during S phase and a high turn-over, relative to the number of molecules, of acetate in the tetra-acetyl form, ac_4H4 in G2 phase (107).

Cell-cycle changes in H4 acetate.

Figure 17 shows an example of the separation of *Physarum* H4 into the 5 forms, ac_nH4 (n = 0, 1, 2, 3, 4), by gel electrophoresis and illustrates the computer procedure that determines the proportion of H4 in each form, fully corrected for overlapping of the peaks. Histone H4 was isolated from *Physarum* plasmodia at various stages of the cell-cycle and analyzed. Figure 18A and 18B shows the changes in overall level of acetate through the cell-cycle and also the total changes in more highly acetylated forms, ac_nH4 (n = 2, 3, 4), through the eell-cycle. Figure 19 shows the changes in the individual forms of H4, with ac_4H4 shown on an expanded scale (106).

Inverse correlation with H1 phosphate and chromosome condensation.

The most striking feature of Figures 18 and 19 is the sharp drop in acetate content in mitosis. This has also been observed in mammalian cells (108). In *Physarum*, the proportion of ac_4H4 falls almost to zero (0.36% compared with 3.6% in S phase) in prophase, just when the H1 phosphate content reaches its maximum value. Consequently, acetylation of H4 appears to be inversely correlated with H1 phosphorylation. There is not yet any other evidence linking the two histone modifications, nor linking de-acetylation with chromosome condensation. Such evidence may be forthcoming from studies of histone de-acetylase activity. Waterborg (unpublished) in our laboratory has prepared a synthetic substrate for histone H4 de-acetylase by isolating the N-terminal 23 residue peptide from calf thymus H4 and chemically acetylating it with ^{3}H-acetic anhydride. This was suggested to us by V.G. Allfrey. The N-terminal peptide contains lysines only at positions 5, 8, 12 and 16 (and methyl-lysine at position 20) which are the positions acetylated *in vivo* so chemical acetylation of this peptide is limited to the *in vivo* sites. Moreover, this N-terminal region is not thought to be part of the globular protein core of the nucleosome (30) so the isolated peptide may be a good substitute for chromatin as far as assays for de-acetylase are concerned. *Physarum* nuclei possess enzyme(s) that de-acetylate this substrate (Waterborg, unpublished). It will be interesting to see how the cell-cycle dependence of this enzyme and acetyltransferase (109, 110) compares with that of H1 histone kinase.

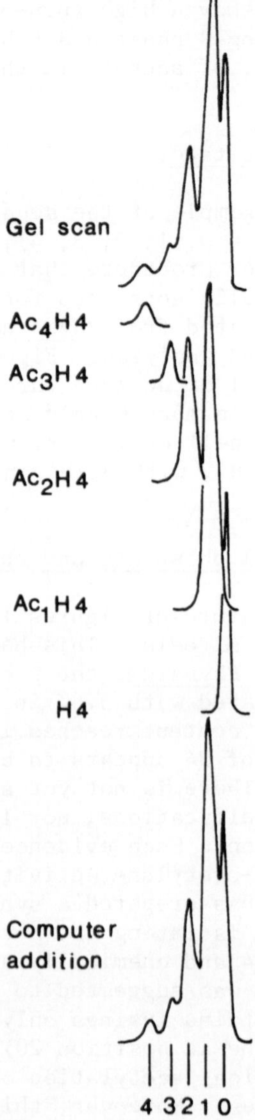

Figure 17. Computer analysis of histone H4 acetylation. Physarum H4 histone from S phase was isolated by chromatography and then fractionated by gel electrophoresis. The top profile is a densitometer scan of the stained gel. Electrophoresis is from left to right. The middle 5 profiles show the individual Gaussian components determined by computer-aided analysis and the bottom profile shows the sum of these components which compares with the experimental scan at the top.

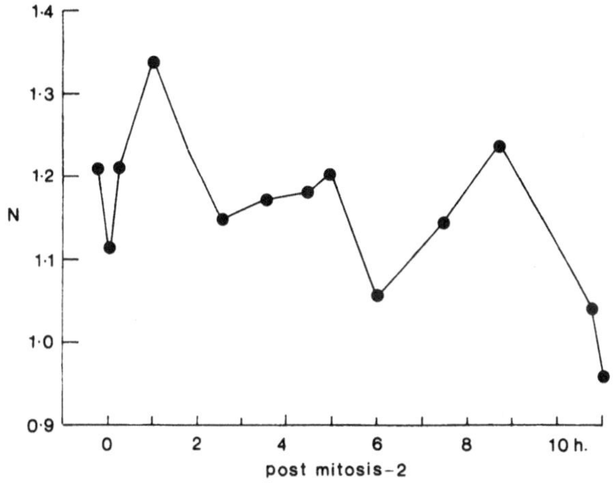

FIGURE 18A. Acetate content of Physarum histone H4 in the cell-cycle. The average number of acetates per H4 molecule was calculated from the data of Chahal et al. (106) as

$$\sum_{n=1}^{4} n.P(n) / \sum_{n=0}^{4} P(n)$$

where P(n) is the precent of total H4 occurring as the species with n acetates. Mitosis occurred at 0 H and again at 11 H. Notice the dip in mitosis and the S phase and G2 phase peaks.

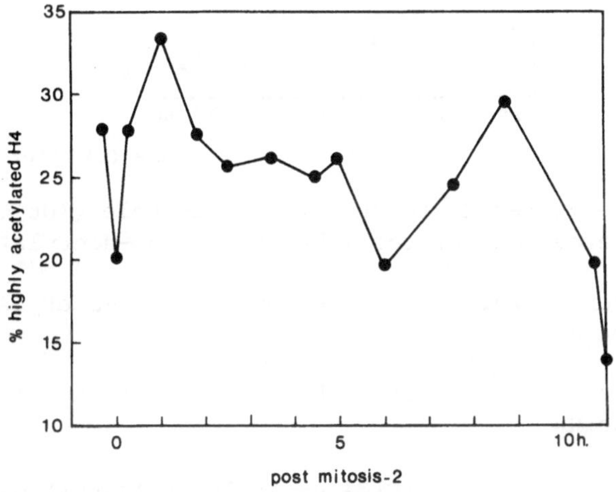

FIGURE 18B. Highly acetylated H4 in the cell-cycle. The percent of H4 in the highly acetylated forms was calculated as:

$$\sum_{n=2}^{4} P(n)$$

where P(n) is the percent of total H4 occurring as the species with n acetates. Notice the dip in mitosis and the S and G2 phase peaks.

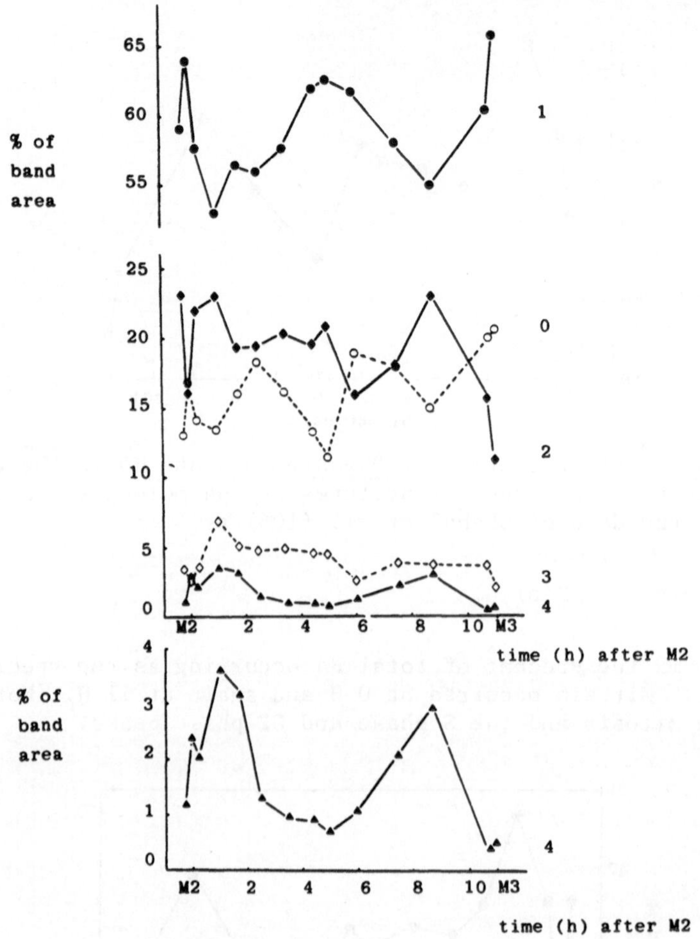

FIGURE 19. Acetylated H4 through the cell-cycle (106).

(a) o -- o Percentage of total H4 in the non-acetylated form.

o — o Percentage of total H4 in the mono-acetylated form, ac_1H4

— Percentage of total H4 in the di-acetylated form, ac_2H4

-- Percentage of total H4 in the tri-acetylated form, ac_3H4

— Percentage of total H4 in the tetra-acetylated form, ac_4H4

(b) is the percentage of ac_4H4 shown on a larger scale.

Correlation of ac_4H4 with transcription.

Transcription in the Physarum plasmodial cell-cycle is biphasic with a minimum in mitosis and a second minimun in early G2 phase (111). There is a variety of evidence that the types of RNA synthesized during S phase and during G2 phase are different and most of the evidence supports the hypothesis that transcription during S phase produces predominantly heterogeneous nuclear RNA (hnRNA) which gives rise to messenger RNA (mRNA) while transcription during G2 phase produces predominantly ribosomal RNA (discussed in Reference 81). The overall acetate content of H4 also shows a biphasic pattern through the cell-cycle (Figure 18) but the maxima and minima are not very marked. However, the amount of ac_4H4 alone does show a very marked biphasic pattern (Figure 19) that is positively correlated with transcription (106). The data suggest that the previously observed correlations of H4 acetylation with transcription (102) reflected specific changes in ac_4H4.

Since G2 phase transcription, in Physarum, is probably concentrated on the genes for ribosomal RNA (rRNA) (112) the level of ac_4H4 on these genes should be normal, or low, in S phase and strikingly high in G2 phase. This can be tested since ribosomal chromatin is localized in nucleoli and, in one experiment so far, has been found to be correct (107).

Correlation of ac_2H4 turn-over with DNA replication.

If ac_4H4 is correlated with transcription where does this leave Dixon's proposal that H4 acetate is associated with chromosome replication? Dixon's (104, 113) proposal involved turn-over of acetate groups on H4 so the effects might be difficult to see when looking at acetate content. However, pulse labelling with ^3H-acetate shows high turn-over of ac_2H4 specifically in S phase and the acetate content data are consistent with this. These results correlate with data from duck erythroid cells (114) and from hepatoma cell cultures (67) which suggest that H4 enters the nucleus in the di-acetyl form and is then de-acetylated after binding to DNA during chromosome replication.

NON-HISTONE CHROMOSOMAL PROTEINS

HMG proteins.

Many eukaryotes possess a small group of proteins bound to chromatin by weak ionic linkages. They can be isolated by washing nuclei or chromatin with 0.35 NaCl and fractionating the extract by precipitation with trichloracetic acid. Several HMG proteins have been extensively characterized (115) and they are of particular interest because they are released from chromatin by mild digestion with DNase-1 (116) in parallel with the digestion of active genes and it has been reported that HMG proteins are necessary for the expression of DNase-1 sensitivity by active genes (117, 118). Physarum contains a group of HMG-like proteins (119) but they have not yet been extensively characterized.

Chromosome structural transitions.

During the mitotic cycle in Physarum, as in other eukaryotes, there are at least 4 different metastable chromosome structures: metaphase chromosomes; replicating chromatin; non-replicating chromatin active in transcription (maybe approximately equated to euchromatin); and non-replicating chromatin inactive in transcription (maybe approximately equated to heterochromatin) (120). Replicating chromatin passes through a stage where it has incomplete nucleosomes and an incomplete complement of nucleosomes. Metaphase chromosomes and non-replicating chromatin both have a nucleosome structure, as far as micrococcal nuclease digestion goes (121, 122) but actively transcribing chromatin has a more open nucleosome structure as judged by its nuclease sensitivity (123-125) and tendency to be isolated as "peak A" in gradients containing nucleosomes (123, 126). Transitions between these structures probably involve non-histone proteins such as polymerases, "scaffold" proteins (127, 128) and HMG proteins (115) but the foregoing data on histone modification suggest it may be a major factor in initiating the structural transition to metaphase chromosomes and be closely involved in all the other transitions. Chahal et al. (106) have proposed a scheme summarizing the involvement of histone modification in chromosome structural transitions. A slightly modified scheme is shown in Figure 20 which brings out the probable involvement of ac_4H4 in transcription and ac_2H4 in replication. A major feature of the scheme is the central role of the "30 nm" coil of nucleosomes in which the histones are essentially unmodified. This "30 nm" coil can be further coiled or condensed by phosphorylation of H1. Conversely, it can be opened out for DNA processing by acetylation of H4. Further work is necessary to characterize the different structural states in more detail and to identify the other factors involved in the structural transitions.

REFERENCES

1. D.R. Hewish and L.A. Burgoyne, Chromatin sub-structure. The digestion of chromatin DNA at regularly spaced sites by a nuclear deoxyribonuclease, Biochem. Biophys. Res. Comm. 52:504-510 (1973)
2. L.A. Burgoyne, D.R. Hewish and J. Mobbs, Mammalian chromatin substructure studies with the calcium-magnesium endonuclease and two dimensional polyacramide gel electrophoresis, Biochem. J. 143:67-72 (1974).

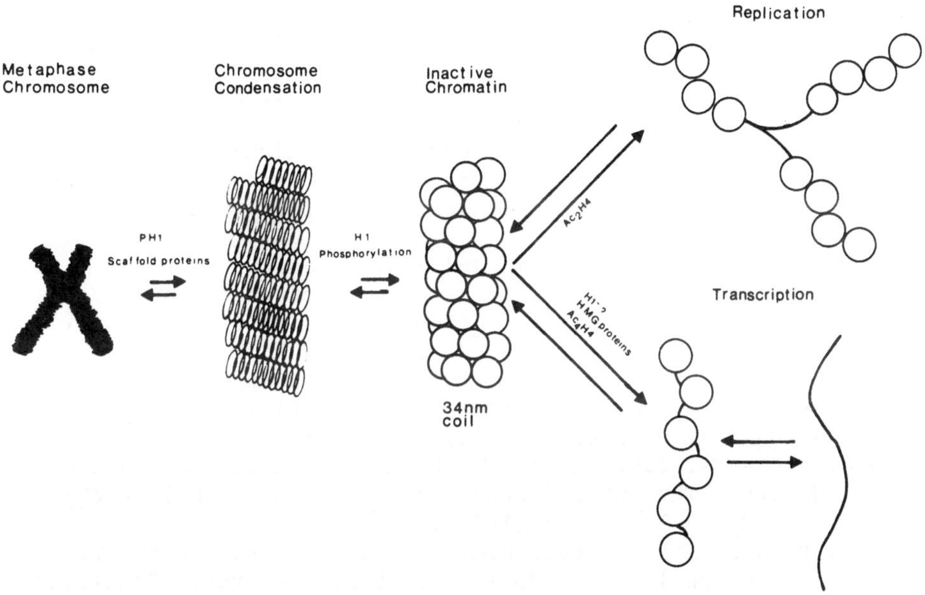

FIGURE 20. Chromosome structural transitions. The various structural states are shown in diagrammatic form only.

3. J.D. McGhee and G. Felsenfeld, Nucleosome structure, Ann. Rev. Biochem. 49:1115-1156 (1980).
4. J.L. Compton, M. Bellard and P. Chambon, Biochemical evidence of variability in the DNA repeat length in the chromatin of higher eukaryotes, Proc. Nat. Acad. Sci. U.S.A. 73:4382-4386 (1976).
5. M. Noll and R. Kornberg, Action of micrococcal nuclease on chromatin and the location of histone H1, J. Mol. Biol. 109:393-404 (1977).
6. K.E. Van Holde and I. Isenberg, Nucleosome chromatin structure, Accts. Chem. Res. 8:327 (1975).
7. L.C. Lutter, Precise location of DNase I cutting sites in the nucleosome core determined by high resolution gel electrophoresis, Nucleic Acids Res. 6:41-56 (1979).
8. E.M. Bradbury, J.P. Baldwin, B. Carpenter, R.P. Hjelm, R. Hancock and K. Ibel, Neutron scattering studies of chromatin, in: "Brookhaven Symp. Biol. 27 IV" B.P. Schoenborn, ed., pp. 97-117 (1975).
9. P. Suau, G. Kneale, G.W. Braddock, J.P. Baldwin and E.M. Bradbury, A low resolution model for the chromatin core particle by neutron scattering, Nucleic Acids Res. 4:3769-3786 (1977).
10. G.W. Braddock, J.P. Baldwin and E.M. Bradbury, Neutron scattering studies of the structure of chromatin core particles in solution, Biopolymers (in press, 1980).
11. J.F. Pardon, D.L. Worcester, J.C. Wooley, K. Tatchell, K.E. Van Holde and B.M. Richards, Low angle neutron scattering from chromatin subunit particles, Nucleic Acids Res. 2:2163-2176 (1975).
12. B.M. Richards, J.F. Pardon, D. Lilley, R. Cotter and J.C. Wooley, The sub-structure of nucleosomes, Cell Biol. Int. Rep. 1:107-116 (1977).
13. J.F. Pardon, R.I. Cotter, D.M.C. Lilley, D. L. Worcester, R.M. Campbell, J.C. Wooley and B.M. Richards, Scattering studies of chromatin subunits, Cold Spring Harbor Symp. Quant. Biol., XLII:11-23 (1978).
14. J.T. Finch, L.C. Lutter, D. Rhodes, R.S. Brown, B. Rushton, M. Levitt and A. Klug, Structure of nucleosome core particles of chromatin, Nature, 269:29-36 (1977).
15. J.T. Finch and A. Klug, X-ray and electron microscope analyses of crystals of nucleosome cores, Cold Spring Harbor Symp. Quant. Biol., XLII:1-11 (1978).
16. G.G. Kneale, J.P. Baldwin and E.M. Bradbury, Neutron scattering of biological macromolecules in solution, Quant. Rev. Biophys. 10:485-527 (1977).
17. R.P. Hjelm, G.G. Kneale, P. Suau, J.P. Baldwin, E.M. Bradbury and K. Ibel, Small angle neutron scattering studies of chromatin subunits in solution, Cell 10:139-151 (1977).
18. E.M. Johnson, V.C. Littau, V.G. Allfrey, E.M. Bradbury and H.R. Matthews, The sub-unit structure of chromatin from

Physarum polycephalum, Nuclei Acids Res. 3:3313-3329 (1976).
19. D. Lohr, J. Cordieu, K. Tachell, R.T. Kovak and K.E. Van Holde, A comparative subunit structure of HeLa yeast and chicken erythrocyte chromatin, Proc. Natl. Acad. Sci. U.S.A. 74:79-83 (1977).
20. H. Ris, Chromosomal structure as seen by electron microscopy, Ciba Found. Symp. 28:7 (1975).
21. P. Suau, E.M. Bradbury and J.P. Baldwin, Higher order structures of chromatin in solution, Euro. J. Biochem. 97:593-602 (1979).
22. L. Klevan, M. Hogan, N. Dattagupta and D.M. Crothers, Electric dichroism studies of the size and shape of nucleosome particles, Cold Spring Harbor Symp. Quant. Biol., XLII: 207-214 (1978).
23. J.D. McGhee, D.C. Rau, E. Charney and G. Felsenfeld, Orientation of the nucleosome within the higher order structure of chromatin, Cell (in press, 1980).
24. F. Thoma, Th. Koller and A. Klug, Involvement of histone H1 in the organization of the nucleosome and of the salt-dependent superstructure of chromatin, J. Cell Biol. 83:403-427 (1979).
25. E.M. Bradbury, G.E. Chapman, S.E. Danby, P.G. Hartman and P.L. Riches, Studies on the role and mode of operation of the very-lysine rich histone H1 (F1) in eukaryote chromatin. The properties of the N-terminal and C-terminal halves of histone H1, Eur. J. Biochem. 57: 521-528 (1975).
26. G.E. Chapman, P.G. Hartman and E.M. Bradbury, Studies on the role and mode of operation of the very-lysine-rich histone H1 in eukaryote chromatin. The isolation of the globular and non-globular regions of the histone H1 molecule. Eur. J. Biochem. 61:69-75 (1976).
27. P.G. Hartman, G.E. Chapman, T. Moss and E.M. Bradbury, Studies on the role and mode of operation of the very-lysine-rich histone H1 in eukaryote chromatin. The three structural regions of the histone H1 molecule, Eur. J. Biochem. 77:45-51 (1977).
28. G.E. Chapman, P.G. Hariman, P.D. Cary, E.M. Bradbury and D.R. Lee, A nuclear-magnetic-resonance study of the globular structure of the H1 histone, Eur. J. Biochem. 86:35-44 (1978).
29. C. Crane-Robinson, personal communication, (1980).
30. P.D. Carey, T. Moss and E.M. Bradbury, High resolution proton magnetic resonance studies of chromatin core particles, Eur. J. Biochem., 89:475-482 (1979).
31. J. Whitlock and R.T. Simpson, Localization of the sites along the nucleosome DNA which interact with NH-terminal histone regions, J. Biol. Chem. 252:6516-6520 (1977).
32. R.T. Simpson, J.P. Whitlock, M. Binn-Stein and A. Stein, Histone-DNA interactions in chromatin core particles, Cold Spring Harbor Symp. Quant. Biol., 42:127-136 (1977).

33. B.G. Carpenter, J.P. Baldwin, E.M. Bradbury and K. Ibel, Organization of subunits in chromatin, Nucleic Acids Res. 3:1739-1746 (1976).
34. J.P. Baldwin, B.G. Carpenter, D. Nixon and E.M. Bradbury, unpublished, (1980).
35. M. Renz, P. Nehlis and p. Mozier, Histone H1 involvement in the structure of the chromatin fiber, Cold Spring Harbor Symp. Quant. Biol., XLII:245:252 (1977).
36. A.D. Olins, Nu-bodies are close-packed in chromatin fibers, Cold Spring Harbor Symp. Quant. Biol., XLII:325-329 (1977).
37. C. Von Holt, W.N. Strickland, W.F. Brandt and M. Strickland, More histone structures, FEBS. Lett., 100:201-218 (1979).
38. R. Kornberg and J.O. Thomas, Chromatin Structure: oligomers of the histones, Science 184:865-868 (1974).
39. R.I. Kelly, Isolation of a histone IIb1-IIb2 complex, Biochem. Biophys. Res. Comm., 54:1588-1595 (1973).
40. J.R. D'Anna and I. Isenberg, A histone cross-complexing system, Biochem. 13:4992-4997 (1974).
41. T. Moss, P.D. Cary, C. Crane-Robinson and E.M. Bradbury, Physical studies on the H3/H4 histone tetramer, Biochem. 15:2261-2267 (1976).
42. L. Bohm, H. Hayashi, P.D. Cary, T. Moss, C. Crane-Robinson and E.M. Bradburg, Sites of histone/histone interactions in the H3-H4 complex, Eur. J. Biochem. 77:487-493 (1977).
43. T. Moss, P.D. Carey, B.D. Abercrombie, C. Crane-Robinson and E.M. Bradbury, A pH-dependent interaction between histone H2A & H2B involving secondary and tertiary folding, Eur. J. Biochem. 71:337-350 (1976).
44. R.J.P. Williams, The conformational properties of proteins in solution, Biol. Rev. 54:389-437 (1979).
45. P.G. Boseley, E.M. Bradbury, G. Butler-Brown, B.G. Carpenter and R.M. Stephens, Physical studies of chromatin. The recombination of histones with DNA, Eur. J. Biochem. 62: 21-31 (1976).
46. T. Moss, R.M. Stephens, C. Crane-Robinson and E.M. Bradbury, A nucleosome-like structure containing DNA and the arginine-rich histones H3 and H4, Nucleic Acids Res. 4:2477-2485 (1977).
47. B. Sollner-Webb, R.D. Camerini-Otero and G. Felsenfeld, Chromatin structure as probed by nucleases and proteases. Evidence for the central role of histones H3 and H4, Cell 9:179-193 (1976).
48. R.D. Camerini-Otero, B. Sollner-Webb and G. Felsenfeld, The organization of histones and DNA in chromatin: evidence for an arginine-rich histone kernal, Cell 8:333-347 (1976).
49. S. Ichimura and M. Zama, The interaction of 8-anilino-1-naphthalenesulfate with polylysine and polyarginine, Biopolymers 16:1449-1464 (1977).
50. S. Ichimura, K. Mita and M. Zama, Conformation of poly (L-arginine): I. Effects of anions, Biopolymers 17:2769-2782 (1978).

51. K. Mita, S. Ichimura and M. Zama, Conformation of poly (L-arginine): II. Complexes with polyanions, Biopolymers 17:2783-2798 (1978).
52. K. Mita, S. Ichimura and M. Zama, Conformation of poly (L-Homoarginine), Biopolymers 19:1123-1135 (1980).
53. E.M. Bradbury, R.J. Inglis, H.R. Matthews and N. Sarner, Histone H1 phosphorylation in Physarum polycephalum correlation with chromosome condensation. Eur. J. Biochem. 33: 131-139 (1973).
54. L.R. Gurley, J.A. D'Anna, S.S. Barham, L.L. Deaven and R.A. Tobey, Histone cells, Eur. J. Biochem. 84:1-16 (1978).
55. L.R. Gurley, R.A. Tobey, P.A. Walters, C.E. Hilderbrand, P.G. Hohman, J.A. D'Anna, S.S. Barham and L.L. Deaven, in: "Cell Cycle Regulation", J.R. Jeter, I.L. Cameron, G.M. Padilla and R.M. Zimmerman, ed., Academic Press, New York (1978).
56. S.G. Fischer and U.K. Laemmli, Cell cycle changes in Physarum polycephalum histone H1 phosphate: relationship to deoxyribonucleic acid binding and chromosome condensation, Biochem. 19:2240-2246 (1980).
57. J. Mohberg and H.P. Rusch, Nuclear histones in Physarum polycephalum during growth and differentiation, Arch. Biochem. Biophys. 138:418-432 (1978).
58. L.R. Gurley, R.A. Walters and R.A. Tobey, Sequential phosphorylation of histone sub-fractions in the Chinese hamster cell cycle, J. Biol. Chem. 250:3936-3944 (1975).
59. R.S. Lake and H.P. Salzman, Occurence and properties of a chromatin-associated F-1 histone phosphokinease in mitotic Chinese hamster cells, Biochem. 11:4817-4826 (1972).
60. R.S. Lake, J.A. Goidl and N.P. Salzman, F1 histone modification at metaphase in Chinese hamster cells, Exp. Cell Res. 73:113-121 (1972).
61. R.K. Lake, Further characterization of the F1 histone phosphokinase of metaphase-arrested animal cells, J. Cell Biol. 58:317-333 (1973).
62. R. Balhorn, O. Riecke and R. Chalkley, Rapid electrophoretic analysis for histone phosphorylation. A reinvestigation for phosphorylation of lysine-rich histone during rate liver regeneration, Biochem. 10:3952-3959 (1971).
63. R. Balhorn, M. Balhorn, H.P. Morris and R. Chalkley, Comparative high-resolution electrophoresis of tumor histones variation in phosphorylation as a function of cell replication rate, Cancer Res. 32:1775-1784 (1972).
64. R. Balhorn, J. Bardwell, L. Sellers, D. ;ranner and R. Chalkley, Histone phosphorylation and DNA synthesis are linked in synchronous cultures of HTC cells, Biochem. Biophys. Res. Comm. 46:1326-1333 (1972).
65. D. Sherod, G. Johnson, R. Balhorn, V. Jackson, R. Chalkley and D. Granner, The phosphorylation region of lysine-rich

histone in dividing cells, Biochem. Biophys. Acta 381:337-347 (1975).
66. V. Jackson, A. Shires, R. Chalkley and D.K. Granner, Studies on highly metabolically active acetylation and phosphorylation of histones, J. Biol. Chem. 250:4856-4863 (1975).
67. V. Jackson, R. Shires, N. Tanphaichitr and R. Chalkley, Modification to histones immediately after synthesis, J. Mol. Biol. 104:471-483 (1976).
68. N. Tanphaichitr, K.C. Moore, D. Granner and R. Chalkley, Relationship between chromosome condensation and metaphase lysine-rich histone phosphorylation, J. Cell Biol. 69: 43-50 (1976).
69. T.A. Langan, Isolation of histone kinases, in: "Methods in Cell Biology" G.S. Stein and J. STein, Ed., Academic Press, New York 19:143-152 (1978).
70. T.A. Langan, Methods for the assessment of site-specific histone phosphorylation, in: "Methods on Cell Biology" G.S. Stein and J. Stein, ed., Academic Press, New York 19:127-142 (1978).
71. T.A. Langan and P. Hohman, Phosphorylation of threonine and serine residues of lysine-rich histone in growing cells, Fed. Proc. 33:1597 (1974).
72. K. Ajiro, T. Borun and L. Cohen, Phosphorylation sites of histone 1 (F1) in relation to the cell cycle, Fed. Proc. 34:581 (1975).
73. T.W. Dolby, K. Ajiro, T. Borun, R.S. Gilmour, A. Zweidler, L. Cohen, P. Miller and C. Nicolini, Physical properties of DNA and chormatin isolated from G1 and S phase HeLa S-3 cells. Effects of histone H1 phosphorylation and stage specific non-histone chromosomal proteins on the molar ellipticity of native and reconstituted nucleoproteins during thermal denaturation, Biochem. 18:1333-1343 (1979).
74. Y. Matsumoto, H. Hasuda, S. Mita, T. Marunouchi and M. Yamada, Evidence for the involvement of H1 histone phosphorylation in chromosome condensation, Nature 284:181-183 (1980).
75. A.J. Adler, B. Shaffhaussen, T.A. Langan and G.D. Fasman, Altered conformational effects of phosphorylated lysine-rich histone (f-1) in f-1 deoxyribonucleic acid complexes. Circular dichroism and immunological studies, Biochem. 10:909-913 (1971).
76. A.J. Adler, T.A. Langan and G.D. Fassman, Complexes of deoxyribonucleic acid with lysine-rich (F1) histone phosphorylated at two separate sites: circular dichroism studies, Arch. Biochem. Biophys. 153:769 (1972).
77. H.W.E. Rattle, T.A. Langan, S.E. Danby and E.M. Bradbury, Studies on the role and mode of operation of the very lysine-rich histones in eukaryote chromatin. Effect of A and B site phosphorylation on the conformation and interaction of histone H1, Eur. J. Biochem. 81:499-505 (1977).

78. T.M. Fasy, A. Inoue, E.M. Johnson and V.G. Allfrey, Phosphorlation of H1 and H5 histones by cyclic AMP dependent protein kinase reduces DNA binding, Biochem. Biophys. Acta 564:322-334 (1979).
79. A.J. louie and G.H. Dixon, Kinetics of phosphorylation of tests histones and their possible role in determining chromosomal structure, Nature New Biol. 243:164-168 (1973).
80. A. Jerzmanowski and K. Staron, Mg as a trigger of condensation-decondensation transition of chromatin during mitosis, J. Theoret. Biol. 82:41-46 (1980).
81. H.R. Matthews, Chromatin proteins and progress through the cell cycle, in "The Cell Cycle", P. John, ed., Cambridge University Press, England (1980).
82. H.R. Matthews and E.M. Bradbury, The role of histone H1 phosphorylation in the cell cycle: turbidity studies of H1-DNA interaction, Exp. Cell Res. 111:343-351 (1978).
83. S. Corbett, E.M. Bradbury and H.R. Matthews, Histone H1 from prophase aggregates DNA better than histone H1 from S phase, Exp. Cell Res. 128:127-132 (1980).
84. E.M. Bradbury, R.J. Inglis and H.R. Matthews, Control of cell division by very lysine-rich histone phosphorylation, Nature 247:257-261 (1974).
85. K. Mitchelson, T. chambers, E.M. Bradbury and H.R. Matthews, Activation of histone kinase in G2 phase of the cell cycle in Physarum polycephalum. FEBS. Lett. 92:339-342 (1978).
86. H.R. Matthews, Chromosome condensation in mitosis, J. Theoret. Biol. 83:367-368 (1980).
87. T. Chambers, "Physarum histone kinase", Ph.D. Thesis, CNAA, Portsmouth, England (1980).
88. D.G. Hardie, H.R. Matthews and E.M. Bradbury, Cell-cycle dependence of two nuclear histone kinase enzyme activities, Eur. J. Biochem. 66:37-42 (1976).
89. J. Schlepper and R. Knippers, Nuclear protein kinases from murine cells, Eur. J. Biochem. 60:209-220 (1975).
90. T.A. Langan, Phosphorylation of liver histone following the administration of glucagon and insulin, Proc. Nat. Acad. Sci. USA, 64:1276-1283 (1969).
91. J.J. Tyson, G. Carcia-Herdugo and W. Sachsenmaier, Control of nuclear division in Physarum polycephalum. Comparison of cycloheximide pulse treatment, UV irradiation and heat shock, Exp. Cell Res. 119:87-98 (1979).
92. B. Chin, P.D. Friedrich and I.A. Bernstein, Stimulation of mitosis following fusion of plasmodia in the myxomycete Physarum polycephalum. J. Gen. Microbiol. 71:93-101 (1972).
93. H.P. Rusch, W. Sachsenmaier, K. Behrens and V. Griter, Synchronization of mitosis by the fusion of the plasmodia of Physarum polycephalum, J. Cell Biol. 31:204-209 (1966).
94. E.N. Brewer and H.P. Rusch, Effect of elevated temperature shocks on mitosis and on the initiation of DNA replication in Physarum polycephalum, Exp. Cell Res. 49:79-86 (1968).

95. C.P. Prior, C.R. Cantor, E.M. Johnson and V.G. Allfrey, Incorporation of exogenous pyrene-labeled histone into Physarum chromatin: a system for studying changes in nucleosomes assembled in vivo, Cell 20:597-608 (1980).
96. E.M. Bradbury, R.J. Inglis, H.R. Matthews, and T.A. Langan, Molecular basis of control of mitotic cell division in eukaryotes, Nature 249:553-556 (1974).
97. R.J. Inglis, T.A. Langan, H.R. Matthews, D.G. Hardie and E.M. Bradbury, Advance of mitosis by histone phosphokinase, Exp. Cell Res. 97:418-425 (1976).
98. I.N. Trakht, I.D. Grozdova, N.N. Gulyaev, E.S. Severin and N.V. Gnuchev, Effect of some protein kinases, cyclic nucleotides and specific phosphorylation inhibitors on the onset time of mitosis in the mycomycetes Physarum polyceph, Biokhimiya 45:783-793 (1980).
99. K. Zanker, R.J. Inglis, H.R. Matthews, and E.M. Bradbury, Cross-reacting nuclear non-histone antigens from Physarum polycephalum and Ehrlich ascites cells, Biochem. Soc. Trans. 5:953-957 (1977).
100. H.R. Matthews, D.G. Hardie, R.J. Inglis and E.M. Bradbury, The molecular basis of control of mitotic cell division, in: "The molecular basis of circadian rhythms", J.W. Hastings and H.G. Schweiger, ed., Dahlem Konferenzen, Berlin, pp. 395-408 (1976).
101. W. Sachsenmaier, Control of synchronous nuclear mitosis in Physarum polycephalum, in: "The molecular basis of circadian rhythms" J.W. Hastings and H.G. Schweiger, ed., Dahlem Konferenzen, Berlin, pp. 409-420 (1976).
102. V.G. Allfrey, Post-synthetic modifications of histone structure, in: "Chromatin and Chromosome Structure", Li and Eckhardt, ed., Academic Press, New York, p. 167 (1977).
103. V.G. Allfrey, R.M. Faulkener and A.E. Mirsky, Acetylation and methylation of histones and their possible role in the regulation of RNA synthesis, Proc. Nat. Acad. Sci. USA 51:786-794 (1964).
104. M.T. Sung and G.H. Dixon, Modification of histones during spermiogenesis in trout: A molecular mechanism for altering histone binding to DNA, Proc. Nat. Acad. Sci. USA, 67: 1616-1623 (1970).
105. S. Corbett, S. Miller, V.J. Robinson, H.R. Matthews and E.M. Bradbury, Physarum polycephalum histones, Biochem. Soc. Trans. 5:943-946 ((1977).
106. S.S. Chahal, H.R. Matthews and E.M. Bradbury, Acetylation of histone H4 and its role in chromatin structure and functions, Nature 287:76-79 (1980).
107. S.S. Chahal, H.R. Matthews and E.M. Bradbury, unpublished.
108. J.A. D'Anna, R.A. Tobey, S.S. Barham and L.R. Gurley, A reduction in the degree of H4 acetylation during mitosis in Chinese hamster cells, Biochem. Biophys. Res. Comm. 77:187-202 (1977).

109. I. Sures and D. Gallwitz, Histone-specific acetyltransferases from calf thymus. Isolation, properties and substrate specificity of three different enzymes, Biochem. 19:943-951 (1980).
110. D.G. Cousens and D. Gallwitz, Different accessibilities in chromatin to histone acetylase, J. Biol. Chem. 254:1716-1723 (1979).
111. C. Mittermeyer, R. Braun and H.P. Rusch, RNA synthesis in the mitotic cycle of Physarum polycephalum, Biochem. Biophys. Acta 91:399-405 (1964).
112. W.D. Grant, The effects of alpha-amanitin and (NH4)2804 on RNA synthesis in nuclei and nucleoli isolated from Physarum polycephalum at different times during the cell cycle, Eur. J. Biochem. 2:94-98 (1972).
113. A.J. Louie, E.P.M. Candido and G.H. Dixon, Enzymatic modifications and their possible role in regulating the binding of basic proteins to DNA and controlling chromosome structure, Cold Spring Harbor Symp. Quant. Biol. 38:803-319 (1973).
114. A. Ruiz-Carillo, L.J. Haugh and V.G. Allfrey, Processing of newly synthesized histone molecules: nascent histone H4 chains are reversibly phosphorylated and acetylated, Science 190:117-128 (1975).
115. G.H. Goodwin, J.M. Walker and E.W. Johns, The high mobility group (HMG) chromosomal non-histone proteins, in: "The cell nucleus IV A," H. Busch Ed., Academic Press, New York (1978).
116. B. Levy-W, N.C.W. Wong and C.H. Dixon, Selective association of the trout-specific H6 protein with chromatin regions susceptible to DNaseII: possible location of HMG-T in the spacer region between core nucleosomes, Proc. Nat. Acad. Sci, USA 74:2810-2814 (1977).
117. S. Weisbrod and H. Weintraub, Isolation of a sub class of nuclear proteins responsible for conferring a DNase I-sensitive structure on globin chromatin, Proc. Nat. Acad. Sci. USA, 76:630-635 (1979).
118. S. Weisbrod, M. Groudine and H. Weintraub, Interaction of HMG 14 and 17 with actively transcribed genes, Cell 19:289-302 (1980).
119. H.R. Matthews, S.S. Chahal, S. Miller, R.J. Inglis and E.M. Bradbury, Physarum chromatin, in: "Current research on Physarum" W. Sachsenmaier ed., University of Innsbruch, Austria, pp. 51-58 (1979).
120. N. Maclean, E.M. Bradbury and H.R. Matthews, "Chromosome structure and function" Blackwells, Oxford, England, in press (1980).
121. E.M. Johnson, V.C. Littau, V.G. Allfrey, E.M. Bradbury and H.R. Matthews, The sub-unit structure of chromatin from Physarum polycephalum, Nucleic Acids Res. 3:3313-3329 (1976).
122. V. Vogt and R. Braun, Repeated structure of chromatin in metaphase nuclei of Physarum, FEBS. Lett. 64:190-192 (1976).

123. E.M. Johnson, V.G. Allfrey, E.M. Bradbury and H.R. Matthews, Altered nucleosome structure containing DNA sequences complementary to 19S and 26S ribosomal RNA in Physarum polycephalum, Proc. Nat. Acad. Sci. USA, 75:1116-1120 (1978).
124. J. Stalder, T. Seebeck and R. Braun, Degradation of the ribosomal genes by DNase I in Physarum polycephalum, Eur. J. Biochem. 90:391-395 (1978).
125. J. Stalder, T. Seebeck and R. Braun, Accessibility of the ribosomal genes to micrococcal nuclease in Physarum polycephalum, Biochem. Biophys. Acta 561:452-463 (1979).
126. E.M. Johnson, H.R. Matthews, V.C. Littau, L. Lothstein, E.M. Bradbury and V.G. Allfrey, The structure of chromatin containing DNA complementary to 19S and 26S ribosomal RNA in active and inactive stages of Physarum polycephalum, Arch. Biochem. Biophys. 191:537-550 (1978).
127. K.W. Adolph, S.M. Cheng and U.K. Laemmli, Role of non-histone proteins in metaphase chromosome structure, Cell 12:805-816 (1977).
128. J.R. Paulson and U.K. Laemmli, The structure of histone-depleted metaphase chromosomes, Cell 12:817-828 (1977).

HISTONES OF TRANSCRIPTIONALLY ACTIVE AND INACTIVE CHROMATIN OF MOUSE CELLS

Franco Gabrielli[+] and Ronald Hancock*

[+]Istituto di Chimica Biologica
Universita di Pisa
Pisa, Italy

*Institut Suisse de recherches
experimentales sur le Cancer
1066-Epalinges, Switzerland

INTRODUCTION

Transcriptionally active chromatin has properties different from those of inactive chromatin, which have been exploited to separate these two fractions of the genome (1,2). The most widely used fractionation procedures are based on the selective sensitivity of transcriptionally active chromatin towards nucleases (1,2). By these fractionation procedures, preparations enriched in transcriptionally active chromatin have been obtained and their properties and macromolecular constituents analysed (1,2,9,10).

We report here data obtained by two-dimensional electrophoretic analysis of proteins associated with transcriptionally active and inactive chromatin fractions of mouse cells. Polynucleosomal chromatin was fractionated by a non-enzymatic method which separates transcriptionally active from inactive chromatin on the basis of their different partition in a two-phase aqueous polymer system (3). We find that transcriptionally active chromatin has a higher content of non-histone proteins, a reduced amount of H1, and a different spectrum of H3 variants than inactive chromatin. These observations are consistent with others obtained transcriptionally active chromatin prepared from other cell types by other fractionation procedures (1,2 and references therein).

MATERIALS AND METHODS

P815 mouse cells (4) were grown with 3H-lysine (2-10 uC/ml) or ^{3}H-tryptophan (6.6 uC/ml) for 22-26 hours. Chromatin was prepared (4) and fractionated by partition in two-phase aqueous polymer systems (3). Proteins were extracted from chromatin by phenol (5), precipitated by addition of 6 volumes of cold acetone, washed 3 times with cold acetone, and resuspended in 0.8 M urea, 10% mercaptoethanol and stored at -70°C. Histones and other acid-soluble proteins were extracted from total chromatin proteins by the following procedure: the urea-B-mercaptoethanol solution was made to 0.4 N H_2SO_4 and then dialyzed for 4 h at 0-4 ^{o}C against 1000 volumes of 0.4 N H_2So_4. Acid insoluble proteins were then removed by centrifugation at 10.000 x g for 30 min and acid-soluble proteins precipitated from the supernatant by addition of 10 volumes of cold acetone. The acid-soluble proteins were dissolved in urea-B-mercaptoethanol and stored at -70°C. Histones were obtained from total chromatin by direct extraction with 0.4 N H_2SO_4.

Proteins were analyzed using two-dimensional polyacrylamide gel electrophoresis (6); the first dimension gel was modified to contain 15% acrylamide, 0.08% bis-acrylamide, 0.36% triton N-101 (Sigma) and 7.5 M urea. Proteins were stained with 0.1% Napthol Blue Black in 45% methanol and 10% acetic acid, and then separated in the 2nd dimension, using a sodium dodecyl sulphate polyacrylamide gel (7) with a 15% acrylamide separating gel. Gels were finally stained with Napthol Blue Black. Radioactive proteins were detected by flurography of dried gels (8). For quantitation of labelled histones, the appropriate areas of gel were excised and extracted with 0.2 M ammonium carbonate buffer (pH 8.5) containing 0.1% SDS and 20 ug/ml of Pronase. Aquasol (New England Nuclear) was added, and radioactivity measured in a liquid scintillation counter.

RESULTS

The acid-soluble proteins of total (unfractionated) chromatin from cells grown with 3H-lysine, analysed by two-dimensional electrophoresis and flurography, are predominantly histones (Fig. 1a). After very long exposure times, other minor histone components, HMG proteins, and other unidentified non-histone proteins become detectable (Fig. 1b).

<u>Proteins of transcriptionally active and inactive chromatin</u>

Chromatin was fractionated into transcriptionally active and inactive fractions by partition in a two-phase polymer aqueous system (3,12) after gentle mechanical shearing. The separation was monitored by following the distribution of nascent RNA chains associated with the transcriptionally active fraction. Among other properties of this

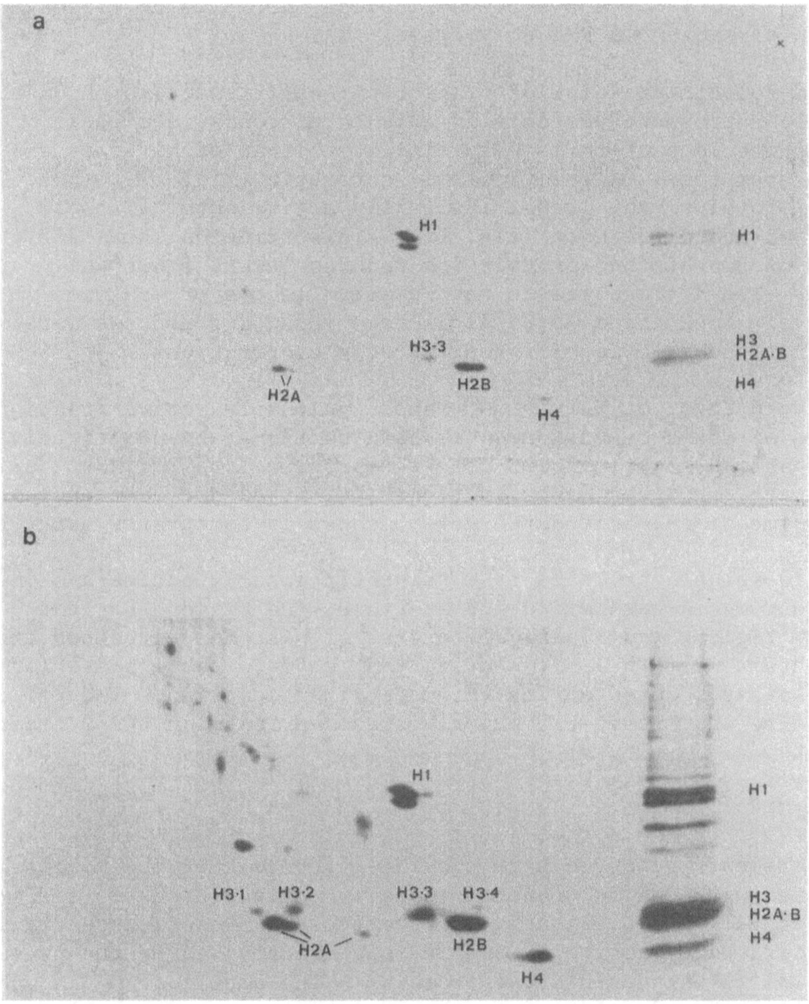

Fig. 1. Fluorogram of a two-dimensional separation of 3H-lysine-labelled histones from total chromatin of P815 mouse cells.

In this and following figures, the 1st dimension (left to right) is in triton-urea, and the 2nd dimension (top to bottom), in SDS. The right side of the fluorograms shows the same proteins separated in the SDS gel used for the 2nd dimension; histones were identified by their migration in this SDS gel. Calf-thymus histones analyzed in an identical two-dimensional systems were used as reference markers (not shown). Variants were numbered so that variant 1 shows the slowest migration in triton-urea.

Exposure times: 1 week (a) and 4 weeks (b).

fraction (3, 12), it shows transcriptional activity in vitro without addition of exogenous RNA polymerase (13).

The acid-soluble proteins of transcriptionally inactive chromatin (fig. 2b) correspond essestially only to histones, and their electrophoretic pattern is very similar to that of histones extracted from total (unfractionated) chromatin (fig. 1). The acid-soluble proteins from transcriptionally active chromatin show several marked difference (fig. 2a). This fraction shows a high content of non-histone proteins, a reduced level of both H1 variants, and differences in the relative levels of proteins tentatively identified as H3 variants. These differences were highly reproducible in different protein preparations.

The reduced level of H1 in the transcriptionally active fraction relative of the core histones has been confirmed by quantitative determinations (14). In SDS gels the protein labelled H3-4 comigrates with the other H3subfractions, but it cannot be detected in one dimensional tritonurea gels because it comigrates with H2B.

When acid-soluble proteins from transcriptionally active and inactive chromatin were mixed and analyzed, the histone variant pattern (fig. 3) was identical to that of histones extracted from total chromatin (fig. 1b), showing that no detectable losses or modifications occured during chromatin fractionation. The gel in fig. 3 also shows that all major histone variants in the 2 fractions comigrate.

DISCUSSION

The experiments reported here provide evidence that the histones associated with DNA in transcriptionally active chromatin differ quantitatively and qualitatively from those in inactive chromatin. Transcriptionally active chromatin is characterized by the presence of many different non-histone proteins. The amount of H1 is substantially reduced relative to the core histones. Our analyses also show a quantitative difference in the relative amount of proteins tentatively identified as H3 variants between transcriptionally active and inactive chromatin. In triton-urea gels, the pattern of mouse H3 variants is very similar to that of other mammalian species (11), which have been shown to differ in primary structure (11); thus the mouse H3 variants we observe may also be primary structure variants. Our analyses whow that the protein labelled H3-4 is found exclusively in transcriptionally active chromatin, while H3-2 is found preferentially in inactive chromatin. Since each of these fractions of chromatin is itself heterogeneous with request to H4 variant content, nucleosomal heterogeneity must exist both between and within the transcriptionally active and inactive fractions.

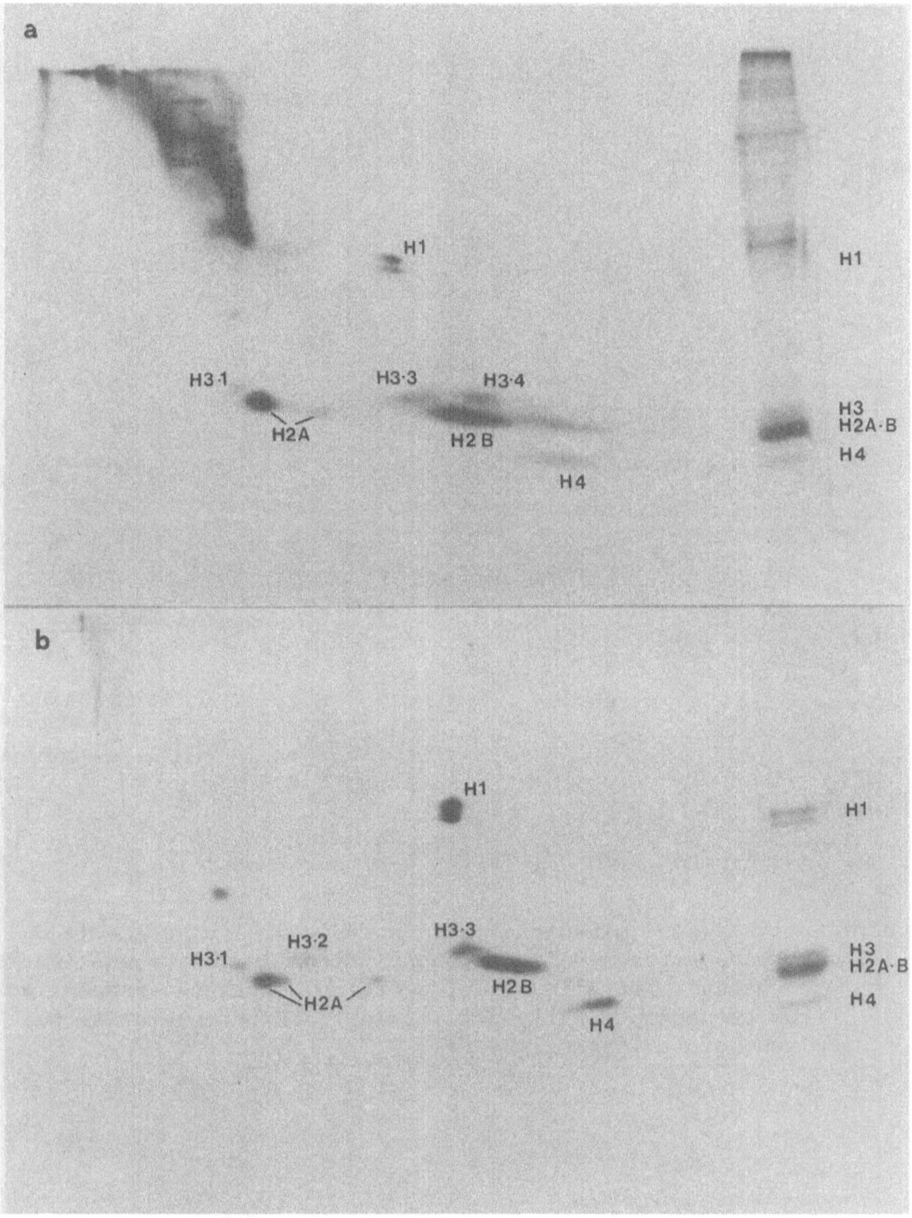

Fig. 2. Acid-soluble 3H-lysine-labelled proteins of transcriptionally active (a) and inactive chromatin (b).

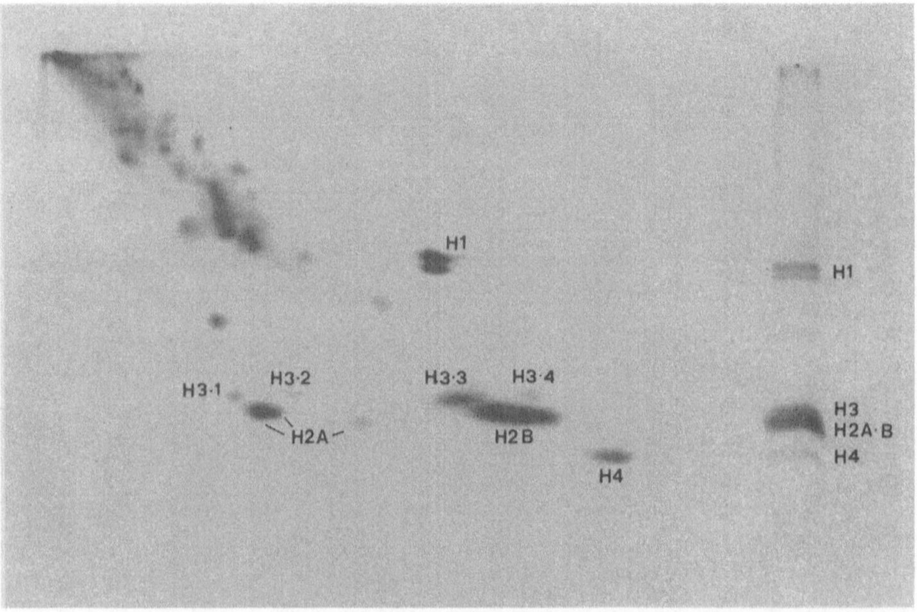

Fig. 3. Analysis of a mixture of acid-soluble 3H-lysine-labelled proteins extracted from transcriptionally active and inactive chromatin. Two aliquots of proteins, extracted respectively from transcriptionally active and inactive chromatin, were mixed before loading the gel.

We believe that the lower content of H1 which we observe in transcriptionally active chromatin is unlikely to be a result of selective losses in the fraction during experimentel manipulation. The strongest evidence for this conclusion comes from quantitative determination of the different histones (14), which shows that the summed quantities of each histone in the two fractions agreed quite precisely with published values for the global histone content of P815 cell chromatin (15).

References

1. T. Pederson (1970). 55, 1-21.
2. D. Mathis, P. Oudet and P. Chambon (1980). Prog. Nucl. Acids Res. and Mol. Biol., 24, 1-100.
3. A.J. Faber (1978). 30, 447-457.
4. R. Hancock, A.J. Faber, S. Fakan (1968). Methods in Cell Biology, 15, 127-147.
5. W.M. LeStourgeon and A.L. Beyer (1975). Methods in Cell Biology 16, 387-406.
6. C. Unger-Ullman and S. Modak (1979). Differentiation 12, 135-144.
7. U.K. Laemmli (1970). Nature 227, 680-685.
8. W.M. Bonner and R.A. Laskey (1974). Eur. J. Biochem, 46, 83-88.
9. D. Levy-Wilson, R.A. Gjerset, B.J. McCarty (1977). Biochim. Biophys. Acta 475, 168-175.
10. J.R. Davie and E.P. Candido (1978). Proc. Natl. Acad. Sci. USA 75, 3574-3577.
11. S.C. Franklin and A. Zweidler (1977). Nature 266, 278-275
12. A.J. Faber, S. Fakan and R. Hancock (1981). In preparation.
13. A.J. Faber (unpublished)
14. F. Gabrielli and R. Hancock (1981). In preparation
15. S. C. A-lright, P.P. Nelson and W.T. Garrard (1979), J. Biol. Chem. 254. 1065-1073.

This work was supported by the Swiss National Scientific Foundation and by an EMBO Fellowship to F. Gabrielli.

MINOR COMPONENTS OF THE CHROMATIN AND THEIR ROLE IN THE RELEASE OF TEMPLATE RESTRICTION

F.A. Manzoli, S. Capitani and N.M. Maraldi

Institutes of Human Anatomy,
Universities of Bologna and Ferrara
Institute of Histology and General Embryology
University of Ancona (Italy)

INTRODUCTION

Many aspects of the expression of the genome informational content have been clarified by using cell-free systems. However, the nucleus is far from being a container of a collection of molecules; on the contrary, it must be considered as one of the most complex exemples of the close relationship existing between the structural arrangement of macromolecules and their metabolic expression. These correlations can be considered as a window open on the cell metabolism, so that different moments of the cell cycle and of the cell differentiation can be monitored by analyzing the morphological changes which occur inside the nucleus. In this respect the nuclear volume enlargement and the chromatin dispersion are targets of metabolic activation. On the other hand, the transitions which occur at the five levels of organization of the chromatin (1) are not the unique factors affecting the nuclear morphology, which depends also on two other structures, the nuclear envelope (NE) and the nuclear matrix.

The close relationships between the NE and chromatin suggest that any modification which occurs at the NE level should influence the chromatin arrangement and functions (2).

Also the nuclear matrix could control the principal nuclear activities, like the nuclear volume and shape variations (3), the transitions of the chromatin structure, the chromosome assembly (4,5), the initiation of replication and the organization of the splicing of the transcripts (6-8) and the transport of nuclear products (9).

Many substances, like steroid hormones, carcinogens, antigens and antibiotics, are known to influence the gene expression. Among them, nonhistone chromosomal proteins (NHP) are thought to play a central role (10). However, a significative amount of NHP is not primarily associated with DNA and is involved in structural functions, like the binding with RNAs and the formation of the nuclear matrix (11,12), while only in few cases particular NHP classes have been demonstrated to be gene specific (13). Therefore the concept that NHP are responsible of the gene control appears an oversimplification, since a variety of other molecular species are also present into nuclei and can interfere with the genome expression.

MINOR CHROMATIN COMPONENTS

Chromatin is currently defined as made up of DNA, proteins and small amounts of RNAs. However, other components are found more loosely bound to the chromatin, once isolated with the present available methods.

In this section, the main evidences on the presence of minor chromatin components, showing some effects on the genome activities, are reported (Table 1).

Polynucleotides

i) Poly-(ADP-ribose)

Nuclei contain a small polynucleotide, the poly-adenosine diphosphate ribose. This molecule is formed by the enzymic transfer of the ADP-ribose portion of NAD to histones and other nuclear proteins (14,15), through the ADP-ribose polymerase, whose activity is mainly associated to the chromatin (16). It has been shown that the ADP-ribosilation increases the DNA synthesis in HeLa cells, probably owing to a removal of histones, due to the anionic nature of this factor (17).

ii) cRNA

Chromosomal RNA (cRNA), constituted by about 40 nucleotides and with a sedimentation coefficient of 3S, is capable of hybridizing with repetitive DNA sequences (18). cRNA has been suggested to be involved in gene activation (19), though the prevalent association with heterochromatin has been interpreted as a probe of its role in maintaining the structure of condensed chromatin (20,21).

iii) SnRNA

In the nucleus of a variety of cells a class of low molecular weight RNAs (Sn RNA) has been identified, whose size ranges between 80 and 320 nucleotides (22-26). The SnRNAs are nuclear in origin and a high percentage of them is associated with the chromatin, though they can shuttle between nucleus and cytoplasm (27-29). These molecules have been interpreted to affect in some way the in vivo gene transcription, since their amount correlates with the differentiation processes (30).

Proteins and peptides

The detection of particular proteins and peptides as components of the chromatin, not derived from degradation of the proteins of the chromosomal pool, has been reported.

Fibronectin, a serum protein, has been found in chromatin preparation from human fibroblasts (31).

Some peptides have been revealed in different cell types and do not appear products of degradation of histones or other chromosomal proteins. These peptides, called "deprimerones", have inhibitory effect on transcription and are considered as endogenous regulatory molecules, since they are localized in chromatin as one molecule every 200 base pairs of DNA, i.e. every nucleosome (32, 33).

The protein A2A has been found in the chromatin of all species and in all cell types (34,35). This protein

Table 1. The major evidences on the presence of minor components in isolated chromatin or in chromatin fractions are listed.

Component	Source	References
Polynucleotides		
Poly(ADP-ribose)	Hen liver	(14)
	Rat liver	(15,16)
cRNA	All cell types	(18-21)
SnRNA	HeLa cells	(22)
	Mouse myeloma	(23,27)
	Rat liver	(24-26)
	Chick embryo	(27)
Proteins and peptides		
Fibronectin	Human fibroblasts	(31)
Deprimerones	Calf thymus	(32)
	Rat liver	(33)
A2A	All cell types	(34-36)
Carbohydrates		
Glycosaminoglycans	Sea urchin	(39-40)
	Rat liver	(41)
Glycoproteins	HeLa cells	(42)
	Rat liver	(43)
Lipids		
Phospholipids	Calf thymus	(44,45)
	Rat liver	(47,48)
	Human lymphocytes	(49,50)
	Chick embryo	(51)
	Ehrlich ascites	(52,54)
Neutral lipids	Calf thymus	(45)
	Ehrlich ascites	(54)
Cholesterol	Rat liver	(53)

is constituted by histone H2A and ubiquitin, linked by an isopeptide bond. Since A2A is present in chromatin in a smaller amount than histones, it has been suggested that it associates with specialized subset of nucleosomes probably involved in the control of chromatin supercoiling(36).

Carbohydrate-containing molecules

The presence of carbohydrate-containing molecules in nuclei has been widely reported (37,38). Evidences have been provided that these molecules are also associated with chromatin. Chromatin fractions from sea urchin embryo contain hyaluronic acid and a protein-mucopolysaccharide complex, while glycosaminoglycans have been found in rat liver chromatin (39-41). Glycoproteins and glycosaminoglycans have been found also in chromatin preparations from HeLa cells (42). Similarly, PAS-positive bands have been identified in the 0.6 M NaCl extract of chromatin from normal liver and Novikoff hepatoma, though it has not been excluded that some glicoproteins may derive from the nuclear envelope (43).

Lipids

The nuclei of mammalian cells contain lipids, mostly confined in the NE. However various evidences have been provided that lipids are associated with chromatin, and that they play some role in determining its structure and function.

Active chromatin has been reported to contain larger amounts of phospholipids in comparison to repressed chromatin. It has been suggested that chromatin lipids could be in the form of lipoproteins, and that in this form they could play a role in regulating the RNA byosinthesis (44-46).

In rat liver it has been found that chromatin contains various phospholipids, whose amount differs from that of the whole nucleus. The main difference concerns the quantity of acidic phospholipids, which are about two-fold increased in chromatin, while sphyngomyelin is largely reduced (47).

Hepatoma nuclei contain increased amounts of phospholipids compared to normal hepatocytes. After nuclei fractionation, a considerable amount of phospholipids is associated with the chromatin fractions (48).

The presence of phospholipids has been revealed by means of two dimensional TLC in chromatin preparations from normal and chronic lymphocytic leukemia (CLL) lymphocytes (49,50). Only NHP appear bound with phospholipids, whose amounts and relative ratios are different compared to nuclear membranes. Significative variations in CLL chromatin-associated phospholipids have been revealed in comparison to normal B lymphocytes (50), as well as in rapidly proliferating embryonic tissues compared to resting ones (51).

Chromatin-associated phospholipids have also been found in Ehrlich ascites tumor cells in a study on the dietary modifications of nuclear lipids, These alterations concern the fatty acid composition of the nuclear membrane phospholipids, while a small quantity of phospholipid, associated with the chromatin, does not reveal appreciable alterations (52).

Also neutral lipids have been found in chromatin preparations from interphase lymphocytes (45) and cholesterol is present in rat liver chromatin with variable ratios of free to esterified forms according to the circadian rhytm of its biosynthesis (53).

Nuclei and chromatin isolated from Ehrlich ascites tumor cells contain lipids, with a neutral lipids/phospholipids ratio higher than in normal cells (54).

RELEASE OF TEMPLATE RESTRICTION IN CELL-FREE SYSTEMS

In this section the investigations carried out with minor chromatin components thought to be active on intact cells as template restriction releasers, or substances presenting structural analogies, are reviewed. The experimental models employed are different nucleotide-polymerizing systems, chromatin or chromatin fractions and isolated nuclei (Table 2).

Polynucleotides

Poly-(ADP-ribose) has been demonstrated to be effective in releasing template restriction by generating new binding sites for the DNA polymerase in isolated chromatin (55).

SnRNAs appear essential for maintaining the transcriptional activity in reconstituted chromatin (56).

In the attempt of interpreting the mode of action of these molecules, natural and synthetic ribonucleic acids have been tested on isolated chromatin and nuclei. In general, it has been demonstrated that purine homopolymers are the most active in releasing DNA template restrictions, while pyrimidine homopolymers and mixed copolymers are less effective (57). Among natural polyribonucleotides the ribosomal RNAs are strong template activators on soluble chromatin, while total nuclear RNAs are effective on isolated repressed chromatin (57,58).

Carbohydrates

During fertilization and embryonic development,gene activity is affected by charged sulfated polysaccharides, like heparin, which is considered as a key substance in the mechanism of nucleo-cytoplasmic interactions and a trigger of the embryonic development (59-61). In cell-free systems heparin has been demonstrated to increase the DNA synthesis and transcription, both in chromatin and in isolated nuclei. Since heparin inhibits the free RNA polymerase activity on naked templates, it probably activates the gene activities in whole nuclei and chromatin by causing a partial displacement of histones (62-71).

Other acidic carbohydrates, like condroitin sulfates, hyaluronic acid and dextrane sulfate are active in promoting gene activation, though to a smaller extent in comparison with heparin (61). In particular, it has been shown that a protein-mucopolysaccharide complex extracted from sea urchin embryos increases the RNA synthesis in isolated nuclei (40). Similarly, proteoglycans are thought to induce chromatin changes during development. They have been shown to exist in the cytoplasm of embryonic cells, from where they are transported in the nucleus at the onset of gene activation. Their effect, as seen by thermal denaturation, occurs at the level of chromatin structure rather than to be of indirect nature (72).

The evidence that acidic polysaccharides have inhibitory effects on the in vitro RNA synthesis by using naked templates has suggested that these molecules interact only with free and non-engaging RNA polymerase, and have no effect in the middle of RNA synthesis. In fact it has

Table 2. In this Table are reported the principal effects of some factors, evaluated by means of thermal denaturation on DNA and chromatin (Chrom.) and of the <u>in vitro</u> levels of replication and transcription by using naked templates (NT), chromatin (Chrom.) and isolated nuclei (Nu.).

	Thermal stability		Replication			Transcription			References
	DNA	Chrom.	NT	Chrom.	Nu.	NT	Chrom.	Nu.	
Polynucleotides									
Poly (ADP-ribose)				+					(55)
SnRNA							+		(56)
Natural/artificial RNAs			−	+	+				(57,58)
Carbohydrates									
Heparin		−	−	+	+		+	+	(63-66,69-72)
Proteoglycans								+	(40,63)
Artificial hydrocarbon polymers and peptides		−		+	+		+	+	(44,57,74,77)
Polyamines						+	+	+	(78,82,83)
Proteins									
Nucleolar basic protein						+			(85,86)
Deprimerones	+					−	−		(32,33)
Lipids									
CMH								+	(87,88)
Gangliosides	−								(89,90)
*Phospholipids									

+ : increased; − : decreased.

*The effects induced by phospholipids are reported in Table 3.

been shown that the inhibitory effect is partly released when employing chromatin as template, and further reduced by using isolated nuclei (73).

Artificial hydrocarbon polymers and polypeptides

The polyanionic compound NSC 46015 pyran copolymer is active in releasing the template restrictions for DNA synthesis both in intact and membrane deprived nuclei, probably by binding to the cationic nuclear proteins which mask the template (74).

Other artificial acidic hydrocarbon polymers, like polyethylene sulfonate (44), polystyrene sulfonate (75) and polyacrylic acid (76), as well as artificial acidic polypeptides, such as polyaspartic acid and polyglutamic acid have been demonstrated to affect both DNA replication and transcription (57,77).

Polyamines

Polyamines occur in essentially all plants and animals and are also found in DNA and RNA viruses. They increase the RNA synthesis with RNA polymerase isolated from several organisms, by enhancing the binding of the enzyme to DNA templates and displacing the transcripts (78). Polyamines are supposed to play an important role also in DNA replication and cell division, being probably required for the G2 phase progression (79). Their cellular levels increase in various stages of tissue development and embryonic growth, suggesting that they could have a regulatory function in RNA metabolism (80). Polyamines interact with anionic metabolites and bind to nucleic acids and could compete with Mg ions for the binding with nucleotides, playing some regulatory function on the nucleotide-dependent reactions (81). Putrescine stimulates the RNA synthesis of rat liver chromatin directed by E. coli RNA polymerase, by increasing the elongation rate and favouring a more efficient use of the initiation sites (82). Spermine and spermidine increase the RNA synthesis in isolated liver nuclei, chromatin and DNA (83).

Proteins

A series of protein factors, isolated from nuclei or

chromatin fractions, have also been found to affect transcription in cell-free systems. Many of these molecules are NHP and affect either the enzyme or the template (84). However some of these factors are not acidic in nature.

A basic protein has been purified from rat liver nucleoli, which stimulates the RNA synthesis directed by RNA polymerase I (85,86).

On the contrary the 'deprimerones', the small peptides associated with chromatin, stabilize the DNA molecule and inhibit DNA-directed or chromatin-directed RNA polymerase activities (32,33).

Lipids

A lipid factor, cholesteryl-14-methylexadecanoate (CMH), found in different animal tissues, appears involved in the control of transcription. In fact it plays a positive role in RNA polymerase activity, by restoring the function of the purified enzyme after extraction with organic solvents (87,88).

Gangliosides are anionic glycolipids capable of forming complexes with histones (89). This ability is expressed also when basic proteins are bound to DNA, indicating that they can compete with DNA for the association with histones. In fact, the thermal stability of nucleohistones is strongly reduced when they are denatured in the presence of gangliosides (90).

Phospholipids affect DNA and RNA synthesis in a different way, depending on the use of naked templates or isolated nuclei.

a) Naked templates (Table 3)

Phospholipids of different origin behave as necessary factors for the optimal activity of E. coli DNA polymerase III. Crude phospholipids or purified PE and DPG, are equally active in supporting the maximal synthesis by the enzyme. It has been suggested that these lipids provide some structural substitute for the complex in which the holoenzyme is present in the cell (91).

The activity of DNA polymerase I from E. coli has been shown to be sensitive to the presence of various phospholipids. The use of different templates indicates that SM, PE and PC induce a stimulation with native and

Table 3. The effects of various phospholipids on duplication and transcription, by using different naked templates, are reported.

Template	Enzyme	Effect +	Effect none	Effect -	Ref.
DNA polymerase					
RNA-primed øX177 SS DNA	E.coli DNA pol. III	DPG PE,Phos.			(91)
Native and denat. CT DNA	E. coli DNA pol. I	SM,PE	PS		(92)
M-band DNA	S.pneumoniae DNA pol.	Gly.		Phos.	(93)
activated CT DNA	S.pneumoniae DNA pol.	Gly.			(93)
poly d (A-T).d(A-T)	Mouse myeloma DNA pol. C I	PE		CL	
	CII	PE,CL			
	CIII		PE	CL	(94)
activated CT DNA	Calf thymus DNA pol.	DOPC	PE	PS PI,SM,	(96)
$d(pA)_{50}$	Calf thymus TdT	SM	PE	PS,PI	(97)
RNA polymerase					
T4 DNA and CT DNA	E.coli RNA pol.		PE,LPE PC	CL,PG	(98)
CT DNA	Mouse myeloma RNA pol. A	PE,PC		CL,PA	(99)

Abbreviations: CT:(calf thymus); Gly:(Glycolipids); Phos:(Phospholipids); DPG:(Diphosphatidylglycerol); PE:(Phosphatidyl ethanolamine); SM:(sphingomyelin); PS:(Phosphatidylserine); CL:(Cardiolipin); DOPC:(Dioleilphosphatidylcholine); PI:(Phosphatidylinositol); LPE:(Lisophosphatidylethanolamine) PG:(Phosphatidylglycerol); PA:(Phosphatidic acid).

heat-denatured DNA, while PS has almost no effect (92).

Various lipids are capable to affect the DNA polymerase activity associated with the M-band extracted from bacteria. Glycolipids and phospholipids have been tested for their ability to influence the DNA polymerizing activity associated with the DNA-membrane fraction isolated from S. pneumoniae. Glycolipids stimulate the DNA synthesis, while phospholipids have almost no effect (93).

It has been demonstrated that the activity of three fractions of the HMW DNA polymerase from mouse myeloma is enhanced by neutral phospholipids and decreased by negatively charged ones (94). Similarly, the activity of DNA polymerase-α, purified from calf thymus, is modified by phospholipids. Evidence has been provided that the action of PC is greatly dependent upon the presence of Mg ions, stressing the role of the polar interactions between lipids and template (95). With other phospholipids, like PE, PS, SM and PI, the DNA polymerase activity is reduced to a variable extent (96).

Comparable results have been obtained by studying the interaction between lipid vesicles and the terminal transferase-directed dATP elongation. Among the phospholipids employed, the acidic PS and PI produce a deep inhibition of the polymerization rate, while the neutral SM and PE vesicles give rise to a certain stimulation and to a slight inhibitory effect, respectively. Neutral vesicles are supposed to increase the chance of binding between the polymer initiator and the enzyme, while the negatively charged vesicles probably compete with the initiator for the binding with the enzyme (97).

Lipid vesicles obtained by sonication of PE, PG, CL, PC and LPE have been tested for their ability of modifying the incorporation of ATP, directed by RNA polymerase from E. coli, by using purified DNA as template. Neutral vesicles have no effect on the RNA polymerase activity, while PG and CL have an inhibitory action proportional to the concentration used (98).

PC vesicles have been shown to increase the RNA synthesis in both active and repressed chromatin, isolated from interphase lymphocytes. This effect has been interpreted as depending on the capability of the phospholipid to antagonize the histone-DNA association (44).

The activity of RNA polymerase from mouse myeloma is affected by sonicated phospholipids. Millimolar concentrations of PC enhance the enzyme activity up to four-fold in comparison to control, while CL inhibits the transcription, suggesting a different regulative role according to the neutral or acidic nature of the lipid molecule (99).

The common feature of these data is that, when polynucleotides or naked DNAs are used as primer-templates, neutral lipids have a general positive effect on the DNA polymerizing activity of E. coli DNA polymerase I and III, mouse myeloma HMW DNA polymerase CI and CII and calf thymus TdT. In some cases, PE produces almost no effect when working with eukaryotic enzymes, while SM, though it is neutral, inhibits the calf thymus DNA polymerase activity. On the contrary, negatively charged phospholipids give rise to an inhibition of such activities except for PS when using DNA polymerase I from E. Coli, and CL which activates the CII fraction of the HMW DNA polymerase from mouse myeloma. The negatively charged PA, PG and CL are inhibitory also on the activity of RNA polymerase from E. coli, and of RNA polymerase extracted from mouse myeloma, except when working with heat-denatured calf thymus DNA as template. The neutral lipids employed act otherwise as stimulatory factors, except for the activity of the bacterial enzyme, which is almost not affected by PE, LPE and LEC.

b) Isolated nuclei

The RNA synthesis in isolated rat liver nuclei is positively affected by PS vesicles, and inhibited to a variable extent by PI, PC, PE and SM vesicles (Table 4).

The rate and the extent of the RNA synthesis occurring in the presence of PS vesicles is dose dependent and reach the maximum level at 5 mM lipid. Almost no effect is present below 0.1 mM, while inhibition takes place at 50 and 100 mM (Table 5). In the presence of optimum amounts of PS vesicles, the RNA synthesis attains levels considerably higher than the control over at least 150 minutes of incubation (100).

Under the conditions in which PS stimulates the RNA synthesis, also the nuclear morphology is deeply altered with respect to controls. The nuclear volume is enlarged and the chromatin undergoes a structural rearrangement consisting in a diffuse decondensation (Figs. 1-4).

Table 4. DNA-dependent RNA polymerase activity in rat liver isolated nuclei, in the presence of different phospholipid vesicles. The data are the mean of at least three determinations.

Lipid	^{3}H-UMP incorporated (pmole/min/mg DNA)	% with respect to the control
Control	30.1	100
PS	55.4	184
SM	21.7	72
PE	23.8	79
PC	14.1	47
PI	20.5	68

Table 5. Influence of the concentration of the phospholipid vesicles on the RNA synthesis in rat liver isolated nuclei.

PS concentration (mM)	% of incorporated ^{3}H-UMP with respect to control
Control	100
0.0005	97
0.001	94
0.005	96
0.01	98
0.05	101
0.1	103
0.5	98
1.0	142
5.0	207
10.0	198
50.0	79
100.0	61

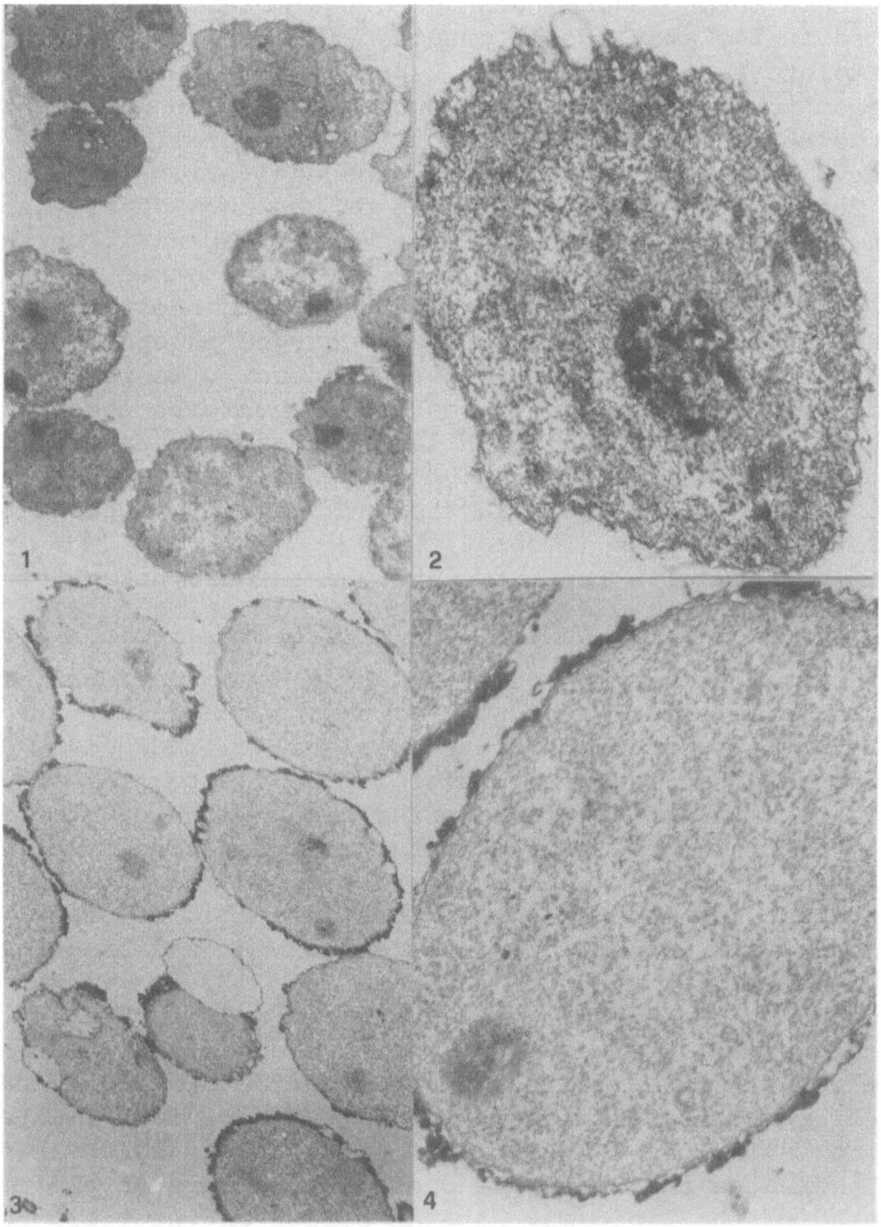

Figs. 1-4. Control rat liver isolated nuclei (1,2) show condensed chromatin and evident nucleoli, while PS treated nuclei dispaly, after two minutes of incubation, a volume enlargement and the diffuse dispersion of the chromatin (3,4).

These morphological changes are similar to that described in isolated nuclei treated with other releasers of template restriction, like heparin and polynucleotides and seem to indicate that the nuclear volume enlargement and the chromatin decondensation are prerequisites for, rather than consequence of, the gene activation (57)

The effects of phospholipids on DNA and RNA polymerizing activities appear therefore quite dependent on the use of a given template. In fact, by using naked templates, neutral lipids generally increase both replication and trancription, while negatively charged phospholipids inhibit these activities. On the contrary, in isolated nuclei, only the negatively charged PS vesicles cause an increase in RNA synthesis, while all the other neutral phospholipids are inhibitory. These effects are in agreement with the variations in thermal stability induced by the different lipids (Table 6).

These results indicate that the anionic nature of some phospholipids may be responsible for the removal of the histones, both through a competition for the same binding sites with the DNA or by inducing conformational changes in the secondary structure of the histones (96).

The decrease of template availability induced by PE, PC and SM on isolated nuclei may be due to the neutral net charge of the lipid polar head, which may decrease their binding affinity for the histone charged residues. Therefore when histones are not removed from DNA, the presence of phospholipids gives rise to a reduced transcription, probably by interfering with the binding of the enzyme, as reported for heparin (71), or by stabilizing the ligand-DNA complex through hydrophobic interactions (51). However the net charge of the lipid vesicle is not sufficient for explaining the different action of the phospholipids in the case of acidic PI, which does not stimulate RNA synthesis in isolated nuclei as PS. This indicates that also the overall structure of the polar head of the lipid may have an effect on the lipid-nucleohistone interaction. In the case of PI, in fact, the presence of an aromatic ring in the polar head may prevent the formation of bindings between the lipid and the histone opposite charges (100).

The possible mode of action of phospholipids on chromatin components can be deduced also by data obtained with different techniques.

Table 6. The prevailing actions of the different lipids, employed at concentrations between 5 and 0.1 mM, are reported.

Lipid	Thermal stability		Replication	Transcription	
	DNA	Nucleohistone	Naked templates	Naked templates	Isolated nuclei
SM	-	-	+	ND	-
PC	-	-	+	+	-
PE	-	-	+	+	-
PS	-	-	-	ND	+
PI	ND	ND	-	ND	-
CL	ND	ND	-	-	ND

+ : increased; - : decreased; ND : not determined

By means of thermal denaturation it has been shown that the conformation of DNA and nucleohistone complexes is affected to a variable extent by cholesterol and phospholipids or their moieties. The general effect of these molecules is a favouring action towards the DNA strand separation, mainly due to the chaotropic effect on the solvent. In the presence of histone-DNA complexes, the phospholipids are likely to affect also the protein secondary structure, inducing a partial removal of histones from the complex (101-106).

Results in good agreement with this model have been obtained by ultrastructural analyses on the interactions among histones, DNA and phospholipids. The affinity between histones and phospholipids, in fact, prevents the binding of DNA with histones and the formation of reconstituted nucleohistone aggregates (107-109).

Therefore it seems likely that phospholipids affect the DNA template availability by:

-exerting a chaotropic effect and disrupting the local organization of water around the double helix, thus favouring the DNA strand separation

-acting as factors which may belong to the holoenzyme complex and which may be lost during the purification of the enzymes

-functioning as counter-ligands, reducing the histone-DNA interactions.

CONCLUSIONS

Many events which induce the onset of cell and tissue differentiation may involve non specific agents, differently distributed and/or uptaken especially for topological reasons, like cell volume, shape, position and contacts.

Small polynucleotides, polypeptides, polysaccharide-containing molecules and lipids have been demonstrated to affect the release of template restriction in different experimental conditions. These factors must be considered as minor chromatin components. In fact chromatin functions depend, besides the tigtly bound DNA, histones and NHP, also on loosely bound molecules which can be demonstrated to enter the nucleus (110). Moreover genetic activation seems to require the involvement of factors which interact with DNA through a mass action on large and only partly selected regions of the genome (110).

Therefore the molecules which can be demonstrated to enter the nucleus and to bind to the chromatin, with some correlation with the cell cycle phases or with the cell differentiation events, should be more attentively considered as possible effectors of the control of gene expression.

ACKNOWLEDGEMENTS

We wish to thank Drs. L. Cocco, P. Santi, E. Caramelli, R. Jovine and S. Papa for carrying out some of the experimental work and for their skillful assistance in the preparation of the manuscript.

This work was partly supported by grants of the National Research Council of Italy, n. CT 80 01587.96, CT 80.004 97.04, CT 79.01927.

REFERENCES

1. C. Nicolini, J. Submicr. Cytol. 12: 475 (1980).
2. D. E. Comings, in 'The Cell Nucleus', H. Bush ed., Academic Press, N. Y., vol. IV, p. 345 (1978).
3. R. Berezney, and D.S. Coffey, Adv. Enzyme Regul. 14: 63 (1976).
4. U. K. Laemmli, S. M. Cheng, K. W. Adolph, J. R. Paulson, J. A. Brown, and W. R. Baumbach, Cold Spring Harb. Symp. Quant. Biol. 42: 351 (1977).
5. K. W. Adolph, J. Cell Sci. 42: 291 (1980)
6. J. H. Shaper, D. M. Pardoll, S.H. Kaufmann, E. R. Barrack, and B. Volgelstein, Adv. Enzyme Regul. 17: 213 (1979).
7. T. E. Miller, C. Y. Huang, and A. O. Pogo, J. Cell Biol. 76: 692 (1978).
8. L. Cocco, N. M. Maraldi, F. A. Manzoli, R. S. Gilmour, and A. Lang, Biochim. Biophys. Res. Comm. in press.
9. R. Berezney, in 'The Cell Nucleus', H. Bush ed., Academic Press, N. Y., vol. VII, p. 413 (1979).
10. L. J. Kleinsmith, J. L. Stein, and G. S. Stein, Proc. Natl. Acad. Sci. 73: 1174 (1976).
11. S. Gilmour, in 'Acidic Proteins of the Nucleus', I. L. Cameron, J. R. Jeter, jr., eds., Academic Press, N. Y., p. 297 (1974).

12. C. B. Kimmel, S. K. Session, and M. C. Mac Leod, J. Mol. Biol. 102: 177 (1976).
13. G. S. Stein, G. L. Stein, L. J. Kleinsmith, J. A. Thompson, W. D. Park, and R. L. Jansig, Cancer Res. 36: 4307 (1976).
14. P. Chambon, J. D. Weill, and P. Mandel, Biochem. Biophys. Res. Comm. 11: 39 (1963).
15. T. Sugimura, Progr. Necleic Acid Res. Mol. Biol. 13: 127 (1973).
16. K. Ueda, R. H. Reeder, T. Honjo, Y. Nishizuka, and O. Hayaishi, Biochem. Biophys. Res. Comm. 31: 379 (1968).
17. J. Roberts, P. Stark, and M. Smulson, Proc. Natl. Acad. Sci. 71: 3212 (1974).
18. J. E. Mayfield, and J. Bonner, Proc. Natl. Acad. Sci. 68: 2652 (1971).
19. J. E. Mayfield, and J. Bonner, Proc. Natl. Acad. Sci. 69: 7 (1972).
20. I. Bekor, Arch. Biochem. Biophys. 155: 39 (1973).
21. I. J. Paul, and S. Duerksen, Mol. Cell. Biochem. 9: 9 (1975).
22. R. A. Weinberg, and S. Penman, J. Mol. Biol. 38: 289 (1968).
23. W. F. Zapisek, A. G. Saponara, and M. Enger, Biochemistry, 8: 1170 (1969).
24. T. S. Ro-Choi, and H. Bush, in 'The Cell Nucleus', H. Bush ed., Academic Press, N. Y., vol. III, p. 151 (1974).
25. R. Reddy, T. S. Ro-Choi, D. Henning, and H. Bush, J. Biol. Chem. 249: 6486 (1974).
26. H. Shibata, T. S. Ro-Choi, R. Reddy, Y. C. Choi, D. Henning, and H. Bush, J. Biol. Chem. 250: 3909 (1975).
27. W. F. Marzluff, E. L. White, R. Benjamin, and R. C. C. Huang, Biochemistry, 14: 3715 (1975).
28. T. S. Ro-Choi, Y. C. Choi, D. Henning, J. Mc Closkey, and H. Bush, J. Biol. Chem. 250: 3921 (1975).
29. G. Zieve, and S. Penman, Cell, 8: 19 (1976).
30. R. P. Hjem jr., and R. C. C. Huang, Biochemistry, 14: 1682 (1975).
31. L. Zardi, A. Siri, B. Carnemolla, L. Santi, W. D. Gardner, and S. O. Hoch, Cell, 18: 649 (1979).
32. G. L. Gianfranceschi, D. Amici, and L. Guglielmi, Biochim. Biophys. Acta 414: 9 (1975).
33. M. Hillar, and J. Przyjemski, Biochim. Biophys. Acta, 564: 246 (1979).

34. L. R. Orrick, M. O. Olson, and H. Bush, Proc. Natl. Acad. Sci. 70: 1316 (1973).
35. I. L. Goldknopf, and H. Bush, in 'The Cell Nucleus', H. Bush, ed. Academic Press, N. Y., vol. VI, p. 149 (1978).
36. T. H. Eickbush, D. K. Watson, and E. N. Moudrianakis, Cell 9: 785 (1976).
37. V. P. Bahavanandan, and E. A. Davidson, Proc. Natl. Acad. Sci. 72: 2032 (1975).
38. R. K. Margolis, C. P. Crockett, W. L. Kiang, and R. V. Margolis, Biochim. Biophys. Acta 451: 465 (1976).
39. S. Ljiljana, and K. Koviljka, Int. J. Biochem. 4: 345 (1973).
40. S. Kinoshita, Exp. Cell Res. 85: 31 (1974).
41. K. Furukawa, and I. Terayama, Biochim. Biophys. Acta 499: 278 (1977).
42. G. S. Stein, R. M. Roberts, J. L. Davis, J. W. Head, J. Stein, C. L. Thrall, J. Van Veen, and D. W. Welch, Nature 258: 639 (1975).
43. A. H. Goldberg, L. C. Yeoman, and H. Bush, Cancer Res. 38: 1052 (1978).
44. J. H. Frenster, Nature 206: 680 (1965).
45. H. G. Rose, and J. H. Frenster, Biochim. Biophys. Acta 106: 577 (1965).
46. J. H. Frenster, Cancer Res. 36: 3394 (1976).
47. M. Song, M. Ledig, and P. Mandel, Life Sci. 8: 253 (1969).
48. M. L. Coetzee, M. Spangler, H. P. Morris, and P. Ove, Cancer Res. 35: 2752 (1975).
49. F. A. Manzoli, L. Cocco, A. Facchini, A. M. Casali, N. M. Maraldi, and C E. Grossi, Mol. Cell. Biochem. 12: 67 (1976).
50. F. A. Manzoli, N. M. Maraldi, L. Cocco, S. Capitani, and A. Facchini, Cancer Res. 37: 843 (1977).
51. F. A. Manzoli, S. Capitani, N. M. Maraldi, L. Cocco, and O. Barnabei, Adv. Enzyme Regul. 17: 175 (1979).
52. A. B. Awad, and A. A. Spector, Biochim. Biophys. Acta 450: 239 (1976).
53. S. K. Erickson, A. M. Davidson, and R. G. Gould, Biochim. Biophys. Acta 409: 59 (1975).
54. Z. Balint, and L. Holczinger, Neoplasma 25: 1 (1978).
55. M. Smulson, P. Stark, M. Gazzoli, and J. Roberts, Exp. Cell Res. 90: 175 (1975).

56. R. C. C. Huang, and P. C. Huang, J. Mol. Biol. 39: 365 (1969).
57. D. S. Coffey, E. R. Barrack, and W. D. W. Heston, Adv. Enzyme Regul. 12: 219 (1974).
58. D. G. Brown, and D. S. Coffey, Science 171: 176 (1971).
59. S. Kinoshita, Exp. Cell Res. 64: 403 (1971).
60. V. G. Allfrey, and A. E. Mirsky, Proc. Natl. Acad. Sci. 48: 1950 (1962).
61. A. Saiga, and S. Kinoshita, Exp. Cell Res. 102: 143 (1976).
62. M. Skalka, J. Matyasova, and M. Cejkova, Folia Biol. 14: 466 (1968).
63. R. T. Cook, and M. Aikawa, Exp. Cell Res. 78: 257 (1973).
64. M. R. Smith, and R. T. Cook, Biochem. Biophys. Res. Comm. 74: 1475 (1977).
65. D. I. De Pomerai, J. Chesterton, and P. M. W. Butterworth, FEBS Lett. 42: 149 (1974).
66. A. T. Ansevin, K. K. MacDonald, C. E. Smith, and L. S. Hnilica, J. Biol. Chem. 250: 281 (1975).
67. C. E. Hildebrand, L. R. Gurley, R. A. Tobey, and R. A. Walters, Biochim. Biophys. Acta 477: 295 (1977).
68. C. F. Warnick, and H. M. Lasarus, Nucleic Acids Res. 2: 735 (1975).
69. Y. Groner, G. Monroy, M. Jacquet, and J. Hurwitz, Proc. Natl. Acad. Sci. 72: 194 (1975).
70. A. Kitzij, N. Defer, B. Dastugm, M. M. Sabatier, and J. Kruh, FEBS Lett. 66: 336 (1976).
71. B. E. M. Coupar, and C. J. Chesterton, Eur. J. Biochem. 79: 525 (1977).
72. S. Kinoshita, Exp. Cell Res. 102: 153 (1976).
73. Y. Aoki, and H. Koshihara, Exp. Cell Res. 70: 431 (1972).
74. S. J. Mohr, J. G. Massicot, and M. A. Chirigos, Cancer Res. 38: 1610 (1978).
75. W. Ragelson, Adv. Cell Res. 11: 223 (1968).
76. D. G. Brown, and D. S. Coffey, Biochim. Biophys. Acta 294: 74 (1973).
77. M. R. Smith, and R. T. Cook, Exp. Cell Res. 110: 15 (1977).
78. T. T. Sakai, and S. S. Cohen, Progr. Nucleic Acid Res. Mol. Biol. 17: 15 (1976).
79. G. Andersson, and O. Hebay, Cancer Res. 37: 4361 (1977).

80. D.H. Russell, in 'Polyamines in Normal and Neoplastic Growth', Raven Press, N. Y. (1973).
81. C. Nakai, and W. Glinsmann, Biochemistry 16: 5636 (1977).
82. D. A. Pierce, and N. Fausto, Biochemistry 17: 102 (1978).
83. G. Moruzzi, B. Barbiroli, M. S. Moruzzi, and B. Tadolini, Biochem. J. 146: 697 (1975).
84. T. Y. Wang, and N. Kostraba, in 'The Cell Nucleus', H. Busch, ed., Academic Press. N. Y., vol. VI, p. 289 (1978).
85. T. Higashinakagawa, T. Inishi, and M. Muramatsu, Biochem. Biophys. Res. Comm. 48: 937 (1972).
86. M. I. Goldberg, J. C. Perriard, and W. J. Rutter, Biochemistry 16: 1648 (1977).
87. E. Komarowa, and J. Hradec, FEBS Lett. 18: 109 (1971).
88. J. Hradec, Progr. Biochem. Pharmacol. 10: 197 (1975).
89. D. A. Booth, J. Neurochem. 9: 265 (1962).
90. M. M. Meisler, and R. H. McCluer, Science 154: 896 (1966).
91. W. Wickner, and A. Kornberg, J. Biol. Chem. 249: 6244 (1974).
92. F. Novello, J. H. Muchmore, B. Bonora, S. Capitani, and F. A. Manzoli, It. J. Biochem. 24: 325 (1975).
93. A. Zerial, I. Gelman, and W. Firshein, J. Bacteriol. 135: 78 (1978).
94. H. J. Hachmann, and A. G. Lezius, Eur. J. Biochem. 50: 357 (1975).
95. S. Capitani, G. Mazzotti, R. Jovine, S. Papa, N. M. Maraldi, and F. A. Manzoli, Mol. Cell. Biochem. 27: 135 (1979).
96. F. A. Manzoli, N. M. Maraldi, and S. Capitani, Bull. Mol. Biol. Med. 3: 99 (1978).
97. S. Capitani, G. Mazzotti, S. Papa, P. Santi, and F. A. Manzoli, Biochem. Biophys. Res. Comm. 89: 1206 (1979).
98. A. Stevens, Biochem. Biophys. Res. Comm. 65: 442 (1975).
99. A. Lezius, and B. Muller-Lornsen, Hoppe-Seyler's Z. Physiol. Chem. 353: 1872 (1972).
100. S. Capitani, E. Caramelli, M. Felaco, S. Miscia, and F. A. Manzoli, Physiol. Chem. Phys., in press.
101. F. A. Manzoli, J. H. Muchmore, B. Bonora, A. Sabioni, and S. Stefoni, Biochim. Biophys. Acta 277: 251 (1972).
102. F. A. Manzoli, J. H. Muchmore, B. Bonora, S. Capitani, and S. Bartoli, Biochim. Biophys. Acta 340: 1 (1974).

103. S. Capitani, J. H. Muchmore, A. Farulla, S. Kovacs, and F. A. Manzoli, IRCS Med. Sci. 2: 1232 (1974).
104. S. Capitani, J. H. Muchmore, N. M. Maraldi, and F. A. Manzoli, IRCS Med. Sci. 3: 312 (1975).
105. F. A. Manzoli, J. H. Muchmore, S. Capitani, B. Bonora, and S. Bartoli, Mol. Cell Biochem. 10: 153 (1976).
106. S. Capitani, N. M. Maraldi, L. Cocco, P. Santi, and F. A. Manzoli, Mol. Cell. Biochem. 20: 159 (1978).
107. S. Capitani, N. M. Maraldi, L. Cocco, A. Antonucci, and F. A. Manzoli, J. Submicr. Cytol. 8: 248 (1976).
108. N. M. Maraldi, S. Capitani, L. Cocco, P. Santi, R. Jovine, S. Papa, and F. A. Manzoli, J. Submicr. Cytol. 10: 397 (1978).
109. N. M. Maraldi, S. Capitani, L. Cocco, and F. A. Manzoli, in 'Chromatin Structure and Function', part B, C. Nicolini ed., Plenum Publ. Corporation, N. Y., p. 371 (1979).
110. H. Busch, in 'The Cell Nucleus', H. Bush ed., Academic Press, N. Y., vol. VII, p. 1 (1979).

HIGHER ORDER CHROMATIN STRUCTURE, PROTEINS, c-AMP, IONS MODIFICATIONS AND CELL CYCLE PROGRESSION: EXPERIMENTAL RESULTS AND POLYELECTROLYTE THEORY

C. Nicolini and A. Belmont

Department of Biophysics - Physiology
Temple University, USA
and National Research Council, Italy

The traditional view on the cell cycle and the numerous biochemical, structural, morphological changes during the various subphases, particularly for what concerns the initiation of RNA synthesis, DNA replication and mitosis, have been well summarized in previous chapters of this book. The intent of this chapter is to furnish additional experimental evidences related to changes occurring at the level of higher order chromatin structure (1) c-AMP, histone - non-histone proteins, and ions, and to attempt a coherent synthesis of alternative mechanisms determining G1→S, G2→M, G0→G1 transitions, within the analytical framework of polyelectrolyte theory (2). All studies here reported utilized mammalian HeLa S3 cells, synchronized by selective mitotic detachment (3).

RADIOACTIVE LABELING PATTERNS

Synchronized HeLa cells are pulse-labeled with 3H-thymidine (15 μCi/ml) at various time intervals after selective mitotic detachment (Fig. 1). Despite the wide variation in residence time in the G1 phase for each single cell, which causes a wide range of values for total cell cycle time and a progressive loss of synchrony (3), clear clusters around homogenous labeling patterns, unique for each time interval appear evident as the cell progresses through the cycle. As shown in Fig. 1 and at odds with other findings on CHO (4), during S-phase, beginning in a few cells as early as 5 hours after mitosis, HeLa nuclei display unique labeling patterns as they progress from early-S (pattern I) through middle-S (pattern II-III) to late-S (pattern IV): specifically, DNA synthesis appear to begin close to the nuclear border, forming a ring of 3-HT grains quite analogous to the recently reported (6) ring of high O.D.

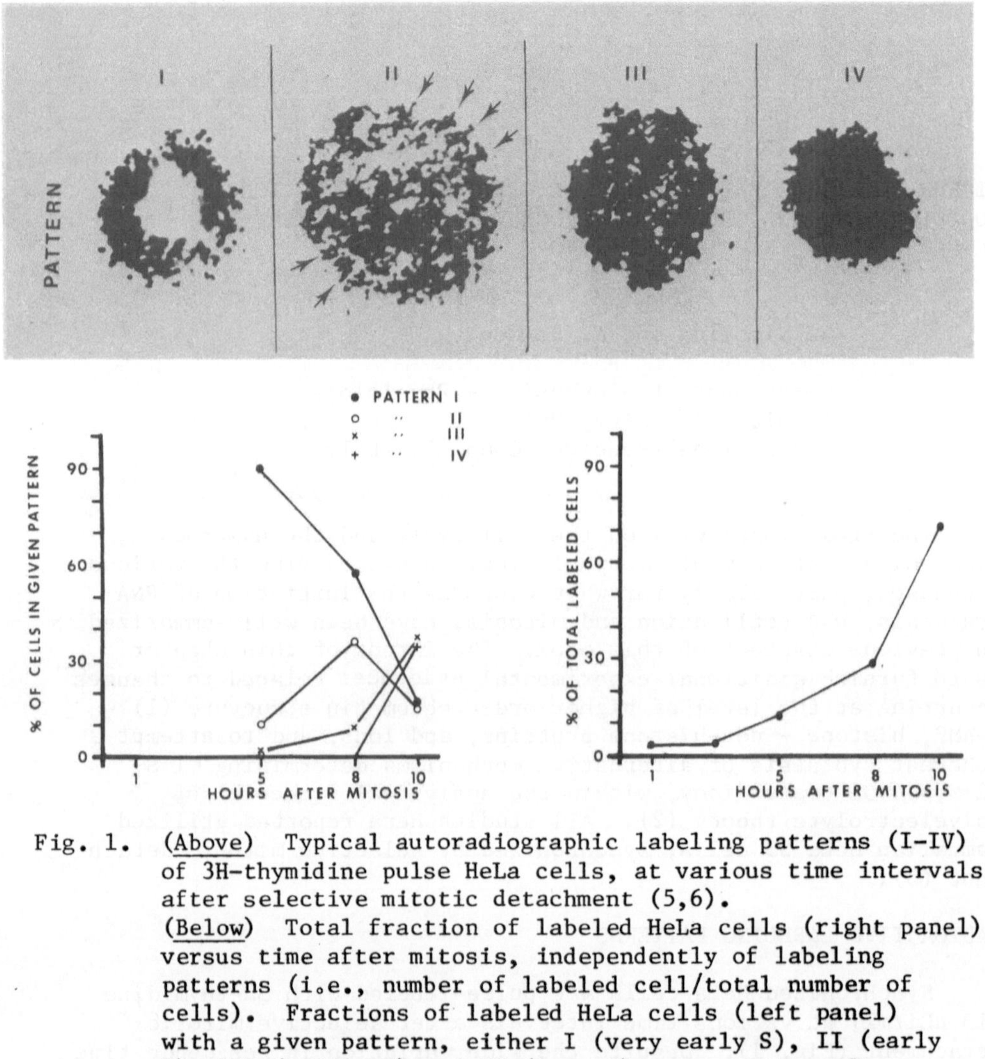

Fig. 1. (Above) Typical autoradiographic labeling patterns (I-IV) of 3H-thymidine pulse HeLa cells, at various time intervals after selective mitotic detachment (5,6).
(Below) Total fraction of labeled HeLa cells (right panel) versus time after mitosis, independently of labeling patterns (i.e., number of labeled cell/total number of cells). Fractions of labeled HeLa cells (left panel) with a given pattern, either I (very early S), II (early S), III (middle S) and IV (late S), versus time after mitosis. The fraction is expressed as a ratio (in percentage) of the number of labeled cells with pattern I divided by the total number of labeled cells (independently of patterns).

elements in corresponding Feulgen-stained nuclei. We will return later to this observation. For the moment, we like only to stress that autoradiographic patterns combined with multiparameter image analysis of Feulgen-stained nuclei (5), in physiologically synchronized HeLa cells permit the monitoring of DNA changes at the level of single cell progressing through the cycle at an accuracy previously unforeseable. Microfluorimetric analysis of nuclear-DNA viably and differentially stained by the proper utilization of acridine-orange (7) may actually permit such analysis on single cells still living.

TERTIARY-QUARTERNARY CHROMATIN-DNA STRUCTURE

The present widely accepted model of the mammalian cell cycle (8) was developed historically from observations based on light microscopic and autoradiographic data, with modifications added to account for so-called "resting cells" (see chapter by Nicolini, Lessin and Abraham). With the advent of more sophisticated biophysical and biochemical techniques, it has been shown that many intracellular substances including chromatin are modulated during the cycle (3,9). Differences in chromatin structure (not only DNA amount) during the cycle have been reported with respect to the amount of actinomycin D bound per unit amount of DNA and to the sensitivity of chromatin to digestion by DNAse (12). Specifically "native" chromatin isolated from the same HeLa cells (accurately probed by spectropolarimetry, where, contrary to spectrofluorimetry, the differences between primary and secondary sites are qualitative and not quantitative) display a number of DNA primary binding sites for an intercalating dye as ethidium bromide very low during mitosis, high in early G1 to decrease progressively during G1 phase and increas again in early S phase (3).

The above physico-chemical data are frequently misquoted as supporting the progressive chromatin decondensation between early and late G1 (11) which, if correct, should be reflected in a progressive increase of primary binding sites. Some authors (11) indeed, to make convincing their case, show only data obtained in "sheared" chromatin by a spectrofluorimetric analysis using actinomycin-D (12) which is quite less accurate and specific in determining primary binding sites than the spectropolarimetric titration using ethidium bromide: indeed, a more direct prove of chromatin structure, as circular dichroism (Fig. 2), do confirm the EB spectropolarimetric data, indicating a small but significant increase in chromatin-DNA super-coiling between early-middle G1 (2.5 hours) and late G1 (5-7 hours). Thermal denaturation studies, coupled to circular dichroism to monitor subtle variations in DNA conformation, have recently (9) more definitely characterized in terms of tertiary-quaternary chromatin-DNA structure (13), the G1-S

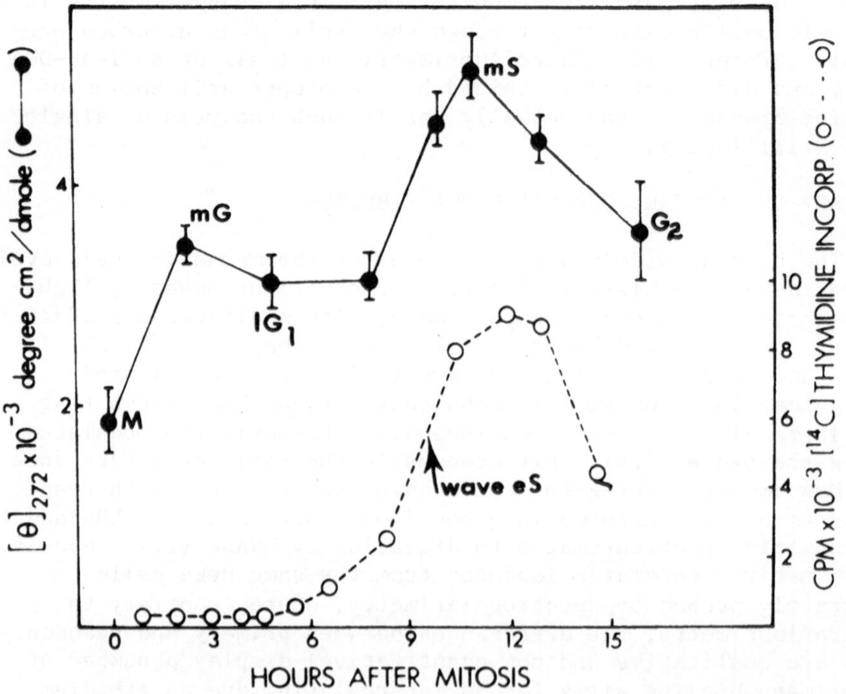

Fig. 2. (●——●) Positive molar ellipticity at 272 nm (DNA tertiary-quaternary structure) of chromatin isolated from synchronized HeLa S3 cells at various times after mitosis. (O---O) Specific activity of HeLa fraction after 3H-thymidine pulse at various times after cells were synchronized and analyzed as previously described (3). Identical time-sequence (3) is obtained with the circular dichroism signal at 308 nm for ethidium bromide-DNA complexes in chromatin isolated from HeLa cells at 0 (mitosis), 2.5 hours (middle G1), 5 hours (late-G1), 8, 10, 11 (middle-S), 14 (G2) and 18 hours after selective mitotic detachment. The data were taken at E.B. saturation and are an exact measure of the number of primary binding sites, since only EB intercalation in the DNA primary sites give a positive signal (optically active), while EB weekly bound to DNA (secondary sites) and free EB are optically inactive (31).

transition, confirming that a decrease in the degree of DNA supercoil at nucleosome and polynucleosome level is associated with the onset of DNA replication.

NUCLEAR-DNA MORPHOMETRY

Recent interest (5,6,14,15,16) in nuclear morphometry based on the Feulgen reaction, seeks to exploit biological image analysis as a tool to quantitatively measure differences in the resulting organization of nuclear DNA space from cells in different functional states. The "chromosome cycle" originally only visualized by light microscopy during mitosis has been resolved during interphase and resting states by such a technique, also at the level of intact nuclei. Data collected from the densitometric-geometric image analysis (15,16) combined to premature chromosome condensation (11) and electron microscopy (17) have indeed been interpreted to corroborate the traditional view of the "oneness" of the cell cycle, reportably showing chromatin decondensation after mitosis and a progressive condensation leading up to mitosis. However, the above interpretation usually relies on the average properties of a population of cells rather than measurements based on individual cells progressing through the cycle. Since imperfect initial synchrony as well as a progressive loss of synchrony (Fig. 1) with time complicates interpretations linking average properties of a population with the response of an individual cell, taking averages over non-identical cells can lead to misinterpretation of results. Indeed, the averaged data presented in (14) and (15) coupled with their heterogenous distributions point towards a pattern not entirely consistent with the traditional view.

For the above reasons, a series of experiments have been conducted recently to overcome some of the pitfalls mentioned (10, 18). The details on the geometric and densitometric image analysis have been given in numerous other publications (6,14,15,18) and for a comprehensive "tutorial" review, the reader is referred to a chapter (19) of a previous two-volume book of this NATO-Life Science Series (1).

In order to circumvent the problems of averaging, we defined cells "typical" of the cells at a given time point (18). The choice of a typical cell was based on examining the multivariate distributions of parameters associated with each cell. We found that three parameters, IOD, Area (or AVOD = IOD/AREA) and Form Factor (FF; Area divided by perimeter square) can characterize the "state" of a cell at any given point. Furthermore, biological interpretations can be given to these parameters (5,19).

At this stage we will refer to IOD/area as a measure of nuclear condensation rather than chromatin condensation (while establishing a direct link between the two later). This is always a difficult matter to address as it is at once a semantic as well as scientific question. Certainly nuclear condensation reflects an overall average condensation of the chromatin on a micron scale but does not necessarily imply a condensation of various lower levels of chromatin organization. Indeed, the term chromatin condensation evokes entirely different phenomena in the minds of histologists, biochemists, or physical chemists.

As an example of how we chose a typical cell, in Figure 3 we display the two dimensional distribution of nuclear-DNA image area versus form factor for cells having 2C DNA content (as selected for by IOD). As we have shown before (1) although DNA synthesis does not begin in most cells for 5-8 hours there is a considerable change in the chromatin (nuclear-DNA) dispersion before this time. We can observe a progressive condensation of the nuclear-DNA of individual cells for the first three time points (indicated by a shift of the area distribution to smaller values) and then a few cells exhibiting larger area at 8 hours. At 1 hour, numerous condensed nuclei are present: visual inspection of the Feulgen-stained cells and time-lapse photography on the same synchronized viable cells (10) reveal indeed that these condensed nuclei (small area) represent cells in telophase or "very early G1", which then abruptly relaxes in early G1. This is shown by the simultaneous presence of disperse nuclei in the same 1 hour smear, which if they were representative of fast-growing cells already in late G1 (11) their average nuclear area should have increased (not decreased, as it occurs!) with time progression during G1 (1-8 hours). Only at 12 hours, when a relatively small number of cells remain in the IOD window, did they all exhibit large areas and several incorporate 3H-thymidine. At 15 hours, when a few cells begin to enter the second cycle (very early G1), as indicated by the large number of telophases and by a drop in 3H-thymidine incorporation (absent in most 2C cells), the nuclear area decreases to increase again abruptly at 18 hours (early G1). As for 1 hour, most of the condensed 2C nuclei (small area) at 15 hours are still in or just existing telophase. This second wave on a few cells confirm the findings on the first wave on most cells (1-12 hours). Interestingly, an increase in nuclear area is always accompanied by a decrease in form factor, confirming that a more disperse nuclear-DNA is also more elongated and convoluted (5). Subsequently, by the multiparameter analysis, the global morphometric parameters for the "typical" cells in each subphase can be obtained (6,18). The modulation of nuclear DNA during the cycle is easily seen in Figure 4. What is new about the figure is that two similar abrupt transitions in chromatin-DNA superpacking (about three-fold decrease in average optical density) occurs between telophase and early G1 (as testified by morphometric and densitometric analysis of the same 1 hour post-mitotic smears) and between late-late G1 and early S (characterized both in the 8 hour post-mitosic smears, by an identical DNA content and differential thymidine uptake). It is interesting that an identical high level of nuclear condensation is achieved both in telophase or very early G1 (56-71 IOD per unit area) and late-late G1 (63 IOD per unit area), just prior to a dramatic increase in nuclear DNA dispersion (24-25 OD per unit area for either early S or early G1). The high level of condensation (and round shape) immediately prior to the two transitions points is identical to the one displayed in other human nuclei as WI-38 fibroblasts (20) and lymphocytes (21) in the "non-cycling" G0 phase. All of the above findings point to a possible role of a "higher order" chromatin-DNA organization in the control of DNA replication and transcription (5). These two abrupt nuclear-DNA decondensations could correspond to the two random transitions recently postulated (22) to account for the responses of quiescent cells to stimulation by growth factors. More strikingly, these two abrupt changes in nuclear chromatin condensation (and volume) correspond exactly to the biphasic changes in the nuclear pore frequency for the same synchronized HeLa, previously shown to form pores at the highest rate in the first hour after mitosis and shortly before S phase (23).

If chromatin is attached to the nuclear membrane (5,6) and indeed to the nuclear pores, as has been recently suggested (5), the parameter form factor should be important in proliferation which is indeed the case (5,6). It has been previously shown that form factor is mainly related to nuclear image convolution (i.e., re-entrances) rather than a mere indication of circular-elliptical shape. If the chromatin is attached at the pores, modulation of chromatin superpacking and frequency of attachment should then be evident as a change in the convolution of the image border, which has indeed a similar biphasic behavior (Fig. 3).

Incidentally, at 8-12 hours, a few HeLa cells with 2C DNA content display as large a nuclear area as do early S cells, but without any thymidine incorporation, suggesting that chromatin decondensation is a prerequisite (but not necessarily a consequence) of DNA synthesis. The progressive condensation in nuclear-DNA between early and late G1 followed by an abrupt relaxation prior to DNA synthesis is confirmed by two successive experiments using two DNA antimetabolites administered 2 hours after mitosis: in fact while ARA-C (5,18) an inhibitor of DNA synthesis through competition with DNA polymerase and ligase, blocks cells in the highly decondensed and convoluted nuclear-DNA at "very late G1" (Fig. 5), theophylline, a phosphodiesterase inhibitor of DNA synthesis through inhibition of H1 phosphorylation (24) blocks cells (even if temporarily compatible with the delay rather than permanent inhibition induced by this drug) in the highly condensed and round "late G1" (Fig. 6). Recent work (Fig. 7) using phase contrast time lapse cinematography clearly shows the abrupt nature of the two nuclear-DNA transitions as viewed through cellular and

2C DNA DURING HeLa CELL CYCLE

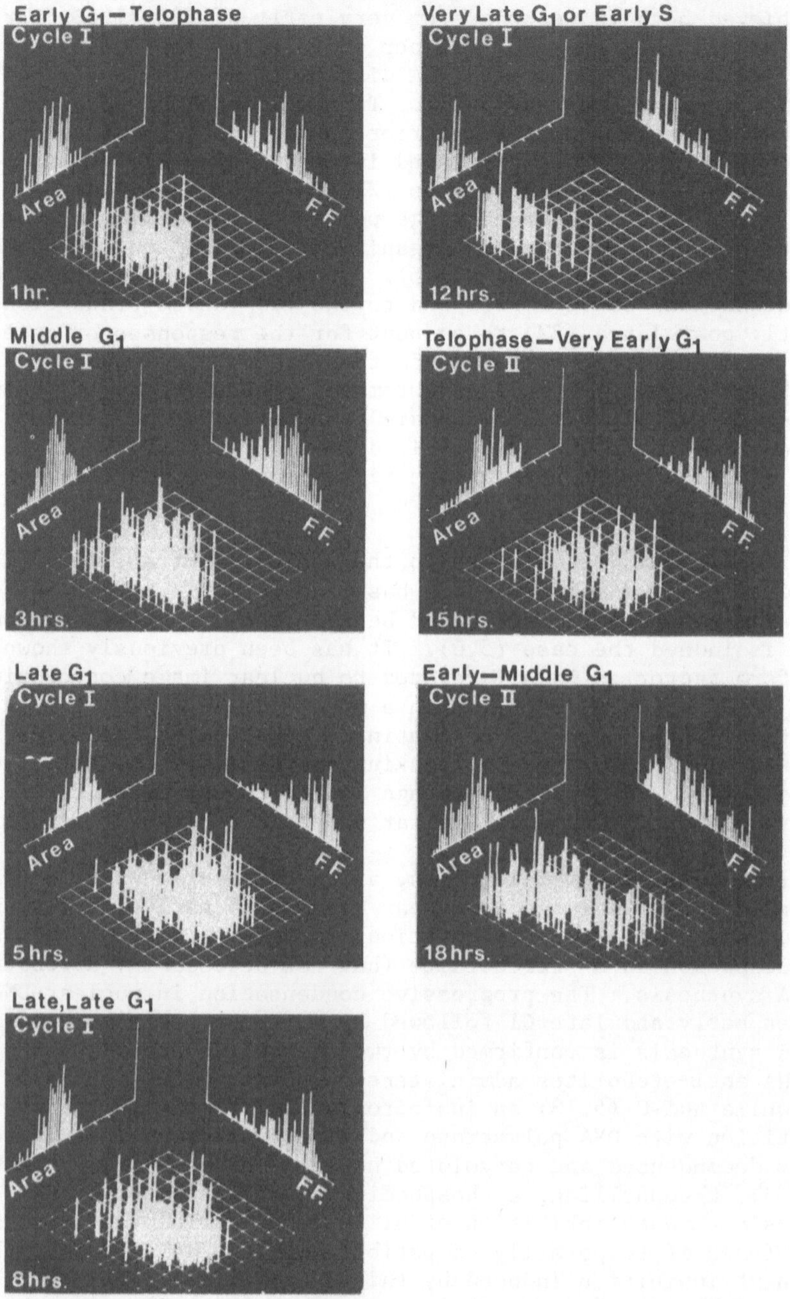

Fig. 3. Two parameters histogram of area versus form factor of "2C DNA" Feulgen-stained synchronized HeLa cells, at 1, 3, 5, 8, 12, 15 and 18 hours after selective mitotic detachment. "2C DNA" cells are selected in terms of their integrated optical density value (18).

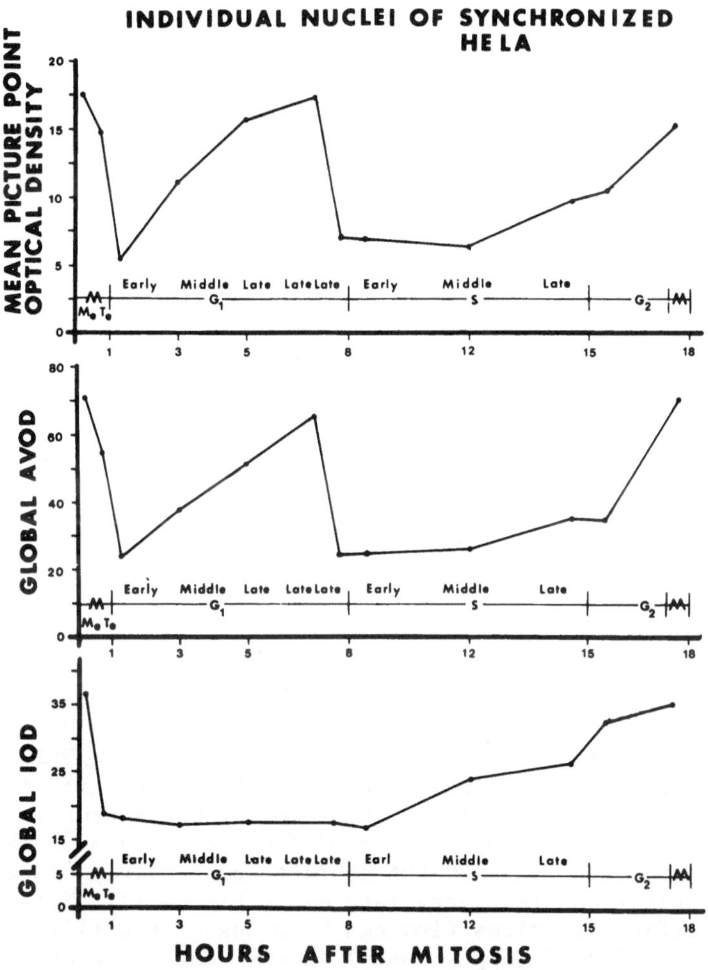

Fig. 4. Integrated Optical Density (lower panel) at the base threshold ("global"), and "global" Average Optical Density (middle panel) of Feulgen stained "typical" nuclei at various hours after selective mitotic detachment of the HeLa cells. The upper panel shows, for direct comparison, the mean optical density of each picture point in the 256 x 256 pp digitized image of the same nuclei (6).

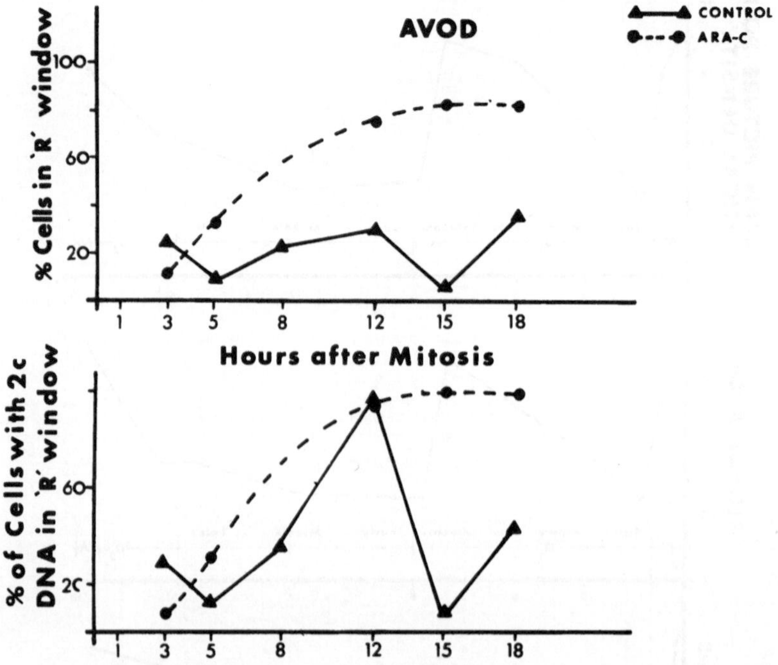

Fig. 5. Synchronized HeLa have been exposed to 40μ g/3x10^5 cells/ml of ARA-C, beginning one hour after mitosis: as a consequence, complete inhibition of DNA synthesis occurs and all cells appear blocked (18) at a given nuclear morphometry with 2C DNA content and a highly disperse nuclear DNA (see very late G1 in Figure 4). This confirms previous findings by means of PCC (11). Percentage of cell (upper panel) in the control group within the "very late G1" window peaks at 12 hours (S phase) and 3 and 18 hours (early-middle G1) while virtually all of the ARA-C treated cells have entered this window at 12 hours and remain there even at 15 and 18 hours. When the percentage is computed by normalizing only to cells with 2C DNA content, the results are as depicted in lower panel.

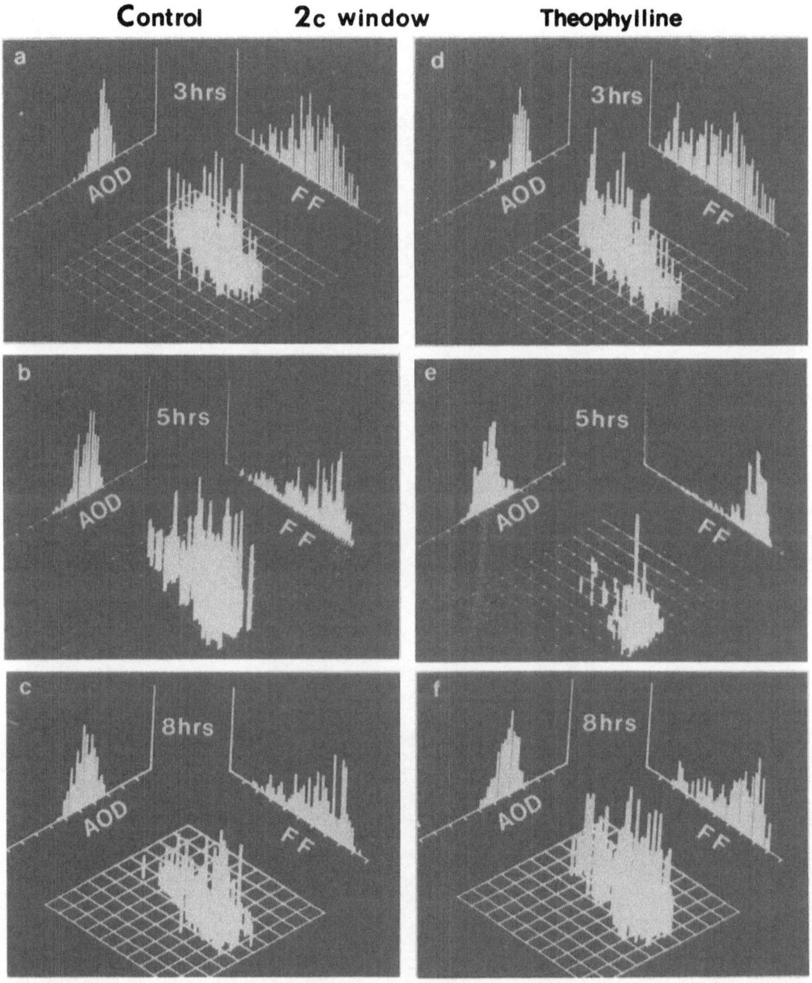

Fig. 6. AOD (Average Optical Density) and FF (Form Factor) histograms for the 2C DNA content subpopulations at 3, 5 and 8 hours after selective detachment. Each sample population represents between 130 and 200 cells. Histograms a, b and c correspond respectively to the control 2C populations at 3, 5 and 8 hours after selective detachment; d, e and f are corresponding histograms for the theophylline treated cells. Note particularly the striking difference at 5 hours (b and e).

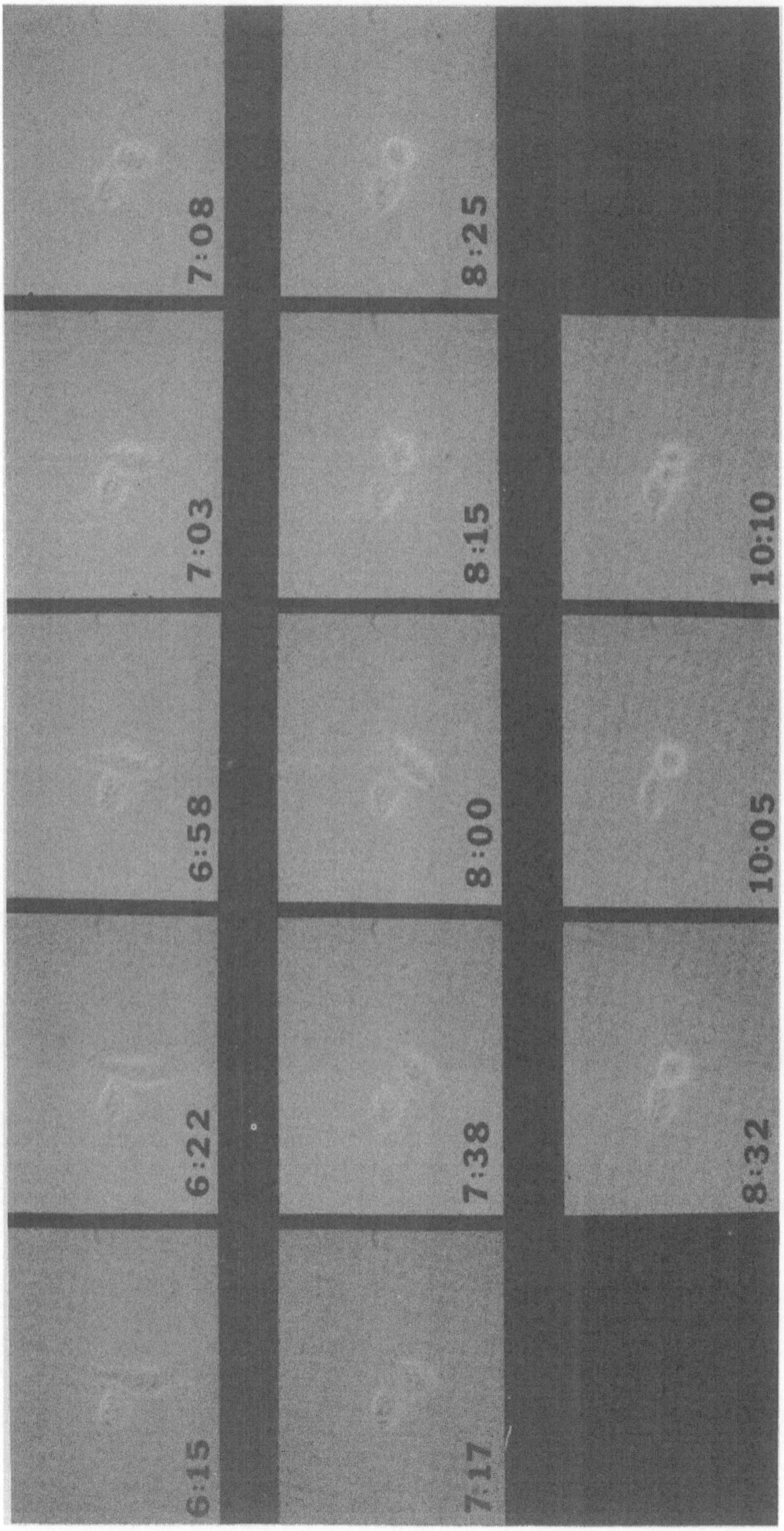
6:15
6:22
6:58
7:03
7:08
7:17
7:38
8:00
8:15
8:25
8:32
10:05
10:10

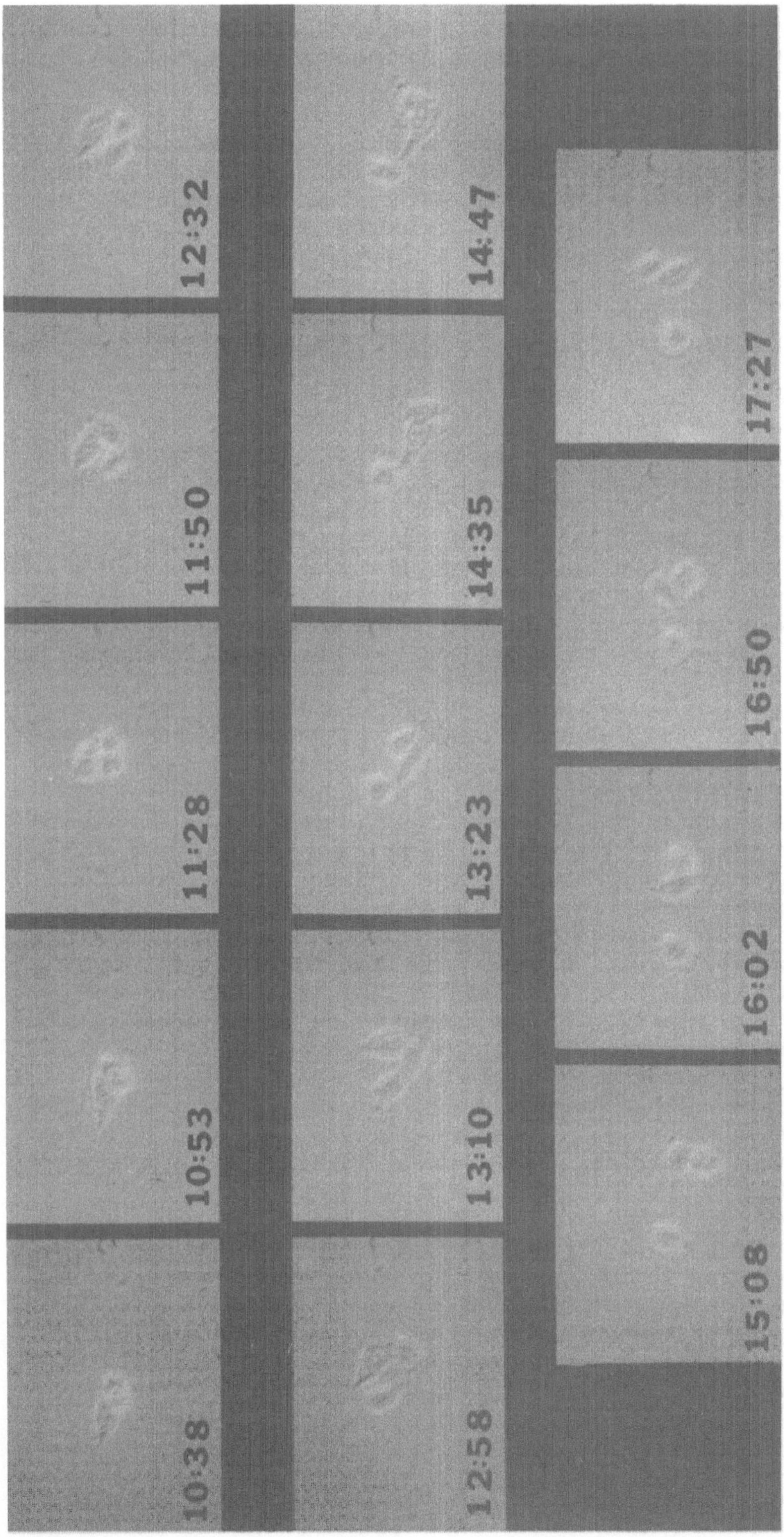

Fig. 7. Phase-contrast time-lapse photography of two daughters synchronized HeLa cells followed for several hours at close intervals after their mitosis (successively occurring in one cell at 6 hours and 10 hours and 20 minutes). The numbers on each panel represent the time (hour:minutes) after the original selective mitotic detachment.

nuclear geometry. In addition, the time of the transitions are in agreement with the data on chromatin morphometry reported above: specifically, immediately after mitosis, cell and nuclei appear round and condensed to abruptly elongate and disperse 1 (cell I) or 2 hours (cell II) later, and to progressively condense and round up until 5 hours after mitosis (late G1). DNA synthesis is then accompanied by a dramatic increase in the area and elongation (see S phase HeLa cells at 6 hours and 15 minutes after selective mitotic detachment). These data on nuclear and cell modulation during the cell cycle were recently confirmed on a large number of cells also in synchronized CHO cells (25), differentially stained for cytoplasmic protein (Napthol Yellow-S) and nuclear DNA (Feulgen).

QUINTERNARY CHROMATIN STRUCTURE

We have alternatively referred so far to nuclear-DNA or to chromatin-DNA condensation for the image analysis data. A three-dimensional reconstruction of individual interphase chromosome and a mathematical analysis of digitized image of intact nuclei do show indeed conclusively that changes in the morphometry of Feulgen-stained nuclei is directly related to changes in the quinternary chromatin-DNA structure (6), and that changes in nuclear-DNA condensation are paralleled by changes in individual chromosome condensation (6,26).

Recently and independently, biophysical studies of native unsheared chromatin-DNA have been suggesting that the nucleosomes (tertiary DNA structure) are folded in a quaternary structure as solenoid or superhelix (13,27) which, as apparent in unstained and unfixed chromatin using wet replica techniques is also folded in a higher order "quinternary" superstructure (28). These putative superstructures collapse into "beads-on-a-string" whenever chromatin IN SITU is fixed and dried in conventional electron microscopy studies or native isolated chromatin is sheared in conventional spectroscopic studies. For the sake of clarity, in this widely accepted terminology (1), the DNA base-pair sequence represents the primary DNA structure, and the Watson-Crick double-helix becomes the secondary DNA structure. Consequently, the folding of DNA double-helix around the octamers, which forms the basic chromatin subunit called nucleosome, becomes the tertiary DNA structure (1). The nomenclature quaternary ("solenoidal" fibers), and quinternary (folding of DNA "solenoidal" fibers) follows naturally from it.

In order to characterize the "quinternary" chromatin structure (and its modulation during the cell cycle) we have transferred the nuclear image point by point into the core of our minicomputer (PDP11/40) by direct memory access under software control (6,19). The image is stored on a mass storage device and is available for display on any display device connected to the Unibus. The image is presented in terms of discrete optical density values in a properly calibrated scale of 63 possible levels for analysis of the patterns of optical density distribution within each nucleus (5,6).

Nuclear DNA images from Feulgen-stained HeLa cells, synchronized at hourly intervals following mitosis, are acquired in a 256 x 256 picture points array (6) and their densitometric-geometric patterns are defined. Unique densitometric properties of individual condensation subsets appear evident for each of the 10 different phases and subphases of the cell cycle (6). In addition to confirm that chromatin undergoes for each cell cycle two distinct identical cycles of condensation followed by abrupt relaxation (see Fig. 7, middle panel), the frequency distributions of picture points gray levels, for the subphases, as seen in nuclear images, show that chromatin assume discrete superpacking configuration, yielding several discrete levels of condensation at the level of optical microscope (Fig. 8). With this work still in progress (6) it is still uncertain whether the OD frequency distribution represent:

a) combination (with occasional overlapping) of two levels of quinternary chromatin fibril organization, one with low degree of superpacking (dominant in very late G1 or early S) and one with high degree (dominant in late G1);

b) several levels of superpacking compatible with the large heterogeneity of native chromatin in a sucrose gradient centrifugation;

c) multiple of a fundamental higher order repeat unit for quinternary chromatin structure.

When the range of DNA condensation (per picture point) is divided in 7 equal intervals (I-VII), most of genome is in the highly relaxed state (I) at early S and in a highly superpacked state (VI-VII) at late G1 (6), suggesting that the structural alteration detected in bulk isolated chromatin (3,9) do involve large fractions (if not all) of the genome, rather than a limited "active" fraction with the majority being an "inactive" spectator.

<u>Nuclease Digestion</u>

In the intact nuclei two chromatin regions have been traditionally identified by classical cytological observations (see chapter by Nagl); one, apparently transcriptionally inert, is seen as highly condensed material (termed heterochromatin), while the other region apparently active in RNA transcription is visualized as extended diffuse fibers (called euchromatin). Efforts to reproduce such dicotomy at molecular level, gave instead contrasting results; only extensive shearing seems capable to produce template active and template inactive chromatin (29) and the template

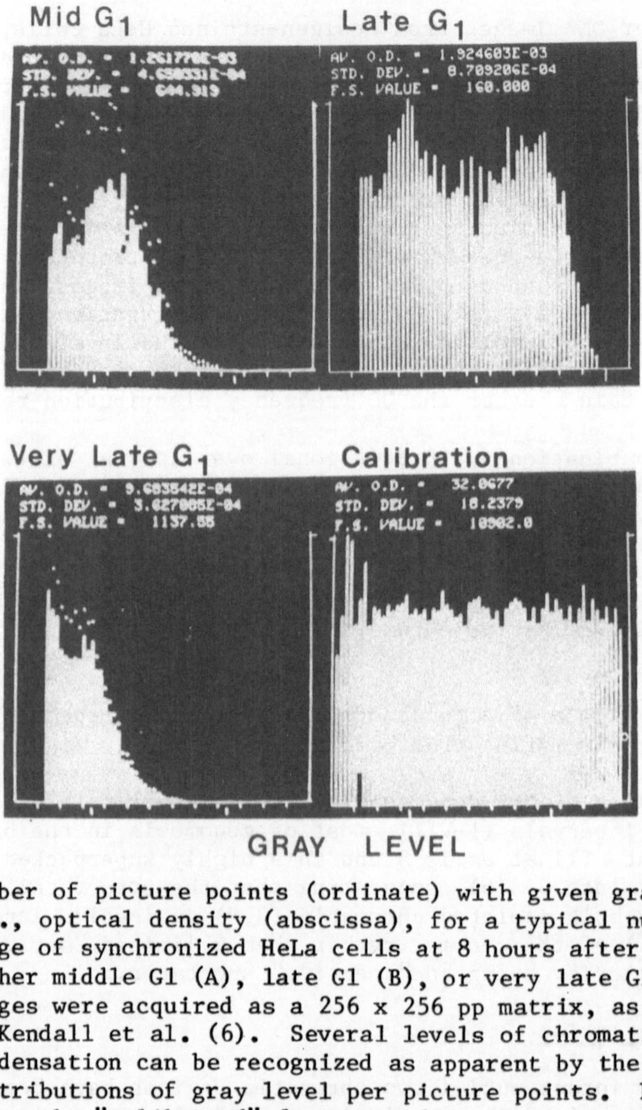

Fig. 8. Number of picture points (ordinate) with given gray level, i.e., optical density (abscissa), for a typical nuclear image of synchronized HeLa cells at 8 hours after mitosis, either middle G1 (A), late G1 (B), or very late G1 (C). Images were acquired as a 256 x 256 pp matrix, as discussed by Kendall et al. (6). Several levels of chromatin condensation can be recognized as apparent by the frequency distributions of gray level per picture points. Panel D shows the "calibrated" frequency distribution generated by a ramp (0 - 1.5 volts) scanning the small frame of 496,676 pp. The linearity of the densitometer is routinely verified using a calibrated wedge filter (6).

activity of putative inactive fraction possess some even if less capacity to support DNA dependent RNA synthesis. To test if the densitometric digitized nuclear images reproduce such dictomy, namely "euchromatin" versus "heterochromatin" we have conducted experiments with micrococcal nuclease (6), which at early times selectively digests active genes (30). A picture-element based image analysis of late G1 and early S Feulgen-stained nuclei, was performed at various times (20 seconds to 20 minutes) after "calibrated" micrococcal nuclease digestion, which at early times preferentially attacks (30) "euchromatic" genes ("disperse" chromatin regions) rather than inactive "heterochromatic" genes ("condensed" chromatin regions). The rate of enzyme digestion (5,6) of the various subpopulations with different degree of chromatin condensation reveals instead that the more condensed ("heterochromatic") picture-elements are preferentially digested at early times: at later times prior to complete disappearance of the entire image, only subpopulations with the lowest absorbance are present. Analysis of the picture elements optical density matrix (6) show that for cells in late G1 (or telophase) the typical ring of high optical density close to the nuclear border is absent already at 30 seconds after nuclease digestion. These data are compatible with the existence of only one fundamental subunit of higher order "superpacking" with given absorbance; a rigorous mathematical model of the rate of enzymatic digestion for the seven DNA condensation levels (Fig. 9) suggests indeed that all subpopulations with different absorbance levels are generated (as the nuclear envelope shrink) though overlapping and/or multiple combination of a "fundamental superpacking" of given absorbance [which modulates along the cycle at the "solenoid" level (3,13,31)]. These data which are not compatible with the traditional concept of euchromatin versus heterochromatin, can be viewed only as preliminary, until similar analysis (Belmont and Nicolini, in preparation) would be confirmed by means of DNAse I and DNAse II enzymes, which according to most authors (29) are more specific in degrading the "euchromatic" expressed regions of the genome. It is interesting moreover to notice that DNA synthesis (as revealed by the 3H-thymidine incorporation in early S HeLa, shown in Fig. 1) begins in the more dense "heterochromatic" regions around the nuclear border (where apparently chromatin fibers are attached and regularly spaced (5,6)).

HISTONES, NON-HISTONES AND c-AMP

Biochemical studies have associated an increase in one type of histone 1 phosphorylation with the onset of DNA synthesis (32,33). Another type of histone 1 phosphorylation, presumably occurring at different sites, has been linked by Bradbury and others (34) to condensation of chromatin from G2 to mitosis. Indeed, as shown experimentally, and indicated on general theoretical grounds (as will be shown later in this chapter), the

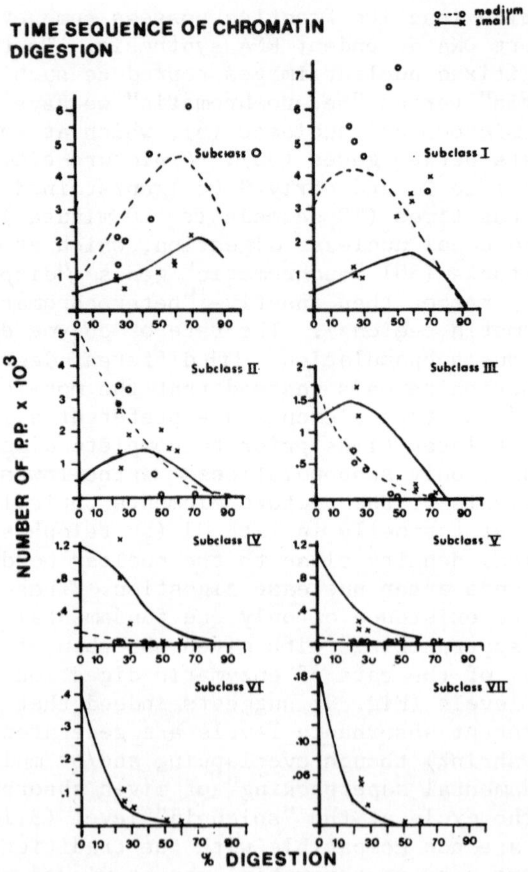

Fig. 9. Experimental [(X) for "late G1" image; (O) for "middle G1" image] and theoretical (.... or ——) time sequence of DNA digestion for seven (I - VII) subclasses of chromatin superpacking (OD level), as classified in ref. 6. Subclass 0 refer to picture points (pp) with very low optical density, mostly due to nuclear edge effects during the acquisition process (19). After fixation, HeLa cells were digested with stephlococcal nuclease enzyme for 1.5, 3, 6, 15 and 30 minutes. Identical nuclear images of identical area were then acquired and from the integrated optical density value the amount of digested DNA computed. For each digestion level (in percentage of total native DNA), the frequency distribution of picture point gray level is then obtained (5,6) and the number of picture points within each subclass computed. The "theoretical" number of pp in each subclass (versus percentage of DNA digested) is obtained with a Poisson distribution, by assuming that all levels of condensation (mean absorbance I through VII) are multiple of only one fundamental superpacking of mean absorbance I, digested at a constant rate.

influence of phosphorylation on histone 1-DNA interactions is strongly site specific. Later circular dichroism measurements coupled with thermal denaturation studies indicated that changes in the interactions of histone 1-DNA and non-histone chromosomal proteins were associated with the phosphorylation of H1 and appeared to be responsible for the relaxation of the highly supercoiled G1 chromatin into a more open or relaxed S-phase configuration (9).

This last conclusion was of great interest in light of the known influence of cyclic nucleotide levels on the phosphorylation state of various proteins and the more recent associations between changing cyclic nucleotide levels and cell proliferation. Specifically, increases in cyclic AMP levels have been observed in many (35) but not all types of cells as they approach quiescence, while increased intracellular levels in cyclic AMP produced either by exogenous dibutyryl cyclic AMP (36), dibutyryl cyclic AMP in conjunction with phosphodiesterase inhibitors (36), or phosphodiesterase inhibitors alone (37), inhibit cell proliferation. Indeed, the action of certain phosphodiesterase inhibitors, including theophylline and caffiene, has been shown to involve a shift of untransformed cells from G1 to G0 (37) as well as a cell cycle arrest in G2 (38). Both elevated concentrations of Ca^{++} and Mg^{++} salts, as well as hormone treatment, have also been shown to reduce mitotic delay after irradiation and in both cases this has been linked to the state of chromatin condensation (38). Thus, in an attempt to determine whether there actually existed a causal relationship between histone 1 phosphorylation and chromatin structure (9), the above associations prompted us to explore the effect of theophylline on progression of synchronized HeLa cells from mitosis to S phase. It appears (24) that 5 mM theophylline, added 2 hours after selective detachment to synchornized HeLa S3 cells, delays the onset and reduces the rate of DNA synthesis while theophylline treatment beginning at 8 hours has no effect on subsequent DNA synthesis. These actions of theophylline are accompanied by an inhibition of histone 1 phosphorylation as monitored by gel electrophoresis (24), and a prevention of the normal relaxation in chromatin structure between G1 and S phases as revealed by image analysis of Feulgen stained nuclei (Fig. 5). In contrast, ARA-C treatment beginning 2 hours after mitosis at a concentration of 40 μg/ml completelyinhibits subsequent DNA synthesis blocking chromatin in a very relaxed state, but not phosphorylation of previously synthesized histone 1 components which also migrated as the three principal peaks (24) even though the apparent level of histone 1 phosphorylation was cut in half (24). This implies that the influence of theophylline on histone 1 phosphorylation and subsequently on chromatin structure (Fig. 5) is not simply a result of the inhibition of DNA synthesis. While histone 1 phosphorylation is reduced in theophylline treated cells, it appears (24) that non-histone phosphorylation at 10 hours after mitosis is increased relative to control cells with marked increases in certain protein species. Thus, it appears that theophylline can potentiate some NHCP phosphorylations while greatly inhibiting histone H1 phosphorylation and the progression of DNA synthesis.

Finally, examination of Fig. 10 reveals that the modulations in higher order chromatin structure toward a more compact state (Fig 5) are not related directly to cyclic AMP levels (24). In agreement with earlier findings (39), measurements on the control HeLa cells showed a small increase in cyclic AMP levels between G1 and S phases about 10-12% increase at the onset of DNA synthesis (∿ 5 hours), decondensation of G1 chromatin and phosphorylation of H1 (9). Exposure of cells to theophylline produced increases in the cyclic AMP levels to values greater than twice the control maximum which peaked depending on the time of administration of the drug between 2 and 4 hours after administration. While it is interesting to note that the dramatic effect produced by theophylline on chromatin at 5 hours after mitosis is coincident with a significant rise in cyclic AMP levels, it must also be noted that even at 10 hours the cyclic AMP level in the treated cells is well above the maximal level reached in the control. Moreover, theophylline treatment at 8 hours produced similar rises in cyclic AMP but did not affect ^{14}C-thymidine incorporation (i.e., DNA synthesis). Although theophylline causes a 230% rise in cAMP levels 2-4 hours after treatment, whether in mid-G1, late G1 or very late G1, or early S, this increase did not obviously produce the onset of DNA synthesis or the decondensation of chromatin.

All above data (ARA-C and theophylline) seem compatible with the hypothesis (5) that the critical event involves modification of the chromatin structure which in turn permits DNA synthesis under those conditions in which the required substrates and enzymes are present and uninhibited, and that theophylline interferes with this normal modification. The observed variations in nuclear morphometry induced by theophylline would now be explained as an expected reflection of interference with this modification. With this hypothesis and above findings, interest in the normal association of S-phase with histone 1 phosphorylation and the possibility that there exists a causal relationship between histone 1 phosphorylation and chromatin structure is greatly enhanced (24).

An explanation of this inhibition of H1 phosphorylation by theophylline may lie in recent observations (40) that histone 1 dephosphorylation can be induced by the activation of an extremely efficient histone 1 phosphatase, which is normally suppressed during S phase by an inhibitory protein with a rapid turnover rate (41). From recent studies (41) it was concluded that passage of cells past the restriction point depended on the synthesis of a labile protein and, if synthesis of this protein was partially inhibited,

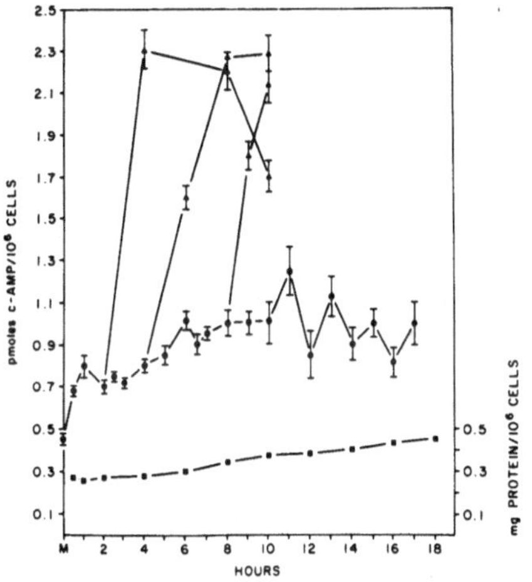

Fig. 10. Levels of intracellular cyclic AMP during the HeLa cell cycle. Cells were synchronized and cAMP determined as described in "Methods". Theophylline was added to parallel cultures in three independent experiments at 5 mM at 2, 4 and 8 hours post mitosis and cAMP determined. Total cellular protein/10^6 cells was also determined and plotted for data conversion to cAMP/mg protein. Data expressed as panel cAMP/10^6 cells $\pm$ 2 S.D. ●——● control cells; △——△ 5 mM theophylline; □——□ mg protein/10^6 cells.

the duration of G1 prior to the restriction point was lengthened, while a stronger inhibition placed cells into a G0 state. As the restriction point has been placed in the middle to late G1, the results of Dolby et al. (24) would suggest that the recently postulated protein (41) is the histone phosphatase inhibitor discussed above (and inhibited by theophylline) and that the necessary event for progression past the restriction point is histone 1 phosphorylation. An alternative explanation could, however, be possible at this stage (see other chapters in this book and ref. 24) with H1 being the effect rather than the cause. An alternative explanation of the results presented above, for instance, may lie in the influence of cyclic AMP on the cellular cytoskeleton and a resulting influence on nuclear morphometry and DNA synthesis (5) (see also the chapter by Nicolini et al., on the coupling between nuclear and cell morphometry).

A UNITARY MODEL OF THE CELL CYCLE

Time-lapse cinematography on single cell culture (22) has shown that the lag time between the stimulus and the onset of DNA synthesis is similar to the minimum intermitotic time in subsequent generations. It we then combine this observation with all data presented in this chapter, it appears (see model in Fig. 11) that the duration of the lag is determined by the same process [first cycle (I) of chormatin condensation during G1] that determines the minimum intermitotic time [second cycle (II) of chromatin condensation between middle S and G2]. It also suggests that normal quiescent cells have to undergo two random transitions (at the two restriction points) separated by a rather lengthy process of progressive chromatin condensation (of similar duration as during S + G_2 phases) before they can enter S phase (22), as indeed expected if normal cells would undergo (as apparently they do) changes similar to the one so far reported for synchronized transformed cells. Experiments with 3T3 cells stimulated to proliferate by serum (22) suggest that cells are not committed to enter S phase more than 5 hours before they actually do, even if their lag time is 14 hours: this would be compatible with the existence of II transition, as shown in Fig. 4. Assuming that the time required for chromatin condensation between early and late G1 is constant [about 4 hours for HeLa, as suggested by the preliminary time-lapse photography data (Fig. 7)] the large variation in G1 residence time is due to the variable time required for the very late (relaxed) chromatin to begin synthesizing DNA. Namely in a slowly growing population as WI-38 fibroblasts with an average 12-24 hours G1 period, most cells with 2C DNA will have in the log-phase a highly decondensed chromatin (G1), while at confluency ("G0") the same fibroblasts, uncapable to undergo the abrupt relaxation, will be arrested at the I restriction point with a highly condensed "2C" chromatin; this indeed occurs, as shown in the chapter by Nicolini, Lessin and Abraham in this volume.

In transformed HeLa (or CHO) cells the level of chromatin condensation is surprisingly identical in both 2C (late-late G1) and 4C (mitosis) nuclear-DNA, and preceeds the similar abrupt dispersion of large fractions of genome which accompanies an increased metabolic activity either in DNA replication (early S phase) or transcription (early G1 phase). Interestingly the level of condensation (and even form factor) of both late and very early G1 nuclei is identical to the one displayed in the WI-38 human fibroblasts (5, 20) and human lymphocytes (21) in the resting G0 phase. It is also useful to remember in further support of the two abrupt transitions experimentally found in synchronized transformed and fibroblast-like cells, and likely generalizable to any mammalian cell system, that quiescent normal cells do begin to enter S phase at 8-18 hours (depending upon cell type) after stimulation by various growth factors, not at a gradually increasing rate throughout the lag, but rather as a rapid abrupt increase at the end of the lag period (22).

Interestingly and also compatible with the model in Fig. 11 transformed CHO cells made quiescent by isoleucine deprivation and blocked apparently in "very late G1" (11) respond promtply to

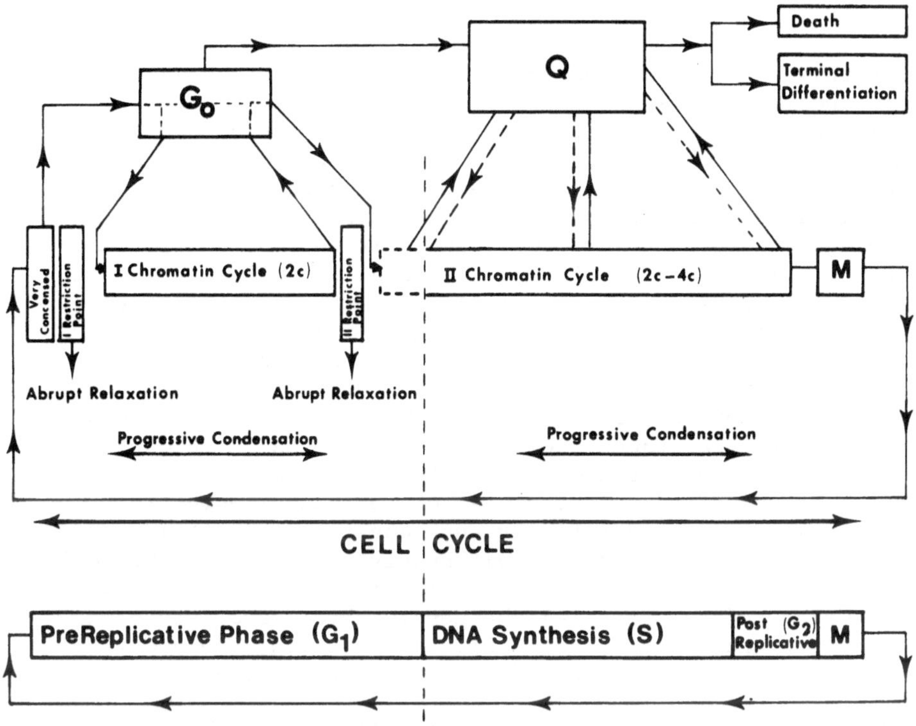

Fig. 11. New model of cell cycle progression for both normal and abnormal mammalian cells which include provision of two equal cycles of chromatin condensation per cell cycle (between two successive mitosis). The traditional cell cycle classification is given in the bottom panel. Variability in the length of pre-replicative phase "G1" is due to the variable (--) residence time in the disperse chromatin state preceeding DNA synthesis and to the random restriction points, where chromatin undergoes an abrupt relaxation from a very condensed state. This condensed state, equivalent to the previously characterized "G0", is a necessary step for cell cycle progression where normal cells may be blocked under proper conditions. Entrance in the non-cycling Q state is discussed in legend. Chromatin condenses at the same rate between early and late G1, and between middle-S and G2 phase. Under the presence of proper enzymatic substrates and chemical environment cells could also bypass the "G1" phase, entering directly into S-phase from mitosis.

administration of isoleucine by entering S phase after a very short lag.

TRANSITION MECHANISMS AND POLYELECTROLYTE THEORY

At this point, some questions may arise. Can such abrupt transitions (as G1 → S) in chromatin structure occur under physiological conditions? Can they quantitatively be explained in terms of simple mechanisms, such as the reported changes in H1 phosphorylation? Histone H1 has been associated as a key, if not essential, ingredient necessary for the stabilization of the "solenoidal" quaternary structure (in addition to the requirement of a minimum ionic strength). Meanwhile at the cellular level, modifications of H1 by phosphorylation at various sites on the protein molecule have been associated with important events of cell cycle progression, namely the G1-S transition in which the chromatin suddenly relaxes to a more dispersed conformation (5, 9, 10) and the condensation of chromatin associated with the G2 → M transition (34); can these two opposite phenomena be reconciled? An elegant analytical treatment recently developed (42), based on Manning's theory of polyelectrolyte counterion screening (2), seem capable to answer positively such questions and to address others (as the role of ions and alternative biochemical changes) through the construction of coherent models in chromatin structural changes which reflect parallel changes in chromosomal proteins (both histone and non-histones, their acetylation, phosphorylation or methylation) and bound ions (monovalent or divalent).

DNA conformation is dependent naturally on the extent of electrostatic repulsion of the negatively charges phosphate groups. Calculated changes in free energy for small angle bending by Manning (2) coupled to DNA persistance length measurements at various ionic strengths indicate that there is an entropy derived negative change in free energy, $\Delta G = 77$ cal/mole - b.p. rad., which would lead to spontaneous bending of DNA if the phosphate charges were neutralized.

Description of the Model

Our objective then is to examine the influence of H1 binding on the conformational properties of a DNA segment of length L using as a framework, general principles of polyelectrolyte theory as outlined by Manning (2). We begin by writing down an expression for "g", the free energy/RT per mole - P of this segment. This expression for g can be considered as the sum of three contributions: an electrostatic term, g^{el}, an entropy related mixing term g^{mix}, and a third term, g^{b}, related to the negative free energy associated with bending in the absence of polyelectrolyte effects.

If we confine ourselves to these ionic conditions in which H1 is known to remain bound ($\lesssim$ .3M), to a first order approximation we can incorporate the effect of H1 binding in Manning's model (2) by changing,

$$g^{el} = -(1-N\Theta_N)^2 \xi \ln(1 - \exp(-kb)) \text{ to}$$

$$\text{eq. 1)}\quad g^{el} = -(1-N\Theta_N - FH)^2 \xi \ln(1 - \exp(-kb))$$

where FH is the fraction of DNA-P charges neutralized by H1 counterion screening.

Here g^{el} refers to that part of the free energy/RT per mole-P arising from the electrostatic potential energy associated with an array of effective charges at the sites of the phosphate groups, where the effective charge, qnet, is equal to the real charge minus the charge of the associated "condensed" counterions (i.e. qnet = $(1-N\Theta_N-FH)q$), N is the valence of the electrolyte counterions, k is the Debeye screening parameter, Θ_N is the number of condensed counterions per phosphate group, and $\xi = q^2/\varepsilon Ktb$ where q is the unit charge, ε the bulk dielectric constant of the solvent, K is Boltzmann's constant, and T the Kelvin temperature. The distance between phosphate groups is given by the parameter b (equal to 1.7A° for the DNA double helix in Manning's linear model).

Assuming uniform curvature of the DNA segment and defining α as the angle between the two radii of curvature which span a uniformly curved segment (42), we then have as the complete expression (2) for g = free energy per mole - P/RT

$$g = g^{el} + g^{mix} + g^{b}$$

$$\text{eq. 2)}\quad g = g^{el} + \Theta_N \ln(\Theta_N V_P^{-1}/C_N) - .0648\,\alpha^2$$

The second term in the expression above refers to the free energy involved in the entropy associated with the mixing of free and condensed counterions where Vp, a function of b and ξ, is the volume to which the condensed counterions are confined in Manning's model (2), and C_N is the concentration of the free electrolyte counterion. Note that the effective b, in both g^{el} and g^{mix} is now a function of α and equal to:

$$\text{eq. 3)}\quad b(\alpha) = (1.7A^{o})\,(2(1-\cos\alpha))^{1/2} /\quad 1.7A^{o}\,(1-\alpha^2/24)$$

(for small α)

The last term g^b in the above expression for g accounts for the negative free energy of bending ($\Delta G = -38.5$ cal/mole-P-rad) in the absence of polyelectrolyte effects.

Predictions of the Model

Minimizing the free energy (eq. 2) numerically with respect to Θ_N and the bending angle, α, yields equilibrium angles of bending as a function of ionic strength for various values of FH as shown in Fig. 12 (for 25°C and a monovalent 1:1 electrolyte). At zero values of FH, corresponding to no H1 protein present, the equilibrium angle of bending is zero as expected from experimental findings (27). In line with our earlier discussion of persistance length (42) this does not preclude random fluctuations in α but due to the large persistance length of DNA these fluctuations will be small for reasonable values of the length of the linker DNA. For instance a linker DNA length of 40 base pairs would have a root mean square angle of about .43 rad given a persistance length of 750A° (corresponding to a concentration of NaCl of .1M) and a ΔG, for a bend of one radian, equal to 2.8 KT. At .01M NaCl, ΔG (α=1 rad) rises to 5.0 KT. Such values would exclude the possibility of solenoid formation except in the presence of very large internucleosome interactions strong enough not only to counter these relatively large values of ΔG but also to impart thermal stability. Even in the presence of such strong nucleosome interactions the final structure would not be one in stable equilibrium and therefore a regular, periodic self-assembling structure could not be expected.

However, with finite values of FH sudden transitions appear in which the equilibrium angle of bending rises to a nonzero value above a given concentration of a 1:1 univalent electrolyte solution. For a value of FH greater than about .13 the critical concentration at which this transition occurs falls below physiological ionic strength (Fig. 12, lower panel). Moreover above this critical FH further concentration increases produce significant equilibrium values for α before a plateau is reached (Fig. 12). Increasing FH increases the final height of this plateau as well as decreases the critical concentration.

Physically these results imply the following sequence of events occuring with increasing electrolyte concentration for a polynucleosome structure whose H1 protein has a value for FH $\gtrsim$ 0.13. At very low concentrations the ΔG associated with bending will still be large (and positive) so as to prevent significant fluctuations in α. As the critical concentration (∿mM) is approached these fluctuations will increase in magnitude as the ΔG of bending approaches zero. Finally above the critical concentration, the ΔG of bending becomes negative up to a given α. Further increases in concentrations increase the α for which $\Delta G(\alpha)$ is a minimum in addition to increasing the absolute magnitude of ΔG. Such a sequence is illustrated in (42) for a value of FH equal to .25 and a 20 base

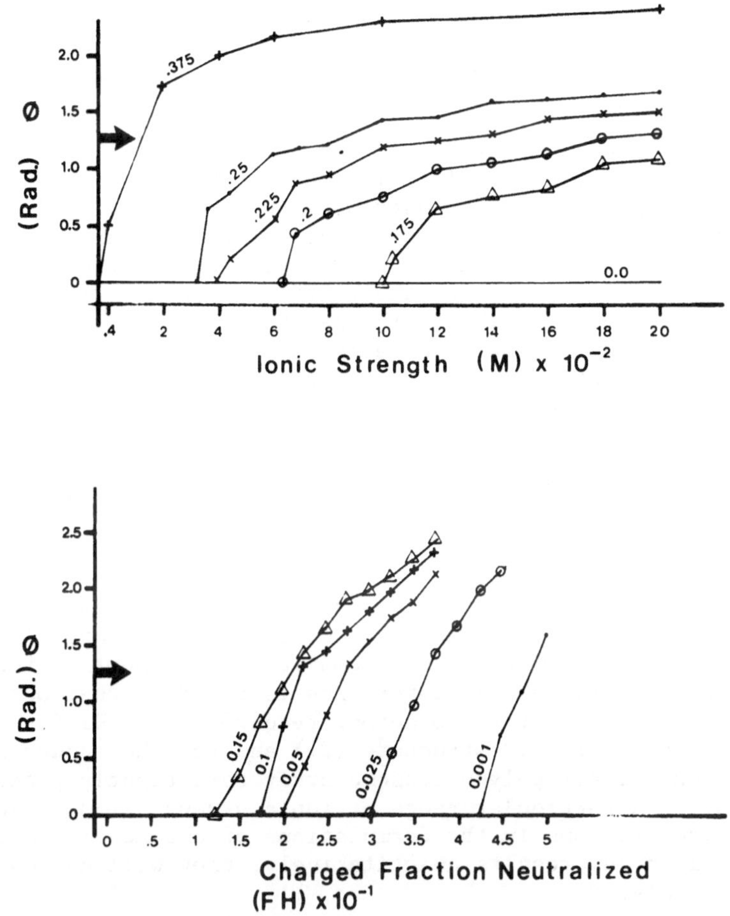

Fig. 12. The equilibrium bending angle ϕ, in radians, is plotted versus ionic strength (assuming a 1:1 electrolyte and $T = 25^{o}C$) for several values of FH, the fraction of DNA-P charges neutralized through counterion screening contributed by histone H1. In the bottom panel, the angle ϕ is plotted versus FH at several ionic strengths.

pair length of DNA interacting with H1.

While detailed modeling (43) of kinetics of enzyme digestion of nucleosomes has indicated that H1 interacts with approximately 20 base pairs of linker DNA, the actual length of DNA involved in the interactions of H1 and DNA has been directly measured in our laboratory by calorimetry (see Appendix I) yielding compatible results of 13 base pairs. Combined with the experimental observation (44) and 10 Na+ ions bound per dissociation of one H1 molecule this produces a FH equal to .25.

Thus at very low concentrations the polynucleosome structure (where H1 is still present) will still be that of an extended "beads on a string" array. In contrast with H1 depleted (FH = 0) polynucleosomes, however, irregularly folded and fluctuating structures will begin to appear at relatively low electrolyte concentrations as $\Delta G(\alpha) \rightarrow 0$ near the critical concentration and the linker DNA becomes flexible. The negative ΔG (for suitable angle) appearing with further increases in electrolyte concentration will permit first a loosely packed superhelical array with some disorder due to thermal fluctuations, which finally with further decreases in $\Delta G(\alpha)$, and in combination perhaps with weak internucleosome (or H1-H1) interactions, will change to a final closely packed solenoid structure whose final dimensions will be limited by contact between nucleosomes. Because of the relatively small absolute magnitude for $\Delta G(\alpha)$, however, this final fiber would be expected to be flexible and to show irregularities due to thermal disorder. All above predictions (42) are strikingly confirmed by independent findings with electron microscopy (45) and light scattering (46) of polynucleosomes arrangement as function of ionic strength. Of greater importance is the fact that our model correctly accounts for the normal requirement of H1 for the development of the solenoid structure (27) and for the sudden shift from an extended linear polynucleosome array to a tightly packed helical array in a particular range of ionic strength (45). For while the approximations in the formulations of our model might be expected to alter our results quantitatively, they will not change them qualitatively.

Discussion

The above calculations dramatize the potential of polyelectrolyte theory to demystify much of the present data in the literature concerning the role of H1 in effecting higher order chromatin structure. For instance although dissociation of H1 at high ionic strength causes condensation of 10 positive monovalent ions, there are roughly 60 positive charges in the entire H1 molecule (44). It is likely therefore that polyelectrolyte effects involving those excess charges in the nonbinding regions of the H1 molecule may alter the conformational dynamics of the H1 molecule allowing more or less interaction of other charges with the DNA molecule (in other words FH would be a function of ionic strength). This might also be predicted due to a cooperative phenomenon in which the spontaneous bending of DNA, occuring at higher ionic strength, by creating a higher negative charge density could induce an increase in FH. These possibilities would explain recent NMR data (34) indicating a reduced degree of binding of the mobile regions of H1 at low ionic strengths relative to an ionic strength of .15M without the need for invoking a hydrophobic component of H1 binding which is not supported experimentally (44). It would also explain how phosphorylation at certain sites in H1 (namely those regions not interacting directly with the DNA but determining the conformation and therefore degree of binding of those regions which do interact directly with the DNA) could cause the observed increase in condensation of chromatin due to an increased degree of binding, rather than a decreased degree of binding and increased cross linking as suggested (34), while phosphorylation at sites normally interacting with the DNA would cause exactly opposite effects. Namely the predicted decrease in binding angle (Fig. 2) and, more clearly in ΔG (42) for an increased phosphorylation (where each incremental FH decrease of 0.025 corresponds to an increase of 1 phosphorylation level per H1 molecule) is compatible with the experimental association between H1 phosphorylation and the abrupt chromatin structure relaxation at the late G1→early S transition (5, 9; see Fig. 4).

Moreover although the biological implications of the solenoid structure are as yet undetermined it is certainly not unreasonable to expect that such numbers as the free energy associated with the bending of a solenoid fiber or the free energy required to open or expand the solenoid may be of key importance in determining chromatin higher order structure and function in vivo. Thus, given such a framework as presented in this paper for considering chromatin structure, it is clear that the exact <u>ionic</u> environment in vivo (not only protein modification, but Mg^{++} or Ca^{++}) could be critical in its effect on chromatin structure and cell function as seen experimentally recently (47).

CONCLUSIONS

The implications and importance of chromatin structure modulation induced by ions, cyclic nucleotides and specific protein modifications (as H1 phosphorylation) during the cell cycle progression and in the control of normal and abnormal cell growth is just beginning to be explored in rigorous and quantiative fashion. The discovery of two condensation cycles during the cell cycle may help, moreover, to unravel some of the inconsistencies of the traditional view of the cell cycle and to develop new insights, which could warrant the departure between pure lengthy semantic-empirical debates (A or L versus G0 or G1 states histone acetylation versus phosphorylation, non-histones versus cyclic nucleotides) and the rigorous physico-chemical approaches here presented.

We are clearly at the beginning of a new fascinating and illuminating avenue, which could lead us to unravel the basic mechanisms regulating cell cycle progression, critically depending by complex and quantitatively predictable interplay among the above named factors; it is our opinion (and this chapter should have proven it) that this goal is likely to be accomplished by an in depth and analytical "multi-probe" characterization of well known mammalian model, rather than by a mere and sometimes confusing reproduction of old "semi-empirical" observations in a wide variety of new biological systems.

APPENDIX

Preliminary Determination of H1-DNA Interaction length by Calorimetry

H1-DNA interactions. The goal of this experiment was to estimate the interaction length of an H1 molecule by a Dupont Differential Calorimeter. The feasibility of such an attempt rested on the assumptions that:

a) there were no significant non-local DNA-H1 interactions, that is the DNA not in contact with the H1 molecule was essentially unchanged in its physical characteristics, and

b) that locally there was a shift of the helix-coil transition to a higher temperature with no significant transition occuring at the normal temperature.

The first measurements were on DNA alone. The total heat exchanged on transition was crudely estimated by measuring the "weight" of the helix-coil transition areas cut out from the graph paper, representing heat capacity ΔH versus temperature (Fig. 13).

The solution conditions were 50 mM NaCl, .7 mM $NaPO_4$ adjusted to pH = 7.1. These conditions were set by dialysis. The final DNA concentration was assayed by molar absorption at 260 nm. The values of entalpy variations ΔH_D in arbitrary units recorded were for three runs 47.8, 31.6, and 35.2 yielding an average of 38.2. Note that the amount of the solution in the pan could not be accurately determined by pipetting due to the high solution viscosity; instead the pan was weighed before and after and the difference in weight used as the measure of sample volume.

The H1-DNA solution was prepared as follows. A solution of .2 ml of H1 was added to 2 ml of the DNA stock solution and the entire dialysis bag immersed in a dialysis solution of .7 mM $NaPO_4$

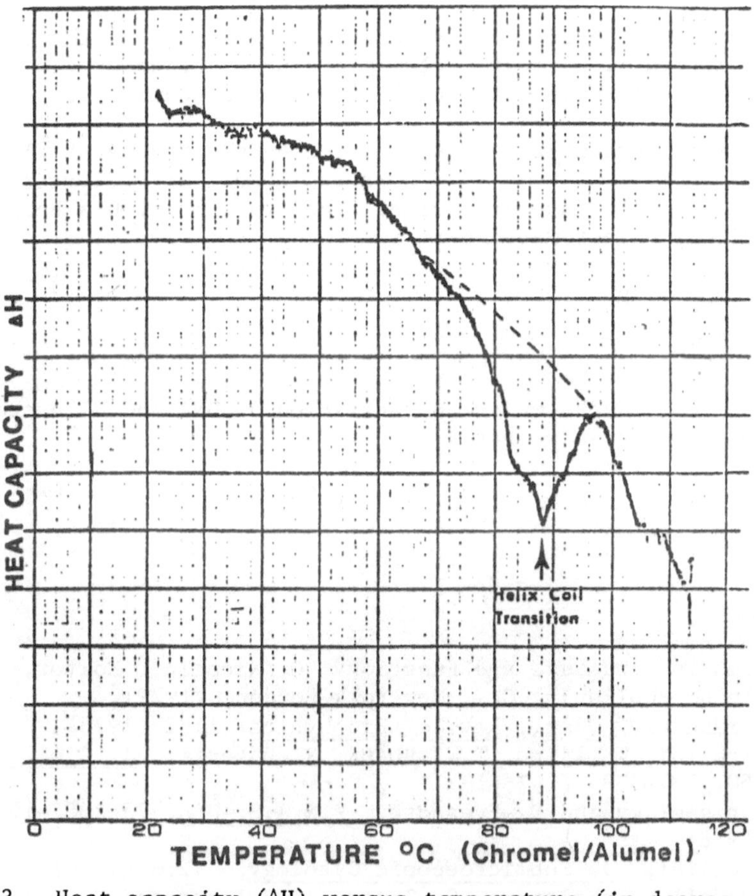

Fig. 13. Heat capacity (ΔH) versus temperature (in degree celsius) for calf thymus DNA, as measured by a Dupont 910 Differential Calorimeter (DSC).

overnight. The preparation was dialyzised again in a solution of 50 mM NaCl, .7mM $NaPO_4$, pH=7.1, before measurement. The heat capacity ΔH_F of the helix-coil transition of free DNA was again determined at ∿.85 °C, with the transition (and remaining heat capacity exchange) of the H1-bound DNA occuring at later temperature. The results of two measurements were 11.5 and 10.8 (again normalized by weight) yielding an average of ΔH_F equal to 11.15 corresponding to ∿29% of the heat capacity for DNA alone.

Spectroscopic absorption measurements were made on the H1-DNA solution, diluted 1,000 fold yielding a concentration of 7.5 x 10^{-3}M and 0.2 x 10^{-3}M for DNA and H1 protein respectively.

H1 protein concentration is computed by absorbance at 219 nm assuming .7 mg/ml/1 OD unit (9). N is the number of nucleotides that one H1 molecule interacts with. The fraction of DNA bound (ΔH_B) to H1 can be estimated by:

$$\frac{\Delta H_B}{\Delta H_D} = \frac{\Delta H_D - \Delta H_F}{\Delta H_D} = \frac{N\ Cp}{C_D}$$

when C_D and Cp are the molar concentration of DNA and H1 protein respectively; and then we get

$$N = \frac{0.71 \times 7.5 \times 10^{-3}}{0.2 \times 10^{-3}} \simeq 26 \text{ nucleotides} = 13 \text{ base pairs.}$$

REFERENCES

1. "Chromatin Structure and Function" C. Nicolini, Editor, Plenum Publishing Co., New York-London, (1979).
2. G.S. Manning, Quant. Rev. of Biophysics 11:2 (1978).
3. C. Nicolini, K. Ajiro, T.W. Borun, R. Baserga, J. Biol. Chem. 250:3381-3385 (1975).
4. R. Yanishelsky and D. Prescott, P.N.A.S.-USA, 75:3307-11 (1978).
5. C. Nicolini, J. Submicroscopic Cytology, 12:3 (1980).
6. F. Kendall, F. Beltrame, S. Zietz, A. Belmont and C. Nicolini, Cell Biophysics, Vol. 2, No. 4 (1980).
7. C. Nicolini, A. Belmont, A. Parodi, S. Abraham and S. Lessin, J. Histo. Cytoche. 27:102-113 (1979).
8. D. Mazia, Scientific American, 230:54 (1974).
9. T.W. Dolby, K. Ajiro, T.W. Borun, R.S. Bilmour, A. Zweidler, L. Cohen, P. Miller and C. Nicolini, Biochem. 18:1333 (1979)
10. C. Nicolini, Cell Biophysics, Vol. 2, No. 4 (1980).
11. P. Rao and S. Hanks, Cell Biophysics, Vol. 2, No. 4 (1980).

12. T. Pederson, P.N.A.S.-USA, p. 2224-2228 (1972).
13. C. Nicolini and F. Kendall, Physiol. Chem. & Physics, 9:265-283 (1977).
14. F. Kendall, R. Swenson, T. Borun, R. Rowinski and C. Nicolini, Science, 196: 1106 (1977).
15. C. Nicolini, F. Kendall, and W. Giaretti, Biophysical Journal, 19:163 (1977).
16. W. Sawicki, J. Rowinski and R. Swenson, J. Cell Physiology, 84:423-428 (1974).
17. G. Setterfield, R. Sheinin, I. Dardick, G. Kiss and M. Bubsky, Journal Cell Biology, 77:247 (1978).
18. C. Nicolini, F. Beltrame and S. Zietz, Science (1980).
19. F. Kendall, F. Beltrame and C. Nicolini, in: "Chromatin Structure and Function", Part A, Editor C. Nicolini, Pelnum Publishing Co., New York-London, pp. 265-292 (1979).
20. C. Nicolini, W. Giaretti, C. Desaive and F. Kendall, Exp. Cell Res., 106:119-125 (1977).
21. E.C. Vonderheid, S. Fang, M.B. Helfrich, S. Abraham, and C. Nicolini, J. Investigative Dermatology (In Press)
22. R. Brooks, D. Bennet, and A. Smith, Cell, 19:493 (1980).
23. G. Maul, H. Maul, J. Scogma, M. Lieberman, G. Stein, B. Hsu and T. Borun, J. Cell Biology, 55:433 (1972).
24. T. Dolby, A. Belmont, T. Borun and C. Nicolini, J. Cell Biology, (1980).
25. S. Parodi, S. Lessin, F. Beltrame, C. Nicolini, Cell Biophysics, 1:271 (1979).
26. M. Grattarola, A. Belmont and C. Nicolini, J. of Cell Science, (In press)
27. J.T. Finch and A. Klug, Proc. Natl. Acad. Sci. USA, 73:1897-1901 (1976)
28. S. Basu, in: "Chromatin Structure and Function", Part A, C. Nicolini, Editor, Plenum Publishing Co., New York-London, 515-540 (1979).
29. J. Bonner, in: "Chromatin Structure and Function" C. Nicolini, Editor, Plenum Press, p. 15-33 (1979).
30. J. Tate and J. Baker, J. Mol. Biol., 118:249-272 (1978).
31. C. Nicolini, in: "Chromatin Structure and Function", C. Nicolini, Editor, Plenum Press, New York-London, 613-616 (1979).
32. D. Marks, W.K. Paik and T.W. Borun, J. Biol. Chem., 248: 5660-5667 (1973).
33. L.R. Gurley, R.A. Walters and R.A. Tobey, BBRC, 50:744-749 (1973).
34. H.W.E. Rattle, G.G. Kneale, J.P. Baldwin, H.R. Mathews, C. Crane-Robinson, P.D. Cary, B.G. Carpenter, P. Suau, and E.M. Bradbury, in: "Chromatin Structure and Function" C. Nicolini, Editor, Plenum Press, New York-London, 451-513 (1979).
35. S. Bannai, and J.S. Shappard, Nature, 250:62-64 (1974).

36. J.E. Froehlich, and M. Rachmeler, J. Cell Biol., 55:19-31 (1972).
37. A. Pardee and L. James, P.N.A.S.-USA, 72:13 (1975).
38. R.A. Walters, L.R. Gurley and R.H. Tobey, Biophysical J., 14:99-118 (1974).
39. J.R. Sheppard, and D.M. Prescott, Exp. Cell Res., 75:293-296 (1972).
40. N. Tanpahaichitr, R. Balhorm, D. Granner and R. Chalkely, Biochemistry, 13:4249-54 (1974).
41. P. Rossow, V. Riddle and A. Pardee, Proc. Natl. Acad. Sci. USA, 76:4446-4450 (1979).
42. A. Belmont and C. Nicolini, J. Theoretical Biol., (1980).
43. K. VanHolde and W. Weischet, in: "The Cell Nucleus", Vol. IV, Harris Busch, Editor, Academic Press, New York (1978).
44. D. Burton, M. Butler, J. Hyde, D. Phyllips, C. Skidmore and L. Walker, in: "Chromatin Structure and Function", Editor, C. Nicolini, Plenum Publishing Co., New York-London (1979).
45. E. Thoma, T. Koller, and A. Klug, J. Cell Biol. 118:248-272 (1979).
46. A. Campbell, R. Cotter and J. Pardon, Nuclei Acids Res., 5:1571-79 (1979).
47. J. Elkington and R. Dulbecco, Proc. Natl. Acad. Sci. USA, 72:1584-1588

ACKNOWLEDGEMENT

This work was supported by NIH Grant CA 20034 (USA), Temple University Research and Study Leave award to C.N. and by the National Research Council, Finalized Project on "Control of Neoplastic Growth", Italy.

THE ORGANIZATION OF GENES IN CHROMOSOMES IN SOME CILIATED PROTOZOA

David M. Prescott, Marshal T. Swanton and Robert E. Boswell
Department of Molecular, Cellular and Developmental Biology
University of Colorado
Boulder, Colorado 80309

Eukaryotic species appear to contain much more DNA in their genomes than can be reasonably accounted for in genetic terms. Mammals, for example, commonly contain DNA sequence complexities of $\sim 10^9$ base pairs (BP). Assuming about 10^3 BP are required to code for an average size polypeptide, then such an organism could contain $\sim 10^6$ genes. This would be at least twenty times more coding capacity than the highest estimates for RNA complexities would require. However, in addition to the amino acid coding capacity many eukaryotic genes have been shown to have one or more transcribed intervening sequences (introns). The gene for ovalbumin, for example, would require only 1,872 nucleotides to code for the structural gene, but the presence of seven intervening sequences expand the "gene" to 7.6 kilobases (KB). The gene for dihydrofolate reductase has been shown to span 42 KB. Thus, at least some of this extra DNA could be accounted for by introns.

Repetitious sequences offer no solution in accounting for the large excess of DNA sequences because repetitious sequences contribute rather little to the total sequence complexity of the genome.

Another phenomenon that might account for large amounts of DNA sequence are the non-transcribed spacers between genes as demonstrated for the genes coding for rRNAs, tRNAs, and histones. These particular spacers separate repeated genes and themselves represent repeated sequences and do not contribute a significant amount of sequence complexity. On the other hand, if single-copy genes, and it is generally assumed that almost all genes are single-copy, were separated from one another by single-copy spacer sequences, a large part of the sequence complexity of a genome might be accounted for by such non-transcribed spacer sequences.

Such a model of the chromosome is, in fact, suggested from studies on Drosophila. Each band of a polytene chromosome in Drosophila contains one, or at most a few genes. The average band contains about 30,000 BP, a length of DNA sufficient to code for about 30 genes. Greater than 85% of the DNA in D. melanogaster is unique sequence DNA. Therefore, the apparent large excess of sequence in each band cannot be accounted for by repetitious sequences. These observations suggest the possibility that most of the DNA in each band of a polytene chromosome is unique sequence DNA that forms spacers between genes.

We present here more direct evidence that much of the unique sequence DNA in eukaryotic chromosomes may form spacers between genes. Our studies exploit the highly unusual structure of the genetic apparatus in those ciliated protozoans known as hypotrichs.

THE MACRO- AND MICRONUCLEI OF CILIATES

Hypotrichous ciliates, like other ciliated protozoa, contain two kinds of nuclei, micronuclei and macronuclei. The micronucleus contains a diploid complement of chromosomes that distribute at division by mitosis. Remarkably, the micronucleus does not carry out detectable RNA synthesis and is genetically inert. The macronucleus contains a much larger amount of DNA but at no stage of vegetative growth are chromosomes perceivable. The macronucleus divides by amitosis, by which it is pinched into two. The mechanism of amitosis is poorly understood; it is not a precise process since the DNA is not distributed exactly equally between daughter nuclei. The macronucleus produces all the RNA (except the trace of RNA produced in mitochondria) needed for vegetative functions, cell growth, and reproduction.

CONJUGATION

The micronucleus does, however, perform a major function in conjugation. In conjugation cells join together in pairs, and the micronuclei in the paired cells undergo meiosis. The cells exchange haploid micronuclei; the exchanged micronuclei fuse with a resident haploid micronucleus (fertilization) to produce a new micronucleus in each cell. The unused haploid micronuclei degenerate, as does the macronucleus, leaving each exconjugant cell with a single, new diploid micronucleus. The new micronucleus then divides by mitosis, and one of the two daughter micronuclei develops into a new macronucleus.

DEVELOPMENT OF THE MACRONUCLEUS AFTER CONJUGATION

Macronuclear development begins with multiple rounds of DNA

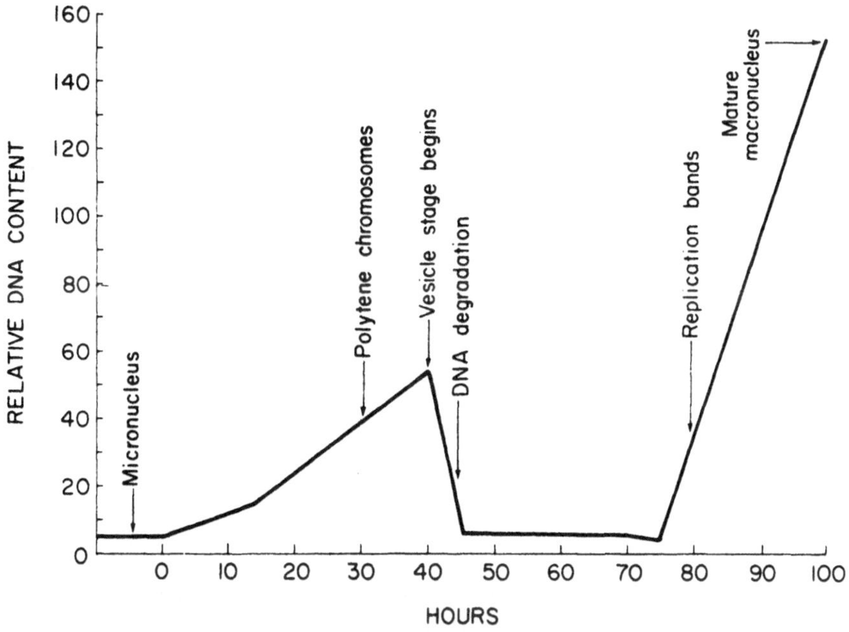

Fig. 1. The changes in DNA during development of the macronucleus in Oxytricha. See text for detailed description.

replication, producing polytene chromosomes (Fig. 1), which are complete by about 40 hours after conjugation (1, 2). The polytene chromosomes are then destroyed by the formation of septa that transect the chromosomes through all of the interbands (3). By this process the chromosomes are cut into short sections, each section corresponding to a chromosome band plus portions of the two adjacent interbands. Each band becomes enclosed within a vesicle, the wall of the vesicle apparently forming from the same kind of material (protein?) that forms the septa through all interbands. In short, the macronucleus is converted from a bag of polytene chromosomes to a bag of vesicles, each vesicle containing the material formerly present in a chromosome band. The number of vesicles is difficult to determine, but it is certainly greater than 10,000.

If the developing macronucleus is isolated and broken open at the vesicle stage, all of the vesicles float free from one another, showing that the polytene chromosomes have truly been transected. This band-by-band transection undoubtedly reduces the molecular weight of the DNA by a large factor, but we have no direct measurement of this.

When the vesicles have formed, most of the DNA in each vesicle

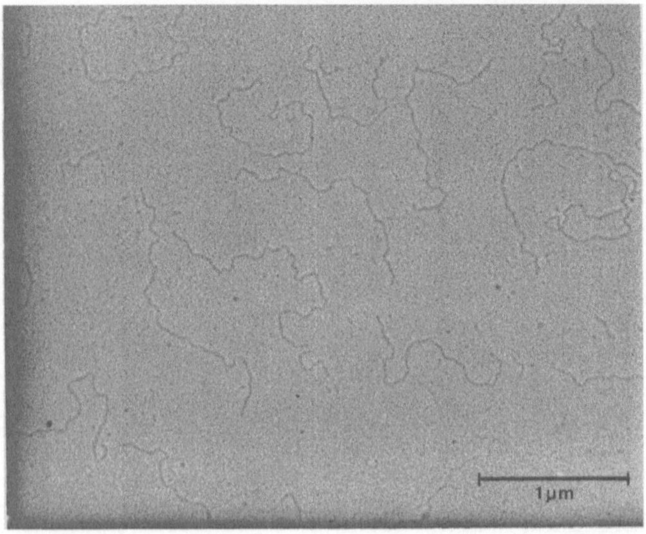

Fig. 2. An electron micrograph of intact, gene-sized DNA molecules from the macronucleus of Oxytricha.

is broken down to nucleotides (Fig. 1), reducing the total amount of DNA in the developing macronucleus by about 95% (2) and creating what is called the DNA-poor stage of macronuclear development. This stage lasts about 24 hours, following which the remaining DNA undergoes five rapid, successive rounds of replication to produce the completed, DNA-rich macronucleus.

All of these post-conjugation events of macronucleus formation require about 4 days at 22°C. When the macronucleus is complete, the cell resumes vegetative growth and may proceed through hundreds of cell cycles before the next conjugation.

THE MOLECULAR WEIGHT OF MICRO- AND MACRONUCLEAR DNAs

The molecular weight of micronuclear DNA is difficult to determine accurately because it is so large. Electron microscope measurements show that the DNA is at least several hundred μm long, corresponding to a molecular weight much greater than several hundred million daltons (4). The size of macronuclear DNA molecules measured by electron microscopy ranges from about 0.14 (∿400 BP) to about 3.0 μm (9,000 BP), with a number average size of about 0.73 μm (2,200 BP) (Fig. 2 and 3). This extremely low molecular weight was verified by extracting intact DNA carefully from macronuclei and separating the molecules by size by electrophoresis in an agarose gel (Fig. 4). When the DNA in the gel is made visible by staining with ethidium bromide, a continuum of DNA molecular sizes ranging from about 400 BP to about 20,000 BP is apparent, although >90% of the molecules are shorter than 5,000 BP. The results are in general agreement with those obtained by electron microscopy.

In addition, a pattern of DNA bands is superimposed on the continuum of DNA in electrophoretic gels. The bands represent DNA molecules of one or another particular size that are present in differentially higher numbers than the other molecules in the size continuum. This banding pattern of macronuclear DNA is constant and highly characteristic for each species of hypotrichous ciliate (5). This confirms that the processing of polytene chromosomes to produce the spectrum of short DNA molecules that make up the genome of the macronucleus is a precise mechanism. Indeed, the data presented in the remainder of this paper confirm that the processing of polytene chromosomes consists in essence in the excision of individual genes or small groups of genes. Most of the DNA molecules, therefore, represent single genes; a small fraction may contain up to several genes, although we have as yet no direct evidence that even the longer molecules (e.g. >10,000 BP) contain more than one gene.

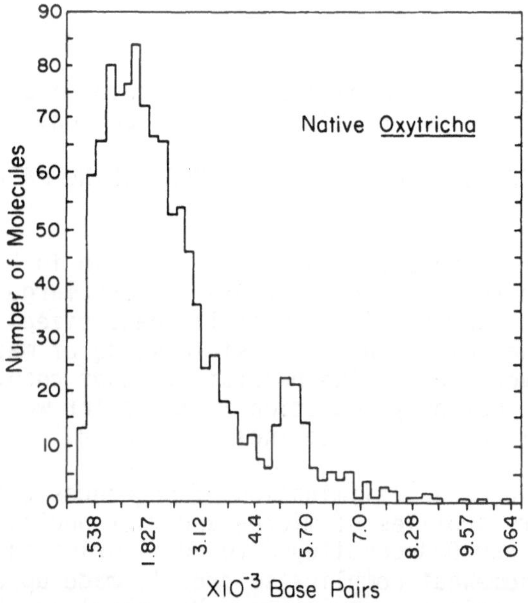

Fig. 3. Histogram showing the size distribution in base pairs of the gene-sized DNA molecules of the macronucleus. DNA of the bacteriophage (small peak) served for size calibration.

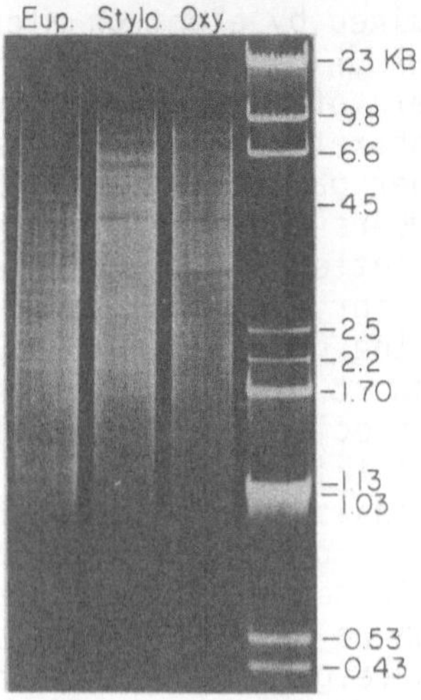

Fig. 4. Size distributions for the DNA molecules from the macronuclei of three species of hypotrichs (Euplotes aediculatus, Stylonychia pustulata, and Oxytricha nova) shown by agarose gel electrophoresis. The column to the right contains calibrating size markers derived by restriction nuclease (Hind III) digestion of λCI-857 and SV40 DNAs.

Characterization of the gene-sized molecules and identification of particular genes is described in a later section.

REDUCTION IN THE SEQUENCE COMPLEXITY OF DNA DURING MACRONUCLEAR DEVELOPMENT

The events of macronuclear development shown in Fig. 1 conform to the observation that micronuclear DNA is very large (chromosome-sized) and macronuclear DNA is very small (gene-sized). The destruction of most of the DNA during the vesicle stage of macronuclear development, however, raises the possibility that changes in the genome may involve not only reduction in molecular weight but also reduction in total nucleotide sequence complexity.

The first evidence of elimination of sequences came from determination of density profiles of micro- and macronuclear DNAs (Fig. 5) (6). Micronuclear DNA certifuged to equilibrium in a CsCl solution generates a somewhat complicated profile made up of at least four different DNA density components. Macronuclear DNA under the same conditions forms a symmetrical peak, indicating the presence of only a single density component. Thus, the formation of the macronucleus involves the elimination from the genome of three or more density components. The elimination of these components occurs in two stages. The DNA of some density components does not replicate during polytenization of the chromosomes (7). Other density components are destroyed during the vesicle stage of DNA breakdown. We assume that the micronuclear diploid copies of density components that do not participate in polytenization of the chromosomes are also eliminated at some point, but we have no independent evidence of this.

Reassociation kinetics of denatured DNA provide a more quantitative means of detecting differences in sequence complexity between micro- and macronuclear DNAs (8). Figure 6 shows the reassociation curves (Cot curves) for E. coli DNA, which serves as a calibration standard, and for macronuclear and micronuclear DNA. The reassociation curve for micronuclear DNA shows that about 30% consists of repetitious sequences and about 70% are unique. Macronuclear DNA reassociates as a single component, showing that at least most of the sequences are present at about the same frequency.

From the reassociation curve the complexity of macronuclear DNA is estimated to be equivalent to $\sim 5.4 \times 10^7$ BP (3.6×10^{10} daltons). The kinetic complexity of the unique sequence DNA in the micronucleus is about 1.26×10^9 BP (8.5×10^{11} daltons). We can estimate from these numbers that about 96% of the unique sequences present in micronuclear DNA are destroyed during formation of the macronucleus. Reasons why repetitious DNA sequences of the micronucleus cannot represent more than a small fraction of the sequences in the macronucleus are presented elsewhere (8). Indeed, it is clear that most of the repetitious sequences are lost during macronuclear development.

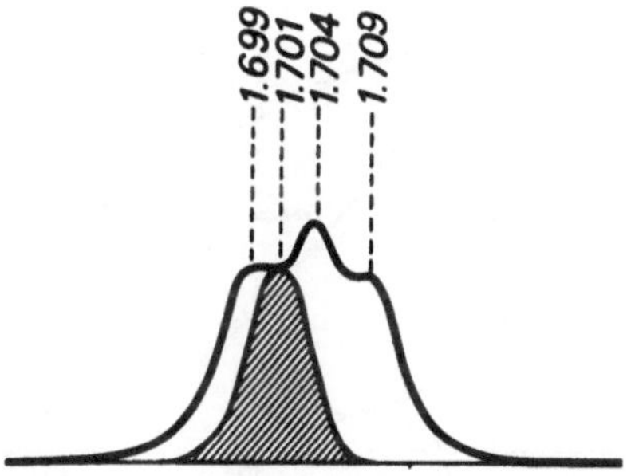

Fig. 5. Density profile of micronuclear and macronuclear DNAs of Oxytricha centrifuged to equilibrium in CsCl. The micronuclear DNA profile is complex; the macronuclear DNA profile (shaded) is symmetrical.

In short, we can conclude that the macronucleus contains only about 4% of the unique sequences present in the micronucleus. This conclusion is important because no detectable RNA synthesis occurs in the micronucleus, and all RNA (except mitochondrial RNA synthesis) needed for cell functions is transcribed from the gene-sized DNA molecules in the macronucleus. We conclude, therefore, that 96% of the unique sequence DNA in the micronuclear chromosomes does not have a genetic coding function, i.e. is not genetic in the traditional sense.

THE ARRANGEMENT OF MACRONUCLEAR GENES IN MICRONUCLEAR CHROMOSOMES

The cytological events of macronuclear development coupled with the changes in DNA during development lead rather forcefully to a model of how macronuclear genes are arranged in micronuclear chromosomes (9). First, we know from the band-by-band transection of the polytene chromosomes and the subsequent destruction of most of the DNA in each band (during the vesicle stage) that the macronuclear sequences must be dispersed throughout the chromosome. It may be reasonably assumed that each band yields one or a few specific macronuclear molecules. The DNA destroyed during the vesicle stage is believed to be nongenetic DNA that separates the macronuclear sequence in one band from the macronuclear sequence in the next band. This nongenetic DNA, moreover, is mostly unique sequence DNA. Thus, as shown in Fig. 7, we conclude that most of the DNA of micronuclear chromosomes is nontranscribed spacer, separating the sequences that give rise to the gene-sized molecules of the macronucleus. (Of course, all of the observations could also be accounted for by a processing model in which all of the eliminated

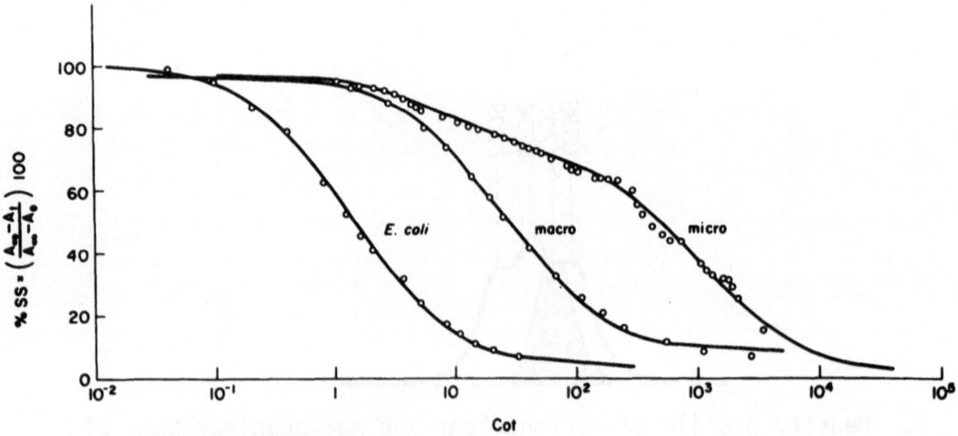

Fig. 6. Reassociation kinetics of macronuclear and micronuclear DNAs of Oxytricha. DNA of E. coli is included for reference.

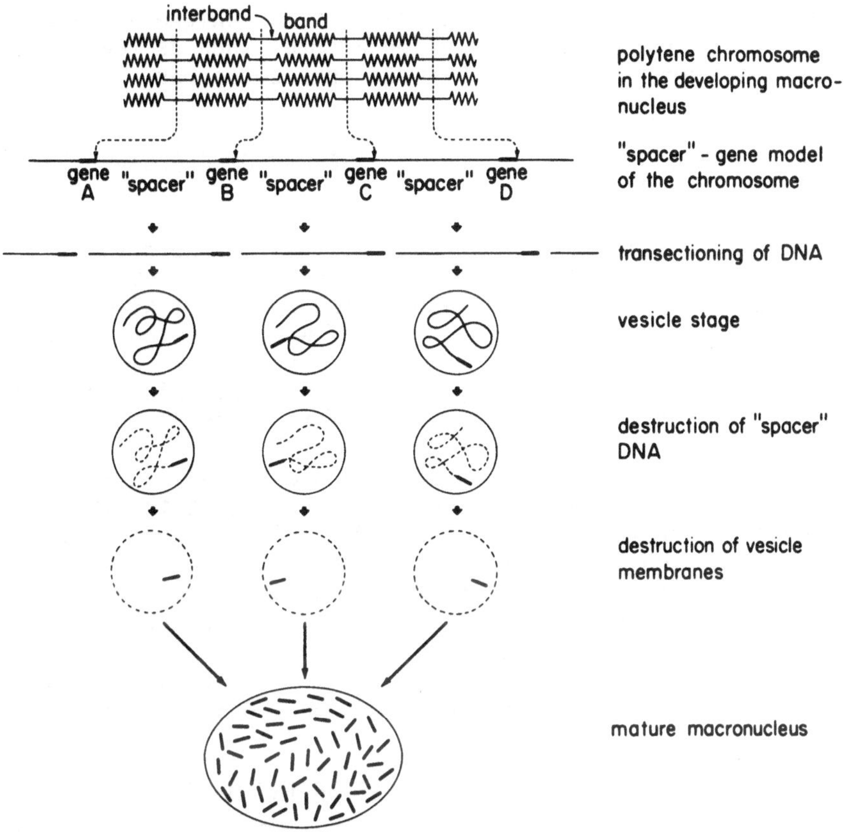

Fig. 7. A model to explain the creation of gene-sized DNA molecules in the macronucleus during the formation of the macronucleus following conjugation.

sequences are introns in genes. According to such a hypothesis 96% of the unique sequences in the chromosomes would make up introns. The model also requires extensive splicing of DNA fragments created by intron excision.)

We have called the nongenetic DNA hypothesized to occur between gene sequences "spacer" in analogy with the nontranscribed spacer DNAs that are present between the successive copies of rRNA, tRNA, and histone genes in other organisms. The function of these nontranscribed spacer sequences is not known, and we can offer no compelling hypothesis for the function of the putative spacers in hypotrichs. Since the chromosomes of the micronucleus do serve in meiosis and the sexual exchange of DNA, it may be that spacers have a role connected with these activities, e.g. serving as regions for crossing over during meiosis.

Finally we suggest that a major part of the unique sequence DNA in virtually all eukaryotes (simpler eukaryotes such as yeast may be exceptions) may serve as spacers between gene sequences.

THE NUMBER OF KINDS OF DNA MOLECULES IN THE MACRONUCLEUS AND THEIR MULTIPLICITY

The kinetic complexity of macronuclear DNA of the hypotrich *Oxytricha nova* is 5.4×10^7 BP and the number average molecular size of the DNA is 2,200 BP. Therefore, the number of kinds of molecules is $\frac{5.4 \times 10^7}{2.2 \times 10^3}$ or about 24,500. If we accept the hypothesis that each molecule is a gene, then we conclude that this hypotrich (*Oxytricha nova*) contains about 24,500 different genes.

The multiplicity of each kind of molecule (gene) can also be calculated. From microspectrophotometry we know that the macronucleus contains about 5.1×10^{10} BP of DNA (8). We know from EM measurements that the number average molecular size of macronuclear DNA is 2,200 BP. The total number of molecules per macronucleus is therefore, $\frac{5.1 \times 10^{10}}{2.2 \times 10^3}$ = about 23×10^6. Since there are about 24,500 different kinds of molecules, each kind of molecule must be present in just under 1,000 copies per macronucleus.

THE STRUCTURE OF THE GENE SIZED MOLECULES OF THE MACRONUCLEUS

The first major clue about the fine structure of macronuclear DNA molecules came from the observation that heat denaturation of purified DNA followed by rapid cooling leads to the formation of single-stranded circles by most of the DNA (10). The single-stranded circles are held together by a short duplex region, or neck, formed by complementary sequences present at the ends of a single-stranded molecule as shown in Fig. 8.

This implies that the duplex (native) DNA molecules have the following configuration.

ABC	cba
abc	CBA

The nucleotide sequences at both the 3' and 5' ends of total macronuclear DNA have now been determined in four hypotrich species. The inverted terminal repeat designated above as ABC..... / abc..... has the following basic sequence.

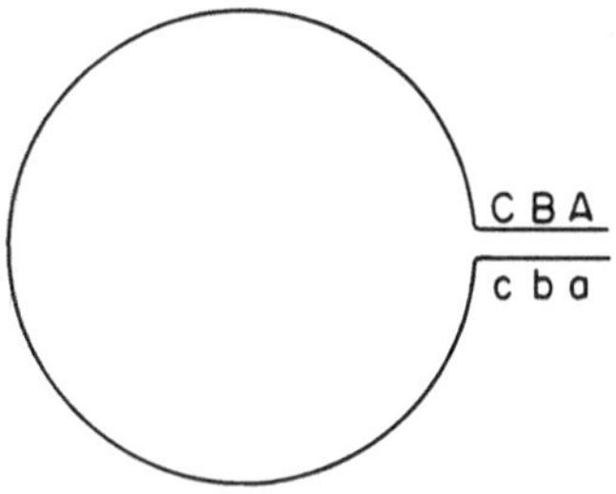

Fig. 8. Diagram of single stranded circular DNA molecule formed from denatured macronuclear DNA. Such circles are held together by an inverted repeated sequence $\left(\frac{ABC}{abc}\right)$.

$$\begin{array}{l} 5'\ C_4A_4C_4A_4C_4\text{--------------------}G_4T_4G_4T_4G_4T_4G_4T_4G_4\ 3' \\ 3'\ G_4T_4G_4T_4G_4T_4G_4T_4G_4\text{--------------------}C_4A_4C_4A_4C_4\ 5' \end{array}$$

In one species (Euplotes aediculatus) the sequence extends into the molecules for 8 more base pairs (A_4C_4). Thus, all or nearly all of the 24,500 different molecules have the same inverted terminal repeat sequences. In addition, all molecules terminate in a 16 base, 3' single stranded protrusion consisting of 3' $G_4T_4G_4T_4$......

The gene-sized molecules therefore all possess the essential property of transposons identified in a variety of eukaryotes, namely inverted terminal repeats. The inverted terminal repeats of transposons are the basis for the frequent translocation of some genes within a genome. In the case of hypotrichs, we believe that the inverted terminal repeats serve a similar function, namely excision of the genes from the polytene chromosome.

The single stranded protrusion of 3' $G_4T_4G_4T_4$ is, on the other hand, probably generated by or in some way involved in the replication of the gene-sized molecules (11).

CODING FUNCTIONS OF PARTICULAR SIZE CLASSES OF MACRONUCLEAR DNA MOLECULES

How genes are encoded within the low molecular weight macronuclear DNA was demonstrated by hybridization of specific gene probes to Southern blot transfers of macronuclear DNA that had been separated by size by gel electrophoresis. Specific gene sequences are found in discrete size classes. This was initially demonstrated for the gene coding for the 19S and 25S ribosomal RNAs. Ribosomal RNA was found to hybridize specifically to a single band in macronuclear DNA of ∿8 KBP, which corresponds to a prominent band that is visible in ethidium stained gels of macronuclear DNA.

Subsequently, the gene coding for 5S RNA has been shown to be encoded by a single size class of 690 BP (12). Not all gene probes have revealed single size classes. For example, using the cloned sea urchin gene repeat (containing all 5 histone genes) as a probe, approximately 20 different sizes of DNA molecules show specific hybridization (13). The bands range in size from 7.6 KBP to approximately 0.73 KBP. If the five sea urchin histone genes are separated from one another and used individually as hybridization probes, each hybridizes to several discrete bands, and no two histone genes hybridize to the same set of bands (14). Many of the macronuclear DNA molecules that hybridize to individual sea urchin histone genes are large enough to encode many copies of that gene. However, it is not known if multiple copies are present on these large molecules.

The genes coding for actin have also been assigned to three discrete sizes of macronuclear DNA (15). These three macronuclear DNA size classes presumably represent a family of actin genes. Each size class could accommodate only a single actin gene sequence.

The conclusion from these hybridizations is clear. Genes in the macronucleus are present as discrete size classes of DNA molecules.

As a first approach to the detailed characterization of macronuclear gene sequences, the intact rDNA gene was cloned into the E. coli plasmid pBR322. The organization of this gene is shown in Figure 9. A long, (1,540 BP) non-coding, leader sequence is followed by the coding sequences for the 19S and 25S rRNAs, which are followed by a non-coding terminal sequence of 620 BP. Those sequences required for the regulation of expression of this gene presumably are contained in the 1,540 BP leader sequence. The terminal 620 BP presumably has specific sequence information required for termination of transcription.

PROSPECTIVES

The study of the genetic apparatus in hypotrichs is continuing along four lines. First, we are now testing the chromosome model in Fig. 7, primarily by analyzing large, cloned segments of micronuclear DNA in lambda bacteriophage. The results so far show that at least some macronuclear DNA sequences are indeed flanked by micronuclear DNA segments (spacers) in a manner that agrees with the model in Fig. 7 (13).

Second, the inverted terminal repeat sequences of macronuclear DNA molecules are probably important in the excision of macronuclear genes from micronuclear chromosomes. Our immediate objective is to determine the nucleotide sequence at the point at which a

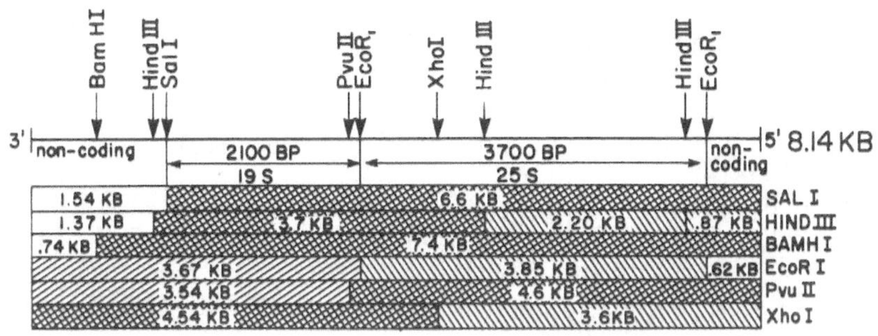

Fig. 9. Organization of coding and non-coding regions of the DNA molecule that codes for 19S and 26S ribosomal RNAs, determined by digestion with various restriction nucleases and hybridization to radioactive rRNA.

macronuclear gene is joined to micronuclear DNA in the chromosome. We have prepared the clones of micronuclear DNA needed to sequence this junction region (13). A major question is whether the sequence in the micronuclear DNA immediately adjacent to the inverted terminal sequence of macronuclear DNA segments contains a further clue to processing, e.g. a micronuclear copy of the inverted terminal repeat arranged as a palindrome.

Third, we continue to analyze cloned copies of gene-sized molecules from the macronucleus to learn about gene structure and function, particularly regulation of initiation and termination of transcription.

Fourth, the single stranded 3' extensions possessed by all gene-sized molecules of the macronucleus may be an important clue about the mechanism of replication of linear DNA molecules, particularly the so-called 5' RNA primer problem. Macronuclear DNA is especially favorable for such study because of the high ratio of ends to molecular weight. An immediate goal is to determine how the 3' tails are generated.

ACKNOWLEDGEMENTS

This work was supported by NIGMS grant #GM19199 to D.M.P.

REFERENCES

1. Ammermann, D. (1964) Naturwissenschaftern 51, 249.
2. Ammermann, D. (1965) Ehrenberg. Arch. Protistenk 108, 109-152.
3. Kloetzel, J.A. (1970) J. Cell Biol. 47, 395-407.

4. Prescott, D.M., Bostock, C.J., Murti, K.G., Lauth, M.R. & Gamow, E. (1971) Chromosoma 34, 355-366.
5. Swanton, M.T., Heumann, J.M. & Prescott, D.M. (1980) Chromosoma 77, 217-227.
6. Bostock, C.J. & Prescott, D.M. (1972) Proc. Natl. Acad. Sci. USA 69, 139-142.
7. Spear, B.B. & Lauth, M.R. (1976) Chromosoma 54, 1-13.
8. Lauth, M.R., Spear, B.B., Heumann, J. & Prescott, D.M. (1976) Cell 7, 67-74.
9. Prescott, D.M. & Murti, K.G. (1973) Cold Spring Harbor Symp. Quant. Biol. 38, 609-618.
10. Wesley, R.D. (1975) Proc. Natl. Acad. Sci. USA 72, 678-682.
11. Klobutcher, L.A., Swanton, M.T., Donini, P. & Prescott, D.M. (1981) Proc. Natl. Acad. Sci. USA, in press.
12. Rae, P.M.M. & Spear, B.B. (1978) Proc. Natl. Acad. Sci. USA 75, 4992-4996.
13. Boswell, R.E., unpublished results.
14. Elsevier, S.M., Lipps, H.J. & Steinbruck, G. (1978) Chromosoma 69, 291-306.
15. Kaine, B.P. & Spear, B.B. (1980) Proc. Natl. Acad. Sci. USA 77, 5336-5340.

CELL CYCLE PHASE-SPECIFIC CHANGES IN RELAXATION TIMES AND WATER CONTENT IN HELA CELLS

Potu N. Rao[1], Carlton F. Hazlewood[2] and Paula T. Beall[2]

[1]Department of Developmental Therapeutics, The University of Texas System Cancer Center M. D. Anderson Hospital and Tumor Institute, Houston, Texas
and
[2]Departments of Pediatrics and Physiology, Baylor College of Medicine, Houston, Texas

The ubiquity of water molecules and their essential role in life processes is well known. Although a number of quantitative methods are now available for the study of water, the structure and function of this simple molecule in solutions and in biological systems is not well understood. Much of the early work with water in biological tissues was confined primarily to simple measurements of water content of different tissues. In 1913, it was noted that the percentage of water in some tumors is higher than that of the host tissues (1). Cramer, in 1916, presented evidence supporting the notion that the percentage of water in tumors is directly related to the rate of tumor growth (2). McEwen et al. (3) later reported that the increased percentage of water in tumors was not due to the loss of dry solids. Evidence demonstrating a systemic effect of tumors on the water content of other organs was reported in 1932 by Schlottman and Rubenow (4). By 1943, this systemic effect of tumors on the percentage of water in organs was well known (1). After the 1940's, interest in the water content of tumors seemed to wane.

An improvement in technology in the 1950's, however, did permit studies of the water content of single cells. Mellors et al. (5) using multiple-beam interference microscopy demonstrated that individual sarcoma cells contained larger amounts of water than embryonic fibroblast cells which were used for comparison. Later, Downing et al. (6), using interferometry, demonstrated the hydration (H_2O/dry mass) of nuclei of normal and cancer cells were the same forcing the conclusion that the increase in hydration of tumor cells must be cytoplasmic and/or extracellular.

Even with these quantitiative methods to determine the amount of water in cells, no real insights were gained as to the role of water in the neoplastic process. Certainly, no general theory of neoplasia involving water as any thing other than a passive entity was ever enunciated. Therefore, the meaning of the elevated percentage of water in malignant cells remained unknown. The development of nuclear magnetic resonance spectroscopy (7) in the forties paved the way for new quantitative studies of the molecular motion of water in living tissues and revived the interest of scientists in water per se and its role in biological tissues.

Damadian's pioneering application of nuclear magnetic resonance to the quantitative study of water molecules in tumors has stimulated a renewed interest in the role of water in neoplasia. Damadian (8) and Hazlewood et al (9) have attributed the lengthened relaxation times of water hydrogens in tumors to a change in physical state of water in those tissues, while others have argued that these changes are due simply to a change in the degree of hydration (10,11,12).

In a attempt to further evaluate the role of water in cell growth and division, we decided to study the changes in water content and relaxation times of water protons during the HeLa cell cycle. As a classic cell line of human cervical carcinoma, it has been well characterized biochemically and can be synchronized in different phases to eliminate many of the variables that plague the interpretation of NMR data. The objective of this chapter is to present comprehensive data on the interrelationships of water and relaxation times as a function of HeLa cell cycle and to discuss the possible mechanisms underlying this phenomenon. A preliminary report of this work has appeared elsewhere (13).

METHODS

Nuclear Magnetic Resonance. A Bruker SXP nuclear magnetic resonance spectrometer (30 MHz) was used to measure the T_1 by a 180-τ-90° pulse sequence, and the T_2 by a Carr-Purcell sequence (14) with the Meiboom-Gill correction at 25°C. The water, culture media, salt and drug solutions were tested for the presence of paramagnetic impurities by comparison to a water standard. The addition of 10% fetal calf serum to Eagle's MEM slightly depressed the T_1 of the medium, which was probably due to the macromolecular content of the serum. All other solutions showed no depression of the T_1 of the distilled water in which they were dissolved.

Cell Culture and Cell Synchrony. HeLa cells were grown as monolayer cultures in Falcon plastic dishes at 37°C in Eagle's minimal essential medium (MEM) supplemented with Eagle's non-

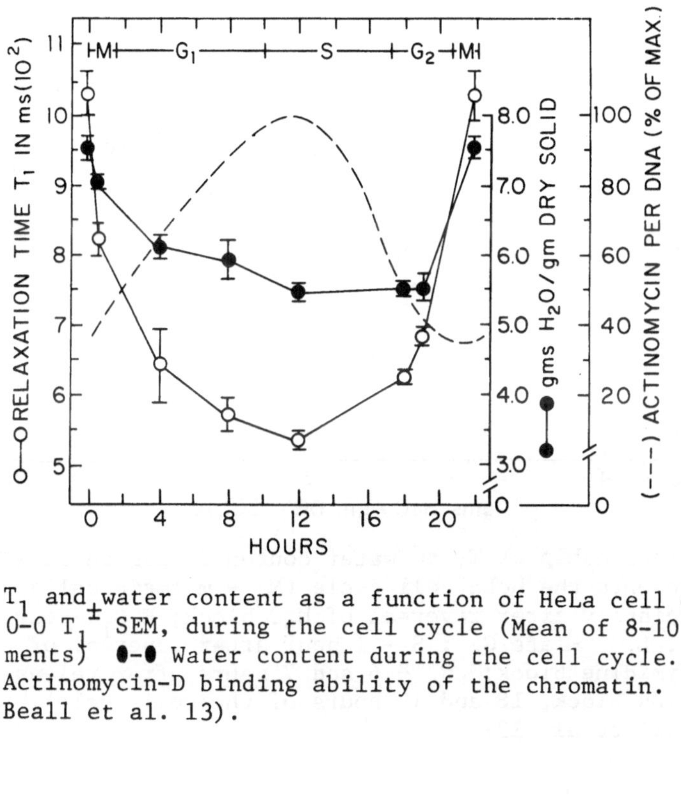

Fig. 1. T_1 and water content as a function of HeLa cell cycle. O-O $T_1 \pm$ SEM, during the cell cycle (Mean of 8-10 experiments) ●-● Water content during the cell cycle. - - Actinomycin-D binding ability of the chromatin. (From Beall et al. 13).

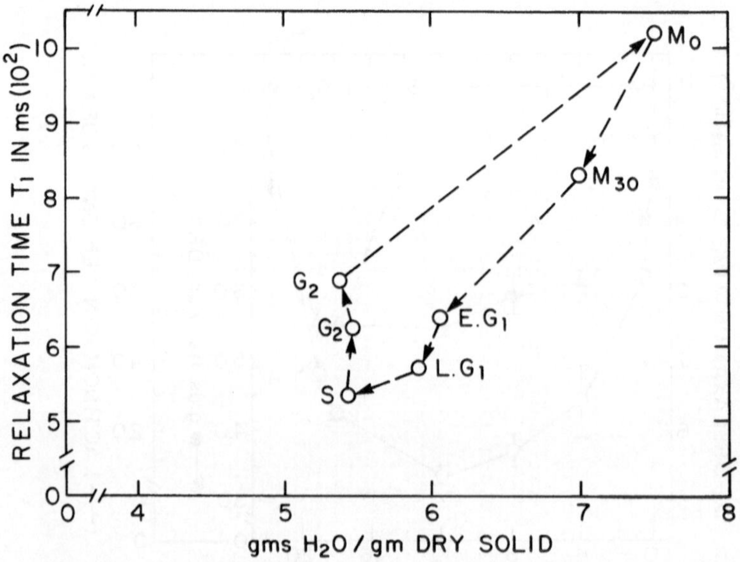

Fig. 2. Relationship of T_1 to water content. Dashed line represents the HeLa cell cycle (M_o - mitotic cells immediately after reversal of N_2O block; $E.G_1$ - early G_1; $L.G_1$ -late G_1 ; S - 1 hour after reversal of a double thymidine block; G_2 - 6 and 7 hours after release of 2xTdR block, 18 and 19 hours of the cell cycle). (from Beall et al. 13).

essential amino acids, heat inactivated fetal calf serum (10%), sodium pyruvate, glutamine and penicillin-streptomycin mixture (15). Cells were synchronized by different methods in the pre-DNA synthetic period (G_1); DNA synthesis period (S); post-DNA synthetic period (G_2); and in mitosis (M). A population of cells, partially synchronized by a single excess thymidine block, were treated with either Colcemid (0.05 mg/ml) or N_2O (80 psi) to obtain mitotic cells. Incubation of N_2O - blocked mitotic cells under regular culture conditions for 4 hr yielded a highly synchronous population of G_1 cells (16). The technique of excess thymidine double block was used to synchronize cells in S and G_2 phases (17). S and G_2 phase cells were harvested at one and six hours, respectively, after the reversal of the second thymidine block. About 2 x 10^7 cells were required for each sample. Samples were prepared for NMR by gently scraping the cells from the dish with a teflon policeman and centrifuging them in the NMR sample tube at 1000 g for 20 minutes. The supernatant medium was removed by suction without disturbing the tightly packed cell pellet, and the samples were sealed and placed on ice until rewarmed to room temperature for NMR measurements.

Determination of the Percentage of Water. Following the NMR measurements, the culture cells were transfered to tared dishes and initial weights determined. The dishes with cells were then placed in an oven at 105°C. The dishes were reweighed at 24 hour intervals until a constant weight was determined.

RESULTS & DISCUSSION

Changes in Relaxation Times During Cell Cycle. The relaxation times T_1 and T_2 and percentage of water for synchronized HeLa cells at various stages of the cell cycle are presented in Table 1. Figure 1 presents the average values for T_1 and and the percentage water as a function of cell cycle. Figure 2 is a plot of the T_1 against the average degree of hydration of the HeLa cells at various stages of the cell cycle. In another study, separate random populations of HeLa and Chinese hamster ovary (CHO) cells were placed in soluations of 0.03-0.5 M NaCl for 15 minutes at 37°C to allow for equilibration and to test the effects of hydration on T_1. The cells were then transfered to NMR tubes and centrifuged at identical g forces for the same periods of time. The pellets were then taken for NMR measurements in the same manner as above. Figure 3 and Table 2 show the effect of hydration on the T_1 of random HeLa and CHO cell populations.

In studies with synchronized cells, the relaxation times were maximum in the M phase and minimum in the S phase (Table 1 and Figure 1). Specifically, mitotic cells had the maximum T_1 value of 1020 ± 84 ms (± SD). The T_1 and T_2 values for mitotic cells

TABLE 1

The HeLa Cell Cycle - Nuclear Magnetic Resonance Parameters and Water Content

Synchronized Stage	Method of Synchrony	Time After Release of Mitotic Block		T_1ms		T_2ms		%.H_2O		$\frac{\text{gms } H_2O}{\text{gm Dry Solids}}$
M_o	(N_2O)			1135		128		---		---
M_o	(N_2O)			1096		137		---		---
M_o	(N_2O)	0 min		1029		142		88.16		7.45
M_o	(N_2O)			917		142		---		---
M_o	(N_2O)			948		115		88.00		7.33
M_o	(2X TdR + Colc.)	24 hrs		995		115		88.60		7.77
			Mean	1020	Mean	130	Mean	88.25	Mean	7.51
			SD ±	84	SD ±	13	SD ±	0.31	SD ±	0.23
			SE ±	34	SE ±	5	SE ±	0.20	SE ±	0.13
M_{30}	(N_2O)			810		107		---		---
M_{30}	(N_2O)			809		104		---		---
M_{30}	(N_2O)			714		114		87.56		7.04
M_{30}	(N_2O)	30 min		894		146		---		---
M_{30}	(N_2O)			736		143		87.45		6.97
M_{30}	(N_2O)			836		144		---		---
M_{30}	(N_2O)			922		131		87.41		6.94
			Mean	817	Mean	127	Mean	87.47	Mean	6.93
			SD ±	76	SD ±	18	SD ±	0.10	SD ±	0.05
			SE ±	29	SE ±	7	SE ±	0.07	SE ±	0.03

TABLE 1 (continued)

The HeLa Cell Cycle - Nuclear Magnetic Resonance Parameters and Water Content

Synchronized Stage	Method of Synchrony	Time After Release of Mitotic Block	T_1ms	T_2ms	%.H_2O	$\frac{\text{gms } H_2O}{\text{gm Dry Solids}}$
S	(2X TdR)		585	105	84.21	5.33
S	(2X TdR)		520	103	84.21	5.33
S	(2X TdR)	12 hrs	462	99	84.53	5.46
S	(2X TdR)		506	---	85.13	5.72
S	(2X TdR)	12 hrs	513	---	85.82	6.05
S	(2X TdR)		509	109	---	---
S	(2X TdR)		511	112	---	---
S	(2X TdR)		585	---	83.39	5.02
S	(2X TdR)		570	89	84.10	5.29
S	(N_2O + 12 hrs)		582	88	84.00	5.25
			Mean 534	Mean 100	Mean 84.42	Mean 5.43
			SD ± 43	SD ± 9	SD ± 0.75	SD ± 0.32
			SE ± 14	SE ± 3	SE ± 0.26	SE ± 0.11
G_2	(2X TdR + 6 hrs)		629	106	---	---
G_2	(2X TdR + 6 hrs)		606	---	84.38	5.40
G_2	(2X TdR + 6 hrs)	18 hrs	585	88	84.71	5.54
G_2	(2X TdR + 6 hrs)		638	94	---	---
G_2	(2X TdR + 6 hrs)		648	97	---	---
			Mean 621	Mean 96	Mean 84.54	Mean 5.47
			SD ± 25	SD ± 8	SD ± ---	SD ± 0.10
			SE ± 5	SE ± 4	SE ± ---	SE ± 0.07

TABLE 1 (continued)

The HeLa Cell Cycle - Nuclear Magnetic Resonance Parameters and Water Content

Synchronized Stage	Method of Synchrony	Time After Release of Mitotic Block	T_1ms	T_2ms	%.H_2O	gms H_2O / gm Dry Solids
G_2	(2X TdR + 7 hrs)		692	---	84.92	5.63
G_2	(2X TdR + 7 hrs)		692	---	84.0	5.25
G_2	(2X TdR + 7 hrs)	19 hrs	692	---	85.0	5.67
G_2	(2X TdR + 7 hrs)		685	---	83.73	5.15
G_2	(2X TdR + 7 hrs)		692	---	84.55	5.47
G_2	(2X TdR + 7 hrs)		685	---	83.51	5.06
			Mean 690	Mean ---	Mean 84.28	Mean 5.37
			SD ± 4	SD ± ---	SD ± 0.63	SD ± 0.26
			SE ± 2	SE ± ---	SE ± 0.26	SE ± 0.10

obtained by Colcemid treatment were identical with those collected by N_2O block. At 30 minutes after the reversal of the N_2O block, when the damage to the mitotic spindle was repaired and the chromosomes were realigned on the metaphase plate (18), the T_1 had decreased rapidly to about 800 ms. The realignment of chromosomes and subsequent decrease in T_1 was prevented if the mitotic cells were held at 4°C immediately after the release of the N_2O block. This suggests that it is not the time interval after N_2O block, but the structural organization following the reversal of the block that is responsible for the decrease in T_1. The values for T_1 decreased throughout G_1 and reached its minimum value of 534 ± 43 ms in S phase. As the cells progressed to G_2, the T_1 started to rise to 621 ± 25 ms in early G_2 and in later G_2 to 690 ms and ultimately returned to the maximum as the cells re-entered mitosis. The reproducible cyclic pattern of T_1 shown in Figure 1 appears to be inversely related to the degree of chromatin condensation, as measured by the amount of actinomycin-D bound to DNA during the cell cycle.

Due to the large nuclear volume of the HeLa cell, we felt this pattern of change in T_1 might be partially due to changes in the physical state of the water molecules associated with the chromatin undergoing condensation and decondensation during the cycle. The other possibility that the pattern was due only to changes in the degree of hydration of the cells was also explored in synchronized and random populations (Figures 2 & 3).

Microscopic examination of random populations exposed to various concentrations of sodium chloride revealed marked changes in the morphology of the cells, including rupture of the cell membranes of HeLa cells outside the range of 0.05-0.25 M NaCl. CHO cells (grown as monolayers in McCoy's modified medium) were able to withstand a greater range of salt concentration, but they too demonstrated shrunken nuclei, crenated nuclear membranes and condensed chromatin at high salt concentrations. In hypotonic solutions of NaCl, the cells were swollen and the cytoplasm was highly vacuolated. In Figure 3, the nonlinearity of water content versus T_1 is shown for both cell lines. It appears that within narrow physiological limits the changing cellular water content and T_1 are linear, but when gross morphological changes become visible, this relationship deviated from linearity. In both the synchronized and random populations, it is obvious that the percentage of water is not always a sensitive determinant of relaxation times (Figures 1-3; Tables 1-3).

Effect of Spermine. It is well known that dramatic changes occur in the nucleus during the cell cycle. The actinomycin-D binding ability of HeLa cells during the cell cycle (Figure 1) gives an indication of the openness of the nuclear chromatin in each phase. Since relaxation times of water protons are at a

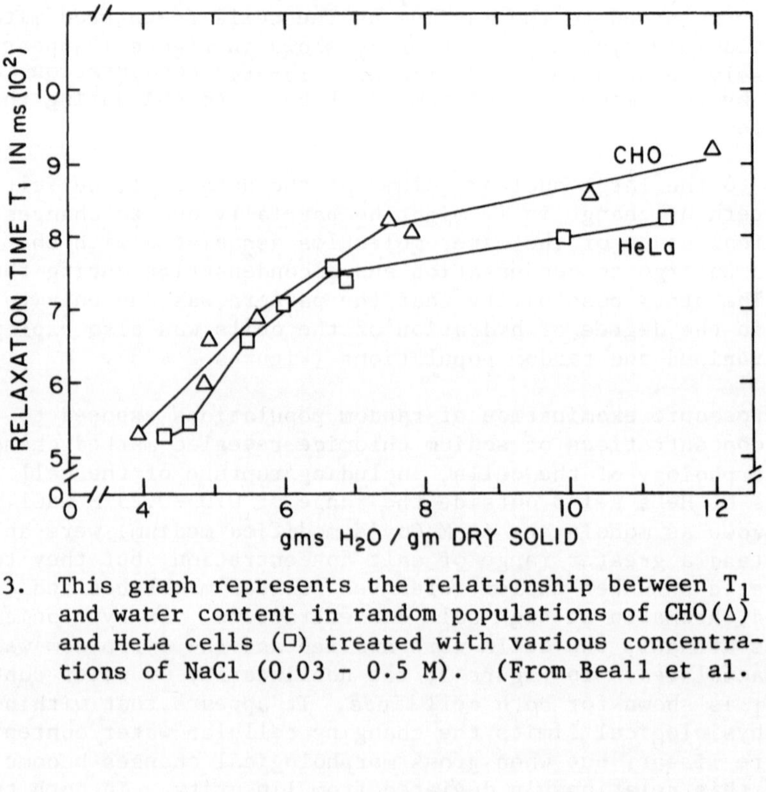

Fig. 3. This graph represents the relationship between T_1 and water content in random populations of CHO(Δ) and HeLa cells (□) treated with various concentrations of NaCl (0.03 - 0.5 M). (From Beall et al. 13)

TABLE 2

THE RELATIONSHIP OF WATER CONTENT AND T_1 AND T_2 OF CELL SAMPLES

Sample		T_1ms	T_2ms	%H_2O	gm H_2O / gm Dry Solids
HeLa Cell Cycle Stages					
M_o	(0 min)	1020	130	88.2	7.51
M_{30}	(30 min)	817	127	87.5	6.98
G_1 early	(4 hrs)	638	110	85.8	6.02
G_1 late	(8 hrs)	570	117	85.5	5.88
S	(12 hrs)	534	100	84.4	5.42
G_2	(18 hrs)	621	96	84.5	5.47
G_2	(19 hrs)	690	---	84.3	5.37
Random HeLa Cells in NaCl					
0.05 M		824	160	91.9	11.36
0.05 M		796	---	90.9	9.94
0.075 M		758	133	87.0	6.67
0.075 M		735	---	87.4	6.92
0.10 M		709	112	85.8	6.06
0.15 M		651	116	84.8	5.56
0.20 M		541	97	82.4	4.68
0.25 M		526	90	81.2	4.31
Random Chinese Hamster Ovary Cells in NaCl					
0.03 M		917	---	92.3	12.00
0.05 M		847	---	91.1	10.27
0.075 M		801	---	88.7	7.82
0.10 M		823	---	88.3	7.51
0.15 M		689	---	85.0	5.65
0.20 M		660	---	83.1	4.97
0.25 M		598	---	83.1	4.91
0.50 M		530	---	79.8	3.96

TABLE 3

EFFECT OF SPERMINE ON THE T_1 OF SYNCHRONIZED S PHASE HELA CELLS

Treatment	Duration of Incubation	T_1 (ms)	$\%H_2O$	$\frac{\text{gms } H_2O}{\text{gm Dry Solid}}$
Control	1 hr	538	84.4	5.523
Spermine (0.02M)	1 hr	638	84.4	5.423
Control	2 hrs	546	84.6	5.493
Spermine (0.02M)	2 hrs	701	85.0	5.667

minimum when the actinomycin-D binding ability of cells is maximum, it is reasonable to suspect that the conformational state of chromatin may exert long range effects on the physical state of water within the nucleus and perhaps in the entire cell.

In order to evaluate the effects of conformational change of chromatin on the relaxation times of water hydrogens, we treated synchronized S phase HeLa cells with spermine (0.02 M in MEM) for 1 and 2 hrs at 37°C. Spermine, which is known to induce chromatin condensation (19), caused an increase in the relaxation times (Table 3). No alteration in the percentage of water was observed, but definite condensation of chromatin occurred. These findings suggest that conformational changes in large macromolecules can be associated with changes in the physical properties of water.

CONCLUSION

Since early in this century, it has been known that the concentration of water in tumors is elevated above that of the normal tissue of origin. By mid-century, it was learned that the concentration of water in the nuclei of cancerous cells is the same as that for normal cells (6). This observation, in view of the fact that the water concentration of tumors is elevated above that of normal tissues, suggests that the changes in water concentration are cytoplasmic and/or extracellular in origin. The volume measurements in HeLa cells (20) and in sarcoma cells (5) coupled with observations of Downing et al. (6) strongly favor the interpretation that the elevated water concentration is cytoplasmic in origin. In our studies, we have evaluated the changes in water concentration in homogeneous cell populations and have found cyclical changes in the percentage of water during the cell cycle. At present, we have no data to determine whether these changes in water concentration are entirely cytoplasmic or nuclear in origin.

We have now learned that certain physical properties (relaxation time of water hydrogen protons) of water are also altered in a cyclical manner throughout the cell cycle. In addition, changes induced in the conformational state of chromatin by treating cells with spermine did result in changes in the physical properties of the cellular water independent of water concentration. More importantly, we have found that prior to mitosis an increase in the relaxation times of water protons precedes the increase in water concentration by a significant time period. This latter finding is consistent with the speculation that a change in the physical state of water preceded cell division.

Ideally, one should separate nuclei from the cytoplasm and determine the role of each cellular component in bringing about alterations. These studies are in progress.

REFERENCES

1. Stern, K. and Willheim, R. (1973) "The Biochemistry of Maligant Tumors" (Reference Press, Brooklyn, N.Y.) p. 75.
2. Cramer, W. (1916) J. Physiol., 50:322-334.
3. McEween, H. D. & Haven, F. L. (1941) Cancer Res. 1:148-150.
4. Schlottmann, H. & Rubenow, W. (1932) Ztschr. f. Krebsforsch 36:120-125.
5. Mellors, R. C., Kupfer, A. & Hollender, A. (1953) Cancer, 6: 372-384.
6. Downing, J. E., Christopherson, W. M. & Broghamer, W. L. (1962) Cancer, 15:1176-1180.
7. Bloch, F. (1946) Phys. Rev., 70:460-474.
8. Damadian, R. (1971) Science, 171:1151-1153.
9. Hazlewood, C. F., Chang, D. C., Medina, D., Cleveland, G. & Nichols, B. L. (1972) Proc. Nat. Acad. Sci. USA 69:1478-1480.
10. Kiricuta, I. C., Demco, D. & Simplaccanu, V. (1973) Arch. Geschwulstforsch, 42:226-228.
11. Inch, W. R., McCredie, A. & Knispel, R. R. (1974) J. Nat. Cancer Inst., 52:353-356.
12. Hollis, D. P., Saryan, L. A., Eggleston, I. C. & Morris, H. P. (1975) J. Nat. Cancer Inst., 54:1469-1472.
13. Beall, P. T., Hazlewood, C. F. & Rao, P. N. (1976) Science 192:904-907.
14. Carr, H. Y. & Purcell, E. N. (1954) Phys. Rev., 94:630-638.
15. Rao, P. N. & Engelberg, J. (1965) Science, 148:1092-1094.
16. Rao, P. N. (1968) Science, 160:774-776.
17. Rao, P. N. & Engelberg, J. (1966) In Cell Synchrony - Studies in Biosynthetic Regulation, I. L. Cameron and G. M. Padilla, editors (Academic Press, New York) p. 332.
18. Brinkley, B. R. & Rao, P. N. (1973) J. Cell Biol., 58:96-106.
19. Rao, P. N. & Johnson, R. T. (1974) In Advances in Cell and Molecular Biology, Vol. 3 (Academic Press, New York.) pp.135-189.
20. Terasima, T. & Tolamch, L. J. (1963) Exptl. Cell Res., 30: 344-362.

DISCUSSION (SECTION III)

OLYHOEK: Since the bacteria have no cytoskeleton, how can we visualize the size sensing system of bacteria in order to assure division at a certain size?

MAZIA: As I understand it, bacteria do not have a size limit. To test for a size limit, the experiment has been done on thymine defective mutants of E.coli; what we call thymine death. There was a case where bacteria needed thymine because they could not make DNA, but they continued growing and grew themselves to death. Whereas an eucaryotic cell under the same circumstances would simply stop growing. You seem to always imply that the cell senses its size and thus gets ready for division. But the bacteria senses the size when its ready for DNA synthesis, and I think that is also the case for the eukaryotic cell.

PRESCOTT: There are now mutants of yeast that have a defect in their size sensing mechanism, so that they divide at a different threshold; so theremay be size sensing mechanisms that is important for initiating mitosis.

ZIETZ: I would like to bring up the data of Dennis Ross where he reversibly blocked cells with excess thymidine, thus decoupling volume growth from DNA synthesis, allowing the cells to grow to 2-3 times their normal volume. After release from the block, cell volume decays back to its normal level in an oscillatory manner consistent with the hypothesis that the cell cycle does not change much. How does this fit into the size scheme?

NANNINGA: When you perturb the system, the cell probably senses concentration of some substance, not amount.

MAZIA: It seems like Nanninga has put his finger on the problem of sensing the size, because this is the ordinary way of thinking in chemistry where one thinks of changes in terms of concentration, but what image can we have of a system responding to an absolute amount?

PRESCOTT: I have struggled with this too. How does the cell count or know what it contains? Perhaps the cell does not sense size, but senses some component of size. Perhaps it senses its protein content, because proteins reflect growth and size more than any other element. You may proceed further and ask what is the protein component that the cell senses? Perhaps it senses its protein synthesis machinery. It counts its ribosomes and it checks its size by counting its ribosomes, and when it has enough ribosomes, than it initiates DNA synthesis. I don't know how the cells can count their ribosomes. One must also consider the ratio between the nucleus and the cell size. A tetraploid cell grows to twice the size of a diploid cell, and therefore there is no absolute size that is sensed, but a ratio between the nucleus and the cell's size.

MAZIA: Again we go back to the oldest hypothesis which express the idea that it all is regulated by the nuclear/cytoplasm ratio. That might be true, but we have to translate it to what we know about molecular genetics, therefore I raised the question of whether there is a relation between transcriptional output, lifetime and proteins.

PRESCOTT: I remember a paper in which the turnover of proteins was much higher in stationary phase compared to log phase. That introduces another factor. Size is not only regulated by protein synthesis, but also by its rate of degradation.

NICOLINI: *Do prokaryotic cells have to pass through restriction points to repeat themselves in time?*

NANNINGA: The point is that when a cell is born, it has accumulated substances from the previous cycle, and the substances from the previous cycle also determine what is happening in the next cycle. There is no fresh start. I'm stressing the start of DNA replication and not the cell division.

MAZIA: In the case of the eukaryotic cell, one can instead say that the cell makes a fresh start at the beginning of each next generation. I think what is coming out of this is the appreciation of the total difference between the procaryotic and the eucaryotic cell. To say it in the most extreme form, namely that they have nothing in common at the level of cellular organization and function. It is two totally different ways of making a cell.

NANNINGA: I don't completely agree with this. I don't think that they are totally different. In these two cases you can say you have some fundamental properties of life; that is duplication of the genome and duplication of the cell. This is what connects the two systems. What I've been trying to convey, is that there are not such big differences between the procaryotic and eukaryotic cells. The problems are the same; the meaning of size, age and what starts the DNA replication.

MAZIA: It seems to me that this is a situation where one distinguishes between words and things. For instance, both have a genome, both genomes have DNA. In other respects they have nothing in common-

RAO: The principal difference in the prokaryotic and the eukaryotic system is the separation of genomes during cell division. There you have the membrane associated mechanism. In the eukaryotic cell where the DNA is very tightly packed in the chromosomes, an entirely different mechanism has to be divised, where the packages of genes are distributed equally between the two daughter cells. The condensation and decondensation processes, which Mazia sketched earlier, may have to do entirely with the distribution of the genomes, which has nothing to do with the prokaryotic cells, and there is no point in saying that they are exactly the same. The basic functions of life are the same, but the distribution of the genomes are the most distinguishing features of the eukaryotic cells.

NAGL: I would just like to make a comment on the dirrerent distribution mechanisms of DNA in prokaryotes and eukaryotic. I do not really think that there are different or opposit mechanisms. If you look at the evolution of mitosis in sheep, for instance, you will find that there is a membrane system for the distribution of chromosomes. In the fungi and algae the chromosomes are fixed to the membrane and are transported by the membrane even if tubules form tracks. They are not linked to the chromosones, so I think in the evolution there is a development from a membrane system to a microtubular system of transport. There is not an either/or situation.

MAZIA: If you look at it from the standpoint that microtubule are the guidance or the mechanism of separation, there is an absolute difference. The other absolute is that the presence or absense of centrosomes, is not a matter of size.

MARALDI: There are evidences that the centrosome is the actual organizer of the cytoskeleton in a cell, since it has been demonstrated that the cytoskeleton reorganization after its destruction caused by VLB, hypothermic treatment and so on, is not associated with the centrosome, but with independent organization sites diffused in the cytoplasm.

MAZIA: It is my impression that there is more than one centrosomal cycle in the cell. But it seems to me that in the image of centrosomes, one concentrates on the centrosome which is physically connected to the nucleus. If this is true, every time you isolate a nucleus, it should also carry a centrosome, i.e. an animal cell should carry a centrosome. In the past, if they saw that, they called it a contamination and subsequently changed their method in order to remove the centrosome; thereby throwing away the most interesting part of their isolation. Now, there are methods of isolation which show that each nucleus carries a centrosome, and people are studying these connections. There is another interesting point related to your question, and that is whether centrosomes is a class among cell components rather than a single unit. There is a recent publication by Burns in Journal of Cellular Biology in which he finds a varying number of centrosomes in fused cells. He can distinguish those which can be mitotic poles from those which are not. So my further proposal is that centrosomes connected to the nucleus are the ones which bipolarize.

ALABASTER: The notion of an initiating event in the cell cycle suffers from the limitation of the philosophical "first cause" argument. Obviously, an argument based on casuality must always acknowledge a preceeding event. Therefore, an "initiating" event is always preceeded by casual events. The question raised by some of your references to casual events is: Should one look for sequential events within a continuum rather than attempt to define initiating events?

MAZIA: You are asking the question that I was asking you, in fact, I was trying to introduce it. The biochemical equivalent to the philosophical view you mentioned - that whatever happens has to be limiting in a complex situation - seems to be casual if the limitation is overcome. That was the purpose of introducing the most untypical case of the Sea Urchin egg where only cytoplasmic pH was limiting, and one would not dream of generalizing from that to the leukocyte where everything is limiting in the first 24 hours or so, - or in the diagram which showed that the wheel could turn in nine minutes, if unlimited, and that regulation could be inserted at any point in the cycle. The limitation is generally inserted somewhere between the chromosome replication and separation, by definition. I gave you at least one other, and rare, example, namely mitosis, where the splitting is the limiting event and would seem to be a casual event for the subsequent sequences of the cycle. I wonder if one will find a sequence of event that is general enough to say that this is the program of the cycle. Someone has suggested that such events have been found.

COOPER: I want to make the suggestion that there is no such program.

PRESCOTT: The question was raised that a universal program of the cell cycle has not been identified. The program has been identified in a good, first approximation in Mazia's diagram of the cycle of chromosome splitting, decondensation, replication, condensation, splitting etc. etc. It remains to determine the molecular basis of this cycle, so that we may understand the cause-effect relationships that underline this chromosome cycle; and the ways in which it may be interrupted.

MAZIA: The stress on the macroscopic discription was only to indicate the direction in which we have to go toward a molecular discription.

PRESCOTT: At the moment we can't give a molecular discription.

BRADBURY: I think in bacteria, there may be something equivalent to condensation and decondensation. The indication of this is the reported findings of histone like proteins in bacteria. I know that Arthur Kornberg has a paper either in PNAS, or is about to come out in PNAS, where he finds a histone 2B - like protein in bacteria. The question I want to ask is related to the fertilized Sea Urchin egg when the pH changes by 0.5 units. We have found, in addition to fertilization following change in pH, that other events may go on. One is that there is a very rapid drop in inorganic phosphate, and secondly, there is a very rapid increase in arginine phosphate. These two events do not take place if the Sea Urchin egg is exposed to an increased pH with weak bases. When you add weak bases, you notice immediately that there is a change in pH, and there is an onset in protein synthesis, but the other two events do not take place. What do you think arginine phosphate do as in the metabolism, which is different in fertilization compared to exposure to weak bases?

MAZIA: Fertilization has an immediate effect on the membrane, while the same events take place after one hour with ammonia. I wonder if someone here is interested in these questions.

BRADBURY: We are interested.

RAO: I have two comments. When we fuse two cells, these cells have their own spindle at mitosis. This we have seen by labelling DNA in one cell, and all the labelled chromosomes are attached to one spindle. The other comment is that when we inject extract from Hela cells in mitosis into a blastomere, the cells stop in metaphase, while the other cells divide. So it looks like the mitotic cells have the same factor as the oocytes. A last comment about the condensation in Sea Urchin eggs. We saw in the cell line we got from Prescott, that they had no G1 or G2 and very rapid decondensation after mitosis. This is in agreement with the Sea Urchin egg pattern.

SECTION IV:
NORMAL VERSUS ABNORMAL CELL GROWTH

REVERSE TRANSFORMATION OF CHINESE HAMSTER CELLS

BY CYCLIC AMP AND HORMONES*

Abraham W. Hsie

Biology Division, Oak Ridge National Laboratory
Oak Ridge, Tenn. 37830

SUMMARY

Treatment with cyclic AMP agents causes morphological conversion of spontaneous transformed epithelioid-like Chinese hamster ovary cells to a fibroblast-like shape. The morphological transformation is accompanied by the reversal of various transformed characteristics to the normal ones, such as the reorientation of the cytoplasmic microtubule-microfilament system through the action of cAMP-dependent protein kinase.

INTRODUCTION

Over the past 15 years, evidence has accumulated which demonstrates that adenosine 3':5'-phosphate (cyclid AMP, cAMP) acts as a second messenger mediating various hormonal actions in eukaryotes (1, 2). Although the molecular basis of cAMP action is not fully understood, it has been suggested that the effects of cAMP in mammalian systems are mediated by cAMP-dependent protein kinase(s) (1, 2). In bacterial systems, where there is no cAMP-dependent protein kinase, cAMP may directly affect transcription of DNA (2, 3).

The effects of cAMP at the cellular level have been documented in several systems, such as the formation of flagella

---

*Research sponsored by the Office of Health and Environmental Research, U.S. Department of Energy, under contract W-7405-eng-26 with the Union Carbide Corporation.

in coliform bacteria (4) and the chemostasis and morphogenesis in the slime mold Dictoidium discoidium (5, 6). The activity of adenylate cyclase, an enzyme which catalyzes the synthesis of cAMP from ATP, was found to be lowered in polyoma-transformed baby hamster kidney cells; methylxanthine compounds, inhibitors of cAMP phosphodiesterase, which degrades cAMP to 5'-AMP, slowed the growth rate of both normal and transformed baby hamster kidney cells (7). In addition, it has been demonstrated that cAMP injection inhibit the growth of transplanted lymphosarcoma in mice (8) and that exogenous cAMP inhibits the proliferation in cultured L and HeLa cells (9, 10). In light of these findings, we were led to examine whether cAMP might modulate mammalian cell growth and associated characteristics.

We used Chinese hamster ovary (CHO) cells, clone K_1 [generally referred to as CHO-K_1 (11)], throughout most of the experiments. When these cells were established in culture over 20 years ago, they had a characteristic fibroblast-like morphology (12; unpublished photomicrographs by T. T. Puck); however, the long-term cultured CHO-K_1 cells, which presumably have undergone spontaneous transformation, assume an epithelioid appearance. The cells were routinely maintained in Ham's F12 medium supplemented with 10% fetal calf serum in a humidified 5% CO_2 incubator at 37°C. For experiments, the F12 growth medium was supplemented with extensively dialyzed, heat-inactivated (56°C for 30 min) serum to provide a better-defined environment for cell growth. The population doubling time of the cells in this medium is 12–14 hr (11, 13).

The Phenomenon of Reverse Transformation by cAMP

Under normal growth conditions for 7 days, each CHO cell develops into a colony which usually contains over 1000 compact and poorly oriented epithelioid-like cells (Fig. 1A,B). When cells are treated with 1 mM Bt_2cAMP for the same period of time, the cells elongate to a spindle shape and line up in parallel to produce a highly oriented colony typical of mammalian fibroblasts (Fig. 1C,D).

The two types of colonies are distinguishable both macroscopically (Fig. 1B,D) and microscopically (Fig. 1A,C). In addition, colonies grown in the presence of dibutyryl cAMP (BT_2cAMP) are contact inhibited, and the cells are confined to growth in a monolayer, whereas in the absence of this cyclic nucleotide the cells grow in multilayers (13).

The morphological transformation is not induced by unsubstituted cAMP, 5'-AMP, adenosine, adenine, or sodium butyrate. The $O^{2'}$-monobutyryl cAMP is also inactive, but

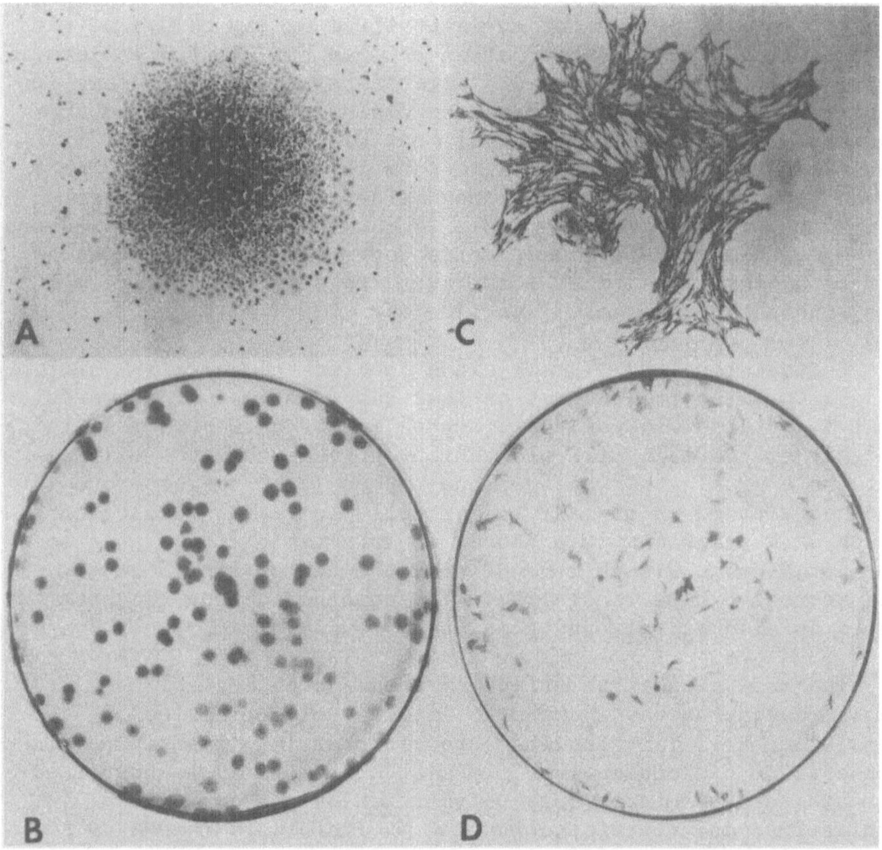

Fig. 1 Effects of Bt_2cAMP on the morphology of CHO cells. (A, B) Colonies grown from single CHO cells on standard medium for 7 days. x50 and x2, respectively. (C, D) Colonies grown in the presence of 1 mM Bt_2cAMP. x50 and x2, respectively. Reproduced from ref. 41 with permission.

N^6-monobutyryl cAMP possesses activity equal to or greater than that of Bt_2cAMP (13). These points will be discussed later.

Time-lapse photomicrograph showed that the morphological transformation induced by Bt_2cAMP is observable within 15 min of treatment, and within 3 hr most of the cells undergo elongation (13–15). Lower concentrations of Bt_2cAMP produce less striking changes, but the effects of concentrations as low as 0.1 mM can be recognized. At a concentration of 1 mM the cloning efficiency of the cells is not appreciably altered, although the population doubling time is slightly increased dependent upon the growth conditions such as medium renewal. At 0.3 mM excellent conversion to the fibroblast-like form is achieved, and neither growth rate nor cloning efficiency is affected. The changes induced by Bt_2cAMP are reversible. When the agent is removed and fresh growth medium is added, cells converted to the elongated form by a 4-hr exposure at 1 mM revert to their original compact form within approximately 1 hr (13–15).

Testosterone (5 – 15 μM) alone does not cause cell elongation at any concentration that has been tested. However, simultaneous addition of testosterone and Bt_2cAMP (0.1 mM) yields a highly exaggerated cell elongation, greater than the maximum achievable with either agent alone. When 15 μM testosterone is combined with 30 μM Bt_2cAMP, clear cell elongation results even though each agent singly produces no detectable effect at these concentrations. With 0.1 mM Bt_2cAMP the action of 1.5 μM testosterone can be recognized. The combined action of these two agents is also rapidly and completely reversible (13).

The cell elongation effect of these morphological transformation agents appears to involve polymerization of preexisting cellular tubulins into microtubules, but not novel synthesis of macromolecules, since this effect can be prevented by inhibitors of microtubule polymerization such as Colcemid, vinblastine, and vincricine but not by inhibitors of DNA, RNA, and protein synthesis (13, 16). Microfilament synthesis is also involved since the morphological transformation does not occur in the presence of cytochalasin B, which prevents microfilament synthesis (13).

Since the change from contact-inhibited to multilayered growth characteristics has been associated with the conversion of normal, differentiated cells to malignant ones, we determine whether the morphologic transformation of CHO cells is accompanied by a reversal of characteristics commonly observed in malignant transformation of mammalian cells in culture. As can be seen from Table 1, in addition to morphological characteristics, various cellular, physiological, and biochemical

Table 1. Characteristics of CHO Cells and Their Modification by Bt_2cAMP

Epithelial-like ($+Bt_2cAMP$ →)	(← $-Bt_2cAMP$) Fibroblast-like
Morphology	
Compact, multipolar, epithelial-like	Stretched, spindle-shaped
Knobbed surface	Smooth surface except for occasional knobs at ends
Knobbed membranes are oscillating	Surface movement limited to slow ruffling
Cells grow randomly without an ordered association and orientation among themselves	Cells associate and orient along their long axis in a "contact-inhibited" fashion
Growth Properties	
Cloning efficiency: ∿100%	∿100%
Population doubling time: 12–14 hr	13–15 hr
Growth in agar: ∿100%	∿1% (delayed colonial growth)
Microtubules	
Fewer, short microtubules randomly distributed within the cytoplasm	More, longer microtubules arranged parallel to long axis of the cell beneath the cell membrane
Collagen Synthesis	
Low	High
Cell Adhesion	
Spontaneously detach from the growth surface	Cells more firmly adherent to growth surface and to each other
Readily dissociation by trypsin or EDTA	Less readily dissociated by trypsin or EDTA
Satellite colonies formed	No growth of satellite colonies
Surface Architecture	
High agglutinability by lectins	Low agglutinability by lectins
Tumorigenicity	
Form undifferentiated sarcoma of low grade malignancy in nude mice	Delayed tumor appearance
Intracellular cAMP Level	
1–1.5 μM	15–20 μM
Protein Kinase Activity	
V_{max}: low	High
Stimulation by cAMP: yes	No
In the complex form consisting of catalytic and regulatory subunits	In the activated form consisting of predominantly catalytic subunits

properties of transformed cells have been reversed as a result of treatment of CHO cells with cAMP agents and hormones. We have termed this phenomenon "reverse transformation" (15), since these compounds induced changes in cellular characteristics in many respects opposite to those seen in normal fibroblasts which were transformed by an oncogenic virus or by carcinogenic chemical or physical agents (17–19).

The Effects of Steroids and Prostaglandins

We examined the synergistic effect of various hormones and 0.1 mM Bt_2cAMP; the latter produces virtually no observable conversion to the fibroblast-like morphology within 18 hr. However, when 15 μM testosterone is also present, maximal cell elongation is observed. We found that besides the androgen testosterone, other compounds such as hydroepiandrosterone, 17α-hydroxypregnenolone, and progesterone, which are in the biosynthetic chains common to both testosterone and 17β-estradiol, display various degrees of activity but usually less than that of testosterone or 5β-dihydrotestosterone (14, 20). Δ1-Testolactone, a synthetic derivative of testosterone used clinically has an effect similar to that of testosterone in causing morphological transformation of CHO cells (21).

In contrast to the synergistic effect of testosterone, high concentrations of 17β-estradiol (30–50 μM) inhibit the morphological transformation (20). Insulin (0.017 – 17 μM), glucagon (0.028 – 28 μM), epinephrine (1 – 100 μM), cholesterol (50 μM), and hydrocortisone (50 μM) were ineffective (14, 20).

Activity similar to or higher than that of testosterone can also be seen with some prostaglandins. In the presence of 0.1 mM Bt_2cAMP, the synergistic action of prostaglandins is demonstrable. Various prostaglandins differ in their ability to exert synergistic action, and the degree of morphological change parallels the ability of each compound to stimulate cAMP synthesis in culture: prostaglandin $E_1 > E_2 > A_1 > F_{2\alpha}$ (14, 20).

An Analysis of Cell-Cycle Effects

It has been reported that Bt_2cAMP has a variety of effects on the cell cycle (22). Most of the studies used synchronized cells to determine cell cycle effects and consequently were limited to the short-term effects of the compound. We observed that continuous exposure to Bt_2cAMP had only minor effects on the cell cycle and cell size when the culture medium was renewed daily (23).

We treated CHO cells in the logarithmic phase of growth to investigate the morphological effects of Bt_2cAMP in all experiments described so far. On close examination of logarithmically growing low-density cells treated with Bt_2cAMP, we found that approximately 30% of the cells are elongated within 3 – 4 hr. The percentage of elongated cells increased to 80% – 90% by 14 – 15 hr. We further observed that in high-density cultures over 75% of the cells become highly elongated within 3 – 4 hr after treatment. Because cells in high-density cultures are mostly arrested in G_1, this phase of the cell cycle appears to be important for morphological change. These observations led us to study the Bt_2cAMP-induced cell elongation in cultures synchronized by a mitotic shake-off technique which yields over 95% mitotic cells. Bt_2cAMP (1 mM) was added to cells at hourly intervals over two cell cycles. Only cells in the first or second early G_1 phases converted to the fibroblast-like form within 2 hr after Bt_2cAMP addition. Cells in other phases were much less affected morphologically, even after long treatment periods. Thus, elements responsible for the cell elongation process are responsive to Bt_2cAMP only during the early G_1 phase of growth (24).

To further delineate the role of cAMP in the control of growth and morphology of CHO cells, we arrested cell growth by a shift to serum-free medium. We observed that during the first 72 hr of arrest there is little change in intracellular cAMP level and cell shape. By 96 hr and thereafter the cAMP level doubles, the cells become more responsive to prostaglandin E_1 (an activator of cellular adenylate cyclase) leading to a higher magnitude of induced cAMP synthesis, and the cells are converted to the fibroblast-like shape. Since these biochemical and cellular transitions are subsequent to growth arrest, cAMP increase is not the cause of growth arrest but the results are consistent with a role for cAMP in regulating cell shape. In addition, these changes point to the importance of the noncycling G_1 cells for initiating cAMP-related events (25).

The Role of Cytoplasmic Microtubules and Microfilaments in Cell Shape

Cytoplasmic microtubules and microfilaments have been known to play an important role in the maintenance of cell shape (26, 27). As demonstrated by inhibitor studies earlier, Bt_2cAMP and hormones may transform CHO cells morphologically by promoting microtubule assembly from precursor tublins already present in the cells (13).

Electron microscopic studies demonstrate that in the presence of Bt_2cAMP the number of microtubules increases and they become arrayed parallel to the long axis of the cell beneath the cell membrane (28, 29). In the absence of Bt_2cAMP a few short microtubules are randomly distributed within the cytoplasm, usually in the centriolar region (28, 29).

These experiments suggest that one manifestation of malignant transformation of fibroblasts is disintegration of the cytoplasmic microtubule-microfilament system (13–15, 20, 28–30); reverse transformation agents restore this disintegrated system to the organized state (13, 15, 20, 28–30).

The development of an antibody to tubulin from bovine brain as an immunofluorescent probe permitted an immediate and rapid analysis of the pattern of cytoplasmic microtubules by light microscopy in a large population of normal and transformed cells (31, 32). It was found that most normal cells have an elaborate system of cytoplasmic microtubules whereas the transformed cells contain a reduced number of microtubules (32). Treatment of sarcoma virus transformed mouse 3T3 cells with Bt_2cAMP was found to restore the microtubule system and morphological appearance of the normal 3T3 cells (32). Recent experiments with a lysed cell microtubule reassembly system showed that these transformed 3T3 cells exhibit a markedly reduced capacity to initiate and elongate microtubules from cytoplasmic microtubule organizing centers; cAMP restores this impaired capacity (33). The distinctive difference in the fluorescent staining pattern of microtubules between normal and transformed cells has led to the consideration of this feature in conjunction with other morphological alterations in distinguishing malignant transformed cells from their normal counterparts (32).

Knobs and Membranal Activities

One morphologic characteristic of CHO cells is the presence of knob-like structures around their cell surface (14, 15, 28, 29) (Fig. 2). When cells grown under normal conditions are examined with a light microscope, knob-like structures of high optical density can be seen extending from the cell periphery into the external medium (13–15). Under the scanning electron microscope the shape is circular from the top, and from the side the knobs appear to have stout stalks (28). Specific histological staining shows that they contain abundant aggregates of polyribosomes, but no other identifiable organelles. However, bands of subplasmalemmal microfilaments are present in the knobs and around the periphery of the cells (28, 29). Time-lapse photomicrography shows the knobs as centers of violent motions of the cell membrane (15), continuously erupting and returning to the main cell body. When cAMP agents are added, the knobs disappear from the cell periphery within a few minutes. A smooth membrane structure can be seen within 15 min, and after 5 – 8 hr the majority of the cells are elongated and fibroblast-like.

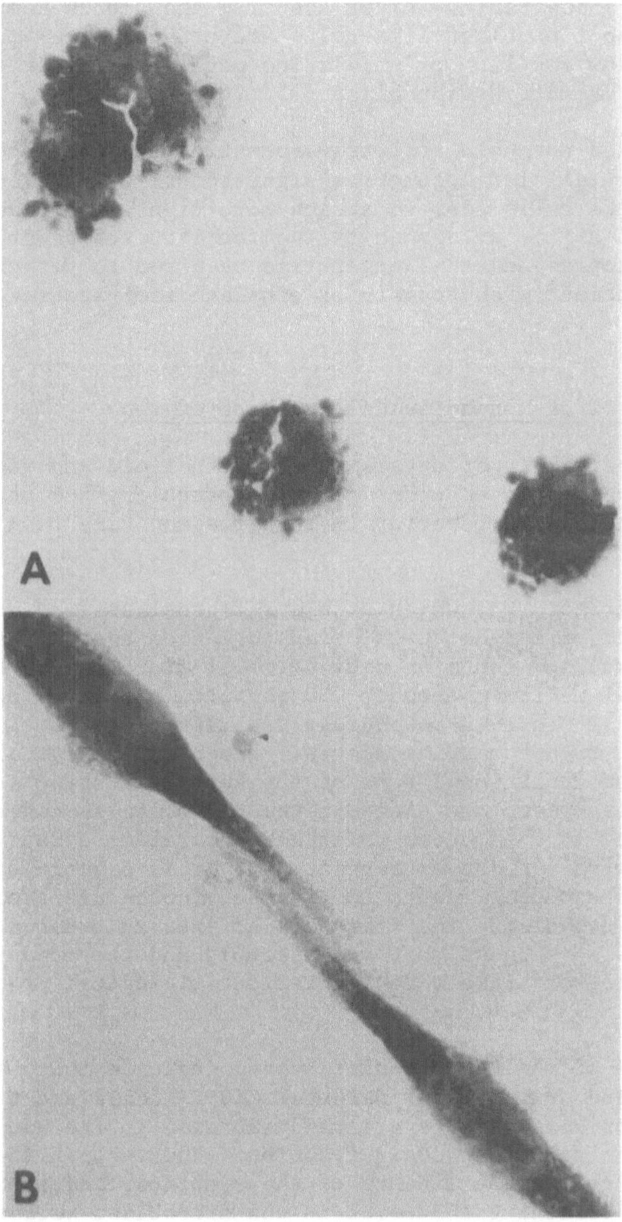

Fig. 2 The presense of knob-like pseudopodal structures in CHO cells grown for 18 hr in basal medium (A), and the disappearance of these structures in similar cultures treated with Bt_2cAMP (0.2 mM) and testosterone (15 M) (B). x1000. Reproduced from ref. 14 with permission.

Thus the processes of knob disappearance and cell elongation appear to be inseparable. After assuming the smooth membrane structure, the fibroblast-like cells are not completely quiescent but have a slow ruffling or undulating movement similar to that of a typical mammalian fibroblast (15, 20).

The marked morphological transformation observed does not seem to alter cellular locomotion significantly. However, the fibroblast-like cells seem to attach more tightly to glass or plastic surfaces, as evidenced by the formation of fewer satellite colonies and the longer time required to detach these cells by treatment with trypsin or ethylenediaminetetraacetic acid (20).

Genetic Control of Morphological Interconversion

Morphological interconversion of epithelioid and fibroblast-like forms appears to be under genetic control, since stable variants with changed behavior in this respect have been isolated (14, 29).

A CHO variant, M6, which displays fibroblast-like morphology in the absence of morphological transformation agents, has a smooth membrane and contains many parallel microtubules even when grown in basal medium. Another CHO variant, M7, which displays a typical epithelioid-like morphology, is refractory to morphological transformation agents. Under morphological transformation conditions, some of the knob-like structures of variant M7 disappear, and although the microtubules appear to increase, they do not become oriented in parallel arrays (29). Experiments with variant M7 suggest that it is capable of responding to hormonal agents to promote microtubule assembly but incapable of organizing the microtubules into an ordered array. The results also suggest that the assembly and the organization of microtubules are likely to be distinct biological processes (29).

Analyses of the microtubular system indicate that when variants M6 and M7 and their parental CHO-K_1 cells are treated there does not appear to be a marked increase in the amount of tubulin in the soluble protein fraction. However, in the treated CHO-K_1 cells colchicine binding of the membrane fraction is increased, approaching the level in the fibroblast variant M6. There is no change in colchicine binding of the membrane fraction of the variant M7 even when treated with Bt_2cAMP. These observations are consistent with the notion that membrane-associated tubulin is associated with microtubule alignment, which confers cell shape (34).

Morpohlogical Transformation of Enucleated CHO Cells

It has been reported that appropriate treatment of cultured mammalian cells with cytochalasin B yields a population of enucleated cells (35, 36). Such enucleates are excellent experimental material for the study of interactions between the nucleus and cytoplasm since certain normal physiological characteristics of the intact cells can be preserved. We enucleated CHO cells and replated them (after recovery) on culture dishes with various morphological transformants. These enucleates underwent a marked change in morphology similar to that of intact cells (37). In the absence of morphological transformation agents, the enucleates assume the compact, epithelial-like morphology characteristic of their nucleated counterparts with knob structures present on their periphery. In the presence of the agent(s), the enucleates elongate to the spindle shape and, like their nucleated counterparts, have smooth edges without knob-like structures, except occasionally at the two pointed tips. As in intact CHO cells, the morphological transformation produced by these agents in enucleates can be prevented by Colcemid and vinblastine (37).

These experiments show that enucleated CHO cells retain the ability to perform three biological processes — hormonal response, membrane alteration, and microtubule assembly; the processes apparently do not require nuclear activity (37).

It has been well documented that the androgenic action of testosterone involves the participation of the nucleus. The morphogenetic activity of Bt_2cAMP plus testosterone in CHO enucleates suggests that the role of testosterone in this case is probably different from that in androgenic action (37).

Mechanism of Bt_2cAMP Action

The observation that Bt_2cAMP and N^6-monobutyryl cAMP are active in inducing the cell elongation while unsubstituted cAMP is inactive has led us to study the properties of these three compounds. Using cell-free extracts we found that these cAMP analogues inhibit cAMP phosphodiesterase and that the inhibition is competitive with respect to cAMP. The N^6-monobutyryl derivative is a more effective inhibitor (38).

We also studied the uptake of radiolabeled cAMP and Bt_2cAMP into CHO cells and their intracellular fate. With cAMP we found uptake of radiolabel and accumulation of AMP, ADP, ATP, and a small amount of adenosine. However, no radiolabeled cAMP was found in the cells. With Bt_2cAMP we found stable intracellular accumulation of Bt_2cAMP and N^6-monobutyryl cAMP at concentrations of 75 and 90 μM, respectively, and of a compound identified as N^6-monobutyryl-AMP. However, no radiolabeled cAMP was found within the cells, although the measured amount of cAMP increased 20-fold in treated cells (38, 39).

These kinetic analyses led us to suggest that the intracellular accumulation of N^6-monobutyryl cAMP is sufficient to inhibit cAMP phosphodiesterase, leading to the observed increase of intracellular cAMP level (38, 39).

Activation of cAMP-Dependent Protein Kinase

We then studied the effect of increased intracellular cAMP on cell behavior. Since it is generally believed that activation of protein kinase(s) is the major mechanism by which cAMP exerts its physiologic function as a second messenger in response to various hormone actions in eukaryotic cells, we investigated the possible intracellular activation of protein kinase after treatment with Bt_2cAMP.

We found that the protein kinase activity in CHO-cell crude extracts is cAMP-dependent and is stimulated three- to fourfold by 1 M cAMP. This stimulation by cAMP occurs with the dissociation of the holoenzyme into regulatory and catalytic subunits, which bind cAMP and catalyze phosphorylation, respectively. In the untreated CHO cells this enzyme exists largely in the high-molecular-weight form of holoenzyme which is cAMP-dependent. However, in the cells treated with Bt_2cAMP this activity was converted to the lower-molecular-weight form which is cAMP-independent (40). Thus, incubation of CHO cells with Bt_2cAMP, which results in an increase in the intracellular level of cAMP, causes the activation of the intracellular cAMP-dependent protein kinase and appears to result in protein phosphorylation through which tubulin is then stimulated to polymerize into microtubules, orient through membrane interaction, and cause the cells to elongate (40, 41).

Several experiments were then carried out to confirm the role of activation of cellular cAMP-dependent protein kinase in converting cells from an epithelioid shape to a fibroblast-like morphology.

First, by treating cells with Bt_2cAMP from 0.1–1.0 mM for 6 hr, we found that Bt_2cAMP caused an increase in intracellular cAMP level, activation of cAMP-dependent protein kinase, and morphological transformation in a dose-dependent fashion. Incubation with low (0.1 mM), intermediate (0.2 mM), and high ($\geq$0.4 mM) concentrations of Bt_2cAMP caused small, intermediate, and obvious effects, respectively (42).

Second, a reversal to the epithelioid shape after removal of 1 mM Bt_2cAMP from the fibroblast-like cells pretreated with 1 mM Bt_2cAMP showed a time-dependent return to pretreatment levels of both cAMP and protein kinase activity (42).

Third, treatment of cells with cholera toxin, an activator of cellular adenylate cyclase, resulted in a dose-dependent and time-dependent change of cAMP level, activation of protein kinase, and morphological conversion (42).

Fourth, incubation of cells with protoglandin E_1, another effective activator for cellular adenylate cyclase, caused a transient increase of cAMP level as well as a transient activation of protein kinase; the cells become more flattened without a clear elongation. It appears that the observed transient activation of protein kinase is not sufficient to cause morphological conversion; prolonged activation is necessary to induce this change (42).

To explore the role of phosphorylation of cell components in the morphologic transformation, the subcellular distribution of the protein kinase was determined (43). Over 80% of both total cAMP-dependent protein kinase (histone as substrate) and cAMP-binding activity is present in the cytosol fraction. The protein kinase activity in this fraction is also the most cAMP-dependent. The other fractions show small amounts of protein kinase activity. When the fractions are assayed in the absence of added histone, a low level of activity representing phosphorylation of endogenous substrate(s) is seen. The cAMP dependence of this activity is difficult to assess because of the limiting amount of substrate present. However, the presence of both protein kinase activity and substrate for phosphorylation indicates that many cell components might be phosphorylated when the intracellular level of cAMP is increased, thus modifying various physiologic characteristics.

Since over 80% of the protein kinase activity is found in the cytosol, the property of the cytosol enzyme was further characterized (44). There appear to be two different protein kinases in this fraction. The two enzymes can be separated by DEAE-cellulose column chromatography and also differ in molecular weight. Kinetically they are similar in their affinity for both ATP and cAMP, but they differ in degree of activation by cAMP as well as in substrate preference (44).

Mechanism of Action of Other cAMP Analogues

Having learned the mode of action of Bt_2cAMP, we then extended our studies of the effects of cyclic nucleotides in

CHO cells to other analogues of cAMP. In addition to Bt_2cAMP and N^6-monobutyryl cAMP, only 8-bromo cAMP and 8-benzylthio cAMP have been found to induce morphologic transformation. As found previously, O^2-monobutyryl cAMP is ineffective, probably due to its rapid degradation within the cell (13, 38). Only Bt_2cAMP and N^6-monobutyryl cAMP cause a net increase in the intracellular level of cAMP, presumably due to phosphodiesterase inhibition. The four analogues which cause cell elongation also result in the activation of the cAMP-dependent protein kinase in the treated cells. Since treatment with 8-bromo cAMP or 8-benzylthio cAMP does not cause any measurable increase in the intracellular level of cAMP, we propose that they activate the protein kinase directly. Treatment with Bt_2cAMP and N^6-monobutyryl cAMP appears to result in kinase activation due to the increase in intracellular cAMP. The activation of the protein kinase in cell extracts is consistent with this notion, since both 8-bromo cAMP and 8-benzylthio cAMP are similar to cAMP in their ability to activate. N^6-monobutyryl cAMP is a less effective activator of the protein kinase (45).

Further Evidence on the Role of Protein Kinase in Morphological Transformation: Study with cAMP-Resistant Variants.

Since activation of cAMP-dependent protein kinase is involved in morphological transformation of CHO cells, one would expect that certain variant CHO cells refractory to such an action of cAMP would have a defective cAMP-dependent protein kinase activity. Several such CHO variants have recently been isolated (46, 47). All these variants were found to be resistant to cAMP agents, and most of them have a defective cAMP-dependent protein kinase activity. These results are additional evidence demonstrating the role of activation of this cAMP-dependent protein kinase in morphological transformation and the associated cellular, physiological, and biochemical modification of CHO cells.

In other mammalian cell systems, variant S49 mouse lymphoma cells which are resistant to cytolysis by Bt_2cAMP have been isolated (48–50). These variants lack cAMP-dependent protein kinase activity (48–50). Consistent with the role of this enzyme in cAMP-mediated cell behavior is the observation that exposure of wild type S49 cells caused a consistent pattern of protein modification, induction, and repressions, none of which are exhibited in the kinase-negative variants (51).

Compatible with the role of cAMP-dependent protein kinase in regulating cell shape is the finding that cAMP stimulates microtubule assembly in the lysed cell microtubule reassembly system of transformed SV3T3 cells; this effect is inhibited by protein kinase inhibitor, and the protein kinase activity toward the endogenous substrate is higher in normal than in the transformed 3T3 cells (33).

ACKNOWLEDGEMENTS

I thank B. R. Brinkley of Baylor College of Medicine, M. M. Gottesman of the National Cancer Institute, and R. A. Steinberg of the University of Connecticut for sending their current reprints and preprints, and my colleagues F. T. Kenney, R. L. Schenley, and R. M. Wallace for reviewing the manuscript.

This paper is dedicated to H. V. Rickenberg of the National Jewish Hospital and Research Center at Denver for his mentorship and friendship.

REFERENCES

1. Robison, G. A., Butcher, R. W. & Sutherland, E. W. (1971) Cyclic AMP (Academic Press, New York).
2. Jost, J. P. & Rickenberg, H. V. (1971) Annu. Rev. Biochem. 40, 741-774.
3. Pastan, I. & Adhya, S. (1976) Bacteriol. Rev. 40, 527-551.
4. Yokota, T. & Gots, J. S. (1970) J. Bacteriol. 103, 513-516.
5. Konijin, T. M., Barkley, D. S., Chang, Y. Y. & Bonner, J. T. (1968) Am. Nat. 102, 225-233.
6. Bonner, J. T. (1970) Proc. Natl. Acad. Sci. USA 65, 110-113.
7. Burk, J. T. (1968) Nature (London) 219, 1271-1275.
8. Gericke, D. & Chandra, P. (1969) Hoppe-Seylers Z. Physiol. Chem. 350, 1469-1471.
9. Ryan, W. L. & Heidrick, M. L. (1968) Science 162, 1484-1485.
10. Heidrick, M. L. & Ryan, W. L. (1970) Cancer Res., 30, 376-378.
11. Kao, F.-T. & Puck, T. T. (1968) Proc. Natl. Acad. Sci. USA 60, 1275-1280.
12. Puck, T. T., Cieciura, S. J. & Robinson, A. (1958) J. Exp. Med. 108, 945-956.
13. Hsie, A. W. & Puck, T. T. (1971) Proc. Natl. Acad. Sci. USA 68, 358-361.
14. Hsie, A. W., Jones, C. & Puck, T. T. (1971) Proc. Natl. Acad. Sci. USA 68, 1648-1652.
15. Puck, T. T., Waldren, C. A. & Hsie, A. W. (1972) Proc. Natl. Acad. Sci. USA 69, 1943-1947.
16. Patterson, D. & Waldren, C. A. (1973) Biochem. Biophys. Res. Commun. 50, 566-573.

17. DiPaolo, J. A. (1974) in Chemical Carcinogenesis, Part B, eds. Ts'o, P. O. P. & DiPaolo, J. A. (Marcel Dekker, New York), pp. 443-455.
18. Sachs, L. (1967) Curr. Top. Dev. Biol. 2, 129-150.
19. Heidelberger, C. (1974) in Chemical Carcinogenesis, Part B, eds. Ts'o, P. O. P. & DiPaolo, J. A. (Marcel Dekker, New York), pp. 457-462.
20. Hsie, A. W., O'Neill, J. P., Schroder, C. H., Kawashima, K., Borman, L. S. & Li, A. P. (1976) in Control Mechanisms in Cancer, eds. Criss, W. E., Ono, T. & Sabine, J. R. (Raven Press, New York), pp. 183-203.
21. Puck, T. T. & Wenger, L. (1973) IRCS Med. Sci. (73-6) 1-8-2.
22. Abell, C. W. & Monahan, T. M. (1973) J. Cell Biol. 59, 549-558.
23. Kimball, R. F., Perdue, S. W. & Hsie, A. W. (1975) Exp. Cell Res. 95, 416-424.
24. O'Neill, J. P., Schroder, C. H., Riddle, J. C. & Hsie, A. W. (1976) Exp. Cell Res. 97, 213-217.
25. O'Neill, J. P. & Hsie, A. W. (1978) J. Cyclic Nucleotide Res. 4, 169-174.
26. Porter, K. R. (1966) in Principles of Biomolecular Organization eds. Wolstenholme, G. E. & O'Conner, M. (Little and Brown, Boston), pp. 308-356.
27. Brinkley, B. R., Fuller, G. M. & Highfield, D. P. (1975) in Microtubules and Microtubule Inhibitors, eds. Borgens, M. & DeBrabander, M. (American Elsevier Publishing Co., New York), pp. 297-312.
28. Porter, K. R., Puck, T. T., Hsie, A. W. & Kelley, D. (1974) Cell 2, 145-162.
29. Borman, L. S., Dumont, J. N. & Hsie, A. W. (1975) Exp. Cell Res. 91, 422-428.
30. Puck, T. T. (1977) Proc. Natl. Acad. Sci. USA 74, 4491-4495.
31. Fuller, G. M., Brinkley, B. R. & Boughter, J. M. (1975) Science 187, 948-950.
32. Brinkley, B. R., Fuller, G. M. & Highfield, D. P. (1975) Proc. Natl. Acad. Sci. USA 72, 4981-4985.
33. Means, A. R., Dedman, J. R., Tash, J. S., Cox, S. M. & Brinkley, B. R. (1979) J. Cell Biol. 83, 349a.
34. Borman, L. S. (1978) Cyclic AMP, Microtubules and Mammalian Cell Shape, Ph.D. Thesis, University of Tennessee—Oak Ridge Graduate School of Biomedical Sciences.
35. Prescott, D., Myerson, D. & Wallace, J. (1972) Exp. Cell Res. 71, 480-485.
36. Wright, W. E. & Hayflick, L. (1970) Exp. Cell Res. 74, 187-194.
37. Schroder, C. H. & Hsie, A. W. (1973) Nat. New Biol. 246, 58-60.

38. Hsie, A. W., Kawashima, K., O'Neill, J. P. & Schroder, C. H. (1975) J. Biol. Chem. 250, 984-989.
39. O'Neill, J. P., Schroder, C. H. & Hsie, A. W. (1975) J. Biol. Chem. 250, 990-995.
40. Li, A. P., Kawashima, K. & Hsie, A. W. (1975) Biochem. Biophys. Res. Commun. 64, 507-513.
41. Hsie, A. W., O'Neill, J. P., Li, A. P., Borman, L. S., Schroder, C. H. & Kawashima, K. (1977) in Cancer Biology, IV (Advances Pathobiol., No. 6), eds. Borek, C., Fenoglio, C. M. & King, D. W., eds. (Stratton Intercontinential Medical Book Corp., New York), pp. 181-191.
42. Li, A. P., O'Neill, J. P., Kawashima, K. & Hsie, A. W. (1977) Arch. Biochem. Biophys. 182, 181-187.
43. Li, A. P. & Hsie, A. W. (1977) Biochim. Biophys. Acta 500, 140-151.
44. Li, A. P. & Hsie, A. W. (1978) Biochim. Biophys. Acta 527, 403-413.
45. O'Neill, J. P., Li, A. P. & Hsie, A. W. (1977) Biochim. Biophys. Acta 497, 35-45.
46. Evian, D., Gottesman, M., Pastan, I. & Anderson, W. B. (1979) J. Biol. Chem. 254, 6931-6937.
47. Gottesman, M. M., LeCam, A., Bukowski, M. & Pastan, I. (1980) Somat. Cell Genet. 6, 45-61.
48. Coffino, P., Bourne, H. R. & Tomkins, G. M. (1975) J. Cell Physiol. 85, 603-610.
49. Steinberg, R. A., O'Farrell, P. H., Friedrich, U. & Coffino, P. (1977) Cell 10, 381-391.
50. Steinberg, R. A., van Daalen Wetters, T. & Coffino, P. (1978) Cell 15, 1351-1361.
51. Steinberg, P. A. & Coffino, P. (1979) Cell 18, 719-733.

CELL CONFORMATION AND GROWTH CONTROL

Scott Wittelsberger and Judah Folkman

From the Department of Surgery, Children's Hospital Medical Center, and the Departments of Anatomy and Surgery, Harvard Medical School, Boston, and the Department of Biology, Massachusetts Institute of Technology, Cambridge, Massachusetts

Studies on the control of growth of mammalian cells have focused primarily on the role of diffusible factors, in the promotion of cell growth. However, a number of experimental results in the last several years suggest that there is another level at which cell growth may be controlled. These studies show that cell conformation plays a crucial role in growth control. The term "conformation" as used here refers both to cell shape and to the nature of the cell-substratum interaction. The experimental studies which will be described here show that cell conformation has a profound influence on (a) cell growth; (b) RNA and protein metabolism; and (c) gene expression. Furthermore, the acquisition of the fully transformed or neoplastic phenotype appears to involve the loss of the conformational-related control of these processes.

CELL SHAPE AND GROWTH

Effect on Growth Rates. The requirement of many untransformed cells for attachment to a surface in order to grow has been a long-recognized phenomenon of cell cultures. The term "anchorage-dependence" was coined by Stoker to refer to this requirement (1). When removed from a surface, anchorage-dependent cells become arrested in the G1 phase of the cell cycle, but resume growth when allowed to reattach (2). Many cells which are completely transformed have lost the requirement for substratum attachment, and indeed, "anchorage-independence" has been frequently used as an assay for the transformed state (3). Anchorage-independence has been found to be one of the most consistant correlates of tumorogenicity *in-vivo* (4).

An obvious difference between attached and non-attached cells is their shape: the attached and growing cells are flat, while the unattached and non-growing cells are spherical. To study the effect of cell shape on growth in a controlled manner, it is desirable to be able to precisely alter cell shape independently of other experimental variables. This has been done by coating standard tissue culture dishes with a thin film of the polymer, poly(2-hydroxyethyl methacrylate) or polyHEMA. Cells on top of polyHEMA-coated surfaces are not able to spread as well as those on normal culture dish surfaces. The greater the thickness of the polyHEMA film, the less the cells can spread. It is thus possible to precisely control the cell shape from a completely flat to a nearly spherical formation by increasing the amount of polyHEMA coating the culture dishes. Cell shape is quantified by measuring one or more of the following parameters: diameter, area, or height. These parameters can be measured rather easily using microscopic techniques. Before any measurements are made, the cells must be allowed enough time to reach "shape equilibrium" after which their average shape does not change. It should be emphasized that both the suitable array of polyHEMA thicknesses and the time for freshly plated cells to reach shape equilibrium vary from one cell type to another. Also, since cells are continually moving and changing their shape, what is measured is actually the average shape of a cell on a given thickness of polyHEMA.

Using polyHEMA-coated plates, the shapes of WI-38 human fibroblasts, A-31 and 3T3 fibroblasts, and bovine aortic endothelial cells were precisely varied from completely spread to a nearly spherical configuration. Cell growth was measured by direct counting of cells, by incorporation of ^{3}H-thymidine and by autoradiography of cells incubated with ^{3}H-thymidine. Regardless of the cell type or the means of measuring growth, a similar result was obtained: the more round the cellular conformation, the lower the rate of growth (5).

DENSITY-DEPENDENT GROWTH CONTROL AND WOUND HEALING

Anchorage-dependent, untransformed cells cease growing in culture dishes when confluency is reached. This property, referred to as density-dependent growth control or "topoinhibition", has been a long-recognized phenomenon of cultured cells. The following experiments by Folkman and Moscona were performed in order to determine if this property of cells is related to cell shape (5).

W1-38 human diploid fibroblasts were grown to confluence and their shape, in terms of cell height, was measured. Cells were then plated at very sparse densities on dishes coated with a polyHEMA thickness at which their shape is identical to that of the cells in the confluent culture. When the incorporation of

^{3}H-thymidine was measured, both the crowded cells in the confluent cultures and the sparse cells on the polyHEMA-coated plates showed identical levels of incorporation. Thus, when cells are forced into an identical shape, whether by the crowding of adjacent cells in a confluent culture or by the use of polyHEMA, the inhibitory effect on cell growth is identical. This result suggests that the more spherical shape cells obtain as they become more crowded is a more important factor in their growth cessation than the presence of adjacent cells or crowding *per se*.

When a confluent culture of cells is scraped or wounded, the cells closest to the wound migrate into it and show an increased labeling index with ^{3}H-thymidine. The labeling index decreases as the distance of cells from the wound increases. That this phenomenon may be shape-related is suggested by the following experiment (1). A wound was made in a confluent culture of A31 mouse fibroblasts. One hour later the height of cells near the wound was measured. The cells in the first, second and third heights were 46, 60, and 70%, respectively, of the initial height. There was virtually no change in the height of cells further removed from the wound. Although ^{3}H-thymidine incorporation was not measured, the expected labeling index of A-31 cells held at these shapes by use of polyHEMA would predict, in agreement with observations from other studies, the highest labeling index for the flattest cells in the first row and decreased labeling indices in the cells further removed.

Serum Sensitivity. Mammalian cells require one or more growth-promoting factors in serum in order to grow. Cells whose rate of growth is dependent upon the concentration of serum in the media are "serum-sensitive". The following experiment was designed to ascertain whether the effect of shape on growth is exerted by changes in serum sensitivity. 3T3 cells were plated on a series of polyHEMA-coated dishes so that their shapes ranged from flat to spherical. The effect of increasing serum concentrations on growth rates of cells maintained at these shapes was then measured. The flattest cells were seen to be the most serum sensitive; whereas nearly round cells undergo little if any increase in growth rate as the serum concentration was increased. It thus appears that the more round a cell is kept, the less it is able to utilize the growth-promoting factor(s) present in serum (6).

Macromolecular Metabolism. The above studies show that shape-related events have a profound influence on cell growth, possibly by mediating the action of one or more serum components. The effect of shape on RNA and protein metabolism has been studied in the laboratory of Penman (7,8). Initial studies used 3T6 cells, an established mouse fibroblast line. In these experiments, 3T6 cells were removed from monolayer cultures with trypsin and suspended in media made viscous with a low percentage of methocel. The shape of the cells was thus changed from flat in monolayer culture, to spherical in suspension culture. The following processes were then measured, using radioactive percursors; protein synthesis, nuclear pre-ribosomal RNA (pre-rNA) synthesis; heterogeneous nuclear RNA (hnRNA) synthesis; and messenger RNA (mRNA) production.

The rate of protein synthesis declined slowly over a period of 72 hours, after an initial lag period of about 20 hours, and reached a minimal level of 20-25% of that in the control monolayer culture. The rate of synthesis of 45S nuclear pre-RNA declined rapidly, reaching a level of about 20% of controls within 12 hours after suspension. The rate of synthesis of hnRNA, however, was unchanged. The production of new mRNA, like pre-RNA, decreased rapidly and reached a level of about 20% of control levels with 12 hours. However, the total cellular content of cytoplasmic mRNA did not change in the suspended cells. This is because the rate of mRNA degradation was decreased. The half-life of mRNA increased from 13 hours in monolayer cells, to about 70 hours in suspended cells. During suspension, the mRNA molecules were slowly withdrawn from translation, and became sequestered as ribonucleoprotein particles. After 72 hours, only about 20% of the mRNA molecules were in the form of polysomes. The remaining polysomes, however, were fullsized.

When 3-day suspended 3T6 were allowed to reattach to and spread out upon culture dishes, protein synthesis resumed rapidly, reaching normal levels within about 5 hours. Cells were also allowed to reattach, but subsequent spreading was prevented by the use of concanavlin A, cytochlasin B, or colchicine. Most interestingly, these cells recovered their protein synthesis in an identical manner to the cells which reattached and spread. This recovery of protein synthesis occured long before the production of new mRNA molecules had resumed. The recovery must, therefore, involved the use of the stabilized mRNA present in the cytoplasm of the suspended cells.

The results of the last experiment imply that for recovery of protein synthesis to occur, only contact with the proper surface is required. The relationship between contact, subsequent spreading, and recovery of particular processes was studied in more detail by Ben-Ze'ev et al. (9). Three-day suspended 3T6 cells were reattached to culture dishes coated with varying thicknesses of polyHEMA. The amount of spreading was thus well controlled, and all cells made contact with a solid surface. Protein synthesis recovered in an identical manner on plastic and on all of the polyHEMA-coated dishes. However, the extent of recovery of mRNA production, pre-RNA synthesis and DNA synthesis was proportional to the extent of cell spreading. Thus, the cytoplasmic process, recovery of protein synthesis, required only contact in order to recover after suspension. The nuclear processes, DNA synthesis, pre-RNA synthesis, and the synthesis processing or export of pre-mRNA, recovered only when cells were allowed to spread.

CELL SHAPE AND GENE EXPRESSION

That shape-related properties may influence the expression of gene products as well as growth and macromolecular metabolism, has been shown in studies in the laboratory of Folkman (10,11). Glowacki, et al., using polyHEMA, studied the effect of cell shape on the expression of glycosaminoglycan (GAG), the cartilage matrix material, by chondrocytes. The synthesis of GAG was found to increase dramatically as cells adopted an increasingly spherical configuration on plates coated with polyHEMA (10).

In another study, Azizkhan et al. examined the effect of shape on the synthesis of gonadotropin by human choriocarcinoma cells (11). The cells that were kept the most round by polyHEMA synthesized the most gonadotropin, even though the growth rate of the choriocarcinoma cells was relatively unaffected by shape changes.

CELL SHAPE AND TRANSFORMATION

One of the most consistent correlates of transformation and tumorogenicity in fibroblasts is loss of the requirement for attachment to a solid surface in order to grow. The term "anchorage-dependence" was coined by Stoker to refer to such a requirement (1). An obvious difference between attached cells and unattached cells is their shape: The attached cells are flat and the suspended ones are spherical. The studies of Folkman and Moscona show that the rate of growth of untransformed cells is tightly coupled to their shape: the more spherical the shape, the slower the rate of growth (5). That the growth rate of completely transformed cells is only slightly affected by shape indicated that the process of transformation involves an escape from at least some of the shape-related controls of cellular functions. A convenient system in which to study the loss of shape-related control mechanisms is that of mouse fibroblasts. This is because cell lines can be derived from embryonic mouse fibroblasts which display varying degrees of the transformed phenotype. Wittlesberger and Penman have studied macromolecular responses to the shape change induced by suspension of the following mouse fibroblast cells: mouse diploid fibroblasts; 3T3; 3T6; high density passage 3T6 (HDP-3T6); and 3T3 cells virally transformed by both SV40 and polyoma viruses, SVPy-3T3 (12). 3T3 cells are transformed in the sense that they are immortal and have an aneuploid chromosome content. However, the conditions used by Todaro and Green for the establishment of this cell line from embryonic cells resulted in cells which retain a flattened morphology and density-dependent inhibition of growth (13). 3T6 cells are similar in origin to 3T3 cells, but were established under conditions which were selected for cells that are less sensitive to crowding than 3T3 cells. 3T6 cells grow to considerably higher densities than 3T3 cells and are also less well spread (13). The HDP-3T6 cells were obtained by growing 3T6 cells to a confluent density of 2×10^5 cells/cm^2 12 times. The HDP-3T6 cells grow to slightly higher cell densities than 3T6 cells, and possess a somewhat rounder morphology. The SVPy-3T3 cells are completely transformed in that they are insensitive to contact-inhibition, anchorage-independent and tumorogenic. The diploid fibroblasts are obtained from embryonic mice and are used after only two in-vitro passages. They are completely normal and untransformed in that they possess an unchanged chromosomal content, are not immortal, and are sensitive to contact-inhibition of growth. The cells studied thus represent a spectrum of transformed phenotypes ranging from the completely untransformed mouse diploid cells to the completely transformed SVPy-3T3 cells. With the exception of the SVPy-3T3, all the cells studied were anchorage-dependent. The experimental study involved placing these cells in suspension culture and measuring protein synthesis, hnRNA synthesis, pre-RNA synthesis; tRNA synthesis, and the production of mRNA. The question asked was: as cells become more transformed, are there accompanying changes in the nature of the response of these macromolecular processes to the shape change which occurs in suspension?

The inhibition of protein synthesis is rapid in diploid fibroblasts, less rapid in 3T3, very slow in 3T6 and HDP-3T6, and absent in SVPy-3T3. In contrast, the inhibition of hnRNA synthesis is lost abruptly as cells become more transformed. Diploid fibroblasts and 3T3 cells shut down hnRNA synthesis to 40% of control monolayer levels after suspension, while 3T6, HDP-3T6 and SVPy-3T3 do not decrease their rates of hnRNA synthesis. The production of mRNA decreases to about 25% after suspension of diploid fibroblasts, 3T3 and 3T6; but is unaffected in HDP-3T6 and SVPy-3T3. Interestingly, 3T6 cells have lost the shape responsiveness to hnRNA synthesis, but not to mRNA production, suggesting that these are independent phenomena. The inhibition of nuclear pre-TRNA synthesis remains unchanged in diploid, 3T3 and 3T6 cells, but is more sluggish in HDP-3T6.

This study shows that during the process of fibroblast transformation there is a progressive loss of shape-responsive metabolic controls. This finding may be related to other studies which have provided evidence that the acquisition of the fully transformed and neoplastic phenotype is a progressive, multi-step process (14,16) -- the "steps" in this case being particular shape-responsive metabolic processes.

DISCUSSION

Conformation, Growth and Gene Expression. The experimental results which have been described here show that cell conformation has a profound influence on growth, macromolecular metabolism, and gene expression. In addition, the process of transformation appears to involve a loss of sensitivity to conformational-related cellular control mechanisms. The mechanism by which conformation influences such important processes is unknown. Indeed the phrases "conformational control" or "shape control" are used because the precise events or processes which are triggered by conformational or shape changes are not known.

It seems that a profitable area in which to search for explanations for some of the experimental observations which have been presented here is in the relationship between cellular structural elements or "cytoskeleton" and the molecular events involved in growth control, gene expression and macromolecular metabolism. This is because the cytoskeletal system must undergo very dramatic changes in arrangement when a cell's shape is changed. The cytoskeletal system is already known to be involved in the process of cell attachment and spreading (17,18). At present, very little is known about the relationships between the cytoskeleton and such molecular events. Experiments in the laboratory of Penman have provided biochemical evidence that mRNA molecules are attached to and translated upon the cytoskeleton (19). The cytoskeleton also appears to be used during viral infection of HeLa cells by poliovirus (20) and vesicular stomatatis virus (21). But the precise nature of the mRNA linkage, the structural elements to which the mRNA is linked, and the means by which conformationally induced changes in cytoskeletal structure affect the mRNA translation, are unknown.

There is a great amount of evidence on the crucial role played by diffusible growth-stimulating factors in the control of cell division. Is it possible that changes in cell conformation or cytoskeletal state also change the nature of the interaction between these factors and the cell? Experimental studies from a number of laboratories suggest that this may be the case. As described earlier, experiments by Robert Tucker in the laboratory of Judah Folkman have shown that changes in cell shape also change the ability of cells to be stimulated by one or more mitogenic serum factors: as cells become more round their response to serum stimulation of growth decreases.

The microtubule disrupting agent colchicine has been reported to both increase (22,23) and decrease (24) the growth stimulating effect of a number of mitogens. This apparent contradiction appears to have been resolved by McClain and Edelman (25). Their studies suggest that the nature of the effect of colchicine on the cellular response to growth factors is dependent upon the cell density. Colchicine increased the effect of growth factors in confluent cells, but decreased the effect in sparse cultures. It is tempting to interpret these findings in terms of cell shape or cytoskeletal organization. The sparse cells are flat and the confluent cells are considerable more rounded; associated with these different shapes are different states of skeletal or microtubule organization which cause the cell to respond to the presence of colchicine and growth factors in different and, in this case, opposite directions. Recently, Selden, Haudenschild and Schwartz have reported on the effect of colchicine and vinblastine on the initiation of division in bovine aortic endothelial cells. These cells, when confluent, form a continuous sheet (26). The only treatment known to re-initiate growth in such cells is wounding. The cells next to the wound are known to flatten out, migrate toward the wound, and concomitantly re-initiate growth. Most interestingly, these workers have also found that confluent, stationary endothelial cells can also be stimulated to grow by brief exposure to colchicine or vinblastine. The cells so treated respond in a manner similar to those adjacent to a wound. There is a retraction of one cell from another and a change in cell shape. Is it possible that the shape and skeletal changes which allow the cells next to the wound to grow are the same, or cause the same stimulating event, as those changes caused by colchicine and vinblastine which allow the confluent cells to grow?

The nature of the substratum to which a cell is attached also appears to play a very important role in cell growth. Experiments in the laboratory of Gospodarowicz have shown that both smooth muscle cells (27) and bovine corneal endothelial cells (28) will grow and reproduce a number of their in-vivo properties only when attached to an extracellular matrix. Since cytoskeletal elements are involved in cell-substrate interactions (17,18), the cytoskeletal arrangement of a cell may also be different on different substrata, and thereby change the growth and functional capacities of a cell.

Some of the experiments in the laboratory of Folkman which have been described here demonstrate that the synthesis of particular gene products is influenced by cell shape. Both chondrocytes and human choriocarcinoma cells synthesize greatly increased amounts of GAG and gonadotrophin, respectively, as their shape becomes more round. In similar studies, using polyHEMA-coated plates, Chen has reported that 3T6 fibroblasts synthesize increased amounts of fibronectin as then shape becomes more flat (29). The results obtained on the synthesis of GAG and fibronectin may be growth-related in that the rate of growth of both chondrocytes and fibroblasts is highly shape-dependent. Such an explanation, however, does not account for the effect of shape on gonadotrophin synthesis by choriocarcinoma cells whose growth rate is only very slightly influenced by shape (11).

These results show that the expression of particular gene products can be greatly changed without simultaneously changing the concentration of any possible gene inducers in the media. These changes in gene expression are the result of changes in cell shape (10,11,29) as well as changes in the substratum (27,28,30), and thus may reflect alterations in the state of the cytoskeleton which influence or control the action of diffusible gene activators in the media.

Cell Conformation and Transformation. Among many cell types, one of the most reliable correlates of the neoplastic state is anchorage-independence. This can be looked upon as an escape from the normal conformational controls upon growth. It was earlier hypothesized that the shape-related control of growth results from cytoskeletal changes which accompany the shape changes and which directly or indirectly are responsible for a cell's progress through the cell cycle. From this point of view, the process of transformation involves a breakage of the coupling between the cytoskeleton and the division process. The frequently reported changed in the cytoskeletal elements of transformed cells (31,32) and the fact that a distortion of the cytoskeletal system is the earliest observable event following expression of the SRC gene may be aspects of an aberrant cytoskeletal-signling system. The findings of Wittelsberger and Penman that the process of transformation involves a progressive loss of shape-sensitive regulatory mechanisms lend very strong support to such a conceptual scheme.

Summary and Conclusion. Experiments to date have shown an intimate correlation between cell conformation and the control of proliferation in fibroblast and endothelial cells. Macromolecular metabolism and gene expression are also profoundly affected by cell conformation. The process of transformation and development toward the neoplastic state appear to involve a progressive diminution in conformation-related control systems. The manner is which conformational control is imposed upon cells may be through altered cytoskeletal states.

REFERENCES

1. Stocker, M., O'Neill, C., Berryman, S., and Waxman, V. (1968). Int. J. Canc. 3, 683-693.
2. Otsuka, H. and Moskowitz, M (1975). J. Cell Phys. 87, 213-220.
3. MacPherson, I. and Montagnier, L. (1964). Virology 23, 291-294.

4. Freedman, V.H. and Shin, S. (1974). Cell 3: 355-359.
5. Folkman, J. and Moscona, A. (1978). Nature 273, 345-349.
6. Tucker, R. and Folkman, J. Unpublished observations.
7. Benecke, B.-J., Ben-Ze'ev, A. and Penman, S. (1978). Cell 14, 931-939.
8. Benecke, B.-J., Ben-Ze'ev, A. and Penman, S. (1980). J. Cell Physiol. 103, 247-254.
9. Ben-Ze'ev, A., Farmer, S.R. and Penman, S. (1980). Cell 21, 365-372.
10. Glowacki, J., Trepman, E. and Folkman, J. (1981). In preparation.
11. Azizkhan, J. and Klagsburn, M. (1981). In preparation.
12. Wittelsberger, A. and Penman, S. (1981). Cell, in press.
13. Todaro, G. and Green, H. (1963). J. Cell Biol. 17, 299-313.
14. Foulds, L. (1975). Neoplastic Development, Vol. 2, (Academic Press, New York).
15. Barrett, J.C. and Ts'O, P.O.P. (1978). Proc. Natl. Acad. Sci. USA 75, 3761-3765.
16. Barrett, J.C. (1980). Canc. Res. 40, 91-94.
17. Abercromb'e, M., Heaysman, J.E. and Pegrum, S.M. (1971). Exp. Cell Res. 59, 393-398.
18. Couchman, J.R. and Rees, D.A. (1979). J. Cell Sci. 39, 149-165.
19. Lenk, R., Ransom, L., Kaufman, Y. and Penman, S. (1977). Cell 10, 67-78.
20. Lenk, R. and Penman, S. (1979). Cell 16, 289,301.
21. Cervera, M., Dreyfuss, G. and Penman, S. (1981). Cell, in press.
22. Otto, A.M., Zunbe, A., Gibson, L., Kubler, A.M. and DeAsua, L.J. (1979). Proc. Natl. Acad. Sci. USA 76, 6435-6438.
23. Friedkin, M., Legg, A. and Rozengurt, E. (1980). Exp. Cell Res. 129, 23-30.
24. McClain, D.A., D'Eustachio, P. and Edelman, G.M. (1977). Proc. Natl. Acad. Sci. USA 74, 666-670.
25. McClain, D.A. and Edelman, G.A. (1980). Proc. Natl. Acad. Sci. USA 77, 2748-2752.
26. Schwartz, S.M. Selden, S.C. and Bowman, P. (1979). In Hormones and Cell Culture, eds. Sato, G.H. and Ross, R. (Cold Spring Harbor Laboratory Press), pp. 593-610.
27. Gospodarowicz, D. and Ill, C.R. (1980). Proc. Natl. Acad. Sci. USA 77, 2726-2730.
28. Gospodarowicz, D., Delgado, D. and Vlodavsky, I. (1980) Proc. Natl. Acad. Sci. USA 77, 4094-4098.
29. Chen, L.B., Summerhayes, I., Hsieh, P. and Gallimore, P.H. (1979). J. Supramolec. Struct. 14, 139-150.
30. Emerman, J.T., Burwen, S.J. and Pitelka, D.R. (1979). Tissue and Cell 11, 109-119.

31. Goldman, R.D., Yerna, M. and Scioso, J.A. (1976). J. Supramolec. Struct. 5, 155-183.
32. Tucker, R.W., Sanford, K.K. and Frankel, F.R. (1978). Cell 13, 629-642.

Supported by grant #RO1-CA14019 and grant #NIH-5-RO1-CA08416 from the U.S. Public Health Service, National Cancer Institute, and by a grant to Harvard University from Monsanto Company.

COUPLING OF NUCLEAR MORPHOMETRY TO CELL GEOMETRY. ITS ROLE IN THE CONTROL OF NORMAL AND ABNORMAL CELL GROWTH

C. Nicolini, M. Grattarola, F. Beltrame and F. Kendall

Department of Biophysics - Physiology
Temple University, Philadelphia, PA USA
and National Research Council, Italy

INTRODUCTION

Recent work has related nuclear DNA morphometry to cell function during the cycle of synchronized mammalian cells (1) (see also chapter by Nicolini and Belmont) during the G0-G1 transition of confluent human diploid fibroblasts stimulated to proliferate (2), as well as during virus transformation (3) and aging (4). However, those studies did not yet establish a causal relation. It is still unclear whether a change in the functional state in the cell results in altered nuclear morphology or whether an altered nuclear morphometry results in an altered functional state (and it is a requirement for the initiation of this altered functional state). At the same time cell geometry appears causally correlated with cell proliferation in normal cells but not in transformed cells (5). We have therefore recently explored the nuclear DNA morphometry changes with respect to cell geometry during various functional states to establish the role of their coupling (or uncoupling) in this control of normal and abnormal cell growth.

DENSITOMETRIC AND GEOMETRIC IMAGE ANALYSIS

In all these studies the stochiometric staining of nuclear chromatin-DNA is accomplished by the Feulgen-basic Fuchsin reaction at the optimal hydrolysis of 5NHCl, at 25°C for minutes. To obtain the geometric densitometric properties of the overall cell including its cytoplasm (hereby referred to as "cell morphometry") cells are double or triple-stained (7) using also napthol yellow-S (protein) in combination with periodic-acid (polysaccharides). By proper filter configuration and taking advantage of the differential absorbance of the dyes utilized (435 nm for naphtol yellow-S;

540 nm for basic Fuchsin), we may then obtain for the same cell its nuclear morphometry (Feulgen-DNA) and its cell morphometry (protein-naphtol yellow-S). In these studies to enhance cell border definition, the triple stain (which include polysaccharides absorbance) was utilized (7). Occasionally nuclear and cell morphometry, exclusive of densitometry were obtained from the same cells, being Wright-stained.

The "global" geometric and densitometric properties of the Feulgen-stained cell (hereby referred to as "nuclear morphometry") are obtained by means of a Quantimet 720-D Image Analyzer (Cambridge Co., Monsey-N.Y.) on line with a PDP11/40 digital computer (1,2,3,8).

Nuclear (or cellular) images were acquired by a Zeiss microscope equipped with a 100 X oil immersion planar achromat of 1.25 NA and illuminated via a condenser of 1.3 NA by means of a 100 W tungsten halogen light source, equipped with a 540 nm filter with a half bandwidth of 40 nm. The image was registered on a plumbicon scanner by means of a Reichert high quality magnification changer. Total magnification was routinely 1250 X. The scanner area is divided into 800 x 588 picture elements whose optical density can be digitized into 64 levels (8). The linear dimension of each approximately square picture element was determined to be 0.08 μm by means of a stage micrometer (American Optical). A blank area of each slide was used to load the shade corrector and to calibrate the densitometer by means of neutral density filters (Kendall, et al., 1979a). The linear response of the densitometer (with respect to OD) was verified by means of an optical density wedge filter (ESCO, Oak Ridge, N.J.) calibrated between 0 and 2 OD. A threshold of 0.03 - 0.04 OD was used to define the nuclear border. Field uniformity tests on a single nucleus positioned in 9 locations around the field yielded for all slides coefficients of variations of less than 2.5% for integrated optical density and 1.0% for area. Variation of either parameter was less than .5% for ten measurements of a single image in the center of the field.

The following parameters were measured as previously described (1,2,3,8) for each nuclear (or cellular) image: integrated optical density (IOD, proportional to DNA content), area, perimeter, horizontal and vertical Feret diameter, and horizontal and vertical complex projections. The horizontal and vertical Feret diameters (FD) are the shadow projections of the longest dimensions onto horizontal and vertical lines. The vertical and horizontal complex projections (CP) are the sums of the shadow projections of all lagging edges on horizontal and vertical lines. An excess projection (EP) is computed as the numerical difference between a complex projection and its correspond-Feret diameter. Thus $EP_1 = CP_H - FD_V$ and $EP_2 = CP_V - FD_H$. These relationships were verified by measuring the excess projections of squares and rectangles created by an electronic frame generator. In these cases EP is zero. Physically EP is a measure of border re-entrance. Euclidean norms of EP and CP are used because this procedure makes the resulting values independent on the physical orientation of each feature with respect to direction of scan (8).

From these parameters the following derived parameters were calculated: average optical density (IOD/Area), form factor (4π (Area/Perimeter Square)), and convolution factor (euclidean norm of the excess projection, EP, divided by the euclidean norm of the complex projection, CP, where EP = CP - Feret diameter). Measurements of IOD, Area and Perimeter are theoretically and practically independent of feature orientation (global values) as are derived parameters computed from them (average OD-AVOD and form factor-FF). The convolution factor is a dimensionless ratio which is quasi-global and serves as a measure of border re-entrance for each feature. The convolution factor is necessary to interpret the geometric significance of form factor (which is only an index of circularity and which is less than unity for features which are completely non-re-entrant but not circular or approximately circular but re-entrant).

TRANSFORMED VERSUS FIBROBLAST-LIKE CHO DURING CELL CYCLE

As previously shown (3,10) administration of 10^{-3}M cAMP to transformed CHO causes the appearance of a normal fibroblast-like cell morphometry and growth characteristics. These changes are also accompanied by a 2 hr lengthening of cell cycle duration: specifically a 2 hr increase in the G_1 phase duration with a subsequent delay in entering S-phase (7) is created. At the level of the intact cell (triple-staining), for every time interval, the decreased form factor and increased perimeter and area in the face of similar integrated optical density (reflecting mostly proteins and polysaccharides and partly DNA) is a quantitative assay of the dramatic increase in CHO cell elongation to a fibroblast-like shape, in the presence of cAMP (7). Both transformed and reverse-transformed cells do progress and undergo similar modulation in their nuclear and cellular morphometry during the cell cycle (as reported for HeLa previously), with the reverse-transformed cells displaying only quite larger convolution dispersion and elongation.

The upper panel of Figure 1 shows that for every cycle subphase either early S (6 hours after mitosis), or middle-S (10 hours after mitosis) a similarly striking linear correlation exists even at the level of intact cells, between changes in cytoplasm condensation (average optical density) and cytoplasm elongation (form factor). Specifically, the increased elongation (lower form factor) of CHO cells (readily apparent as spindle-shape by visual

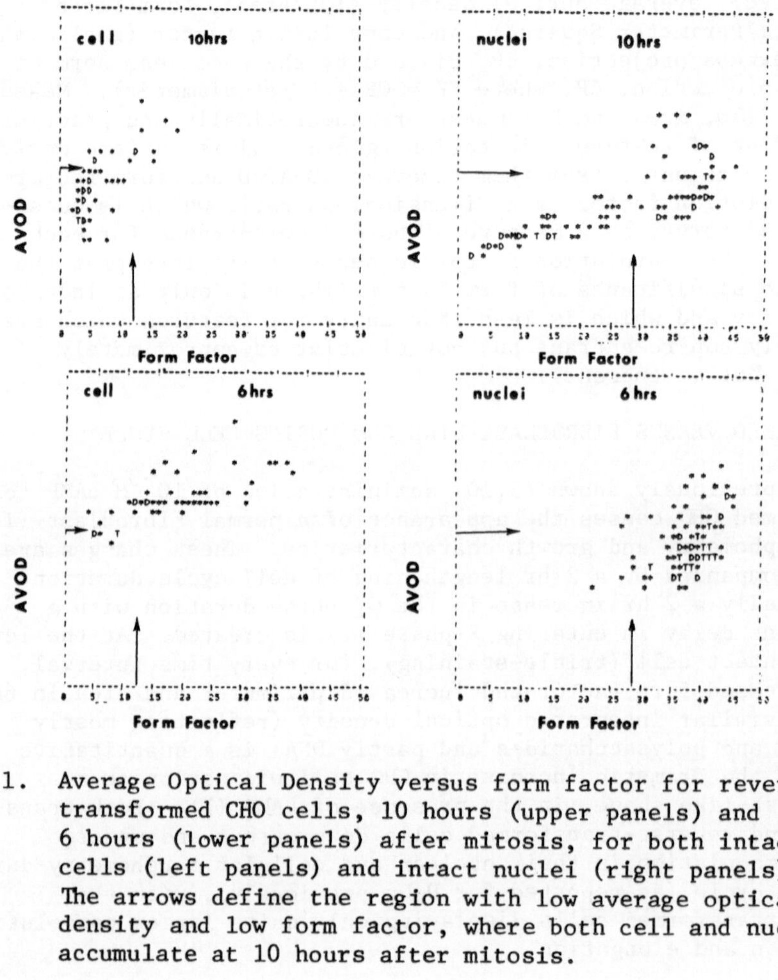

Fig. 1. Average Optical Density versus form factor for reverse transformed CHO cells, 10 hours (upper panels) and 6 hours (lower panels) after mitosis, for both intact cells (left panels) and intact nuclei (right panels). The arrows define the region with low average optical density and low form factor, where both cell and nuclei accumulate at 10 hours after mitosis.

inspection) is paralleled by an increased cytoplasm dispersion (lower IOD per unit area). The above correlation previously shown (7) between mean values of geometric and densitometric parameters of bulk population clearly reflects properties of individual cells or nuclei (Fig. 1).

Namely, as shown in Fig. 1 during the cycle of reverse-transformed cells, a compact and round cell is accompanied by a round and compact nucleus (at 6 hours after mitosis), while a disperse and elongated nucleus (at 10 hours after mitosis) is accompanied by a disperse and elongated cell. As shown in Fig. 2 (upper panel), it appears however that during the cell cycle the correlation between changes in nuclear geometry and cell geometry is lacking for transformed cells (correlation coefficient r = 0.04) and good (r = 0.91) for their reverse-transformed fibroblast-like counterparts. Similarly, the average optical density of both nuclei and cell (lower panel of Fig. 1) indicates that during the cycle, reorganization of protein and polysaccharides space (mostly cytoplasm) strikingly parallel reorganization of chromatin-DNA. Indeed (14) the degree of cytoplasm condensation correlates well with the degree of chromatin condensation for fibroblast-like (r = 0.91), but not for transformed CHO cells (r = 0.04).

It appears that nuclear morphometry, which reflects the higher order chromatin structure (13, 16), not cell geometry is coupled with cell growth and progression during the cycle in both cell types, suggesting the existence of a direct causal relationship between higher order chromatin structure and cell function. This is particularly enlightning since previous findings (5, 15) have indicated that in transformed cells cell shape and cell growth are uncoupled, being coupled in normal cells.

When cells are in suspension culture they acquire a round shape with eventual loss of plating efficiency for normal untransformed cells but not for the transformed ones. Contrary to normal, transformed cells can also grow as rapidly on the thickest substrate as on the thinnest despite that their cell shape varied with substrate adhesiveness in the same proportion as for untransformed cells (5).

EFFECT OF POLY-HEMA SUBSTRATE

Pursuing the line of investigation on the relationship between cell geometry and cell growth, it has been recently proposed by Folkman and Moscona that density dependent inhibition of cell growth is mediated through cell geometry (5). By changing the chemical nature of the tissue culture substrate through the addition of increasing amounts of the polymer poly (HEMA) (poly-2-hydroxyethylmethacrylate) it has been possible to progressively increase the cell height of plated cells, non-transformed and transformed

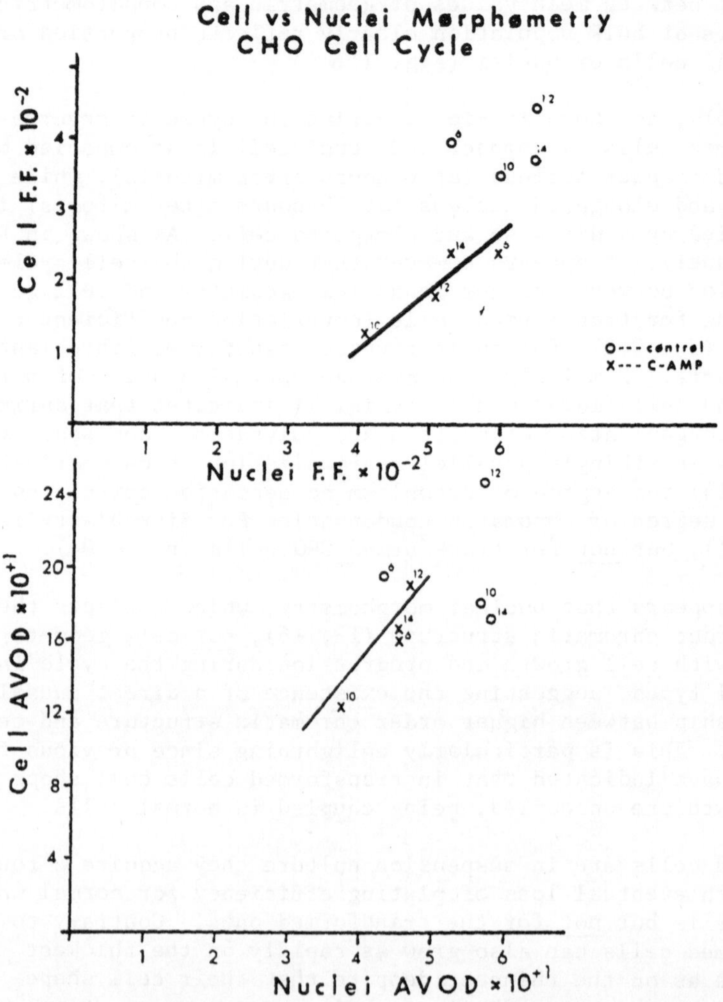

Fig. 2. (Above) Mean values of cell form factor vs. nuclei form factor of transformed (0) and cAMP reverse-transformed (X) synchronized CHO cells at 6, 10, 12 and 14 hours after mitosis. Form factor is here defined as the ratio of area divided by perimeter square, without multiplying by 4π.

(Below) As above, but mean values of cell cytoplasm average optical density vs. nuclear-DNA average optical density.

alike (5). Increase in cell height and cell spherical shape correlated with a decrease in DNA synthesis for nontransformed cells (5). Moreover nontransformed cells with equivalent cell heights and shape, whether produced by crowding in dense cultures or as low density growth on a chemically modified substrate, had equivalent rates of DNA synthesis (5). At present, it is unclear which are the important variables likely associated with rounded cells that inhibit proliferation, such as transport of certain nutrients or ions and/or nuclear DNA organization. It is then interesting to find that, when we assay for the nuclear morphometry of WI-38 cells plated on increasing thicknesses of the polymer, Poly-HEMA (11), changes in nuclear morphometry (Fig. 3) are found to parallel changes in cellular morphometry: namely a transition to a more rounded and compact conformation, with increasing thicknesses of the polymer (11). Such changes in nuclear morphometry induced by mere changes in cell geometry (with all nutrition required still present) are opposite to those already observed after stimulation of proliferation by serum addition and similar to those observed with increasing confluency (12). All above findings confirm the high correlation between changes in cell geometry and changes in nuclear morphometry.

EXPRESSION OF METASTATIC POTENTIAL

Immunological, biophysical and biochemical differences have been described between transformed "neoplastic" cells and their normal counterparts. Several criticisms however can be raised when a tumor is compared to its tissue of origin since differences may reflect cell type rather than manifestation of cell transformation. Similarly, analogous comparative studies <u>in vitro</u> (with transformed cells frequently aneuploid) may detect differences related to cell proliferation rather than transformation.

When comparing quiescent WI-38 fibroblasts to its stationary SV-40 virus transformed counterparts (2RA), the findings (3) are apparently contrary to the common expectation that 2RA cells are in a more active state. Nuclear morphometry did show, in fact, that the nuclear chromatin of the stationary 2RA cells is more condensed than that of confluent WI-38 cells, quite compatible with the surprising fact that the template activity of chromatin from stationary 2RA cells is less than that of quiescent WI-38 cells (see Fig. 2). Reduced template activity has also been observed in Walker tumor chromatin when compared to rat liver and mammalian chromatin.

More recently (see Fig. 4) the same increase in nuclear chromatin condensation has been reported when comparing (I) fibroblast-like cAMP reverse-transformed to transformed CHO-K1 cells (7), (II) low metastatic cell variants (F1) to high metastatic variants (F10) from the same B16 tumor in mice (12). It then appears that regardless of the mode of cell transformation (either spontaneous, viral or chemical) a dramatic increase in nuclear chromatin

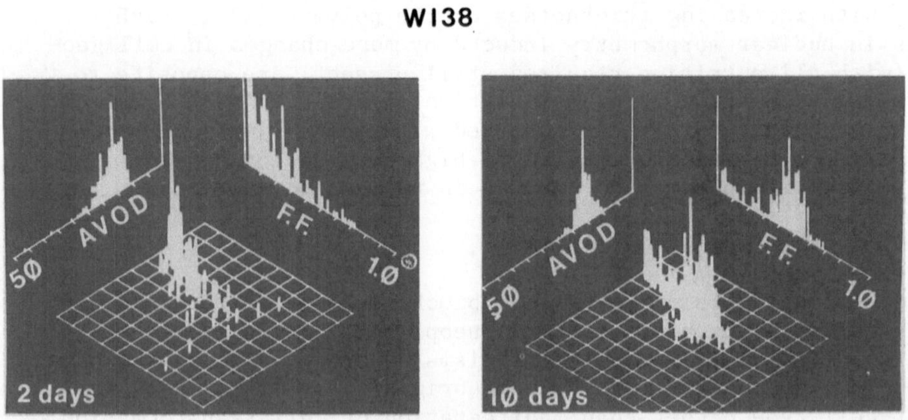

Fig. 3. Two parameter histograms of average optical density (AVOD) versus form factor (FF) of Feulgen-stained nuclei from 2C DNA WI-38 human diploid fibroblasts at 2 days on plastic (cycling G1; upper left panel), 7 days on plastic (confluent G0; upper right panel), and 2 days on 5 x 10^{-3} Poly-HEMA (lower panel), after plating. Confluent G0 cells display clustering at significantly larger form factor and average optical density (with respect to cycling G1 cells). For details see ref. 2, 11.

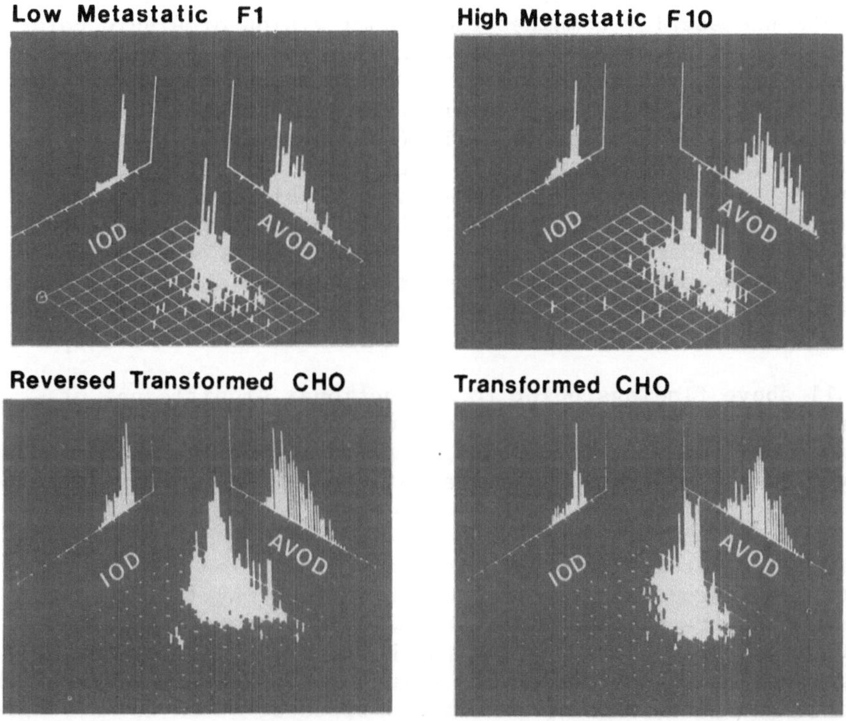

Fig. 4. (Above) Integrated optical density (IOD; i.e. DNA content) versus average optical density of low metastatic F1 cell variants (left panel) and high metastatic F10 cell variants (right panel), cloned from B16 melanoma tumor as previously shown (17). These subpopulations selected by proper manipulation appear to possess unique characteristics that determine their capacity to successfully metastatic to the lung (12, 17).

(Below) As above, but Feulgen-stained transformed (right panel) and cAMP reverse-transformed (left panel) CHO-K1 cells. For further details see ref. 7.

condensation characterizes the expression or enhancement of any "abnormal growth" behavior, consistent with the idea that a limited transcription of the genoma is necessary to maintain the transformed phenotypes (13). This finding did not emerge earlier by traditional biochemical studies, maybe because any technique which measures any chemical or physical parameter of bulk preparations necessarily combines the tunnel vision of single parameter observation with a loss of that distributional information that would exist if each contributor could be observed independently (as with multiparameter image analysis). At the same time, as per transformed versus normal cells (see Table I and Fig. 5), when we compare (12) highly metastatic cell variants (F10) and low metastatic cell variants (F1) from the same B16 melanoma tumor in vivo, the coupling between variations in nuclear form factor and cell form factor are higher in F1 ($r = 0.37$) than F10 ($r = 0.09$). Similarly high and low correlation exist between changes in nuclear area and cellular area when comparing F1 and F10 metastatic variants (Fig. 5).

All above findings point to a functional significance of nuclear morphometry and namely its coupling (or lack of it) with cell morphometry, as unique probe (not only of cell transformation) but even of the expression of metastatic potential in animal tumor (12).

The lack (or gradual reduction) of correlation between cell geometry and chromatin geometry (which in turn may influence transcription and translation) for increasingly metastatic (as for transformed) cells, could be due to the previously reported (10) disorganization of macromolecular structures, such as microtubules and microfilaments, whose presence and orientation in normal cells may instead allow transfer of modulation in cell geometry into modulation of nuclear morphometry, linked to the higher order chromatin structure, which in turn determines the metabolic activity of the cell. It still remains to be explained whether the coupling between cell and nuclear morphometry is direct, initially depending on mechanical forces transmitted from membrane to nucleus by properly oriented microtubule-microfilament, and/or indirect, depending on ionic fluxes, transport mechanism or activity changes of membrane enzymes induced by changes in physical parameters of cell membranes.

CELLULAR AGING

There have been several reports in the literature indicating that normal diploid mammalian cells have a finite life span.

Table I. Correlation Coefficents (4) Between Nuclear and Cell Morphometry

A. Low (F1) versus High (F10) Metastatic Cell Variants from B16-Melanoma Tumor

F1	r = 0.39
F10	r = 0.17

For each of 100 individual cells in log-phase growth (12) nuclear and cellular form factor were measured and the correlation coefficient computed by a least square fit.

B. cAMP Reverse-Transformed (Fibroblast-Like) versus Transformed CHO-K1 Cells

Fibroblast-like	r = 0.92
Transformed	r = 0.08

The mean values of nuclear form factor and cell form factor, respectively, were computed for "bulk" synchronized cells at different times after mitosis (14) and the correlation coefficient computed by a least square fit. The correlation coefficients, at individual cell levels would be lower for both cell types, maintaining however a similarly striking difference.

These studies have shown that human fibroblasts progress through three definite stages. Phase I corresponds to the development of a primary culture; Phase II cells grow exponentially to form confluent monolayers upon sub-cultivation; while Phase III shows a decreased proliferative capacity, ultimately leading to complete growth failure upon sub-cultivation. Several biochemical (alteration in DNA metabolism, chromosomal aberration, histone acetylation) and

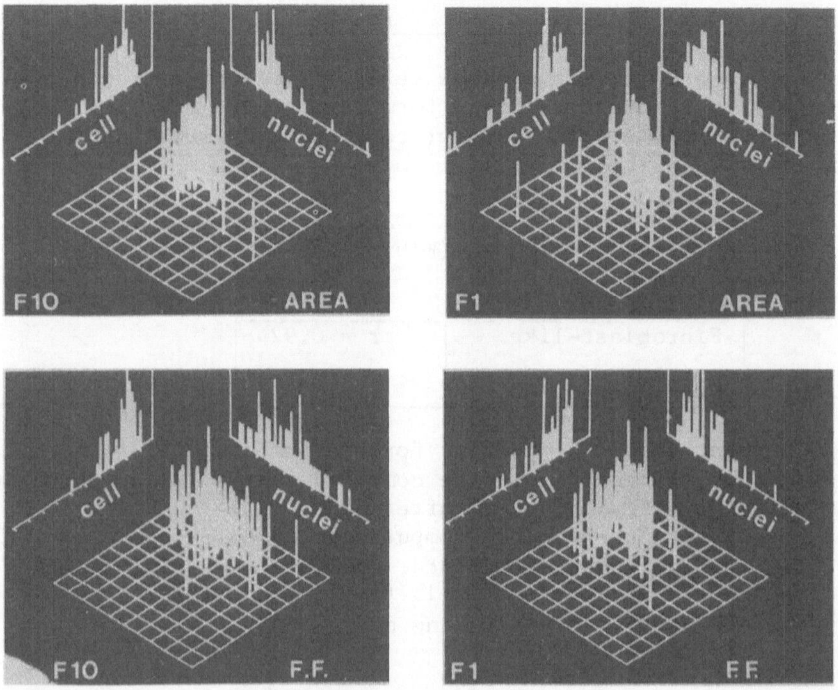

Fig. 5. (Above) Cell versus nuclear area for F1 (low metastatic variants) and F10 (high metastatic variants) from B16 melanoma primary tumor growing in mice (12, 17).

(Below) As above, but cell versus nuclear form factor.

biophysical (increased on cell volume, increase in number of lysosomes and autophagic vacuoles) changes have been shown to occur between Phase I and Phase III, which have been interpreted as an expression of cellular aging (18, 19). The apparent decreased proliferative capacity with age has been associated in WI-38 with a progressive increase in the nuclear-DNA circularity and compactness (4), accompanied by an increase of polyploidy. The increase with age, for each cell cycle phase (4) in chromatin condensation in situ agrees well with the decrease in primary binding sites (20) and template activity (18) of chromatin isolated from the same cells.

Multiparameter analysis (Figs. 6 - 7) confirms these findings on the average properties of bulk WI-38 preparations, but reveals wide variations in nuclear shape and condensation at any age of cell growth. Large heterogeneity (21) has been observed also in overall cell size, shape and behavior, partly explained by the large cell-cycle related variations but partly suggesting that the "so-called" senescence in vitro is a gradual phenomena differentially affecting cells within a population. Furthermore, by comparing fibroblast at different ages, nuclear convolution (or circularity) and nuclear condensation appear highly correlated within each phase of growth (see Fig. 8), being instead uncorrelated between phases. Namely (Fig. 7) considering G0 and G1 cells as defined by a 2C DNA content and proper clustering in the two-parameter histograms of cycling and non-cycling cells (as shown in Fig. 3), within either Phase I, II or Phase III. A decrease in form factor between G0→G1 is always accompanied by a corresponding decrease in average optical density; conversely, cells in the same proliferative state (either G0 or G1) of any age have similar form factor, but an average optical density linearly increasing with the number of passages in vitro (nearly doubling between Phase I and Phase III). This would indicate that while nuclear chromatin condensation is a parameter equally depending from both cell cycle phase and cell type (transformed versus normal, old versus young), nuclear chromatin form factor (related to nuclear convolution) is primarily dependent on cell proliferation. If we then use these morphometric criteria (2C window and clustering in the two-parameter histogram of form factor versus average OD) to define the G1 cycling versus non-cycling G0 cells at different ages, preliminary data indicate that the growth fraction G1/G1+G0 versus time after plating (Fig. 7) is equal (or even larger) in the older cells, suggesting an invariant proliferating capability with increasing age (22). This finding would be then compatible with a recent explanation for the finite life-span of WI-38 in culture, in terms of cell differentiation or of an increasing number of cells "committed to die" (22) in culture. Nuclear chromatin condensation (which could be an expression of cell differentiation) increases linearly between Phase I and Phase III, for any cell cycle phase, indicating that significant alterations occur in the genome during the "senescence" in vitro: the stability and irreversibility of such chromatin changes during "aging", has been proved by the fusion of diploid

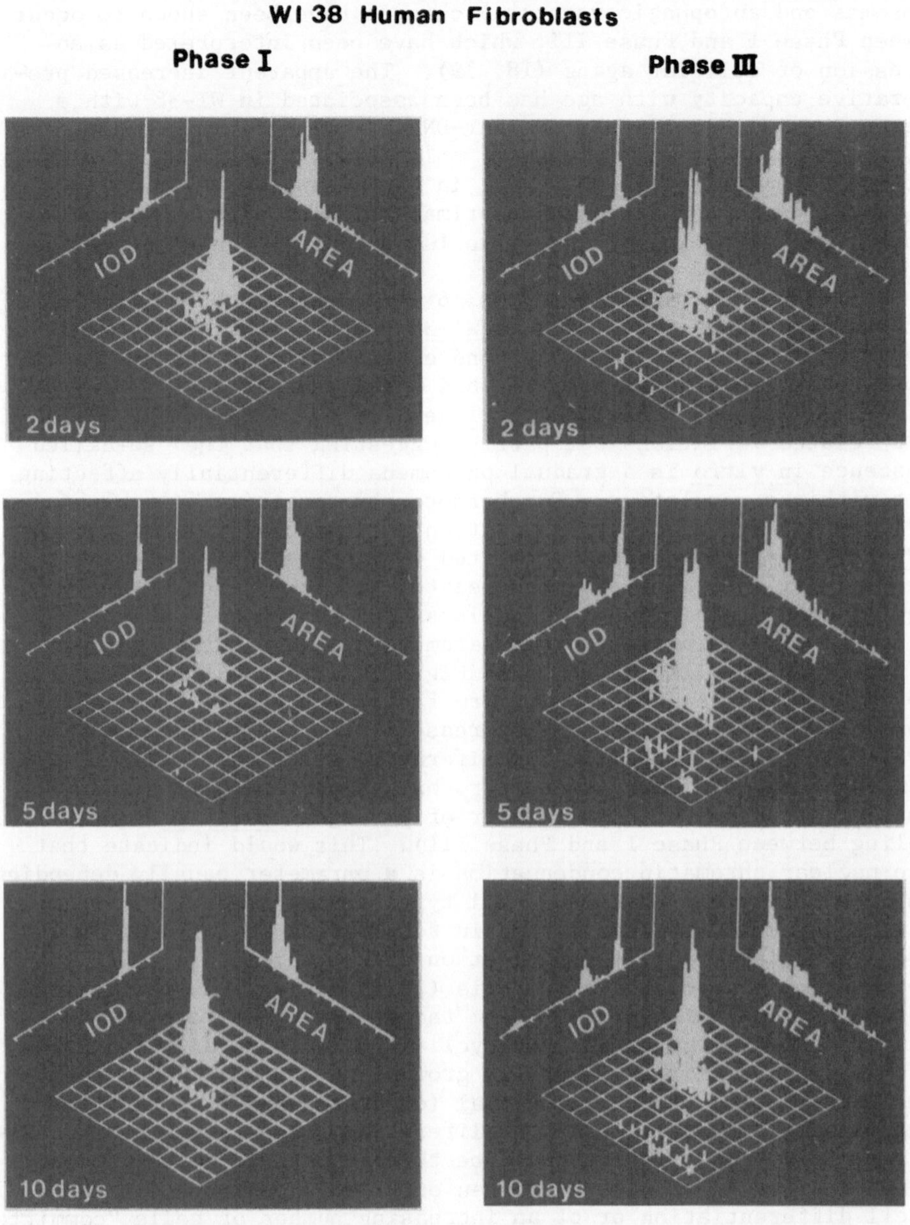

Fig. 6. Integrated optical density (DNA content) versus area for late Phase I (left panels). WI-38 fibroblasts at 2 days (cycling), 5 days confluent and 10 days (deeply confluent). Similar data are shown for late Phase III WI-38 in the right panels.

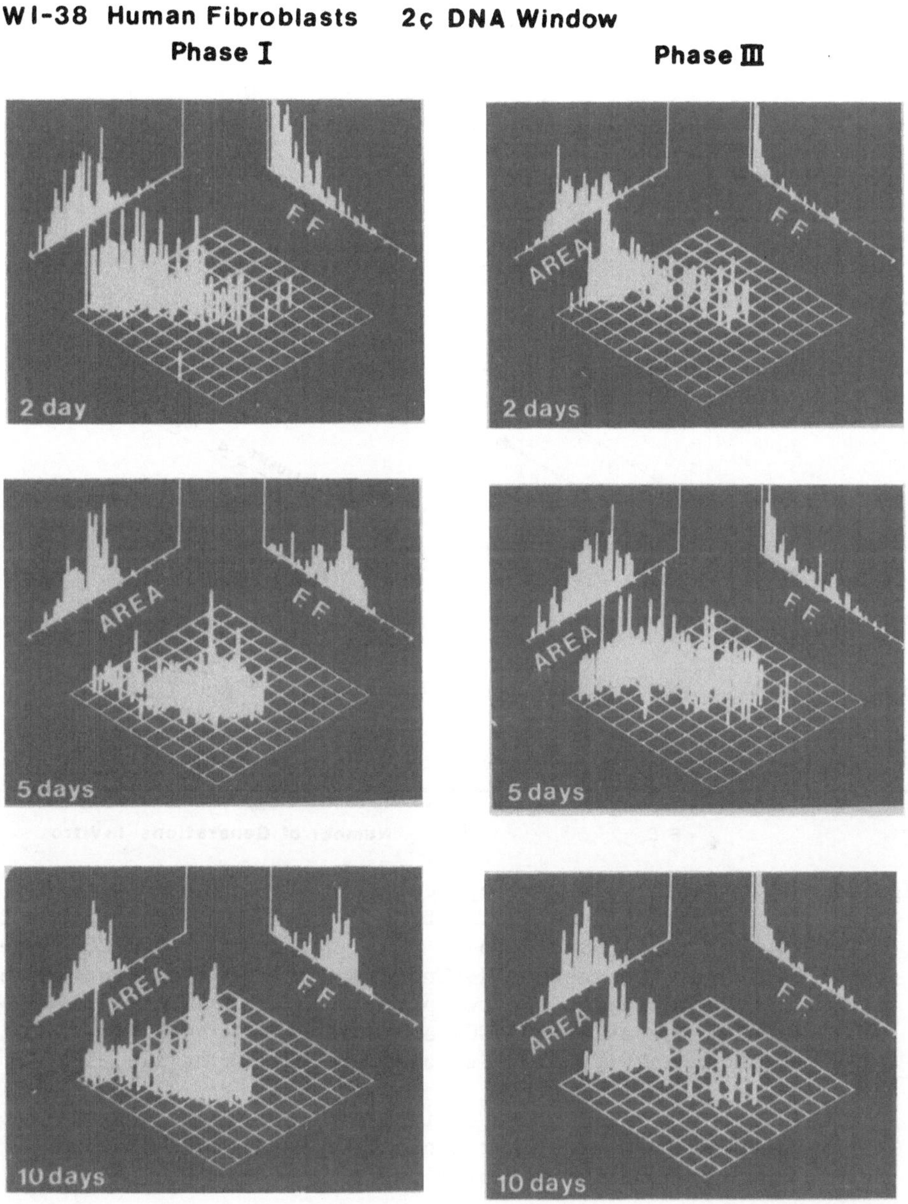

Fig. 7. As Figure 6, but area versus form factor (FF) of "2C-DNA" WI-38 fibroblasts.

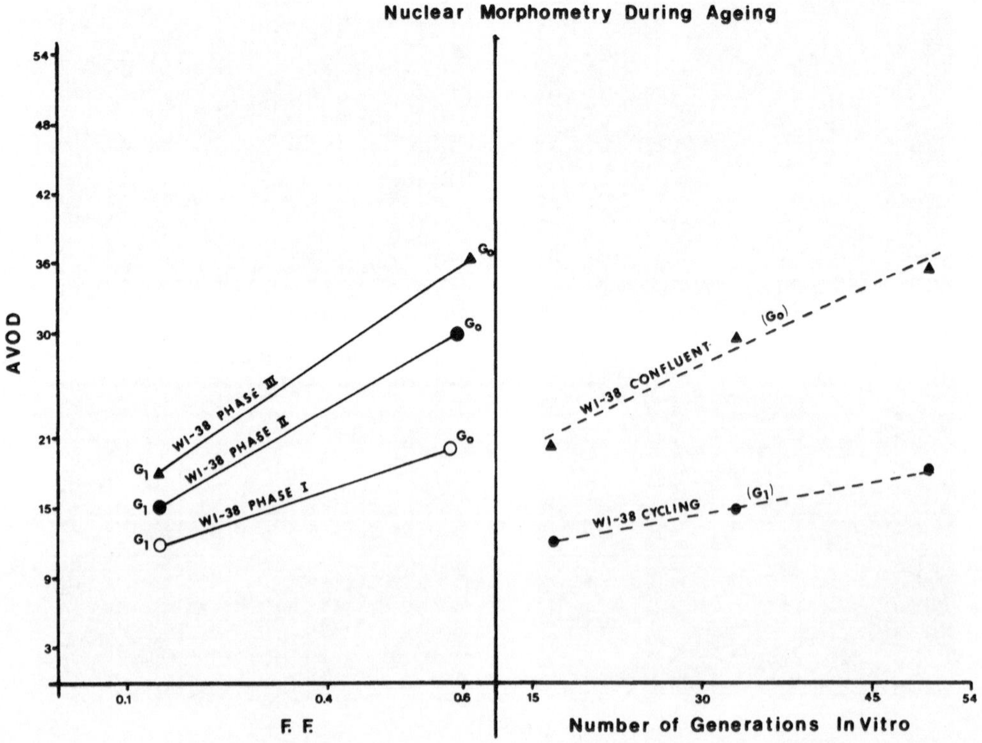

Fig. 8. (Left Panel) Average optical density versus form factor of Feulgen-stained nuclei from Phase III (——), Phase II (•——•), and Phase I (o–o) WI-38 cells in G1 and G0 phase. Cell cycle phases are defined by the 2C DNA content and by the clustering in the two parameters distribution respectively at confluency (G0) and during log-phase growth (G1), as shown in Fig. 5. (See also Grattarola et al., 1980).

(Right Panel) Average optical density versus number of passages in culture for cycling (•---•) and confluent G0 (——) WI-38 human fibroblasts.

"old" cells into "young" cells (19). More interestingly (Fig. 9) is the correlation coefficient between changes in nuclear area and cytoplasmic cell area for over 200 individual cells during the cell cycle of WI-38 human fibroblasts, is 0.72 in Phase II (middle age cells) and 0.89 in Phase III (old cells).

Interestingly enough, the observed relationship between cell and nuclear morphometry in WI-38 cells increasing with age, is apparently confirmed by independent investigators reporting with aging both a decreased spreading capability and an increased cell form factor (23) which parallels the increased form factor and AVOD nuclei (Fig. 7). In analogy with the poly (HEMA) experiments (Fig. 3 and ref. 11) such findings suggest that plating senescent WI-38 on a substrate with increased spreading (24) might increase their proliferative capabilities. Such a suggestion has been supported tentatively by recent preliminary experiments involving the growth of Phase III WI-38 cells on polylysine (see Abraham, personal communication).

CONCLUSION

As shown previously (1,2,13) cell cycle phases (including the non-cycling G0 and Q compartments discussed in previous chapters) regardless of cell type, are uniquely characterized by form factor and integrated optical density (DNA content); at the same time the average optical density displays large variations (13) gradually related to cell transformation (reflecting even differences in the degree of metastatic potential) and to cellular aging (as between old and young fibroblasts), which obscure and overlap to the changes in chromatin condensation occuring during cell cycle (1,13). It would then seem that a higher order chromatin superpacking and a reduced template activity (13) are a prerequisite for the expression of both the transformed phenotype or of the cellular senescence. This apparent paradox is however resolved when we measure the degree of coupling between changes in nuclear morphometry and changes in cellular morphometry: as cells progress through the cycle, the correlation between nuclear form factor and cellular form factor is high for young fibroblast or fibroblast-like cells, very low in transformed cells, being increased but still poor in tumor cells with lower metastatic potential, but becomes nearly perfect in old fibroblasts. The mechanisms by which cell geometry and growth are respectively coupled in normal and uncoupled in transformed cells (5) could indeed be the physically or chemically-induced coupling between cell morphometry and nuclear morphometry (11, 13, 14). However, only in old fibroblast cells, where certain conditions (as rigid properly-oriented microtubule-microfilament assembly) warrant propagations of an increasingly round and compact structure from the nucleus to the cells, constant increase in chromatin condensation would cause ultimately the cell to become round and eventually detaeh from the culture flask (even if its

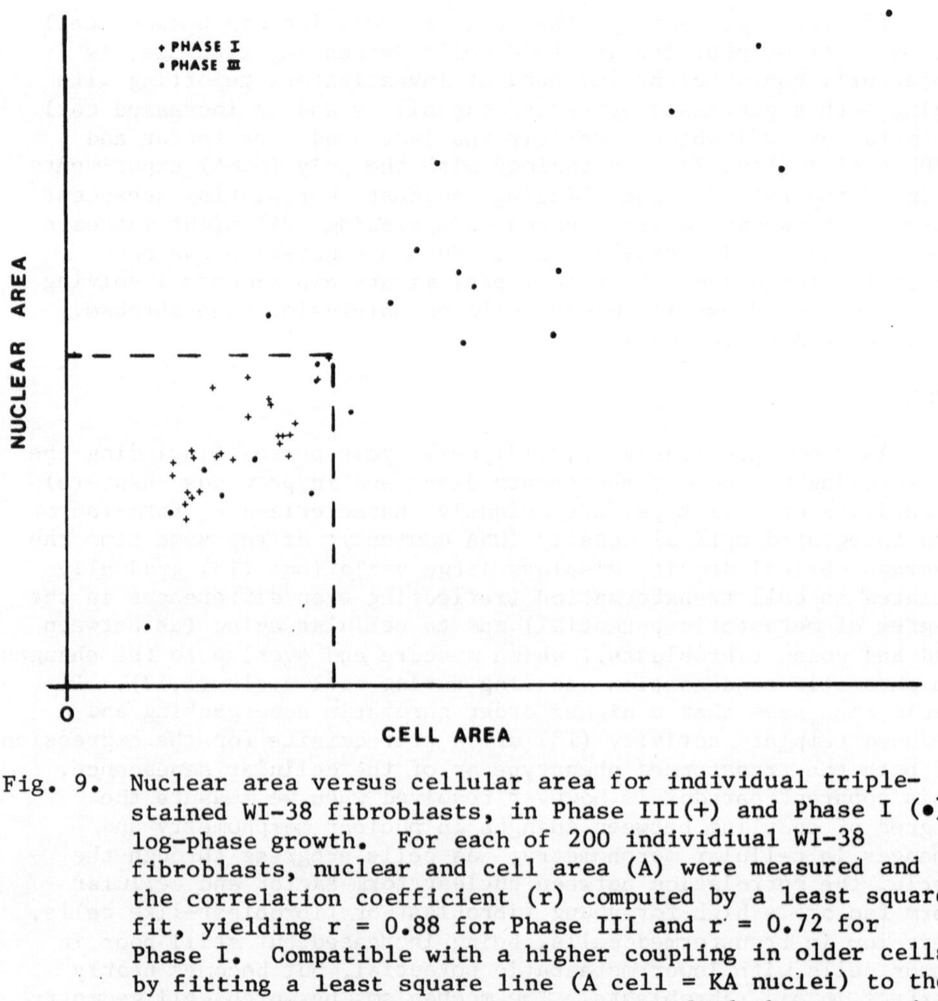

Fig. 9. Nuclear area versus cellular area for individual triple-stained WI-38 fibroblasts, in Phase III(+) and Phase I (•) log-phase growth. For each of 200 individiual WI-38 fibroblasts, nuclear and cell area (A) were measured and the correlation coefficient (r) computed by a least square fit, yielding r = 0.88 for Phase III and r = 0.72 for Phase I. Compatible with a higher coupling in older cells, by fitting a least square line (A cell = KA nuclei) to the data and passing through the origin we find that per unit change in nuclear area, the cellulae area changes 50% more in the old cells (K = 11.5) than in young cells (K = 7.8); both fits exhibiting the same relative dispersion.

proliferating capability would be increased). On the contrary, in transformed cells the absence of any coupling would let a highly condense chromatin to progress through its cycles of condensation and decondensation (apparently with a higher growth rate related to its higher fiber superpacking), regardless of the shape assumed by intact cell, including a round one as in suspension or as induced by cell-cell interaction, nutritional deprivation or cell-substrate interaction. Within this framework a gradual change in the degree of coupling would then imply a gradual expression of malignancy, as indeed occurs between low and high metastatic variants (12). The provocative conclusions and hypothesis were recently introduced (13) and have resummarized, do strictly follow a rigorous inductive process, triggered only by the need to offer a consequential and coherent interpretation of the large amount of image analysis data collected in recent years from a wide variety of cell systems both in vivo and in vitro. Other more imaginative and accurate pictures could likely be devised linking the data here presented to the other numerous morphological, biophysical and biochemical data already in the literature. Our primary intention is indeed to attempt a zero-approximation synthesis, to aid in the design of further probatory experimentations.

Recent experimentation (26, 27) has furthermore shown that the disruption of the microtubules-microfilaments in fibroblast, induced by agents as colchicine, did cause either stimulation (26) or inhibition (27) of cell growth, depending on the stage of the cycle (or chromatin condensation) in which the cell resides at that time, [as expected within a recently proposed analytical model (25)]. Namely colchicine will act as a stimulator, alone or in synergism with other mitogen (26), when normal fibroblasts are confluent and therefore in a highly condensed state (see Fig. 3) by causing chromatin relaxation and cycle progression with removal of the nuclear geometry constraints imposed by the round cell shape. At the same time, if the abrupt relaxation (at II restriction point) has indeed a critical role in creating favorable thermodynamic conditions for DNA replication (25), when sparse fibroblasts are blocked by serum starvation (likely in a relaxed state, as G1 of Fig. 3), colchicine should instead inhibit (27) any mitogenic stimulation acting at the level of cell surface-geometry: the lack of coupling, due to microtubule-microfilament disruption, would in this case prevent the needed chromatin condensation and subsequent abrupt transition.

It still remains to explain whether the coupling between cell and nuclear morphometry, during various phases of the cycle in normal cells is direct, initially depending on mechanical forces transmitted from membrane to nucleus by microtubule-microfilament (see chapter by Brinkley & Hsie) and/or indirect, depending on ionic fluxes, transport mechanism or activity changes of membrane enzymes induced by changes in physical parameters of cell membranes. What is clear at this time is that this coupling, i.e., present in fibroblasts-like cells but absent at varying degrees in transformed cells, is a reproducible observation which is compatible with previous concepts (23) on the mechanism controlling cell transformation, and which demands further investigation as a means of furthering our understanding of abnormal cell growth.

REFERENCES

1. C. Nicolini, F. Kendall and W. Giaretti, Biophysical J. 19:163-177 (1977).
2. C. Nicolini, F. Kendall, W. Giarretti and C. Desaive, Experimental Cell Res., 106:118-125 (1977)
3. F. Kendall, F. Beltrame, and C. Nicolini, IEEE Trans. Biomed. Eng. 3:172-175 (1979).
4. W. Sawicki, in: "Chromatin Stucture and Function", Edited by C. Nicolini, Plenum Publishing Co., New York-London, pp. 667-682 (1979).
5. J. Folkman and A. Moscona, Nature 273:345-352 (1978).
6. C. Nicolini, in: "Chromatin Structure and Function", Part B. Editor, C. Nicolini, Plenum Publishing Co., New York-London, pp. 612-666 (1979).
7. S. Parodi, S. Lessin, F. Beltrame, and C. Nicolini, Cell Biophysics, Vol. 1 (1979).
8. F. Kendall, F. Beltrame and C. Nicolini, in: "Chromatin Structure and Function", Part A. Editor, C. Nicolini, Plenum Publishing Co., New York-London, pp.265-292 (1979).
9. A. Hsie, J. Puck, P.N.A.S. - USA, 68:358 (1971).
10. J. Puck, C.A. Waldren and A. Hsie, P.N.A.S.-USA , 69:1943-48 (1972).
11. A. Belmont and C. Nicolini, Cell Biophysics, Vol. 2, No. 2, (1980).
12. S. Lessin, S. Abraham and C. Nicolini Nature (1980).
13. C. Nicolini, J. Submicroscopic Cytology (1980).
14. C. Nicolini and F. Beltrame, P.N.A.S.-USA (1980).
15. J. Folkman and H. Greenspan, Biophysical Biochemia Acta 217:211-236 (1975).
16. F. Kendall, F. Beltrame, S. Zietz, A. Belmont and C. Nicolini Cell Biophysics, Vol. 2, No. 3 (1980).
17. I. Fidler, Nature New Biology, 242:148-149 (1973).
18. J. Ryan and V. Cristofalo, Expt. Cell Res. 90:456-458 (1975).
19. A. Muggleton-Harris and L. Hayflick, Expt. Cell Res. 103:321-330 (1976).
20. A. Maizel, C. Nicolini and R. Baserga, Expt. Cell Res. 96:351-359 (1975).
21. J. Wolosewick and K. Proter, Am. J. Anatomy, 149:197-226 (1977).
22. R. Holliday, L. Huschtscha, G. Tarrant and T. Kirkwood, Science 198:366-373 (1977).
23. T. Grusberg, B. Hoskins and R. Widdas, Expt. Cell Res. 118:39 (1979).

24. D. Mazia, G. Schalten and N. Sale, J. Cell Biol. 66:198 (1975).
25. C. Nicolini Cell Biophysics, Vol. 2, No. 3 (1980).
26. M. Tang, J. Bartholomew and M. Bissel Nature, 268:739 (1977).
27. D. McClain and A. Edelman, P.N.A.S.-USA, 77:27-48 (1980).

ACKNOWLEDGEMENT

This work was supported by NIH Grant CA 20034 (USA), Temple University Research and Study Leave award to C.N. and by the National Research Council, Finalized Project on "Control of Neoplastic Growth", Italy.

INFORMATIONAL MACROMOLECULES IN STATIONARY AND DIVIDING HEPATOCYTES AND HEPATOMAS

J. Paul, H. Jacobs, R. Shott, P. Wilkes and G. D. Birnie

Beatson Institute for Cancer Research
Garscube Estate, Switchback Road
Bearsden, Glasgow, G61 1BD, Scotland

It is a fundamental tenet of cell biology that the phenotype of the cell is a manifestation of the proteins of which it is composed and that the protein spectrum results from modulated expression of the genes coding for them. It is now clear that regulation at translational level, if it plays a part at all, plays a relatively minor part in determining the amounts of proteins made, and this leaves the implication that the phenotype of a cell is mainly determined by the population of messenger RNA molecules within it. Less than 10 years ago we had no idea of how this population varied in cells of different phenotypes or in different physiological states, and we did not know how the population was comprised. However, the development of analytical methods using nucleic acid hybridization has enabled us to establish a number of general principles and more recently the availability of techniques to prepare DNA libraries has made it possible to start to work out some details.

Until the introduction of these analytical methods there could only be speculation about the number of genes expressed in cells, particularly eukaryotic cells. It was known that many cells, for example mammalian cells, contained enought DNA for over a million genes big enough to code for globin messenger RNAs, but genetic reasons had been advanced many years before to suggest that because of the mutational load the actual number of functioning genes was unlikely to exceed about 5×10^4. In Drosphila the correlation of bands in the giant chromosomes with lethal mutation sites suggested that the number of genes might be on the order of 5,000 or slightly more (Judd et al, 1972). One of the first fruits of the application of hybridization analyses was the establishment of good estimates of the number of messenger RNAs and hence the number of functioning genes in a cell.

The analysis of the kinetics of reannealing of DNA by Britten and Kohne (1968) established the principle of measuring the complexity of nucleic acid molecules. In this context complexity can be defined as the number of nucleotides in unique sequences in a sample of nucleic acid. The first method of determining RNA complexity was based on knowing the complexity of genomic DNA (Davidson and Hough, 1971). These invesigators prepared a sample of unique sea urchin DNA by denaturing total labelled DNA, allowing it to reanneal to a point at which all repetitive sequences were in double-stranded form but unique sequences were in single-stranded form, and fractionating these. The highly labelled unique DNA was then incubated with a vast excess of RNA such that all complementary DNA sequences formed hybrid molecules. This gave an estimate of the total fraction of DNA complexity which could be accounted for by molecules which were transcribed into RNA. With this approach Davidson and his colleagues estimated that the messenger RNA of sea urchins had a complexity of between 15 and 22 x 10^6 nucleotides depending on developmental stage (Anderson et al, 1976, Hough-Evans et al, 1977). This is roughly equivalent to 10,000 to 15,000 messenger RNAs. The second technique described simultaneously by Bishop et al (1974) and Birnie et al (1974) requires the isolation of messenger RNA and its transcription with reverse transcriptase to obtain a preparation of cDNA which is representative of the messenger RNA population. By following the kinetics of hybrid formation a direct estimate of complexity can be obtained. This latter method provides a further piece of information in that the relative abundance of different RNA molecules can be measured from the rates of hydridization, since the most abundant species form hybrids more rapidly than the least abundant according to second order kinetics. When applied to a wide variety of cells, this method gives estimates of the order of about 20,000 different kinds of messenger RNA in each cell.

These two methods give complementary information because the saturation method gives much more precise estimates of total complexity, whereas the kinetic method gives more meaningful estimates of abundance.

The availability of these methods enables some questions to be answered immediately. The generality of the rule that in nearly all mammalian cells between 10,000 and 20,000 different mRNAs are present has held up well with one exception which has not yet been fully explained. The exception is mammalian brain in which the messenger RNA population seems to be at least twice as complex.

Comparison of the complexity of nuclear RNA and cytoplasmic messenger RNA led to the discovery that the former is 5-10 times more complex than the latter. Hence, 80-90% of all the RNA transcripts in the nucleus do not reach the cytoplasm, and although this discrepancy can in part be ascribed to processing of transcripts from introns, it is by no means clear that this accounts for all of it.

DO DIFFERENT CELLS USE DIFFERENT SETS OF GENES?

It has been assumed for many years that differentiation may be caused by gene switching and, although it was obvious that certain "housekeeping genes" will be expressed in different tissues, the view has been widely held that a large number of genes may be tissue specific. With the availability of methods for analyzing complexity of RNA molecules, it has become possible to tackle this question and two different answers have been obtained which at first sight appear mutually contradictory. Studies of sea urchin development by Hough-Evans et al (1977) yielded evidence for different RNA complexities at different stages of development. For example, they found that about 14,000 genes appeared to be expressed in the oocyte but that this number diminished progressively until at the pluteus stage only about 10,000 messengers were present. However, comparisons of mature differentiated mammalian cells seemed to provide a different answer in that, with the exception of brain which has already been referred to, essentially the same range of messenger RNAs appeared to be present in a variety of cell types (Young et al, 1976, Hastie and Bishop, 1976). The main differences these authors noticed between tissues were not so much in the absolute complexity of the messenger RNA populations as in the abundance of a relatively small number of molecules which might be present in low concentration in one cell and high concentration in another. This discrepancy may reflect a real difference in the extent of gene switching in early embryogenesis (as exemplified by the sea urchin) and in mature differentiated cells. However the discrepancy is quite likely to be one of degree rather than an absolute difference because in the mammal it is known that certain genes such as the globin and ovalbumin genes are expressed at a high level in some tissues but are apparently not expressed at all in others. In mature reticulocytes, for example, there are about 150,000 copies of globin mRNA, whereas it is not possible to measure globin mRNA in fibroblast cells. Hence, in mature mammalian cells, as in sea urchin embryonic cells, there may be real differences in the complexity of the messenger RNA, but these do not form such a large proportion of the total complexity. By contrast, these experiments show that in differentiated mammlian cells the most striking difference is the relative concentrations of some RNA molecules as compared with others.

RNA POPULATIONS IN GROWING AND NON-GROWING CELLS

Before a complete understanding of nucleic acid complexity in eukaryotic cells had been obtained, some hybridization experiments were undertaken on regenerating and non-regenerating liver which seemed to show rather striking changes (e.g. Church and McCarthy, 1967). In these experiments newly synthesized RNA was hydridized to DNA immobilized on nitrocellulose filters in conditions which we now know would give estimates only of transcripts from highly repetitive DNA. In more recent experiments designed to look at messenger RNA sequences by complexity analysis in growing and resting mouse fibroblasts, very little difference was found (Williams and Penman, 1975). In these experiments, 3T6 mouse fibroblasts were grown either in high serum or in low serum, the latter conditions being adjusted so that no mitosis was visible although the cells remained viable. When the kinetics of hybridization of messenger RNA to cDNA were measured, they appeared identical in cells from the different growth states. On performing cross-hybridizations between the two cDNAs and the messenger RNAs, it appeared that the sequences were virtually the same although some evidence was obtained that 3% of the messenger RNA in both stages was not present in the other stage. This kind of experiment therefore revealed surprisingly little difference in the composition of the RNA populations. However, experiments of this kind can always be objected to on the grounds either that tissue culture cells are not normal or that the method of growth control is artificial. Experiments have therefore been undertaken to compare normal liver with liver undergoing regeneration following partial hepatectomy. Hepatocytes in normal liver provide a classical example of cells in G_0. Moreover, they are highly differentiated cells performing a number of specialized functions. In experiments in which we compared the complexity and abundance of RNAs in normal and regenerating liver, we found very little difference in complexity but were able to show that abundant sequences in normal liver became much less abundant when the liver began to regenerate, possibly reflecting a reduction in specialized syntheses (Wilkes et al, 1979). These experiments also suggested a slight compensatory increase in the abundance of some sequences in regenerating liver as compared with non-regenerating liver. There was very little evidence for differences in the complexity of polysomal messenger RNA but nuclear RNA in regenerating liver was 10-15% more complex than in non-regenerating liver, suggesting that during regeneration additional transcription may have occurred of RNA which remained confined to the nucleus. Similar results were reported by Krieg et al (1979).

RNA POPULATIONS IN NORMAL AND TUMOR CELLS

The main chracteristics which distinguish tumor cells from normal cells are that they respond inadequately to normal growth regulation such as is exhibited in tissue culture by contact or density inhibition, that they often exhibit losses of differentiated characteristics which may be manifested by morphological changes, e.g. as in cell transformation, and that at least in tissue culture they seem to be able to grow indefinitely, whereas primary cells often have a restricted lifetime. It is usually postulated that these changes reflect changes in gene expression and this can be investigated in part by studies similar to those already described. A study by Grady and Campbell (1973) reported that a very significantly higher fraction of mouse DNA was saturated by total RNA from polyoma transformed cells than from normal fibroblasts from which these were derived. A similar result was reported for SV40 transformed cells as compared with Balb 3T3 cells. However, two other studies in which the kinetics of hybridization between RNA and cDNA was studies revealed less striking results. Williams et al (1977) studied messenger RNA complexity in normal human fibroblasts and in the same fibroblasts transformed with SV40 virus. They found a high degree of homology between these and concluded that at most 3% of messenger RNA in a transformed cell had sequences not present in the normal parental cell. Rolton et al (1977) compared both nuclear and cytoplasmic RNAs in Balb 3T3 cells and the same cells transformed with mouse sarcoma virus to give a non-producer transformed line. They were unable to reveal any significant difference in the nuclear RNA of the normal and transformed cells. They were also unable to measure any highly significant difference in the overall complexity of the messenger RNAs but showed that transformation of 3T3 cells were accompanied by an increase in the abundance of certain RNA sequences which were present in low concentration in normal cells and by reduction in the abundance of some other sequences. These experiments did not exclude the possibility that there was some slight difference in overall complexities measured by saturation, but certainly a difference of the order observed by Grady and Campbell (1973) was not seen.

We also undertook experiments to compare RNA populations in normal rat liver and in HTC cells derived from a minimum deviation rat hepatoma (Jacobs et al, 1980). Saturation hybridization of RNA to non-repetitive DNA showed that the polysomal poly(A) RNAs of normal rat liver and of minimum deivation rat hepatoma were very similar. A study of the kinetics of hydridization revealed that a proportion of the liver messengers were at very much reduced abundance in the hepatoma but that the converse was not true.

USE OF A cDNA LIBRARY TO STUDY mRNA ABUNDANCE CHANGES IN NORMAL AND REGENERATING LIVER AND HEPATOMA CELLS

The experiments described are of limited usefulness because, although these give useful general information, resolution is inadequate to recognize quite substantial and potentially important differences in individual messengers. Complexity differences corresponding to 100 genes are difficult to measure and differences in abundance of some messengers can be concealed by compensatory changes in others. To carry the analyses further it would be desirable to obtain data on the behaviour of individual gene products and this has become possible with the development of genetic recombinant methods. These experiments are based on isolation of individual cDNAs for each of a large number of mRNAs. This can be achieved by creating a cDNA library.

A preparation of total mRNA is first transcribed to give cDNA. This is again transcribed to give double-stranded DNA which can be linked to a plasmid vector. In this way a random population of DNA molecules corresponding to the original RNA population is obtained. Bacteria transformed with recombinants made in this way were plated out and colonies screened by hybridization to cDNAs transcribed from mRNA from homologous or heterologous sources. Some colonies which scored as strongly positive with normal liver cDNA, scored negative or very weakly positive with HTC cell cDNA (Figure 1).

In preliminary studies four colonies have been studied further, two of which were present at different abundances in liver and HTC cells, and two of which appeared to be at similar abundances in both. These were designated pRR 83, pRR 5B, pRR 133 and pRR 117.

The cultures were grown in bulk and DNAs prepared from them. These were bound to nitrocellulose filters and these were reacted with a large excess of high specific activity cDNAs transcribed from nuclear and polysomal poly(A)+ RNAs from normal and regenerating liver and from HTC cells in conditions in which the cDNA bound is proportional to its concentration.

The results of these experiments (performed in replicate, usually triplicate) are shown in Table 1.

All these clones represent highly abundant RNAs in normal liver. Sequences complementary to pRR 83 and pRR 5B are abundant in both normal and regenerating liver mRNA but are present at much lower concentration in hepatoma mRNA. These differences do not reflect differences of similar degree in HnRNA and therefore suggest a post-transcriptional control. In contrast, sequences complementary to pRR 133 behave rather similarly in all cases. pRR 117 represents a moderately abundant species in normal liver but is elevated somewhat in regenerating liver and cultured hepatoma.

Grunstein-Hogness screening of cDNA clones constructed from rat liver polysomal poly A(+)

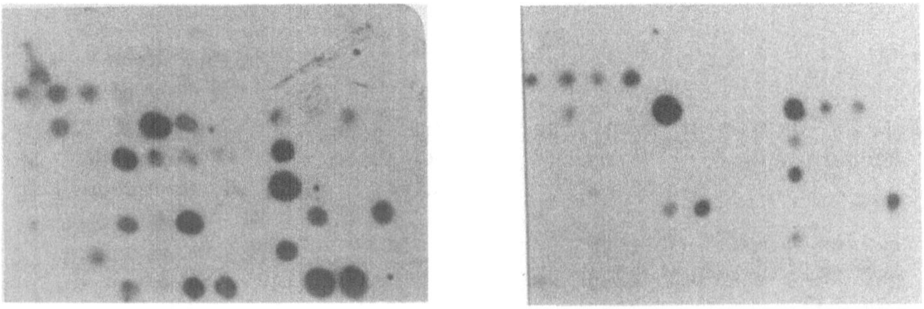

cDNA probe for rat liver polysomal poly A(+) mRNA

cDNA probe for hepatoma polysomal poly A(+) mRNA

Figure 1 Grunstein-Hogness Colony screening to show differing abundances of various mRNAs in rat liver and hepatoma tissue culture cells. ds cDNA transcribed from rat liver polysomal polyA(+) mRNA was inserted into the Bam HI site of pAT 153 by blunt end ligation. Recombinant clones obtained after transformation of E. Coli HB 101 were grown and fixed to nitrocellulose filters.

Duplicate filters were hybridized with ^{32}P labelled cDNA transcribed from polysomal polyA(+) mRNA from (a) rat liver (1.2×10^7 cpm per filter); (b) hepatoma tissue culture cells (0.7×10^7 cpm per filter).

TABLE 1. Mass Fractions (x 10^{-2} of cDNAs Hybridizing to Cloned DNAs from a Rat Liver Polysomal cDNA Library

Recombinant DNA	Source of cDNA Preparation					
	Normal Liver		Regenerating Liver		Cultured Hepatoma	
	HnRNA	mRNA	HnRNA	mRNA	HnRNA	mRNA
pRR 83	3.7	44.	1.8	23	1.4	.72
pRR 5B	6.7	150.	15.	28	3.4	1.8
pRR 133	.94	44.	1.2	32	3.3	63.
pRR 117	.79	6.7	14.	13	4.2	27.

TABLE 2. Ratios of Mass Fractions (mRNA/HnRNA) From TABLE 1.

Recombinant	Normal Liver	Regenerating Liver	Cultured Hepatoma
pRR 83	11.9	12.8	0.5
pRR 5B	22.4	5.2	0.53
pRR 133	46.8	26.7	19.
pRR 117	8.5	0.92	6.43

These are preliminary experiments and the data are too meagre to draw hard and fast conclusions about the roles of the mRNAs represented here, although one could speculate that the behaviour of pRR 83 and pRR 5B might be expected of luxury gene products, that of pRR 133 of a "housekeeping" gene and of pRR 117 of a gene concerned with replication. The data themselves are compatible with the conclusions to be drawn from kinetic experiments and indicate that extension of this technology is likely to yield detailed information about the behaviour of whole RNA populations. The most striking feature to emerge from the data is the evidence for post-transcriptional regulations of the level of individual mRNAs.

This work was supported by grants from MRC and CRC.

References

1. Anderson, D.M., Galan, G.A., Britten, R.J. and Davidson, E.H., Devel. Biol., 51:138-145, 1976.
2. Birnie, G.D., Macphail, E., Young, B.D, Getz, M.J. and Paul, J., Cell Diff., 3:221-232, 1974.

3. Bishop, J.O., Morton, J.G., Rosbash, M. and Richardson, M., Nature, 250:199-204, 1974.
4. Britten, R.J. and Kohne, D.E., Science, 161:529-540, 1968.
5. Church, R.B. and McCarthy, B.J., J. Mol. Biol., 23:477-486, 1967.
6. Davidson, E.H. and Hough, B., J. Mol. Biol. 56:491-506, 1971.
7. Grady, L.J. and Campbell, W.P., Nature New Biology, 243:195-198, 1973.
8. Hastie, N.D. and Bishop, J.O., Cell, 9:761-774, 1976.
9. Hough-Evans, B.R., Ernst, S.G., Britten, R.J. and Davidson, E.H., Dev. Biol., 69:258-269, 1977.
10. Jacobs, H. and Birnie, G.D. Nucl. Acids. Res., 8:3087-3103, 1980.
11. Judd, B.H., Shen, M.W. and Kaufman, Z.C. Genetics, 71:139-152, 1972.
12. Krieg, L., Alonso, A. and Volm, M. Eur. J. Biochem., 96:77-85, 1979.
13. Rolton, H.A., Birnie, G.D. and Paul, J. Nucl. Acids Res., 6:25-39, 1977.
14. Wilkes, P.R., Birnie, G.D. and Paul, J. Nucl. Acids Res., 6:2193-2208, 1979.
15. Williams, J.G., Hoffman, R. and Penman, S., Cell 11:901-907, 1977.
16. Williams, J.G., and Penman, S., Cell, 6:197-206, 1975.
17. Young, B.D., Birnie, G.D. and Paul, J., Biochem, 15:2823-2829, 1976.

CELL GROWTH AND NUCLEAR DNA INCREASE BY ENDOREDUPLICATION AND DIFFERENTIAL DNA REPLICATION

W. Nagl

Department of Biology
University of Kaiserslautern
Federal Republic of Germany

INTRODUCTION

Somatic variations in the amount of nuclear DNA due to somatic polyploidization and differential DNA replication are events, which occur much more frequently than thought by many biologists and biochemists. This might be, in part, because there is particular interest in the genetic information of the DNA rather than on its "nucleotypical" (1) or "nucleoskeletal" effects (2). Actually, it is widely believed that the genome is an extremely stable constant, and that the genetic information stored in the nucleus is identical in all cells of an individual (dogma of DNA constancy). Classical cytologists, however, discovered already in 1939 the process of endomitosis (3), and somewhat later that of DNA endoreduplication (4). Recently, it was shown in a review that some kind of somatic polyploidy can be found in nearly every taxon investigated (5), and that often up to 70% of the cells of an organism may be polyploid (6). Hence, we should no longer ignore this fact, but search for the biological significance of somatic DNA increase.

There is another phenomenon which is only badly understood, but which may be important for the understanding of complex organisms: <u>morphogenesis</u>, i.e. growth and differentiation of an individual according to the species-specific construction ("<u>Bauplan</u>"). It seems that thinking in linear causality ("differentiation is the consequence of differential gene activity") does not adequately describe living systems (7,8). Therefore, since the discovery of endomitosis, suggestions were elaborated that somatic polyploidization should be seen in connection to differentiation and morphogenesis (3,5,9-14). Most scientists, however, unfortunately still neglect nuclear dynamics. In this essay, I shall therefore give evidence for the occurence of DNA endoreduplication (and other kinds of somatic polyploidy) and differential DNA replication in various animals and plants. Then, the mechanisms will be discussed which lead to the increase of nuclear DNA, and finally some ideas on its control and its significance for cell growth and morphogenesis will be proposed.

OCCURENCE OF NUCLEI AND MULTIPLE GENOMES

One of the fundamental features of cell nuclei of eukaryotes, from protists and algae to man, is the high variability of their DNA content. Table 1 shows a few examples taken from the animal and plant kingdoms; a more complete list of species from which somatic DNA increase is known was recently published (5).

Cell and tissue growth often occurs by nuclear DNA increase and not by cell proliferation. Particularly the postembryonic development of insects and angiosperms is predominately characterized by somatic DNA variation. Figure 1 shows the main strategems of growth with respect to the nuclear behavior.

MECHANISMS OF SOMATIC DNA INCREASE

The term "somatic polyploidization" covers various kinds of cell cycles which are curtailed before cytokinesis (cell division) (Figs. 2 and 3). The result is an increase in nuclear DNA content and the number of chromosome complements in a geometrical order. The evolution and regulation of such cell cycles in eukaryotes can be best understood, if they are seen as steps of an evolutionary strategy which leads to cell-specific DNA increase by step-wise reduction and curtailment of the mitotic cell cycle with its complex mitotic machinery (5). The first step which evolved by means of this strategy lies in the omission of cytokinesis only, leading to a multinucleate or polyenergid cell. Such a cell is functionally polyploid. The last step that can be found at present is the omission of all stages, except DNA replication of a specific DNA sequence, e.g. a single gene (= DNA amplification).

Polyenergid Cells

Polyenergid organization is frequently found in some taxa of algae and fungi and, in higher plants, in endosperm, at least during early differentiation. Binucleate and multinucleate cells arise regularly in the anther tapetum of many angiosperms (15), and the liver of mammals (16). In several legums, the embryo-suspensor is made up of multinucleate chambers, e.g. in *Pisum* and *Lathyrus*. While lysis of the cell wall and subsequent cell fusion, whose result is also the multinucleate state, occurs only rarely in plants (e.g. in galls caused by nematodes), cell fusion is the

Table 1. Examples of Maximum Endopolyploidy Levels in Various Organisms*

Taxon	Species	Tissue	Degree*
Sarcodina: Foraminifera	Cibicides lobatulus	Somatic cells	30 C
Cnidaria: Hydrozoa	Hydra canuliculata	Digestive & epidermal cells	6 C
	Carymorpha palma	Endoderm	16 C
Aschelminthes: Nematoda	Ascaris lumbricoides	Uterine epithelium	163C
	Ascaris suum	Esophagal gland	262 C
Annelida: Polychaeta	Ophryotrocha puerilis	Nurse cells	256 C
Mollusca: Gastropoda	Aplysia californica	Giant neurons	200,000 C
	Helix pomatia	Salivary gland	64 C
Echinodermata: Asteroidea	Asterias amurensis	Stomach epithelium	32C

Table 1. (Continued)

Taxon	Species	Tissue	Degree**
Arthropoda:			
Crustacea	Artemia salina	Optical ganglion	8 C
Arachnida	Amaurobius ferox	Fat body	16 C
Collembola	Bilobella massoudi	Salivary gland	1,024 C***
Coleoptera	Cassidia viridis	Malpigh. tubules	64 C
Diptera	Drosophila melanog.	Salivary glands	2,048 C
		Trophocytes	512 C
	Chironomus tentans	Salivary gland	32,768 C
Hemiptera	Gerris lateralis	Malpighina tubules	64 C
		Testis wall	16 C
	Lygaeus saxatilis	Salivary gland	2,048 C
Hymenoptera	Apis mellifera (worker)	Pharyngeal gland	256 C
		Thoracic gland	256 C
	Melipona quadrifasciata	Silk gland	64 C
Lepidoptera	Bombyx mori	Silk gland	524,288 C

Taxon	Species	Tissue	Degree**
Orthoptera	Chorthippus hammastroemi	Testis follicles	8 C
Chordata:			
Pices	Scorpaena porcus	Nervus terminalis	256 C
Aves	Gallus domesticus	Cerebellum Purkinje cells	8 C
Mammalia	Microtus arvalis	Trophoblast	2,048 C
	Mus musculus	Trophoblast	1,024 C
	Rattus rattus	Trophoblast	4,096 C
	Homo sapiens	Bone marrow megakaryocytes	128 C***
Phycophyta:			
Chlorophyceae	Chara contraria	Rhizoid	32 C
Bryophyta:			
Musci (gametophytes)	Polytrichum formosum	Hydroids	8 C

Table 1. (continued)

Taxon	Species	Tissue	Degree**
	Drepanocladus exannulatus	Paraphyses	8 C
Magnoliatae (dicots):			
Ranunculales	Aconitum neomontanum	Antipodal cells	128 C
Papaverales	Papaver rhoeas	Antipodal cells	128 C
	Corydalis cava	Elaiosome	512 C
Urticales	Urtica caudata	Stinging hairs	256 C
Fabales	Phaseolus coccineus	Suspensor (embryo)	8,192 C
	Phaseolus vulgaris	Suspensor (embryo)	2,048 C
	Phlomis viscosa	Endosperm	384 C
Myrtales	Trapa natans	Suspensor (embryo)	256 C
Geraniales	Geranium phaeum	Integument	512 C
	Tropaeolum majus	Suspensor (embryo)	2,048 C
Violales	Viola declinata	Elaiosome	16 C

Taxon	Species	Tissue	Degree**
Cucurbitales	Bryonia dioica	Anther hairs	256 C
	Cucumis sativus	Anther hairs	64 C
	Cucurbita pepo	Endosperm	284 C
	Echinocystis lobata	Endosperm	2,072 C
Caryophyllales	Cucubalus baccifer	Suspensor (embryo)	128 C
	Dianthus sinensis	Suspensor (embryo)	128 C
	Echinodorus tenellus	Suspensor (embryo)	128 C
	Gymnocalycium quehlianum	Elaiosome	256 C
Scrophulariales	Bartsia alpina	Endosperm haustorium	768 C
	Melampyrum pratense	Endosperm haustorium	1,536 C
	Plantago atrata	Endosperm haustorium	1,536 C
Liliatae (monocots): Alismatales	Alisma plantago-aquatica	Suspensor (embryo)	512 C

Table 1. (Continued)

Taxon	Species	Tissue	Degree**
Liliales	Allium triquetrum	Elaiosome	512 C
	Crocus suaveolens	Antipodal cells	64 C
	Scilla bifolia	Elaiosome	4,096 C
Orchidales	Cymbidium hybridum	Protocorm	128 C
Cyperales	Carex hirta	Tapetum (anther)	8 C
	Cyperus alternifolius	Tapetum (anther)	8 C
Poales	Hordeum distichum	Antipodal cells	256 C
	Triticum aestivum	Antipodal cells	196 C
Arales	Arum maculatum	Endosperm haustorium	24,576 C

* References and further examples in (5).
** Values which are not multiples of 2C (or 3C in endosperm) indicate differential DNA replication.
*** Estimated from figures, no measurements published.

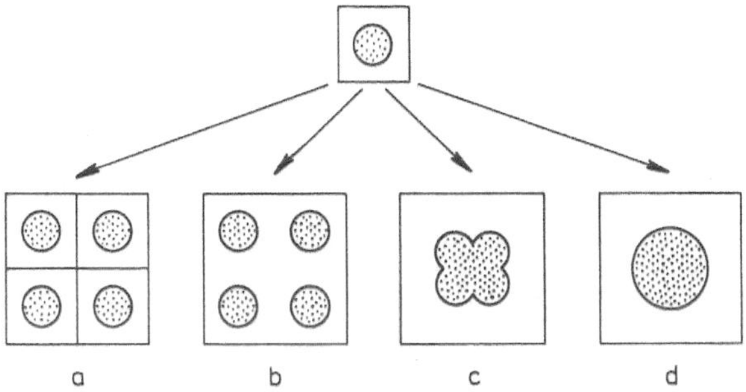

Fig. 1. The various ways of cell and tissue growth starting with a diploid cell: (a) growth by mitotic cell cycles and cell divisions, (b) growth by mitotic cell cycles, but failure of cell division (or secondary cell fusion), (c) growth by restitution cycles (due to failure of mitotic spindle), and (d) growth by endo-cycles (endoreduplication and endomitosis). The total DNA content is the same in all differentiated tissues shown, but the distribution of the genomes is different.

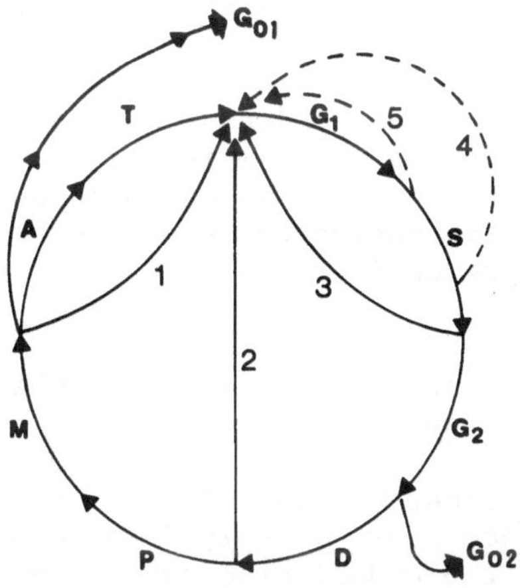

Fig. 2. Diagram to illustrate the course of the mitotic cycle and the curtailed cell cycles: G_{01} and G_{02} = non-cycling cells with the 2C and 4C DNA content, respectively, S = DNA synthetic period, D = "dispersion stage" of plant mitosis and endomitosis, P = prophase, M = metaphase, A = anaphase, T = telophase; 1 = restitution cycle, 2 = endomitotic cycle, 3 = endoreduplication (and polytenization cycle, 4,5 = differential DNA replication cycles: 4 = DNA underreplication, 5 = DNA amplification.

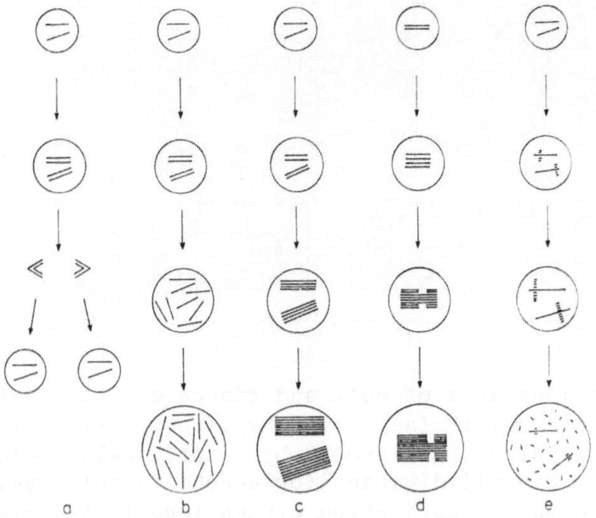

Fig. 3. The result of various cell cycles occurring in higher organisms (modified from 5): (a) mitotic cycle - the result is two nuclei which are genetically identical to the ancestor nucleus, (b) endomitotic cycle - the result is a polyploid nucleus; the genomes are duplicated during each cycle, (c) endoreduplication or polytenization cycle - although the same number of genomes are present as in the nuclei shown under (b), the endochromosomes do not become separated, because no condensation phase occurs, (d) underreplication cycle - in this case underreplication of a certain DNA sequence is shown in an endoreduplication cycle, (e) amplification cycle - a certain gene or regulatory sequence is repeatedly extra replicated.

common way of syncytium formation in mammals (e.g. syncytiotrophoblast formation in humans).

The following short-cuts of the mitotic cell cycle affect mitotic stages, and hence they directly lead to somatic polyploidy: the restitution cycle, the endomitotic cycle, and the endoreduplication cycle.

Nuclear Restitution Cycles

Nuclei in the anther tapetum, the endosperm, the embryo-suspensor insect galls of many angiosperms, and in different tissues of mammals, often undergo restitution, that is, the chromosomes enter mitosis but do not complete anaphase but re-enter an interphase state within one and the same nucleus. Hence the chromosomes undergo a consendation - decondensation cycle, but karyokinesis is omitted, mostly due to failure of the spindle mechanism. Restitution may take place as early as in prophase, clearly showing the continual transition of the mitotic cell cycle to the endo-cycle. Moreover, it demonstrates the independency in the control of chromosome condensation and decondensation, break-down and reconstruction of the nuclear envelope, and formation and destruction of the spindle apparatus. Nuclear restitution takes also place after poisoning of the spindle by a tubulin-binding agents, e.g. colchicine.

Restitution nuclei are normally identifiable by their irregular, often bizarre shape due to the intimate attachment of the nuclear envelope around the randomly distributed chromosomes, and due to the minimal motion of the chromatin domains after reconstruction of the interphase state. Such nuclei, particularly dumb-bell-shaped anaphase restitution nuclei, have been often mis-interpreted as amitotic figures.

Endo-cycles

In angiosperms, like in insects, endocycles predominate. The term "endo-cycles" has been introduced by Nagl (5,12,17) to designate a DNA replication cycle within the nuclear envelope and without spindle formation, i.e. the endomitotic and the endoreduplication cycle. Both cycles irrevocably lead to endopolyploidy and are evidently under strict genetic control. Endomitosis (3) and endoreduplication (4) differ from each other in only one aspect: in an endoreduplication cycle, no mitosis-like structural changes can be seen in the nucleus, and G and S phases follow each other. On the other hand, as the term suggests, structural changes comparable with those seen in mitosis do occur in a nucleus during an endomitotic cycle. Chromosome condensation and separation during endomitosis is, however, visible in certain insect taxa only. In angiosperms, a structural change is visible if heterochromatin is present. There occurs a stage called "Z phase" (Zerstaubungsstadium; 18) or dispersion stage, during that the heterochromatin undergoes decondensation to the level of euchromatin (Fig. 4). The Z stage can also be observed in the mitotic cycle, immediately before onset of prophasic chromosome coiling. This dispersion stage is the only structural feature by which an endomitotic cycle can be recognized in plants (5,10,18,19).

Because Z phase cannot be seen in those species which lack sufficient masses of heterochromatin, many authors refer all kinds of endo-cycles in plants as to DNA endoreduplication. Moreover, there exists an uncertainty as to whether the Z phase coincides with the late S period (5,20,21; see also the accompanying paper by Nagl in this volume).

Nuclei which have passed through endo-cycles are called to be endopolyploid. Their structure may be similar to the diploid

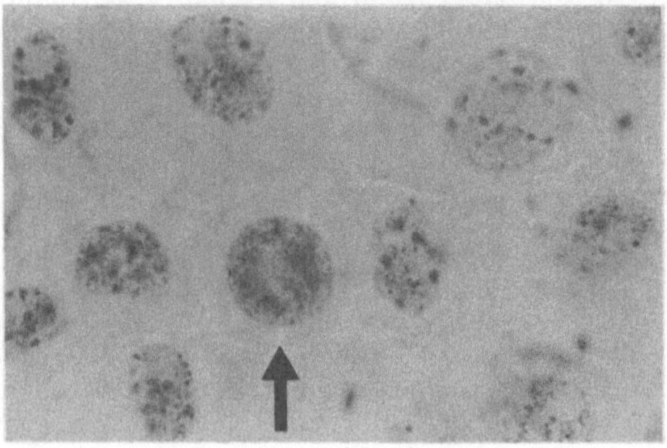

Fig. 4. Micrograph of nuclei in the protocorm of *Cymbidium*; note presence of chromocenters at interphase, and absence of chromocenters at "dispersion phase" (arrow; approx. x 1,000).

ancestor nuclei, or different. As the sister chromatids do not undergo any coiling in an endoreduplication cycle, they normally remain close together thus forming bundles of sister chromatids (or endochromosomes). The endoreduplication cycle (or polytenization cycle), therefore, consequently leads to polytene nuclei. Z phase in the plant endomitotic cycle causes some loosening of the chromatid bundles, so that the polytene (or giant) chromosomes normally are not banded. Bands can, however, be induced by experimental inactivation of RNA synthesis resulting in extreme condensation (Fig. 5d; references in 5,12). Functionally, no difference could be found between nuclei with polyploid and polytenic organization of the chromatin.

Figure 5 shows endopolyploid/polytenic nuclei from animals and plants. It is evident that the nuclei with the highest levels of somatic polyploidy predominantly are found in gland cells, tissues involved in embryo nutrition, and other highly active cells (Table 2, and the chapter on biological significance of DNA variation).

DIFFERENTIAL DNA REPLICATION

Differential DNA replication implies the endoreduplication of part of the genome only (= DNA underreplication), or the extra replication of a certain gene of DNA sequence (= DNA amplification), Heitz (18) suggested already in 1929 that α-heterochromatin is underrepresented in polytene nuclei of salivary glands of *Drosophila*. Differential underreplication of respectively heterochromatin and satellite DNA was later evidenced by various techniques

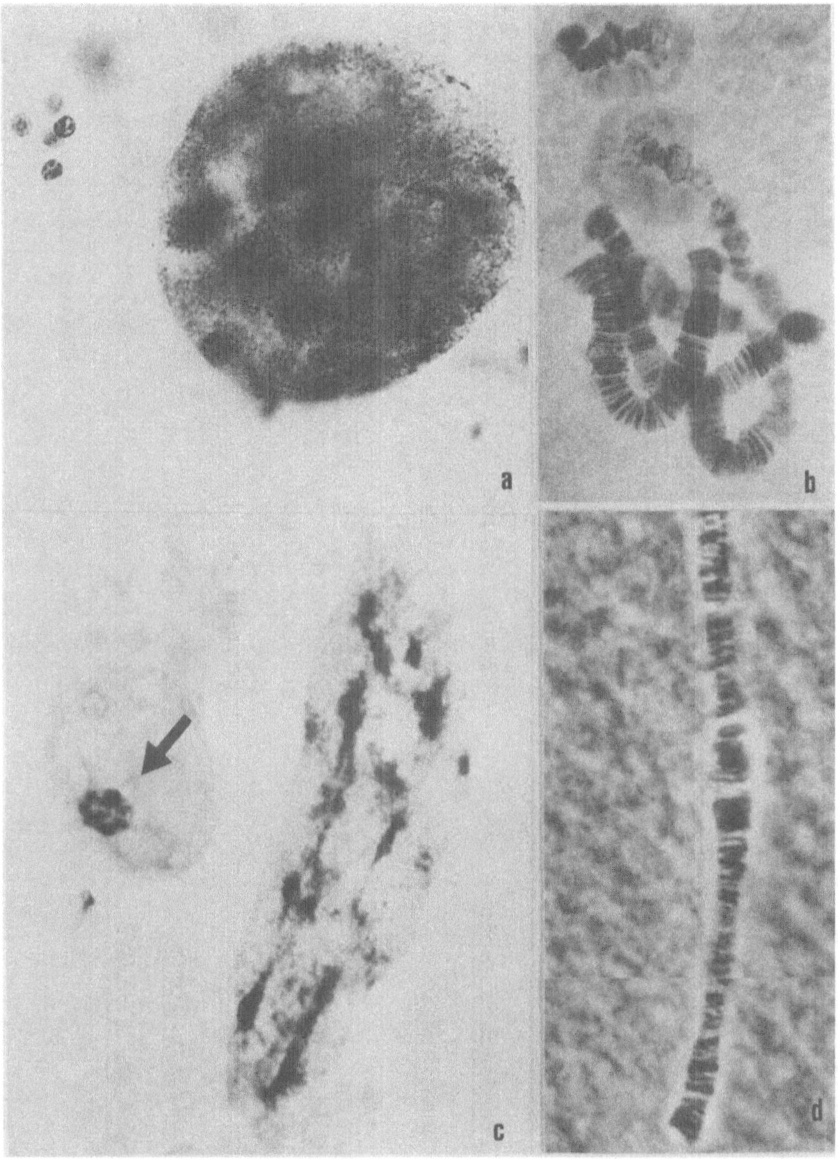

Fig. 5. Examples of polyploid and polytenic nuclei in animals and plants. (a) Rat trophoblast giant cell (note also diploid nuclei in vicinity; Feulgen, x 500), (b) polytene chromosomes in the salivary gland of *Chironomus* (x 540), (c) low-polyploid nucleus (arrow) and highly polyploid nucleus from the *Tropaeolum* suspensor (toluidine blue, x 1,400), (d) Polytene chromosome from a highly endo-polyploid nucleus of the *Phaseolus* suspensor (phase contrast, x 2,000).

Table 2. A Few Examples of Differential DNA Replication*

Taxon	Species	Tissue/Stage	Event
Algae	Acetabularia mediterranea	Meiosis	rDNA amplification
	Chlamydomonas reinhardii	Cell cycle	rDNA amplification
Gymnospermes	Pinus glauca	Seed	repet. DNA amplification
Angiospermes	Crepis capillaris	Tissue culture	sat. DNA amplification
	Cucumis melo	Fruit	sat. DNA/heterochromatin underreplication and amplification
	Cybidium hybridum	Protocorm differentiation	heterochromatin/AT-rich DNA amplification
	Daucus carota	Cell culture	different. DNA replicat.
	Hedera helix	Phase change	different. DNA replicat.
	Nicotiana glauca	Pith culture	rDNA/sat. DNA amplific.
		Flowering stems	DNA amplification
	Phaseolus coccineus	Suspensor etc.	rDNA underreplication, sat. DNA amplification

Table 2 (Cont'd)

Taxon	Species	Tissue/Stage	Event
Angiospermes (Cont'd)	Pisum sativum	Seedling	Heterochromatin/repet. DNA underreplication
	Rhoeo discolor	Flower buds	DNA amplification
	Sinapis alba	Seedling	Heterochromatin amplific.
	Triticum aestivum	Seedling	rDNA amplification
	Tropaeolum majus	Suspensor	AT-rich DNA/heterochromatin underreplicat.
Ciliates	Oxytricha sp.	Macronucleus	repet. DNA underreplicat. gene amplification
	Tetrahymena sp.	Macronucleus	Same events
Coleoptera	Dermestes maculatus	Soma	Sat. DNA/heterochromatin underreplication
Diptera	Drosophila melan.	All tissues	Differential sat. DNA/ heterochromatin underreplication
	Calliphora sp.	Salivary gland	Pulff-DNA amplification
Amphibia	Triturus vulgaris	Embryogenesis	Repet. DNA amplification

Table 2 (Cont'd)

Taxon	Species	Tissue/Stage	Event
Aves	Gallus domesticus	Cartilage and retina differentiation	Repet. DNA amplification
Mammalia	Mus musculus	Cultured cells	Dihydrofolate reductase gene amplification
	Rattus rattus	Liver	rDNA amplification
		Hepatoma cells, suprarenal gland, cerebal cortex neurons	DNA amplification
	Davia cobaya	Retina	DNA amplification
	Homo sapiens	Lymphocytes	DNA amplifications
All Animals	(nearly)	Oocytes	rDNA amplification

*References and further examples are given in (5). Abbreviations: rDNA = ribosomal DNA (ribosomal genes and spacers), sat. DNA = satellite DNA, repet. DNA = repetitive DNA. Most of the examples are not unequivocally accepted by the scientific community.

(cytophotometry, CsCl density gradient centrifugation, melting curves and reassociation kinetics, filter hybridization, etc.) in many tissues of Drosophila and other insects (Fig. 6; Table 2). More recently, DNA underreplication was also detected in plants, e.g., in epicotyl cells of Pisum sativum, in leaves and fruits of Cucumis species, in the embryo suspensor of the nasturtium, Tropaeolum majus (Figs. 7 and 8), and other systems (Table 2; for reviews see 5, 22 - 24).

Amplification of the ribosomal RNA genes is well known to occur in the oocytes of many organisms. It can be understood as a local and transitory polytenization of ribosomal DNA (25, 26). The extra-DNA is soon degraded. Amplification is the only way for germ line cells to increase the number of DNA templates, because endopolyploidy would lead to meiotic disturbances. In insects with meroistic ovaries, the supply of the maturing egg cell with RNP material used during cleavage divisions and early embryogeny is taken over by abortive oogonia, which become highly polyploid nurse cells. In this instance the oocytes do not undergo DNA amplification, or only to a small extent (Fig. 9). These examples clearly show the close relationship between variation in DNA content and composition at the one hand, and cell size and function at the other hand.

The structure of the amplified gene copies at the light and electron microscopic level show considerable variation among organisms. In amphibians and the unicellular alga, Acetabularia, extra loops are synthesized during a lampbrush stage of the chromosomes, while the extra copies of the ribosomal genes in beetles and flies form a condensed chromatin body which later disperses.

Somatic DNA amplification has not yet been so intensely studied as meiotic amplification. In the orchid Cymbidium, heterochromatin amplification was recognized in several cell types (27, 28; Fig. 10). It could be shown that nucleolus organizers are not involved in this process (29), but that an AT-rich DNA fraction is extra replicated (30; Fig. 11). It is not yet understood, why heterochromatin is synthesized in excess in certain cells and at specific development stages of the protocorms in this orchid, but probably the event can be seen in the light of cell growth and differentiation (see following chapter).

Somatic DNA amplification has also been reported to occur in sciarid flies and other insects. In the polytene salivary gland cells of Sciaridae, DNA puffs occur (31). The non-ribosomal DNA copies produced in DNA puffs are either not transcribed (32), or if transcribed, the RNA is not transferred into the cytoplasm (33). The transcripts of the DNA puffs in Rhynchosciara have been found to be translated into proteins; the RNA is polyadenylated, migrates rapidly into the cytoplasm, and is unstable (35). Thus, the extra DNA, or the RNA and proteins transcribed thereof, may be of

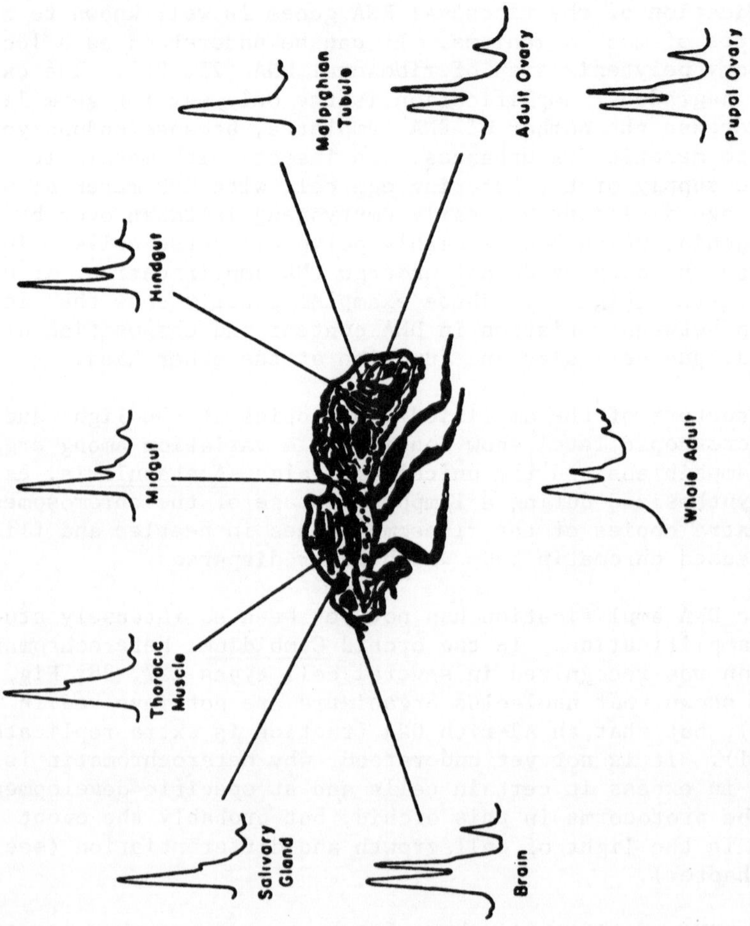

Fig. 6. Diagram illustrating differential underreplication in various organs of Drosophila virilis as found by analytical CsCl density gradient centrifugation (modified from 52).

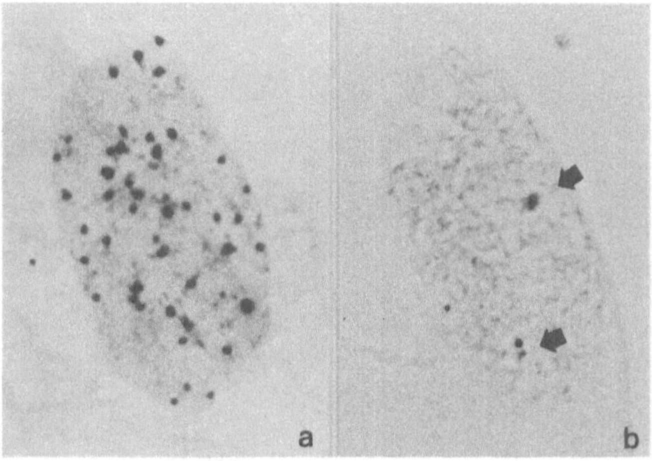

Fig. 7. Indication of heterochromatin underreplication in certain cells of the Tropaeolum suspensor.

(a) Endopolyploid nucleus with heterochromatin amount (chromocenters) proportional to the ploidy level;

(b) endopolyploid nucleus with the diploid amount of heterochromatin (arrows; Feulgen, x1,=400). Modified from (47).

regulatory character in the sense of Britten and Davidson (34), or of nucleoskeleton character (2).

Among several other reports on somatic DNA amplification in plants and cell (and tissue) cultures (Table 2), the amplification of repetitive DNA was noted in differentiating neural retina cells and in cartilage cells of the chicken (36, 37). In general, the coincidence of somatic DNA amplification and certain steps of differentiation, and the transitory nature of the amplified sequences indicate some role of the extra DNA in the complex control system of cell differentiation.

Recently, a clear example of amplification of a mRNA gene could be induced in cultured murine cells. After treatment with methotrexate some of the cells become resistent against this toxic agent. This resistance could be shown to be a consequence of amplification of the dihidrofolate reductase gene (38-40). The some hundred-fold amplification of this gene is visible at the light microscopic level as a large euchromatic (Giemsa-negative) chromosome band (40). Evidently, also other mRNA genes are capable

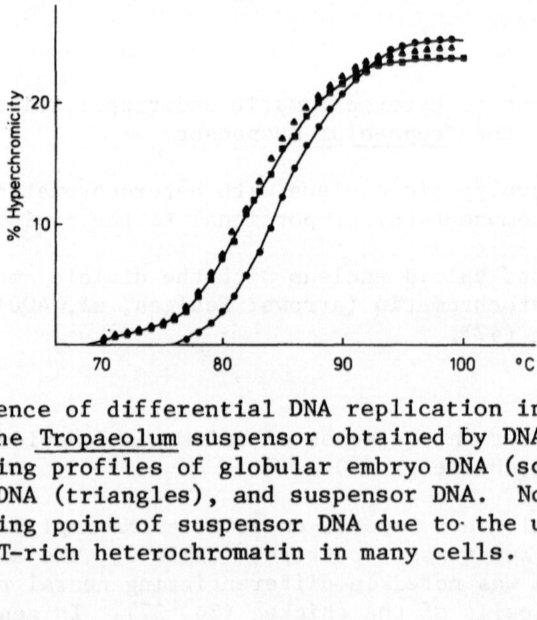

Fig. 8. Evidence of differential DNA replication in the nuclei of the Tropaeolum suspensor obtained by DNA denaturation: melting profiles of globular embryo DNA (squares), leaf bud DNA (triangles), and suspensor DNA. Note the higher melting point of suspensor DNA due to the underreplication of AT-rich heterochromatin in many cells.

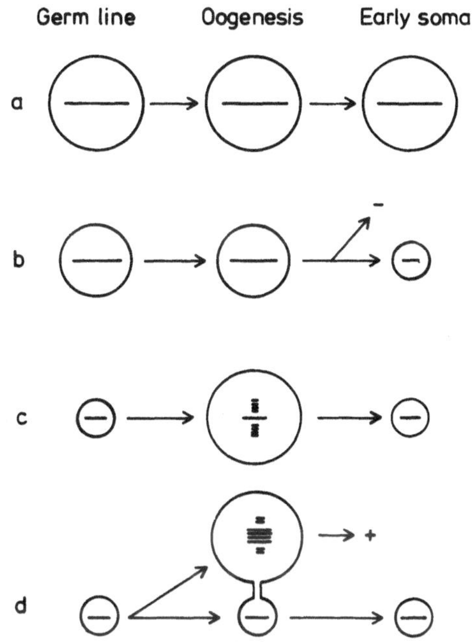

Fig. 9. Endopolyploidy and differential DNA replication during insect oogenesis as a consequence of genome size (C value) reduction. Under "early soma" elimination events during the cleavage divisions are indicated, but not the later endopolyploidization steps.

(a) Species with large basic nuclear DNA amount showing DNA constancy during oogenesis (it cannot yet be decided whether such species really exist).

(b) Species which show chromosome elimination or chromatin diminution during early somatic development, probably because of a negative nucleotypic effect of the DNA mass (e.g. Sciara).

(c) Transient DNA amplification (extra replication of cistrons) in the oocytes (many species).

(d) The oogonia differentiate into oocytes and trophocytes. The trophocytes become endopolyploid and undergo, in addition, rDNA amplification. All the RNA-synthesizing activity is undertaken by the trophocytes and transferred to the oocytes via cytoplasmic bridges or channels (+ = the trophocytes do not survive with the oocyte). Modified from (47).

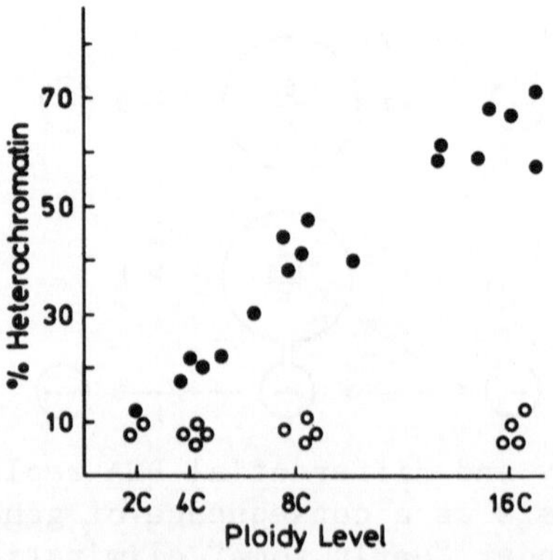

Fig. 10. Diagram illustrating heterochromatin amplification in protocorms of Cymbidium as obtained by cytophotometry of Feulgen-stained nuclei. The proportion of heterochromatin-DNA is between 7 and 10% in "standard nuclei" (o) of each ploidy class, while it is increased to more than 70% during endopolyploidization of the "heterochromatin-rich" nuclei (•); each point represents the mean of 10-20 measurements. Modified from (14).

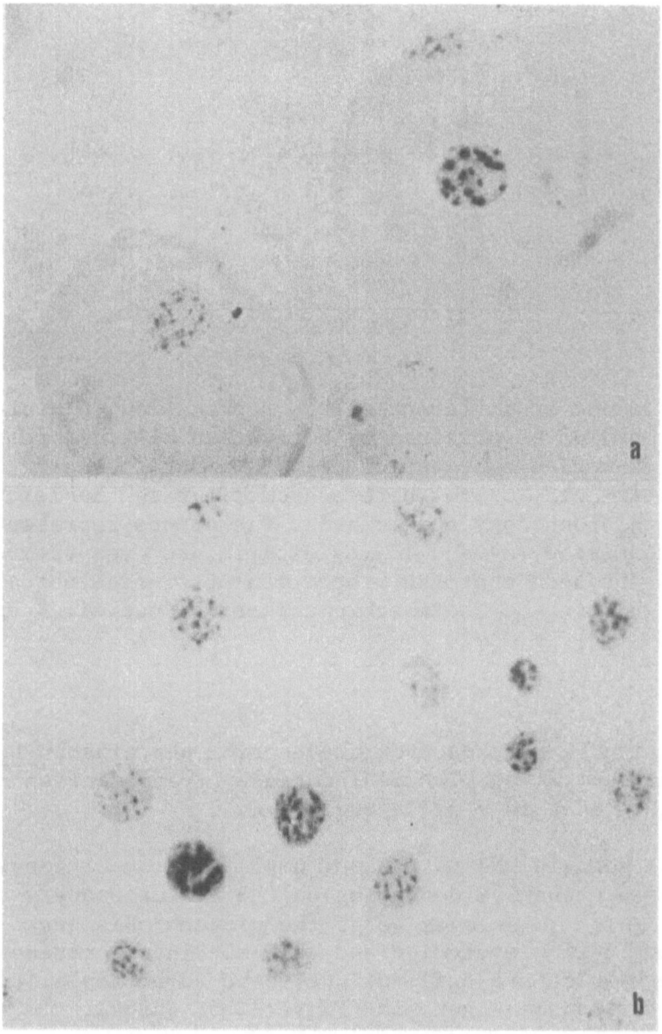

Fig. 11. Nuclei of Cymbidium of various ploidy which have, in part, undergone heterochromatin amplification.

(a) Squash preparation showing one cell with a "standard nucleus" (at left), and a cell with a "heterochromatin-rich" nucleus (at right).

(b) isolated nuclei with considerably different amounts of heterochromatin (Feulgen x 400).

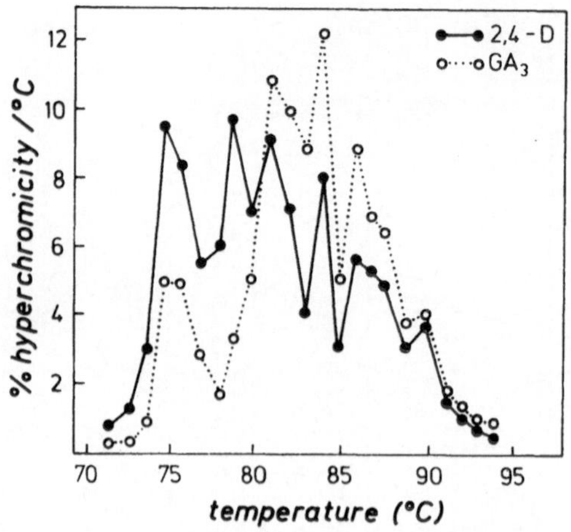

Fig. 12. Evidence of differential DNA replication in protocorms of Cymbidium as obtained by derivative melting profiles. Cultures were either enriched in "standard nuclei" by gibberellic acid, or in "heterochromatin-rich" nuclei by 2,4-dichlorophenoxy acetic acid. The points represent the means of 6 (o) and 12 (o) experiments, respectively. It can be seen that the proportion of AT-rich sequence is higher in the "heterochromatin-rich" nuclei. Redrawn from (30).

of amplification (e.g. those for the chorion proteins in Drosophila, 41). Possibly, amplification is induced by translocation of a transposon, but this idea is still speculative.

Like all somatic DNA variation, amplification of genes may be understood as a mechanism depending on the evolutionary state of the genome. While some genes (e.g. the histone-mRNA genes) have been amplified during evolution and tandemly incorporated into the genome, others which are still unique, have to be amplified "laterally" during somatogenesis (Fig. 13; 5, 42-44).

BIOLOGICAL SIGNIFICANCE OF SOMATIC DNA VARIATION

The biological significance of endopolyploidy and differential DNA replication is under discussion ever since their detection. While some authors don't see any role for them in development (45, 46), others envisage them as "switches" (coordinating control events) for differentiation and morphogenesis (for reviews see 5, 8, 10, 11, 13, 23, 47). I think there is evidence for a twofold role of somatic DNA variation in development: the one is directed to a general or specific increase of the RNA and protein synthesizing capacity, the other one is directed to high speed differentiation, and to growth and morphogenesis in general (organ modeling, etc.).

Active Cells Have Increased DNA Contents

As Fig. 3 shows, any polyploidization cycle and cell cycle with differential DNA replication leads to excess DNA in addition to the diploid genome. The genetic effects of such cell cycles can be estimated from the DNA composition in the mature cell.

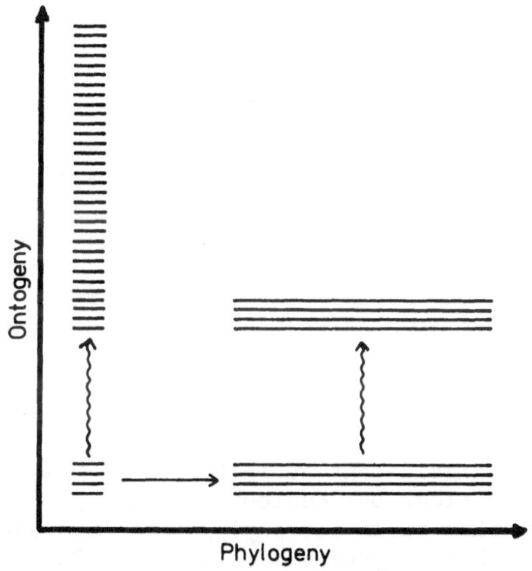

Fig. 13. Diagram to illustrate the possible strategem which underlies the occurrence and degree of endopolyploidy (exemplified as polyteny).

(a) The bars represent the genome (chromosome complement) of a species with little DNA (at left) and a species whose genome increased during evolution by tandem addition of DNA.

(b) It is assumed that the highly active and specialized cells of eukaryotes need a defined mass of DNA of both types, coding and regulatory sequences. In the DNA-poor species this DNA amount is synthesized by endo-cycles; the DNA-rich species do not need "lateral" multiplication of the genome. Redrawn from (5).

Giant cells with polyploid and polytene nuclei dispose of a great potentiality of gene activity, because they are supplied with multiple templates (up to 500,000 per nucleus; Table 1). It is well known from developmental-specific and cell-specific puffing in dipters that qualitatively and quantitatively different patterns of gene activity are induced in such nuclei, and that this allows a wide range of specialization in cell function. Therefore, high degrees of somatic polyploidy are found in cells and organs which reach high functional activity, but which often produce only a few gene products in abundance. Well known examples are the salivary and silk gland cells of insect larvae, the nurse cells of oocytes in meroistic insect ovaries, suspensor haustoria of plant embryos, antipodal cells, endosperm nuclei, giant cells in the trophoblast of rodents, and other specialized, highly active, but normally shortlived cells. Fig. 14 shows ultrastructural features of the highly polyploid suspensor cells of the bean, Phaseolus coccineus. which are suggestive for high functional activity.

Lower degrees of polyploidy are, however, found in many cells of nearly all organisms, normally in a tissue-specific pattern of certain degrees of polyploidy (reviews: 5, 49). In angiosperms, estimates of the proportion of polyploid cells have been made in Beta vulgaris (49) and Scilla dicidua (6), and a percentage of 70-80% could be established. These frequently occurring low levels of endopolyploidy are to be interpreted in a different functional sense.

Cell Size Depends On Nuclear DNA Content

One of the most significant nucleotypic effects is the control of cell size by the nuclear DNA content (1, 2, 17). This nucleo-cytoplasmic relationship holds for phylogenetic and ontogenetic variation in nuclear DNA. Species with low 2C values have small diploid cells, species with high 2C values have large cells. As a certain cell size seems to be a prerequisite for certain functions, species with low 2C values show a higher trend to higher somatic polyploidy levels than species with high 2C values (42, 44; Fig. 13). Somatically, the pattern of endopolyploidy is reflected by the pattern of cells of various size. Therefore, somatic polyploidy can be envisaged as an important tool in organ modeling. For instance, the differently sized scales of a butterfly wing exhibit corresponding differences in endopolyploidy. The size of plant hairs is dependent on the level of endopolyploidy, although in plants also the vacuole may increase the size of a cell. Further examples of plants and animals are listed by Nagl (5).

The role of non-coding DNA in the control of cell size is an important aspect of morphogenesis. This fact rules out to speak about "selfish DNA" or "sense-less DNA" (50,51). As living matter responds to the laws of thermodynamics, it it very unlikely that any

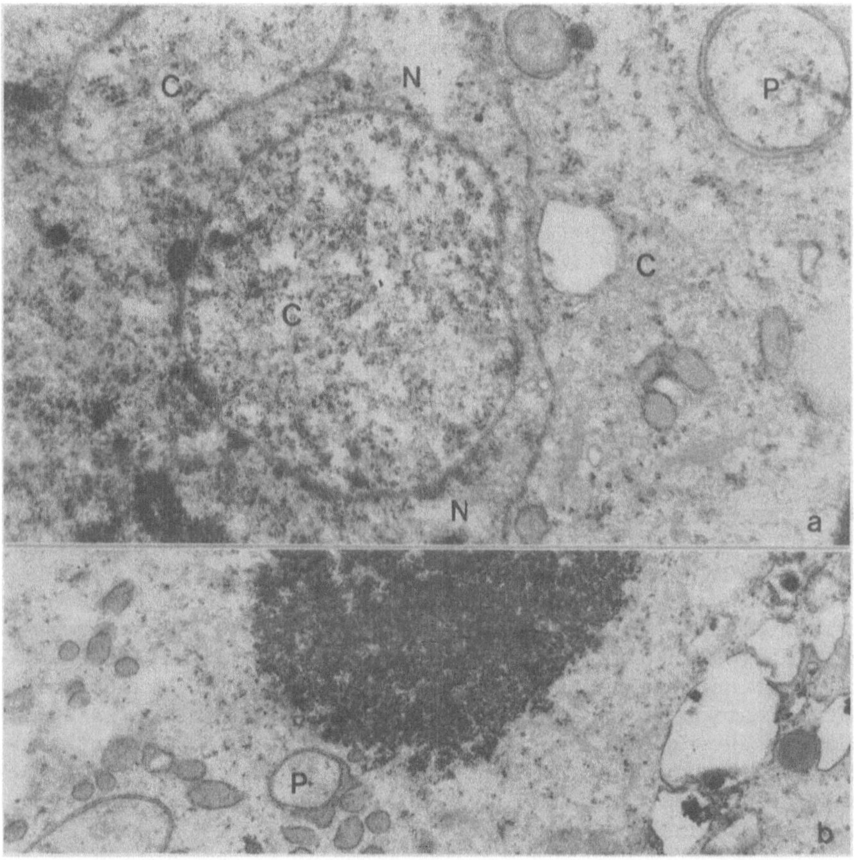

Fig. 14. Electron micrographs indicating the high synthetic activity of endopolyploid suspensor cells in Phaseolus coccineus.

(a) Intimate interdigitation between cytoplasm and nucleoplasm (N = nucleoplasm, C = cytolasm; x 15,000).

(b) "Free cytoplasmic nucleolus, which was exported from the nucleus (x 12,000). Note also the plastolysomes (P) (53) which are involved in the autolysis of the cells.

DNA without biological significance is replicated and stored. Statements on "selfish DNA" etc. are, in my mind, dangerous, because they keep many scientists from searching for the function of non-coding DNA, which in fact may play the critical role in the control of speciation, differentiation and transformation. But it must be kept in mind that nothing in biology can be generalized to all organisms. While somatic polyploidy is the mechanism of morphogenesis in plants and insects, changes in chromatin condensation are apparently the main mechanism in vertebrates (a physiological example for diverse mechanisms leading to the same effect is the C3, C4 and CAM way of photosynthesis).

The Economy of Differential DNA Replication

Differential DNA replication is, in spite of many methodical difficulties and sources of errors, proven for a number of cases (5, 22, 23, 47). It is the consequent continuation in the evolution of curtailed cell cycles. Its selective advantage can be seen in its energetic economy: only those DNA sequences are replicated, which are actually needed by a given cell.

Underreplication always affects heterochromatin and highly repetitive DNA (e.g. satellite DNA). This portion of the genome is possibly of some significance for sexual, meiotic and speciation events (see accompanying paper by Nagl in this volume), and for mitotic spindle attachment, as the centromeric regions are normally heterochromatic. Endopolyploid, mature somatic cells, particularly highly polyploid ones, will not need material for meiosis nor for spindle attachment, because they never will divide. Therefore, non-replication or underreplication of heterochromatin/satellite DNA saves energy that can be used for tissue-specific syntheses. The ultimate evolutionary step of this economic strategem can be seen in elimination of "meiotic" heterochromatin from somatic cells during the first cleavage divisions.

Gene amplification represents an easily understood strategem. Those genes, which are not reiterated in the genome, but whose products are needed in huge amounts within a short time, are extra replicated. Amplification of ribosomal genes in oocytes seems to be necessary in order to supply the egg cell with the high amount of ribosomes which are involved in translation during early cleavage divisions, as long as the embryonal genome is inactive. In methotrexate-resistant murine cells, the dihydrofolate reductase gene is amplified (38-40). In highly active somatic cells, the ribosomal genes or certain mRNA genes (e.g. those for the chorion proteins) may be amplified.

Besides structural genes, also regulatory sequences may be amplified. Heterochromatin amplification in *Cymbidium* can be seen in this direction. Strom et al. (36, 37) reported the amplification of a middle-repetitive DNA sequence in differentiating cartilage and retina cells of the chicken. Such amplification of non-coding DNA sequences may influence the development of an organ or tissue by altering the pattern of active genes via determination of the pattern of chromatin domain condensation (see the accompanying article by Nagl in this volume), or by altering the growth parameters (nucleotypic effects). Nicolini (55) reached a similar conclusion and stated:

"The critical role in the control of cell proliferation and cell transformation could be linked not to the activation of specific genes (with most parts of the genome being permanently switched off), but to the overall periodic geometry of the genome at the three interrelated levels, as determined by the chemical-electrostatic environment, mostly in terms of proteins or ions modifications, and partly in terms of the viscosity - diffusion properties per se of the highly condensed chromatin-DNA and its surrounding aqueous medium."

CONCLUDING REMARKS

Figure 15 shows the "DNA optimization model (5, 43), in which the interrelationships of DNA changes in phylogenesis and ontogenesis are unified to a single hypothesis. In addition, I would like to indicate the factors which drive the DNA changes: predefined by the nucleotide sequence, and according to the laws of thermodynamics and non-linear thermodynamics far away from equilibrium (56), the genome responds to environmental stimuli via the membrane system and ion changes, in a predictable manner. The main response in higher animals is determination of the chromatin domains to be condensed and permanently inactivated, the main response in plants and insects, but also visible in mammals, is DNA endoreduplication and differential DNA replication (curtailment of cell cycles). These and other events shown in Fig. 15 (e.g. chromosome rearrangements) are the intrinsic ability of DNA, and are only induced, but not controlled by the environment, and may lead via instabilities, fluctuations (56) and the going-on of hyper-cycles (57) to the predefined genomic response. The following changes in chromatin condensation and DNA content/composition determine the relative spatial relationship and arrangement among the genes and their regulators. This spatial relationship was suggested to be informational and specific for cell function (55, 58).

The mystery of diversification of bio-matter may be solved in the near future not by the mutation-selection theory of genes (because genes have much less diverged during evolution than organisms), but by increasing understanding of the huge changes in non-coding DNA in phylogeny and ontogeny, and by the changes in the three-dimensional structure of chromatin-DNA. The realization of the species-specific construction during ontogeny, i.e. cell

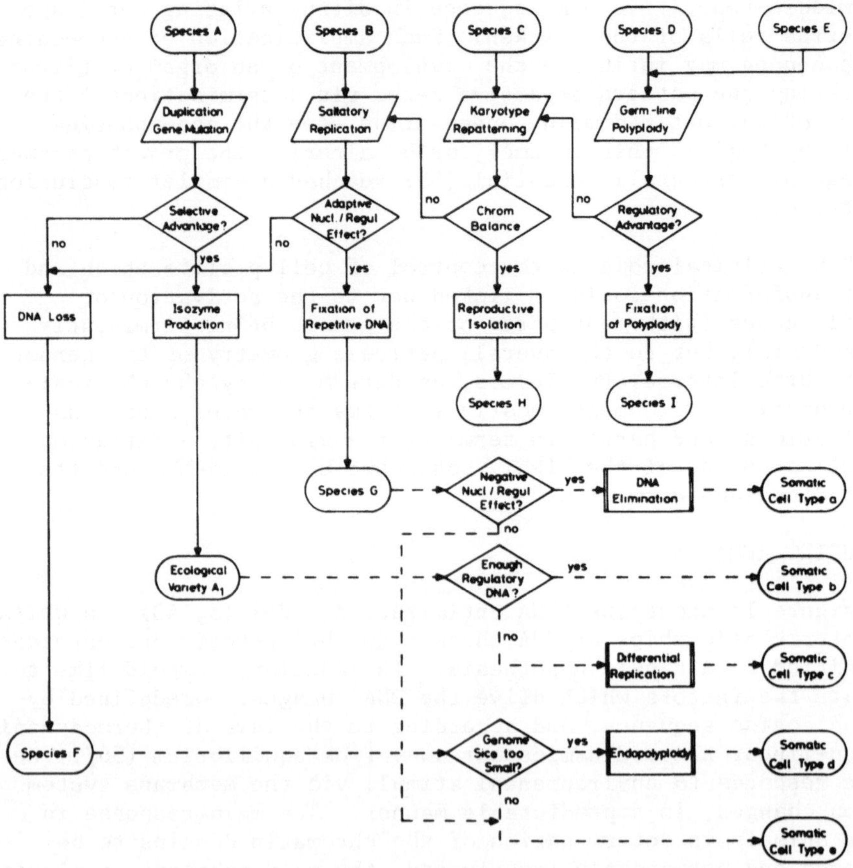

Fig. 15. The DNA optimization model (flow diagram) as an attempt to understand the quantitative and qualitative changes in nuclear DNA as the molecular basis of biological diversification in phylogenesis and ontogenesis. The model shows that somatic DNA variation such as differential DNA replication may be a predefined process, dependent on the evolutionary state of the basis genome in the zygote. The symbols have the following meanings:

oval = terminal interrupt; paralleologram = input, output; rhomb = decision; rectangle = process; rectangle with double lines = predefined process.

Details of the model will not be discussed here, but it should be realized that differential DNA replication may be a necessary step of cytodifferentiation in certain species. The content of a symbol is just an example; for instance, the rhomb (decision) before differential DNA replication may also include the questions: Enough ribosomal genes?, etc. Redrawn from (47).

differentiation and morphogenesis. are predefined in the DNA nucleotide sequence not by the genetic code, but by "instability sequences" and "condensing protein-binding sequences", which answer environmental stimuli according to the laws of thermodynamics. Carcinogenesis may be based on an erroneous response or over-fluctuation; compared with the innumerable cells which differentiate normally, such errors can be neglected from a biophysical point of view.

REFERENCES

1. M.D. Bennett, Brookhaven Symp. Biol., 25:344-366 (1973).
2. T. Cavalier-Smith, J. Cell Sci., 34:247-278 (1978).
3. L. Geitler, Chromosoma, 1:1-22 (1939).
4. A. Levan and T.S. Hauschka, J. Natl. Cancer Inst. 14:1-43 (1953).
5. W. Nagl, "Endopolyploidy and Polyteny in Differentiation and Evolution", North-Holland, Amsterdam (1978).
6. B. Frisch and W. Nagl, Plant Syst. Evol., 131:261-276 (1979).
7. B. Wright, Trends Biochem. Res., 4:N110-N111 (1979).
8. W. Nagl, in: "Genome and Chromatin: Organization, Evolution, Function", W. Nagl, V. Hemleben and F. Ehrendorfer, editors, p. 3-25, Springer, Vienna-New York (1979).
9. F. D'Amato, Caryologia, 17:41-52 (1964).
10. E. Tschermak-Woess, in: Handbuch der Allgemeinen Pathologie", H.W. Altmann, et al., editors, p. 569-625, Springer, Berlin-Heidelberg-New York (1971).
11. F. D'amato, "Nuclear Cytology in Relation to Development" Cambridge University Press, New York (1977).
12. W. Nagl, Ann. Rev. Plant Physiol., 27:39-69 (1976).
13. V. Ya Brodsky, I.V. Uryvaeva, "Cell Polyploidy: Its Relation to Growth and Differentiation" Academic Press, New York (1977).
14. W. Nagl, in: "Encyclopedia of Plant Physiology", Vol. 16, B. Parthier and D. Boulter, editors, Springer, Berlin-Heidelberg-New York (in press)
15. K. Carniel, Oesterr.Bot.Z., 110:145-176 (1963).
16. V.Ya. Brodsky and I.V. Uryvaeva, Intern. Rev. Cytol. 50: 275-332 (1977).
17. W. Nagl, Zelkern und Zellzyklen. Ulmer, Stuttgart (1976).
18. E. Heitz, Ber. Deutsch. Bot. Ges., 47:274-284 (1929).
19. L. Geitler, Ergebn. Biol., 18:1-54 (1941).
20. P.W. Barlow, Protoplasma, 90:381-391 (1976).
21. W. Nagl, Protoplasma, 91:389-407 (1977).
22. M. Buiatti, in: "Plant Cell, Tissue, and Organ Culture", J. Reinert and S. Bajaj, editors, p. 358-374, Springer, Berlin-Heidelberg-New York (1977).
23. W. Nagl, V. Hemleben and F. Ehrendorfer, editors "Genome and Chromatin: Organization, Evolution and Function" Springer, Vienna-New York (1979).

24. W. Nagl, Progr. Bot. 39:132-152 (1977).
25. A. Lima-de-Faria, Cold Spring Harb. Symp. Quant. Biol., 38:559-571 (1974).
26. W. Nagl, in: "Tissue Culture and Plant Science 1974", H.E. Street, editor, p. 19-42, Academic Press, New York, (1974).
27. W. Nagl, Cytobios, 5:145-154 (1972).
28. W. Nagl, J. Hendon and W. Rucker, Cell Diff., 1:229-237 (1972).
29. D. Schweizer and W. Nagl, Exptl. Cell Res., 98:411-423 (1976).
30. W. Nagl and W. Rucker, Nucleic Acids Res., 3:2033-2039 (1976).
31. A. Brito-da-Cunha, C. Pavan, J.S. Morgante and M.C. Garrido, Genetics Suppl., 61:335-349 (1969).
32. L. Walter, Chromosoma, 41:327-360 (1973).
33. J. Balsamo, J.M. Hierro, and F.J.S. Lara, Cell Diff., 2: 119-130 (1973).
34. R.J. Britten, E.H. Davidson, Science, 165:349-357 (1969).
35. M.F. Bonaldo, R.V. Santelli, and F.J.S. Lara, Cell, 17: 827-833 (1979).
36. C.M. Strom, and A. Dorfman, Proc. Natl. Acad. Sci. US, 73:3428-3432 (1976).
37. C.M. Strom, M. Moscona, and A. Dorfman, Proc. Natl. Acad. Sci. US, 75:4451-4454 (1978).
38. F.W. Alt, R.E. Kellems, J.R. Bertino and R.T. Schimke, J. Biol. Chem. 253:1357-1370 (1978).
39. R.J. Kaufman, P.C. Brown, and R.T. Schimke, Proc. Natl. Acad. Sci. US, 76:5669-5673 (1979).
40. B.J. Dolnick, R.J. Berenson, J.R. Bertino, R.J. Kaufman, J.H. Nunberg and R.T. Schimke, J. Cell Biol., 83:394-402 (1979).
41. A.C. Spradling, and A.P. Mahowald, Proc. Natl. Acad. Sci. US, 77:1080-1100 (1980).
42. W. Nagl, The Nucleus, 20:10-27 (1977).
43. W. Nagl, Chromosomes Today, 6:151-152 (1977).
44. W. Nagl, Nature, 261:614-615 (1976).
45. P.W. Barlow, Protoplasma, 90:381-391 (1976).
46. L.S. Evans, J. Van't Hof, Amer. J. Bot., 62:1060-1064 (1975).
47. W. Nagl, Z. Pflanzenphysiol., 95:283-314 (1979).
48. C.R. Partanen, Intern. Rev. Cytol., 15:215-243 (1963).
49. Th. Butterfass, Mittlg. Max-Planck-Ges., 1:47-58 (1966).
50. W.F. Doolittle, and C. Spaienza, Nature, 284:601-603 (1980).
51. L.E. Orgel and F.C.H. Crick, Nature, 284:604-607 (1980).
52. S.A. Endow, and J.G. Gall, Chromosoma, 50:175-192 (1975).
53. W. Nagl, Z. Pflanzenphysiol., 85:45-51 (1977).
54. P.J. Gartner, and W. Nagl, Planta xx,xx-xx (1980).
55. C.A. Nciolini, in: Chromatin Structure and Function", C.A. Nicolini, editor, p. 613-666, Plenum Press, New York (1979).
56. I. Prigogine, "Thermodynamics of Irreversible Processes" Wiley-Interscience, New York (1955).

57. M. Eigen, and P. Schuster, "The Hypercycle" Springer, Berlin-Heidelberg-New York (1978).
58. P.O.P. Ts'o, "The Molecular Biology of the Mammalian Genetic Apparatus" North-Holland, Amsterdam (1979).

CELL TRANSFORMATION BY RNA SARCOMA VIRUS

Heinz Bauer

Institut für Virologie der Justus-Liebig-Universität

D-6300 Giessen

INTRODUCTION

Autonomous cell proliferation appears to be an essential characteristic of most neoplastic cells. Present knowledge of the biochemical and biological properties by which the phenotype of tumor cells may be defined, is largely based on experimental studies in vitro, in which normal cells are converted to a transformed state by virus infection. This transformed state in vitro appears to correlate with the phenotype of tumor cells in vivo. Numerous viruses with the capacity to transform cells have been identified and genetic analyses have revealed that many of these bear genetic information, generally referred to as an onc-gene, which is directly responsible for transformation. These viral onc-genes have proven to be powerful tools in the study of the molecular processess which initiate and maintain malignant cell growth (see CSH Symp. 44, 1980).

Considerable progress has been made recently by the biochemical identification of proteins which are coded for by the onc-genes of the various tumor viruses. Investigations of their functions, however, seem to confirm earlier genetic studies indicating that the onc-genes or their products are multifunctional and that initiation of cell proliferation in these systems is a more complex phenomenon than one may have previously hoped. On the other hand, metabolic pathways in transformed cells do not appear to differ from those in normal replicating cells (Preskott, 1976). One might therefore assume that the processess which initiate proliferation in transformed cells are not unique but are similar or identical to those which occur when normal cells are stimulated to proliferate under physiological conditions. This fundamental biological question concerning the intiation of cell proliferation, is not understood in normal cells and it is at this dilemma in which tumor virology and cell biology meet.

The purpose of this article is to briefly review some of the current information on the biological properties of cells transformed by Rous sarcoma virus and to relate these to the possible functions of the viral onc-gene.

THE ROUS SARCOMA VIRUS (RSV)

Structure and Biology of the Virus

As with other retroviruses, the core of RSV consists of a single stranded diploid RNA genome associated with a basic protein and reverse transcriptase which is surrounded by a protein capsid. During the maturation process which occurs by budding through the plasma membrane, this nucleocapsid of about 70 nm diameter acquires part of the cell membrane as a lipid envelope in which virus-coded glycoproteins are inserted. These glycoproteins are required for the specific absorption of virions to a target cell, a primary step in infection. After virus infection, the RNA is transcribed by the viral polymerase into a double-stranded DNA (reverse transcription) which is then integrated into the cellular DNA (see CSH Symp. Vol. 44, 1980). RSV will transform chicken embryo cells (CEC) within less than 24 hours post-

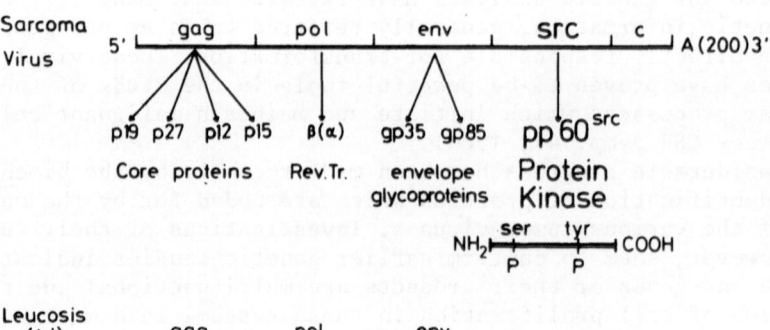

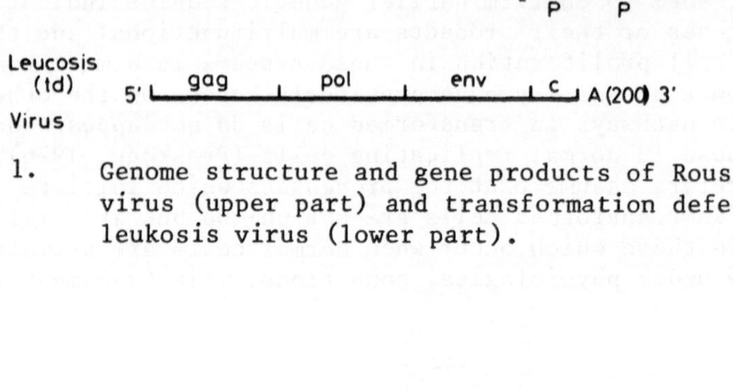

Fig. 1. Genome structure and gene products of Rous sarcoma virus (upper part) and transformation defective (td) leukosis virus (lower part).

infection in vitro, or cause visible fibrosarcomas in vivo within a few days. Within the group of rapidly transforming retroviruses, RSV is unique in its ability to replicate independently of a helper virus since it contains all the genes necessary for virus replication. These are gag, env, and pol, in the order from the 5'- to 3' end of the RNA (Fig. 1). The gag gene codes for a 76,000 MW precursor protein (pr76) that is cleaved to yield the core and nucleocapsid protein p12, p15, p15, and p27; the pol gene codes for the RNA-dependent DNA polymerase (reverse transcriptase); and the env gene codes for the envelope glycoproteins gp85 and gp37.

The Src Gene and its Gene Product $pp60^{src}$

The fourth gene, called src, which is located between the env gene and the 3' end of the genome codes for a phosphoprotein of about 60,000 MW, $pp60^{src}$ (Brugge and Erikson, 1977; Brugge et al., 1978), which is believed to induce all of the transforming functions. The $pp60^{src}$ is phosphorylated by a cAMP- dependent kinase in a serine residue at the N-terminal part of the molecule and by a cAMP-independent kinase in a tyrosine residue at a C-terminal site (Erikson et al., 1979; Brugge et al., 1978; Hunter and Sefton, 1980). The tyrosine is phosphorylated by an unusual kinase that is found to be closely associated with $pp60^{src}$, and there is genetic and biochemical evidence though not final proof that this kinase activity is a property of the $pp60^{src}$ (Erikson et al., 1978; Levinson et al., 1978; Rübsamen et al., 1979). The kinase was originally detected and is usually assayed for by its capacity to phosphorylate the heavy chain of IgG after immunoprecipitation of $pp60^{src}$ with tumor-bearing rabbit (TBR) sera (Collett and Erikson, 1978; Levinson et al., 1978).

There is ample evidence for the transforming function of the src gene and its product (see Friis, 1978; Bishop, 1978). Genetic studies have shown that a temperature sensitive (ts) lesion in src such that the gene is defective at 42° C will affect the transforming capacity while leaving viral replication intact. Furthermore, deletion mutants which lack part or all of the src gene retain the capacity to replicate but no longer transform CEC in vitro.

The availability of ts-mutant viruses as well as transformation-defective viruses (the latter called Rous-associated viruses (RAV) or lymphoid leukosis viruses (LLV), depending on their origin) provides further controls in addition to normal uninfected cells in the study of transformation-associated cellular events in vitro. Furthermore, they provide the opportunity to shorten the lag period between infection and the onset of the transformed phenotype. A cell culture fully infected at the restrictive temperature with a ts transformation-defective (td) mutant virus will convert rapidly to the transformed phenotype upon shift to the permissive temperature.

THE PHENOTYPE OF RSV TRANSFORMATION

The most apparent modification occurring in the transformation of fibroblasts is a change from the flat to a rounded morphology. This change is detectable by light microscopy (Fig. 2). Another transformation-induced phenomenon is the fact that transformed cells, in contrast to normal fibroblasts, are less adherent to a substrate and are capable of growing in multilayers. This substrate-independent growth is best documented by the formation of large colonies when transformed cells are grown in semiliquid medium. Furthermore, transformed cells grow to a higher density than normal cells and replicate in about 1/10 the calf serum concentration that is required for normal cell division, apparently the result of reduced serum growth factor requirements.

All of these properties may be of great importance for the growth and movement of a tumor cell in vivo. Since they would appear to be greatly dependent upon the structure and function of the plasma membrane, one can assume that this membrane is largely involved in important structural, physiochemical and biological alterations characteristic for the transformed phenotype. This assumption is further verified by scanning electron microscopic investigations showing dramatic changes in the cell surface morphology (Fig. 3).

In fact, most of the biochemical and biological alterations to the transformed fibroblast involve the plasma membrane or structures which are closely associated with it, as is summarized in the following section.

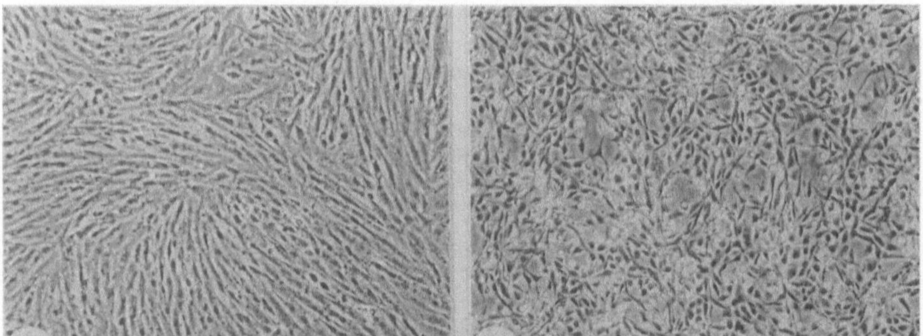

Fig. 2. Light microscope photography of a normal (a) and a RSV-transformed (b) chick embryo cell.

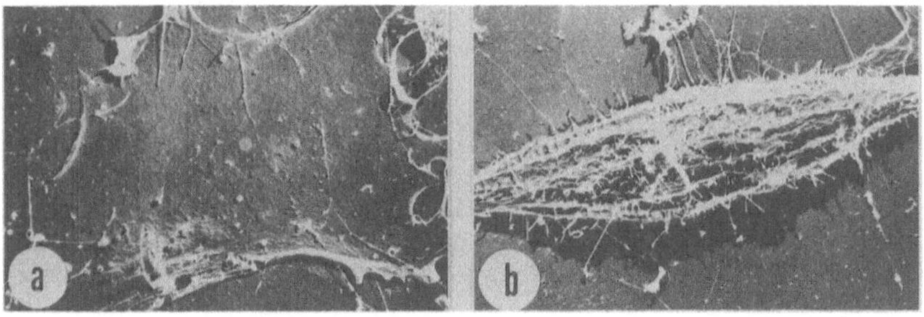

Fig. 3. Scanning electron micrograph of a normal (a) and a RSV transformed (b) chick embryo cell.

ALTERATIONS ASSOCIATED WITH THE CYTOPLASMIC MEMBRANE OF TRANSFORMED FIBROBLASTS

The plasma membrane consists of a liquid phase lipid bilayer containing proteins, most of which are glycosylated. The proteins are loosely associated either with the inner or with the outer membrane surface or they are actually embedded within the membrane (see Singer, 1974; Nicolson, 1976a). There is little doubt that the function of this membrane in transporting ions and macromolecules as well as mediating signals to the inside or outside of the cell are of decisive importance for the biological behaviour of the cell, and certainly also influence cell proliferation. Accordingly, numerous reports have described structural and functional changes in the plasma membrane of transformed cells. Those which currently appear most important are summarized in Fig. 4.

RSV-transformed cells release macromolecules which have been shown to enhance transformation by RSV (Kryceve et al., 1976), and to stimulate cell growth (Rubin, 1970) or hexose uptake (Lawrence and Jullien, 1980) upon addition to normal cells. These factors thus resemble a family of so-called growth factors such as those released by murine sarcoma virus transformed cells (Todaro and De Larco, 1978). This may be the reason that in some tumor cell systems the growth-factor receptors appear to be reduced or blocked (Todaro and De Larco, 1976; Cohen, 1976; Blomberg et al., 1980), though this phenomenon has not yet been documented in RSV-transformed cells. One might speculate that the function of growth factors released from transformed cells is to react with receptors on one and the same cell; i.e. the transformed cell may continuously release factors which stimulate its proliferation from without. This would also help explain the reduction of serum requirements for division of transformed cells.

The significance of the increased agglutinability of transformed cells by lectins is unclear. It is known, however, that the local density of the respective receptors is increased (see Nicolson, 1976b).

Unknown mechanisms cause an alteration in the passive diffusion or active transport of ions and macromolecules in transformed cells. After RSV transformation, cells take up 2-deoxyglucose five to ten times faster (Weber, 1973) and the intracellular Na^+ and K^+ concentrations are elevated (Johnson and Weber, 1979). The intra-

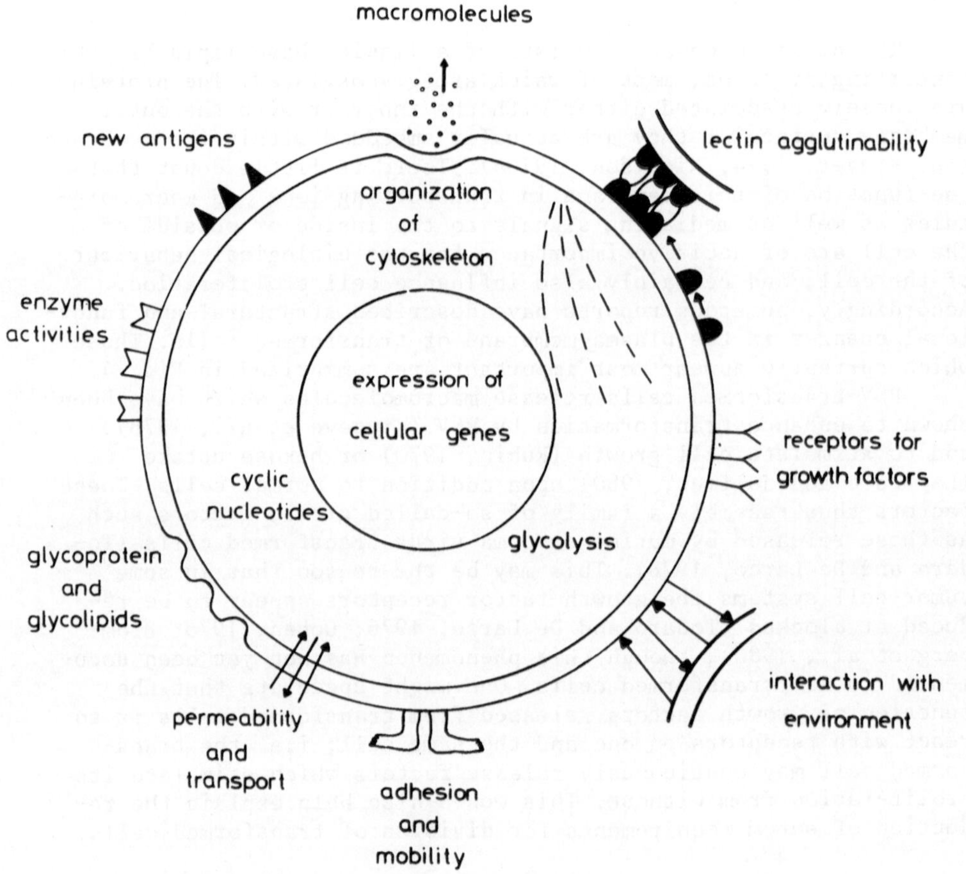

Fig. 4. Constituents and functions of the plasma membrane which are altered after transformation.

cellular Ca^{++} concentration is increased after RSV transformation, and probably as a consequence of this, the uptake of radioactive calcium is lower than in normal cells (Barnekow, Rose, and Bauer, unpublished). Interestingly, RSV-transformed cells require much lower extracellular Ca^{++} concentration for proliferation than do normal cells (Balk et al., 1979).

Changes in ion concentrations in particular for Ca^{++}, appear to be of utmost importance for the regulation of cell metabolism and proliferation. Calcium or calcium-activated calmodulin has been shown in various cells to influence either directly as a second messenger or by regulating the second messenger function of cyclic nucleotides, the fluxes of other ions like K^{+}, to react with constituents of the cytoskeleton, as well as to affect glycolysis and protein and nucleic acid synthesis (see Durham, 1978; Means and Dedman, 1980). The ability of epidermal growth factors to stimulate glycolysis is markedly dependent on K^{+} (Carpenter and Cohen, 1979), and by alteration of the extracellular Na^{+}concentration normal CEC express some characteristics of RSV-transformed cells (Moyer et al., 1980). Thus a change in the intracellular kation concentration may be a key event in transformation.

In regard to the alterations in glycolipid and glycoprotein glycosylation, it is assumed that these mainly reflect the changed metabolism of transformed cells. On the other hand, the function of molecules which mediate cell-cell interactions may be influenced by the degree of glycosylation. For instance glycolipids serve as receptors for growth inhibiting factors, which upon binding to their respective substrates affect cyclic nucleotides (see Critchley, 1979).

Highly characteristic for transformed fibroblasts is the marked decrease in fibronectin, which is assumed to be responsible for adhesion of normal cells (see Hynes, 1979).

A plasminogen activator protease which increases as much as 100-fold upon RSV-transformation (see Quigley, 1979) has gained much interest. One may speculate that proteases are necessary for the invasiveness of tumor cells in metastases.

New antigens can be detected at the cell surface by various immunological methods which occur either as a consequence of the expression of viral genes, by the derepression of cellular genes or by the alteration of cell membrane structures. In addition to the viral envelope glycoprotein found at virus budding and nonbudding sites (Kurth and Bauer, 1972; Gelderblom et al., 1972), two further kinds of transformation-associated cell surface antigens have been described. One, tentatively called onco-fetal antigen (OFA), is expressed early in embryogenesis in presumably undifferentiated cells (Yoshikawa et al., 1979) and in RSV as well as methylcholanthrene-transformed cells, but not in normal CEC grown after several passages in vitro or in the tissue of older embryos or adult birds. A second antigen termed TSSA (tumor-specific cell surface antigen) is only found in RSV-transformed cells, but not in MC-transformed, in avian leukemia virus-transformed or in normal cells (Ignjatovic et al., 1978). Since the latter antigen does not seem to represent one of the major structural virus proteins, and since in contrast to OFA it is antigenically expressed even at the nonpermissive temperature in ts-RSV-infected cells (Bauer et al., 1979) it is tempting to speculate that this antigen is identical to or part of the src gene product, $pp60^{src}$ (see below).

Finally, some of these structural and biochemical modifications of the plasma membrane may cause the substrate independent mobility and division of transformed cells.

WHAT ARE THE FUNCTIONAL CHARACTERISTICS OF $PP60^{SRC}$?

Pleiotropism

Studies with ts-transformation-defective RSV mutants have shown that the cellular alterations briefly described above (and probably others still unknown) are all dependent on the function of the src gene. This has raised the question as to the interrelationships between such alterations, whether they represent individual steps in a cascade which has been initiated by the interaction of $pp60^{src}$ with a single cellular component, or whether $pp60^{src}$ is pleiotropic and affects several cellular targets which then in a concerted action lead to cell transformation.

The investigation of the transformation phenotype of a variety of ts transformation-defective RSV mutants has been helpful in studying this question and has led to the conclusion that the src gene of RSV is multifunctional (Caloty and Pessac, 1976; Becker et al., 1977; Friis et al., 1977; Weber and Friis, 1979), because it was found that a src gene defect does not necessarily affect all transformation-associated phenotypic changes. Investigation of revertants from RSV-transformed vole cells led to a similar conclusion (Collett et al., 1979). An intriguing result was the finding that a ts-defect in RSV mutants as shown in fibroblast transformation is not observed in the chorioallantoic membrane, i.e. here ts mutants induce invasive cell proliferation at both temperatures, permissive and nonpermissive (Poste and Flood, 1979). This can be explained upon the assumption that the src gene has a pleiotropic effect and that a ts src gene function is compensated for in a possibly fully permissive environment in the chorioallantoic membrane.

The Cellular Localization of $pp60^{src}$

The precise determination of the cellular compartments where $pp60^{src}$ accumulates may help to identify the molecular structures with which $pp60^{src}$ interacts. Most reports dealing with this question more or less agree that significant amounts of $pp60^{src}$

are in the cytosol and in addition close to or associated with the cytoplasmic membrane (Willingham et al., 1979; Courtneidge et al., 1980; Krueger et al., 1980a), while only two reports describe part of the protein in the periphery of the cell nucleus (Rohrschneider et al., 1979; Krueger et al., 1980b). Because of the requirement for detergent in the extraction of $pp60^{src}$ it has been suggested that it is at least in part embedded in the plasma membrane through hydrophobic interactions. In this laboratory significant amounts of $pp60^{src}$ have been detected at the cell surface and found to be released by intact cells into the supernatant (Barnekow et al., 1980; 1981). Thus it is possible that $pp60^{src}$ exerts some or all of its functions in close proximity to the plasma membrane.

Possible Targets for $pp60^{src}$

The only known enzymatic function to be associated with $pp60^{src}$ is the cAMP-independent tyrosine-specific kinase activity mentioned before. It has been shown with ts-td viruses that the $pp60^{src}$-associated kinase is temperature sensitive as is the phosphorylation of the tyrosine residue in $pp60^{src}$. Therefore protein phosphorylation appears to be a key event in the transformation process and in fact the cellular levels of phosphotyrosine in proteins is increased by a factor of ten upon RSV-transformation (Sefton et al., 1980). Much work has been done to identify cellular substrates for that kinase and the unusual tyrosine specificity of the kinase has facilitated the search for target phosphoproteins. Thus, a protein of 36,000 mol. weight has been identifired which is present in normal cells and is phosphorylated in tyrosine upon RSV-transformation (Radke et al., 1980; Erikson and Erikson, 1980). The function and cellular localization of this protein is unknown.

Another candidate target for the kinase activity is the pyruvate kinase which is a key enzyme in glycolysis and which is phosphorylated and inactivated by a cAMP-independent kinase (PKK). It has been speculated that the $pp60^{src}$-associated kinase interacts with pyruvate kinase because its enzymatic and biophysical properties cannot be clearly distinguished from those of PKK (Presek et al., 1980). In fact, aerobic glycolysis is specifically increased in RSV-transformed cells as has been shown with ts-td mutants (Carroll et al., 1978). Interestingly, a direct effect on glycolysis by a transforming protein has also been discussed recently in the murine retrovirus system (Anderson et al., 1979).

A marked change to be observed 6 to 12 hours after the onset of RSV-transformation concerns the cytoskeleton and consists in a severe disorganization of the microfilament bundles (Edelmann and Yamara, 1976; Wang and Goldberg, 1976). More refined studies in this laboratory have indicated that the microfilament proteins are affected very early in transformation. Ruffle-like evaginations of the plasma membrane, also named flowers, which occur early after downshift to the permissive temperature of ts-td RSV-infected cells (Ambros et al., 1975) seem to be the consequence of a re-arrangement of microfilament proteins. Fifteen minutes after temperature shift such "flowers" can be observed to contain actin, α-actinin, myosin, and tropomyosin in a re-organized form (Fig.5) (Boschek et al., 1979; 1981). None of the other transformation-associated parameters so far observed occur as rapidly as flower formation, and together with the loss of microfilament bundles it is the only known effect that appears after temperature downshift when new protein synthesis is blocked by cycloheximide (Friis et al., 1980). This seems of particular interest because other studies have shown the prompt reversibility of the ts-defect of $pp60^{src}$ by demonstration that kinase activity and phosphorylation of $pp60^{src}$ itself, both ts properties of $pp60^{src}$, re-occur within 15 minutes after shift to the permissive temperature in the absence of protein synthesis (Friis et al., 1980; Ziemiecki and Friis, 1980). Therefore, one is inclined to speculate, that a microfilament protein may be one of the targets affected by $pp60^{src}$. This is also supported by the finding of Rohrschneider (1980) that $pp60^{src}$ accumulates in substrate adhesion plaques, the points of microfilament bundle attachment.

As an alternative, the microfilament disarrangement could be the rapid result of another $pp60^{src}$ effect, for instance the change of cation concentrations. Ca^{++} as well as Ca^{++}-activated calmodulin have been shown to affect cytoskeleton structures. It is unclear at present by which mechanism the ion concentrations are altered following RSV-transformation. Of the various possibilities, two seem to be especially attractive: One mechanism speculates that $pp60^{src}$ changes the effectiveness of ion pumps by direct sterical or enzymatical interaction. Since $pp60^{src}$ is shed by the cell (Barnekow et al., 1980; 1981) another possibility could be an effect of $pp60^{src}$ from without, by reaction with cell surface receptors in a similar way as it is known for growth factors. It has also been shown that one of the effects exerted by growth factors may be an alteration of intracellular ion concentrations.

Concluding Remarks

Cell division requires not only the replication of the cellular genome by DNA synthesis, but also cell growth, i.e. increased synthesis of other essential constituents of the cell. Which of these events occur first and how these two principle mechanisms interact with each other is not known. Considering the multiple evidence for a pleiotropic effect of the src gene as well as the various metabolic and structural targets for the functions of $pp60^{src}$, one may hypothesize therefore that the $pp60^{src}$ acts in

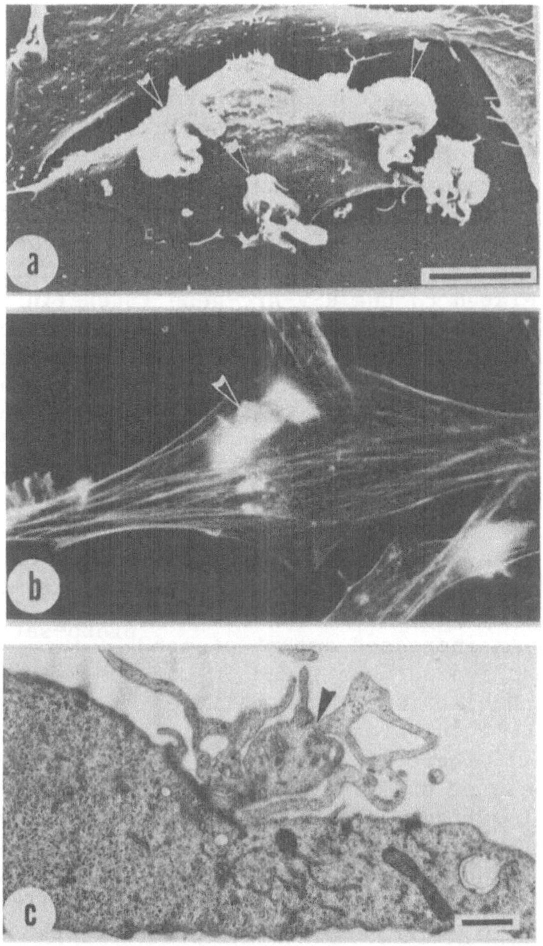

Fig. 5. Ruffle-like flowers (indicated by arrows) on dorsal surface of RSV-infected cells as demonstrated by a) scanning electron microscopy b) immunofluorescence after staining with antibody to actin and c) by transmission electron microscopy of thin section. Marker a: 10 μm; c: 1 μm

two manners, possibly independently from each other, one which would induce cell growth and the other DNA synthesis.

It would be premature, however, to draw a detailed model of the molecular events induced by such a transforming gene and which lead to unrestricted cell proliferation. Not only are the functions of the transforming protein poorly understood but the present knowledge about the signals and events that promote restricted proliferation of normal cells is likewise limited. This is a serious disadvantage, because it seems apparent that the metabolic events accompanying tumor cell proliferation are similar to or identical with those occuring in normal cell replication. This, on the other hand, raises the question whether one can relate the transformed state of a tumor cell to a physiological state of the normal cell counterpart, implying that the normal cell bears genetic information that is similar to the transforming virus gene with respect to its biological effects. Indeed, the normal chicken cell possess a gene, called sarc, that is biochemically and functionally closely related to the RSV-src-gene (Stehelin et al., 1976; Wang et al., 1980). In normal cells one can detect a phosphoprotein of about 60,000 dalton with similar properties to the viral $pp60^{src}$ (Collett et al., 1978) and which is called $pp60^{proto-sarc}$ (Courtneidge et al., 1980). Since the amount of endogenous $pp60^{proto-sarc}$ in fibroblast cultures is only about 1/50 to 1/100 of the amount of exogeneous $pp60^{src}$ (Karess et al., 1979), cell proliferation induced by this kind of protein may be dosis dependent.

An elevated rate of cell proliferation is physiologically normal during embryogenesis and one might therefore assume an increased expression of the proto-sarc gene during that period of organism development. An increased amount of $pp60^{proto-sarc}$-associated kinase activity has indeed been found in mesenchymal tissue of 5 day old embryos and also in primary cultures of chick embryo cells when compared to cell cultures after several subcultivations (Barnekow et al., 1981). There are other similarities between transformed and embryonic or undifferentiated cells. One, the expression of common cell surface antigens (OFA), has been mentioned above. A specific function of fibroblasts, the synthesis of collagen which is not found in the early stages of embryogenesis is distinctly reduced after transformation of chicken fibroblasts by RSV (Kamine and Rubin, 1977; Rowe et al., 1978). Other target cells for RSV-transformation, such as myoblasts, melanoblasts, and chondroblasts, have likewise after RSV-transformation been described to lose functions which are specific for their final differentiation stage (Holtzer et al., 1975; Roby et al., 1976; Pacifici et al., 1977; Fiszmann and Fuchs, 1975; Boettiger et al., 1979). Further, a metabolic property of embryonic cells, increased aerobic glycolysis, is also characteristic for RSV-transformed fibroblasts (Carroll et al., 1978).

Embryonic and transformed cells in this system have not been investigated with respect to other parameters, however, those mentioned above allow one to speculate, that the chick fibroblasts upon RSV-transformation, acquire metabolic and growth properties similar to those which are under the control of the endogeneous proto-sarc gene in an early stage of development and differentiation. Therefore, the mechanism of virus transformation may lead to a shift back in the differentiation stage of the cell (Bauer and Yoshikawa, 1980). The essential difference between the RSV-transformed cell and the embryonic cell could be then that the endogeneous proto-sarc gene is under the physiological control of the organism and the microenvironment of the cell, causing a limited proliferation capacity of the embryonic cell. The viral src gene, on the other hand, would not be restricted by any cellular control mechanism due to its association with another genetic structure, the virus genome, allowing continued expression.

More than a dozen transforming viral genes have been identified in the past which all have their homologues in normal cells. Interestingly, related genes are not only found in the animal species from which the virus probably has originated, but also in other species indicating that they are highly conserved in evolution (see CSH Symp. Vol. 44, 1980). Therefore, one can assume, that the cellular genes corresponding to viral onc-genes are of great importance for the physiological development and differentiation of the organism and it is suggested, that such genes, upon genetic mutation by biochemical or biophysical agents may be activated in tumors of non-virus origin and cause unlimited cell proliferation. Such a hypothesis can now be experimentally tested, and there is in fact one report showing that cell transformation can be induced by transfection of normal cellular DNA (Cooper et al., 1980).

ACKNOWLEDGMENT

The help of Valerie Bosch and of Bruce C. Boschek with the preparation of the manuscript is gratefully acknowledged.

The own experimental work has been supported by the Deutsche Forschungsgemeinschaft (SFB 47, Virologie).

REFERENCES

Ambros, V. R., Chen, L. B., and Buchanan, J. M., 1975, Surface ruffles as markers for studies of cell transformation by Rous sarcoma virus. Proc. Natl. Acad. Sci.USA 72:3144.

Anderson, G. R., Marotti, K. R., and Whitacker-Dowling, P. A., 1979, A candidate rat-specific gene product of the Kirsten murine sarcoma virus. Virology 99:31.

Balk, S. D., Plimeni, P. I., Hoon, B. S., LeStourgeon, D. N., and Mitchen, R. S., 1979, Proliferation of Rous sarcoma virus-infected, but not of normal, chicken fibroblasts in a medium of reduced calcium and magnesium concentration. Proc. Natl. Acad. Sci. USA 76:3913.

Barnekow, A., Boschek, C. B., Ziemiecki, A., and Bauer, H., 1980, Detection of the src-gene product $pp60^{src}$ and its associated protein kinase on the surface of Rous-sarcoma virus-transformed cells.Biochem. Soc. Transactions 8:735.

Barnekow, A., Bauer, H., Boschek, C. B., Friis, R. R., and Ziemiecki, A., 1981, Rous sarcoma virus transformation: Action of the src gene product, in: "International Cell Biology 1980/81" (H. G. Schweiger, ed.), Springer-Verlag, Berlin, Heidelberg, New York, pp. 457.

Bauer, H., Hayami, M., Ignjatovic, J., Rübsamen, H., Graf, T., and Friis, R. R., 1979, On the origin and in vivo immunogenicity of avian sarcoma cell surface antigens. in: "Avian RNA-Tumor Viruses" Piccin-Medical Books. (S. Barlati, C. De Guili-Morghen, eds.), pp. 252.

Bauer, H., and Fleischer, B., 1980, The immunobiology of avian RNA tumor virus-induced cell surface antigens, in: "Mechanism of Immunity to Virus-Induced Tumors" Marcel Dekker, Inc., New York, in press

Bauer, H., and Yoshikawa, 1980, Oncofetal antigens as markers for retrodifferentiation in malignant transformation. in: "Cold Spring Harbor Conferences on Cell Proliferation" Vol. 7, pp.1231.

Becker, D., Kurth, R., Critchley, D., Friis, R. R., and Bauer, H., 1977, Distinguishable transformation defective phenotypes among temperature sensitive mutants of Rous sarcoma virus. J. Virol. 21:1042.

Bishop, J. M., 1978, Retroviruses. Ann. Rev. Biochem. 47:35.

Blomberg, J., Reynolds jr.. F. H., Van de Ven, W. J. M., and Stephenson, J. R., 1980, Abelson murine leukemia virus transformation involves loss of epidermal growth factor-binding sites. Nature 286:504.

Boettiger, D., Roboy, J., Brumbaugh, J., Biehl, J., and Holtzer, H., 1977; Transformation of chicken embryo retinal melanoblasts by a temperature-sensitive mutant of Rous sarcoma virus. Cell 11:881.

Boschek, C. B., Jockusch, B. M., Friis, R. R., Back, R., and Bauer, H., 1979, Morphological alterations to the cell surface and cytoskeleton in neoplastic transformation. Beitr. elektronenmikroskop. Direktabb. Oberfl. 12:47.

Boschek, C. B., Jockusch, B. M., Friis, R. R., and Back, R., 1981, Early changes in the distribution and organization of microfilament proteins during cell transformation. Cell 24: in press

Brugge, J. S., and Erikson, R. L., 1977, Identification of a transformation-specific antigen induced by an avian sarcoma virus. Nature 269:346.

Brugge, J. S., Erikson, E., Collett, M. S., 1978, Peptide analysis of the transformation-specific antigen from avian sarcoma virus-transformed cells. J. Virol. 26: 773.

Calothy, G., and Pessac, B., 1976, Growth stimulation of chick embryo neuro-retinal cells infected with RSV: relationship to viral replication and morphological

transformation. Virology 71:336.
Carpenter, G., and Cohen, S., 1979, Epidermal growth factor. Ann. Rev. Biochem. 48:193.
Carroll, R. C., Ash, J. F., Vogt, P. K., and Singer, S. J., 1978, Reversion of transformed glycolysis to normal by inhibition of protein synthesis in rat kidney cells infected with temperature-sensitive mutants of Rous sarcoma virus. Proc. Natl. Acad. Sci. USA 75:5015.
Cohen, S., 1976; Transformation by murine and feline sarcoma viruses specifically blocks binding of epidermal growth factor to cells. Nature 264;26.
Collett, M. S., and Erikson, R. L., 1978, Protein kinase activity associated with the avian sarcoma virus src gene product. Proc. Natl. Acad. Sci. USA 75:2021.
Collett, M. S., Brugge, J. S., and Erikson, R. L., 1978, Characterization of a normal avian cell protein related to the avian sarcoma virus transforming gene product. Cell 15:1363.
Collett, M. S., Brugge, J. S., Erikson, R. L., Lau, A. F., Kryzek, R. A., and Faras, A. J., 1979, The src gene product of transformed and morphologically reverted ASV-infected mammalian cells. Nature 281:195.
Cold Spring Harbor Symposis on Quantitative Biology Vol. 44, 1980.
Cooper, G. M., Okenquist, S., and Silverman, L., 1980, Transforming activity of DNA of chemically transformed and normal cells. Nature 286:418.
Courtneidge, S. A., Levinson, A. D., and Bishop, J. M., 1980, The protein encoded by the transforming gene of avian sarcoma virus ($pp60^{src}$) and a homologous protein in normal cells (pp60 proto-sarc) are associated with the plasma membrane. Proc. Natl. Acad. Sci. USA 77:3783.
Critchley, D. R., 1979, Glycolipids as membrane receptors important in growth regulation, in: "Surfaces of Normal and Malignant Cells" (R. O. Hynes, ed.), John Wiley and Sons, Chichester, 1979, pp.63.
Durham, A. C. H., 1978, The roles of small ions, especially calcium, in virus disassembly, takeover, and transformation. Biomedicine 28:307.
Edelmann, G. M., and Yahara, I., 1976, Temperature-sensitive changes in surface modulation assembles of fibroblasts transformed by mutants of Rous sarcoma virus. Proc. Natl. Acad. Sci. USA 73:2047.
Erikson, E., Collett, M. S., and Erikson, R. L., 1978, In vitro synthesis of a functional avian sarcoma virus transforming-gene product. Nature 274:919.
Erikson, R. L., Collett, M. S., Erikson, E., and Purchio, A. F., 1979, Evidence that the avian sarcoma virus transforming product is a cAMP-independent protein kinase. Proc. Natl. Acad. Sci. USA 76:6260.

Erikson, E., and Erikson, R. L., 1980, Identification of a cellular protein substrate phosphorylated by the avian sarcoma virus-transforming gene product. Cell 21:829.

Fiszman, M. Y., and Fuchs, P., 1975, Temperature-sensitive expression of differentiation in transformed myeloblasts. Nature 254:429.

Friis, R. R., Schwarz, R. T., and Schmidt, M. F. G., 1977, Phenotype of Rous sarcoma virus-transformed fibroblasts. An argument for a multifunctional src gene product. Med. Microbiol. Immunol. 164:155.

Friis, R. R., 1978, Temperature-sensitive mutants of avian RNA tumor viruses: A review. Current Topics in Microbiol. and Immunol. Vol. 79:261.

Friis, R. R., Jockusch, B. M., Boschek, C. B., Ziemiecki, A., Rübsamen, H., and Bauer, H., 1980, Transformation-defective, temperature-sensitive mutants of Rous sarcoma virus have a reversibly defective src-gene product. Cold Spring Harbor Symp. on Quant. Biol. Vol. 44:1007.

Gelderblom, H., Bauer, H., and Graf, T., 1972, Cell surface antigen induced by avian RNA tumor viruses: Detection by immunoferritin technique. Virology 47:416.

Holtzer, H., Biehl, J., Yeoh, G., Meganathan, R., and Kaji, K., 1975, Effect of oncogenic virus on muscle differentiation. Proc. Natl. Acad. Sci. USA 72:4051.

Hunter, T., Sefton, B. M., 1980, Transforming gene product of Rous sarcoma virus phosphorylates tyrosin. Proc. Natl. Acad. Sci. USA 77:1311.

Hynes, R. O., 1979, Proteins and glycoproteins, in: "Surfaces of Normal and Malignant Cells", (R. O. Hynes, ed.), John Wiley and Sons, Chichester, 1979, pp. 103.

Ignjatovic, J., Rübsamen, H., Hayami, M., and Bauer, H., 1978, Rous sarcoma virus-transformed avian cells express four different cell surface antigens that are distinguishable by a cell-mediated cytotoxicity-blocking test. J. Immunol. 120:1663.

Johnson, M. A., and Weber, M. J., 1979, Potassium fluxes and ouabain binding in growing, density-inhibited and Rous sarcoma-virus transformed chicken embryo cells. J. Cell. Physiol. 101:89.

Kamine, J., and Rubin, H., 1977, Coordinate control of collagen synthesis and cell growth in chick embryo fibroblasts and the effect of viral transformation on collagen synthesis. J. Cell. Physiol. 92:1.

Karess, R. E., Hayward, W. S., and Hanafusa, H., 1979, Cellular information in the genome of recovered avian sarcoma virus directs the synthesis of transforming proteins. Proc. Natl. Acad. Sci. USA 76:3154.

Krueger, J. G., Wang, E., and Goldberg, A. R., 1980a, Evidence that the src gene product of Rous sarcoma virus is membrane associated. Virology 101:25.

Krueger, J. G., Wang, E., Garber, E. A., and Goldberg, A. R., 1980b, Differences in intracellular location of $pp60^{src}$ in rat and chicken cells transformed by Rous sarcoma virus. Proc. Natl. Acad. Sci. USA 77:4142.

Kryceve, C., Vigier, P., and Barlati, S., 1976, Transformation enhancing factor(s) produced by virus-transformed and established cells. Int. J. Cancer 17:370.

Kurth, R., and Bauer, H., 1972, Cell surface antigens induced by avian RNA tumor viruses: Detection by a cytotoxic microassay. Virology 47:426.

Lawrence, D. A., and Jullien, P., 1980, Hexose uptake enhancing factor released from Rous sarcoma cells. J. Cell. Physiol. 102:245.

Levinson, A. D., Oppermann, H., Levintow, L., Varmus, H. E., and Bishop, J. M., 1978, Evidence that the transforming gene of avian sarcoma virus encodes a protein kinase associated with a phosphoprotein. Cell 15:561.

Means, A. R., and Dedman, J. R., 1980, Calmodulin - an intracellular calcium receptor. Nature 285:73.

Moyer, M. P., Garry, R. F., Moyer, R. C., and Waite, M. R. F., 1980, Intracellular ion concentrations and the transformed phenotype. Europ. J. Cell Biol. 22:526.

Nicolson, G. L., 1976a, Transmembrane control of the receptors on normal and tumor cells. I. Cytoplasmic influence over cell surface components. Biochim. Biphys. Acta 457:57.

Nicolson, G. L., 1976b, Trans-membrane control of the receptors on normal and tumor cells. II. Surface changes associated with transformation and malignancy. Biochim. Biophys. Acta 458:1

Pacifici, M., Boettiger, D., Roby, K., and Holtzer, H., 1977, Transformation of chondroblasts by Rous sarcoma virus and synthesis of sulfated proteoglycan matrix. Cell 11: 981.

Poste, G., and Flood, M. K., 1979, Cells transformed by temperature-sensitive mutants of avian sarcoma virus cause tumors in vivo at permissive and nonpermissive temperatures. Cell 17:789.

Preskott, D. M., 1976, The cell cycle and the control of cellular reproduction. Adv. Genetics 18:99.

Presek, P., Glossmann, H., Eigenbrodt, E., Schoner, W., Rübsamen, H., Friis, R. R., and Bauer, H., 1980, Similarities between a phosphoprotein (pp60src)-associated protein kinase of Rous sarcoma virus and a cylic adenosine 3':5'-monophosphate-independent protein kinase that phosphorylates pyruvate kinase type M_2. Cancer

Res. 40:1733.

Quigley, J. P., 1979, Proteolysis enzymes of normal and malignant cells, in: "Surface of Normal and Malignant Cells" (R. O. Hynes, ed.), John Wiley and Sons, Chichester, 1979, pp.247.

Radke, K., Gilmore, T., and Martin, G. S., 1980, Transformation by Rous sarcoma virus: a cellular substrate for transformation-specific protein phosphorylation contains phosphotyrosine. Cell 21:821.

Roby, K., Boettiger, D., Pacifici, M., and Holtzer, H., 1976, Effects of Rous sarcoma virus on the synthesis programs of chondroblasts and retinal melanoblsts. Am. J. Anat. 147:401.

Rohrschneider, L. R., Eisenman, R. N., Leitch, C. R., 1979, Identification of a Rous sarcoma virus transformation-related protein in normal avian and mammalian cells. Proc. Natl. Acad. Sci. USA 76:4479.

Rohrschneider, L. R., 1980, Adhesion paques of Rous sarcoma virus-transformed cells contain the src gene product. Proc. Natl. Acad. Sci. USA 77:3514.

Rowe, D. W., Moen, R. C., Davidson, J. M., Byers, P. H., Bornstein, P., and Palmiter, R. D., 1978, Correlation of procollagen mRNA levels in normal and transformed chick embryo fibroblasts with different rates of procollagen synthesis. Biochem. 17:1581.

Rubin, H., 1970, Overgrowth stimulating factor released from Rous sarcoma virus cells. Science 167:1271.

Rübsamen, H., Friis, R. R., and Bauer, H., 1979, Src gene product from different strains of avian sarcoma virus: Kinetics and possible mechanism of that inactivation of protein kinase activity from cells infected by transformation-defective, temperature-sensitive mutant and wild-type virus. Proc. Natl. Acad.Sci. USA 76:967.

Sefton, B. M., Hunter, T., Beemon, K., and Eckhart, W., 1980, Evidence that the phosphorylation of tyrosine is essential for cellular transformation by Rous sarcoma virus. Cell 20:807.

Singer, S. J., 1974, Molecular biology of cellular membranes with applications to immunology. Adv. Immunol. 19:1.

Stehelin, D., Varmus, H. E., Bishop, J. M., and Vogt, P. K., 1976, DNA related to the transforming gene(s) of avian sarcoma viruses is present in normal avian DNA. Nature 260:170.

Todaro, G. J., De Larco, J. E., and Cohen, S., 1976, Transformation by murine and feline sarcoma viruses specifically blocks binding of epidermal growth factor to cells. Nature 264:26.

Todaro, G. J., and De Larco, J. E., 1978, Growth factors produced by sarcoma virus-transformed cells. Cancer Res.

38:4147.

Wang, E., and Goldberg, A. R., 1976, Changes in microfilament organization of surface topography upon transformation of chick embryo fibroblasts with Rous sarcoma virus. Proc. Natl. Acad. Sci. USA 73:4065.

Wang, L.-H., Snyder, P., Hanafusa, T., and Hanafusa, H., 1980, Evidence for the common origin of viral and cellular sequences involved in sarcomagenic transformation. J. Virol. 35:52.

Weber, M. J., 1973, Hexose transport in normal and in Rous sarcoma virus-transformed cells. J. Biol. Chem. 248: 2978.

Weber, M., and Friis, R. R., 1979, Dissociation of transformation parameter using temperature-conditional mutants of Rous sarcoma virus. Cell 16:25.

Willingham, M. C., Jay, G., and Pastan, I., 1979, Localization of the ASV src gene product to the plasma membrane of transformed cells by electron microscopic immunocytochemistry. Cell 18:125.

Yoshikawa, Y., Ignjatovic, J., and Bauer, H., 1979, Tissue-specific expression of onco-fetal antigens during embryogenesis. Differentiation 15:41.

Ziemiecki, A., and Friis, R. R., 1980, Phosphorylation of $pp60^{src}$ and the cycloheximide insensitive activation of the $pp60^{src}$-associated kinase activity of transformation-defective temperature-sensitive mutants of Rous sarcoma virus. Virology 106:391.

MOLECULAR MECHANISMS OF THE CONTROL OF CELL GROWTH IN CANCER

Arthur B. Pardee
Department of Pharmacology
Harvard Medical School

and

Laboratory of Tumor Biology
Sidney Farber Cancer Institute
44 Binney Street
Boston, Massachusetts 02115

ON THE NATURE OF CANCER

Most cells *in vivo* are in a quiescent state. They metabolize and perform various functions. In adults they do not generally divide unless they are to be replaced. Some cells practically never divide (brain cells) or are terminally differentiated and cannot divide (red cells). Others divide quite frequently (skin, intestine, bone marrow cells); some divide after an insult such as wounding. Cells thus contain regulatory mechanisms that control their capacity to divide.

Cancer exhibits itself *in vivo* as one or more masses of cells growing improperly for host function (1). Cancer cells divide at inappropriate times, responding even when the appropriate growth stimuli are not presented. They do not necessarily grow unusually rapid but rather seem to have relaxed their growth controls. Since cancer cells are generally clonal, coming from one original cell, some modification of an inhereited control mechanism seems to be at the basis of uncontrolled growth. Malignancy of course involves other further changes in cell properties, especially ability to grow in wrong places (see Pitot (2) for an overview of the fundamentals of cancer). The general supposition is that several fundamental changes have occurred to create the original tumor cell, permitting abnormal growth in time and space (1). Tumors can be investigated *in vivo* for these changes by methods such as histology, measurement of mass, biochemistry, and lethality. Difficulties of controlling conditions of growth in vivo, among other factors, have made many investigators turn to cells in culture to seek a fundamental basis of cancer (3, 4, 5, 6).

For many years studies were carried out on tumor growth in animals. Spontaneous tumors are hard to study because their production is not predictable, and each tumor tends to differ from others. Transplantable tumor lines were developed, by mincing pieces of existing tumors and injecting them into other animals of the same genetic background. These tumors could grow and be transplanted for indefinite periods of time; one such tumor, the Erlich Ascites, has been passaged from mouse to mouse continually since the late 1800's.

Transplantable tumors will kill the host, in a time that depends on the tumor type. For example, Morris rat hepatoma lines differ widely in the time required to kill the host, from weeks to years (7). Tumor lines are used extensively for testing the efficacy of drugs for chemotherapy, usually with mice as hosts. With characterized lines such as L1210 leukemia, a standard intraperitoneal inoculum of 10^6 cells is lethal in about 7 days (8). Many studies have been carried out on the metabolic properties of established tumor lines. Correlations have been developed between the growth rate of tumor lines and the activity of various metabolic pathways such as the glycolytic pathway (9). With respect to normal tissue, definite enzyme changes have been discovered in tumor lines; however, these enzymatic changes have not as yet provided useful leverage for chemotherapy protocols.

One of the problems with transplantable tumor lines is that the faster growing cells are constantly being selected as the tumor is passed from one animal to the next. Eventually the line is metabolically stripped down, and all non-essential properties are lost; the line bears little resemblance to a cell in a tumor. Examples include the human HeLa cell, and most of the classical transplantable rat lines (7). An important priority is research on cancer thus is to decide which cancer to study. Spontaneous tumors differ from one another in many ways, including growth rates, immunology, morphology, histology, metastastic ability, and lethality. The tissue and cell of origin may greatly influence the properties of a tumor as well as of the cells of which it is composed. The same tissue can produce tumors of very different properties as in Morris hepatomas (7), which differ in vivo in growth rate, karyotype, biochemistry, and so forth.

We will limit this chapter to a discussion of basic ideas about cell growth control, and its defects in cancer cells. The work to be described has been mostly done with cell lines in tissue culture. This is because growth conditions in culture are readily controlled. These lines possess regulatory mechanisms that can be lost upon "transformation" to the state which permits them to form tumors in animals. They must, however, be considered only as models for tumor cells since their culturing in vitro changes many of their properties (1). Until very recently most of this work has been with fibroblastic cells.

Cells can be re-implanted only into hosts that do not reject them. The "nude" mouse has been developed as a general host with very low immunological activity against foreign cells. It is possible to inject foreign cells from humans, chickens, etc. into these animals, and learn whether they will grow (as do tumor cells) or not (normal cells). The nude mouse test appears to provide one of the best criteria to determine whether or not a given cell is malignant (10).

CANCER AS A CELLULAR DISEASE

Tumors have been shown to be clonal. That is, malignant cells within a tumor are the progeny of a single cell. This original cell apparently had lost the property of being controlled by its environment. These fundamental considerations lead one to consider cancer as a disease that has its basis in the properties of individual cells. We are thus interested in questions such as:

a) How does a cell grow?

b) By what mechanism is cell growth controlled?

c) What changes in growth pattern and control do we observe in cancer vs. normal cells?

d) Can we take advantage of such differences for treatment?

Cell Growth Properties in Culture

The inability to manipulate the growth conditions of cells within an animal, host, and lack of convenience have led to development of in vitro systems for studying the growth of cells (tissue culture). Many types of cells cannot grow in culture, but more and more varieties are being cultured as methods are developed for them. Most tissue culture work is done with fibroblasts, cells that grow out of a tissue sample and take over the culture. In early passages, fibroblasts appear to retain their original properties and, in general, lose differentiated functions. Nevertheless these lines show properties of growth control, and tumor cells derived from them have a variety of properties that are different from those of normal cells (4,5; Fig. 1).

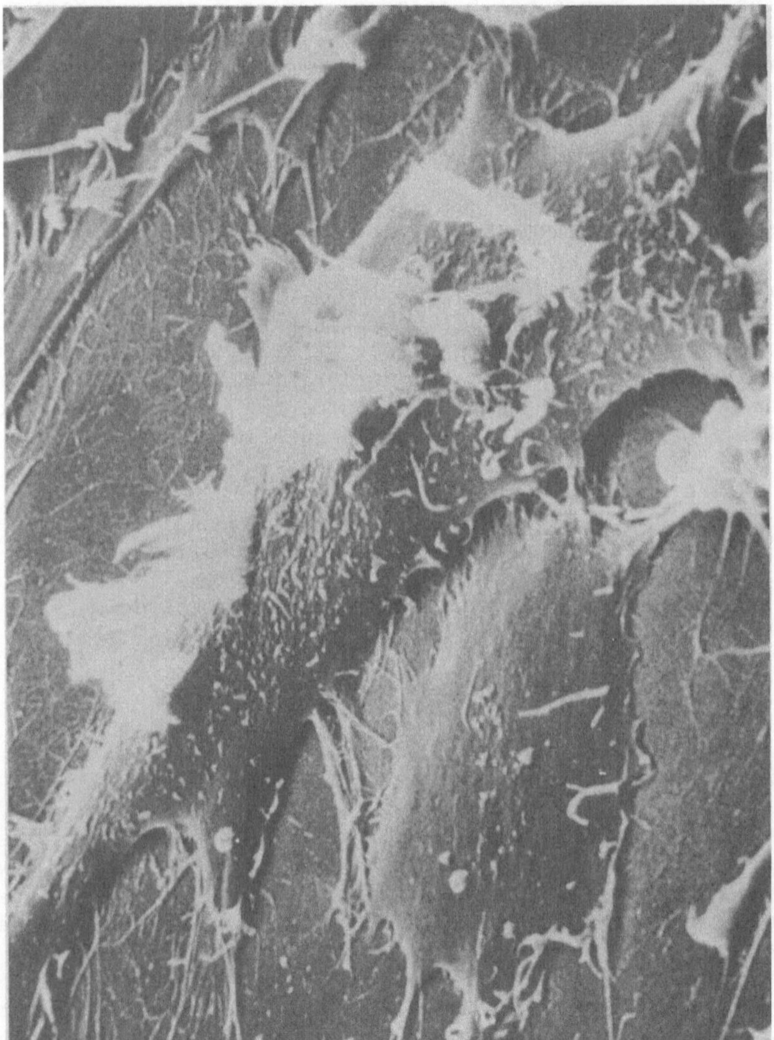

Fig. 1. Cells in culture. This is a scanning electron micrograph of chick cells transformed by Rous virus, growing on a plastic surface (courtesy of Dr. L.B. Chen).

Fibroblasts in culture (chick, rat, mouse, hamster, and human are all used) generally require a substratum of glass or treated plastic on which to grow. Normal (non tumor-forming) cells generally cannot grow unless they are able to attach to a solid surface and spread out. By contrast, tumor-forming cells that have been obtained by transformation of normal cells with viruses or chemicals usually can grow in suspension. In addition to differences in this so-called "anchorage" requirement for growth, the morphology of normal and transformed cells also usually differs strikingly. The normal cells grow as a monolayer until they cover the available surface, and then stop growing. Transformed cells continue to grow after confluence is reached, and can form multilayers in which the cells criss-cross over one another. Thus, transformation often is seen to result in loss of a "density-dependent" inhibition of growth.

A confluent monolayer of normal cells can be scratched to remove some cells locally. Cell proliferation occurs in this "wound", but at the same time cells in the remainder of the culture are density inhibited. This experiment shows that growth control is local and not a function of the entire culture.

In order for cells to grow they need to be supplied with a complex medium. The medium contains essential amino acids, vitamins, minerals, glucose, and often other factors such as purines and pyrimidines. Importantly, the cells require serum, obtained as a rule from clotted calf blood, although other sera are used. Normal cells need a high serum concentration (10%) for optimal growth, but transformed cells usually can grow well on much less serum (1% or less). Serum serves as a source of growth factors. Some of these factors are known, such as insulin and transferrin, but most are unknown. Several new factors have been discovered in the past few years, including fibroblast growth factor, epidermal growth factor, and a factor from platelets. All of these are proteins of rather small size (a few 10-thousands molecular weight). So far, only a few established cell lines grow with serum totally replaced by growth factors. Apparently tumor cells in culture lose some of their growth factor requirements.

There are a few cases where drugs differently affect the growth of normal and transformed cells. For example, caffeine was shown to arrest the growth of baby hamster kidney cells (BHK) reversibly, but not to affect the cells that had been transformed with polyoma, a DNA tumor virus (11).

The Cell Cycle

The well-known cell cycle consists of 4 parts. Following production of new cells by division, rapidly growing cells enter the G_1 period during which protein and RNA synthesis occurs. Then after a few hours they usually start DNA synthesis, and by definition enter the S period (S for "synthesis"). After the S period of about 6-8 hours they stop DNA synthesis and enter the G_2 period which lasts a few hours. Finally, the cells go into mitosis (M period), which usually lasts less than an hour, and culminates with separation of the daughter cells. It is clear that this description is very primitive. Many different metabolic events occur during each of these period, because the entire cell must replicate itself. Each period contains events which seem to be necessary for the next. Thus, if DNA synthesis is blocked, as with hydroxyurea, the cells do not proceed to mitosis; or if protein synthesis is blocked in G_1 with cycloheximide, cells do not reach the beginning of S period (Fig. 2).

Most practical chemotherapy depends on drug susceptibility of cells at particular stages of the cycle. Of the 20 or so widely used anticancer drugs, most affect DNA synthesis and hence are active only on cells that are in the S period. A few, such as vincristine and vinblastine, affect the cells in M. Some drugs damage non-replicating DNA (bleomycin, adriamycin, alkylating agents). These cells continue DNA synthesis if they are stimulated, but do not go on to mitosis. The cycle-specific drugs are effective against growing cells because, as will be discussed below, resting cells are generally in G_1 (or G_0), not in S or M. The drugs that damage DNA show the most promise for killing slowly growing tumors, since they are not dependent on the cycle and on cell growth.

Under suboptimal conditions in culture or at high density, cells stop growing. Normal cells grown to high density or to arrest with low serum, or in the presence of some drugs (see below) or placed in suspension (12) are mainly arrested with a G_1 DNA content. (See Gelfant (13) for a discussion of other arrest points.) There has been considerable discussion over the years as to whether this arrested state is different from some part of G_1 (4, 5). G_0 cells have recently been distinguished from cells in phases of the cycle using flow microfluorimetry (14, 14a, 14b) image analysis (14c). For a complete and updated review on the non-cycling cells see also the chapter by Nicolini, et al. in this book. This specificity of arrest suggests that there is a growth control mechanism responsive to the environment that acts at some part of G_1. If conditions are appropriate for growth, the control mechanism allows the cells to proceed for another round. If not, the cells go into a holding state, generally called G_0. In this state the cells are not susceptible to the cycle-dependent antineoplastic drugs.

Signals for growth come from outside the cell. There are three main ideas about density-control, not necessarily exclusive. 1) Crowded cells have difficulty taking up some necessary growth factors or nutrients. 2) Crowded cells have difficulty secreting

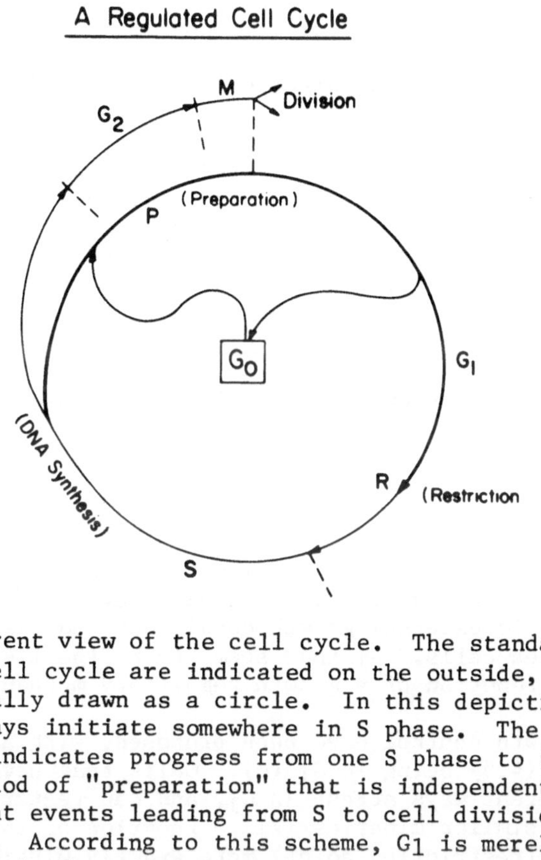

Fig. 2. A current view of the cell cycle. The standard phases of the cell cycle are indicated on the outside, and are generally drawn as a circle. In this depiction, parallel pathways initiate somewhere in S phase. The interior line indicates progress from one S phase to the next through a period of "preparation" that is independent of the concurrent events leading from S to cell division (the exterior line). According to this scheme, G_1 is merely the time interval after division and before DNA synthesis starts again. The restriction point for control is located before S.

inhibitory substances. 3) Cell contact directly inhibits cell growth (true density dependent inhibition). Recent experiments show that isolated membranes in contact with cells can block growth. Cultures arrest in low-serum medium; they can be made to grow by adding serum or, for some lines, purified hormones. There is great interest in studying hormones that regulate growth.

Emergence of cells from the G_0 state is more sensitive to some drugs including histidinol (15) and propargyl glycine (16) than is continued growth of an exponential cell population. This result, together with a time longer than G_1 for cells to go from G_0 to S, and the special requirement of platelet factor to permit cells in G_0 to emerge into the cycle (17), all suggest that special biochemical events are required to leave the quiescent state. The requirement for platelet factor is abolished by infection with SV40 (17). Little work seems to have been done on the effects of RNA virus or chemical transformation on emergency of cells from the resting state. This process could be very important in cancer, viewed as excess entry of new cells into growth relative to replacement of lost cells (18).

A Growth Control Point in the Cell Cycle

Dividing cells go through the classical cycle stages G_1, S, G_2 and M; for the 3T3 mouse line these stages take 5, 9, 3 and 0.8 hours respectively (19; Fig. 3). The last three stages are fairly constant under different conditions, but G_1 varies from cell to cell in a culture over a 2-3 fold range (20). Furthermore, the average G_1 is greatly extended under adverse growth conditions, but the rest of the cycle is not changed much. The basis for variability of G_1 is not clear. We surmise a biochemical event in G_1 that is highly responsive to external growth conditions and also to differences between cells. G_1 thus seems to be the likely part of the cycle for function of a growth regulatory mechanism.

Points of growth control have been proposed, with a major growth regulatory event (4,5) being in G_1 (6). Cells that have been arrested in different ways arrest in G_1, and can restart their growth with very similar kinetics (21). Location of this putative regulatory event (restriction point) more exactly within G_1 is difficult because there are not "landmarks" to subdivide this part of the cycle. One approach suggests that it is a couple of hours before the beginning of DNA synthesis (S phase) (19).

The G_1 phase of the animal cell cycle, as viewed from a biochemical standpoint, is not well understood. Yet it may well be the most important determinant affecting regulation of growth. The single major event which causes cells to continue to grow (when they are faced with a choice) occurs in G_1. This control event, in general efficient for normal cells, appears ineffective for at least some neoplastic cells.

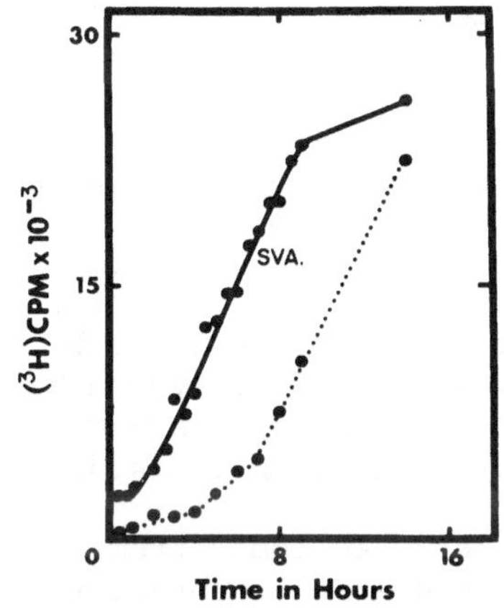

Fig. 3. Inhibition of emergency of cells from G0. This experiment shows that Chinese hamster ovary cells (CHO) are blocked from initiating DNA synthesis when streptovitacin A (SVA), an inhibitor of protein synthesis, is added initially, but not if the inhibitor is added later at the time DNA synthesis commences. Cultures were put into the G0 state by starving them for isoleucine for 36 hr. Growth was then initiated by putting them into complete medium containing ^{3}H-thymidine. One set of cultures was assayed at intervals for ^{3}H-thymidine incorporation to measure DNA synthesis. S started around 6 hr (•••●•••). The other line (——●——) represents effects of adding a low concentration of SVA at the times shown and then collecting all cultures at 25 hrs and determining the amounts of ^{3}H-thymidine incorporated. DNA synthesis was almost completely blocked if SVA was added at 0 hr. But if SVA was added later, e.g. at 8 hr., there was almost no inhibition. Thus, early events after G0 are particularly sensitive to modest inhibition of protein synthesis.

We conceive that release from growth restriction of animal cells depends upon: 1) an event in G_1; 2) a diffusible substance in the cytoplasm; and 3) synthesis of a protein with a fairly rapid turnover rate. We will summarize experiments which relate to the proposed model. Some background information in support of this hypothesis is presented first.

The regulatory event's location in the cycle. Many observations have been made regarding that part of the cycle in which cells are found at rest; these data have recently been discussed in books and reviews (6). Most results report that arrested cells have the G_1 complement of DNA; but a few find arrested cells in G_2. In general, no notice was taken as to whether the cells were normal or transformed. As commented on below, the restriction event can be lost in at least some transformed cells (11); hence these cells would be arrested in parts other than G_1 of the cycle. The cells arrested with G_1 DNA content are proposed to be in a resting state G_0 (4) different from any stage of G_1. There has been a long controversy as to whether resting cells are really in a different state or are simply slowly traversing G_1 (4,5).

Evidence for the latter view comes from observations that all the cells of a quiescent population eventually become labeled with tritiated thymidine (^{3}H-TdR) (22). Smith and Martin (23) have proposed a model in which all cells during G_1 go into a state A in which they remain for an indefinite time. They emerge probabilistically, that is, with a constant probability per unit time. Conditions that permit only very slow proliferation are, according to this model, ones that make the probability of emerging very low, and so cells stay in state A for long periods. They therefore equate the G_0 state (or rather replace it) with an A state of very long duration (on the average), and one which is common to all cells that enter G_1.

We know that a resting cell differs in a dozen ways from the average cell traversing G_1. These differences are well summarized by Baserga (4). Even in the Smith and Martin model (23), quiescent A cells have a very low probability per unit time of progressing to the B condition (the rest of the cell cycle). In contrast, growing cells have a high probability of passing from A to B. Therefore, there must be biochemical differences during the A states between quiescent and proliferating cells which are responsible for these different transition probabilities. The ability of all quiescent cells to take up thymidine eventually could be attributed to repair of DNA damage and to cell death and replacement over long intervals.

At present one cannot identify any portion of G_1 in which the cells have the same biochemical properties as resting cells. This is partly because synchrony of G_1 cells is so poor that a restricted sub-portion of G_1 cells cannot be isolated, and because one does not know what to look for in the relevant subpopulation in G_1. As a practical matter, we as biochemists maintain that resting cells have different biochemical properties than do growing cells at any measurable stage of the cycle. We identify the state of these cells as "G_0". Cells in G_0 are considered to have left the cycle at some point in G_1. When restored to growth conditions, they re-enter the cycle in G_1. This G_0 is much like the G_{1B} state of Temin (24).

Augenlicht and Baserga (25) have shown that as cells are kept in non-proliferation (low serum) conditions, they sink into states of greater quiescence as measured by the time required for them to recover. Therefore, G_0 is probably not one condition; rather, the cells might initially enter a first "G_0" and then gradually change further, perhaps because of protein and RNA degradation. If so, then the question recurs as to whether this initial G_0 is indeed different from the state of any cells in G_1. We consider an initial G_0 to exist because kinetic and transport differences appear shortly after cells stop growing, and persist for several days (26).

This may seem semantic in character. Yet an alternative quiescent state (G_0) is important in depicting the restriction point model that we use. As a basis for thinking about cell cycle control, we visualize a single growth event which decides between alternative possibilities -- growth or quiescence.

The restriction point's properties. It is thus important to study the biochemical nature of this regulatory event in the cycle. We propose this regulatory event occurs at a specific "metabolic place" in G_1; i.e. each cell carries out certain early G_1 processes and then must undergo a special event at the restriction point in order to carry out further biochemical processes leading to DNA synthesis (21). If the cell cannot carry out this restriction point event, it passes off into the G_0 state. This event can be considered a sort of "switch" at which a choice is made between proliferation and quiescence. A variety of conditions that control growth appear to accomplish the same restriction event. That is, only one control point exists and is responsive to many growth-restrictive conditions.

Some sort of crucial G_1 event was also suggested earlier (see Baserga (4) and Prescott (5)). Temin's experiments (24) showed serum to be dispensable to the growth of chick fibroblasts about halfway through G_1; he suggested that all serum-dependent events were accomplished at that time. Hershko, et al. (27) proposed a pleiotypic control, meaning that when cells are arrested in a variety of ways, they always show several characteristic changes including slower transport of some nutrients, diminished macromolecular syntheses, and enhanced protein breakdown. By analogy with the stringent response of bacterial RNA synthesis to amino acid deficiences, Hershko, et al. proposed that these effects were mediated by a pleiotypic modulator compound of low molecular weight. This putative compound has yet to be identified. Smith and Martin's concept (23) comes close to the idea of a restriction point; they propose a point (in G_1) in which all cells under all conditions become arrested and from which they escape probabilistically, either rapidly or slowly, depending on growth conditions.

Lower organisms also show a cell cycle specific control event prior to DNA synthesis initiation. Hartwell, et al. (28) did outstanding work in showing that yeast are arrested at a specific point, under conditions of poor nutrition or when they are exposed to mating factor peptide made by the opposite mating type. A temperature sensitive mutant arrested at this event has been isolated. In Bacilli there is also a critical event at which the decision is made as to whether the bacteria continues vegatative growth or forms a spore (29). Shilo, et al. has done careful -inetic studies with yeast to show that the variability in reaching S from the resting state is not after the startup process, but occurs in it (30).

A strong indication of some special cycle control event in G_1 is seen in studies of variability of cycle duration. Several workers have demonstrated that most of the cycle variability lies in G_1 (see Baserga (4) and Prescott (5)). This variability is observed with exponentially growing cells (31), with cells synchronized so as to start cycling at mitosis (M) (32), and is similarly variable (32) with cells started from G_0 (20, 33). Growth is slower and more variable in low serum, mostly because the G_1 period is extended through greater variability beyond a minimal time (20; Fig. 4).

There are two main explanations for G_1 variability. One is the notion of cell differences. Newly born cells in a population are different; their G_1 durations could reflect this difference. This notion gains support from observations that cells differ in size several-fold (34). Although in the two-thirds majority of larger cells, cycle times are independent of size, in the smaller third of the population G_1 durations are increasingly longer as smaller cells are observed (35, 36). More subtle changes than gross size could be responsible for variabilities of the majority of cells. The alternative model is that proposed by Smith and Martin (23): that all cells of a population are basically identical under any given condition of growth, and that G_1 variability is probablistic. In other words, prior to the event there is no way of predicting the duration of G_1 for any given cell. In either case, these questions focus on G_1 as the part of the cycle within a population most subject to variability.

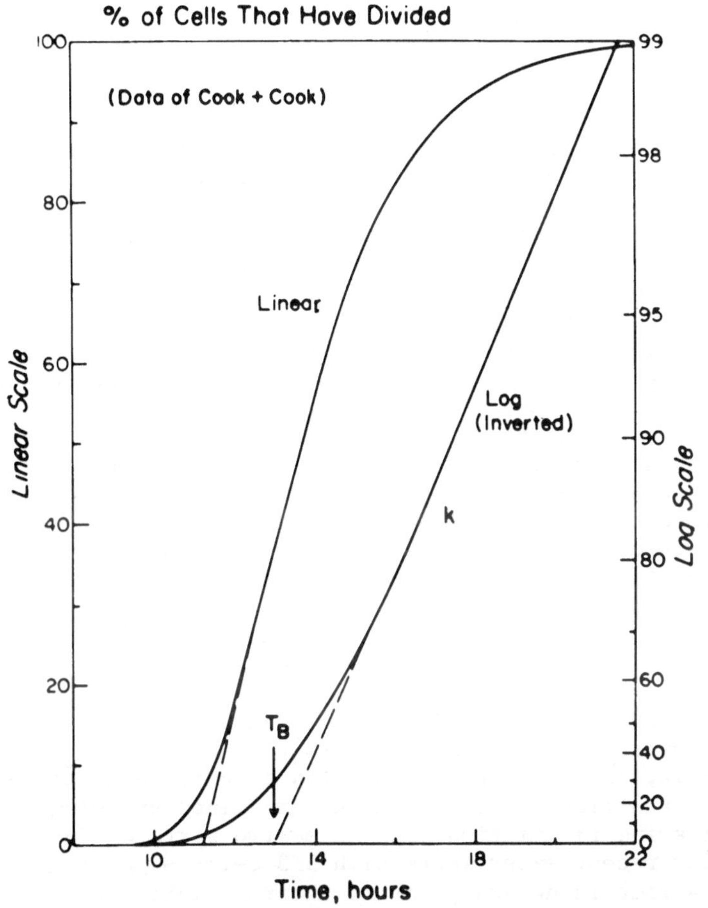

Fig. 4. Distribution of durations of the cell cycle. Cell cycle durations of a large number of Euglena cells were determined by Cook and Cook (see 38). The number of cells that had divided by the time shown on the abscissa is on the ordinate. Note that the shortest cycle duration was about 11 hr and the longest nearly twice that. The same data are plotted on a log scale (an inversion of the α plot of Smith and Martin, 23). Note linearity of most of this log plot, from which Smith and Martin inferred that variability is due to a probablistic event. But the initial 50% of the cells do not lie on this line, and extrapolation indicates a minimal time of about 13 hr, vs. 11 hr for the linear plot.

Various kinetic data are quantitatively fairly consistent with Brooks, et al's newer model for animal cells (37) and for yeast (Shilo, et al., 30). However, the data are equally well fitted by other than probablistic distribution functions such as the Normal and Log Normal (38). New approaches are needed to choose between the bases for variability of G_1 (38a).

Our proposal depends on the idea that specific biochemical changes occur at the restriction point if the cell is to continue to proceed through the cycle; they do notoccur when the cell is destined to become quiescent. To identify these changes experimentally we must identify as precisely as possible the time in G_1 at which these changes should occur. The experiments of Temin (24) cited above suggest it to be near the middle of G_1 in chick cells initiated from a serum-starved G_0 state. It has, however, been objected that the stickiness of serum prevents its rapid removal. Hence the effects of serum could persist for some time after a change to serum-free medium (22). We used several drugs that appear rapidly and quite selectively to arrest cells at the restriction point, and showed that BHK cells become insensitive to these drugs several hours before starting DNA synthesis (11). This result is consistent with Temin's conclusion. But Martin and Stein (39) propose that the restriction point is actually located at the beginning of DNA synthesis, based on postulated effects of SV40 virus T-antigen in temperature-sensitive mutant infected cells. Thus even the general section of G_1 within which to look is debated, let alone determining the exact time distribution of the restriction points in a population of cells. Furthermore, more hours are required for cells to cycle from G_0 to S than from M to S (4, 32). Whether this extra time is required before or after the restriction point can be determined by doing kinetic studies comparing restriction with these two sorts of initiation systems. Our recent experiments with 3T3 cells suggest that 4 hours' extra time is needed prior to the restriction point, which itself is 2 hours before S in the average cell (19).

One of the most interesting consequences of the restriction point idea is that at least some transformed cells are not arrested in the same way as their untransformed counterparts. It has long been known that transformed cells are not arrested in low serum medium (see, for example, 40), and also they are not blocked by some drugs at low concentrations (streptovitacin A (41) and puromycin aminonucleoside (42)) which arrest normal cells. Transformed cells that do not stop in G_0 proceed slowly around the cycle (43) and they do not survive (11, 44).

It has now been demonstrated in our laboratory and in several others (11, 45, 46) that normal cells can be protected preferentially against neoplastic agents by moving them out of cycle with reversible G_1-arresting drugs.

Location of Regulatory Molecules Within the Cell

Several sorts of evidence indicate that regulation of the cell cycle occurs in the cytoplasm rather than in the nucleus. Whether or not a cell will continue to proliferate depends on external factors such as substratum, serum, nutrients, or neighboring cells (4, 5). But these factors must interact with the membrane or be transported across it. Then, they or their effects (as is the case with neighboring cells) must pass through the cytoplasm, eventually to trigger DNA synthesis.

That diffusible cytoplasmic factors control nuclear events was indicated early by experiments of Vasiliev, et al. (47), who showed that adjacent ascitic hepatoma cells were generally coordinated in both DNA synthesis and mitosis. Dewey, et al. (48) have demonstrated improved synchrony when CHO cell populations became denser. Nuclear transplantation and cell fusion studies have shown that the regulation of DNA initiation closely involves interactions between nuclei and cytoplasmic factors. The cytoplasm from S cells contains factors which stimulate nuclei from non-S cells to synthesize DNA. Graham (49) showed that when "inactive" nuclei from mouse liver, brain or blood cells are injected into Xenopus eggs, the inactive nuclei begin to replicate their DNA. Similarly, Gordon and Cohen (50) found that fusing actively dividing melanocytes with macrophages caused activation of DNA synthesis in the macrophage nuclei. The signal for DNA synthesis originated from the melanocyte component of the heterokaryon, and an influx of proteins from the melanocyte to the macrophage nuclei was observed. Rao and Johnson (51) demonstrated a rapid induction of DNA synthesis in nuclei of G_1-phase HeLa cells following fusion with S-phase HeLa cells. Although these and similar studies have helped establish that DNA synthesis is under positive cytoplasmic control in animal cells, these systems do not allow the identification of the factor(s) responsible for inducing DNA synthesis. The two or more nuclei of lower eukaryotes such as Physarum (52) or in binucleate rodent cells initiate DNA synthesis at the same time. In contrast, the two mitoses in binucleate DNA virus-transformed cells occur at different times; but in binucleate RNA virus-transformed cells, the two mitoses occur at the same time (as with untransformed cells) (53).

These results suggest that one should look in the cytoplasm for agents involved in normal S phase initiation. It is not known how close in sequence these G_1 events in the cytoplasm are to initial events at the membrane or to DNA initiation.

Chemical Nature of the Regulatory Molecule(s).

Protein synthesis has been strongly implicated in the progression of cells through G_1 (4,5). Thus, when inhibitors such as cycloheximide were added for a few hours to G_1 cells and then removed, there was a corresponding delay in the initiation of DNA synthesis. A more detailed analysis was carried out by Schneiderman et al. (54) who at different times added cycloheximide (to inhibit protein synthesis totally) to CHO cells synchronized by mitotic shake at M and removed the inhibitor after various intervals. They found the duration of the delay in S initiation to be greater than the period the cycloheximide had been present. The duration of this extra delay depended both on the time in G_1 of addition of the inhibitor and on duration of exposure. Their results were explained in terms of a protein that has to build up during 4¼ hours following M and is needed at a high concentration until S has started. Furthermore, this protein was proposed to be fairly labile, with a half-life of about 2 hours. We have been able to confirm these very striking experiments with CHO cells, and to extend them to mouse cells.

Inhibitor concentrations that only slow the rate of protein synthesis arrest normal cells much as they are arrested in G_1 by low serum, in A state (20), restriction point (21). The simplest way to explain a special requirement for new proteins at this point is based on Schneiderman, et al's model (54). When all proteins are synthesized more slowly (in the presence of the inhibitor at low concentration), an unusually labile "restriction" protein cannot be built up to the required concentration because it is too rapidly degraded.

Several temperature-sensitive mutants have been isolated which stop at various points in G_1 at around 39°C (55). A plausible explanation is that their modified genes code for temperature sensitive proteins that are required for various G_1 processes, possibly including growth control functions.

Several laboratories have searched for specific protein synthesis changes in the cell cycle, and particularly in G_1 relative to other phases of the cycle (4, 56, 57; Fig. 5). For example, one of the earliest results, obtained in this laboratory (58), showed that S-phase CHO cells make proteins that bind to DNA which are not made by G_1 cells. We have also demonstrated with synchronized Escherichia coli that a DNA-binding protein of 80,000 M.W. is produced in a burst about 15 minutes before DNA synthesis begins (59). Ley (60) has recently obtained some striking results with synchronized CHO cells, showing that a protein of 80,000 M.W. Is synthesized in the cytoplasm, beginning at the start of G_1; it reaches a peak near the initiation of DNA synthesis. We have confirmed his results with CHO but not 3T3 cells (61). The connection of Ley's result to the model of Schneiderman, et al. (54) is striking.

Proteins in nuclei, both histones and non-histone chromosomal proteins (4) have been extensively studied. Kinetics of formation of the latter suggest a connection to DNA initiation (62).

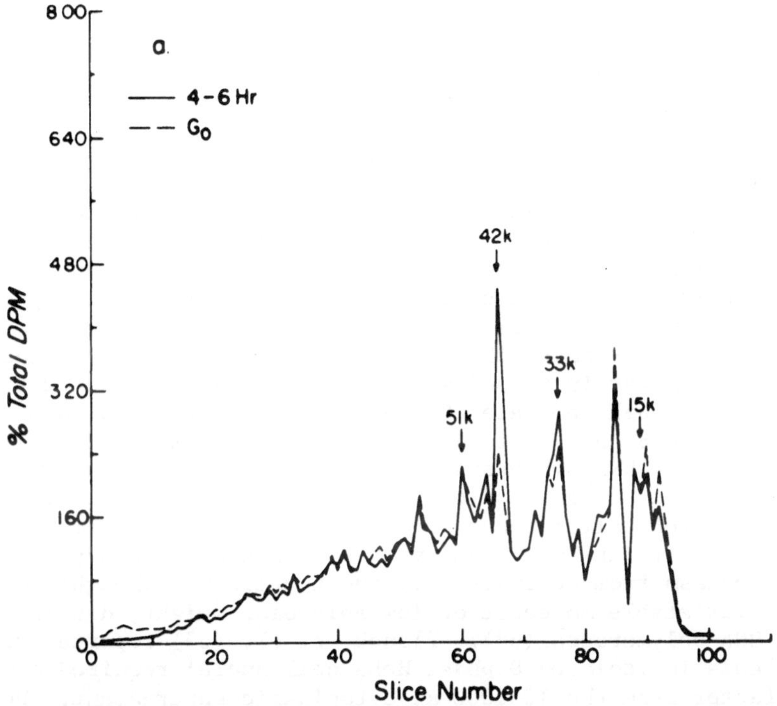

Fig. 5. Protein synthesized by 3T3 cells. Two cultures of 3T3 mouse cells were prepared and put into G0 in a low serum medium (61). One was labeled with radioactive amino acids while still in G0, and the other 4 to 6 hr after growth was stimulated. The proteins were run on acrylamide gels, which were sliced and counted. The labeled protein patterns were quite similar, with a few exceptions particularly the peak at 42,000 daltons, which is the position of actin.

Compounds that had extensive consideration as cytoplasmic regulators are cyclic nucleotides. But in spite of many experiments, there is not strong evidence for their having a specific role in DNA initiation (63). A newly discovered nucleotide, AP4A, is a candidate for DNA initiation (64, 65). We have looked for cyclically produced low molecular weight compounds that are labeled with adenine or phosphate. Although we have identified at least two such compounds that change cyclically, we do not think they are regulatory for DNA startup since the changes occur mainly in the S period.

Studies With Subcellular Systems

The eventual, definite identification of factors for regulation of DNA and RNA synthesis will be to add them to isolated nuclei or to other subcellular preparations and achieve new macromolecular synthesis. The importance of cytoplasmic factors comes from observations that the most active isolated nuclear preparations still have some cytoplasm associated with them. Nuclei stripped of all cytoplasm are generally poor DNA synthesizers. Kumar and Friedman (66) demonstrated that the number of G_1-phase HeLa nuclei synthesizing DNA was increased when S-phase HeLa cytoplasm was introduced, although the rate of ^{3}H-TTP incorporation was not increased by the S-phase cytoplasm. Thompson and McCarthy (67) found the rate of DNA synthesis to increase 15-20 fold when mouse liver nuclei were combined with cytoplasm from actively dividing Tapor liver tumor cells; mouse liver cytoplasm had no stimulating effect. The stimulation of DNA synthesis in hen erythrocyte nuclei by the cytoplasm from actively dividing L cells was thought to be due to a heat stable molecule of low molecular weight, distinguishable from DNA polymerases (68). Planck, et al. (69) reported that DNA synthesis in isolated S-phase HeLa cell nuclei required a protein factor from the 105,000 xg cytoplasmic supernatant. However, the cytoplasm from S-phase HeLa cells was unable to stimulate DNA synthesis in nuclei from cells not in S.

In one report of a cytoplasmic preparation from mammalian somatic cells which can initiate DNA synthesis, Jazwinski, et al. (70) has stimulated DNA initiation in Xenopus oocytes with proteins in extracts from proliferating higher animal cells and from yeast. However, very little DNA was synthesized. The nuclear preparations of Krokan, et al. (71) and Goulian and Tseng (72) apparently can perform ligation of small pieces of animal cell DNA (2-4s) to larger pieces (30s). Hershey and Taylor (73) and Friedman (74) were unable to demonstrate ligation of DNA in their nuclear systems.

Another approach is to use nuclei in the presence of cytoplasm. Lazarus (75) has made a preparation of nuclei in a gel of cell sap, and reported incorporation of TdR into DNA for several hours. The amounts of material he worked with were very limited. Klein and

Bonhoeffer (76) reported on a somewhat similar gel of disrupted bacteria (*E. coli*) capable of making DNA. Seki, et al. (77) permeabilized HeLa cells to nucleotides by hypotonic swelling. This preparation was capable of converting ribonucleoside triphosphates and deoxyribonucleoside triphosphates into acid insoluble material; only a limited incorporation of TdR into acid insoluble material was observed (less than 25% of above). Approximately 50% of the -ellular protein was lost during the permeabilization procedure. Billen and Olson (78) have described a procedure which renders CHO cells permeable to nucleotides, and they have demonstrated both RNA synthesis and semiconservative DNA synthesis. Berger and Johnson (79) have recently reported a promising permeabilized system capable of DNA synthesis for 30 minutes. Cells permeabilized with high salt or lysolecithin have been used to study DNA synthesis (80).

CANCER CELL VS. NORMAL CELLS

Carcinogenesis

General remarks on transformation. Tumors can be induced by several sorts of carcinogens: chemicals, DNA tumor viruses, RNA tumor viruses, radiation (x-ray and ultraviolet light), and "spontaneously", that is, by unidentified agents. The properties of tumor cells differ according to the carcinogen, as will be discussed below. Furthermore, even tumor cells that are produced from one kind of normal cell by one carcinogen can have very different properties, as with fibroblasts transformed *in vitro* with SV40 virus (see 81).

Cancer and cancer-forming cells thus exist in much greater variety than can be readily studied. Choices have to be made as to the kinds that deserve intensive investigation. One factor is the frequency of occurrence of cancers of various sorts. Most human tumors (80% or more) are thought to be caused by chemical carcinogens (82). It would seem relatively important to study them, as compared to tumors produced by viruses; which in humans are proven to produce only a few kinds of tumors (by Herpes). The study of cells transformed to tumorigenicity by viruses has greatly increased knowledge of molecular biology and has offered insights into the processes of uncontrolled cell growth (3, 83). However, balance among studies of cells transformed by chemicals vs. viruses seems somewhat askew.

Another often cited fact is that most human tumors are carcinomas that arise from epithelial cells. Thus studies of epithelial tumors would appear to be most valuable. Whether this is a wise choice at this time is open to question, since culture of epithelial cells is difficult compared to culture of fibroblasts, which produce sarcomas. The difference in frequency of tumor appearance in humans might reflect merely greater frequency of exposure of epithelial cells to carcinogens rather than some fundamental cellular difference.

These introductory remarks suggest that more studies on tumor-forming properties of cells transformed by chemicals are very worthwhile. We need to know more about whether chemical transformation produces tumor cells (from one original cell type) with properties in common to each other, and different from those of non-tumor forming cells. Are classes of chemically transformed cells distinguishable? (An extension of these questions is whether different chemical carcinogens produce tumor cells with similar or different traits). Assuming some properties are revealed to be in common, we should study them more carefully to learn more about the biochemical nature of cancer.

Determination of tumorigenicity. The most fundamental property of tumor cells is their tendency to grow into a tumor when injected into a suitable host (2, 3). A considerable number of properties in vitro have been correlated with tumor-forming ability in vivo (84) (see below). However, it has become clear in the past few years that there are exceptions to all of these correlations, and in both directions. Cells with property X usually form tumors but there are tumor cells lacking this property. Also some non-tumor forming cells do have property X. For example, hybrids formed between human fibroblast and HeLa cells have a set of properties generally ascribed to tumors, but they do not form tumors in the nude mouse (85). A set of properties of DNA (SV40) transformed cells has been correlated with tumor formation in nude mice. These properties are growth without anchorage, changes in cytoarchitecture, and production of plasminogen activator (81). It can be objected that nude mice are not optimal hosts for all cells which might form tumors in another host. Yet studies by Shin, et al. (8, 86) and Stiles, et al. (10, 87) have shown that nude mice support growth of tumors from a wide variety of animal cells including those of humans. We consider the ability of cells to form tumors in nude mice the most useful criterion of tumorigenicity. By using this technique for isolating tumor lines, one can retain all traits in culture unselected, as indicators for study of phenotypes of the tumor lines.

Tumor selection in nude mice becomes more powerful and quantitative if a procedure of coinjection of 10^7 non-tumor forming cells with the tumor cells is used (88, 89). For example, 10^4 or more HeLa cells alone had to be injected to produce a tumor, but only 10 HeLa cells plus 10^7 normal, non-tumor forming irradiated cells formed a tumor. By this technique tumorigenicity can be quantitated in terms of how few transformed cells will form a tumor upon coinjection.

The property of fibroblasts in vitro that most often correlates with tumor formation in vivo is growth without anchorage (see 81,

89, 90, 91). If one isolates unanchored cell clones, as well as cells from tumors in nude mice, one will increase the likelihood of getting a sufficient number of tumor-forming clones. Anchorage-independent growth can be determined by several methods. One is growth in methocel over an agar base (92). Another is growth on soft agar (93). A third is growth on a film of Hydron (94). These anchorage independent clones can then be passed through nude mice to select truly tumorigenic cells for study in subsequent experiments. Since passage through the mouse can alter properties of the cells (95), there is apparently a selection for properties required for tumorigenicity. Hence comparisons of transformed lines' properties beofre and after passage through the mouse should provide additional information.

Transforming agents. Cancer production by treatment with chemicals has a vast literature. Of particular interest to us here is carcinogenesis *in vitro*, a subject reviewed by Heidelberger fairly recently (82). Initial work on chemical carcniogenesis *in vitro* was to a large extent carried out with the cell lines BALB/c-3T3 (96), BHK (97), and C3H/10T½ (98). A variety of properties have been used by various workers to identify transformed cells. In general it was not clearly shown that the cells so identified will actually form tumors. Barrett and Ts'o (99) have shown that transformation is progressive, and early changes after treatment (morpho-ogy, for instance) are not sufficient to indicate tumorigenicity. Somatic mutation has many properties in common with transformation, particularly ability of the same agents to act as mutagens and transformers (97, 100, 101). However, Barrett and Ts'o (102) have shown that the relation between mutation and transformation is not simple. Transformation seems to be a multi-step process, the details of which are far from being understood (99). As one example, we refer to the sequential steps of initiation and promotion by different chemicals studied for a quarter of a century (103, 104).

Most studies of transformation *in vitro*, such as those cited above, were done with rodent cells -- usually mouse and Syrian hamster. Sager and Kovac (89) have chosen to do genetic investigations with Chinese hasmter cells, owing to the desirable properties of these lines with respect to karyotype stability, ease of cell fusion, non-tumorigenicity of one line (CHEF/18), and growth in defined medium. This line has been transformed to tumorigenicity with EMS, as measured by growth in the nude mouse and growth in methocel. There is a tumor-forming line (CHEF/16) that arose spontaneously in the original culture.

Transformation of human cells has been difficult. Freedman and Shin (105) treated human fibroblasts with MNNG and obtained clones that were anchorage independent but did not form tumors in nude mice. Kakunaga (106) treated human fibroblasts with 4-NQO or MNNG to obtain lines that grew in soft agar and nude mice. Numerous cell divisions were required to establish these lines (see also 99, 102). Milo and DiPaolo (107) transformed human fibroblasts with six carcinogens, using the same criteria as above. A complex procedure was used, including treating the population in S phase.

About a dozen chemical transforming agents have been used often. They fall into the category of compounds that react with DNA and cause mutations. Whether carcinogens that alter DNA in different ways lead to different phenotypes is a subject of interest for the future. Choice of a particular carcinogen for routine study would seem at present to depend on its efficacy, convenience of use, and importance in the etiology of human cancer. To provide a concrete example of an interesting carcinogen, we will discuss benzo(a) pyrene briefly. It is a major carcinogen in nature (82, 108). Benzopyrene has been used for in vitro carcinogenesis of rodents (99, 109, 110, 111). It is converted to a diol epoxide which is the active form (112) and causes strand scission of DNA in vitro (D'Andrea and Haseltine, personal communication). It has not been used to transform Chinese hamster cells, to our knowledge.

Most studies in which characteristics of mammalian transformed cells were compared to untransformed cells have been carried out with a few DNA virus-transformed lines, and most of our present ideas about the loss of growth control after transformation were obtained with these cells (3, 4, 5, 6).

Transformation of hamster cells can also be accomplished in vitro with Herpes viruses (113). These tumor cells grow rapidly and metastasize in synergeneic hamsters. Since little has been done on the in vitro properties of these transformed cells, a long-range problem would be to investigate them as possible models for human virus-induced cancer, along the lines discussed here. The same can be done with cells transformed by RNA tumor viruses (114). (See below for references to studies on comparisons of growth characteristics of some DNA virus and RNA virus-transformed cells.)

Characterization of Transformed Cells

Properties of transformed cells. Several properties have been proposed to be correlated with tumorigenicity. They have been discussed very often by us as well as by others (see 3, 4, 5, 6). Tumorigenic cells may differ from non-tumorigenic ones by: 1) having more abnormal karyotypes, 2) having abnormal cytology, 3) requiring lower serum concentrations for optimal growth in culture, 4) growing to higher density in culture, 5) growing into multilayers in culture, 6) not arresting in a state with G_1 DNA content (G_0) at high density or in low serum, 7) growing without anchorage (see above), having altered surface properties, including 8) higher transport rates of several nutrients, 9) changed chemical composition, less large, extracellular trypsin-sensitive protein (LETS or fibronectin), or 10) producing a proteinase plasminogen activator. Most of these differences have been observed by comparisons of normal and DNA virus-transformed cells, although there have been a few studies with RNA virus transformed, and a very few with chemically transformed cells (see below).

There are other in vitro properties correlated with transformation (81, 99, 102). Barrett and Ts'o (99) have shown that some of these properties such as altered morphology and fibrinolytic activity appear earlier and with considerably higher frequency than does ability to grow without anchorage. They suggest that carcinogens are generally mutagens, and thus their application creates changes in the genetic material aside from those required for tumor formation. Hence cells that have been selected for alteration of one property are likely to be altered in others as well. The modifications that are needed for tumor formation either must be more numerous or less likely (smaller target) than those required for ordinary gene mutations, which might be able to change morphology. None of these transformed characteristics has been shown to be diagnostic for tumor-forming ability. However, any cell that possesses several of these properties can be suspected of being tumorigenic. Some of these tests are not at present amenable to interpretation at a biochemical level. Others, such as reduced serum requirement, non-arrest in G_0, or surface transport alterations have led to current ideas as to how growth control might be modified in cancer. Any of these properties that are altered frequently in chemically transformed lines can be worthy of detailed study.

Chemotherapy. Chemotherapy, a subject in itself (115), is a triangle involving drug, normal cells of all types in the host, and tumor cells (16). It is not hard to find drugs that kill mammalian cells, but the difficulty is to find one with any reasonable therapeutic index. Aside from selection against growing cells by cycle-dependent drugs, there are few further bases for rational chemotherapeutically useful distinctions between normal and tumor cells. Trial-and-error has produced perhaps two dozen clinically useful drugs, of several hundred-thousand screened. Small modifications can lead to useful drugs (117).

The problem is made more difficult by the need to kill nearly all tumor cells; otherwise, the survivors are likely to produce a new tumor. A large tumor (10 cm in diameter) contains about 10^{12} cells (1 kilo). There has to be a reduction of probably 9 logs in tumor cell number without host lethality. Drugs used singly seem at best to give 2 logs' kill of tumor cells. Repeated application at intervals can in theory give a successful series of 2 log kills; but time is required between doese to permit recovery of the hosts' cells, and during this time the tumor cells also proliferate. There is a limit to how many times a drug can be applied, partly because drug-resistant mutant cells either are originally present or are induced by the drug. These mutants eventually will form a drug-resistant tumor. Consequently, multiple drug therapy is in common use, both to increase the kill per application and to present the appearance of drug-resistant mutants. Much of modern cancer chemotherapy is devoted to optimizing the schedule of drug delivery (18, 118).

Some cells are constantly being formed in the adult. These include blood cells that largely come from stem cells in the bone marrow. The stem cells are stimulated in such a way as to keep the concentrations of leukocytes and red cells at physiological levels. Antineoplastic drugs kill many of the bone marrow cells, as well as intestinal cells that are likewise proliferating. The lethal effects of the drugs and their low therapeutic index can largely be attributed to depletion of the bone marrow or intestine (116).

Cells can be transformed so that they wholly or in part lose their responsiveness to external growth control. The morphology of the culture changes, seen primarily as cells growing over one another, and growth beyond confluence (4, 5). The requirement for serum is diminished as well. The cells transformed by DNA tumor viruses do not arrest in G_1 the way untransformed cells do, but are found in all parts of the cycle under conditions that slow down their growth (43). These cells do not recover well when put back into complete medium. Following a week or so of serum or amino acid deprivation they die, unlike untransformed cells that have gone into G_0, the quiescent state.

This continued growth of tumor cells under conditions which make normal cells quiescent is the basis for much of chemotherapy, since drugs that are toxic to cells making DNA or that are in M do not affect G_0 cells. The difference can be amplified by using a second drug that puts normal cells into G_0 but does not arrest growth of transformed cells (11). Such protective drugs include caffeine and weakly acting inhibitors of protein and RNA synthesis. If added first to put the normal cells into quiescence, S phase (or M phase) specific drugs show a large differential kill of tumor vs. normal cells in culture.

<u>Changes in growth control after transformation</u>. We divide cell growth control into three main steps. First there is action of an external agent on the cell membrane. Second, this signal is transmitted through the cytoplasm, either by the original agent or a "second messenger". Third is action in the nucleus, to start up DNA synthesis. Surface changes of many sorts follow

transformation (119, 120). Such changes include alterations of surface proteins and other membrane-related components, changes in ability of the cells to be agglutinated by proteins from plants (known as lectins), and changes in transport of small molecules into the cell (121). None of these has a known causative role in growth control.

We suggested as early as 1964 that the principal differences of cancer cells vs. normal cells might reside in changes in the cell membrane that altered the cell's interaction with growth controlling chemical factors (122). Serum is now known to contain growth factors that are required for transit through the cell cycle (40, 123, 124, 125). At that time these factors were unknown. We proposed that they had to be transported through the membrane, between cell cytoplasm and environment.

The ability of transformed cells to grow on medium containing low serum suggests that transformation reduces or eliminates the cells' requirement for limiting serum factors (4). One interesting dissection of these serum requirements is based on the ability of transformed cells to grow with plasma. Non-transformed cells require serum, which contains a factor from platelets. They do not grow with plasma. Addition of this platelet-derived growth factor (PDGF) to plasma allows growth of 3T3 cells (17). Transformation with DNA tumor virus abolishes the requirement for PDGR (18). PDGR is needed by quiescent cells, prior to the other factors provided in plasma, to allow the cells to re-enter the cycle and initiate DNA synthesis. PDGF requirements of chemically transformed tumor cell lines have not been studied.

Defined media, containing known growth factors rather than serum, have been worked out for some transformed cells (125, 126, 127, 128). Of particular interest is a medium devised by Dr. Gordon Sato, et al., that allows BHK cells to grow rapidly when supplemented with only the four growth factors: epidermal growth factor (EGF), fibroblast growth factor (FGF), insulin and transferrin. We have slightly modified this medium so that it will permit growth of CHEF/18 cells at more than half the rate permitted by serum-containing medium (129; Fig. 6). Transformation of BHK cells with polyoma virus abolished the growth factor requirements except for transferrin. Transformation of CHEF/18 with EMS moderately diminished the EGF requirement, whereas cells from a tumor formed in nude mice had a larger decrease. There was definitely a lower EGF requirement by transformed cells as compared to the original 18 cells.

Altered membrane receptors could modify the cells' response to growth factors and thereby alter growth control. The relation of growth factors to transformation has been studied using measurements of the binding of EGF to the cell surface (130, 131).

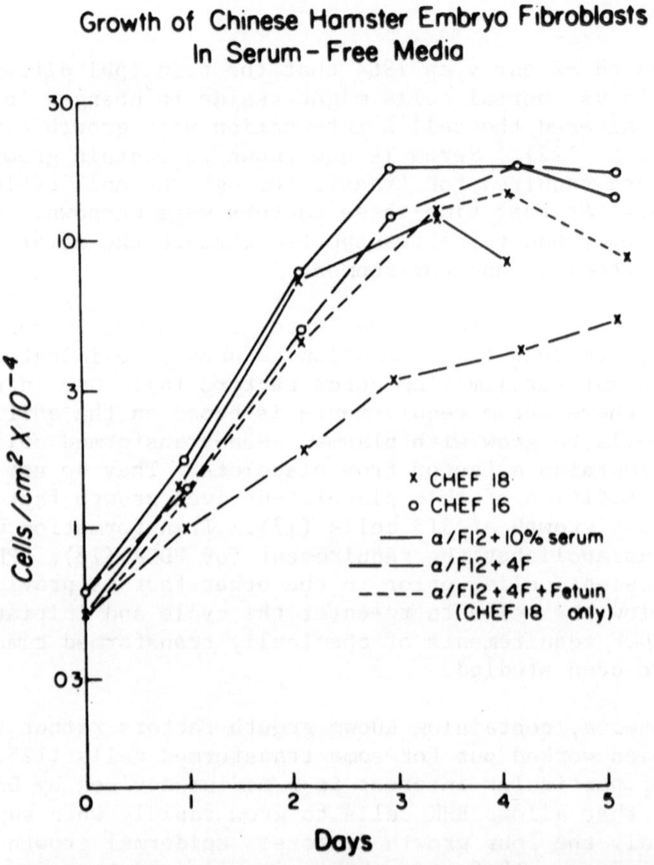

Fig. 6. Growth of cells in serum-free medium. Chinese hamster embryo fibroblasts were grown in media containing serum and in media lacking serum but supplemented with four known growth factors (transferrin, insulin, epidermal growth factor, and fibroblast growth factor). Growth was equally good for several days for the CHEF-16 (transformed) cells, but growth of the untransformed CHEF-18 cells was further aided by unknown factors provided by a crude preparation of fetuin (129).

Receptor site determinations have also been made for insulin (132). Whether there is requirement for such bound hormones to enter the cell and function internally (133) or whether they can function solely externally (134) by producing a "second messenger" within the cell is uncertain (135).

Effects in the cytoplasm are poorly understood. That signals can be transmitted through the cytoplasm to both nuclei of a binucleate cell was shown by cell fusion experiments, where both nculei of an untransformed cell (but not of DNA virus-transformed cells) start to make DNA at the same time (53). In the early 1970's the idea arose that cAMP and cGMP are regulators of animal cell growth. By the mid-1970's the data became confusing, and this idea is now less popular (63). More recently, emphasis has been placed on protein kinases and phsopho-protein in relation to growth and physiological control (136, 137). The overall function perceived in control is that low serum raises cAMP which increases protein kinase activity; this enzyme phosphorylates proteins so as to stop growth. Other enzymes are involved in this control system (e.g. adenylate cyclase, phophodiesterase, and protein phosphatases). GTP, cGMP and Ca^{++} also appear to be involved.

A role of protein kinases in growth control can offer a biochemical marker for the transformation process. Modifications of control in tumor cells could be placed before or after the kinase-dependent process. Less cAMP binding protein (kinase subunit) was reported in a tumor line (138). Mutants defective in protein kinase have been isolated. Coffino, et al. (139), have shown a close correspondence between loss of protein kinase activity and loss of growth control (G_0 arrest) by db-cAMP. Similar results were obtained by Simantov and Sachs (140, 141). But Wigglesworth, et al. (142) did not see such changes between 3T3 and SV40-transformed cells. Note that this DNA virus transformation might merely bypass the G_0 control (uncouple it) without affecting the kinase. 3T3 has only one cAMP-binding protein, although it might have more than one kinase. SV40 infection creates an additional kinase (142, 143). Note too that the src gene of RNA sarcoma virus (Rous) is a protein kinase (144). The most striking observation is that loss of protein kinase prevented db-cAMP-resistant mutant cells from entering G_0, though poorly blocked in low serum (142). The original (tumor) cells probably had a defect in growth control at a step prior to the kinase step. Interestingly, cells without cAMP dependent kinase continued to cycle normally (139). With regard to the role of protein kinase in growth, we observed a higher phosphorylation of proteins *in vitro* by quiescent BHK cells, especially with cAMP present, relative to growing cells (145). Protein phosphorylation patterns *in vivo* and in extracts were altered by transformation (146).

A very relevant study is by Costa, et al. (146, 147). Mitotic shake synchronized CHO cells show a dramatic increase of kinase II in late G_1 and a drop of kinase I after mitosis. Also, cAMP tended to rise in G_1. Protein kinase II may be needed to allow transit of G_1. By comparison Coffino, et al.'s work suggests that entry into G_0 requires protein kinase I (139).

There are also numerous studies on phosphoproteins in relation to growth and growth control. For instance, phosphorylation of nuclear acid proteins, and their turnover during the cell cycle, have been reported (148). Numerous studies have been performed on phosphorylation of histones and membrane proteins. A relevant example of a protein kinase under the influence of EGF is membrane phosphorylation (149). One would like to have comparisions of protein kinase activities of various chemically transformes lines.

The nuclear events of DNA initiation are unknown. They can be classified into changes of DNA structure, changes of synthesis and location of enzymes of DNA synthesis, and changes of availability of DNA precursors. There is not much evidence regarding the former two possibilities. Recent work with animal cells that have been made permeable to small molecules has shown that these cells can make DNA normally at the rates of intact cells (80), and that the non-availability of precursors (deoxyribonucleoside triphosphates) is not limiting the ability of cells to make DNA during the G_1 period (150). We have recently suggested that assembly of enzymes and other proteins into a complex on the DNA may be the activating step (151).

<u>Comparison of transformation by different agents</u>. Most of the ideas about growth control were obtained from studies of DNA-virus transformed cells. Similar studies have recently been done with RNA-virus transformed cells, and a few with chemically transformed cells. These experiments suggest that absolute loss of abilityof transformed cells to enter G_0 under adverse conditions may be found only in DNA virus transformed cells (21, 43). RNA virus and chemically transformed cells seem merely to have relaxed their growth control so that they accumulate in G_0 under suboptimal conditions; part of the population remains in cycle, however. Holley, et al. (152) studied serum-dependent growth of a benzopyrene tranformed 3T3 line. The culture (BP3T3) grows to about 5X higher density than 3T3 with high serum, but its growth is greatly restricted below 1% serum. The cells then accumulate with a G_1 DNA content, as with 3T3, but unlike SV-40 transformed 3T3 (43). BP3T3 cells grow in suspension. They have a lower requirement for the growth factor FGF, possibly because they destroy it less rapidly than do 3T3 cells. Moses, et al. (153) have similar results with two 3-methylcholanthrene-transformed mouse lines. The growth of these lines is arrested because nutrients rather than serum factor (as for 3T3) are depleted from the medium. This indicates that

low molecular-weight nutrients might have a role in growth control of some transformed cells.

O'Neill (154) compared properties of DNA and RNA virus-transformed cells of hamster, rat and mouse. In general, the DNA virus-transformed cells completely lost growth control as measured by their ability to grow in agar, to grow to high density beyond confluence, and to grow in medium containing low serum. In contrast, the RNA virus-transformed cells showed much less escape from growth control by these criteria. O'Neill, et al. (155) demonstrated that cells prevented by cytochalasin B from dividing show "controlled nuclear division" - arrest of nuclear division after one or two rounds (at most). By contrast, DNA virus-infected cells have uncontrolled nuclear division and give rise to multinucleate cells under the same conditions.

The responses of transformed and normal cells to various drugs also differ depending on the agent used for transformation. DNA-virus transformed cells growth was less strongly inhibited by several drugs including caffeine (11), protein synthesis inhibitors at low concentrations (11), cytochalasin B (155), puromycin aminonucleoside (11, 42, 45), high cAMP (46), and picolinic acid (156). Picolinic acid was also tested on a series of RNA virus-transformed cells, and arrested the cultures in different parts of the cell cycle (156).

Studies of cells transformed by different agents indicated that the various cells are altered differently with respect to their growth properties (157, 158). Results of this sort with different lines have been extended in our laboratory (159). DNA virus-transformed mouse and hamster cells in low serum medium were distributed around the entire cycle; but cells transformed with any of three RNA viruses or two chemically carcinogens (benzopyrene or 7, 12-dimethyl-benzathracene) were mainly arrested in G_1 (G_0). Furthermore, the drugs listed above had very different effects on growth of the differently transformed cells, with no clear pattern emerging.

These few studies make it clear that our current views about changes in growth control derived largely from studies with DNA tumor viruses, will be modified when we study chemically transformed cells. The nature of control, therefore, will probably also be different for actual human tumors.

MOST RECENT DEVELOPMENTS

The last part of this Chapter describes our most recent work on the problem. Our principal objective is to identify, isolate, characterize, and study molecules having primary importance in regulating the animal cell cycle and their derangements in cancer.

Animal cells reach a crucial control (restriction) point during the G1 phase of the cycle from which they either continue to proliferate or detour to a quiescent state G0. The control event is probably cytoplasmic. Protein synthesis is critical in influencing this decision to grow or not to grow. We have done many kinetic experiments, showing that inhibitors can arrest cells at this restriction point. A variety of conditions that control growth involve the same restriction event. We propose that this event occurs at a specific "metabolic place" in G1; i.e. each cell must pass this restriction point in order to initiate biochemical processes leading to DNA synthesis. Cells that cannot carry out this restriction point event pass into the G0 state. The event can thus be considered a "switch" at which a choice is made between proliferation and quiescence.

Growth factors and growth requirements

Non-tumor forming cells generally require higher concentrations of serum than do tumor forming cells. The main factors that become growth limiting for "normal" 3T3 cells at high density or deprived of serum were shown to be insulin and epidermal growth factor (Rossow and Pardee, unpublished). Differences between transformed cell lines were studied by flow cytofluorimetry and autoradiography (159). Cell Cycle distributions of Balb/3T3 and seven transformed lines after three days growth in low serum were a function of the mode of transformation. DNA virus transformed cells were spread around the cycle, whereas RNA virus and chemically transformed cells had accumulated G1 to various degrees. The changes in cell number during a two day growth period were used to measure the growth responses of the Balb/3T3 series to the drugs: rotenone, antimycin A, azide, oligomycin, arsenate, chloroquine, ouabain, tetracaine, actinomycin D, lucanthone, 6-azauridine, 3-acetylpridine, picolinic acid, caffeine, methylgyozal bis-(guanylhydrazone) and streptovitacin A. Some of these drugs had affected transformed and untransformed cells differently. For no drug was every transformed line either more or less sensitive than the corresponding untransformed line, and for most drugs there was variation in growth responses among lines transformed by the same agent. However, growth responses to several drugs were observed which were characteristic of a class of transformants. This study demonstrates that the behavior of a single transformed line cannot be taken to represent the behavior of all transformed lines, or even of all lines transformed by the same agent.

Totally defined medium (125) (α/F12, 1:1) plus four known growth factors--insulin (10 μg/ml) epidermal growth factor (10 ng/ml), fibroblast growth factor (10 ng/ml, transferrin (5 μg/ml) and $FeSO_4$ (2.5 mM) gave rapid growth of BHK cells. We optimized this medium for growing Chinese hamster embryo fibroblast (CHEF/18) cells (89). We studied the effects of deprivation of one or more of these factors upon growth of several cell lines, and the points in the cycle at which the cells were arrested (129). Transferrin was essential throughout the cycle, particularly in G2; the other three factors were important for passage through G1.

Polyoma (DNA virus) transformation greatly reduced requirements for all the G1-related factors. We isolated a highly tumorigenic variant of BHK cells, by passage through nude mice and isolation of clones from the ensuing tumor. It has lost most of its growth control. Transformation in vitro of BHK and CHEF/18 cells was accomplished using mutagens. A pair of very similar Chinese hamster cells, CHEF/18 and CHEF/16 isolated from the same embryo culture, and chemically transformed lines were also investigated. In all cases, the major factor no longer required for growth in Sato's defined medium was EGF. Little is known about the further functioning of any of these factors after they attach to surface receptors.

Thrombin at ng/ml concentrations was found to stimulate growth of CHEF/18 in the difined medium. It replaced and was superior to EGF (160). Criteria for deciding which factors might be regulatory were reviewed (161). One such criterion is that media containing an insufficient supply of a regulatory factor (such as EGF) arrest non-tumor forming cells in G1 (G0). A second criterion is that some tumor forming cells should have diminished requirements for this factor.

Studies on growth kinetics

Our experiments derive from the idea if a cell is to continue through the cycle that specfic biochemical changes must occur at or up to the restriction point. If they are not able to occur the cell is destined to become quiescent. To identify these changes experimentally we must define as precisely as possible their timing in G1. The experiments of Temin (24) suggest this time to be near the middle of G1, in chick cells initiated from a serum-starved G0 state. It has, however, been objected that the stickiness of serum prevents its rapid removal; hence the effects of serum could persist for some time after a change to serum-free medium (22). As counter to this possibility, we used several drugs that rapidly and quite selectively act and arrest cells at the restriction point; we showed that BHK cells become insensitive to these drugs several hours before they start DNA synthesis (11) (19). This result in consistent with Temin's conclusion. But Martin and Stein (39) proposed that the restriction is actually located at the beginning of DNA synthesis, based on postulated effects of SV40 virus T-antigen in temperature-sensitive mutant infected cells.

A method combining Flow cytofluorimetry and shift-down of medium was developed to determine at what cycle time exponential cells pass beyond their growth requirements for serum factors (19). The technique showed that depreivation of serum arrests those 3T3 cells situated about 2 hrs before the start of DNA synthesis. Deprivation for isoleucine prevented all the cells that had not yet started to make DNA from progressing into S. The isoleucine arrested cells, after being resupplied with isoleucine, still required serum to reach S phase even though they had passed beyond the point (2 hrs before S) where they require serum (161).

More hours are required for cells to cycle from G0 to S than from M to S (32,4). Amino acid analogs including histidinol can block entry of cells into the cycle from G0, but they do not at the same concentrations affect cycling cells (15). These and other data (16) indicated that special biochemical events are required for cells to emerge from quiescence; e.g. that G0 is a distinct cell state.

The labile protein model of growth control

The experiments to be described support the idea that progress of a cell from G0 or M to G1 requires synthesis of specfic, quite labile protein(s). Protein synthesis rates were studied in synchronized cells, particularly during G1 phase using one dimensional gel electrophoresis following pulse labeling with amino acids (61). A major protein, actin, was made very rapidly for a few hours after quiescent cells were stimulated to proliferate, and other protein synthesis were more gradually changed. In contrast to cells emerging from quiescence, cells going through G1 did not have a peak of actin synthesis (162). Thus, dramatic changes of specific protein synthetic rates in the cycle can result from emergence from quiescence, as opposed to their being periodically occurring cyclic events. The role of actin in emergence from the quiescent state is not known; it may relate to current views on the cytoskeleton.

A most interesting set of results led to the hypothesis that completion of the restriction point process requires the synthesis of an adequate amount of some labile protein. Its net accumulation is proposed to require a high rate of synthesis needed to counterbalance its degradation (56, 163). With growing 3T3 cells low levels of CHM slow the cell doubling time in proportion to inhibition of protein synthesis. The cycle extension is almost all in G1, and prior to the serum dependent restriction point (2 hrs before S, independent of rate of growth). Inhibitory CHM concentrations that slow the rate of protein synthesis up to 70% arrest normal calls in G1 much as they are arrested by low serum, etc., i.e. at the restriction point (21). Our work with other inhibitors of protein or RNA synthesis support this. The simplest way to explain a special requirement for protein synthesis at this point is based on a labile protein model (54). All proteins are synthesized more slowly in the presence of CHM, but an unusally labile "restriction" protein would not be accumulated as rapidly to a required concentration, because it is too rapidly degraded for synthesis to raise its level (56).

We have done similar experiments with yeast, which has a "start" signal very similar to the restriction point of animal cells. Hartwell and his colleagues (28) have investigated the role of proteins in "Start", basing their ideas mainly on a stringent-type effect as found in E. coli for regulation of RNA synthesis by amino acid supply. Studies of the kinetics of the start event and emergence of yeast cells from the block at "start", show similarly to animal cells a marked sensitivity to CHM, and to low concentrations of essential nutrients that affect protein synthesis (30,164). The results are consistent with the requirement that a labile protein (half-life only about 6 minutes) be made in adequate amounts before yeast can go on beyond "Start" (165).

As a test of the labile protein hypothesis we have examined effects of CHM and other inhibitors on cell cycle kinetics of various untransformed and transformed cells obtained from mice, hamsters and humans. In all cases low CHM caused the non-tumor cells to accumulate in G1. But none of the tumor forming cells were nearly so strongly delayed in G1 (166). These results indicate a relaxation in tumor cells of the requirement for rapid protein synthesis to permit passage through G1. They strongly support the importance of our mechanism (labile protein model) in growth control.

Cell cycle events possibly involved in growth regulation

Cells in G0 population differ in nuclear size. The cells with larger nuclei are more likely to enter S phase earlier, a result suggesting that variability of the cycle is not purely probabilistic (167). A novel finding regarding cycle changes is formation on centrioles of a cilium that contains tubulin. A "centriole cycle" that must be coordinated with the usual DNA-division cycle was proposed (168).

Since the first, yet delayed event that can be observed following the serum-and protein synthesis-depent restriction point is DNA synthesis, attention was focused on biochemistry of DNA at S phase initiation. Studies have been carried out on proteins made during S phase, and changes in their rates of synthesis were identified (169).

Enzymes are found in the cytoplasm that need to be translocated to the nucleus at the time of DNA replication (170). Moreover, proteins of unknown function migrate in and out of L cell nuclei depending on the stage of the cell cycle (171). It is important to discover if the quantitative localization of DNA synthesizing enzymes at the nuclear site of their utilization will correlate with changing DNA synthesis during the cell cycle. The specific periodic localization of DNA synthesizing enzymes, including those associated with DNA precursor synthesis, with reference to the cell cycle enables us to propose a novel controlling element associated with the induction of cell growth and proliferation.

In order to evaluate the functional distribution of enzymes associated with DNA metabolism, locations of thymidine kinase, thymidylate synthetase, dCMP kinase, NDP kinase, and dihydrofolate-reductase along with DNA polymerase were initially compared in quiescent and logarithmically growing CHEF/18 cells (151). On the average only about 15% of these enzyme activities were present in the nuclear component (karyoplasts) of quiescent cells. By contrast, about 50% of these activities were recovered in the nuclear component of log-growing cells. These figures correspond well with the proportion of the cells making DNA in the respective cultures. In accordance with these observations of increased activities in karyoplasts of growing cells, concommitant decreases were reflected in the corresponding cytoplasmic compartments (cytoplasts). These results suggest that the enzymes of DNA metabolism are located in the cytoplasm of quiescent cells and are actually moved into the nucleus of cultures that are replicating DNA, possibly at the beginning of S phase.

For a detailed investigation of specific localization and association of these enzymes in synchronized populations, we have devised techniques to synchronize CHEF/18 cells. Our data clearly demonstrate that when DNA synthesis increase (about 20 fold) as the cells progressed from G0/G1 to S, there were only modest activity increases (about 4-5 fold in thymidine kinase and a smaller increase if DNA polymerase; the total cell activities of dihydrofolate reductase and NDP kinase remained the same. Further, when the cells progressed to S to G2, these total enzyme activities apparently remained unchanged in the cells. Therefore, we believe that it is not the total enzyme activities existing in the cell but the specific fraction that is located at the site of their function (muclei), possible as an integral part of a complex, that may be responsible for determining ability of cells to make DNA.

Consistent with the above speculation, our studies indicate that a significant fraction of DNA polymerase, thymidine kinase, dTMP kinase, NDP kinase and dihydrofolate reductase, but not adenylate deaminase and hypoxanthine phosphoribosyl-trans-ferase, in lysates of karyoplasts co-sedimented rapidly on sucrose density gradients. However in lysates prepared from cytoplasts these enzyme activities (when present) sedimented slowly, as expected for free enzymes. The rapidly sedimenting enzyme activities are unique to cells that are in the S phase. In lysates of cells obtaines 12 hours after releasing from isoleucine block a major fraction of thymidine kinase rapidly sediments on a sucrose gradient. Our in vitro assay system for DNA polymerase is sensitive to N-ethylmaleimide and so includes polymerase α.

In vivo kinetic studies with regenerating rat liver demonstrated the association of newly replicating DNA with the nuclear protein matrix (172). We inquired if our putative "multienzyme complex" isolated on sucrose density gradients contains any such bound DNA. The DNA polymerase activity in the rapidly sedimenting fraction did not show any requirement for exogenous "activated" DNA. These results indicated that the "multienzyme complex" possibly contain template and newly replicating DNA.

An alternative approach to the existence of multienzyme complexes is to evaluate kinetics of metabolite incorporation when enzymes are retained in a physiological state inside the cell and made accessible to exogenous compounds. Two different methods were devised to make animal cells permeable to different degrees and to molecules of different size (173,174). Under conditions that permit entry of only small molecules such as nucleoside triphosphates the cells are able to reseal and continue to grow. These techniques have been applied to about 20 cell lines (175) and have been used to study enzymology of DNA repair after damage by bleomycin (176) as well as for preliminary studies on transport defects as a basis of drug resistance. We showed that permeabilized late G1 cells could not synthesize DNA (although S phase cells did so) when supplied with all four precursors. Therefore, production of these DNA precursors (dNTPs) is not limiting for initiation of DNA synthesis at the G1/S boundary. We have not so far been able to accomplish our primary aim of getting permeabilized cells to make DNA by supplying them with extracts of S phase cells. We have earlier reported that permeabilized animal cells supplied with deoxyribonucleoside diphosphates as substrates synthesized DNA at a rate similar to intact cells (174). Under these conditions, substituting rNDPs for dNTPs decreased the DNA synthesis rate to 24% (173). However, in our recent experiments the incubation mixture was modified to activate ribonucleoside diphosphate reductase by provision of formate, THF and dithiothreitol; we then observed that rNDPs are incorporated into DNA 20 to 25% more effectively than were dNTPs in their original incubation mixture. Further, in this modified incubation mixture the synthesis of DNA from dNTPs was decreased to about 20%. This ability to synthesize DNA was increased with dNTPs by adding hydroxyurea (a compound which inhibits ribonucleoside diphosphate reductase). Our interpretation of these results is that the complex in a physiological state channels rNDPs directly to the site of their incorporation; the channel is poorly accessible to the "more proximal" dNTP precursors. Functional integrity of the complex can be altered by inactivating one of its essential components, to allow the more proximal precursors (dNTPs) to participate in DNA replication directly at the replication fork.

ACKNOWLEGEMENTS

We thank E. Fingerman for preparing the manuscripts. This work was supported by U.S. Public Health Service Grant, GM 24571.

REFERENCES

1. J. Poten, Biochem. Biophys. Acta, 458:397 (1976).
2. H.C. Pitot, "Fundamentals of Oncology", Marcel Dekker, Inc. N.Y. (1978).
3. B. Clarkson and R. Baserga (ed.) "Control of Proliferation in Animal Cells", Cold Spring Harbor, New York (1974).
4. R. Baserga, "Multiplication and Division of Mammalian Cells" Marcel Dekker, Inc., N.Y. (1976).
5. D.M. Prescott, "Reproduction in Eukaryotic Cells", Academic Press, N.Y. (1976).
6. A.B. Pardee, R. Dubrow, J. Hamlin and R.F. Kletzein, Ann. Rev. Biochem., 47:715 (1978).
7. V.R. Potter, Br. J. Cancer, 38:1 (1978).
8. A. Schmid, D.J. Hutchinson, G.M. Otter and C.C. Stock, Cancer Treatment Reports, 60:23 (1976).
9. G. Weber, New England J. Med., 296:486 & 541 (1977).
10. C.B. Stiles and A. Kawahara, <u>in</u>: "The Nude Mouse in Experimental and Clinical Research, J. Fogh and B. Giovanella, editors, Academic Press, N.Y. (1978).
11. A.B. Pardee and L.J. James, Proc. Natl. Acad. Sci. USA, 72:4994 (1975).
12. O.I. Epifanova, Int. Rev. Cytol., 49 (Suppl): 303 (1977).
13. S. Gelfant, Cancer Res., 37:3845 (1977).
14. C. Nicolini, F. Kendall, C. Desaive, B. Clarkson and J. Fried, Cancer Treatment Rep., 60:1818-27 (1976).

14a. C. Nicolini, S. Parodi, A. Belmont, S. Abraham and S. Lessin, J. Histoch. Cytochem., 27:102-113 (1979).

14b. Z. Darzynkiewica, D.P. Evenson, L. Staiano-Coico, T. Sharpless, and M.R. Melamed, J. Cell. Physiol., 100:425 (1979).

14c. C. Nicolini, F. Kendall, C. Desaive and W. Giarretti, Expt. Cell Res., 106:118-127 (1977).

15. A. Yen, R. Warrington, A.B. Pardee, Exp. Cell Res., 114: 458 (1978).
16. V.G.H. Riddle, P. Rossow, R. Boorstein, M. Addonizio, and A.B. Pardee, <u>in</u>: "The Molecular Base of Immune Cell Function", J.C. Kaplan, editor, Elsevier/North Holland, Amsterdam (1979).
17. C.S. Sher, W.J. Pledger, P. Martin, H. Antoniades, and C.D. Stiles, J. Cell Physiol, 97:371 (1978).
18. B.T. Hill, Biochem. Biophys. Acta., 516:3879 (1978).
19. A. Yen and A.B. Pardee, Exp. Cell Res., 116:103 (1978).
20. R.F. Brooks, Cell, 12:311 (1977).
21. A.B. Pardee, Proc. Natl. Sci., USA, 71:1286 (1974).
22. H. Rubin and R. Steiner, J. Cell Physiol., 85:261 (1975).
23. J.A. Smith, and L. Martin, Proc. Natl. Sci. Acad. USA, 70:1263 (1973).
24. H.M. Temin, J. Cell. Physiol., 78:161 (1971).
25. L. Augenlicht and R. Baserga, Exp. Cell Res., 89:255 (1974).

26. D.D. Cunningham and A.B. Pardee, Proc. Natl. Sci. Acad. USA, 64:1049 (1970).
27. A. Hershko, P. Mamont, R. Shields, and G.M. Tomkins, Nature New Biol., 232:206 (1971).
28. L.H. Hartwell, Bacteriol. Revs., 38:164 (1974).
29. J. Mandelstam and S.A. Higgs, J. Bacteriol., 120:38 (1974).
30. B. Shilo, V. Shilo, and G. Simchen, Nature, 264:767 (1976).
31. J.E. Sisken and L. Morasca, J. Cell. Biol., 25:179 (1965).
32. G. Sander, and A.B. Pardee, J. Cell. Physiol., 80:267 (1972).
33. K.D. Ley and R.A. Robey, J. Cell. Biol., 47:453 (1970).
34. D. Killander, and A. Zetterberg, Exp. Cell Res., 40:12 (1965).
35. T.O. Fox and A.B. Pardee, Science, 167:80 (1970).
36. A. Yen, J. Fried, T. Kerahara, A. Strife and B.D. Clarkson, Exp. Cell Res., 95:303 (1975).
37. R.F. Brooks, D.C. Bennett, and J.A. Smith, Cell, 19:493 (1980).
38. A.B. Pardee, B. Shilo and A. Koch, in: "Hormones and Cell Culture", Cold Spring Harbor, N.Y., p. 373 (1979).
38a. C. Nicolini, Cell Biophysics, Vol. 2, Number 3, 1-10 (1980).
39. R.G. Martin and S. Stein, Proc. Natl. Acad. Sci. USA, 73: 1655 (1976).
40. R.W. Holley, Nature, 258:487 (1975).
41. B.K. Bhuyan and T.J. Fraser, Cancer Res., 34:778 (1974).
42. G.P. Studzinski and J.F. Gierthy, J. Cell Physiol., 81:71 (1973).
43. J.C. Bartholomew, H. Yokota and P. Ross, J. Cell. Physiol., 88:277 (1976).
44. D. Paul, Biochem. Biophys. Res. Commun., 35:3111 (1973).
45. M.O. Bradley, K.W. Kohn, N.A. Sharkey and R.A. Ewig, Cancer Res., 37:2126 (1977).
46. E. Rozengurt and C.C. Po, Nature, 261:701 (1976).
47. J.M. Vasiliev, I.M. Gelfand, V.I. Guelstein and A.G. Malenkov, Int. J. Cancer, 1:451 (1966).
48. W.C. Dewey, H.H. Miller, and H. Nagasawa, Exp. Cell Res., 77:78 (1973).
49. C.F. Graham, J. Cell Sci., 1:363 (1966).
50. S. Gordon and Z. Cohen, J. Exp. Med., 134:935 (1971).
51. P. Rao and R. Johnson, Nature, 225:159 (1970).
52. J. Tyson, G. Garcia-Herdugo, and W. Sachsenmaier, Exp. Cell Res., 119:87
53. F.J. O'Neill, Cancer Res., 34:107 (1974).
54. M.H. Schneiderman, W.C. Dewey and D.P. Highfield, Exp. Cell Res., 67:147 (1971).
55. C. Basilico, Adv. Cancer Res., 24:223 (1977).
56. P. Rossow, V.G.H. Riddle, A.B. Pardee, Proc. Natl. Acad. Sci., 76:4446 (1979).
57. B.K. Choe and N.R. Rose, Exp. Cell Res., 83:271 (1974).
58. T.O. Fox and A.B. Pardee, J. Biol. Chem., 246:6159 (1971).
59. L.G. Gudas, R. James and A.B. Pardee, J. Biol. Chem., 231:3470 (1976).
60. K.D. Ley, J. Cell Biol., 66:95 (1975).

61. V.G.H. Riddle, R. Dubrow and A.B. Pardee, Proc. Natl. Acad. Sci. USA, 76:1298 (1979).
62. E.W. Gerner, R.E. Meyn, and R.M. Humphrey, J. Cell Physiol. 87:277 (1976).
63. L.I. Rebhun, Int. Rev. Cytol., 49:1 (1977).
64. E. Rapaport and P.C. Zamecnik, Proc. Natl. Acad. Sci. USA, 73:3984 (1976).
65. F. Grummt, G. Waltl, H. Jantzen, K. Hamprecht, U. Heubscher, and C.C. Kuenzle, Proc. Natl. Acad. Sci., USA, 76:6081 (1979).
66. K. Kumar and D. Friedman, Nature New Biol., 239:74 (1972).
67. L. Thompson and B. McCarthy, Biochem. Biophys. Res. Comm., 30:166 (1968).
68. L. Thompson and B. McCarthy, Fed. Proc., 30: Abst. 1177 (1971).
69. S. Planck, S. Seki, M. LeMahieu and G.C. Mueller, Fed. Proc., 33: Abst. 303 (1974).
70. S.M. Jazwinski and G.M. Edelman, Proc. Natl. Acad. Sci. USA, 73:3933 (1976).
71. H. Krokan, L.A. Cooke and H. Prydz, Biochem., 14:4233 (1975).
72. M. Goulian and B.Y. Tseng, J. Mol. Biol., 99:317 (1975).
73. H.V. Hershey and I. Taylor, Exp. Cell Res., 85:79 (1974).
74. D.L. Friedman, Biochem. Biophys. Acta, 353:447 (1974).
75. L.H. Lazarus, F.E.B.S. Letters, 35:166 (1973).
76. A. Klein and F. Bonhoeffer, Ann. Rev. Biochem., 41:301 (1972).
77. S. Seki, M. LeMahieu and G.C. Mueller, Biochem. Biophys. Acta, 378:333 (1975).
78. D. Billen and A.C. Olson, J. Cell Biol., 69:732 (1976).
79. N.A. Berger and E.S. Johnson, Biochem. Biophys. Acta, 425:1 (1976).
80. J.J. Castellot, Jr. in: "Introduction of Macromolecules into Viable Mammalian Cells" R. Baserga, S. Croce and G. Rovera, editors, Allen R. Liss, Inc., New York (1980).
81. S. Shin, V.H. Freedman, R. Risser and R. Pollack, Proc. Natl. Acad. Sci. USA, 72:4435 (1975).
82. C. Heidelberger, Ann. Rev. Biochem., 44:79 (1975).
83. J. Tooze, "The Molecular Biology of Tumor Viruses, Cold Spring Harbor, N.Y. (1973).
84. A. Smets, Biochem. Biophys. Acta. 605:93 (1980).
85. E.S. Stanbridge, and J. Wilkinson, Proc. Natl. Acad. Sci. USA, 75:1466 (1978).
86. V.H. Freedman and S. Shin, Cell, 3:355 (1974).
87. C.B. Stiles, W. Desmond, L.H. Chuman, G. Sato and M.H. Saier, Jr., Cancer Res., 36:3300 (1976).
88. C.B. Stiles, L.M. Chuman and M.H. Saier, Jr., J. Cell Biol., 70:169a (1976).
89. R. Sager and P. Kovac, Somatic Cell Genet., 4:375 (1978).
90. H. Eagle, G.E. Foley, H. Koprowski, H. Lazarus, E.M. Levine and R.A. Adams, J. Exp. Med., 131:863 (1970).

91. R.W. Tucker, K.K. Sanford, S.L. Handleman and G.M. Jones, Cancer Res., 37:1571 (1977).
92. I. MacPherson and L. Montagnier, Virology, 23:291 (1964).
93. T. Kuroki, Meth. Cell Biol., 9:157 (1975).
94. J. Folkman and A. Moscona, Nature, 273:345 (1978).
95. R. Kitchen and R. Sager, Somatic Cell Genet., (In Press)
96. J.A. DiPaolo, K. Takano, and N.C. Popescu, Cancer Res., 32:2686 (1972).
97. N. Bouck, and G. diMayorca, Nature, 264:722 (1976).
98. C.A. Reznikoff, D.W. Brankow and C. Heidelberger, Cancer Res., 33:3231 (1973).
99. J.C. Barrett and P.O.P. Ts'o, Proc. Natl. Acad. Sci. USA, 75:3761 (1978).
100. B.N. Ames, J. McCann and E. Yamasaki, Mutation Res., 31:347 (1975).
101. E. Huberman, P.J. Donovan and J.A. DiPaolo, J. Nat. Cancer Inst. 48:837 (1972).
102. J.C. Barrett and P.O.P. Ts'o, Proc. Natl. Acad. Sci. USA, 75:3297 (1978).
103. I. Berenblum, Cancer Res., 14:471 (1954).
104. S. Mondal, D.W. Brankow and C. Heidelberger, Cancer Res., 36:2254 (1976).
105. V.H. Freedman and S. Shin, J. Nat. Cancer Inst., 58:1873 (1977).
106. T. Kakunaga, Proc. Natl. Acad. Sci. USA, 75:1334 (1978).
107. G.E. Milo, Jr. and J.A. DiPaolo, Nature, 275:130 (1978).
108. National Academy of Science USA Report, "Particulate Polycyclic Organic Matter", Washington, D.C. (1972).
109. J.A. DiPaolo, R.L. Nelson and P.J. Donovan, Science, 165: 917 (1969).
110. E. Huberman, S.L. Sach, S.K. Yang, H.V. Gelboin, Proc. Natl. Acad. Sci. USA, 73:607 (1976).
111. S.K. Yang, D.W. McCourt, P.P. Rollec, H.V. Gelboin, Proc. Natl. Acad. Sci. USA, 73:2594 (1976).
112. T. Meehan and K. Straub, Nature, 277:410 (1979).
113. S. Kimura, V.L. Flannery, B. Levy, and P. Schaffer, Int. J. Cancer, 15:786 (1975).
114. G.M. Cooper and H.M. Temin, Cold Spring Harbor Symp. 39: 1027 (1975).
115. W.B. Pratt and R.W. Ruddon, "The Anticancer Drugs" Oxford Univ. Press (1979).
116. W. Sutow, J. Vietti and D.J. Fernback (Editors) "Clinical Pediatric Oncology", Chapters 5 and 6, St. Louis, The C.V. Mosby Company (1973).
117. M. Israel, J. Modest and E. Frei, III, Cancer Res., 35: 1365 (1975).
118. E. Frei, III, Cancer Res., 32:2593 (1972).
119. G.L. Nicolson, Biochem. Biophys. Acta, 457:57 (1976).
120. G.L. Nicolson, Biochem. Biophys. Acta, 458:1 (1976).
121. A.B. Pardee, Biochem. Biophys. Acta, 417:153 (1975).

122. A.B. Pardee, Nat. Cancer Inst. Monograph, 14:7 (1964).
123. P.S. Rudland and L. Jimenez de Asua, Biochem. Biophys. Acta, 550:91 (1979).
124. D. Gospodarowicz and J.S. Moran, Ann. Rev. Biochem., 45:531 (1977).
125. I. Hayashi and G. Sato, Nature, 259:132 (1976).
126. A. Rizzino and G. Sato, Proc. Natl. Acad. Sci. USA, 75:1844 (1978).
127. K. Mierzejewski and E. Rozengurt, Biochem. Biophys. Res. Commun., 73:271 (1976).
128. H. Bush and M. Shodell, J. Cell Physiol., 90:573 (1977).
129. P.V. Cherington, B. Smith and A.B. Pardee, Proc. Natl. Acad. Sci. USA, 76:3937 (1979).
130. G. Carpenter and S. Cohen, J. Cell Biol., 71:159 (1976).
131. J.E. DeLarco and G.J. Todaro, J. Cell Physiol., 94:335 (1978).
132. M.D. Hollenberg and P. Cuetrecaras, J.Biol. Chem., 250: 3845 (1975).
133. M. Das, T. Miyakawa, C.F. Fox, R.M. Pruss, A. Aharonov and H.R. Her-chman, Proc. Natl. Acad. Sci. USA, 74:2790 (1977).
134. Y. Shechter, L. Hernaez and P. Cuatrecuras, Proc. Natl. Acad. Sci. USA, 75:5788 (1978).
135. M. Das, Proc. Natl. Acad. Sci. USA, 77:122 (1980).
136. P. Greengard, Science, 199:146 (1978).
137. R.A. Jungmann and D.H. Russell, Life Sciences, 20:1787 (1977).
138. C.W. MacKenzie, III and R.H. Stellwagen, J. Biol. Chem. 249:5755 (1974).
139. P. Coffino, H.R. Bourne, U. Friedrich, J. Hochman, P.A. Insel, I. LeMaire, K.L. Melmon, and G.M. Tomkins, Recent Progress Hormone Res., 32:669 (1976).
140. R. Simantov and L. Sachs, J. Biol. Chem., 250:3236 (1975).
141. R. Simantov and L. Sachs, Eur. J. Biochem., 59:89 (1975).
142. N.M. Wigglesworth, A. Mastro, H.R. Bourne, and E. Rozengurt, Arch. Biochem. Biophys., 180:258 (1977).
143. A.J. Gharrett, A.M. Malkinson and J.R. Sheppard, Nature, 264:673 (1976).
144. A.D. Levinson, H. Oppermann, L. Levintow, H.E. Varmus, and J.M. Bishop, Cell, 15:561 (1978).
145. R.F. Kletzien, M.R. Miller and A.B. Pardee, Nature, 270: 57 (1977).
146. M. Costa, E.W. Gerner and D.H. Russell, J. Biol. Chem., 251:3313 (1976).
147. M. Costa, E.W. Gerner and D.H. Russell, Biochem. Biophys. Acta, 538:1 (1978).
148. J. Karn, E.M. Johnson, G. Vidali and V. Allfrey, J. Biol. Chem., 249:667 (1974).
149. G. Carpenter, L. King, Jr., and S. Cohen, Nature, 276:409 (1978).
150. J.J. Castellot, Jr., M. Miller and A.B. Pardee, Proc. Natl. Acad. Sci. USA, 75:351 (1978).

151. P.V. Reddy and A.B. Pardee, Proc. Natl. Acad. Sci. USA, 76:0000 (1980).
152. R.W. Holley, J.H. Baldwin, J.A. Kiernan and T.O. Messmer, Proc. Natl. Acad. Sci. USA, 73:3229 (1976).
153. H.L. Moses, J.A. Proper, M.E. Volkenant, D.J. Well and M.J. Getz, Cancer Res., 38:2807 (1978).
154. F.J. O'Neill, Exp. Cell Res., 117:393 (1978).
155. F.J. O'Neill, Cancer Res., 35:3111 (1975).
156. J.A. Fernandez-Pol, V.H. Bono, Jr. and G.S. Johnson, Proc. Natl. Acad. Sci. USA, 74:2889 (1977).
157. L. Mallucci, G.H. Poste and V. Wells, Nature New Biol., 235:222 (1972).
158. L. Mallucci, V. Wells and M.R. Young, Nature New Biol., 239:53 (1972).
159. R. Dubrow, V.G.H. Riddle and A.B. Pardee, Cancer Res., 39: 2718 (1979).
160. Cherington, P.V. and Pardee, A.B. (1980) J. Cell Physiol. 105: 25-32.
161. Pardee, A.B. Cherington, P.V. and Medrano, E.E. (1980) ICN-UCLA Sumposium on Control of Cell Division and Development. in press.
162. Riddle, V.G.H. and Pardee, A.B (1980) J. Cellular Physiol. 103: 11-15
163. Riddle, V.G.H., Pardee, A.B. and Rossow, P. (1979) J. Supromolecular Structure. 11: 529-538.
164. Shilo, B., Simchen, G. and Pardee, A.B. (1978) J. Cell Physiol. 97: 177-188.
165. Shilo, B., Riddle, V.G.H. and Pardee, A.B. (1979) Exp. Cell Res. 123: 221-227.
166. Medrano, E.E., and Pardee, A.B. (1980) Proc. Natl. Acad. Sci. U.S.A. 77:4123-4126.
167. Yen, A. and Pardee, A.B. (1979) Science 204: 1315-1317.
168. Tucker, R., Pardee, A.B. and Fujiward, K. (1979) Cell 17: 527-535.
169. Hamlin, J.L. and Pardee, A.B. (1978) in Vitro 14: 119-127.
170. Fansler, B. and Loeb, L.A. (1969) Exp. Cell Res. 57: 305-
171. Mitchison, J.M. (1971). The Biology of The Cell Cycle (London: Cambridge U. Press).
172. Brezney, R. and Coffey, D.S. (1977). J. Cell Biol. 73: 616
173. Miller, M.F., Castellot Jr., J.J. and Pardee, A.B. (1978) Biochemistry 17: 1073-1080.
174. Castellot Jr., J.J., Miller, M.R. and Pardee, A.B. (1978) Proc. Natl. Acad. Sci. U.S.A. 75: 351-355.
175. Miller, M., Castellot Jr., J.J. and Pardee, A.B. (1979) Exp. Cell Res. 120: 421-426.
176. Castellot Jr., J.J., Miller, M., Lehtomaki, D. and Pardee, A.B. (1979) J. Biol. Chem. 254: 6904-6908.

GENE MUTATION, QUANTITATIVE MUTAGENESIS, AND MUTAGEN SCREENING IN MAMMALIAN CELLS: STUDY WITH THE CHO/HGPRT SYSTEM

Abraham W. Hsie

Biology Division, Oak Ridge National Laboratory
Oak Ridge, Tennessee 37830

Summary

We have employed CHO cells to develop and define a set of stringent conditions for studying mutation induction to TG resistance. Several lines of evidence support the CHO/HGPRT system as a specific-locus mutatational assay. The system permits quantification of mutation at the HGPRT locus induced by various physical and chemical mutagens. The quantitative nature of the system provides a basis for the study of structure-function relationships of various classes of chemical mutagens. The intra- and interlaboratory reproducibility of this system suggests its potential for screening environmental agents for mutagenic activity.

Introduction

Recent advances in the molecular genetics of bacteria have to a large extent been attributed to the fact that mutants of desirable phenotypes can be readily isolated from haploid microrganisms. Since the successful cloning of a near-diploid Chinese hamster ovary (CHO) cell line over two decades ago (1) there has been interest in utilizing CHO and other cells for studying mechanisms of mammalian genetics. Because the great majority of mammalian genes exist in the diploid state, it has not

---

Research sponsored by the Office of Health and Environmental Research, U. S. Department of Energy, under contract W-7405-eng-26 with the Union Carbide Corporation.

been feasible to isolate recessive mutants, which are the predominant mutant type.

Mutagen-induced cell variants with altered nutritional requirements and drug sensitivity were first demonstrated in CHO (2) and V79 (3) cells. These successful isolations were attributed to, in the case of CHO auxotrophs, the hemizygosity of the affected genes (2), and in V79 drug-resistant variants, localization of the affected gene, hypoxanthine-guanine phosphoribosyl transferase (hgprt), on the functionally monosomic X chromosome (3). Many mutants with other phenotypes have since been isolated and characterized from these and other cells (4,5).

In this review, I will briefly discuss the CHO/HGPRT system (6-9), a system affecting HGPRT activity in CHO cells, to illustrate studies of gene mutation, quantitative mutagenesis, and mutagen screening. For studies of the genetic, biochemical, and molecular basis of gene mutation, isolation of one or a few mutants of the desired phenotypes would be sufficient; however, studies of quantitative mutagenesis require that a mutational protocol select for mutants a great majority (or all) of which are affected at a single gene. For mutagen screening a quantitative single-locus mutational assay must also demonstrate its intra- and interlaboratory reproducibility (6).

The CHO/HGPRT system

For most of our studies, we used CHO-K_1-BH_4 cells (7), a subclone of the near-diploid CHO cell line. CHO cells are genetically well characterized and are readily synchronized by various physical and chemical means. These cells exhibit a high cloning efficiency and a relatively stable karyotype. They grow well either on solid substrate or in suspension with a population doubling time of 12-14 hr. In addition to being used extensively for studying mutagen-induced cytotoxicity and gene mutation, CHO cells have also been favored for cytogenetic studies.

We have previously standardized the experimental procedures for cell culture, treatment with chemicals, and measurement of cytotoxicity and gene mutation. These are presented briefly in Fig. 1. We measure gene mutation by quantifying the frequency of mutants resistant to a purine analogue, 6-thioguanine (TG). The CHO/HGPRT system has been defined in terms of medium, pH, TG concentration, optimal cell density for selection, recovery of the presumptive mutants, and expression time for the mutant phenotype (6-8). A metabolic activation system derived from Aroclor 1254-preinduced male Sprague-Dawley rat livers has been used to determine the mutagenicity of promutagens (6,8,9). Since mutant selection is based on the loss of HGPRT, a phenotypic delay

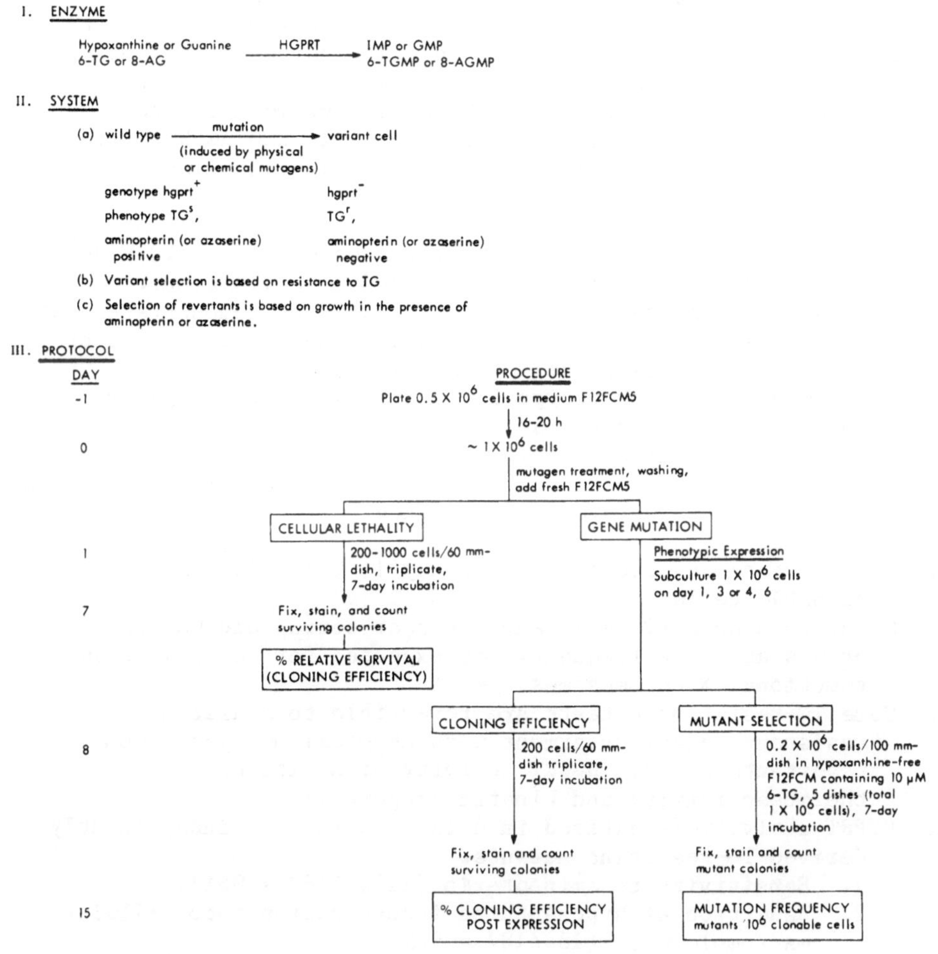
I. ENZYME
Hypoxanthine or Guanine
6-TG or 8-AG
HGPRT
IMP or GMP
6-TGMP or 8-AGMP
II. SYSTEM
(a) wild type
mutation
variant cell
(induced by physical or chemical mutagens)
genotype hgprt+
hgprt−
phenotype TGs,
TGr,
aminopterin (or azaserine) positive
aminopterin (or azaserine) negative
(b) Variant selection is based on resistance to TG
(c) Selection of revertants is based on growth in the presence of aminopterin or azaserine.
III. PROTOCOL
DAY
PROCEDURE
-1
Plate 0.5 X 10^6 cells in medium F12FCM5
16-20 h
0
~ 1 X 10^6 cells
mutagen treatment, washing, add fresh F12FCM5
CELLULAR LETHALITY
GENE MUTATION
1
200-1000 cells/60 mm-dish, triplicate, 7-day incubation
Phenotypic Expression
Subculture 1 X 10^6 cells on day 1, 3 or 4, 6
7
Fix, stain, and count surviving colonies
% RELATIVE SURVIVAL (CLONING EFFICIENCY)
CLONING EFFICIENCY
MUTANT SELECTION
8
200 cells/60 mm-dish triplicate, 7-day incubation
0.2 X 10^6 cells/100 mm-dish in hypoxanthine-free F12FCM containing 10 μM 6-TG, 5 dishes (total 1 X 10^6 cells), 7-day incubation
Fix, stain and count surviving colonies
Fix, stain and count mutant colonies
15
% CLONING EFFICIENCY POST EXPRESSION
MUTATION FREQUENCY mutants/10^6 clonable cells

Figure 1. The CHO/HGPRT system.

is expected; maximum stable expression of the TG-resistance phenotype is reached 7-9 days after mutation induction and remains constant for a diverse spectrum of mutagens such as ethyl methanesulfonate (EMS), N-methyl-N'-nitro-N-nitrosoguanidine, ICR-191, ICR-170, dimethylnitrosamine, benzo(a)pyrene, ultraviolet light, and Cr(VI) (K_2CrO_4) (6,10).

Early studies of mammalian cell mutagenesis have questioned the genetic vs epigenetic origin of phenotypic variants. In the absence of direct evidence of gene mutation through analyses of the nucleotide sequence of the hgprt gene and the amino acid sequence of the HGPRT protein for TG-resistant mutants, the following indirect evidence indicates that the CHO/HGPRT system fulfills the criteria for a specific-locus mutational assay (6,10,11).

A. There is a low spontaneous mutation frequency of 0-20 x 10^{-6} mutants/clonable cell in ~95% of 500 determinations.
B. Fluctuation analyses of spontaneous mutation demonstrate that the TG-resistance phenotype appears randomly at a rate of 2-4 x 10^{7} mutations/cell/generation.
C. The TG-resistance phenotype is stable for over 100 generations among 200 independent isolated mutants.
D. Physical and chemical mutagens induce TG-resistant mutants with a linear dose-response relationship, and mutants with a temperature-sensitive TG-resistance phenotype have been isolated.
E. The spontaneous reversion frequency is 0-1 x 10^{-7} mutants/clonable cells.
F. Mutagen-induced TG resistance in near-tetraploid CHO cells occurs at a low frequency, as expected for cells with two functional X chromosomes.
G. Some TG-resistant mutants are revertible to azaserine resistance after treatment with chemical mutagens; some revertants exhibit HGPRT activity with altered thermosensitivity and kinetic properties.
H. HGPRT activity is altered in a large number of independently derived TG-resistant mutants.
    1. Sensitivity to aminopterin (1170/1180 = 98%).
    2. Reduction of hypoxanthine incorporation into cellular macromolecules (184/187 = 98%).
    3. Reduced or altered HGPRT enzyme activity in cell-free extracts (96/98 = 98%).

Quantitative Mutagenesis with the CHO/HGPRT System

The quantitative nature of this system was evident from mutagenesis experiments performed over 6 years ago employing a direct-acting chemical mutagen, EMS (6-11) and a physical agent, ultraviolet light (12). With EMS we found that mutation induction with a treatment time of 16 hr occurred over the entire survival curve (7,13,14). Further studies using treatment times of 2–24 hr and varying EMS concentrations demonstrated the existence of a limited reciprocity of EMS mutagenesis; that is, when different combinations of EMS concentrations (0.05–3.2 mg/ml) are multiplied by varying treatment times (2–10 hr), the product [(mg/ml)·hr] yields a constant mutation frequency and cytotoxicity (13). Since 1974 our laboratory has studied the mutagenicity of more than 100 chemical and physical agents (6).

The quantitative nature of the system permits studies of structure-mutagenicity of various classes of direct-acting mutagens, and when the system is coupled with a liver S9-metabolic activation system, promutagens can also be studied (6,9). In analyzing mutagenesis data we describe the mutagenic activity as the number of mutants per 10^6 clonable cells induced by 1 μM of chemical tested. These results listed below together with determinations of chemically induced DNA lesions (14,15) provide some insights on the mechanisms of chemical mutagenesis.

A. Analyses of 10 alkylating chemicals reveal that their mutagenicity decreases with increasing size of the alkyl group; methylating agents are 3-6 times more mutagenic than the corresponding ethylating agents (16,17).
B. Studies of 19 heterocyclic mustards (ICR compounds) show that compounds with the tertiary amine side chain are more mutagenic than those with secondary amine side chain (18,19).
C. Of 6 platinum (II) chloroammines studied, <u>cis</u>-Pt$(NH_3)_2Cl_2$ is mutagenic whereas its steric isomer, <u>trans</u>-Pt$(NH_3)_2Cl_2$, exhibits highly reduced mutagenicity (15).
D. The mutagenic activity of quinolines is 4-nitroquinoline-1-oxide > quinoline > 4-aminoquinoline-1-oxide > 8-aminoquinoline > hydroxyquinoline, 8-nitroquinoline, and dichloroquinoline (San Sebastian, J. R., and Hsie, A. W., unpublished results).
E. While diphenylnitrosamine does not appear to be mutagenic, dimethylnitrosamine and diethylnitrosamine exhibit similar mutagenicity (San Sebastian, J. R., and Hsie, A. W., unpublished results).
F. The mutagenicity of haloethanes increases with increasing bromination, i.e., ethylene dibromide > ethylene bromochloride > ethylene dichloride (20).

Mutagen Screening with the CHO/HGPRT System

We have routinely used EMS as a positive control for direct-acting mutagens and found that EMS at 200 μg/ml for a treatment time of 5 hr consistently yields 300 mutants/10^6 clonable cells with a 15% standard error (6–20). Several laboratories such as those at DuPont Co., Chemical Industry Institute of Technology, Allied Chemical Co., Carnegie-Mellon Institute, Inhalation Toxicology Research Institute, etc. have used the CHO/HGPRT system as an integral part of their toxicology research and obtained similar results with EMS. Demonstration of the intra- and interlaboratory reproducibility in studies with standard mutagens such as EMS emphasizes the reliability of the CHO/HGPRT system for quantitative mutagenesis and mutagen screening.

The sensitive, quantitative, and reproducible results obtained from over 80 standard chemical and physical agents encouraged us to employ the CHO/HGPRT system to screen for the mutagenic activity of over 30 chemicals, some of which exist as organic mixtures (21,22). For example, we have shown that the acetone fraction of a coal-liquified crude oil which contains polycyclic aromatic primary amines and polycyclic aromatic nitrogen heterocyclics is the major contributor to the mutagenic activity of synthetic oil (22); thus, the system appears to be useful for determination of mutagenicity of organic mixtures and for corroboration of results from other biological assays.

Acknowledgements

I thank my colleagues P. A. Brimer, D. B. Couch, N. L. Forbes, A. W. Fuscoe, A. P. Li, M. H. Hsie, N. P. Johnson, R. Machanoff, J. P. O'Neill, J. R. San Sebastian, R. L. Schenley, C. H. Schroder, E.-L. Tan, and K. R. Tindall for experimental work and B. S. Hass, C. A. Jones, R. Machanoff, and R. L. Schenley for manuscript review.

References

1. Puck, T. T. and Marcus, P. I. (1955) Proc. Natl. Acad. Sci. USA 41, 432-437.
2. Puck, T. T. and Kao, F.-T. (1967) Proc. Natl. Acad. Sci. USA 58, 1227-1234.
3. Chu, E. H. Y. and Malling, H. V. (1968) Proc. Matl. Acad. Sci. USA 61, 1306-1312.
4. Thompson, L. H. and Baker, R. M. (1973) In Prescott, D. M. (ed.), Methods in Cell Biology, Academic Press, New York, pp 209-281.
5. Siminovitch, L. (1976) Cell 7, 1-11.
6. Hsie, A. W. (1980) In Williams, G. M., Kroes, R., Waaijers, H. W. and Van de Poll, K. W. (eds.), Predictive Value of Short-Term Screening Tests in Carcinogenicity Evaluation, Elsevier/North-Holland Biomedical Press, Amsterdam/New York/Oxford, pp 89-102.
7. Hsie, A. W., Brimer, P. A., Mitchell, T. J. and Gosslee, D. G. (1975) Somat. Cell Genet. 1, 247-261.
8. O'Neill, J. P., Brimer, P. A., Machanoff, R., Hirsch, G. P. and Hsie, A. W. (1977) Mutat. Res. 45, 91-101.

9. Machanoff, R., O'Neill, J. P., San Sebastian, J. R., Brimer, P. A. and Hsie, A. W. (1980) Proc. 11th Annu. Meet. Environ. Mutagen Soc., Mar. 16-19, 1980, Nashville, TN, p 102.
10. Hsie, A. W., O'Neill, J. P., Couch, D. B., San Sebastian, J. R., Brimer, P. A., Machanoff, R., Fuscoe, J. C., Riddle, J. C., Li, A. P., Forbes, N. L. and Hsie, A. W. (1978) Radiat. Res. 76, 471-492.
11. Hsie, A. W., Brimer, P. A., Machanoff, R. and Hsie, M. H. (1977) Mutat. Res. 45, 271-282.
12. Hsie, A. W., Brimer, P. A., Mitchell, T. J. and Gosslee, D. G. (1975) Somat. Cell Genet. 1, 383-389.
13. O'Neill, J. P. and Hsie, A. W. (1977) Nature 269, 815-817.
14. Thielman, H.-W., Schroder, C. H., O'Neill, J. P. and Hsie, A. W. (1979) Chem.-Biol. Interact. 26, 233-243.
15. Johnson, N. P., Hoeschele, J. D., Rahn, R. O., O'Neill, J. P. and Hsie, A. W. (1980) Cancer Res. 40, 1463-1468.
16. Couch, D. B. and Hsie, A. W. (1978) Mutat. Res. 57, 209-216.
17. Couch, D. B., Forbes, N. L. and Hsie, A. W. (1978) Mutat. Res. 57, 217-224.
18. O'Neill, J. P., Fuscoe, J. C. and Hsie, A. W. (1978) Cancer Res. 38, 506-509.
19. Fuscoe, J. C., O'Neill, J. P., Peck, R. M. and Hsie, A. W. (1979) Cancer Res. 39, 4875-4881.
20. Tan, E.-L. and Hsie, A. W. (1980) Proc. 11th Annu. Meet. Environ. Mutagen Soc., Mar. 16-19, 1980, Nashville, TN, p 56.
21. Hsie, A. W., Brimer, P. A., O'Neill, J. P., Epler, J. L., Guerin, M. R. and Hsie, M. H. (1980) Mutat. Res. 78, 79-84.
22. Hsie, A. W., O'Neill, J. P., Machanoff, R., Schenley, R. L. and Brimer, P. A. (1981) In de Serres, F. J. (ed.), International Programs for the Evaluation of Short-Term Tests for Carcinogenicity (in press).

ROUND TABLE DISCUSSION ON MECHANISMS CONTROLLING NORMAL VERSUS ABNORMAL CELL GROWTH

CLAUDIO NICOLINI: Could you, John, summarize the events or steps that take place during cell growth at the molecular level, and during normal or abnormal cell growth?

JOHN PAUL: It is rather clear from the experiments that we and other people have done recently, that the gross macromolecular changes that one can measure during growth are not that great. There are obviously quite specific changes that take place in synthesis but one does not certainly see evidence for the switching on of large number of genes which is one thing we might have possibly expected. There is a paradox here because we know that in viral trans-formation we are introducting new genetic functions into the cell but recollecting that there are only one or two. So, from the point of view of our studies it would seem that there is no strong evidence for the switching on of specific genes, and if there is, it maybe only one or two. As Dr. Bauer has shown there are normal genes rather similar to the characteristics of an onconovirus. It may be that these type of genes play a role in part in initiating the process of cell division. But from our studies, it would seem that some of the most striking observations at the macromolecular level are that changes in the processing of RNA are occuring rather than changes in the synthesis of RNA. These effects may be playing a more important part than de novo synthesis in the control of normal and abnormal cell growth.

CLAUDIO NICOLINI: Could you, Arthur, explain in a few words what are the biochemical mechanisms behind the idea of a restriction point?

ARTHUR PARDEE: My idea of the restriction point is that it is the time at which the cells become independent of normal growth regulatory factors from the outside, i.e. they no longer depend upon their conformation or their serum factors to continue through the cycle. My own bias is that in this preristuction part of the cycle you are building up a labile protein that is needed for the cell to go on to DNA synthesis.

CLAUDIO NICOLINI: Could you then summarize few molecular or cellular events that distinguish normal and transformed cells?

ARTHUR PARDEE: There is a long list of differences between normal cells and transformed cells in culture but none of them correlate really well with the ability to form a tumor in an animal, some correlate better than others but there are exceptions. Growth in suspension is one of these that seems to correlate well with tumor formation. However, none of these are really universal yet, a cell can be transformed by some of the tests but not transformed by others.

CLAUDIO NICOLINI: Could you, Scott, summarize from your laboratory point of view what is the difference between normal and abnormal cell growth, i.e. what could be the critical factor?

SCOTT WITTLESBERGER: From my point of view when a cell is growing abnormally it is not responding to the proper signals that control its growth and its social behavior as the normal cell does. I tend to look upon this as being an alteration in its cytoskelatal structure that somehow it cannot sense certain phenomona like where it is, what the components of the media are, etc.

CLAUDIO NICOLINI: Could you summarize the role of geometry in cell growth?

SCOTT WITTLESBERGER: From our data it is obviously playing some role in cell growth, RNA and protein metabolism and even gene expression. But when we say cell geometry we say that because there is no better word. What exactly it is were looking at I am not sure yet. It is just a rather crude expression because what we alter experimentally is geometry.

CLAUDIO NICOLINI: Do you Barlati have something to add to this point about the extracellular matrix?

JERGIO BARLATI: Yes, concerning how fibronectin degradation products might be related to cell transformation. Fibronectin is a molecule with many functions, i.e. multifunction molecule that interacts with proteins of the extracellular matrix, like collegen and proteins in the cell membrane like proteoglycans. So if fibronectin is intact you can build up the extracellular matrix. If the fibronectin is degraded by plasmin or plasminogen activator derived activities, you have a breakdown of fibronectin into 3 or 4 molecules that compete for the active sites but have different functions. This competition prevents the buildup of the extra cellular matrix by the loss of the multivarient functions of the fibronection.

CLAUDIO NICOLINI: Can you, Abraham, summarwze the differences between normal and transformed cell or fibroblast-like cells and transformed cell in terms of the cytoskeleton network?

DR. HSIE: Transformed cells generally have a lower cAMP level and exhibit a disorganized cytoskeleton as shown also by Dr. Brinkley at this institute. By that, we mean microtubles and microfilaments organization in the cytoplasm. In contrast normal counter-parts have a higher cAMP level and organized cytoskeleton. Unfortunately this is true in a model fibroblast system whether you transform the cell by spontaneous, virons or chemical means. But the situation in epitheal cells does not stand up well. This is unfortunate, since 90% of all human tumors are epithelial in origin.

CLAUDIO NICOLINI: Now Gary has genetic engineering and chromosone mapping helped in illuciating the mechanism of cell growth? What are the major understandings or breakthroughs, and what are the problems still remaining?

GARY STEIN: Gene mapping and genetic engineering are essentially tools that permit us to approach problems at the molecular level. One might search for types of aberrant gene expression as a function of neoplasia or tumor progression. One can think of these techniques being applied ultimately for screening various types of tumors and also for screening birth defects. These are ways these tools could be utilized at a very practical and diagnostic level.

CLAUDIO NICOLINI: Are Gary non-histone chromosonal proteins changing during the cell cycle and are they different in the normal and transformed cells in amount and in terms of enzymetic modifications?

GARY STEIN: There have been a number of cases showing differences in non-histone proteins in normal and transformed cells. I think the real problem is what is the function of these non-histone proteins. It is fine to do a run on one or two gels and say here is a clear cut example of a difference in a protein. But the point is what is that protein doing? To my way of thinking I don't know of any functional role of any chromosonal protein which is unique to the transformed cells. I think there are promising areas at this point but I would not be overly enthusiastic of anything shown to date.

CLAUDIO NICOLINI: There is an abundant literature on H1 phosphorylation, methylation, acetylation and their role in the control of DNA replication. Could you, Morton, summarize from your point of view the hard facts in these data?

MORTON BRADBURY: A few general comments first: I find the complexity of what I heard at this meeting rather horrifying. I see cycles within cycles, changes between the cytoplasm and the nucleus parameters being measured for which I cannot see where the experiments are leading to.

If one extends our knowledge of bacteria, then some of the questions asked are based on our ideas which come from our understanding of bacteria. I am just wondering in general whether after all these years if we have been looking at the problem of cell growth in the right way. I think there are several points I can make in relation to this.

In terms of the question of the relavancy of the G1 period David Prescott came to the view that G1 was really a growth period and if there was a switch then it probability occured at the G1-S boundary. I think that Arthur Pardee is not very far from that although he may bring his restruction point further into G1 than the G1-S boundary. Clearly he is talking in terms of events of growth, i.e., because of unequal division cycle less privileged cells are trying to catch up with the more privileged cells to reach a situation where they can undergo division. If we take this sort of general view then what we are coming to the point of view of Mazia, i.e. one has to look at the cell cycle in terms of a G1 which can be variable and which precedes the basic chromosomal cycle. The function of cell cycle is then to replicate the chromosones in that cell in their entirity. By replicating the chromosones, it means replicating the total genome, the total genetic potintial of that cell. It also means replicating those aspects of chromatin structure which are allowing a particular cell type to express its individuality. You have to look at the chromatin for answer about the replication of the genome and the replication of the characteristics of a particular set of chromosomes.

Scientists have been looking at chromatin structure and what is clear is that we have now an outline and understanding of the chromatin subunit, the first order coiling and we really don't have much understanding beyond that point.

When looking at such structures it is important to ask oneself what are the events that take place for chromatin to replicate itself. When comparing the bacterial and eukaryotic chromosones the real difference between them is the histones. The features of eucharyotic histones that are important are histones 3 and 4 which have evolved with eukaryotes. Whatever histones 3 and 4 do, their function is extremely important to chromatin structure and function, as they show extraordinary sequence conservation. Then you must ask oneself why has nature gone to the trouble of setting up a whole system of acetyl transferases, acetylases in order to modify these highly conserved proteins. It becomes quite clear then, that histone modifications like acetylation of the core histones are set up in response to important cell functions. The correlation of histone modification with important cell functions shows that acetylation of histones 3 and 4 are involved in DNA replication though we don't know how or what are the series of events that lead up to this point, but acetylation is involved in the transition from inactive to active chromatin. There is no doubt about these correlations and if one monitors what is the effect of histone modification on structure of chromatin then we can produce plausible models. A first approximation steady-state model would be the physical-chemical conditions of the cell nucleus where the chromatin is not chemically modified by the histones. This structure would be the 30 nanometer coil that one sees in electron micrographs.

If you go below the 30 nanometer coil then acetylation of the histones is an important event but that has not been demonstrated. In my opinion acetylation does destabilize the 30 nanometer coil but clearly it must do something more. We do know that after trypsin digestion of acetylated internal areas of the core particle you still have a particle. So the function of acetylation is not to open the particle. It may destablize the particle by allowing it to slide. We know that histones 3 and 4 are absolutely essential for the nucleosome structure. Histones 2A and 2B you can add to it but 3 & 4 are clearly the fundamental building block protein of the eucharyotic chromosone.

As you proceed up the hierarchy of histones, histones 2A & 2B show more variability and in H_1 there is a substantial variability for there is a whole set of very lysine rich histones which are now found in cells in different conditions. The question is whether or not this variability has a functional significance. It may very well be that the higher order coiling has to be modulated in some way for recognition purposes. It may be that one part of the genome which is to become available, for a whole set of linked genes to be expressed, has a particular type of H_1 subfraction. This particular type of H_1 subfraction could destablize the higher order coiling which then allows it under certain conditions to be expressed, like a puffed gene, like a loop coming out which is then available to the fine control of the cell itself.

In terms of H1 phosphorilation there is no exception to the general rule that there is a H_1-like molecule in eurcharyote and that H1 molecules are heavily phosphorilated in G_2, i.e. 6 phosphates per molecule and peaking before metaphase. It is quite clear that if nature has gone to the trouble of setting up a whole system of cAMP independent phosphor-kinases and phosphatases then it is there for a very important function.

In terms of the strategic location of H_1 in chromatin that is near the linker DNA, H_1 is a 3 domains protein, one domain attached to the core particle and the other 2 domains which are totally different in amino acid composition which are the modified regions of the molecule and can extend over large distances. It becomes quite easy to think of models whereby this type of mechanism can cause chromatin condensation.

Therefore histone modifications are important events which we must understand first in order to establish a sound foundation of chromosone structural and functional relationship. With such a sound foundation one can then ask oneself questions like what HMG proteins do to structure? Where do they bind? When do they bind? Is the acetylation of HMG protein important to their function? But you must have this very firm basis from which to work, if you don't have it, then you are just fishing. You must have a framework to base your basis thoughts upon.

CLAUDIO NICOLINI: Let me go a step further on this argument by recalling recent experimental observations, which may clearify the functional significance of chromatin structure and, by adding further complexity, points to a need of rigorous physicochemical models on which to base our thoughts and future experimentation.

Observations have been made, that chromatin structure, both lower and higher order structure, does change during cell growth, namely does undergo two abrupt transitions per cell cycle, when a cell is preparing either to replicate or to transcribe its DNA. These alterations of DNA structure may be determined and are indeed causally correlated with changes in H_1 phosphorilation, with changes in primary cilia, nuclear envelop organization and even with changes in bound and freeions. This yields a complex picture by which chromatin structure is the target, the key event, whose modification may lead to different functional events and that alteration maybe controlled not simply by a modification of one protein component but may happen at the electrostatic level and may have concurrent contributions of ionic changes, phophorilation, methylation and acetylation which could all be candidates for the same structural alteration.

Not all cells need to have those two transitions, in fact cells which do not have a G1 period do only go through one abrupt transition, directly from mitosis to S phase. So that is one event which may clearly define a G1 period, i.e. during G1 the chromatin has to undergo a cycle of condensation, that is an extra cycle of condensation, and then abrupt relaxation in order to have an increase in accessible templates in the chromatin DNA needed for the cell to express S phase functions. Cells which already possess the needed substrates and enzymes initiate DNA replication immediately after mitosis, going through only one cycle of chromatin condensation and decondensation.

Chromatin structure is affected not only by changes at the molecular level but maybe affected also by changes occuring at the level of cell geometry and shape, which are able to propagate through the microtubles-microfilaments network or any other suitable mechanism thru to the chromatin structure in normal cells, but cannot propagate to chromatin structure in the transformed cells, which indeed lack microtuble-microfilament organization and experience uncontrolled growth.

E. GENDEL: What are the capabilaties of gene sequencing and gene hybridization, i.e. genetic engineering to obtain antenatal diagnosis for conditions like sickle cell disease and thalassemia.?

JOHN PAUL: Hybridization techniques can be used for antenatal diagnosis and thalasemia is a case where a number of attempts are being made to develop probes which will make it possible to identify genetic defects in utero. Although I said we have not been able to pick up any significant evidence for genes being switch on or off, I think there is sufficient evidence to make us suspect that this does happen particularly in the transition of non growing tissues to growing tissues. We lack the probes, but if there is a breakthrough in finding probes which enable us to identify genes which are characteristic of growing versus non-growing tissues I think it is possible to appreciate that these would be a very useful kind of probes in looking at and looking for potential cancer cells. Again, certain cancers do produce specific proteins or are alleged to do so, or produce them in larger amounts than what is to be expected from normal cells. These again if we can obtain a probe for the appropriate messenger RNA, there is a good chance we maybe able to use these diagnostically. These are all methods wherein I think we can see possibilities of using genetic engineering techniques but I would emphasize that it is very early yet and it has been only 2 or 3 years that we have been able to manipulate genes in this way.

GARRY STEIN: I would like to expand on Paul statements concerning proliferation, neoplasia and genetic disorders, utilizing recombinant DNA technology.

As long as one could go ahead and identify a gene product, even gene products present in relatively small quantities in a cell and whose messenger RNA's are in small quantities, utilizing some very high powered immunological techniques one can isolate those messenger RNA's. It doesn't require very much messenger to go ahead and prepare a clone or to use that coupled with other techniques to be able to isolate genomic sequences. I think there is a tremendous hope in developing high resolution probes that could be used for studying these diseases at the molecular level.

Another application is that in certain situations within the cell cycle there is no "yes or no" situation in terms of the absolute presence or absence of a particular protein or enzyme, but differences are at the quantitative level. With respect to gene expression we are talking about qualitative differences in gene expression but we are talking about quantitative differences in the amount of messenger present. Here recombinant DNA technology is a very powerful tool in being able to tritiate the quantities present.

The last point is that I can forsee in 10 or 15 years in the future the possibility of clinically utilizing recombinant DNA technology for gene replacement therapy. I don't think it is totally unrealistic but it is important to realize that there is a lot of technology that has to be developed in order for that to take place. Clinically, I think there is a great deal that is yet to come.

CLAUDIO NICOLINI: Could you elaborate on the hard facts of the G0-G1 transition?

JOHN PAUL: I am certainly no expert on the hard facts behind the G0-G1 transition except that it is a very common observation and that most of the tissues in our bodies don't normally grow. Now this has been a condition which has been very difficult to reproduce in vitro for study purposes. I think perhaps the nearest thing that we have to this is 3T3 cells and primary fibioblasts. I am not even sure whether that reflects the situation that one sees in liver and in muscle. I think this is an area of complete and utter mystery that is fundamentally important. The G0-G1 transitions seem to me a rather fundamental step, the kind of step that may indeed involve the activation of specific genes.

CLAUDIO NICOLINI: How does the restriction point relate to the G0-G1 transition?

ARTHUR PARDEE: The difference between these cells I think depends on the available growth factors and the effects of neighboring cells. Platelet derived growth factor which is essential for G0 cells to be exposed to first in order for them to go on into the cycle. If you don't provide that factor the substances in serum don't work.

FRANCO GABRIELLI: If I have understood correctly, the following characteristics are what should be possessed by a cell component in order for it to be classified as a gene controller.

1. Bound to DNA of all eukaryotic cells.
2. Their synthesis should be coupled to DNA synthesis.
3. They should be heterogenous enough to control $5X10^6$ genes.
4. They should repetitively be bound along DNA.
5. They should not have turnover to stabilize differentiation.
6. They should have at least two post-translational modifications with relatively high turnover. Their post-translational modifications should be related to chromatin condensation and decondensation that take place during the cell cycle and to DNA transcription.
7. At least one of these components should be tissue specific and reversibly bound to DNA.
8. Change when the cell differentiates.
9. Once knowing their composition we should be able to tell if the cells are able to differentiate and determine cell histocompatability to distinquish tumor cells from normal cells. We have such a component and they are the histones!

GARRY STEIN: First even accepting as a model that histones are regulatory, I am not comfortable with the idea that there is sufficient heterogenity to take into account the fact that in human cells you have 50 copies of the genes and you have some post-translational modifications. But what you have to do to be able to alter the transcription of the gene is to alter using your model the histone DNA interaction. So your regulatory molecule would not actually be a histone but a histone modifying enzyme. Secondly, is there enough heterogenity amongst the histone modifying enzymes? Are there enough acetylases, kinases, phosphatases and deacetylases to bring that about. But my principle reservation is that I still don't see sufficient heterogenity to be able to have specificity even for large groups of genes.

CLAUDIO NICOLINI: Furthermore by only changing the geometry you induce an altered genome expression as previously shown and this is done without directly affecting the histones. The fact that histones may have a role doesn't imply that everything else is not there.

SECTION V:
CELL KINETICS AND CLINICAL APPLICATIONS

CLINICAL APPLICATIONS OF FLOW CYTOMETRY

Walfried A. Linden

Institute of Biophysics and Radiation Biology, University of Hamburg, Martinistraße 52, D-2000 Hamburg 20, FRG

INTRODUCTION

During the last decade appreciable effort has been undertaken to elucidate the impact of cell kinetic concepts for clinical oncology. As flow cytometry (FCM) has become a valuable and reliable research tool in cell cycle kinetics (for a review see (1,18) and as many of the changes in cell transformation from normal to malignant at least principally can be measured in flow systems, this method has been applied both in cancer diagnostics (as well as estimating of prognosis) and cancer treatment monitoring. It has been demonstrated in previous chapters that flow cytometry is a rapid method for the quantitative determination of biologically important substances in single cells in suspension. In clinical applications of FCM the most important quantitative parameters are nucleic acid content, protein content, cell volume and light scattering and cell surface antigens. The aim of this chapter is to present some typical FCM determinations of those parameters in different fields of interest. A critical discussion of these examples should help to get an answer to the question concerning the present and potential benefit of FCM data for the clinician.

CANCER DIAGNOSTICS

Prescreening for Cervical Cancer Detection

The well established method for the early detection of cancers of the uterine cervix is the microscopic inspection of vaginal-cervical cell material, obtained by scraping and stained with the traditional Papanicolaou technique. This method affords high qualification and experience of the cytotechnologist and pathologist, and on the other hand is very tedious and has appreciable error rates. Thus, there has been the desire to automate this procedure, using suitable instrumentation. Flow cytometry could serve as a prescreening device selecting from the big number of definitely normal cytology specimens a small number of suspicious ones for further microscopic investigation, thus performing the work now done by the cytotechnologist.

It has been demonstrated, that measurement of a single parameter is inadequate for FCM studies in cervical cytology (2, 3). Thus, the more recent work has been using a combination of several measuring parameters, generally nuclear DNA content and a cell volume correlated parameter, like small angle scattering or protein content.

A typical example is shown in Fig. 1. It presents fluorescence distributions of a benign and a malignant cervical smear. The two samples display characteristic differences. The benign smear (Fig.1a) shows a narrow distribution on the DNA axis, which on the protein scale extends from small to very high protein content. This is the distribution of the different normal cells in a cervical smear, which have the same DNA content, but different protein content, the latter increasing from basal to superficial cells. The malignant smear on

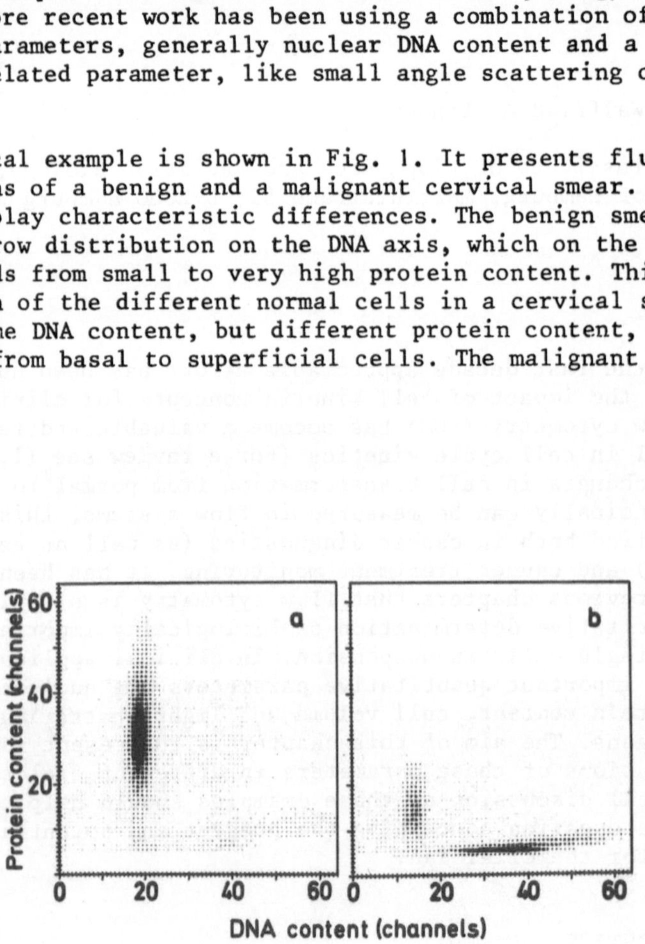

Fig. 1. Fluorescence distributions of human cervical smears stained with 4'-6-diamidino-2-phenylindole (DAPI) for DNA and sulforhodamine (SR 101) for total cellular protein. The graphs are scatter-plots; the number of cells per channel is represented, using increasing grey levels. a: normal specimen; b: malignant smear (carcinoma in situ).

the contrary (Fig. 1b) shows a very broad DNA distribution shifted to smaller protein content. This reflects the increased DNA content of tumor cells and their higher nuclear to cytoplasmic ratio. From such differences of benign and malignant smears we have derived criteria for their discrimination and applied to preliminary measurements of about 100 cases.

A comparison of our FCM data with the cytologic diagnoses obtained after Papanicolaou staining is set out in Table 1. The results look promising. We did not obtain false negative FCM diagnoses, i.e. the 5 cytologically malignant specimens of the Papanicolaou grades IV and V were all detected by FCM and classified as positive. Also the fraction of false positive diagnoses (normal specimens, classified as positive by FCM and thus requiring further analysis) is low, amounting to 11%. Of course the number of cases is too low for a statistically significant statement.

A major problem of FCM applied to cervical cytology is the occurence of artefacts leading to false alarms. Some of the sources of false alarms, causing a false positive classification of the specimen, have been identified. The most frequent false alarms are those from aggregates of normal cells. An example of another type of false alarm is shown in Fig. 2. It is a FCM distribution of a benign specimen with adherent bacteria. In contrast to the symmetrical ridge, obtained from an uncontaminated sample (Fig. 1a), a

Table 1. Comparison of Flow Cytometric Results after DAPI/SR 101 Staining with the Cytologic Diagnoses after Papanicolaou Staining[a].

Cytologic Diagnosis	Benign	Equivocal or Suspicious	Malignant	Total of FCM Diagnoses
Papanicolaou Grading	I and II	III and IIID	IV and V	
Flow Cytometry				
Negative	87% (85)	100% (1)	0% (0)	83% (86)
Positive	11% (11)	-	100% (5)	15% (16)
Unsatisfactory	2% (2)	-	-	2% (2)
Total of microscopic diagnoses	100% (98)	100% (1)	100% (5)	100% (104)

[a]Results are given in percentages and absolute count in parentheses

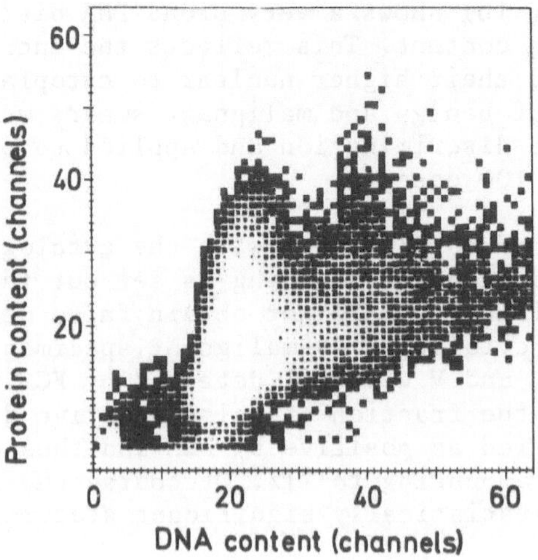

Fig. 2. FCM distribution of a sample contaminated with hemophilus vaginalis bacteria, as determined by cytology. The scale of the vertical axis (Number of cells per channel) has been set in such a way that the lower 10% of the plot at the foot of the peak are resolved (4).

"bump" to the right of the ridge of normal cells can be observed. Quite a number of cells appear at elevated DNA fluorescence, thus being measured outside the region of normal cells in the so-called plain of dysplasia, where neoplastic cells are expected.

Another problem to be solved is that of data analysis of multiparameter flow cytometry. We are working on a mathematical procedure to identify normal as well as abnormal cells by deconvolution (5).

On the whole, when critically judging the present results of our group as well as those published in the literature, the question, if a flow cytometric prescreening for cervical cancer will at any time be feasible, has to remain undecided. On the other hand we believe, as was mentioned also by Herman et al. (6), that flow cytometry may provide a quantitative definition of the spectrum of premalignant changes in epithelial cells, thus extending our capabilities in cytopathological research.

Assessment of Ploidy and Proliferation in Solid Tumors and Leukemia

Flow cytometric measurements of solid tumors and leukemia at present yield two criteria that might be correlated with the degree of malignancy: the ploidy level and the fractions of cells in the various phases of cell cycle, which indicate the proliferation capacity of the malignant system. Examples are shown in Figs. 3 and 4. Figure 3 presents FCM DNA distributions (histograms) of two samples taken from the colon of a patient with a colorectal adenocarcinoma. The first histogram (Fig. 3a) shows the distribution obtained for the normal mucous membrane with a first peak corresponding to the G_o/G_1 fraction of the normal cells with a diploid DNA content of $2C = 6$ pg, and with a second peak consisting of G_2+M cells. The second histogram gives the distribution obtained for the center of the tumor. In addition to the 2C peak which comprises diploid normal and/or diploid tumor cells in G_o/G_1 phase a second peak at elevated DNA content, corresponding to hyperdiploid tumor cells in G_o/G_1 phase can be recognized. By comparing the fluorescence intensity at which this second peak appears with that of the normal cells in G_1 phase a measure of the ploidy level of the tumor cell population is obtained. In this case we observe a tumor cell population with a 3C DNA content, which proliferates having its G_2+M peak at 6C DNA content.

A comparison of the ploidy levels, as determined by flow cytometry, of 20 colorectal adenocarcinomas with the histological grading is given in Table 2. We see a good correlation between the tumor differentiation and the ploidy level. Highly differentiated tumors mostly appear at diploid levels (2C), whereas tumors with a medium or low degree of differentiation mainly appear at enhanced ploidy levels (>2C). Apparently the dedifferentiation of tumor cells is accompanied by an increase of DNA content.

Figure 4 shows DNA distributions of two brain tumors, a meningioma and a medulloblastoma. The benign meningioma (Fig. 4a) has diploid DNA content with a G_o/G_1 peak at a DNA value of 2C. It is usual in FCM studies of this kind to take the fraction of cells

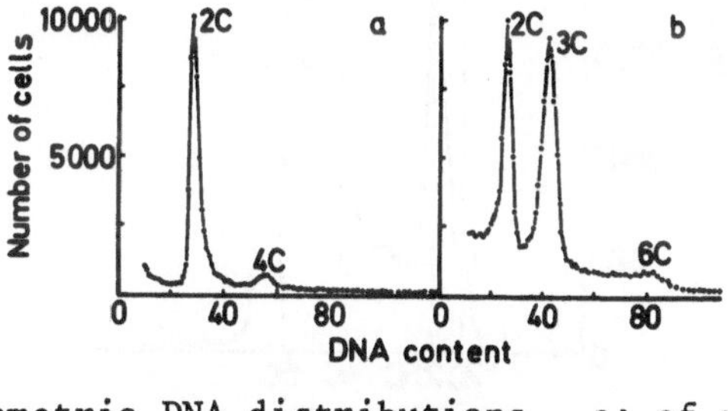

Fig. 3. Flow cytometric DNA distributions, a: of normal mucous membrane (edge of the resection material) and b: of a colorectal adenocarcinoma (center of the tumor) (3).

Table 2. Comparison of the Ploidy Levels as Determined by Flow Cytometry with the Degree of Differentiation of 20 Colorectal Adenocarcinomas (3).

Ploidy level from flow cytometry	Degree of differentiation from histology	
	High	Medium - Low
2C	5	1
More than 2C	2	12
Number of cases	7	13

in S and G_2+M phases as a criterion for cell proliferation, though of course especially in the "G_2+M" fraction of FCM histograms an appreciable amount of resting cells with double the DNA content may be contained. Following this operational definition the proliferation of the meningioma is relatively low with a fraction of cells in S+G_2+M phases of 16%.

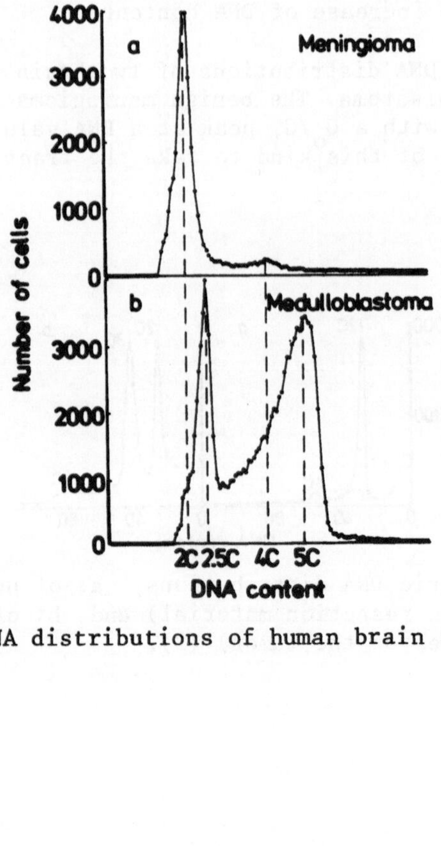

Fig. 4. DNA distributions of human brain tumors (7).

The DNA distribution of the malignant medulloblastoma is quite different. The DNA content is hyperdiploid, the G_o/G_1 peak being at 2.5C, and cell proliferation seems to be highly increased with a fraction of cells in $S+G_2+M$ phases of 73%.

However, these differences between benign and malignant brain tumors could not be confirmed in an extended study. A comparison of ploidy and histological classification of 64 brain tumors is presented in Table 3. Tumors with a G_o/G_1 peak at 2C DNA content, eventual further peaks only in geometrical order of 2C and cell number progressively decreasing with higher DNA content were designated euploid all the others as aneuploid. With this definition all benign brain tumors had euploid DNA content. However, also the majority of semimalignant and malignant tumors (66%) belonged to the euploid group. Thus no correlation between ploidy and histological dignity could be established. The same was true for the fraction of cells in $S+G_2+M$ phases. Though the mean fraction of cells in S phase was higher for the malignant tumors compared to the benign ones the difference was not statistically significant.

Also the results in the literature are rather controversal. In a study of 26 miscellaneous solid tumors Barlogie et al. (9) found that except for one breast and one colonic carcinoma, all tumors had aneuploid DNA contents. But no correlation between ploidy and proliferative cell characteristics was apparent. Collste et al.(10)

Table 3. Comparison of Ploidy as Determined by FCM[a] with the Histological Classification of 64 Brain tumors (8).

Ploidy	Histological Classification: Benign	Histological Classification: Semimalignant and malignant
Euploid	100% (32[b])	66% (21)
Aneuploid	0% (0)	34% (11[c])

[a]Results are given in percentages and absolute count in parentheses.

[b]In this group are contained 6 neurinomas; each of them had a main fraction of cells with 2C DNA content but regularly also smaller fractions with 4C and 8C DNA content.

[c]Containing 5 anaplastic tumors with a higher fraction of cells at 4C than at 2C DNA content, but no irregular DNA content.

reported FCM results of a study of 54 urothelial tumors. The fraction of cells with DNA values in $S+G_2+M$ of the cell cycle varied between 7 and 57%, but no statistically significant differences between different histological groups could be established. On the other hand Roters et al. (11) found in a FCM study of 55 prostatic lesions, that all adenomas had diploid ploidy levels and G_o/G_1 fractions higher than 95%, whereas all the 48 cases of prostatic carcinomas had G_o/G_1 fractions lower 95%; 64% of the carcinomas being diploid, and only 36% tetraploid. One metastatic lesion had G_o/G_1 cells at 3C DNA content. Atkin & Kay (12) demonstrated that the modal DNA value of the tumor tissue has prognostic significance. In a study of 1465 tumors they found that for all tumor sites except the cervix uteri, patients in the low (near-diploid) range of tumor DNA content showed better survival.

Moving to the work performed on leukemia, Fig. 5 shows DNA histograms of peripheral blood cells. Fig. 5a is a DNA histogram of blood from a healthy individual. Polymorphonuclear leucocytes, monocytes and lymphocytes are measured in a peak at 2C DNA content. Figure 5b refers to blood of a patient with chronic lymphatic leukemia. Besides a 2C peak there is a S plateau and a second peak at 4C. Figure 5c is a DNA histogram of blood of a patient with Sezary syndrome. Besides the 2C peak we realize a peak at 3.5C DNA content. Figure 5d gives the DNA histogram of a transformed immunocytoma. The main fraction of cells is recorded at 4C DNA content.

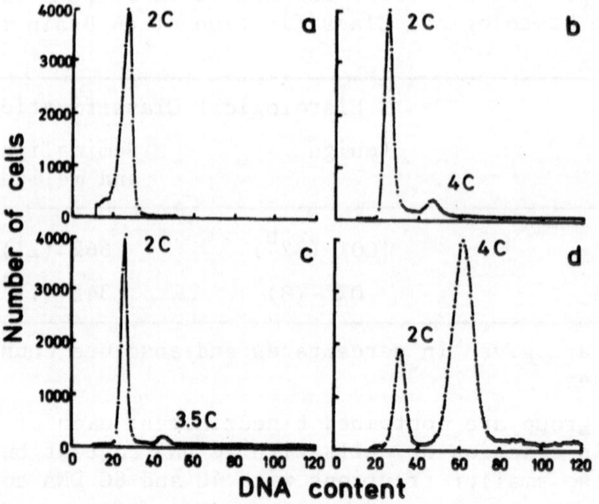

Fig. 5. DNA histograms of peripheral blood cells. a: healthy individual, b: patient with chronic lymphatic leukemia, c: patient with Sézary Syndrome, d: patient with transformed immunocytoma (13). For recent findings on Sézary Syndrome, see ref. 19.

In an investigation of 16 patients with leukemic non-Hodgkin's lymphomas - untreated and in different stages of therapy respectively - we obtained in 13 patients unimodal DNA distributions of peripheral blood at 2C DNA content similar to that in Fig. 5a. Bimodal DNA histograms were obtained in one case of chronic lymphatic leukemia (Fig. 5b) one case of Sézary syndrome (Fig. 5c) and one case of immunocytoma (Fig. 5d). The hyperploid cells in the case of Sézary syndrome and immunocytoma could be identified microscopically. In the case of the immunocytoma these cells appeared 60 months after onset of the lymphoma and 5 months after the last application of alkylating agents and can be interpreted to reflect increased malignancy in a preterminal stage of the disease.

These examples and others published in the literature show, that the determination of cellular DNA content as an additional parameter in the assessment of hematological malignancy may be of assistance in diagnosis, classification and estimation of prognosis. On the other hand the determination of DNA content alone is of very poor diagnostic significance. FCM DNA measurements of bone marrow of 175 patients with leukemia and lymphoma revealed in only 16.6 % of the cases abnormal DNA distributions, whereas concomitant cytogenetic analysis carried out on 160 patients showed karyotype abnormalities in 50% of acute leukemias (14).

Multiparameter techniques may yield better results (20,21) in studies of solid tumors as well as in investigations on leukemia. But at present the diagnostic relevance of flow cytometry in grading of malignancy is highly questionable.

CANCER TREATMENT MONITORING

Flow cytometry has been applied in monitoring the effects of chemotherapy and radiation therapy. The most experience has been obtained so far for studies in chemotherapy of acute leukemia. From FCM DNA histograms of leukemic bone marrow cells the effects of vincristine, prednisone, adriamycin, daunomycin and cytosine arabinoside on the distribution of cells in the cell cycle are known (for a review see 15). However, the relationship of these cell kinetic effects to lethal cell damage of leukemic blast cells has not yet been established. On the other hand in a FCM study of 115 adult patients with acute leukemia performed by Dosik et al. (16) pretreatment S phase compartment size correlated positively with the degree of bone marrow blast cytoreduction during the first 8 days of induction treatment. But there was no correlation with complete remission or with response duration. The study suggests that the percentage of pretreatment bone marrow cells in S phase is predictive for rate of cell kill and number of courses necessary for remission but not for attainment of complete remission. On the whole the therapeutic significance of cell kinetic studies under therapy of leukemia has still to be defined.

Another and perhaps more promising approach is the assessment of the leukemic blast cell compartment during therapy. This can obviously not be done by DNA measurements for, as has been demonstrated, leukemic blast cells very often have diploid DNA content. In the past years the development of leukemia specific antisera has made much progress (17). It is assumed that during remission there still may be a tumor burden in the bone marrow as high as 10^9 leukemic cells, which cannot be detected microscopically. There is now reasonable hope that using immunofluorescent techniques with a fluorescence activated cell sorter it will be possible to assess the fraction of leukemic blast cells during remission; thus eventually detecting relapse before clinical recurrence. This would of course be a very effective means for treatment monitoring giving also a rational basis for continuing or stopping of chemotherapy.

An example obtained in our institute is given in Fig. 6. Low angle scattering and fluorescence of bone marrow cells was measured of a patient in relapse of a chronic myeloic leukemia. The antiserum

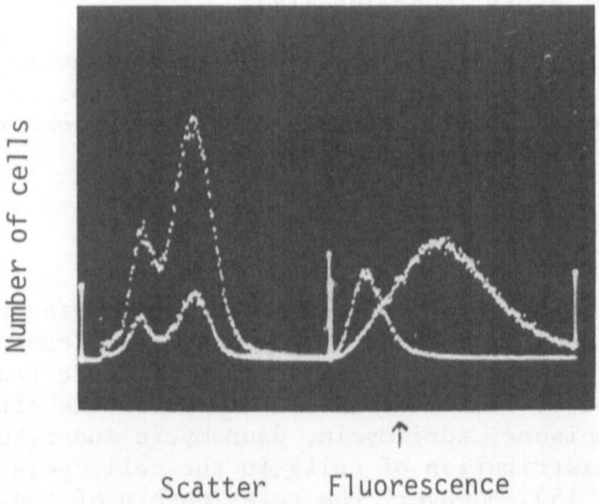

Fig. 6. Low angle scattering and fluorescence intensity of bone marrow cells of a patient in relapse of a chronic myeloic leukemia. The fluorescence distribution to the left was measured with unlabeled cells reflecting the autofluorescence of the cells, the fluorescence distribution to the right with cells labeled with an antibody against the common-ALL (cALL) antigen, coupled with fluorescein isothiocyanate. The arrow marks the border for the cALL positive cells. In the left part of the figure the respective scatter distributions are shown, the smaller one corresponding to the unlabeled cells. The measurements were performed using a FACS IV cell sorter (H. Baisch, unpublished results).

applied contained an antibody against the so-called common-ALL (cALL) antigen, labeled with fluorescein isothiocyanate. From a comparison with the autofluorescence distribution of the unlabeled cells a border can be established for the cALL positive cells (arrow). With this border a fraction of 79% cALL positive cells is obtained, which result perfectly agrees with the 80% of leukemic blast cells scored microscopically. Of course this example shall demonstrate only the potential of this method, which also needs further development.

CONCLUSIONS

When critically judging the present state of the art of clinical use of flow cytometry, the following conclusions can be drawn:

(i) Though flow cytometry is a powerful research method, in most present applications it cannot supply data, which are of substantial relevance for clinical decision making.

(ii) In cancer diagnostics detailed studies concentrating on very few oncologic diseases like prostatic lesions, colorectal carcinomas or acute leukemia should be performed under close collaboration of clinicians, pathologists and biophysicists, to establish the diagnostic and prognostic relevance of FCM distributions. In any tumor system some 50 values are without significance; on the contrary the studies have to be performed for several years trying to get a correlation with the clinical development. Furtheron in many neoplasias like leukemia measurement of DNA content has to be supplemented by the determination of other parameters like cell volume, RNA content etc.

(iii) In cancer treatment monitoring - perhaps apart from the studies on untreated leukemia - determination of cell cycle phase distributions of cancer cells are of very limited value. They can only be interpreted, when having data on the colony forming ability of the tumor cells. Thus those studies should mostly be confined to experimental tumor systems like animal tumors or human xenografts in nude mice. A very promising approach in treatment monitoring is the assessment of tumor burden in acute leukemia using leukemia specific antisera and a fluorescence activated cell sorter.

ACKNOWLEDGEMENTS

I would like to thank Mrs. D. Lensch for carefully typing the manuscript and Mrs. B. Prosch for preparing the figures.

REFERENCES

1. Gray, J.W., Dean, P.N. & Mendelsohn, M.L. (1979) in Flow Cytometry and Sorting, eds. Melamed, M.R., Mullaney,P.F. & Mendel-

sohn, M.L. (John Wiley & Sons, New York), pp. 383-407.
2. Linden, W.A., Ochlich, K., Baisch, H., Scholz, K.-U., Mauss, H.-J., Stegner, H.-E., Joshi, D.S., Wu, C.T., Koprowska, I. & Nicolini, C. (1979) J. Histochem. Cytochem. 27, 529-535.
3. Linden, W.A., Beck, H.-P., Baisch, H., Gebbers, J.-O., Heienbrok, W., Junghanns, P., Roters, M., Scholz, K.-U., Stegner, H.-E., Winkler, R. & Wöllmer, W. (1980) in Flow Cytometry IV, eds. Lindmo, T., Thorud, E. & Laerum, O.D. (Norwegian University Press, Oslo), pp. 408-412.
4. Gieseking, F., Baisch, H., Scholz, K.-U., Stegner, H.-E. & Linden, W.A. (1980) Analytical Quant. Cytol. in press.
5. Scholz, K.-U. (1980) Analytical Quant. Cytol. in press.
6. Herman, C.J., Bunnag, B. & Cassidy, M. (1979) in Flow Cytometry and Sorting, eds. Melamed, M.R., Mullaney, P.F. & Mendelsohn, M.L. (John Wiley & Sons, New York), pp. 559-572.
7. Heienbrok, W., Roters, M. & Linden, W.A. (1978) in Advances in Neurosurgery 5, eds. Frowein, R.A., Wilcke, O., Karimi-Nejad, A., Brock, M. & Klinger, M. (Springer, Berlin), pp. 285-288.
8. Borchers, H. (1981) Thesis, University of Hamburg.
9. Barlogie, B., Göhde, W., Johnston, D.A., Smallwood, L., Schumann, J., Drewinko, B. & Freireich, E.J. (1978) Cancer Res. 38, 3333-3339.
10. Collste, L.G., Darzynkiewicz, Z., Traganos, F., Sharpless, T.K., Devonec, M., Claps, M.L.K., Whitmore Jr., W.F. & Melamed, M.R. (1979) Br. J. Cancer 40, 872-877.
11. Roters, M., Lämmel, A., Kastendieck, H. & Becker, H. (1980) in Flow Cytometry IV, eds. Lindmo, T., Thorud, E. & Laerum, O.D. (Norwegian University Press, Oslo), pp. 397-401.
12. Atkin, N.B. & Kay, R. (1979) Br. J. Cancer 40, 210-221.
13. Düllmann, J., Wulfhekel, U., Linden, W.A., Beck, H.-P. & Hausmann, K. (1980) Blut 324, 1-9.
14. Göhde, W., Schumann, J., Büchner, Th., Otto, F. & Barlogie, B. (1979) in Flow Cytometry and Sorting, eds. Melamed, M.R., Mullaney, P.F. & Mendelsohn, M.L. (John Wiley & Sons, New York), pp. 599-620.
15. Arlin, Z.A., Fried, J. & Clarkson, B.D. (1979) in Flow Cytometry and Sorting, eds. Melamed, M.R., Mullaney, P.F. & Mendelsohn, M. L. (John Wiley & Sons, New York), pp. 583-597.
16. Dosik, G.M., Barlogie, B., Smith, T.L., Gehan, E.A., Keating, M. J., McCredie, K.B. & Freireich, E.J. (1980) Blood 55, 474-482.
17. Jannossy, G., Roberts, M.M., Capellaro, D., Greaves, M.F. & Francis, G.E. (1978) in Immunofluorescence and Related Staining Techniques, eds. Knapp, W., Holubar, K. & Wick, G. (Elsevier/North Holland Biomedical Press, Amsterdam), pp. 111-122.
18. C. Nicolini, in Advances in Neuroblastoma Research, ed. A. Raven, Press. New York, 271-286 (1980).

19. Vonderheid, E. S. Fang, Helfrich, M., Abraham, S., and Nicolini, C., Journal of Investigative Dermatology, 76, 28-37 (1981) .
20. Nicolini, C., Linden W., Zietz, S. and Wu, C., Nature, 270, 163-176 (1977).
21. Zietz, S., Grattarola, M. Desaive, C., and Nicolini, C., Cell Tissue Kinetics, 13, 473-484 (1980).

THE RELEVANCE OF CELL KINETICS IN DETERMINING DRUG ACTIVITY IN VITRO

Benjamin Drewinko and Barthel Barlogie

The University of Texas System Cancer Center
M.D. Anderson Hospital and Tumor Institute
6723 Bertner
Houston, Texas 77030

(Supported by Grant CA16763 through the National Large Bowel Cancer Project and Grants CA14528 and CA23272 through the National Cancer Institute, NIH, DHEW.)

Cell cultures provide a rapid, efficient and economic system for initial cytotoxicity screenings of antitumor agents, allowing elucidation of the mode of action of a drug in a controlled, systematic fashion with a high degree of resolution. The main assumption for _in vitro_ studies with chemotherapeutic agents is that the survival response of cultured cells will reflect that of _in vivo_ cells once the drug reaches the neoplastic elements, thus circumventing the pharmacokinetics determinants of tumor response (i.e. absorption, transportation, combination, transformation and degradation). While a direct translation from _in vitro_ to _in vivo_ systems is not possible, many survival responses of proliferating mammalian cells are reasonably similar to the two situations (1-4) provided that strict criteria are established to define cell survival. Drug-induced cell killing is the result of an interplay between the type, extent, and duration of the damaging effect caused by a drug to critical biosynthetic pathways or subcellular structures and the capacity of living elements to bypass or repair such damage. Because of this interplay, a lethally damaged cell may divide several times before the entire progeny perishes from the damage inherited from the single ancestor (5-8). Furthermore, antitumor agents generate two types of unrelated effects in treated cells: lethality and delayed transit through the mitotic cycle (9-11). Therefore, cell proliferation may be decreased because of actual cell kill, cell cycle progression delay, a reversible or irreversible block in a particular stage of the cycle or, more commonly, a combination of these factors. This temporary lag in the multiplication rate may increase the doubling time of the treated population and produce results indistinguishable from those obtained in populations whose cells are actually killed (12).

For these reasons, cellular death in proliferating populations is best evaluated by the permanent loss of the reproductive integrity of the individual cells (12-14). In vitro, this permanent loss is reflected in the inability of cells to proliferate indefinitely, forming colonies (clonogenic capacity) under the appropriate experimental conditions (15). In vitro clone formation implies a minimum of 5-6 (but usually 8-10) cell divisions which represent at least a 32-64 (and up to 1,000-fold) increment in the number of cells initially exposed to the drug. Under these circumstances, colony formation is independent of the intitial biochemical or transit-delay effects caused by the agent on the treated cell (for example, a colony may be small but will still represent the progeny of a surviving cell) and the assessment will be more accurate but the lethal efficacy of the agent may appear decreased (15, 16). Applying the colony formation method to cell culture systems, dose response effects can be analyzed quantitatively, since the exact concentration of the drug bathing cells and the duration of exposure is known, and these quantitative responses can be used to compare the efficacy of different agents on a given cell type or the activity of a specific drug on different cell classes.

A major limitation of experiments involving in vitro cells resides in the difficulty of extrapolating information detailed on cultured exponentially growing cells to the expected responses of tumor cells in vivo, where large fractions of the population may be in the quiescent state (G_0 cells) (17-19). Quiescent cells are usually considered less sensitive than proliferating cells to the lethal activity of most antitumor drugs (19-27). However, recent investigations suggest the usefulness of utilizing stationary phase cultures as an adequate in vitro model for the biological behavior of in vivo neoplasms with low fractions of proliferating cells (low growth fractions) (20, 22, 23, 28-32). Cultures in stationary phase can be obtained by a variety of methods, all of which lead the exponentially growing cells to a state where net increments in cell number can no longer be demonstrated. By comparing the survival response of cells treated in these two phases of in vitro growth, a more clinically relevant evaluation of cell-drug interactions can be obtained (20, 24, 31).

For the exponentially growing populations, the position in the cycle occupied by a cell at the time of drug administration is an important determinant for survival. Most antitumor drugs demonstrate significant differences in efficacy at each stage of the cell cycle (age-dependent response) (11, 33, 34). These age-dependent differences may have significant impact in the development of improved clinical chemotherapeutic regimens. The kinetics of in vitro cell killing may indicate dose and time manipulations of clinical scheduling for different combinations of drugs given simultaneously or in sequence. Thus, it may be possible to accumulate a large population of cells in a given stage of the cell cycle and utilize a second drug (or combination of drugs), the main killing effect of which occurs in that specific stage to sterilize the tumor more efficiently. On the other hand, agents with widely different activities in distinct stages of the cycle could be combined in a paired delivery to attack all of the cells composing the tumor population. Therefore, in vitro studies on the mode of action of various drugs in conjunction with cytokinetic data may provide alternative approaches to combination therapy which could prove more effective than the largely empirical methods utilized to date.

This review presents the experience gathered over the past decade in our studies on the influence of cellular proliferation kinetics on the cytotoxic activity of antitumor drugs.

MATERIALS AND METHODS

Cell Lines

Two established human cell lines were used in this investigation. One was an IgA-producing lymphoid cell line (T_1 cells) derived from a patient with lymphocytic lymphoma (35). The other was a carcinoembryonic antigen-producing line (LoVo cells) derived from the metastatic nodule of a patient with colon carcinoma (36). Both lines are maintained as monolayer cultures in Ham's F-10 medium (Grand Island Biological Company, Grand Island, New York) supplemented with 20% fetal calf serum, 1% glutamine, 1% MEM vitamins and antibiotics, LoVo cells can generate glandular structures in vitro when grown as monolayers (37) and in vivo when propagated as xenografts in nude mice (38).

The doubling time of exponentially-growing T_1 cells is 44 hours and the generation time is 31.1 hours. Transit times through each stage of the cycle are as follows: G_1 = 14.2 hours; S = 9.9 hours; and G_2+M = 7.0 hours (39). The corresponding values for LoVo cells are: doubling time = 37 hours; generation time = 29.3 hours; G_1 phase = 14.7 hours, S phase = 10.7 hours and G_2+M phases = 4.8 hours (37). Exponential growth of LoVo cells lasts for about five days, after which time cells enter stationary phase. The plating efficiency of T_1 cells ranges from 35 to 70 percent.

Cell Synchronization

Cells accumulated in S phase were obtained with a single treatment of 24 hour duration with 3 mM Thd for T_1 cells, and

7 mM Thd for LoVo cells. After removing Thd, cell cycle transit was monitored with the labeling and the mitotic index, and with flow cytometry (FCM). For labeling index determinations, the cells were pulse labeled for 30 minutes with {^{3}H}Cdr before harvesting. Cytocentrifuge preparations were processed for radioautography using a 50% solution of Ilford emulsion (Polysciences, Inc., Warrington, Pennsylvania) in distilled water, exposed for two weeks and developed in Kodak D-19 (Eastman Kodak, Rochester, New York). Labeled cells were identified by the presence of 5 or more grains overlying the nucleus. FCM studies were conducted on mithramycin-ethidium bromide stained cells using a Phywe ICP 11 pulse cytophotometer (Phywe, A.G., Gottingen, Germany) with methods previously described (37, 39).

At the end of the synchronization procedures, 85%-90% of T_1 cells were in S phase. This level was maintained for the next 6 hours, after which the proportion of S phase cells declined rapidly; at 13 hours about 78% of the cells were in the G_2 phase. A mitotic peak ranging from 5-12% occurred between 15 and 17 hours and cells subsequently entered G_1 phase composing 84% of the population at 28 hours. In the case of the LoVo cells, about 80-85% of the population was accumulated in S phase at the end of the Thd block. Synchrony was maintained in G_2, with 80% of the cells accumulated in that stage, but rapidly decreased when cells reached G_1. Twenty four hours after release from the block, the compartment distribution was similar to that observed for asynchronous LoVo cells.

Survival Assays

Stock cultures were harvested and counted with the aid of a Coulter Counter Model ZBI electronic counter (Coulter Electronics, Hialeah, Florida). Cell suspension aliquots were seeded into 60 mm petri dishes (5 x 10^5 cells/dish). The cells were incubated without medium replenishment at 37°C in a 5% CO_2 atmosphere in air for 48 to 72 hours to achieve exponential growth and, in the case of LoVo cells, for 8 days to achieve stationary phase of growth. The medium was discarded and the cells were exposed in increasing drug concentrations for exactly one hour at 37°C. The drug was decanted and the cells were washed twice in Hanks' balanced salt solution, harvested as a monodispersed suspension and counted. Known aliquots of the cell suspension were dispensed into 60 mm petri dishes so that 50-100 colonies would appear after 21 days of incubation. The colonies were stained with 2% crystal violet in 95% ethanol. Viability was defined as the ability of single cells to give rise to a colony of greater than 50 cells. In each experiment, the plating efficiency of at least 6 control cultures was assessed simultaneously. Control cultures consisted of cells treated in exactly the same manner as the test cells but without receiving drug. The survival fractions for the different drug concentrations were normalized with respect to the individual controls for each experiment. All experiments were repeated at least twice with triplicate samples for each drug concentration.

Survival curve patterns were classified in 5 types: simple exponential (type A); biphasic exponential (type B); threshold exponential (type C); exponential plateau (type D); and ineffectual (type E) as previously described (31). Because in many instances the shape of the survival curve obtained for LoVo cells in stationary phase differed substantially from that determined for the exponentially growing counterpart it was impossible to quantify differences in the degree of cell kill by conventional parameters (13). Therefore, we arbitrarily calculated differences in efficacy on these 2 classes of cells as the ratio of the survival levels determined at the mid-range point of concentrations used for each drug.

Age-dependent cell survival was investigated by incubating synchronized cells with increasing concentrations of drug at regular intervals throughout the cell cycle. Plating efficiency controls were obtained at each time point.

Drugs

All drugs (listed in Table 1) were obtained from the Cancer Chemotherapy Branch, Division of Cancer Treatment, National Cancer Institute, U.S.A. Drug solutions were always prepared in growth medium immediately before each experiment and the pH was adjusted, if necessary, to 7.2 - 7.4. Water soluble drugs were first dissolved in saline solution or in distilled water. Lipid soluble drugs were first dissolved in pure ethanol or in 20% ethanol-80% propylene glycol. In some cases (VP-16, AMSA, etc.), the drug was dissolved first with the solvent furnished by the manufacturer. At the final concentrations used in our experiments, such solvent used alone failed to affect the viability of either exponential or stationary phase cells.

RESULTS

Survival was compared for LoVo cells in exponential growth (2-3 days after subculture) and in stationary phase (8 days after subculture). Cis-DDP was more effective (efficacy ratio = 2.3) on cells in stationary phase than on exponentially growing ones (Figure 1). This effect resulted from the abrogation of the shoulder region of the type C survival curve, while the slope was maintained essentially intact. VDS was also more effective on cells in stationary phase (efficacy ratio = 18.9), but in this case the type B pattern of the survival curve was similar for both cell classes.

Methyl-GAG, hycanthone and VBL had a similar low killing effect (less than 90%) on both exponentially growing and stationary

TABLE 1

LIST OF ANTITUMOR AGENTS

Common Name	Chemical Name	Abbreviation	NSC #
Adriamycin	14-hydroxydaunorubicin	ADR	123127
AMSA	Methanesulfon-M-anisidide, 4'	AMSA	24992
Anthracenedione	1,4-dihydro-5,8-bis{{2-{2-hydroxyethyl) amino}-ethyl}amino}-9,10-anthracenedione dihydrochloride	DHAQ.Cl	301739
Bleomycin	--	BLEO	125066
Camptothecin	--	CS	100880
Carmustine	1,3-bis(2-chloroethyl)-1-nitroso-urea	BCNU	409962
Cis-acid	4-(3-(2-Chloroethyl)-3-nitroso-ureido)-cis-cyclohexanecarboyxlic acid	cis-acid	153174
Cisplatin	cis-diamminedichloroplatinum	cis-DDP	119875
Cytosine arabinoside	1-β-D-arabinofuranosylcytosine	ara-C	63878
Epipodophyllotoxin	4'demethylepipodophyllotoxin, 9-(4,6-O-ethylidene-β-D-glucopyranoside)	VP-16-213	141540
5-Fluorouracil	--	5-FU	19893
Hycanthone	--	--	142982

TABLE 1 (continued) - LIST OF ANTITUMOR AGENTS

Common Name	Chemical Name	Abbreviation	NSC #
Hydroxyurea	--	HU	32065
Lomustine	1-(2-chloro-ethyl)-3-cyclohexyl-1-nitrosourea	CCNU	79037
Maytansine	--	MAYT	153858
Melphalan	3-(p-(bis(2-chloroethyl)amino)phenyl)-L-alanine	L-PAM	8806
Methotrexate	glutamic acid,N-(P-((2,4-diamino-6-pteridinyl) methyl)methylamino)benzoyl)	MTX	740
Methyl-GAG	Methylglyoxal bis-guanyl-hydrazone	Methyl-GAG	32946
Mitomycin C	--	Mito C	26980
Peptichemio	m-(di(2-chloroethyl)amino)-L-pheylalanine, multipeptide complex	PC	247516
Prednisolone	pregna-1,4-diene-3,20-dione,11β,17,21-trihydroxy	--	9900
Rubidazone	daunorubicin benzoylhydrazone hydrochloride	RUB	164011
Semustine	1-(2-chloroethyl)-3-trans-(4-methyl-cyclohexyl)-1-nitrosourea	MeCCNU	95441
Transplatin	trans-diamminedichloroplatinum	trans-DDP	--

TABLE 1 (continued) - LIST OF ANTITUMOR AGENTS

Common Name	Chemical Name	Abbreviation	NSC #
Vinblastine	--	VBL	49842
Vincristine	--	VCR	67574
Vindesine	desacetyl vinblastine amide sulfate	VDS	245467
Yoshi 864	1-propanol, 3,3'-iminodi-dimethanesulfonate (ester), hydrochloride	--	102627

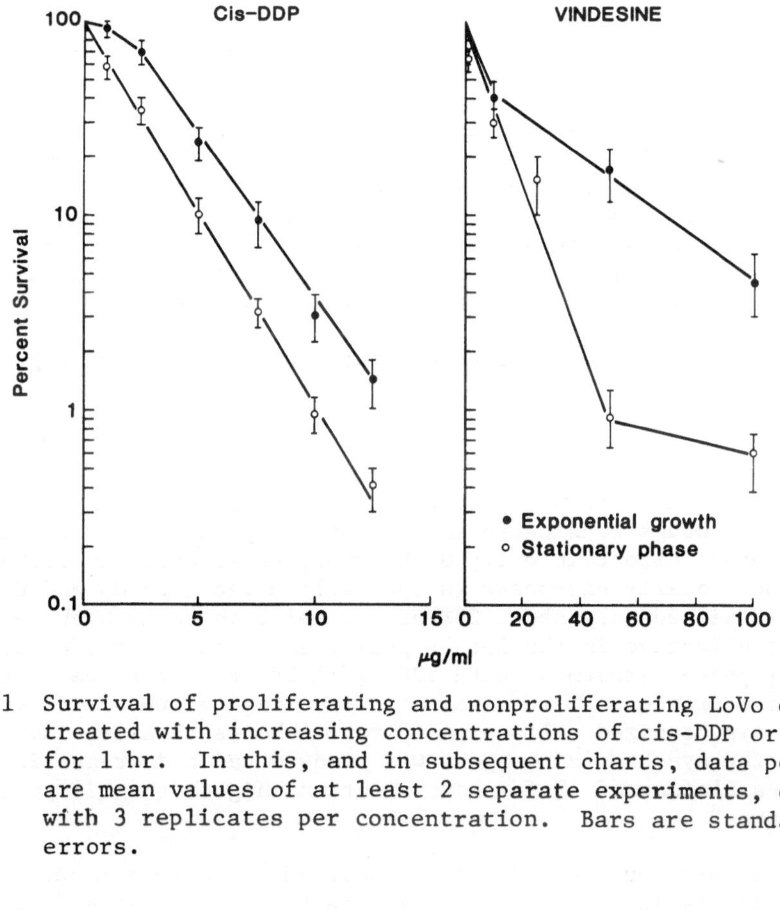

Fig. 1 Survival of proliferating and nonproliferating LoVo cells treated with increasing concentrations of cis-DDP or VDS for 1 hr. In this, and in subsequent charts, data points are mean values of at least 2 separate experiments, each with 3 replicates per concentration. Bars are standard errors.

phase cells; radiomimetic agents (mitomycin C, BCNU and cis-acid) displayed the same powerful effect (greater than 90% cell kill) and the same type C survival curve on both cell classes.

Agents considered to be cell cycle stage-sensitive and to act primarily on cells positioned in S phase (5-FU, MTX and HU) were considerably less lethal on stationary phase cells than on exponentially growing cells (Figure 2 ; Table 2). In fact, MTX completely failed to kill cells in stationary phase of growth. Mitotic inhibitors (VCR, MAYT and VP-16) all had similar type B survival curves and in all instances their efficacy on stationary phase cells was about 2.5 to 3-fold less than on exponentially growing cells.

Similar differences in efficacy between exponential and stationary phase cells were observed for DNA intercalating agents such as anthracycline derivatives (ADR and RUB) and AMSA.
The ineffective trans isomer of diammine-dichloroplatinum was even less active when used on stationary cells and the antibiotic BLEO also showed markedly less activity (efficacy ratio = 8) on stationary phase cells than on exponentially growing ones.

When synchronized cells were treated with antitumor drugs for one hour at different stages of the cell cycle, a variety of age-dependent response patterns were observed. Both prednisolone and camptothecin had similar quantitative efficacies in G_2 and G_1 phases. However, while camptothecin displayed its greatest efficacy on cells in S phase, such cells were completely refractory to prednisolone. HU and 5-FU, drugs considered to be S phase-specific, exerted their maximal lethal effect in early S. However, their efficacy decreased but did not totally disappear as the cells moved into G_2 and G_1 phase. MTX killed only about 20% of the cells in early S phase, was even less effective in the late S phase, and totally ineffectual in G_2 and G_1 phase. Treatment with 500 μg/ml of ara-C for one hour failed to decrease cell survival at every stage of the cell cycle. However, synchronized cells exposed for 14 hours to ara-C immediately after the synchronizing agent was removed had a decrease in survival to a plateau of 4% for concentrations higher than 25 μg/ml. (See figure 3.)

There were considerable differences in the age-dependent response to alkylating agents. While mephalan was most effective in the early G_1 phase, peptichemio, its congener, had a greater efficacy in G_2 phase. Both agents were considerably less effective in S phase. In contrast, Yoshi 864 and cis-acid, the latter an alkylating and carboamylating water-soluble nitrosourea derivative, had their greatest efficacy in early S phase. Their efficacy persisted well into mid-S phase, decreased during late S phase, and increased again in late G_1 phase for Yoshi 864 and in G_2 phase for cis-acid.

TABLE 2

DIFFERENTIAL EFFICACY OF ANTITUMOR DRUGS

ON PROLIFERATING AND NONPROLIFERATING LOVO CELLS

Drugs	Stationary/Exponential*
5-FU	1.3
MTX	1.9
HU	1.3
VCR	2.5
MAYT	2.6
VP-16	3.2
ADR	2.5
RUB	3.3
AMSA	3.8
trans-DDP	3.0
BLEP	8.0

*Ratio of Survivals at mid-range dose point.

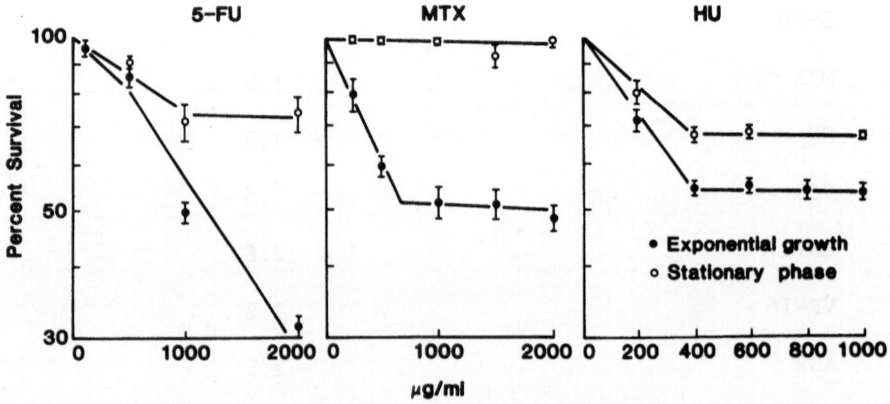

Fig. 2. Survival of proliferating and nonproliferating LoVo cells treated for 1 hour with S-phase sensitive agents.

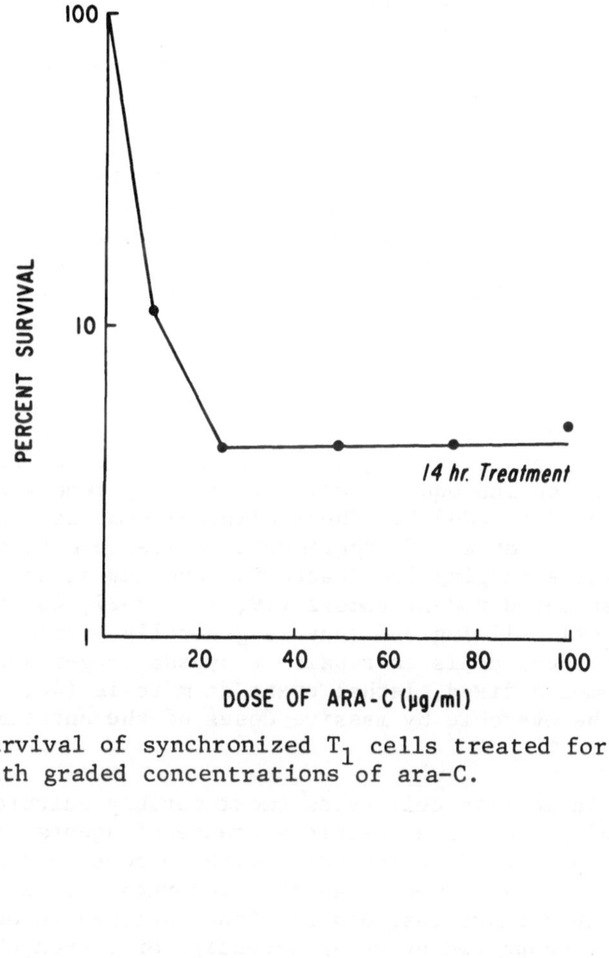

Fig. 3. Survival of synchronized T_1 cells treated for 14 hours with graded concentrations of ara-C.

Three nitrosourea derivatives killed cells in all stages of the cell cycle but each exerted preferential killing effects at different phases: BCNU in early S and G_2, methyl-CCNU in early S, and CCNU in early S and G_1 (not shown). In the case of BCNU, age-dependent changes were reflected in modifications of both the extent of the shoulder region and the slope of the exponential part of the type C dose-dependent survival curve (not shown). Only the length of the shoulder region varied for cells treated with CCNU and only the slope changed for cells treated with methyl-CCNU.

VCR and VDS were equally effective in all stages of the cell cycle of synchronized LoVo cells (Figure 4). In contrast to the marked age-dependent cytotoxicities observed for nitrosourea derivatives on T_1 cells, LoVo cells exposed to BCNU and cis-acid, showed no significant increase in cell killing as a function of position in the cycle.

In many instances, the particular age-dependent pattern defined for a given drug did not change regardless of the concentration level employed in the analysis (not shown). For other agents, cell cycle stage sensitivity was either intensified or attenuated as the concentration of the agent was increased.

DISCUSSION

There is generalized agreement concerning the increased sensitivity of rapidly proliferating tumors to most anticancer regimens with respect to the inadequate response shown by tumors with low growth fractions (25, 40-42). These clinical conclusions are supported by a vast array of experimental evidence obtained in a variety of systems ranging from bacterial and mammalian cell cultures to transplanted rodent tumors (19, 21, 23-27, 43-49). The difference in cell killing efficacy is generally attributed to the ability of quiescent cells to repair or bypass drug-induced damage before it becomes a fixed, lethal event at mitosis (46, 50-54) and can sometimes be overcome by massive doses of the antitumor agent (55).

However, in certain cell types (most notably cultured tumors of rodent origin) and for a restricted range of agents, increased sensitivity of nonproliferating cells with respect to that of exponentially growing elements has been reported (22, 27, 29, 56-58). In most such instances, observations obtained in one cell system were not supported by other investigators, even when using similar cell types. Thus, the reported increased efficacy of BCNU (22, 29) BLEO (29) or HU (57) on nonproliferating cells was not confirmed by others in different or even similar cultured cell systems (23, 26, 27, 30, 57, 59, 60). These discrepancies can be attributed in part to species-related biological differences.

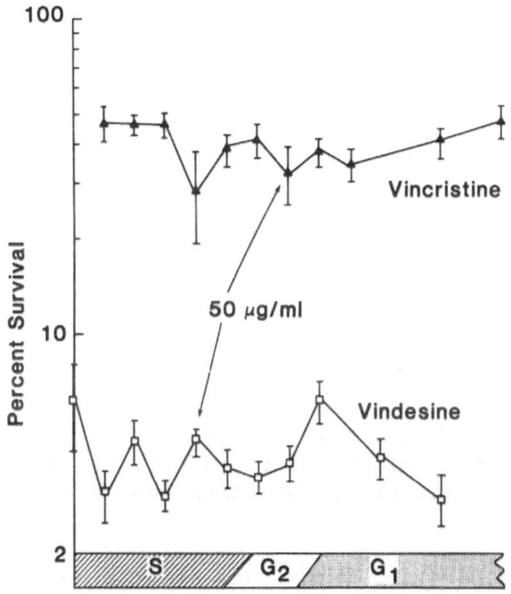

Fig. 4. Survival of synchronized LoVo cells treated for 1 hour with vinca alkaloids at different stages of the cycle.

More important, perhaps, is the diversity in methodology used to obtain nonproliferating cultures (20, 23, 28, 31, 32, 61, 62). This diversity can generate significant differences in physiological properties of the treated cells, and in their metabolic disposition of the noxious agent (31, 32, 43, 63, 64) and also lead to differences in sensitivity and clonogenicity as a function of the duration of the stationary growth (32, 65). Another equally important possibility resides in the fact that the stationary phase cultures used by different investigators differ significantly in their cell cycle compartment distribution; thus, it is possible that purported differences in drug sensitivity of stationary versus proliferating cells may actually reflect the sensitivity of cells accumulated in different stages of the cell cycle. For instance, Tobey and Ley (62) and Madoc-Jones and Bruce (66) respectively demonstrated that Chinese hamster and L cells were arrested in the G_1 stage of the cycle during stationary phase, while Thatcher and Walker (59), Drewinko, et al.(67), Macieira-Coelho (68) and Ross and Sinclair (69) showed considerably accumulation in G_2 phase for hamster embryo, human myeloma, human embryo fibroblasts and Chinese hamster cells respectively.

For our LoVo cells in stationary phase of growth, the cell cycle stage compartment distribution analyzed by FCM showed a considerable proportion of cells with S phase (15%) and G_2+M (16%) DNA content, although the labeling index was only 1% (37). These results indicate that mammalian cells in the stationary phase of growth do not necessarily accumualte in G_1 phase and that failure to incorporate (^{3}H)Thd cannot be considered evidence of non-S-stage position. Apparently, many cells in stationary phase can be delayed in the G_2 phase of the cycle (8, 59, 68, 70), while others are stopped in the midst of synthesizing DNA or synthesize DNA very slowly.

As shown by others for occasional agents, our cultured human cells also disclosed two drugs (cis-DDP and VDS) that were more effective on stationary phase cultures than on exponentially growing cells. In a manner similar to that observed by Hagemen et al. for BCNU on mouse plasmacytoma cells (50), the increased effectiveness of cis-DDP on the human tumor cells resulted from the abrogation of the shoulder region of the survival curve while the slope was retained intact. Such changes in the shape of the survival curve can also be observed in exponentially growing LoVo cells if cis-DDP is administered at 41° (71), suggesting that the capacity to absorb drug-induced damage without expressing a lethal effect is thermosensitive and modulated by the proliferative status of the LoVo cells.

VDS was almost 20 times more effective on stationary phase cells than on exponentially growing ones. It is difficult to reconcile this observation with the lethal mode of action ascribed to vinca alkaloids (i.e., inhibition of mitotic spindle). Thus, it is possible that either VDS has an additional killing mechanism not shared by its cogeners or that the activity of this particular agent persists long enough to inhibit division when the treated cells in stationary phase are allowed to proliferate exponentially for colony-formation. In direct contrast to the findings of Olah, et al for Chinese hamster cells (30), vincristine was less sensitive on LoVo cells in stationary phase than on exponentially growing ones and both, our findings and those of Olah et al, vary from the observations of Hill (72) for Syrian hamster ovary cells where either vincristine and vindesine were totally ineffective against cells in stationary phase. Thus, it appears that species differences mediate important modifications of the interactions between antitumor drugs (at least vinca alkaloids) and the target cells under various proliferative conditions.

In the manner described by others for alkylating agents (59, 73, 74), the efficacy of nitrosourea derivatives and mitomycin C was quite similar for both exponentially growing and stationary phase LoVo cells. This similar efficacy could originate from an intrinsic property of the interaction of alkylating agents and the target cells, or result from the inability of stationary phase cells to repair potentially lethal damage before they are manipulated into exponential growth (52).

Of great clinical importance, most antitumor drugs displayed decreased efficacy on LoVo cells in stationary phase of growth. The decreased efficacy was noted for cell cycle stage-sensitive agents, for mitotic inhibitors, and for DNA-intercalating agents. These findings reiterate the significance of quantifying cellular proliferation kinetics for determing potential sensitivity to tumor therapy and demonstrate that, in contrast to some cells of rodent origin, nonproliferating human tumor cells have decreased sensitivity to most antitumor agents. Although a rare drug may show increased activity on nonproliferating cells, in general only those agents that display alkylating properties can be expected to, at most, be as effective on nonproliferating as on proliferating cells.

As indicated before, position in the cycle at the time of exposure to a drug is an important determinant for survival of proliferating cells. Some drugs (i.e., MTX) may only kill cells positioned in the S phase of the cycle, while virtually every drug exhibits preferential kill in a particular stage. Because most drugs exhibit preferential rather than specific cell kill in a given stage of the cycle, it is perhaps more appropriate to describe such agents as "cell cycle phase-sensitive" rather than with the commonly employed term "phase-specific". For instance, our studies with HU and 5-FU, agents considered to kill cells by inhibiting DNA synthesis, also displayed substantial killing in G_2 and G_1 phases, as also observed by Bhuyan et al. for DON cells (34). Conversely,

T_1 cells synchronized in S phase and exposed for 1 hour to MTX showed a mere 20% decrease in survival, indicating that either all cells positioned in the cycle stage purportedly sensitive to an agent are not equally susceptible to its lethal action or that cell age alone is not sufficient to predict a lethal outcome. This fact is further emphasized by our results with ara-C. This agent completely failed to kill S phase cells when administered for only one hour but readily decreased survival when treatment was extended for an interval equal to the length of S phase. This finding has been previously observed on other cell lines (75, 76) and its explanation can be ascribed, at least in part, to the lethal mode of action of the antitumor agent. Drugs that inhibit DNA synthesis kill cells by the mechanism of "unbalanced growth" (76, 77). To achieve lethality by this mechanism, inhibition of DNA synthesis must be maintained for a period longer than the generation time (78), lest the cells resume their growth, synchronized but still capable of unlimited proliferation. For drugs that bind irreversibly to their macromolecules, degrade slowly, and are exposed to cells which have a limited capacity to synthesize new macromolecules, continuous DNA inhibition may be achieved even after brief exposure to the drug. If one or more of these conditions are not satisfied, lethality can only be obtained if the cells are incubated continuously with that antitumor agent.

Although realizing that the sensitivity in a particular stage of the cycle may not always be related to the particular biochemical event responsible for cell death, Mauro and Madoc-Jones (33) and Bhuyan et al. (34) analyzed the age-dependent response of different cell lines to a great number of antitumor agents with the intention to extract generalizations useful in the classification of cell-drug interactions. For some agents, their observations on cell cycle stage sensitivity were similar. But for many drugs, their conclusions disagreed not only with each other but also with the age-dependent patterns reported by other investigators employing the same drug (or their congener) on different cell lines. Our present report serves to emphasize these discrepancies and the inadequacy of extrapolating results obtained on a given cell line to a generalized statement applicable to the response of all mammalian cells. For instance, both Madoc-Jones and Mauro (11) and Bhuyan et al. (34) considered that alkylating agents were characterized by a distinct preferential cell kill in G_1 phase or at the G_1/S transition. In our studies, this pattern was observed only for melphalan. Its congener, peptichemio, in addition to killing cells in G_1 phase, exhibited the greatest activity in G_2 phase. Other alkylating agents such as Yoshi 864 also killed cells in early S phase, while cis-acid was most active in early S and G_2 phase. Vinca alkaloids considered more efficacious in S phase cells by other authors (30, 79) showed almost equal activity in all stages of the cycle in our studies.

Minor structural modifications of a drug may generate profound changes in the age-dependent patterns. In the studies of Mauro and Madoc-Jones, vincristine and vinblastine, congeners whose molecular structure differs by only two H atoms, showed distinct concentration-dependent activities and different age-dependent survival patterns, the latter characterized by particular modifications of the shape of the survival curve (79). These observations are similar to our present findings where three closely related nitrosourea congeners interacting with the same cell type (T_1 cells) originated each a unique age-dependent pattern of survival and a distinct modification of the capacity to absorb damage and/or sensitivity of the target molecule.

Several factors may be responsible for the disparity in results among different investigators. To analyze cell age-dependent cytotoxic effects, so-called synchronized cells are exposed to the antitumor agents at regular intervals throughout the cell cycle transit. Strictly, a synchronized population of cells is that in which all elements pass through a specific point of the cycle at the same time. Current available methods impose several technical limitations in the achievement of this purpose and no techinque provides more than partial synchrony. Thus, although cells in one stage of the cycle may predominate, the population is actually spread at various points throughout that stage and even in other stages of the cycle. Therefore, the operational term "synchronization" does not imply the phasing of all cellular processes, but refers to the simultaneous manifestation of some distinct marker, such as the synthesis of DNA, the presence of a mitotic apparatus, etc. While the cell population may appear phased with respect to one biochemical process, other biochemical activities, which should normally occur in subsequent stages, may have already taken place during the interval required to obtain the "synchronous" population. In addition, synchronization is an ephemeral event since mammalian cells have random rates of transit through the cycle stages and synchrony rapidly decays, preventing clear-cut interpreation of results obtained after the release of the synchrony induction. Hence, just as discrepancies in the sensitivity of proliferating and nonproliferating cells can be sometimes ascribed to the variety of methods used by different investigators, diversities in age-dependent response patterns may only reflect an assortment of factors unrelated to actual cell cycle sensitivity (i.e., methods used to synchronize cells, degree of synchrony achieved, rate of synchrony decay, etc.).

Conceivably, dissimilar cell cycle stage-sensitivity patterns could be sometimes related to variations in cell species origin. For instance, the age-dependent pattern observed for our human lymphoma cells treated with nitrosoureas differed substantially from that reported by Bhuyan, et al. for DON cells (34) and by Barranco and Humphrey for CHO cells (80). The pattern observed for Vinca alkaloids on our human colon carcinoma cells is totally unrelated to that observed by Mauro and Madoc-Jones and by Hill respectively for Chinese and Syrian hamster cells (72, 79). Yet, even within a particular species, diversity in histological origin may originate distinct age-dependent response patterns. For instance, our human lymphoma cells showed limited sensitivity (<20%) to MTX while the human colon carcinoma cells showed greater than 50% cell kill, a proportion greater than that of cells present in S phase during the treatment interval. A similar age-dependent survival pattern was obtained for both BCNU and cis-acid when the target cells were colon carcinoma, a pattern that differed significantly from that obtained on lymphoma cells. However, the response of lymphoma cells to both agents were largely similar. These findings explain, in part, the distinct responses of different tissues elicited by a particular drug and could acquire significant value in treatment strategies requiring precise scheduling of antitumor agents.

It must also be considered that differences in age-dependent patterns among investigators may actually result from the dissimilar selection of concentration levels and unequal treatment intervals used to examine the sensitivity of the synchronized cells. Even if the treatment interval is maintained uniform, it may not be biologically equivalent. Different cell classes may have different generation and cell cycle stage transit times, and, therefore, during a given exposure interval, more cells of a faster cell type may pass through the sensitive stage of the cell cycle.

In spite of these shortcomings, considerable information has been accumulated on the cell cycle stage-dependent effects of a variety of antitumor agents. Unfortunately, this information has been inappropriately utilized by some clinical oncologists to develop "kinetic" chemotherapeutic regimens which have thus far met with varying, mostly unsatisfactory, degrees of success in terms of improved clinical responses. This implementation of the *in vitro* information has obviously been too premature, not only because of the inadequacy of present *in vitro* knowledge, but also because of the vagaries of our current information regarding the cytodynamics of *in vivo* tumors. A clinical chemotherapeutic strategy based on a cell kinetics rationale requires accurate knowledge of the fraction of proliferating cells and of the durations of the phases of the cell cycle, features that change continously during the natural history of disease (17, 81-83), and as a consequence of the perturbation induced by the therapeutic maneuvers (84, 85). Such variations require alterations in the scheduling tactics of drugs in accordance with the fluctuating growth kinetics properties of neoplastic cells. Most "kinetic" protocols have thus far adhered to rigid schema of drug scheduling and were based on insufficient information on the growth kinetics properties of human tumors. Development of new and better techniques for studying these parameters (86, 87) and acquisition of a vast data bank on human tumor growth dynamics could alter this situation in the not too distant future. It is then when the full impact of the in vitro information on cell cycle stage-dependent drug effects may be finally attained.

REFERENCES

1. B. K. Bhuyan, Cancer Res. 30:2013-2017 (1970).
2. H. Madoc-Jones and F. Mauro, J. Nat. Cancer Inst. 45:1131-1143 (1970).
3. T. Kaneko and G.A. LePage, Cancer Res. 38:2084-2090 (1978).
4. M. L. Rosenblum, K. T. Wheeler, C. B. Wilson, M. Barker and K. D. Kuebel, Cancer Res. 35:1387-1391 (1975).
5. C. G. Whitmore and J. E. Till, Annual Rev. Nucl. Sci. 14:347-374 (1964).
6. L. H. Thompson and H. D. Suit, Int. J. Rad. Biol. 13:347-362 (1969).
7. S. Peel and D. M. Cowen, Brit. J. Cancer, 26:304-314 (1972).
8. U. K. Ehmann and K. T. Wheeler, Eur. J. Cancer, 15:461-473 (1979).
9. B. Drewinko and B. Barlogie, Cancer Treat. Rep. 60:1707-1717 (1976).
10. B. Barlogie and B. Drewinko, Eur. J. Cancer, 14:741-745 (1978).
11. H. Madoc-Jones and F. Mauro, in: "Handbook of Experimental Pharmacology", A. C. Sartorelli and D. G. Johns, editors, Springer, Germany, pp. 205-219 (1975).
12. P. Roper and B. Drewinko, Cancer Res., 36:2182-2188 (1976).
13. B. Drewinko, P. R. Roper and B. Barlogie, Eur. J. Cancer, 15:93-99 (1979).
14. B. K. Bhuyan, B. E. Loughman, T. J. Fraser, and K. J. Day, Expt. Cell Res., 97:275-280 (1976).
15. T. T. Puck and P. I. Marcus, Proc. Natl. Acad. Sci., 41:432-437 (1955).
16. A. W. Nias and M. Fox, Br. J. Radiol. 41:468-474 (1968).
17. G. G. Steel, in: "Growth Kinetics of Tumours", Clarendon Press, Oxford, (1977).
18. O. I. Epifanova and V. V. Terskikh, Cell Tissue Kinet. 2:75-93 (1969).
19. M. F. Rajewsky, in: "Recent Results in Cancer Research, Grundmann and R. Gross, editors, Springer, Germany, pp. 156-171 (1975).
20. S. C. Barranco and J. K. Novak, Cancer Res. 34:1616-1818 (1974).
21. W. R. Bruce, B. E. Meeker and F. A. Valeriote, J. Natl. Cancer Inst., 37:233-245 (1966).

22. B. K. Bhuyan, T. J. Fraser and K. J. Day, Cancer Res. 37: 1057-1063 (1977).
23. O. I. Epifanova, I. N. Smolenskaya and V. A. Polunovsky, Br. J. Cancer, 37:377-385 (1978).
24. F. Valeriote and L. van Putten, Cancer Res., 35:2619-2630 (1975).
25. L. M. van Putten, Cell Tissue Kinet. 7:493-504 (1974).
26. L. M. van Putten and P. Lelieveld, Eur. J. Cancer, 7:11-16 (1971).
27. L. M. van Putten, P. Lelieveld and L. K. J. Kram Idsenga, Cancer Chemoth. Repts., 56:691-700 (1972).
28. S. C. Barranco, W. E. Bolton and J. K. Novak, 12:11-16 (1979).
29. S. C. Barranco, J. K. Novak and R. M. Humphrey, Cancer Res. 33:691-694 (1973).
30. E. Olah, I. Palyi and J. Sugar, Eur. J. Canc. 14:895-900 (1978).
31. G. M. Hahn and L. B. Little, Current Topics Radiation Res. 8:39-83 (1972).
32. F. Mauro, B. Falpo, G. Briganti, R. Elli and G. Zupi, J. Natl. Cancer Inst. 52:705-713 (1974).
33. F. Mauro and H. Madoc-Jones, Cancer Res., 30:1397-1408 (1970).
34. B. K. Bhuyan, L. G. Scheidt and T. J. Fraser, Cancer Res. 32:398-407 (1972).
35. J. M. Trujillo, B. Drewinko and M. J. Ahearn, Cancer Res. 32:1057-1065 (1972).
36. B. Drewinko, M. M. Romsdahl, L. Y. Yang, M. J. Ahearn and J. M. Trujillo, Cancer Res., 36:467-475 (1976).
37. B. Drewinko, L. Y. Yang, B. Barlogie, M. M. Romsdahl, M. Meistrich, M. A. Malahy and B. Giovanella, J. Natl. Cancer Inst., 61:75-83 (1978a).
38. J. J. Stragand, J. P. Bergerat, R. A. White, J. Hokanson and B. Drewinko, Cancer Res., (In Press).
39. B. Drewinko, B. Bobo, P. R. Roper, M. A. Malahy, B. Barlogie and B. Jansson, Cell Tiss. Kinet. 11:191-197 (1978).
40. B. D. Clarkson in: "Control of Proliferation in Animal Cells" B. Clarkson and R. Baserga, editors, Cold Spring Harbor Press, Maryland, p. 945 (1974).
41. M. Tubiana, M. Guichard and E. P. Malaise, in: "Growth Kinetics and Biochemical Regulation of Normal and Malignant Cells" B. Drewinko and R. M. Humphrey, editors, William and Wilkins, Baltimore, pp. 827-842 (1977).
42. G. C. Zubrod, Can. Conf., 8:31-39 (1969).
43. O. I. Epifanova, Int. Rev. Cytol. 5:303-335 (1977).
44. H. Ian, Cancer Res., 33:1716-1720 (1973).
45. R. F. Pittillo, F. M. Schabel, Jr. and H. E. Skipper, Cancer Chemoth. Rept., 54:137-142 (1970).
46. G. R. Ray, G. M. Hahn, M. A. Bagshaw and S. Kurkjian, Cancer Chemoth. Rept. 57:473-475 (1973).
47. F. M. Schabel, Jr., Cancer Res., 29:2384-2389 (1969).

48. F. M. Schabel, Jr., H. E. Skipper, M. W. Trader and W. D. Wilcox, Cancer Chemoth. Rept., 48:17-30 (1965).
49. L. J. Wilkoff, H. H. Lloyd and F. A. Dulmadge, Chemotherapy, 16:44-60 (1971).
50. R. F. Hagemann, L. L. Schenken and S. Lesher, J. Natl. Cancer Inst. 50:467-474 (1973).
51. G. M. Hahn, Nature, 247:741-742 (1968).
52. G. M. Hahn, G. R. Ray, D. E. Gordon and R. F. Kallman, J. Natl. Cancer Inst. 50:529-533 (1973).
53. B. Papirmeister, C. L. Davidson, Biochem. Biophys. Acta, 103:70-92 (1965).
54. J. J. Roberts, T. P. Brent and A. R. Crathorn, Eur. J. Cancer 7:515-524 (1971).
55. F. A. Valeriote, D. Vieth and S. Tolen, Cancer Res., 33: 2658-2661 (1973).
56. G. M. Hahn, D. E. Gordon, S. D. Kurkjian, Cancer Res., 34: 2373-2377 (1974).
57. F. Mauro, B. Falpo, G. Briganti, R. Elli and G. Zupi, J. Natl. Cancer Inst. 52:715-722 (1974).
58. R. M. Sutherland, Cancer Res., 34:3501-3503 (1974).
59. C. J. Thatcher and I. G. Walker, J. Natl. Cancer Inst. 42:363-368 (1969).
60. P. R. Twentyman and N. M. Bleehen, Brit. J. Cancer, 28: 500-507 (1973).
61. A. B. Pardee, Proc. Natl. Acad. Sci. 71:1286 (1974).
62. R. A. Tobey and K. D. Ley, J. Cell Biol., 46:151-157 (1970).
63. A. D. Glinos and R. J. Werrlem, J. Cell Physiol. 79: 79-90 (1972).
64. A. Kouns, Expl. Cell Res., 94:15-22 (1975).
65. P. R. Twentyman and N. M. Bleehen, Br. J. Cancer, 31:68-74 (1975).
66. H. Madoc-Jones and W. R. Bruce, Nature 215:302-303 (1967).
67. B. Drewinko, B. Barlogie, W. Mars, M. A. Malahy and B. Jansson Cell Tissue Kinet. 12:675 (1979).
68. A. Macieira-Coelho, Proc. Soc. Exp. Biol. Med., 125:548-552 (1967).
69. D. W. Ross and W. K. Sinclair, Cell Tissue Kinet. 5:1-14 (1972).
70. S. Gelfant, Cancer Res., 37:3845-3862 (1977).
71. B. Barlogie, P. Corry and B. Drewinko, Cancer Res., 40: 1165-1168 (1980).
72. B. T. Hill in: "Current Chemotherapy and Infectious Disease" J. D. Nelson and C. Grassi, editors, American Society of Microbiology, Washington, pp. 1571-1573 (1980).
73. N. M. Blackett and K. Adams, Br. J. Haematol. 23:751-758 (1972).
74. S. K. Lahiri, Cell Tissue Kinet. 6:509-514 (1973).
75. M. Karon and S. Shirakawa, Cancer Res., 29:687-696 (1969).
76. J. H. Kim, A. G. Perez and B. Djordjevic, Cancer Res., 28: 2443-2447 (1968).

77. W. C. Lambert and G. Studzinsky, Cancer Res., 27:2364-2369 (1967).
78. S. Brachetti and G. F. Whitmore, Cell Tiss. Kinet. 2:193-211 (1969).
79. H. Madoc-Jones and F. Mauro, Cell Physiol. 185-196 (1969).
80. S. C. Barranco and R. M. Humphrey Cancer Res., 31:191-195 (1971).
81. I. Tannock and E. Frei, Natl. Cancer Inst. Monograph, 34: 19-23 (1971).
82. B. Drewinko and R. Alexanian, J. Natl. Cancer Inst. 58: 1247-1253 (1977).
83. H. E. Skipper, Cancer, 28:1479-1492 (1971).
84. L. A. Dethlefsen, Int. J. Rad. Oncology Biol. Phys. 5:1197-1203 (1979).
85. L. A. Dethlefsen, S. P. Sorensen and R. M. Riley, Cancer Res. 35:694-699 (1975).
86. J. J. Stragand, B. Drewinko, P. G. Braunschweiger, H. R. Jacob and L. M. Schiffer, Eur. J. Cancer, 16:293-295 (1980).
87. B. Barlogie, B. Drewinko, G. Dosik and E. J. Freireich, Acta Path. Microbiol. Scand. (In Press)

CELL KINETICS IN CLINICAL ONCOLOGY*

Barthel Barlogie, Benjamin Drewinko, Martin N. Raber
and Douglas E. Swartzendruber

The University of Texas System Cancer Center
M.D. Anderson Hospital and Tumor Institute
Houston, Texas 77030, U.S.A.

While the etiology of most human cancers is still unknown and may be multifactorial, a number of epiphenomena discriminating normal from malignant cells have been discovered, including genetic (1, 2), biochemical (3, 4, 5), and cytokinetic parameters (6, 7). Focusing on the kinetics of tumor growth in this chapter, we will: (1) delineate how measurements of cell kinetics parameters can serve as quantitative descriptors of malignant disease potentially determining response to cytotoxic therapy; (2) review strategies to either differentially enhance tumor cell kill or to establish a viable symbiosis between tumor and host via tumor cell differentiation; and (3) address the feasibility of cytokinetic methods to predict drug responsiveness in vitro and in vivo.

CELL KINETICS AS DESCRIPTOR OF MALIGNANT DISEASE

First, we would like to demonstrate the ethical rationale supporting evaluation of cell kinetics properties within a clinical setting. At a time of stagnating results in the treatment of disseminated neoplasms, full attention should be given not only to the development of new agents and new treatment concepts, but also to extensive quantitiation of tumor biologic parameters such as tumor growth characteristics. It is not inconceivable that the information provided will aid in a more prudent use of currently available treatment armamentarium, particularly since similar tactics have been successful in eradicating and curing advanced animal tumors (8).

*Supported in part by Grants CA5831, CA11520, CA14528 from the Natl. Cancer Inst., Natl. Inst. of Health, Bethesda, MD 20205

Tumor cell kinetics encompass both the events occurring within individual cells from mitosis to mitosis ("microkinetics") and the dynamics of the cell population as a whole ("macrokinetics"). Macrokinetics deals with the description of measurable tumor growth, usually expressed as doubling time of tumor volume, which is intimately related to a variety of microkinetics parameters, such as magnitude of the growth fraction, length of cell generation time, degree of cell differentiation and extent of cell loss.

Rapidly growing tumors such as choriocarcinoma, oat cell carcinoma, acute leukemia and undifferentiated large cell lymphoma generally have a high response rate to chemotherapy compared to well differentiated epithelial tumors with limited tumor expansion over time. However, on occasion a remarkably dramatic regression is observed in slowly growing tumors which can be partially explained by eminent drug sensitivity, but may also reflect a high proliferative activity balanced by a high spontaneous cell loss factor. Also, there is an abundance of reports demonstrating, as well as arguing, the prognostic significance of pretreatment measurement of microkinetics (predominately by way of the tritiated thymidine labelling index) (9).

We therefore recognize the need to explore further both macrokinetics and microkinetics as essential descriptors of tumor growth, which may account to a considerable extent for the variability of treatment efficacy noted among tumors of similar if not identical histologic and possibly biochemical makeup.

Macrokinetics

Because of the seemingly inexhaustable availability of powerful antitumor agents, early and aggressive treatment of marginally responsive tumors has perhaps been over emphasized. While there are some neoplastic diseases that demand prompt institution of treatment at the time of diagnosis (e.g. promyelocytic leukemia, Burkitt's lymphoma, oat cell carcinoma), there is a larger number of both hematologic and solid tissue neoplasms which are not immediately life-threatening and for which curative treatment is not really yet available. Thus, short-term follow-up could be clinically feasible and possibly therapeutically rewarding. It is an established fact that histologically identical tumors can demonstrate vastly different responses to identical therapy, thus escaping even the most sophisticated prognostic factor models (10). Thus, it is quite conceivable that careful follow-up of unresectable disease in terms of changes in tumor dimensions and humoral markers would provide new insight into its biology, capturing a brief segment of the natural tumor-host relationship. Such information may prove to be of great prognostic importance and hence could be crucial to the development of new treatment strategies.

Examples of such approaches include DeVita's proposal to await dedifferentiation of non-Hodgkin's lymphomas with favorable histology to more aggressive forms of the disease (large cell lymphoma) before instituting chemotherapy (11). In the area of multiple myeloma, Alexanian et al demonstrated the usefulness of withholding chemotherapy from patients with indolent myeloma and to institute treatment-free periods for patients achieving remission of their overt myeloma (12). Another example is the strategy of approaching smouldering or oligoblastic leukemia. In this disease, early institution of aggressive cytotoxic treatment has been detrimental, whereas careful observation of the natural progression of the disease has not compromised survival and in fact has provided important insight into the biology of this disorder (13, 14, 15). For instance, in a recent clinical trial of lithium carbonate, we were able to identify a number of pre-treatment characteristics of patients with oligoleukemia that predicted outcome with regard to either marrow failure or leukemic progression (16, 17).

Microkinetics

Cell Cycle Compartment Distribution. Microscopic determination of the mitotic index and autoradiographic evaluation of the labelling index have been the backbone of most clinical kinetics investigations until recently (18). Observation of a low mitotic index in leukemias lead to questioning of the old concept that malignant cells have a growth advantage over normal tissue counterparts (19). However, since mitosis is the shortest cell cycle phase, the mitotic index is invariably very low and hence has limited discriminatory power to characterize the overall proliferative activity of human neoplasms. The usefulness of labelling index determination as an indicator of the kinetics behavior of tumors is likewise limited. The labelling index does not accurately reflect the growth fraction of a tumor mass, a parameter of importance in defining its potential growth, and it depends on other kinetics parameters such as the length of the cell cycle and the length of S phase. Thus, proliferating cells with a longer generation time may have low labelling index values, whereas cells with shorter generation times may show high labelling index values; yet, the fraction of proliferating cells in both populations may be the same.

Both mitotic and labelling index determinations have largely been replaced by DNA flow cytometry, which is a more expendient and objective technique of high statistical accuracy (20, 21, 22). However, while technically far superior to autoradiographic examination of the labelling index, the added discrimination of G_2 cells by DNA flow cytometry does not add significant usefulness in the analysis of unperturbed tumors because of the generally low (G_2+M) compartment. Thus, we and others have accumulated data in leukemia and solid tumors indicating that the major fraction of tumor cells has a G_1 content (18, 23). Since the DNA compartment composition is determined by the relative cycle phase durations, any predominant interference with cell progression through one particular cycle phase will result in cell reassortment. Thus, DNA flow cytometry has been found useful to describe drug and radiation effects on cell cycle progression (24), which might be exploited for the design of combination chemotherapy with phase-specific agents.

A disadvantage of exclusive evaluation of DNA content relates to the inability to distinguish per se normal and malignant subpopulations as well as cell heterogeneity in terms of cell differentiation. It is for this reason that many investigators still prefer autoradiographic procedures in order to perform differential cytokinetic measurements on cell subpopulations of commonly heterogeneous human tumor cell specimens. However, in corroboration of cytogenetic data, we have recently demonstrated that some 90% of all solid tumors and a great majority of lymphoid malignancies have a frankly abnormal DNA content, which can be readily determined in reference to a normal diploid standard (23, 25, 26, 27). Thus, DNA flow cytometry not only provides prompt cell cycle analysis but also permits quantitation of an objective cellular marker of malignancy. Investigations are underway in several laboratories to identify additional probes of cell differentiation and malignancy, so that automated cytology should soon provide adequate means of quantitative differential cytologic evaluation. Along these lines, we have been interested in adapting the nucleolar antigen assay to flow technology. The exclusive association of the nucleolar antigen with malignant cells represents a major advancement in objective tumor cell recognition (5). Preliminary data in our laboratory would suggest that this cellular feature can be readily measured by flow cytometry. Several investigators have also reported on the use of combined DNA/RNA or DNA/protein analysis to distinguish tumor vs. normal cells, demonstrating a general increase in RNA and protein content in tumor cells (21, 23, 28). This is particularly striking in the area of acute non-lymphocytic leukemia, where a high RNA content has been found to be characteristic of immature myeloid cells (28). Thus, in the case of acute leukemia with a low incidence of DNA content abnormality of approximately 17% (25). Cellular markers such as RNA and nucleolar antigen content may effectively discriminate the various stages of myeloid differentiation and malignant transformation, respectively. In the case of multiple myeloma with a frequency of generally low-degree hyperdiploid DNA content abnormality in the range of 80% for active disease (29,30,31), RNA content was found to be particularly high exceeding mean values of normal marrow and lymphocytes by 5 and 15-fold, respectively (23,32). Other approaches have also been reported in the literature (see also chapter by Nicolini et al. in this volume), which advocate by means of flow (115) and scanning (116) cytometry the possibility to discriminate cycling and non-cycling cells (117,118), normal and tumor cells (120), in terms of chromatin-DNA structure (41) and RNA amount (119), and even high versus low metastatic variants (121), in terms of membrane properties (122).

Growth Fraction. The classical means for determining the tumor growth fraction has been the percent-labelled-mitoses (PLM) technique (33), which has been applied to study a limited number of human neoplasms (34). This technique involves in vivo administration of tritiated thymidine, is tedious and plagued with statistical inaccuracies, and is valid only during steady state conditions (6). Similar considerations apply to the use of prolonged in vivo administration of tritiated thymidine in an effort to exclusively label all potentially cycling cells (35, 36). While this technique, in contrast to the PLM method, does not encounter problems associated with expression of a well-defined second wave of labelled mitoses, it is subject to bias through expansion of the labelled cell compartment via cell division. A number of new approaches have been recently advocated including the primer DNA polymerase (PDP) assay (37), the modification of DNA-specific fluorescence by BUdR incorporation (38), and the morphology of prematurely condensed chromosomes in G_1 phase (39, 40). Furthermore, cytochemical probes have been demonstrated to be useful for the distinction of G_0 and G_1 cells (41, 42, 43). The acridine orange method described by Darzynkiewicz et al takes advantage of the greater susceptibility of chromatin to denaturation in non-proliferating vs. proliferating cells (43). This property can be visualized by the metachromatic fluorescent dye acridine orange, which yields green fluorescence upon intercalation with double-stranded DNA and red fluorescence upon dye-stacking to single stranded-denatured DNA. Thus, the extent of DNA denaturation is reflected in a progressive shift in the ratio of red:green fluorescence. Specifically, in lymphocyte cultures, G_0 and mitotic cells have been demonstrated to be more susceptible to DNA denaturation by heat or acid treatment than G_1 and G_2 cells, respectively (43). Alternative rigourous physico-chemical explanation in terms of mass action law has been given of the above observations(119).

We have applied the acridine orange technique to study tumor cells in culture and clinical specimens. In a human multiple myeloma cell line (ARH-77), we were able to distinguish G_0 and G_1 cells as well as mitotic and G_2 cells. The transition of ARH-77 cells from exponential into plateau phase of growth was associated with significant increase in the proportion of cells with high red and low green fluorescence in the $G_{1/0}$ area (Figure 1). Very similar results were obtained in CHO, V-79, human breast (MCF 7) and human colon cancer (LoVo) cells, where the proportion of G_0 cells identified by acridine orange technology was in close agreement with results obtained by PLM and PDP technology. However, when human tissue samples (bone marrow, 25; solid tumors, 8) were investigated, discrimination of 2 distinct G_1 and G_0 subpopulations was noted in only 2 instances (not shown). Thus, the acridine orange technique has yet to identify distinct cell subpopulations with G_0 and G_1 staining characteristics in fresh human samples. This may be a reflection of true differences between cultured cells and in vivo tissue, with the latter showing a wide range of cycle transit times rather than a true non-cycling G_0 compartment (6).

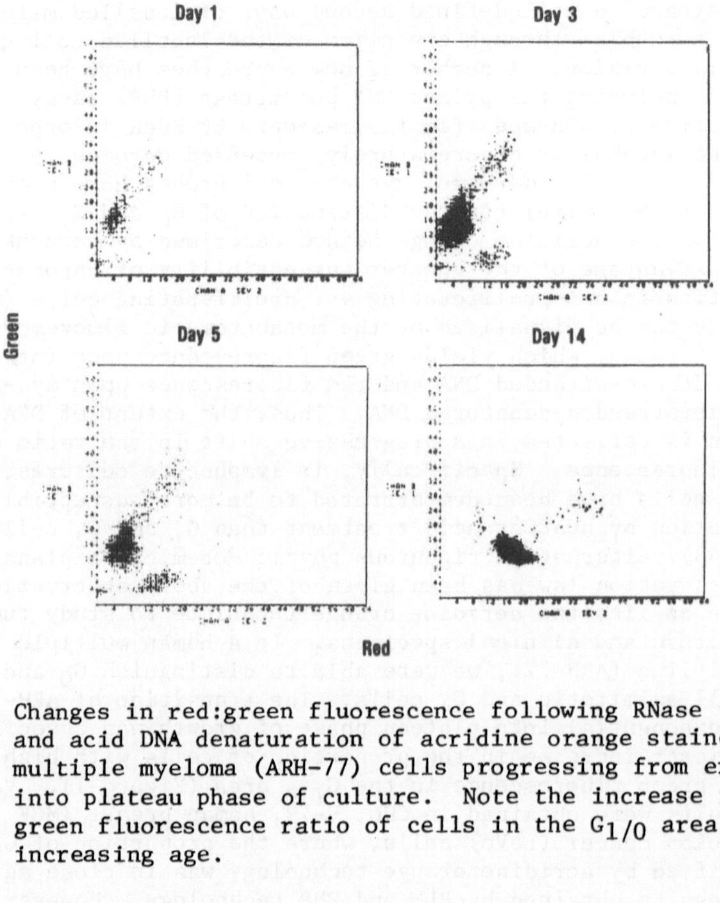

Fig. 1 Changes in red:green fluorescence following RNase treatment and acid DNA denaturation of acridine orange stained human multiple myeloma (ARH-77) cells progressing from exponential into plateau phase of culture. Note the increase in red: green fluorescence ratio of cells in the $G_{1/0}$ area with increasing age.

We have also investigated the feasibility of in vitro exposure of human bone marrow cells to BUdR to quench acridine orange green fluorescence as a means of distinguishing in vitro cycling and non-cycling cells (38). Although in vitro culture may alter the intrinsic proliferative characteristics of cells, it is conceivable that the degree of in vitro recruitability of potentially non-cycling cells into proliferation is systematically related to and hence simply amplifies in vivo conditions. Thus far, we have noted considerable variation both among individuals with normal marrow (6) and among patients with leukemia (7); both from the point of view of the time course of fluorescence quenching and concerning the ultimate proportion of the residual cell population with original fluorescence properties (representing non-cycling cells) (44). Perhaps, the time course of quenching can also be utilized to measure in vitro generation times (45).

Cell Cycle Traverse Rate. The classical means for cycle transit time determination has been the percent-labelled-mitoses (PLM) technique (33). As in the case of growth fraction determination and for the same reasons, the clinical application of PLM technology for analysis of cytodynamic parameters has been very limited. Hence, the information on cycle time parameters of human tumors is extermely sparse. Similar technical and feasibility problems are inherent to in vivo stathmokinetic methods (46, 47, 48, 49). Time lapse cinematography (40) involves prolonged in vitro culture with its pitfalls of significantly altering intrinsic cytokinetic properties, even if appropriate conditions for long-term growth could be provided. Only few cells at any one time can be observed. In view of considerable variation in cycle times noted for established cell lines between successive cell generations and among daughter cells from the same parent cell, time lapse photography is too much of a detailed individual cell microkinetic method to yield meaningful information on average cycle traverse rates (51).

Some investigators have advocated the use of double labelling autoradiography to determine the S phase duration (52, 53, 54). It is conceivable that through the combined use of double labelling autoradiography, DNA flow cytometry and a method to determine the tumor growth fraction (such as the PDP assay), a rather complete description of cytokinetic parameters can be obtained (18).

Recent advances in flow cytometry and sorting may enable measurements of cell cycle time durations by serial cell sorting through a defined "window" in the cycle following pulse labelling with tritiated thymidine (55). Thus far, such RCS_i or RCG_1 (18) methods have been limited to in vitro investigations in experimental systems, and are plagued with at least some of the problems of the PLM technique (i.e. in vivo administration of radioisotopes, dampening of second wave, etc.) It is possible that the use of non-radioactive heavy isotopes may make this technique useful for human *in vivo* work in the future.

We have already alluded to the possibility of employing BUdR-induced quenching of DNA specific fluorochromes for *in vitro* generation time analysis (45). Like any long-term in vitro procedure, this approach is complicated by possible changes in intrinsic proliferative characteristics as a result of *in vitro* growth conditions.

Recovery Kinetics. In contrast to cytodynamic measuremetns of unperturbed systems, similar investigations following cytotoxic injury ("recovery kinetics") (56, 57, 58, 59) are at once facilitated and further complicated. They are facilitated by virtue of induction of a parasynchronous recovery readily demonstrable by serial analysis of DNA content in conjunction with measurements of DNA synthetic activity (58, 59); they may be complicated by cell inactivation and problems inherent in identifying the cytodynamics of the surviving "clonogenic" cells.

Clonogenic Assays. Clonogenicity designates the unlimited proliferative propensity of cells and can be measured by colony formation (60). Tumor stem cells are the real targets to be eradicated by cytotoxic therapy, while a significant proportion of normal stem cells is required to repopulate the various normal tissue components. The universally low *in vitro* plating efficiency observed in most tumors poses the question whether the proportion of stem cells grown *in vitro* might not significantly underestimate the *in vivo* representatives. Nonetheless, tumor stem cell assays have gained considerable interest from the point of view of *in vitro* prediction of clinical drug responsiveness (61, 62) and for the identification of new agents for clinical phase I investigations (63).

Cell Loss. The measurement of cell loss requires knowledge of actual and potential tumor cell doubling times, the latter being dictated by growth fraction and cell cycle time (64). Hence, cell loss can only be determined indirectly and concerns population kinetics. We do not have detailed knowledge about the time course and cell cycle stage of either metabolic or clonogenic cell inactivation under unperturbed or perturbed conditions (65). We already alluded to the practical clinical consequence of a high spontaneous tumor cell loss rate accounting for the occasional rapid tumor regression following cytotoxic treatment of slowly growing tumors.

Tumor Cell Differentiation. Abrogation of tumor clonogenic potential can also come about by way of tumor cell differentiation, either spontaneously or as a result of treatment. The importance of this cellular pathway is probably largely underestimated. Most tumors display some degree of spontaneous differentiation. Specific clinical examples concern the occasional differentiation of neuroblastomas into neurogangliomas (66). In the case of germ cell tumors, aggressive combination chemotherapy has not infrequently been associated with the persistence of mass lesions consisting of histologically benign teratomas (67). Similarly, in the case of oligoblastic leukemias, there is evidence from karyotype analysis of prematurely condensed chromosomes, DNA flow cytometry, and the nucleolar antigen assay, that at least some of the normal appearing mature myeloid cells carry malignant cell markers and therefore represent differentiated neoplastic cells (68). There has been considerable interest recently to identify and apply non-cytotoxic differentiation-inducing agents such as retinoids (69) and low dose chemotherapy (70). Preliminary clinical results of such "differentiating" cancer therapy are very promising. Thus, remarkable antitumor effects have recently been observed with thioproline, an inducer of "reverse transformation" (71). There was a 100% objective response rate with 6 complete remissions among 10 patients with head and neck cancer. Of similar importance as the identification of differentiation-inducing agents is the establishment of differentiation markers so that the mechanism of spontaneous and induced cell differentiation can be further elucidated. Automated cytology will be of great benefit in addressing this problem in a quantitative fashion by making use of malignant cell markers such as abnormal DNA content and the nucleolar antigen in conjunction with antigenic or enzymatic determinants of cell differentiation.

Recommendation for Cytokinetic Studies in Human Malignancies.

A major problem in cell kinetics research related to human tumors in the past has been the use of only one pet technology by any one investigator. The only extensive data available relates to labelling index and DNA flow cytometric measurements (18). In addition, this information has not been gathered under standardized laboratory conditions. Furthermore, clinical details are generally sparse, so that data comparison is severely compromised. We emphasize the need to interface cytokinetic information with all other clinical and laboratory data, (a) to assure comparability of patient groups in terms of disease stage, sample site, treatment status and histopathology; and (b) to discover tumor-host interactions. Because of the variability of human tumors and cytokinetic data already available, an extensive data bank should be established, which contains pertinent disease-and host-related features.

In the following section, we will elaborate on the rationale for selection of both cytokinetic parameters and methods to be employed in clinical cell kinetics research, in order to overcome the deficiencies of past studies and help to resolve the continuing controversy about the clinical importance of tumor cell kinetics.

The selection of cytokinetic parameters should consider the cell kinetics determinants for drug-and radiation-induced cell kill in experimental systems. These include the growth fraction (72), the cell cycle stage and the rate of cycle traverse (18). Most tumors contain a significant proportion of nonproliferating (G_0) cells, which retain clonogenic potential (73). For the most part, such G_0 cells are rather resistant to chemotherapeutic and radiation cytocidal damage (74, 75), although non-ionizing radiation in the form of hyperthermia may be a distinct exception (76). There is an abundance of investigations on the cell cycle phase sensitivity of antitumor agents (77, 78, 79), which is more readily appreciated than the underlying biochemical mechanism, e.g. fluctuating enzyme activities during cell cycle progression (80). Other data demonstrate the marked dependence of cell kill on the rate of cycle traverse (81, 82). Since clonogenic cells are most threatening to the tumor-bearing host, the above parameters should be performed on such viable cells, and the proportion of spontaneous or treatment-induced tumor cell differentiation must be appreciated. If there are several methods available, the information value and feasibility of such measurements have to be considered in the clinical setting, where there are limitations with regard to tissue quantity and sampling times.

Without doubt, DNA flow cytometry has a superior role in determining cell cycle distribution for its practicality, speed and accuracy. Its shortcomings relate to the identification of cell heterogeneity (tumor vs. normal cells, degree of differentiation) and appreciation of cell function (stem cells, cycle traverse rate). The introduction of multi-parameter flow cytometry using additional markers of malignancy (nucleolar antigen) and of cell differentiation (RNA, surface antigens) will permit differential cell cycle distribution measurements of normal and various tumor subpopulations in generally heterogeneous clinical samples. Figure 6 illustrates how DNA/RNA differentiation by acridine orange can serve to distinguish high RNA content myeloma tumor cells from normal hemopoietic bone marrow cells with lower RNA content. We have preliminary data to indicate that the presence and density of estrogen receptor expression can be measured by flow cytometry (Figure 2). Concurrent measurement of a tumor cell marker such as abnormal DNA content or the nucleolar antigen will allow quantitation of estrogen receptor expression in relationship to individual tumor cells rather than to quantity of breast tissue containing an unknown proportion of normal cells. Moreover, the flow cytometric assessment of estrogen receptor content as a function of cell cycle characteristics in aneuploid breast cancer will permit in the individual patient a direct evaluation of the purported inverse relationship between ER^+ expression and proliferative activity, as demonstrated by population statistics measurements (83) (Figure 8). Finally, the question of co-existence of ER^+ and ER^- tumors in the same individual can be addressed.

Tumor growth fraction analysis remains a significant challenge

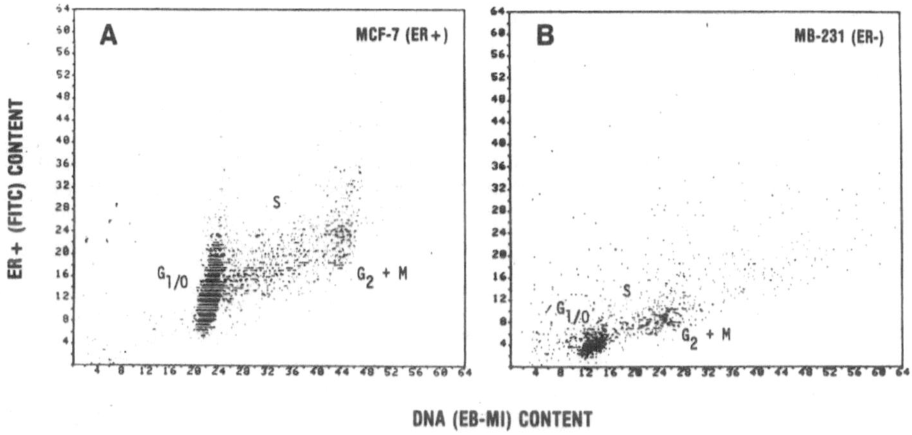

Fig. 2. Biparametric analysis of estrogen receptor (ER) and DNA content in human ER^+ (MCF 7, A) and ER^- (MB 231, B) breast cancer lines. Cells were incubated for 4 hr with 10^{-5}M estradiol-FITC. The FITC conjugated estradiol was kindly provided by Dr. George Barrows. Note the dispersion in ER content of G_1 cells and a linear increase in average ER content during S phase of MCF 7 cells (A). When the same technique was applied to poorly differentiated ER^- MB 231 cells, only non-specific background fluorescence was noted.

(104), for cyclophosphamide in neuroblastoma with a significant increase in tritiated thymidine labelling index and mitotic index on day 7 for responders (89), and for a number of combination chemotherapy programs with diverse changes in labelling index at various times after treatment (for review, see (9)).

In vitro prediction of drug responsiveness using kinetics principles has rested on two major assays. The already cited tumor stem cell assay developed by Hamburger and Salmon has gained considerable interest and is currently in use in most cancer centers (61). Similar tumor stem cell assays have also been developed for acute myelogenous leukemia (105, 106, 107). We and others have investigated the feasibility to predict antitumor effect on the basis of drug-induced suppression of tritiated thymidine incorporation into DNA *in vitro* (108, 109, 110, 111). Our assumption was that lethal cell sensitivity should be reflected in the dose-dependent suppression of DNA synthesis. In fact, we were able to demonstrate that, in adult acute leukemia, the combined resistance index for adriamycin and Ara-C completely separated patients obtaining complete remission from those failing to respond to anthracycline/Ara-C combination chemotherapy (Figure 3) (112).

CONCLUSION

While the biological importance of cytokinetic determinants for tumor cell kill has been well established in experimental systems, clinically feasible methods to determine such cytokinetic parameters have just recently become available. For a complete cytokinetic description of tumor and normal tissue, we recommend DNA flow cytometry for cycle stage distribution analysis, the primer dependent DNA polymerase (PDP) assay for growth fraction determination, and *in vitro* double labelling autoradiography for S phase transit time determination. Automated cytology for both growth fraction analysis and cycle time determination may soon become feasible for application to human tumor cell kinetics. A broad data base of cytokinetic parameters needs to be established for correlation with clinical and other laboratory data of known prognostic importance. Population kinetics measurements should be carried out whenever clinically justifiable, i.e. the case of slowly progressive tumors for which curative cytotoxic therapy is not yet available. Considering the influence of both macrokinetics and microkinetics on tumor growth, there are multiple strategies to re-establish or mimick the normal equilibrium of cell proliferation, differentiation and cell death. Cell kinetics strategies to enhance the differential cell kill of tumor vs. normal cells should be employed according to the baseline cytokinetic measurements. Of particular practical interest is the concept of normal tissue protection by cell cycle freezing or blocking agents. Another concept of tumor growth control employs tumor cell differentiation with the aim to establish a viable

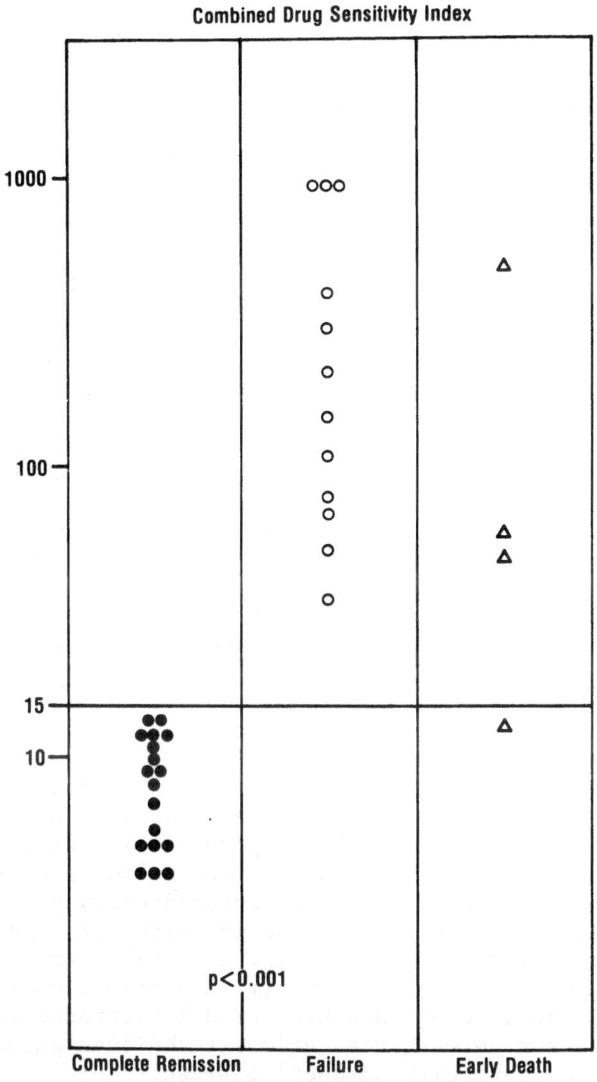

Fig. 3. In vitro prediction of complete remission in adult patients with acute leukemia receiving anthracycline/ARA-C combination chemotherapy. Bone marrow cells were cultured for 24 hrs in RPM1 1640 medium without growth stimulator. At this time, cell aliquiots were exposed for 4 hr to increasing concentrations of either adriamycin or ARA-C, and the relative suppression of tritiated thymidine incorporation during the last hour of exposure was determined. Using Mahalanobis D^2 test (114), a resistance index was computed, which separated patients achieving remission from patients failing to respond to this induction chemotherapy program. Early deaths were excluded from analysis.

Given today's technology, it appears that the autoradiographic examination of the primer DNA polymerase (PDP) activity is probably best suited to approximate the fraction of actively proliferating cells. This technique has already been successfully applied to the study of hematologic malignancies and few solid tumor specimens (84, 85). The existence in cultured cells of different chromatin susceptibility to denaturation in G_0 and G_1 cells as visualized by acridine orange metachromasia was not consistently apparent in our own studies on human tumor samples. This may reflect technical problems of resolution or true differences between in vivo and in vitro conditions. We are presently evaluating whether time-dependent "melting" profiles of chromatin may give further insight into the growth fraction. To this end, acid or heat denaturation of chromatin will be measured by an increase in acridine orange red fluorescence over time. The differential stainability of non-stimulated vs. stimulated lymphocyte cultures using subsaturation concentrations of Hoechst 33342 dye might be applicable to the distinction of cycling vs. non-cycling cells in general (42). This assay explores potential differences in fluorochrome binding in quiescent vs. actively proliferating cells, which we documented in the case of cultured human myeloma cells. The use of BUdR to quench fluorescence intensity from various DNA specific dyes including acridine orange (green fluorescence) involves in vitro culturing of tumor cells and hence perturbation of in vivo conditions, so that the maximum proportion of cells with quenched fluorescence may not be a true measure of in vivo growth fraction.

Cell cycle transit time determination for the time being still rests on PLM technology. For ethical reasons, such measurements can not be conducted in patients with potentially curable malignancies. In order to gain insight into the validity of potential substitutes, such as double labelling autoradiography for S phase transit time determination (in conjunction with cell cycle compartment distribution by DNA flow cytometry and growth fraction determination) or in vitro cycle transit time determination from the time course of BUdR-induced quenching of DNA fluorescence intensity, extensive comparison between the various techniques should be carried out in patients with advanced disease.

A fundamental problem relates to the identification of stem cells through means different from functional assessment of colony formation (60), thus avoiding problems of underestimation due to low in vitro plating efficiency and time delay (61). As demonstrated by Swartzendruber, et al., a number of quantitative cellular probes have been identified to be associated with cellular differentiation in the case of a teratocarcinoma model (86). It is quite possible that the search for unique differentiation or stem cell enzyme activities, cell surface markers, etc. will be successful. Likewise, the chromatin structure of stem cells may differ from that of doomed cells resulting in different DNA melting profiles, which can be measured by the acridine orange method.

Observation of macrokinetics is feasible and ethically justifiable in patients with advanced but not immediately life-threatening malignancies, for which no curative treatment exists. Documentation of tumor doubling time together with microkinetic measurements would provide a rather complete description of tumor growth parameters, which have been so instrumental for the understanding and treatment of experimental animal tumors.

From the point of view of potential exploitation of tumor cell kinetics for the systemic treatment of malignant disease, knowledge is also required about normal host cell kinetics, particularly of those tissues which are frequent targets of dose-limiting host toxicity from cancer therapy, such as bone marrow and gut. Such studies would also permit evaluation of potential growth interactions between tumor and normal host tissue, as have been demonstrated in the case of human leukemia between leukemic blast and normal hemopoietic cells (87).

EXPLOITATION OF CELL KINETICS FOR THE TREATMENT OF MALIGNANT DISEASE

In case of confirmation of autoradiographic and flow cytometric data consistent with generally lower proliferative activity in tumor vs. normal host cells, in vivo manipulations of tumor and/or host cell kinetics may be suitable to increase tumor and reduce normal tissue lethal injury from systematically administered cytoxic chemotherapy. Such manipulation may follow one of several general concepts, i.e., tumor cell recruitment into the proliferative cycle, tumor cell synchronization, normal tissue protection, or induction of tumor differentiation (18).

Tumor cell recruitment describes the treatment-induced triggering of non-proliferating cells into the division cycle and seems to occur as a result of effective tumor cell kill. Past cytokinetic studies have postulated the existence of tumor cell recruitment by inference from serial observation of the tritiated thymidine labelling index in leukemias and solid tumors (88, 89, 90). Burke et al., have demonstrated the generation of a humoral stimulator following high dose cytotoxic chemotherapy, which was common to normal and malignant hemopoiesis and solid tumors (90, 91). Verification of such cytokinetic events relies on the availability of growth fraction measurements of unperturbed and perturbed cell systems. Both the PDP assay and cytochemical assessment of chromatin function using a-ridine orange metachromasia following DNA denaturation may be suitable for detection of such short-term changes in growth fraction.

While cell recruitment is an appealing strategy for low growth fraction tumors, cell synchronization maneuvers should be limited to high growth fraction tumors, in which a significant cycle perturbation effect can be achieved. At this time, we do not have effective means to distinguish between truly synchronized cells retaining their full clonogenic potential and those cell populations whose viability is severely compromised as a result of cycle perturbation effects (92). In extensive in vitro studies, we have demonstrated that G_2 block is a common end point of cycle perturbation from a large number of chemotherapeutic agents (84). The magnitude and time course of such G_2 block was shown to be a complex function of drug concentration, incubation time and cell cycle stage of treatment (79). Of particular importance was the notion that the lack of cycle perturbation of treated cells could still be followed by an effective block of cycle progression in daughter cells. The time of manifestation of G_2 block during the life span of treated cells or during the daughter cell generation was determined by the cycle stage of drug exposure (93, 94). Comparative studies of lethal and cycle perturbation effects did not yield a systematic relationship (79, 95). In order to escape the dilemma of synchronization-associated cell kill, we have recently designed a clinical trial evaluating the synchronizing property of cytosine arabinoside in patients with relapsed acute leukemia resistant to the lethal effects of Ara-C. We were able to demonstrate in 8 patients that an Ara-C-induced cytokinetic effect with S phase accumulation (96) can still be universally provoked in the absence of lethal cell damage (97). Such reassortment in cell cycle distribution was then utilized for more effective cell kill by 3-deazauridine, an S phase specific agent with limited antileukemic efficacy. Our data to date indicate that even in the presence of Ara-C resistance in all and 3-deazauridine resistance in half of the patients, significant antileukemic efficacy could be promoted from such kinetically oriented combination chemotherapy (97).

The concept of normal tissue protection can be pursued by observation of differential recovery kinetics (56, 57) as well as by differential induction of cell cycle block in normal vs. tumor cells. As for the latter case, methyl GAG has been demonstrated to differentially block normal cells in G_1 phase of the cell cycle while transformed cells continue to cycle under the experimental condition employed (98). Differential protection against the S phase specific drug hydorxyurea was noted in non-transformed cells blocked in G_1 phase. Similarly, anguidine at non-cytotoxic concentrations has been found to effectively freeze the entire cycle of both human colon cancer (LoVo) (99) and CHO cells (100). In drug combination experiments in CHO cells, anguidine was noted to afford marked protection by 10 to 100 fold against cytotoxic damage from Ara-C and adruamycin (101). It is conceivable that in the case of preferential block of normal cells in vivo, significant normal tissue protection against cytotoxic injury can be accomplished. From a practical technical standpoint, the strategy of normal tissue protection by either methyl GAG or anguidine can be readily verified by DNA flow cytometry in conjunction with measurements of radio-labelled DNA precursor uptake.

The in vivo induction of cell differentiation is just now being explored in the treatment of human disease. In murine leukemia, Sachs, et al. have demonstrated the induction of differentiation by steroidal hormones and by chemotherapeutic agents such as Ara-C and actinomycin D (70). Early clinical trials are now underway to assess the in vivo differentiation capacity of retinoic acid (69). The promising results with thioproline in squamous cell carcinoma of the head and neck have stimulated further clinical investigations in the entire area of "biological response modifiers" (71). More extensive research at the experimental and clinical levels is warranted to identify agents capable of and specific diseases most likely sensitive to differentiation induction. Clinical investigators should become more aware of the potential options of establishing a viable balance between host and tumor. Such approaches have been taken in non-malignant diseases, where pallative as well as causative therapy may achieve long-term clinical benefits.

PREDICTION OF TUMOR CELL SENSITIVITY IN VIVO AND IN VITRO

Like other oncologic laboratory disciplines, cell kinetics has amplified our understanding of the biology of a given disease and provided a number of treatment concepts to improve the differential tumor-host tissue toxicity. However, ti would be most desirable to accurately predict the efficacy of a given treatment for every patient and thus avoid administration of potentially hazardous and ineffective therapy. Many investigators have addressed the feasibility of such in vitro drug sensitivity tests focusing on the prediction of tumor responsiveness. An optimal assay, in addition, should include the individual patient's normal tissue response as well. Predictive assays reported to date have employed cytokinetic, molecular pharmacologic and cytogenetic principles. There is experimental evidence of a linear correlation between in vitro clonogenic survival and the extent of DNA-cross linking following cis-platinum treatment (102). Likewise, the degree of chromosomal damage in interphase cells has been demonstrated to correlate with adriamycin-induced lethal cell injury (103).

Among the cytokinetic parameters employed to predict tumor cell sensitivity, both in vivo and in vitro tests have been pursued. Several investigators including ourselves have evaluated the relationship between drug-induced early detectable interference with cycle progression in vivo and subsequent clinical response. To date, positive data exist for AMSA in relapsed acute leukemia with a significant steeper increase in the bone marrow S phase compartment size by day 7-10 in clinical responders symbiosis between tumor and host. Promising leads exist to employ cytokinetic principles to predict tumor cell sensitivity in vitro, so that an era of individualized treatment with highly effective drugs rather than empirical therapy based on population statistics can be anticipated in the near future.

REFERENCES

1. P. C. Nowell, "Chromosomes changes and the clonal evolution of cancer" in: Chromosomes in Cancer, edited by J. German, New York, Wiley and Sons, (1974).
2. J. D. Rowley and D. Potter "Chromosomal banding patterns in acute non-lymphocytic leukemia" Blood, 47:705-721 (1976).
3. R. Baserga, "Biochemical, molecular and pharmacologic Approaches to large bowel cancer. An Overview" Cancer, 45:1168-1171 (1980).
4. H. Busch, N. R. Ballel, R. K. Busch, Y. C. Choi, F. Davis, I. L. Goldknopf, S. I. Matsui, M.S. Rao and L. I. Hoffbloom, "The nucleolus. A model for analysis of chromatin controls" Cold Spring Harbor Symposium, Vol. 42 (1978).
5. H. Busch, F. Gyorkey, R. K. Busch, F. M. David, P. Yorke and A. K. Smetana, "A nucleolar antigen found in a broad range of tumor specimens" Cancer Res. 39:3024-3030 (1979).
6. G. G. Steel, Growth kinetics of tumors; cell population kinetics in relation to the growth and treatment of cancer, Clarendon Press, Oxford (1977).
7. R. Baserga, The cell cycle and cancer, Marcel Dekker, Inc., New York (1971).
8. F. M. Schabel, Jr., D. P. Griswold, Jr., T. H. Corbett, W. R. Laster, Jr., J. G. Mayo and H. H. Lloyd "Testing therapeutic hypothesis in mice and man: observation on the therapeutics activity against advanced solid tumor of mice treated with anticancer drugs that have demonstrated potential clinical utility for treatment of advanced solid tumor in man" In: Cancer Drug Development-Methods in Cancer Research, edited by H. Busch and V. DeVita, Jr., Vol. 17, New York Academic Press, Inc. (1979).
9. R. B. Livington and J. S. Hart "The clinical application of cell kinetics in cancer therapy" Ann. Rev. Pharmacol. Toxicol. 71:529-543 (1977).
10. E. A. Gehan, T. O. Smith, E. J. Freireich, G. Body, V. Rodriguez, J. Speer and K. McCredie "Prognostic factors in acute leukemia" Seminars in Oncology, 3:271-281 (1976).
11. V. T. DeVita, Jr., E. R. Glatstein, R. C. Young, S. M. Hubbard, R. M. Simon, J. L. Ziegler and P. H. Wiernik, "PH: Changing Concepts: the lymphomas" in: Adjuvant Therapy of Cancer II, edited by S. E. Jones and S. E. Salmon, Grune and Stratton (1979).

12. R. Alexanian and B. Drewinko, "Multiple myeloma and related plasma syndromes" in: Cancer patient care at M.D. Anderson Hospital and Tumor Institute, edited by R. L. Clark and C. D. Howe, Year Book Medical Publishers, Chicago (1976).
13. C. Jacquillat, V. Izrael, M. Weil, C. Chastang, N. Boiron, and J. Bernard, "A study of 120 patients with oligo-blastic leukemia (OBL)" American Association for Cancer Research, Abstract #412, pp 103 (1975).
14. J. R. Cohen, W. P. Creger, P. L. Greenberg and S. L. Schrier, "Subacute myeloid leukemia, a clinical review" The American Journal of Medicine, 66:959-966 (1979).
15. P. L. Greenberg and B. Mara, "The preleukemic syndrome, correlation of in vitro parameters of granulopoiesis with clinical features" The American Journal of Medicine, 66:51-958 (1979).
16. B. Barlogie, G. Spitzer, G. Dosik, D. A. Johnston, K. B. McCredie and E. J. Freireich, "Lithium Carbonate (Li) for the treatment of oligoblastic leukemia" XVII Congress of the International Society of Hematology, Paris, France, (1978).
17. B. Barlogie, D. A. Johnston and L. Smallwood "Clinical evolution of oligoblastic leukemia following lithium carbonate treatment" Submitted to Blood.
18. B. Barlogie, B. Drewinko, G. Dosik and E. J. Freireich, "Cell Kinetics and the management of malignant disease" Acta Pathologica Microbiologica Scandinavia (In press)
19. H. Begemann and W. Hemmerle "Die Mitosetatigkeit des menschlichen Knochenmarks unde ihre Beeinflussung durch cytostatische Substanzen Klin" Wochenschr. 27:530-537 (1949).
20. W. Gohde and W. Dittrich "Impulsfluorometrie-ein neuartiges Durchflussverfahren zur ultraschnellen Mengenbestimmung von Zeillinhaltsstoffen" Acta Histochem. 10:429-437 (1971).
21. W. Gohde, J. Schuman, T. Buchner, F. Otto and B. Barlogie "Pulse cytophotometry: Its application to tumor cell biology and clinical oncology" in: Flow Cytometry and Sorting, M. R. Melamed, P. F. Mullaney and M.L. Mendelsohn, editors, New York, John Wiley and Sons, Inc., (1979).
22. B. Barlogie, G. Spitzer, J. S. Hart, D. A. Johnston, T. Buchner, J. Schumann and B. Drewinko "DNA-histogram analysis of human hemopoietic cells" Blood 48:245-258 (1976).
23. B. Barlogie, J. Latreille, C. T. Fu, J. Franco, M. Meistrich, and M. Andreeff, "Characterization of Hematologic Malignancies by Flow Cytometry" Blood Cells (In press).
24. B. Barlogie and B. Drewinko "Cell Cycle stage dependent induction of G_2 phase arrest by different antitumor agents" in: Third International Symposium on Pulse Cytophotometry, D. Lutz, editor, European Press Medikon, Gent, Eur, J. of Cancer 14:7 (1978).

25. B. Barlogie, W. Hittelman, G. Spitzer, J. S. Hart, J. M. Trujillo, L. Smallwood and B. Drewinko "Collreation of DNA distribution abnormalities with cytogenetic findings in human adult leukemia and lymphoma" Cancer Res. 37:4400-4407 (1977).
26. B. Barlogie, W. Gohde, D. A. Johnston, L. Smallwood, J. Schumann, and B. Drewinko, "Determination of ploidy and proliferative characteristics of human solid tumors by pulse cytophotometry. Cancer Res. 38:3333-3339 (1978).
27. B. Barlogie, B. Drewinko, J. Schumann, W. Ghode, G. Dosik, J. Latreille, D. A. Johnston and E. J. Freireich, "Cellular DNA content as a marker of human neoplasia" The American Journal of Medicine, 69:196-203 (1980).
28. M. Andreeff, Z. Darzynkiewicz, T. Sharpless, B. Clarkson and M. R. Melamed "Discrimination of human leukemia subtypes by flow cytometric analysis of cellular DNA and RNA" Blood 55:282-293 (1980).
29. J. Latreille, B. Barlogie, G. Dosik, D. A. Johnston, B. Drewinko and R. Alexanian "Cellular DNA content as a marker of human multiple myeloma" Blood, 55:403-407 (1980).
30. J. Latreille, B. Maxwell, D. Johnston, J. Franco, B. Barlogie, T. Loo and R. Alexanian "Ploidy analysis of multiple myeloma bone marrow by DNA flow cytometry and mitotic karyotyping" Annual Meeting American Association for Cancer Research, Proceedings Abstract #C-640 (1980).
31. J. Latreille, B. Barlogie, D. A. Johnston, B. Maxwell, B. Drewinko and R. Alexanian "Ploidy and proliferative characteristics in monoclonal gammopathy" Submitted to Blood.
32. J. Latreille, B. Barlogie, R. Lefebvre, R. Alexanian, "Acridine orange flow cytometry of DNA/RNA content multiple myeloma" Submitted Fifth International Symposyum on Flow Cytometry, Rome, Italy (1980).
33. M. L. Mendelsohn "Autoradiographic analysis of cell proliferation in spontaneous breast cancer in C3H mouse. II. Growth and Survival of Cells labelled with Tritiated Thymidine" Journal of National Cancer Institute 25:45-500 (1960).
34. I. Tannock, "Cell kinetics and chemotherapy: a critical review" Cancer Treatment Rep. 62:1117-1133 (1978).
35. R. Baserga, W. E. Kisieleski and K. Haldorsen "A study on the establishment and growth of tumor metastates with tritiated thymidine" Cancer Res. 20:910-917 (1960).
36. B. Drewinko, R. Alexanian, H. Boyer, B. Barlogie and S. I. Rubinow "The growth fraction of human myeloma cells" Blood (In press).
37. J. S. R. Nelson and L. M. Schiffer "Autoradiographic detection of DNA polymerase containing nuclei in sarcoma 180 ascites cells" Cell and Tissue Kinetics 6:45-54 (1973).

38. Z. Darzynikiewicz, M. Andreef, F. Traganos, T. Sharpless, and M. Melamed "Discrimination of cycling and non-cycling lymphocytes by BUdR-suppressed acridine orange fluorescence in a flow cytometric system" Exp. Cell Res. 115:31-35 (1978).
39. P. Rao, B. Wilson and T. Puck "Premature chromosome condensation for cell cycle analysis" Journal Cell Phisiol. 91: 131-142 (1977).
40. W. Hittleman and P. Rao "Mapping G_1 phase by the structural morphology of the prematurely condensed chromosome" Journal of Cell Physiol. 95:333-342 (1978).
41. C. Nicolini, S. Ng and R. Baserga "Effect of chromosomal protein extractable with low concentrations of NaCl on chromatin structure of resting and proliferating cells" Proceedings National Academy of Sciences, USA 72:2361-2365 (1975).
42. M. E. Lalande, and R. C. Miller "Fluorescence flow analysis of lymphocytes activation using Hoechst 33342 dye" Journal of Histochem and Cytochem 27:394-397 (1979).
43. Z. Darznkiewicz, F. Traganos, M. Andreef, T. Shárpless and M. R. Melamed "Different sensitivity of chromatin to acid denaturation in quiescent and cycling cells revealed by flow cytometry" Journal Histochem and Cytochem 27:478-485 (1979).
44. L. Teodori, A. M. Maddox, B. Barlogie, J. J. Stragand "5-Bromodeoxyuridine (BUdR) quenching of acridine orange (AO) in human bone marrow as a potential probe for growth fraction determination in vitro" Submitted Fifth International Symposium on Flow Cytometry, Rome Italy, (1980).
45. R. M. Bohmer "Flow cytometric cell cycle analysis using the quenching of 33258 Hoechst fluorescence by bromodeoxyuridine incorporation" Cell and Tissue Kinetics 12:101-110 (1979).
46. O. J. Eigsti and P. Dustin "Colchicine" Iowa State College Press, AMES, Iowa (1955).
47. T. T. Puck and J. Steffin "Life cycle analysis of mammalian cells" Biophys. J. 3:379-397 (1963).
48. E. Frei, III, J. Whang, R. B. Scoggins, E. J. van Scott, D. P. Rall and M. Ben "The stathmokinetic effect of vincristine" Cancer Res. 24:1918-1925 (1964).
49. G. M. Dosik, B. Barlogie, R. Allen, W. Gohde and B. Drewinko "A rapid automated stathmokinetic method for determination of in vitro cell cycle transit times" Cell and Tissue Kinetics (In press).
50. J. E. Sisken and L. Moraska "Intrapopulation kinetics of the mitotic cycle" Journal Cell Biol. 25:179-189 (1965).
51. B. Drewinko, B. Bobo, P. Roper, M. A. Malahy, B. Barlogie and B. Jansson "Analysis of the growth kinetics of a human lymphoma cell line" Cell and Tissue Kinetics 11: 177-191 (1978).

52. D. E. Wimber and H. A. Quastler, "A 14C- and 3H-thymidine double labelling technique in the study of cell proliferation in Tradescantia, Root Tips" Exp. Cell Res. 30:8-22 (1963).
53. S. E. B. Harris and D. Hoelzer "An evaluation of various double labelling and autoradiogrpahic techniques for the measurement of DNA synthesis times in leukemic cells" Journal Microsc. 96:205-217 (1972).
54. B. Schultze, W. Maurer and H. Hagenbusch "A two emulsion autoradiographic technique and the discrimination of 3 different types of labelling after double-labelling with 3H- and 14C-thymidine. Cell and Tissue Kinetics 9:245 (1976).
55. J. W. Gray, J. H. Carver, Y. S. George and M. L. Mendelsohn, "Rapid cell cycle analysis by measurements of the radioactivity per cell in a narrow window in S phase (RCS_i)" Cell and Tissue Kinetics 10:97-110 (1977).
56. S. H. Rosenoff, F. Bostik and R. C. Young "Recovery of normal hemopoietic tissue and tumor following chemotherapeutic injury from cytophosphamide (CTX). Comparative analysis of biochemical and clinical techniques" Blood 45:465 (1975).
57. S. H. Rosenoff, J. M. Bull and R. C. Young "The effects of chemotherapy on the kinetics and proliferative capacity of normal and tumor tissues in vivo" Blood 45:107-118 (1975).
58. W. R. Vogler, Z. H. Israili, A. G. Zoliman, S. Moffitt and B. Barlogie "Marrow Cell kinetics in patients treated with methotrexate and citrovorum factor" Cancer (In press).
59. S. B. Howell, A. Drishan and E. Frei, III "Cytokinetic comparison of thymidine and leukovorum rescue of marrow in humans after exposure to high-dose methotrexate" Cancer Res. 39: 1315-1320 (1979).
60. T. T. Puck and P. Marcus "A rapid method for viable cells tritration and clone production with HeLa cells and tissue culture: The use of x-irradiated cells to supply conditioning factors" Proceedings National Academy of Science USA, 41: 432-438 (1955).
61. A. W. Hamburger and S. E. Salmon "Primary bioassay of human tumor stem cells" Science 197:461-463 (1977).
62. S. E. Salmon, A. W. Hamburger and B. Soehnlen "Quantitation of different sensitivity of human tumor stem cells to anticancer drugs" New England Journal of Medicine 298:1231-1327 (1978).
63. S. E. Salmon "Human tumor stem cells in adjuvant therapy of cancer" in: Adjuvant therapy of cancer, II, S. E. Jones and S. E. Salmon, editors, Grune and Stratton, Inc. New York, (1979).
64. G. G. Steel "Cell loss from experimental tumor" Cell and Tissue Kinetics 1:193-207 (1968).
65. D. Price Dynamics of proliferating tissues University of Chicago Press, Chicago (1958)

66. P. C. Dyke and D. A. Mulkey "Maturation of ganglioneuroblastoma to ganglioneuroma" Cancer 20:1343-1349 (1967).
67. L. H. Einhorn and J. P. Donohue "Combination chemotherapy in disseminated testicular cancer: The Indiana University Experience" Seminars in Oncol. 6:87-93 (1979).
68. W. Hittleman, B. Barlogie and F. Davis Unpublished observations.
69. N. Levine and F. L. Meyskens "Topical vitamin-A-acid therapy for cutaneous metastatic melanoma" Lancet 2:224-226 (1980).
70. L. Sachs "Control of normal cell differentiation in leukemic white blood cells" in: Cell differentiation and neoplasia, G. S. Saunders, editor, Raven Press, New York (1978).
71. A. Brugarolas and M. Gosalvez "Treatment of cancer by an inducer of reverse transformation" Lancet 1:68-70 (1980).
72. M. L. Mendelsohn "The growth fraction: A new concept applied to tumors" Science 130:1496 (1960).
73. R. Kallman, C. Combs, A. Franko, B. Furlong, S. Kelley, H. Kemper, R. Miller, B. Rapacchietta, D. Schoenfeld, and M. Takahashi "Evidence for the recruitment of non-cycling clonogenic tumor cells" in: Radiation Biology in Cancer Res., R. E. Meyn and H. R. Withers, editors, Raven Press, New York, (1980).
74. B. K. Bhuyan, T. J. Fraser and K. J. Day "Cell proliferation kinetics and drug sensitivity of exponential and stationary populations of cultures L1210 cells" Cancer Res., 37:1057-1063 (1977).
75. B. Drewinko, M. Patchen, L. Y. Yang and B. Barlogie "Differential killing efficacy of 20 antitumor drugs on proliferating and non-proliferating human tumor cells" Cancer Res. (In press).
76. B. Barlogie, P. M. Corry and B. Drewinko "In vitro thermo-chemotherapy of human colon cancer cells with cis-platinum and multomycin-C" Cancer Res. 40:1165-1168 (1980).
77. W. R. Bruce, B. E. Meeker and F. A. Valeriote "Comparison of the sensitivity of normal hemopoietic and transplanted lymphoma colony forming cells to chemotherapeutic agents administered in vivo" Journal of the National Cancer Institute 37:233-245 (1966).
78. F. Mauro and H. Madoc-Jones "Age responses of cultured mammalian cells to cytotoxic agents" Cancer Res. 30:1397-1408 (1970).
79. B. Barlogie and B. Drewinko "Cell cycle related induction of cell progression delay" 29th Annual Symposium on Fundamental Cancer Research, Houston, TX March 10-12, 1976.
80. L. K. Persohn, B. Drewinko and T. L. Loo "Uptake of 1-β-Darabinofuranosyl cytosine by synchronized human lymphoma cells" Med. Pediatric Oncology, 2:291-297 (1976).
81. S. S. Ford and S. E. Shackney "Lethal and sublethal effects of hydroxyurea in rélation to drug concentration and duration of drug exposure in sarcoma 180 in vitro" Cancer Res. 37:2628-2637 (1977).

82. S. E. Shackney, B. W. Erickson and C. E. Lengel "Schedule optimization of cytosine arabinoside (CA) in hydroxyurea (HU) in sarcoma 180 in vitro" AACR Abstract #900 (1978).
83. M. Raber, B. Barlogie, H. Fritsche and C. Bedrossian "Cytokinetic, cytogenetic and hormonal assessment of human breast cancer" American Association for Cancer Research, Proceedings Abstract #348, May (1980).
84. L. M. Schiffer, P. G. Braunschweiger, J. J. Stradand and L. Poulakoa "The cell kinetics of human mammary cancers" Cancer 43:1707-1719 (1979).
85. G. L. Wantzen, H. Karle and S. A. Killman "Nuclear DNA polymerase estimation in human keukemic meyloblasts" British Journal of Hematology 33:329-334 (1976).
86. D. E. Swartzendruber, K. Z. Cox and M. E. Wilder "Flow cytoenzymology of the early differentiation of mouse embryonal carcinoma cells" Differentiation 16:23-39 (1980).
87. H. E. Broxmeier, N. Jacobson, J. Kurland, H. Mendelsohn and M. A. Moore "In vitro suppression of normal granulocytic stem cells by inhibitory activity derived from human leukemia cells" J. National Cancer Institute 60:497-511 (1978).
88. B. C. Lampkin, T. Nagao and A. M. Mauer "Synchronization and recruitment in acute leukemia" J. Clin. Invest. 50:2204-2214 (1971).
89. F. A. Hayes, A. A. Green and A. M. Mauer "Correlation of cell kinetics and clinical response in chemotherapy in disseminated neuroblastoma" Cancer Res. 37:3766-3770 (1977).
90. P. J. Burke, J. E. Karpe, H. G. Braine and W. P. Vaughan "Timed sequential therapy of human leukemia based upon the response of leukemic cells to humoral growth factors" Cancer Res. 37:2138-2146 (1977).
91. P. J. Burke, Personal communication.
92. L. M. van Putten "Are cell kinetics relevant for the design of tumor chemotherapy schedules?" Cell and Tissue Kinetics 7:493-504 (1974).
93. B. Barlogie and B. Drewinko "Lethal and kinetic response of cultured human lymphoid cells to melphalan" Cancer Treatment Rep. 61:415-436 (1977).
94. B. Barlogie, B. Drewinko, W. Gohde and G. P. Bodey "Lethal and kinetic effects of peptichemio on cultured human lymphoma cells" Cancer Res. 37:2583-2588 (1977).
95. W. Gohde, M. Meistrich, R. Meyn, J. Schumann and B. Barlogie "Cell cycle phase dependence of drug induced cycle progression delay" J. Histochem. Cytochem. 27:470-473 (1979).
96. B. Barlogie, T. Buchner, U. Asseburg, D. Kamanabroo "Zellkinetischer effekt von cytosin-arabinoside im klinischen test" Med. Welt. 25:1532 (1974).
97. B. Barlogie, W. Plunkett, M. Rabert, J. Latreille, M. Keating and K. McCredie "In vivo cellular kinetics and pharmacology studies of Ara-C and 3-DAU chemotherapy for relapsing acute leukemia. Cancer Res. (In press).

98. H. T. Rupniak and D. Paul "Selective killing of transformed cells by exploitation of their defective cell cycle control by polyamines" Cancer Res. 40:293-297 (1980.
99. G. M. Kosik, B. Barlogie, D. A, Johnston, W. K. Murphy and B. Drewinko "Lethal and cytokinetic effects of anguidine on a human colon cancer cell line" Cancer Res. 38:3304-3309 (1978).
100. L. Reodori, C. T. Fu and B. Barlogie "In vitro cell cycle and lethal effects of anguidine" Cell Kinetics Society, Abstract #48 (1980).
101. L. Teodori, B. Barlogie, B. Drewinko, D. Swartzendruber and F. Mauro "Reduction of Ara-C and adriamycin cytotoxicity following cell cycle arrest b y anguidine" Submitted to Cancer Res.
102. P. C. Raich "Prediction of therapeutic response in acute leukemia" Lancet 1:74-76 (1978).
103. W. Hittleman and P. Rao "The nature of adriamycin-induced cytotoxicity in chinest hamster ovary cells as revealed by premature chromosome condensation" Cancer Res. 35:3027-3030 (1975).
104. M. Haq, J. Latreille, M. Keating and B. Barlogie "In vivo cell cycle effect of AMSA in relapsing adult acute leukemia" Cell Kinetics Society, Abstract #51 (1980).
105. R. N. Buick and E. A. McCullouch "A clonogenic assay for leukemic blast cells used to measure anthracycline sensitivity in culture" Proceeding AACR, Abstract #360 (1978).
106. R. M. Buick, H. A. Messner, J. E. Till and E. A. McCullouch "Cytotoxicity of adriamycin and daunorubicin for normal and keukemia progenitor cells of man" Journal of National Cancer Institute 62:249-255 (1979).
107. C. H. Parks, M. Amare, M. A. Savin, J. W. Goodwin, M. M. Newcomb and B. Hoogstraten "Prediction of chemotherapy response in human leukemia using an in vitro chemotherapy sensitivity test on the leukemic colony-forming cells" Blood 41:595-601 (1980).
108. M. J. Cline and E. Rosenbaum "Prediction of in vivo cytotoxicity of chemotherapeutic agents by their effects on leukocytes from patients with acute leukemia" Cancer Res. 28:2516-2521 (1968).
109. J. E. Byfield and J. J. Stein "A simple radiochemical assay of inhibition by chemotherapeutic agents of precursor incorporation into biopsy samples, effusion and leukocyte preparations" Cancer Res. 28:2228-2231 (1968).
110. P. C. Raich "Prediction of therapeutic response in acute leukemia" Lancet 1:74-76 (1978).
111. G. M. Dosik, B. Barlogie, D. A. Johnston, D. Mellard and E. J. Freireich "In vitro drug sensitivity test to predict clinical response in acute myeloblastic leukemia" European Journal of Cancer (In press).

112. M. Haq, B. Barlogie, G. M. Dosik, K. McCredie and M. Keating "In vitro drug-induced deoxyribonucleic acid (DNA) synthesis inhibition as a predictive test for clinical response in acute leukemia" The American Society of Clinical Oncology C-459 (1980).

113. F. Traganos, Z. Darzynkiewicz, T. Sharpless and M. R. Melamed: "Simultaneous staining of ribonucleic and deoxyribonucleic acids in unfixed cells using acirdine orange in a flow cytofluorometric system" J. Histochem. Cytochem. 25:46 (1977).

114. A. M. Kshirsayer, in: Multivariate Analysis, New York, Marcel Dekker (1972).

115. Nicolini, C., Parodi, S., Lessin, S., Belmont, A., Abraham, S., Zietz, S., and Grattorola, M. (1979); Chromatin study in situ. II. Static and flow microfluorometry. In: Chromatin Structure and Function, edited by C. Nicolini, pp. 293-323. Plenum Press, New York.

116. Nicolini, C., Desaive, C., Kendall, F.M., and Giaretti, W. (1977): The G0-G1 transition of WI-38 cells. II. Geometric and densitometric texture analysis. Exp. Cell Res., 106:118-127.

117. Nicolini, C., Linden, W.A., Zietz, S., and Wu, C.T. (1977): Objective identification of non-proliferating cells in a melanoma B16 tumor in vivo. Nature, 270:163-176.

118. Abraham, S., Lessin, S., Vonderheid, E. and C. Nicolini. "Reversible (G0) and irreversible (Q) non cycling cells in human peripheral blood: immunological, biological and structural characterization". Cell Biophysics, 1981, 2,4.

119. Nicolini, C., Parodi, S., Belmont, A., Abraham, S., and Lessin, S., (1979): Mass action and acridine orange staining. J. Histochem. Cytochem., 27:102-113.

120. M. Grattarola, S. Zietz, S. Lessin, C. Desaive and C. Nicolini, Early Detection of Micrometastases Via Flow Microfluorimetry, Cancer Biochem, Biophys., (1979), Vol. 4, pp. 13-18.

121. C. Nicolini, FMF Analysis of the Cell Cycle: Membrane and Nucleic Acid Stainings, Advances in Neuroblastoma Research edited by Audrey E. Evans. Raven Press, New York (c) 1980. 271-286.

122. Lessin, S., Abraham, S., and Nicolini, C. (1981): "High versus low metastatic potential melanoma cells: biophysical cytological studies."Histochemical Journal.

ROUND TABLE DISCUSSION ON
"ARE CELL KINETICS USEFUL IN CANCER CHEMOTHERAPY"

WALFRIED LINDEN: Probably most of you know that the two main achievements in cancer therapy during the last ten years were: 1) the radiation therapy of Hodgkin's disease, (using very large radiation doses), and 2) polychemotherapy of acute childhood leukemia. This is the point that I wish to make: all these new therapeutic regimens were designed especially by physicians. I think this presents a challenge for cell biologists trying to apply their knowledge to the problems of oncology. One of our questions might be: will this devleopment occur in the near future? Becoming more specific: I would like to recall the symposium of this afternoon and to try to find the points on which all of us agree and those on which we have problems and questions.

I think, we all agree, that cell kinetics can only be applied to cancer therapy if we get a very comprehensive set of cell kinetic data, using either autoradiography or flow systems to supply the information needed. The main kinetic parameters of a malignant system have already been mentioned this afternoon: distribution of cells throughout the cell cycle, rate of progression through the cycle, growth fraction, fraction of cells that are capable of more or less unlimited growth, and cell loss. I have the impression that we have now the techniques to determine most of these parameters in vitro and this leads from the point of general consenses to our question: *is there any evidence, that, if we have all this information on an individual tumor, we can use this knowledge for better regimens of therapy?* Or to formulate the question in a different way: *We have animal tumor systems where we have all the necessary cell kinetic information, can we treat them better than by using the empirical schemes?*

ARTHUR PARDEE: I am very pessimistic about this problem. The tumor is only one half of the problem and the patient is the other half. The object of chemotherapy is obviously to eliminate the tumor and not to eliminate the patient. What has to be realized is that the patient is composed of many different kinds of cells (like intestine or bone marrow) and those tissues are very sensitive proliferative populations. One not only has to perform such tricks as killing cells specifically in S phase but one also has to not kill too many S phase cells in the intestine. We all know that antineoplastic drugs are highly lethal and the major margin of safety is often said to be a factor of two in concentration (i.e. with twice the dose you kill the patient). So the simplistic idea of killing only cells in S phase can work in cases like leukemia where you have a relatively simple system with a large number of cycling cells, but when you come to real solid tumors most of the cells are quiescent most of the time. Beside the tumors are not homogenous and it has been shown that different parts of the tumor can have different degrees of drug resistance. Therefore, I think this idea of using S phase or mitotic properties of cycling cells is lucky if it works with leukemia but we have to come up with something different now.

BENJAMIN DREWINKO: I could answer in five seconds: I don't know. But I'd like to express my one piece of philosophy. I sounded very negative in my lectures on the application of cell kinetics to oncology, and it could be construed that cell kinetics has nothing to contribute to oncology. That is not true. What I was trying to point out is that we have to be very humble in our claims how cell kinetics can help clinical oncology. It is by human nature that we would like to build up gods or heros and when they do not deliver what we expect, we also like to destroy them with great vigor. Cell kinetics went the route of what virology and immunology went a few years ago in promising to cure cancer. It has not then nor will it do so in the near future because we still do not know anything. The problem is that anytime we say that kinetics can be applied to clinical oncology it can also be said: it may work but if it does not work it is only because we do not know. That opens the field for anybody who wants to do studies. But we must still realize that we are at the level of empirical description, not even of interpretation.

STANLEY ZIETZ: I feel that most of the frustration that has been voiced concerning the utilization of cell kinetics in cancer chemotherapy is due to the fact that cell kinetics has until recently mostly been used as an empirical tool. Global unperturbed population parameters, such as labeling index and mitotic index, although sometimes used to help stage a patient, are clearly too gross to be able to help design treatment rigimens or even predict response to therapy. The concept of growth fraction, over 20 years old, has been an elusive one, since until recently we could only indirectly measure the non-cycling compartment before treatment, and the measurement of recruitment was always inferential. Attempts to synchronize either the tumor or normal tissues met with little success partly because of the great heterogeneity of the populations in question.

Recent advances in automated cytology (the chapter by Nicolini etal in this volume) however, have given us powerful tools to monitor response to treatment. We can now quickly (less than one hour) and objectively measure the non-proliferating population and thus directly quantitate the recruitment of cells into the cycle after treatment. Also, multiparameter techniques have helped us characterize and monitor subpopulations in the tumor which have enhanced metastatic capability. Thus we may overcome the problems of heterogenous populations and focus on cell populations of interest.

However, to maximally utilize this advanced technology, we must develop in parallel quantitative mathematical tools capable of utilizing the experimental data to help predict effects of therapy on the tissues of interest (both tumor and normal). Such a quantitative approach, encompassing both cell and drug kinetics, as well as dosage affects and drug interactions has been developed at Temple during the last 7 years (as can be seen from the numerous literature). This quantitative approach is needed in order to help avoid the quasi-random (and here to fore frustrating) searches for the best therapy. Once the quantitation of the dose response of single agents has been accomplished, we may then utilize optimal control theory, or more precisely Cooperative Differential Game Theory, in order to attempt to predict optimal multi-agent therapy. Again, I must emphasize that the value of this obviously long approach lies in the continued feedback between experiments and predictions of models. We are begining to see its fruition in developing better therapies against the metastatic murine tumor and we are looking forward to bringing this approach to the clinic to attack the human disease. We are now starting this effort.

WALFRIED LINDEN: Just a short comment: Well, Dr. Pardee, of course we are quite well aware of the fact that we are looking for differential effects. We would like to kill the tumor and save the normal bone marrow and normal intestine. And that was just the idea behind proposed cell kinetics applications in the first place. They should serve for purposes just like that. Now I will ask a more specific question: Is there any example where cell kinetics ever worked to give an improved therapeutic response? Are we aware of any cases in the literature of animal tumor systems for some evidence of this?

BRIGITTE SCHULTZ-MAURER: I think I can not answer this second question. On the whole I would say that I am also pessimistic but not so pessimistic apart from our lack of knowledge of the proliferation of the individual human tumor. There is very little known about the effects of cytotoxic drugs on the proliferation of the tumor cells and the normal cells. I think that is a better insight into their action on the proliferation of both normal and tumor cells might provide the oncologist with better knowledge concerning the choice of the various drugs in combination therapy and perhaps also for the timing of these drugs. Furthermore, I see another problem: I think it will be more difficult in the future to carry out randomized studies with the various drug combinations in humans so we will have to check the drugs more carefully in animals.

CLAUDIO NICOLINI: I will like to address the first original question. I think if cell kinetics has to be used the way it has been used up to now, and in the way the large majority of people use it, cell kinetics is useless. There is no purpose in doing this on animals, monkeys, or humans. You see, again we come back to the usual problem. We forget that if you give a drug to a human body, (this is what Dr. Pardee was saying) the drug goes through various parts of the body, gets metabolized, takes time to be catabolized, and these events depend on the transport of the durg through the body, tissues and various cellular membranes. These events also depend on how the overall macromolecules in the cells, cycling versus noncycling, tumor versus normal, are organized. So the overall end point is quite complex and the way it has been probed up until now has been very primitive and very empirical. Thus people are now very discouraged about the usefulness of cell kinetics. I would be very discourage. I would not spend even one minute of my time on this. Take for instance the way flow systems are utilized today. They are utilized as pieces of equipment yielding no more information than standard autoradiography. So let me say when you talk about cell kinetics you are referring to the traditional cell kinetics, which I think is useless. We are just wasting our time. However there is the possibility now, as Dr. Zietz has said, to discriminate within the bulk tumor the population capable of metastasis, which is a very small fraction of the overall total and which to our surprise this population is also nearly all cycling. But you have to model this system and monitor it by double labeling and by the proper multiparameter analysis. If you do that on real time, using flow systems in the proper ways, it will allow you to take samples of solid tumor or certain normal tissues whenever you can bring then in usupension, and have the response in real time while you can still utilize it, rather than having to inject thymidine and monitor sequentially as has been done in the past. You see some people forget that when we sent a rocket to the moon we used optimal control theory and not randomized empirical trails Optimal control theory, or other mathematical approaches combined with new observables monitored by flow and scanning cytometers, is a way by which you can try to optimize a regimen; for instance, the timing and dosage by which the clinician tries to optimize the maximum killing of tumor cells compatible with the minimum killing of normal cells. There is no way, since there is a different response of each tissue, that synchrony or recruitment will provide a simple answer. Synchrony even in cell populations with high growth fraction will not solve the problem because you cannot generalize the response of the drug even between one tissue and another tissue in the very same mouse. So it is hopeless if you do not approach things analytically. Furthermore, we have to focus the chemotherapy not on the overall bulk tumor which is highly differentiated and poorly cycling, but to the small fraction which is capable to metastasize. This is a difficult approach but all challenges are difficult and complex. Some of the problems involved in realistically optimizing chemotherapy would challenge the first rate system engineer. The cells in various tissues are the most complex system you can get. So even if I am very pessimistic for the present about the way cell kinetics and pharmaco-enzyme kinetics are done in clinical oncology, I am more hopefull for the future since we have finally developed the needed tools, both mathematical and physico-chemical, which led us to explain and simulate in mice even the most complex drug-cell and drug-macromolecules interactions regulated by feedback mechanisms.

Even if our major emphasis should undoubtedly lie in the search of the molecular mechanisms determining cell proliferation and transformation, we have still to attempt to help people with cancer now. We have then to approach the problem analytically and not empirically as done with little or no success up to now, and try to solve it or at least optimize it. I am still pessimistic but I must say few people have seriously tried the quantitative approach. Ask me again in several years and I may tell you this is hopeless, but first we have to make a serious effort along this line before we reach a negative conclusion.

PARTICIPANTS AT ERICE ASI ON CELL GROWTH, 1980
(I Course of the International School of Biostructure)

Sitting From Left To Right,First Row:S. Alberti,G. Starace,L. Teodori,A. Yen,M. Grattarola,F. Julemont, S. Lessin. Sitting From Left To Right,Second Row:H. Bauer,P. Rao,E.M. Bradbury,D. Mazia,W. Schreil,C. Nicolini,W. Linden,H. Halvorson,S. Avrameas. Standing From Left To Right,Front Row: F. Gabrielli,D. Statakis,K. Rao,M. Numes,V. Barlati,A. Suau,M. Moreno,E. Mantegani,M. Robert-Gero,M. Enginum,M. Milla, O. Alabaster. Standing From Left To Right,Back Rows:D. Prescott,B. Maurer-Schultz,G. Algan,D. Lohang, S. Svetina,N. Maraldi,W. Wallmer,M. Zerbini,E. Gendel,F. Kendall,F. Beltrame,H. Reinhardt,P. Reynolds, T. Bradbury,S. Cooper,V. De Boni,C. Leung,S. Stragand,E. Aarholt,C. Lucke-Huhle,R. Van Meeteren,S. Elci, F. Lawrence,S. Ayvali,M. Vasconcelos,W. Nagl,S. Abraham,F. Toker,S. Colombini,S. Zietz,L. Walder,B. Sells,G. Zimmerman,A. Olojhoeck,J. Domingo,K. Konig,D. Skagen,P. Steck. Persons Missing: R. Baserga, B. Drewinko,B. Hill,P. Omedeo,G. Stein,J. Stein,A. Pardee,R. Tucker,T. Brinkley,H. Hsie,N. Nanninga,J. Paul,A. Guialis,M. Lis,M. Di Pasquale,A. Barnekow,M. Purello,I. Young.

LIST OF PARTICIPANTS

Aarholt, E.	Dept. Electrical Engineering, University of Salford Salford M5-4WT, UK
Abraham, S.	Temple University - H.S.C. Philadelphia, PSA, USA
Alabaster, O.	The George Washington University, Medical Center - Division of Ematology and Oncology, Washington, DC, USA
Alberti, S.	Istituto Mario Negri, Via Eritrea, 62, 20157 Milano, Italy
Algan, G.	Genaral Botany Department, Faculty of Science University of Ankara, Turkey
Avrameas, S.	Unite d' Immunocytochimique, Departement de Biologie Moleculaire, Institute Pasteur 75 Paris, France
Ayvali, C.	Department of Zoology, Ankara University Ankara, Turkey
Badaracco, G.	Istituto Tumori Regina Elena, Viale Regina Elena, 291, Roma, Italy
Barlati, S.	Consiglio Nazionale delle Ricerche, Laboratorio di Genetica e Evoluzionistica 27100 Pavia, Italy
Barnekow, A.	Institut fur Virologie, Fachbereich Humanmedizin, D-6300 Giessen, FRG
Baserga, R.	Department of Pathology, Temple University School of Medicine, Broad and Ontario Street, Philadelphia, PA 19140, USA
Bauer, H.	Department of Tumor Virology, D-6300 Giessen, FRG
Belmont, A.	Temple University - H.S.C., Philadelphia, PA 19140, USA
Beltrame, F.	Istituto di Elettrotecnica, University of Genova, Genova, Italy
Bologna, M.	Istituto di Patologia Generale, Facolta di Medicina e Chirurgia, 67100 L'Aquila, Italy
Bradbury, E.M.	Department of Biological Chemistry, Medical School, University of California, Davis, CA, USA

Cooper, S. Department of Microbiology, The University of Michigan, Medical School, Ann Arbor, Michigan, USA

Colombini, S. Istituto di Clinica Medica 1, Cattedra di Ematologia-Policlinico, Via del Pozzo 71 41100 Modena, Italy

De Boni, U. Department of Physiology, Medical Sciences Building, Toronto, Canada, M5S 1A8

Di Pasquale, M. CNR - Lafbil, Via Buonarroti, 56100 Pisa, Italy

Domingo, J. Catedra de Histologia y Anatomia Patologica Facultad de Medicina, Barcelona, Spain

Drewinko, B. Anderson Hospital and Tumor Inst., Dept. of Laboratory Medicine, Houston, Texas 10021, USA

Elci, S. Faculty of Sciences, University of Firat, Elagiz, Turkey

Ellwart, J. Institut fur Hematologie, Landwehrstrasse 61, 8000 Munchen, 2, FRG

Enginum, M. Istanbul University, Analitik Kimya Kursusu Istanbul, Turkey

Gabrielli, F. Istituto di Chimica Biologica, Via Roma, 55 56100 Pisa, Italy

Gendel, E. Cytogenetics Laboratory, Metropolitan Hospital Center, 1901 First Avenue, New York, NY 10029, USA

Giannoulis, N. Saint Savvas Cancer Hospital, 171 Alexandras Avenue, Athens 603, Greece

Grattarola, M. Istituto di Elettrotecnica, Viale Francesco Causa, 13, Genova, Italy

Guialis, A. The National Hellenic Research Foundation Biological Research Center, 48 Vassileos Constantinou Avenue, Athens 501/1, Greece

Hill, B. T. Laboratory of Cellular Chemotherapy, Imperial Cancer Research Fund Laboratories, P.O. Box n. 123, London, WC2A 3PX, UK

Huot, J. Department of Pharmacology, Faculty of Medicine, Lavac University, Quebec G1K 7P4, Canada

Julemont, F. X. Faculte des Sciences Agronomique, Place Croix du Sud, 1348 Louvain la Neuve, Belgium

Kendall, F. Dept. Physiology-Biophysics, Temple University - H.S.C., 3223 North Broad Street Philadelphia, PA, USA

Konig, K. Universitat Hamburg, Institut fur Biophysik und Strahlembiologie, 2 Hamburg 20, FRG

Lawrence, F. Centre National de la Recherche Scientifique Institut de Chimie des Substances Naturelles 91190 Gif-sur-Yvette, France

Lessin, S. — Temple University, Medical School - H.S.C. Philadelphia, PA 19140, USA

Leung, C. — Department of Physiology, Faculty of Medicine University of Manitoba, Winnipeg, Manitoba, Canada

Linden, W. — Institut fur Biophisics, and Radiation Biology, Krankenhaun Eppendorf, 2 Hamburg 20, FRG

Lis, M. — Laboratory of Compared Endocrinology Institute de Recherches Clinique, 110 Avenue des Pins Ouest, Montreal H2W1R7, Quebec, Canada

Lovhaug, D. — Norwegian Defence Research Establishment Division for Toxicology, P.O. Box 25 N-2007 Kjeller, Norway

Lucke-Huhle, C. — P.O. Box 3640, Nuclear Research Center D-75 Karlsruhe, FRG

Maraldi, N. M. — Istituto di Istologia ed Embriologia Generale Universita, Ancona, Italy

Maurer-Schultze, B. — Institute of Medical Radiology, University of Wurzburg, 80700 Wurzburg, FRG

Martegani, E. — Catte dra di Biochimica Comparata, Facolta di Science, via Saladini, 50, 20133 Milano, Italy

Milla, M. — Tarlac College of Technology, Tarlac, 2101, Republic of the Philippines

Moreno, M.L. — C.S.I.C., Instituto de Biologia Celular Velasquez 144, Madrid 6, Spain

Nagl, W. — Department of Biology, The University of Kaiserlautern, P.O. Box 3049, Kaiserlautern, FRG

Nicolini, C. — Temple University, Department of Biochemistry and Physiology, Philadelphia, PA 19140, USA

Numes, M. — Instituto Portogues de Oncologia, Franciso Gentil, Palhave, Lisboa, Portugal

Olijhoek, A.J.M. — Department of Electron Microscopy, University of Amsterdam, Amsterdam, The Netherlands

Omodeo, P. — Istituto di Biologia Umana, Via Loredan Padova, Italy

Pardee, A.B. — Sidney Farber Cancer Institute, Department of Pharmacology, Division of Cell Growth and Regulation, Boston, MA 02115, USA

Paul, J. — The Beatson Institute for Cancer Research Wolfson Laboratory for Molecular Pathology Garscube Estate, Glasgow, C61 1BD, UK

Principe, P. — Istituto Superiore di Sanita, Viale Regina Elena 299, 00161 Roma, Italy

Purrello, M. — Istituto di Biologia Generale, Via Androne, 81, 95124 Catania, Italy

Rao, K.N. — Department of Pathology, School of Medicine Pittsburg, PA 15261, USA

Rao, P.N.	Section of Cellular Pharmacology, Department of Developmental Therapeutics, M.D. Anderson and Tumor Institute, Texas Medical Center, Houston, Texas 77030, USA
Reinhardt, R.	Department of Biology, Universitatstrasse 10 775 Konstantz, FRG
Reynolds, T.L.	The Ohio State University, Department of Botany, 1735 Neil Avenue, Columbus Ohio, USA
Robert-Gero, M.	Institute de Chimie des Substences Naturelles Centre National de la Recherche Scientifique, 91190 Gif-sur-Yvette, France
Schreil, W.	Istituto Internazionale di Biofisica e Genetica, Via Guglielmo Marconi, 10 80125 Napoli, Italy
Sells, B.H.	Faculty of Medicine - H.S.C., Memorial University of New-Foundland, St. John, A1B-3V6, Canada
Skagen, Dankert	School of Medicine, Medical Department A 5016 Haukaland Sykehus, Bergen, Norway
Starace, G.	CNR, c/o I.M. E/R.S., Via Luigi Bodio 58 00191 Roma, Italy
Statakos, D.	Department of Biology, Nuclear Research Center "Demokritos", Athens, Greece
Stein, G.S	University of Florida, Dept. Immunology and Medical Microbiology, Gainesville, FL. 32610, USA
Stein, J.L.	University of Florida, Dept. Immunology and Medical Microbiology, Gainesville, FL 32610, USA
Steck, P.	Department of Biochemistry, Michigan State University, East Lansing, Michigan 48824, USA
Stragand, J.	MD Anderson Hospital and Tumor Institute Texas Medical Center, Houston, Texas 77030 USA
Suau, P.	Departamento de Bioquimica, Faculttad de Ciencias, Universidad Autonoma Bellaterra Barcelona, Spain
Svetina, S.	Institute of Biophysics, Medical Facukty and J. Stefan Institute, University of Ljubljana Ljubljana, Yugoslavia
Teodori, L.	IMERS - CNEN, Via Luigi Bodio 58, 00191 Roma, Italy
Toker, C.	General Botany Department, Faculty of Science Ankara, Turkey

Tucker, R.W.	John Hopkins Oncology Center, 600 North Wolfs Street, Baltimore, MD, USA
Van Meeteren, R.	Rijksuniversiteit Utrecht, Vakgroep Molecular Cell Biologie, State University 3584 CH Utrecht, The Netherlands
Van Wijk, R.	Department of Molecular Cell Biologie, State University, 3584 CH Utrecht, The Netherlands
Vasconcelos, M.	Av. Almirante Jaao, Azevedo Coutinho Murtal, S. Pedro do Estoril, Portugal
Walder, L.	Institut fur Biologie III, Schanzlestrasse 1 7800 Freiburg, FRG
Wollmer, W.	Institut fur Biophysik, Martinistrasse 52 2000 Hamburg, FRG
Yen, A.	Memorial Sloan Kettering Cancer Center 1275 York Avenue, New York, NY 10021, USA
Young, I.E.	Department of Biology, York University 4700 Keele Street, Downsview, Ontario M3 J-1P3 Canada
Zerbini, M.L.	Istituto di Microbiologia, Policlico S. Orsola, Via Massarenti, 9, Bologna, Italy
Zietz, S.	Drexel University, Dept. Mathematical Science, Philadelphia, PA 19140, USA
Zimmerman, G.J.	Institute for Experimental Surgery, Technical University Munich, 800 Munich, 80, FRG

CONTRIBUTOR INDEX

SUBJECT INDEX